CASE STUDIES IN
INFECTIOUS
DISEASE

CASE STUDIES IN INFECTIOUS DISEASE

Peter M. Lydyard

Michael F. Cole

John Holton

William L. Irving

Nino Porakishvili

Pradhib Venkatesan

Katherine N. Ward

GS Garland Science
Taylor & Francis Group

NEW YORK AND LONDON

Vice President: Denise Schanck
Editor: Elizabeth Owen
Editorial Assistant: Sarah E. Holland
Senior Production Editor: Simon Hill
Typesetting: Georgina Lucas
Cover Design: Andy Magee
Proofreader: Sally Huish
Indexer: Merrall-Ross International Ltd

Peter M. Lydyard, Emeritus Professor of Immunology, University College Medical School, London, UK and Honorary Professor of Immunology, School of Biosciences, University of Westminster, London, UK. **Michael F. Cole**, Professor of Microbiology & Immunology, Georgetown University School of Medicine, Washington, DC, USA. **John Holton**, Reader and Honorary Consultant in Clinical Microbiology, Windeyer Institute of Medical Sciences, University College London and University College London Hospital Foundation Trust, London, UK. **William L. Irving**, Professor and Honorary Consultant in Virology, University of Nottingham and Nottingham University Hospitals NHS Trust, Nottingham, UK. **Nino Porakishvili**, Senior Lecturer, School of Biosciences, University of Westminster, London, UK and Honorary Professor, Javakhishvili Tbilisi State University, Tbilisi, Georgia. **Pradhib Venkatesan**, Consultant in Infectious Diseases, Nottingham University Hospitals NHS Trust, Nottingham, UK. **Katherine N. Ward**, Consultant Virologist and Honorary Senior Lecturer, University College Medical School, London, UK and Honorary Consultant, Health Protection Agency, UK.

ISBN 978-0-8153-4142-0

Library of Congress Cataloging-in-Publication Data

Case studies in infectious disease / Peter M Lydyard ... [et al.].
    p. ; cm.
  Includes bibliographical references.
  SBN 978-0-8153-4142-0
  1. Communicable diseases--Case studies. I. Lydyard, Peter M.
  [DNLM: 1. Communicable Diseases--Case Reports. 2. Bacterial
  Infections--Case Reports. 3. Mycoses--Case Reports. 4. Parasitic Diseases--
  Case Reports. 5. Virus Diseases--Case Reports. WC 100 C337 2009]
  RC112.C37 2009
  616.9--dc22
                        2009004968

Published by Garland Science, Taylor & Francis Group, LLC,
an informa business
270 Madison Avenue, New York NY 10016, USA,
and 2 Park Square, Milton Park, Abingdon, OX14 4RN, UK.

Printed in the United States of America

15 14 13 12 11 10 9 8 7 6 5 4 3 2 1

Visit our web site at http://www.garlandscience.com

Preface

The idea for this book came from a successful course in a medical school setting. Each of the forty cases has been selected by the authors as being those that cause the most morbidity and mortality worldwide. The cases themselves follow the natural history of infection from point of entry of the pathogen through pathogenesis, clinical presentation, diagnosis, and treatment. We believe that this approach provides the reader with a logical basis for understanding these diverse medically-important organisms.

Following the description of a case history, the same five sets of core questions are asked to encourage the student to think about infections in a common sequence. The initial set concerns the nature of the infectious agent, how it gains access to the body, what cells are infected, and how the organism spreads; the second set asks about host defense mechanisms against the agent and how disease is caused; the third set enquires about the clinical manifestations of the infection and the complications that can occur; the fourth set is related to how the infection is diagnosed, and what is the differential diagnosis, and the final set asks how the infection is managed, and what preventative measures can be taken to avoid the infection.

In order to facilitate the learning process, each case includes summary bullet points, a reference list, a further reading list and some relevant reliable websites. Some of the websites contain images that are referred to in the text. Each chapter concludes with multiple-choice questions for self-testing with the answers given in the back of the book.

In the contents section, diseases are listed alphabetically under the causative agent. A separate table categorizes the pathogens as bacterial, viral, protozoal/worm/fungal and acts as a guide to the relative involvement of each body system affected. Finally, there is a comprehensive glossary to allow rapid access to microbiology and medical terms highlighted in bold in the text. All figures are available in JPEG and PowerPoint® format at www.garlandscience.com/gs_textbooks.asp

We believe that this book would be an excellent textbook for any course in microbiology and in particular for medical students who need instant access to key information about specific infections.

Happy learning!!

The authors
March, 2009

Acknowledgments

In writing this book we have benefited greatly from the advice of many microbiologists and immunologists. We would like to thank the following for their suggestions in preparing this edition.

William R. Abrams (New York University College of Dentistry, USA); Abhijit M. Bal (Crosshouse Hospital, UK); Keith Bodger (University of Liverpool, UK); Carolyn Hovde Bohach (University of Idaho, USA); Robert H. Bonneau (The Pennsylvania State University College of Medicine, USA); Dov L. Boros (Wayne State University, USA); Thomas J. Braciale (University of Virginia Health Systems, USA); Stephen M. Brecher (VA Boston Healthcare System USA); Patrick J. Brennan (Colorado State University, USA); Christine M. Budke (Texas A&M University, USA); Neal R. Chamberlain (A.T. Still University of Health Sciences/KCOM, USA); Dorothy H. Crawford (University of Edinburgh, UK); Jeremy Derrick (University of Manchester, UK); Joanne Dobbins (Bellarmine University, USA); Michael P. Doyle (University of Georgia, USA); Sean Doyle (National University of Ireland); Gary A. Dykes (Food Science Australia); Stacey Efstathiou (University of Cambridge, UK); Roger Evans (Raigmore Hospital, UK); Ferric C. Fang (University of Washington School of Medicine, USA); Robert William Finberg (University of Massachusetts Medical School, USA); Joanne Flynn (University of Pittsburgh School of Medicine, USA); Scott G. Franzblau (University of Illinois at Chicago, USA); Caroline Attardo Genco (Boston University School of Medicine, USA); Geraldo Gileno de Sá Oliveira (Oswaldo Cruz Foundation, Brazil); John W. Gow (Glasgow Caledonian University, UK); Carlos A. Guerra (University of Oxford, UK); Paul Hagan (University of Glasgow, UK); Anders P. Hakansson (SUNY at Buffalo, USA); Tim J. Harrison (University College London, UK); Robert S. Heyderman (Liverpool School of Tropical Medicine, UK); Geoff Hide (University of Salford, UK); Stuart Hill (Northern Illinois University, USA); Stephen Hogg (University of Newcastle, UK); Malcolm J. Horsburgh (University of Liverpool, UK); Michael Hudson (University of North Carolina at Charlotte, USA); Karsten Hueffer (University of Alaska Fairbanks, USA); Paul Humphreys (University of Huddersfield, UK); Ruth Frances Itzhaki (University of Manchester, UK); Aras Kadioglu (University of Leicester, UK); A. V. Karlyshev (Kingston University, UK); Ruth A. Karron (Johns Hopkins University, USA); Stephanie M. Karst (Louisiana State University Health Sciences Center, USA); C. M. Anjam Khan (University of Newcastle, UK); Peter G.E. Kennedy (University of Glasgow, UK); Martin Kenny (University of Bristol, UK); H. Nina Kim (University of Washington, USA); George Kinghorn (Royal Hallamshire Hospital, UK); Michael Klemba (Virginia Polytechnic Institute and State University, USA); Brent E. Korba (Georgetown University Medical Center, USA); Awewura Kwara (Warren Alpert Medical School of Brown University, USA); Jerika T. Lam (Loma Linda University, USA); Robert A. Lamb (Northwestern University, USA); Audrey Lenhart (Liverpool School of Tropical Medicine, UK); Michael D. Libman (McGill University, Canada); David Lindsay (Virginia Technical University, USA); Dennis Linton (University of Manchester, UK); Martin Llewelyn (Brighton and Sussex Medical School, UK); Diana Lockwood (London School of Hygiene & Tropical Medicine, UK); Francesco A. Mauri (Imperial College, UK); Don McManus (Queensland Institute of Medical Research, Australia); Keith R. Matthews (University of Edinburgh, UK); Ernest Alan Meyer (Oregon Health and Science University, USA); Manuel H. Moro (National Institutes of Health, USA); Kristy Murray (The University of Texas Health Science Center, USA); Tim Paget (The Universities of Kent and Greenwich at Medway, UK); Andrew Pekosz (Johns Hopkins University, USA); Lennart Philipson (Karolinska Institute, Sweden); Gordon Ramage (University of Glasgow, UK); Julie A. Ribes (University of Kentucky, USA); Alan Bernard Rickinson (University of Birmingham, UK); Adam P. Roberts (University College London, UK); Nina Salama (Fred Hutchinson Cancer Research Center and University of Washington, USA); John W. Sixbey (Louisiana State University Health Sciences Center-Shreveport, USA); Deborah F. Smith (York Medical School University of York, UK); John S. Spencer (Colorado State University, USA); Richard Stabler (London School of Hygiene & Tropical Medicine, UK); Catherine H. Strohbehn (Iowa State University, USA); Sankar Swaminathan (University of Florida Shands Cancer Center, USA); Clive Sweet (University of Birmingham, UK); Clarence C. Tam (London School of Hygiene & Tropical Medicine, UK); Mark J. Taylor (Liverpool School of Tropical Medicine, UK); Yasmin Thanavala (Roswell Park Cancer Institute, USA); Christian Tschudi (Yale University, USA); Mathew Upton (University of Manchester, UK); Juerg Utzinger (Swiss Tropical Institute, Switzerland); Julio A. Vázquez (National Institute of Microbiology, Institute of Health Carlos III, Spain); Joseph M. Vinetz (University of California, San Diego, USA); J. Scott Weese (University of Guelph, Canada); Lee Wetzler (Boston University School of Medicine, USA); Peter Williams (University of Leicester, UK); Robert Paul Yeo (Durham University, UK); Qijing Zhang (Iowa State University, USA); Shanta M. Zimmer (Emory University School of Medicine, USA); Prof G. Janossy (University College, London, UK).

Table of Contents

Pathogens by type and body systems affected

Guide to the relative involvement of each body system affected by the infectious organisms described in this book: the organisms are categorized into bacteria, viruses, and protozoa/fungi/worms.

Organism	Resp	MS	GI	H/B	GU	CNS	CV	Skin	Syst	L/H
Bacteria										
Borrelia burgdorferi						4+		1+	1+	
Campylobacter jejuni			4+			2+				
Chlamydia trachomatis				2+	4+	2+				
Clostridium difficile			4+							
Coxiella burnetti	4+						4+			
Escherichia coli	4+				4+	4+			4+	
Helicobacter pylori			4+							
Leptospira spp.				4+	4+	4+				
Listeria monocytogenes			2+			4+	2+		4+	
Mycobacterium leprae						2+		4+		
Mycobacterium tuberculosis	4+								2+	
Neisseria gonorrhoeae					4+				2+	
Neisseria meningitidis						4+			4+	
Rickettsia spp.							4+	4+	4+	
Salmonella typhi			4+						4+	
Staphylococcus aureus	1+	1+	2+				1+	4+	1+	
Streptococcus mitis						1+	4+			
Streptococcus pneumoniae	4+					4+				
Streptococcus pyogenes		3+						4+	3+	
Viruses										
Coxsackie B virus	1+	1+				4+		1+		
Epstein-Barr virus									2+	4+
Hepatitis B virus				4+						
Herpes simplex virus 1					2+	4+		4+		
Herpes simplex virus 2					4+	2+		4+		
Human immunodeficiency virus			2+						2+	4+
Influenza virus	4+					1+	1+			
Norovirus			4+							
Parvovirus	2+	3+						4+	2+	
Respiratory syncytial virus	4+									
Varicella-zoster virus	2+					2+		4+		

Protozoa/Fungi/Worms	Resp	MS	GI	H/B	GU	CNS	Skin	Syst	L/H
Aspergillus fumigatus	4+						1+	2+	
Echinococcus spp.	2+			4+					
Giardia lamblia			4+						
Histoplasma capsulatum	3+					1+		4+	
Leishmania spp.							4+	4+	
Plasmodium spp.				4+					4+
Schistosoma spp.			4+	4+	4+				
Toxoplasma gondii						2+			4+
Trypanosoma spp.			4+			4+		4+	
Wuchereria bancrofti									4+

The rating system (+4 the strongest, +1 the weakest) indicates the greater to lesser involvement of the body system.

KEY:

Resp = Respiratory: MS = Musculoskeletal: GI = Gastrointestinal

H/B = Hepatobiliary: GU = Genitourinary: CNS = Central Nervous System

Skin = Dermatological: Syst = Systemic: L/H = Lymphatic-Hematological

Case 1

Aspergillus fumigatus

A 68-year-old Caucasian man was diagnosed with B-cell chronic lymphocytic **leukemia** (B-CLL) and received various regimens of chemotherapy. As a patient with chronic leukemia he attended the CLL clinic regularly. Ten years later the patient presented with **pneumonia** symptoms and was examined by chest **CT scan**. The results were suggestive of aspergillosis and additional laboratory tests were done. Positive *Aspergillus* serology allowed the doctors in the clinic to give a diagnosis of *A. fumigatus* pneumonia. The patient was not neutropenic and his condition improved following an 8-month course of itraconazole followed by voriconazole for 6 months.

Two years later the patient was diagnosed with pulmonary aspergillosis. The diagnosis was based on a CT scan, cytology results, and a history of prior infection (Figure 1). He was treated with amphotericin B, monitored by radiography, followed by caspofungin for 9 days, but he died 2 days later of drug discontinuation. An autopsy was performed and the diagnosis of invasive pulmonary aspergillosis was confirmed.

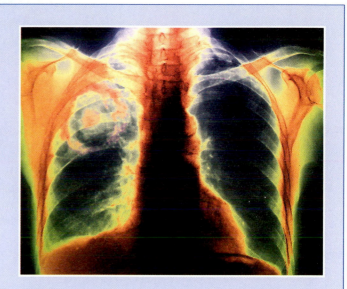

Figure 1. Chest X-ray showing that the fungus has invaded the lung tissue. There is a large cavity in the upper left lobe of the lung, with a fungus ball within the cavity.

1. What is the causative agent, how does it enter the body and how does it spread a) within the body and b) from person to person?

Causative agent

Aspergillosis is caused by *Aspergillus*, a saprophytic, filamentous fungus found in soil, decaying vegetation, hay, stored grain, compost piles, mulches, sewage facilities, and bird excreta. It is also found in water storage tanks (for example in hospitals), fire-proofing materials, bedding, pillows, ventilation and air conditioning, and computer fans. It is a frequent contaminant of laboratory media and clinical specimens, and can even grow in disinfectants!

Although *Aspergillus* is not the most abundant fungus in the world, it is one of the most ubiquitous. There are more than 100 species of *Aspergillus*. Although about 10 000 genes have been identified in the *Aspergillus* genome, none of the gene sets is shared with other fungal pathogens.

The cell wall of *A. fumigatus* contains various polysaccharides (Figure 2). Newly synthesized β(1-3)-glucans are modified and associated to the other cell wall polysaccharides (chitin, galactomannan, and β(1-3)-, β(1-4)-glucan)

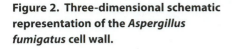

Figure 2. Three-dimensional schematic representation of the *Aspergillus fumigatus* cell wall.

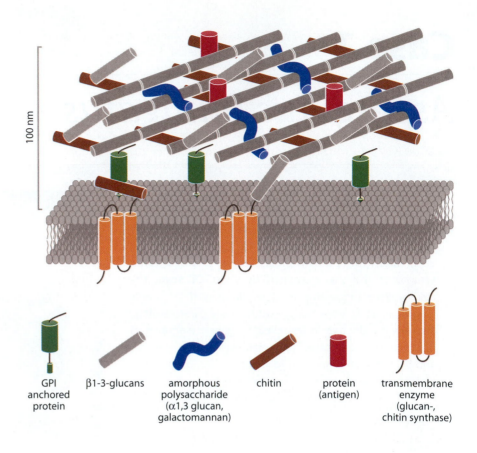

100 nm

GPI anchored protein	β1-3-glucans	amorphous polysaccharide (α1,3 glucan, galactomannan)	chitin	protein (antigen)	transmembrane enzyme (glucan-, chitin synthase)

leading to the establishment of a rigid cell wall. Glycosyltransferases bound to the membrane by a glycosylphosphatidyl inositol (GPI) anchor play a major role in the biosynthesis of the cell wall. Fungal cell composition affects its **virulence** and susceptibility to immune responses.

Of over 100 species of *Aspergillus* only a few are pathogenic, most of all *A. fumigatus*, but other species, including *A. niger*, *A. terreus*, *A. flavus*, *A. clavatus*, and *A. nidulans*, have also been implicated in pulmonary allergic disorders. However, other than *A. fumigatus*, *Aspergillus* species do not normally grow at temperatures higher than 37°C, and therefore do not cause invasive disease (see Section 3). *A. nidulans* can cause occasional infections in children with **chronic granulomatous disease**.

A. fumigatus is a primary pathogen of man and animals. It is characterized by thermotolerance: ability to grow at temperatures ranging from 15°C to 55°C, it can even survive temperatures of up to 75°C. This is a key feature for *A. fumigatus*, which allows it to grow over other aspergilli species and within the mammalian respiratory system.

A. fumigatus is a fast grower. It reproduces by tiny spores formed on specialized conidiophores. *A. fumigatus* sporulates abundantly, with every conidial head producing thousands of conidia. The conidia released into the atmosphere range from 2.5 to 3.0 μm in diameter and are small enough to fit in the lung alveoli. The spores are easily airborne both indoors and outdoors since their small size makes them buoyant. There is no special mechanism for releasing the conidia, it is simply due to the disturbances of the environment and air currents. When inhaled the spores are deposited in the lower respiratory tract (see below, Section 2).

A. fumigatus can be identified by the morphology of the conidia and conidiophores. The organism has green-blue echinulate conidia produced in chains basipetally from greenish phialides (Figure 3). A few isolates are nonpigmented and produce white conidia. *A. fumigatus* strains with colored conidia indicate the presence of accumulated metabolites and are more **virulent** than nonpigmented strains since they give the fungus an advantage to adapt to its environment.

In the past *A. fumigatus* was considered as an "asexual" fungus. Recent studies, however, have indicated the existence of a fully-functional sexual reproductive cycle.

Entry and spread within the body

We normally inhale 100–200 *Aspergillus* conidial spores daily, but only susceptible individuals develop a clinical condition. The spores enter the body via the respiratory tract and lodge in the lungs or sinuses. Once inhaled, spores can reach distal areas of the lung due to their small size. Very rarely other sites of primary infection have been described such as the skin, peritoneum, kidneys, bones, eyes, and gastrointestinal tract, but these are not clinically important. Usually the invasion of other organs by *A. fumigatus* is secondary and follows its spread from the respiratory tract.

When *Aspergillus* spores enter the human respiratory system at body temperature they develop into a different form, thread-like hyphae, which absorb nutrients required for the growth of the fungus. Some enzymes, particularly proteases, are essential for this fungal pathogen to invade the host tissue. Proteases are involved in the digestion of the lung matrix composed of elastin and collagen. In the case of infection of respiratory tissues, this contributes to the pathogenesis (see Section 3). Together the hyphae can form a dense mycelium in the lungs. However, in the case of healthy immunocompetent individuals the spores are prevented from reaching this stage due to optimal immune responses (see Section 2), and there is some colonization but limited pathology.

Aspergillus species are essentially ubiquitous in the environment worldwide, and no geographic preference of the exposure to airborne conidia or spores has been reported.

Person to person spread

This organism is not spread from person to person.

Until recently, *A. fumigatus* was considered a causative factor of mainly allergic conditions such as **farmer's lung**. However, over the past 20 years, due to the increase in aggressive immunosuppressive therapies and the AIDS pandemic, the number of immunocompromised patients developing aspergillosis has grown significantly. Thus severe and often fatal invasive infections with *A. fumigatus* have increased several times in developed countries and it has become the dominant airborne fungal pathogen.

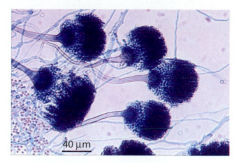

Figure 3. Microscopic morphology of *Aspergillus fumigatus* showing typical conidial heads. Conidiophores are short, smooth-walled, and have conical shaped terminal vesicles, which support a single row of phialides. Conidia are produced in basipetal succession forming long chains. They are green and rough-walled to echinulate. Chains of spores can be seen emerging from phialides surrounding the head.

2. What is the host response to the infection and what is the disease pathogenesis?

When the fungi colonize the respiratory tract the clinical manifestation and the severity of the disease depend on the efficiency of the immune responses. In susceptible hosts, *Aspergillus* conidia germinate to form swollen conidia and then progress to hyphae, its invasive form. The main goal of the immune system is to recognize and kill *Aspergillus* conidia and to prevent its transition to the hyphal form.

Mechanical barriers and innate immunity

In immunocompetent hosts, innate immunity to the inhaled spores begins with the mucous layer and the ciliated epithelium of the respiratory tract. The majority of the conidia are normally removed from the lungs through the ciliary action. However, *A. fumigatus* can produce toxic metabolites such as gliotoxin, which inhibit ciliary activity, and proteases which can damage the epithelial tissue.

Because of the site of the infection by *A. fumigatus*, bronchoalveolar macrophages – the resident phagocytic cells of the lung, together with recruited neutrophils, are the major cells involved in the **phagocytosis** of *A. fumigatus*. While macrophages mostly attack conidia, neutrophils are more important for elimination of developing hyphae.

Bronchoalveolar macrophages sense *A. fumigatus* through **pathogen-associated membrane patterns** (**PAMPs**) on the conidia via their **Toll-like receptors** TLR2 and TLR4, followed by engulfment and phagocytosis. Inhaled conidia via galactomannan bind some soluble receptors such as pentraxin-3 and lung surfactant protein D. This enhances phagocytosis and inflammatory responses. Phagosomes containing conidia fuse with endosomes followed by activation of **NADPH oxidase-dependent killing**. Nonoxidative mechanisms are also essential for the digestion of phagocytosed conidia by macrophages. Swelling of the conidia inside the macrophage appears to be a prerequisite for fungal killing. Conidial swelling inside macrophages or in the bronchoalveolar space alters cell wall composition and exposes fungal β-glucan. This further triggers fungicidal responses via mammalian β-glucan receptor dectin-1. However, the killing is delayed and quite slow and a total distruction of inhaled conidia by alveolar macrophages has never been reported. *A. fumigatus* is often able to block phagocytosis by producing hydrophobic pigments – melanins such as conidial dihydroxynaphthalene-melanin. Melanins are expressed on the conidial surface and protect the pathogen by quenching **reactive oxygen species (ROS)**.

In the cases when resident bronchoalveolar macrophages fail to control the fungus, conidia germinate into hyphae. Neutrophils and monocytes are then recruited from the circulation to phagocytose and kill hyphae.

Neutrophils adhere to the surface of the hyphae, since hyphae are too large to be engulfed. They are often seen clustered around fungal hyphae. This triggers a respiratory burst, secretion of ROS, release of lysozyme, neutrophil cationic peptides, and degranulation of neutrophils. NADPH oxidase-independent killing through the release of lactoferrin, an iron-sequestering molecule, is important. In contrast to the slow and sub-efficient killing of conidia by macrophages, hyphal damage by neutrophils is rapid, possibly through a release of fungal cell wall glycoproteins with the help of polysaccharide hydrolases produced by the neutrophils. **Defensins** may also play a role in responses to *A. fumigatus* hyphae. A defense mechanism of the pathogen is that it produces oxidoreductases, which could neutralize phagocytic ROS.

The importance of neutrophils in protection against *Aspergillus* is illustrated by a development of invasive aspergillosis in immunodeficient patients with chemotherapy-induced **neutropenia**. Corticosteroid-based treatment, purine analogs (fludarabine) and some monoclonal antibody treatment (Campath 1H – anti-CD52) and immunosuppression lead to neutropenia and/or neutrophil dysfunction. **Corticosteroids** reduce the oxidative burst and superoxide anion release by neutrophils, thereby inhibiting hyphal killing.

Platelets also play some role in protection against *Aspergillus*. They attach to the cell walls of invasive hyphae and become activated, thus enhancing direct cell wall damage of *A. fumigatus* and neutrophil-mediated fungicidal effect. Hence **thrombocytopenia**, which is associated with prolonged neutropenia during chemotherapy, increases the risk of infection by *A. fumigatus*.

Invasion with *A. fumigatus* enhances the levels of serum fibrinogen, C-reactive protein, and other **acute-phase proteins**. Resting conidia activate the **alternative complement pathway**, and induce neutrophil chemotaxis and deposition of complement components on the fungal surface. To combat this, *A. fumigatus* produces a specific lipophilic inhibitor of the alternative complement pathway and enhances proteolytic cleavage of C3 complement component bound to conidia by molecules present in the outer wall.

Antigen-presenting **dendritic cells (DCs)** exposed to hyphae in the lung migrate to the spleen and draining lymph nodes where they launch peripheral T helper (Th)-cell responses.

Adaptive immunity

T-cell responses

T-cell responses against *A. fumigatus* are mostly confined to the **CD4+ T cells**. To a certain extent their efficiency resides in the ability of **Th1** cells to further enhance neutrophil-mediated killing. A Th1 response, associated with a strong cellular immune component and increased levels of **IFN-γ**, granulocyte and granulocyte-macrophage colony-stimulating factors (**G-CSF** and **GM-CSF**), **TNF-α, interleukin-1 (IL-1)**, IL-6, IL-12, and IL-18, provides resistance to mycotic disease. Th1 proinflammatory signals recruit neutrophils into sites of infection. TNF-α enhances the capacity of neutrophils to damage hyphae; G-CSF, GM-CSF, and especially IFN-γ enhance monocyte and neutrophil activity against hyphae; while IL-15 enhances hyphal damage and IL-8 release by neutrophils. IL-8 recruits more neutrophils to sites of inflammation and mediates release of **antimicrobial peptides**.

On the other hand, a **Th2** response, which is associated with a minimal cellular component and an increase in antibody production, and secretion of IL-4, IL-5, and IL-10, appears to facilitate fungal invasion. Production of IL-4 by CD4+ T lymphocytes impairs neutrophil antifungal activity and IL-10 suppresses oxidative burst. Pathogenicity of allergic bronchopulmonary aspergillosis (ABPA, see Section 3) is associated with a pulmonary **eosinophilia** – the result of production of Th2 **cytokines**. *Aspergillus*-specific CD4+ T-cell clones isolated from ABPA patients have a Th2 phenotype.

It appears that a balance between beneficial Th1 and damaging Th2 types of immune responses is dependent upon the nature of antigens that prime DCs. Exposure of conidia to DCs leads to the activation of Th1 CD4+ T cells, while priming of DCs with hyphae enhances Th2-mediated pathways of CD4+ T-cell responses. **Regulatory T cells** may be also involved in determination of the Th1/Th2 bias.

A role of **CD8+ T cells** in the resistance to *A. fumigatus* was found to be very limited since fungal gliotoxin suppresses granule **exocytosis**-associated cellular cytotoxicity.

Antibody responses

Humoral immunity to *Aspergillus* species is poorly characterized. Although even in severely immunocompromised patients the production of specific antibodies has been described, their protective role, if any, remains unclear. The **antibody isotypes** produced are **IgG**1, IgG2, and IgA (particularly in bronchial lavage) but not IgG3, a pattern associated with a Th2 response. Immune serum did not enhance phagocytosis of conidia *in vitro*, but did induce macrophage-mediated killing. Neutralizing antibodies to proteases or toxins may also be beneficial to the host. Serum antibodies to *A. fumigatus* are often found in the absence of disease as a result of environmental exposure.

Serum samples from patients with ABPA contain elevated levels of antigen-specific circulating antibodies, mainly of IgG and **IgE** isotypes, which participate in pathogenesis of ABPA. B cells secrete IgE spontaneously as a result of IL-4 production, while IL-5 recruits eosinophils. Eosinophilic infiltration and basophil and mast cell degranulation in response to *A.fumigatus* antigens and IgE complex releases **pro-inflammatory mediators**. This leads to further chemotaxis of eosinophils and activated CD4+ lymphocytes to the site of the infection. In patients with ABPA, immediate skin reactions are mediated mainly by **type I hypersensitivity** and IgE antibody. The late reaction (**Arthus reaction**) to *Aspergillus* antigens is the result of IgE-mediated mast cell activation or immune complex formation (**type III hypersensitivity**) with complement activation. **Immune complexes** of specific IgG and *A. fumigatus* antigens trigger the generation of leukotriene C4 by mast cells, which in turn promotes mucus production, bronchial constriction, **hyperemia**, and **edema**. **Granuloma** formation in the lung has also been reported since some patients have granulomatous **bronchiolitis**.

Patients with **aspergilloma** (see Section 3), particularly those who recover from **granulocytopenia**, have increased levels of specific IgG and **IgM**, mostly against fungal carbohydrates and glycoproteins.

Generally the efficiency of host immune responses to *A. fumigatus* is a result of a dynamic interaction between fungal cell wall components and immune cells. Redundancy of host defense mechanisms may lead to the tissue-damaging inflammation favoring the invasive potential of the fungal cells and development of aspergillosis.

3. What is the typical clinical presentation and what complications can occur?

The spectrum of pulmonary diseases caused by *A. fumigatus* is grouped under the name of aspergillosis. These conditions vary in the severity of the course, pathology, and outcome and can be classified according to the site of the disease within the respiratory tract, the extent of fungal invasion or colonization, and the immunological competence of the host. There are four main clinical types of pulmonary aspergillosis : allergic bronchopulmonary aspergillosis (ABPA), chronic necrotizing *Aspergillus* pneumonia (CNPA), invasive aspergillosis (IA), and pulmonary aspergilloma (Figure 4).

Allergic bronchopulmonary aspergillosis (ABPA)

The main allergic condition caused by *A. fumigatus* is ABPA, which develops as a result of a hypersensitivity reaction to *A. fumigatus* colonization of the tracheobronchial tree. Estimating the frequency of ABPA is difficult due to the lack of standard diagnostic criteria (see Section 4). It often appears not as a primary pathology, but as a complication of other chronic lung diseases such as **atopic asthma**, **cystic fibrosis**, and **sinusitis**. It occurs in approximately 0.5–2% of asthmatic patients (and in up to 15% of asthmatic patients sensitized to *A. fumigatus*) and in 7–35% of cystic fibrosis patients.

The clinical course often follows as classic asthma, but can also lead to a fatal destruction of the lungs. IgE- and IgG-mediated type I hypersensitivity and type III hypersensitivity based on immune complexes are the leading causes of pathology (described in more detail in Section 2).

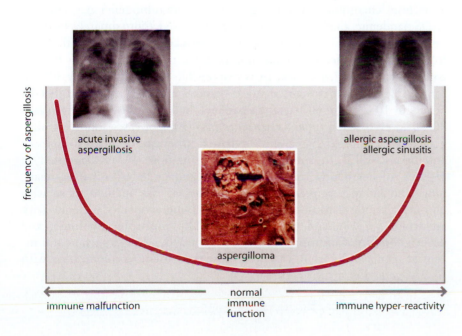

acute invasive aspergillosis

allergic aspergillosis allergic sinusitis

aspergilloma

frequency of aspergillosis

immune malfunction

normal immune function

immune hyper-reactivity

Figure 4. Relative risk of *Aspergillus* infection. The two chest X-rays show examples of acute invasive and allergic pulmonary aspergillosis. The fungal ball (aspergilloma) that was removed from a lung and measures about 6 cm in diameter is also shown. Hypersensitivity accompanies development of allergic aspergillosis, immunodificiency leads to invasive aspergillosis, whilst aspergilloma can be observed in immunocompetent individuals.

Since ABPA presents as a bronchial asthma the symptoms are similar to asthma and include wheezing, cough, fever, malaise, and weight loss. Additional symptoms include recurrent pneumonia, release of brownish mucoid plugs with fungal hyphae, and recurrent lung obstruction. In the case of secondary ABPA unexplained worsening of asthma and cystic fibrosis is observed. It is essential to diagnose and treat ABPA at the onset of the disease, which can be traced to early childhood or even infancy. ABPA should be suspected in children with a history of recurrent wheezing and pulmonary infiltrates. The outcome of the disease depends on asthma control, presence of widespread **bronchiectasis**, and resultant chronic fibrosis of the lungs (Figure 5). Respiratory failure and fatalities can occur in patients in the third or fourth decade of life.

Chronic necrotizing pulmonary aspergillosis (CNPA)

CNPA is a subacute condition mostly developing in mildly immunocompromised patients, and is commonly associated with underlying lung disease such as steroid-dependent **chronic obstructive pulmonary disease** (**COPD**), interstitial lung disease, previous thoracic surgery, chronic corticosteroid therapy or alcoholism. Patients may have been on long-term treatment with antibiotics or antituberculosis drugs without response, have **collagen vascular disease**, or chronic granulomatous disease.

CNPA presents as a subacute pneumonia unresponsive to antibiotic therapy, which progresses and results in cavity formations over weeks or months. Symptoms include fever, cough, night sweats, and weight loss. Because it is uncommon, CNPA often remains unrecognized for weeks or months and causes a progressive cavitatory pulmonary infiltrate. It is often found at autopsy. The reported mortality rate for CNPA is 10–40% or higher if it remains undiagnosed.

Invasive aspergillosis (IA)

Exposure to *A. fumigatus* in immunocompromised individuals can lead to invasive aspergillosis, which is the most serious, life-threatening condition. Due to the development of immunosuppression in transplantation and anticancer chemotherapy leading to severe immunodeficiency and the AIDS pandemic, the incidence of IA has increased approximately 14 times during the past 10–20 years. IA has even overtaken **candidiasis** as the most frequent fungal infection. Leukemia or bone marrow transplant (BMT) patients are at particular risk. IA is responsible for approximately 30% of fungal infections in patients dying of cancer, and it is estimated that IA occurs in 10–25% of all leukemia patients, in whom the mortality rate is 80–90%, even when treated. It occurs in 5–10% of cases following allogeneic BMT and in 0.5–5% after **cytotoxic therapy** of blood diseases or autologous BMT. In solid organ transplantation, IA is diagnosed in 19–26% of heart-lung transplant patients and in 1–10% of liver, heart, lung, and kidney recipients. Other patients at risk include those with chronic granulomatous disease (25–40%), neutropenic patients with leukemia (5–25%), and patients with AIDS, **multiple myeloma**, and **severe combined immunodeficiency** (about 4%). Drugs such as antimicrobial agents and steroids can predispose the patient to colonization with *A. fumigatus* and invasive disease.

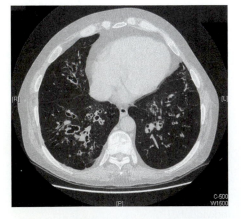

Figure 5. High-resolution CT scan of chest demonstrating remarkable bronchial wall thickening in the context of longstanding ABPA.

Currently four types of IA have been described. Clinical symptoms of the different types of IA depend on the organ localization and the underlying disease.

- Acute or chronic pulmonary aspergillosis (lungs).
- Tracheobronchitis and obstructive bronchial disease (bronchial mucosa and cartilage).
- Acute invasive rhinosinusitis (sinuses).
- Disseminated disease (brain, skin, kidneys, heart, eyes).

IA starts with pneumonia, and then the fungus usually disseminates to various organs causing **endocarditis**, **osteomyelitis**, **otomycosis**, **meningitis**, vision obstruction, and cutaneous infection (Figure 6). *Aspergillus* is second to *Candida* as a cause of fungal endocarditis. *Aspergillus*-related endocarditis and wound infections may occur through cardiac surgery. In the developing world, infection with *Aspergillus* can cause **keratitis** – a unilateral blindness.

Symptoms are usually variable and nonspecific: fever and chills, weakness, unexplained weight loss, chest pain, **dyspnea**, headaches, bone pain, a heart murmur, decreased **diuresis**, blood in the urine or abnormal urine color, and straight, narrow red lines of broken blood vessels under the nails. Patients develop **tachypnea** and progressive worsening **hypoxemia**. IA is accompanied by increased sputum production (sometimes with blood), sinusitis, and acute inflammation with areas of **ischemic necrosis**, **thrombosis**, and **infarction** of the organs involved.

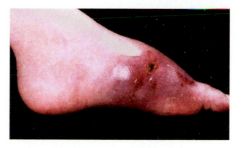

Figure 6. Bone infection caused by invasive aspergillosis.

Aspergilloma

An aspergilloma, also known as a mycetoma or fungus ball, is a clump of fungus which populates a lung cavity. It occurs in 10–15% of patients with pre-existing lung cavities due to the conditions such as tuberculosis, cystic fibrosis, lung abscess, **sarcoidosis**, emphysematous bullae, and chronically obstructed paranasal sinuses. Although *Aspergillus* species are the most common, some *Zygomycetes* and *Fusarium* may also form mycetomas. In patients with AIDS, aspergilloma may occur in cystic areas resulting from prior *Pneumocystis jiroveci* pneumonia infection. The fungus invades, settles, and multiplies in a cavity mostly outside the reach of the immune system. The growth results in the formation of a ball shaped like a half-moon (crescent). It consists of a mass of hyphae surrounded by a proteinaceous matrix, which incorporates dead tissue and mucus with sporulating structures at the periphery. Some cavities may contain multiple aspergilloma (Figure 7).

Patients with aspergilloma do not manifest many related symptoms, and the condition may go on for many years undiagnosed. It is often discovered incidentally by chest X-ray or by CT scans and appears as spherical masses usually surrounded by a radiolucent crescent. The most common, but still rare, symptom is **hemoptysis**. This happens when aspergilloma disrupts the cavity wall blood vessels or bronchial artery supply. The bleeding is not usually life-threatening due to the small amount of blood produced, but rarely hemoptysis may be massive and even fatal. The patients may cough up the fungus elements, and sometimes chains of

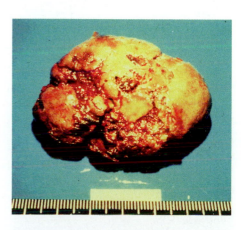

Figure 7. Multiple aspergillomas. Gross pathology showing three fungus balls in one cavity.

conidia can be seen in the sputum. Aspergilloma can lead to **pleural thickening**.

Rarely, in immunocompromised individuals, aspergillomas can be formed in other body cavities. They may cause abscesses in the brain, or populate various face sinuses, ear canals, kidneys, urinary tracts, and even heart valves. Secondary aspergillomas may occur as a result of IA when a solid lesion of IA erodes to the surface of the lung. These lesions can be detected by a chest CT scan and must be taken into account when further immuno-suppressive therapy for relapsed IA is prescribed.

4. How is the disease diagnosed and what is the differential diagnosis?

Diagnosis of ABPA

ABPA is a particularly difficult syndrome to diagnose since the symptoms are not specific. The disease presents with bronchial asthma with transient pulmonary infiltrates and at later stages proximal bronchiectasis and lung **fibrosis**. Chest radiography is also not specific and shows various transient abnormalities: consolidation or collapse, thickened bronchial wall, peripheral shadows.

The following criteria are currently used for diagnosis of ABPA: asthma, a history of pulmonary infiltrates, and central bronchiectasis. This is supplemented by laboratory tests: peripheral blood eosinophilia (>10% or 1000 mm^{-3}), immediate skin reactivity to *A. fumigatus* antigenic extracts within 15 ± 5 min, detection of precipitating IgG and IgM antibodies in >90% of cases, and elevated levels of total IgE in serum (>1000 ng/ml). Specific IgE antibodies against *A. fumigatus* are normally measured by IgE **RAST**.

Isolation of *A. fumigatus* from sputum (Figure 8), expectoration of brown plugs containing eosinophils and **Charcot-Leyden crystals**, and a skin reaction occurring 6 ± 2 h after the application of antigen are used as a complementary diagnosis.

Since the majority of the features are not specific and may appear at different stages of the course of the disease, not all the criteria are present simultaneously. The diagnostic value of some criteria, such as radiographic findings, eosinophilia, or the detection of precipitating immunoglobulins, should be taken in conjunction with an existing primary condition (cystic fibrosis, asthma). There is even a concept of 'silent' ABPA when none of the diagnostic criteria appear, but there is damage to the respiratory mucosa in response to *Aspergillus* conidia.

Diagnosis of IA

IA in the early stages is also difficult to diagnose. A safe diagnosis can only be made at autopsy with the histopathological evidence of mycelial growth in tissue. Differential diagnosis from the invasion of hyphae of other filamentous fungi such as *Fusarium* or *Pseudallescheria* is often difficult and requires **immunohistochemical** staining or *in situ* **hybridization** techniques. Clinical symptoms are usually nonspecific and require further laboratory tests.

Figure 8. Microscopy of sputum. A typical example of a wet mount of a sputum sample from a patient with ABPA.

Criteria currently used for the diagnosis of IA are: a positive CT scan (see Figure 1), culture and/or microscopic evaluation, and the detection of *Aspergillus* antigens in the serum.

Radiographic pictures of pulmonary IA can vary from single or multifocal nodules, with and without cavitation, to widespread often bilateral infiltrates. A CT scan is more reliable than radiography and can demonstrate the number and the size of the lesions. However, the appearances are heterogeneous throughout the course of the disease, with the most specific being at the early stages and presenting a 'halo' of hemorrhagic **necrosis** surrounding the fungal lesion or pleura-based lesions. In nonpulmonary forms of the disease such as cerebral aspergillosis, a CT scan together with brain **magnetic resonance imaging** (**MRI**) can detect the extent of the disease and the bone invasion.

The diagnostic value of the microscopic examination of sputum is limited due to the presence of airborne conidia of *Aspergillus* and the possibility of accidental contamination. However, in neutropenic or BMT patients the predictive value of a sputum culture positive for *A. fumigatus* exceeds 70%. The presence of *A. fumigatus* in **bronchoalveolar lavage** fluid (BAL) samples from patients with leukemia and BMT is found in 50–100% of those who have definitive or probable aspergillosis. Nasal swabs of patients also have diagnostic value, although **bronchoscopy** is preferable due to the sterility of the clinical sample. For the same reason percutaneous lung biopsy or aspirated material are the specimens of choice. However, invasive procedures in immunocompromised patients require careful consideration.

Cell culture

After the microscopic identification of *A. fumigatus*, cell culture may be critical in supporting the diagnosis of aspergillosis. The specimen is usually inoculated onto a plate with Sabouraud glucose agar, inhibitory mold agar (IMA) or other appropriate medium with antibiotics – gentamicin or chloramphenicol, but not cycloheximide, which is toxic for *Aspergillus* species. The plates are incubated at 30°C for up to 6 weeks with the cultures being examined at 3-day intervals.

Diagnosis based on cell culture therefore takes a long time. IA is a life-threatening condition and requires the development of early diagnosis methods such as **enzyme-linked immunosorbent assay (ELISA)**, which measures the presence of serum antigens and is both sensitive and specific.

Antigen detection

A highly specific (99.6%) and sensitive (1 ng ml^{-1}) test for detection of *Aspergillus* galactomannan (GM) has been developed for screening and for early diagnosis of IA in serum, bronchoalveolar lavage, and cerebrospinal fluid. GM is a part of the *Aspergillus* cell wall (see Section 1), and can be often released into the patient's bodily fluids. The detection of *A. fumigatus* GM by ELISA becomes possible at an early stage of infection thus allowing timely initiation of therapy. In 65.2% of patients GM can be detected in serum 5–8 days before the development of IA symptoms. In addition, ELISA can be used for monitoring the disease treatment.

Positive results in two consecutive serum samples allows the diagnosis of IA. However, in some cases false positive reactions can be observed due to cross-reactivity with nonspecific antigens derived from other fungi such as *Rhodotorula rubra*, *Paecilomyces varioti*, *Penicillium chrysogenum* and *P. digitatum*.

Diagnosis of aspergilloma

A definitive diagnosis of aspergilloma requires bronchoscopy, lung biopsy or resection, but this is rare. The diagnosis is usually made accidentally or specifically by chest radiography. A pulmonary aspergilloma appears as a solid ball of water density, sometimes mobile, within a spherical or ovoid cavity. It is separated from the wall of the cavity by the air space. Pleural thickening is also characteristic. A chest CT scan can sometimes detect aspergilloma with a negative chest radiograph. The radiographic picture must be differentiated from other conditions such as cavitating neoplasm, blood clot, disintegrating **hydatid cyst**, and pulmonary **abscess** with necrosis.

Clinical analysis should be coupled with serologic tests since a number of other fungi such as *Candida*, *Torulopsis*, *Petriellidium*, *Sporotrichum*, and *Streptomyces* can lead to the development of mycetoma. Since aspergilloma is a condition often observed in immunocompetent individuals, a laboratory diagnosis based on a humoral response is feasible. The most commonly used methods in clinical diagnosis are double immunodiffusion (Figure 9), immunoprecipitation, and counter-immunoelectrophoresis because they are simple, cheap, and easy to perform.

The diagnosis of an aspergilloma is confirmed when radiographic findings are supported by serological tests with >95% sensitivity for aspergilloma. It must be noted that patients undergoing corticosteroid treatment may become seronegative.

Positive sputum cultures are found in >50% of patients with aspergilloma, but this is not a specific diagnostic marker and is seen in many aspergillosis conditions.

Polymerase chain reaction (**PCR**) analysis has recently been introduced for diagnosis. The most reliable and well-characterized antigens of *A. fumigatus* are RNase, catalase, dipeptidylpeptidase V, and the galactomannan.

There is a general consensus in the field of aspergillosis diagnostics that the development of the genetic diagnostic approaches should be pursued. Combined use of PCR and ELISA should result in a fast definitive diagnosis of IA, even without clinical symptoms, and should allow the detection of a transient aspergillosis, which may occur in neutropenic patients.

5. How is the disease managed and prevented?

ABPA

Although ABPA is a chronic condition, acute corticosteroid-responsive asthma can occur and lead to fibrotic end-stage lung disease. The aim of the treatment of ABPA is to suppress the immune reaction to the fungus and to control bronchospasm. For this high doses of oral corticosteroids are used: 30–45 mg/day of prednisolone or prednisone in acute phase and

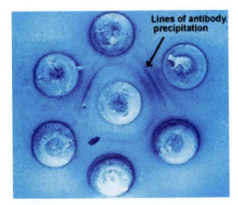

Lines of antibody precipitation

Figure 9. Double diffusion test for aspergillosis. The central well contains *A. fumigatus* antigen and wells at the top and bottom contain control antiserum. The three peripheral wells with precipitin bands contain sera from patients with *A. fumigatus* aspergilloma. More bands present in the upper right case is characteristic of aspergilloma. The well in the bottom left position is negative.

a lower maintenance dose of 5–10 mg/day. Sometimes antifungal drugs such as itraconazole (see below) are used as well to suppress fungal growth, although its eradication is not possible.

The administration of medications can be combined with removal of mucus plugs by bronchoscopic aspiration. Regular monitoring by X-rays, pulmonary function tests, and serum IgE levels are essential. Successive control of ABPA leads to a drop in IgE levels, whilst their increase indicates relapse.

Invasive aspergillosis

It is difficult to achieve a timely diagnosis of IA due to the rapid progression (1–2 weeks from onset to death) and delayed methods of diagnosis (see Section 4). Waiting for the confirmation of diagnosis would put the patients at a greater risk of untreatable fungal burden. The decision to start antifungal treatment has therefore to be empirical and based more on the presence of risk factors such as long-term (12–15 days) severe neutropenia (< 100–500 mm^{-3}).

The antifungal regimen for the treatment of IA includes voriconazole, amphotericin B (deoxycholate and lipid preparations), and itraconazole. Voriconazole is particularly effective against invasive aspergillosis and in reducing mortality. In addition, voriconazole, itraconazole, and amphotericin B exhibit a broad-spectrum activity against *Aspergillus* and the related hyaline molds.

Amphotericin B (AmB) has been used in the antifungal therapy for more than 30 years, mostly under the name of Fungizone® and remains a first-line drug, although the overall success rate of AmB therapy for IA is only 34%.

Despite some progress in antifungal therapy, mortality rates from IA remain very high: 86%, 66%, and 99% for pulmonary, sinus, and cerebral aspergillosis, respectively. As mentioned above, early initiation of antifungal therapy is critical since all untreated patients, or those treated for 1 week only, died. Of those who received antifungal drugs for 2 weeks, only 54% responded. Few patients with severe persistent neutropenia survive, and neutrophil recovery usually is followed by resolution of aspergillosis. The duration of neutropenia is therefore an indication of a possible outcome of IA.

Aspergilloma

Aspergilloma can be prevented by timely and effective management of diseases that increase the risk of its development, such as tuberculosis. Complication of aspergilloma by severe hemoptysis is an indication for surgery to remove it and thus stop the bleeding, although it is associated with high risks of morbidity and mortality. Surgical resection of aspergilloma is one of the most complex procedures in thoracic surgery, since prolonged chronic infection and inflammation lead to thickened fibrotic tissue, induration of the hilar structures, and obliteration of the pleural space. Postoperative complications include hemorrhage, bronchopleural fistula, a residual pleural space, and **empyema**. Surgery is restricted to patients with severe hemoptysis and adequate pulmonary function. The procedure is usually performed under local anesthesia via an incision over the cavity, guided by CT scans.

On the other hand, in the absence of surgical intervention the course of the disease remains unpredictable.

An alternative treatment in patients with severe pulmonary dysfunction comprises topical therapy. This includes intracavitary instillation of an antifungal drug such as AmB, percutaneous injection into aspergilloma cavities of sodium and potassium bromide, and endobronchial instillation of ketoconazole via fiberoptic bronchoscopy. However, topical therapy is labor-intensive and nonapplicable for the cases with multiple aspergilloma.

SUMMARY

1. What is the causative agent, how does it enter the body, and how does it spread a) within the body and b) from person to person?

- *Aspergillus* is a saprophytic fungus, and one of the most ubiquitous. Its natural habitat is soil, but it is also present around construction sites and indoors in water storage tanks, fire-proofing materials, bedding, and ventilation and air-conditioning systems.

- The cell wall of *A. fumigatus* contains various polysaccharides including a galactomannan core.

- Of over 100 species of *Aspergillus* only a few are pathogenic: *A. fumigatus*, *A. flavus*, *A. terreus*, *A. clavatus*, and *A. niger*.

- *A. fumigatus* is a primary pathogen of man and animals and causes all manifestations of aspergillosis. It is thermotolerant and grows at temperatures ranging from 15°C to 55°C, it can even survive temperatures of up to 75°C. This is a key feature that allows it to grow over other aspergilli species and within the mammalian respiratory system.

- *A. fumigatus* sporulates abundantly, with every conidial head producing thousands of conidia. The conidia released into the atmosphere range from 2.5 to 3.0 μm in diameter. We normally inhale 100–200 spores daily, but only susceptible individuals develop a clinical condition. The spores enter the body via the respiratory tract and lodge in the lungs or sinuses.

- Once inhaled, spores at body temperature develop into a different form – thread-like hyphae, which invade the host tissue. Together the hyphae can form a dense mycelium in lungs. However, in the case of healthy immunocompetent individuals the spores are prevented from reaching this stage due to the optimal immune responses, and there is some colonization but limited pathology.

2. What is the host response to the infection and what is the disease pathogenesis?

- In immunocompetent hosts fungal conidia may be cleared by ciliated epithelium of the terminal bronchioles and ingested by tissue macrophages or alveolar macrophages. While macrophages mostly attack conidia, neutrophils are more important for elimination of the next, hyphal form of the fungus.

- Neutrophils adhere to the surface of the hyphae and trigger a respiratory burst, secretion of reactive oxygen species (ROS), release of lysozyme, neutrophil cationic peptides, and lactoferrin.

- In immunodeficient, particularly IA patients, corticosteroid-based treatment, purine analogs (fludarabine) and some monoclonal antibody treatment (Campath 1H, anti-CD52) lead to neutropenia and/or neutrophil dysfunction. Corticosteroids reduce oxidative burst and superoxide anion release by neutrophils, thereby inhibiting hyphal killing.

- Th1 cytokines are important in neutrophil-mediated killing of hyphae. The Th2 response, which is associated with an increase in antibody production, seems to facilitate fungal invasion rather than protection.

- ABPA pathogenesis is associated with elevated levels of antigen-specific circulating IgG and IgE. B cells secrete IgE spontaneously as a result of IL-4 production, while IL-5 recruits eosinophils. Development of type I and type II hypersensitivity leads to inflammation and tissue damage.

- Patients with aspergilloma have increased levels of specific IgG and IgM, mostly against *Aspergillus* carbohydrates and glycoproteins. The protective role of specific antibodies to *Aspergillus* remains unclear.

3. What is the typical clinical presentation and what complications can occur?

- Pulmonary diseases caused by *A. fumigatus* are classified into four main clinical types: allergic bronchopulmonary aspergillosis, chronic necrotizing *Aspergillus* pneumonia, invasive aspergillosis, and pulmonary aspergilloma.

- Allergic bronchopulmonary aspergillosis (ABPA) is the result of a hypersensitivity reaction to *A. fumigatus* colonization of the tracheobronchial tree. It often appears as a complication of other chronic lung diseases such as atopic asthma (0.5–2%), cystic fibrosis (7–35%), and sinusitis. ABPA occurs in up to15% of asthmatic patients sensitized to *A. fumigatus*.

- ABPA symptoms are similar to asthma and include wheezing, cough, fever, malaise, and weight loss. Additional symptoms include recurrent pneumonia, release of brownish mucoid plugs with fungal hyphae, and recurrent lung obstruction. In secondary ABPA there is unexplained worsening of asthma and cystic fibrosis

- Chronic obstructive pulmonary disease (COPD) presents as a subacute pneumonia that is unresponsive to antibiotic therapy. Symptoms include fever, cough, night sweats, and weight loss. It develops mostly in mildly immunocompromised patients and is commonly associated with underlying lung disease.

- Invasive aspergillosis (IA) is the most serious form of aspergillosis and normally occurs in immunocompromised individuals. There are four types of IA: acute or chronic pulmonary

aspergillosis (lungs); tracheobronchitis and obstructive bronchial disease (bronchial mucosa and cartilage); acute invasive rhinosinusitis (sinuses); and disseminated disease (brain, skin, kidneys, heart, eyes).

- IA symptoms are variable and nonspecific: fever and chills, weakness, unexplained weight loss, chest pain, shortness of breath, headaches, bone pain, a heart murmur, blood in the urine, decreased diuresis, and straight, narrow red lines of broken blood vessels under the nails. IA is accompanied by increased sputum production (sometimes with hemoptysis), sinusitis, and acute inflammation with ischemic necrosis, thrombosis, and infarction of the organs.

- Aspergilloma occurs in 10–15% of patients with cavitating lung diseases such as tuberculosis, sarcoidosis, lung abscess, emphysematous bullae, cystic fibrosis, and paranasal sinuses. It is often discovered incidentally by chest X-ray or by computed tomography (CT) scans. The most common symptom is hemoptysis.

4. How is the disease diagnosed, and what is the differential diagnosis?

- ABPA diagnostic criteria include: asthma, a history of pulmonary infiltrates, and central bronchiectasis supplemented by the laboratory tests: peripheral blood eosinophilia, immediate skin reactivity to *A. fumigatus*, precipitating IgG and IgM, elevated levels of total IgE in serum and specific IgE against *A. fumigatus* measured by IgE RAST.

- A safe diagnosis of IA can be only provided at autopsy showing evidence of mycelial growth in tissue. Differential diagnosis from the invasion of hyphae of other filamentous fungi such as *Fusarium* or *Pseudallescheria* requires immunohistochemical staining or *in situ* hybridization techniques. Diagnosis of IA: a positive CT scan, culture and/or microscopic evaluation, and detection of *Aspergillus* antigens in serum.

- *A. fumigatus* cell culture may be critical in the diagnosis of aspergillosis although it takes a long time.

- Since IA is life-threatening, early diagnosis using ELISA and PCR-based techniques is essential. Reliable antigens of *A. fumigatus* include RNase, catalase, dipeptidylpeptidase V, and the galactomannan (GM). GM detection by ELISA has become the basis for the most popular *A. fumigatus* diagnostic tests.

- A definitive diagnosis of aspergilloma requires bronchoscopy, lung biopsy or resection. The diagnosis is usually made accidentally or specifically by chest radiography. Clinical analysis should be coupled with serologic tests. Positive sputum cultures are found in > 50% of patients, but are common for many other aspergillosis conditions.

5. How is the disease managed and prevented?

- The aim of the treatment of ABPA is to suppress the immune reaction to the fungus and to control bronchospasm. High doses of oral corticosteroids are used. Antifungal drugs such as itraconazole are sometimes used. Regular monitoring by X-rays, pulmonary function tests, and serum IgE levels is essential.

- Since it is difficult to achieve timely diagnosis of IA the decision to start antifungal treatment has to be empirical and based on the presence of risk factors, particularly prolonged neutropenia. Antifungal treatments include voriconazole (particularly effective), amphotericin B (AmB), and itraconazole, which exhibit a broad-spectrum activity against *Aspergillus* and the related hyaline molds. Itraconazole is prescribed to patients with AmB-induced nephrotoxicity.

- Aspergilloma can be prevented by the treatment of diseases that increase its risk, such as tuberculosis. In the absence of surgical intervention the course of the disease remains unpredictable. Itraconazole for 6–18 months is recommended for the oral treatment.

FURTHER READING

Barnes PD, Marr KA. Aspergillosis: spectrum of disease, diagnosis and treatment. Infect Dis Clin N Am, 2006, 20: 545–561.

Bennett JE. *Aspergillus* species. In GL Mandell, JE Bennett, R Dolin (eds), Mandell, Douglas and Bennett's Principles and Practice of Infectious Diseases, 4th edition. Churchill Livingstone, New York, 1995: 2306–2310.

REFERENCES

Ascioglu S, Rex JH, de Pauw B, et al. Defining opportunistic invasive fungal infections in immunocompromised patients with cancer and hematopoietic stem cell transplants: an international consensus. Clin Infect Dis, 2002, 34: 7–14.

Boutboul F, Alberti C, Leblanc T, et al. Invasive aspergillosis in allogeneic stem cell transplant recipients: increasing antigenemia is associated with progressive disease. Clin Infect Dis, 2002, 34: 939–943.

Brookman JL, Denning DW. Molecular genetics in *Aspergillus fumigatus*. Curr Opin Microbiol, 2000, 3: 468–474.

Buzina W, Braun H, Freudenschuss K, Lackner A, Habermann W, Stammberger H. Fungal biodiversity – as found in nasal mucus. Med Mycol, 2003, 41: 149–161.

Casadevall A, Feldmesser M, Pirofski L-A. Induced humoral immunity and vaccination against major human fungal pathogens. Curr Opin Microbiol, 2002, 5: 386–391.

Denning DW, Stevens DA. Antifungal and surgical treatment of invasive aspergillosis: review of 2,121 published cases. Rev Infect Dis, 1990, 12: 1147–1201.

REFERENCES

Fujimura M, Ishiura Y, Kasahara K, et al. Necrotizing bronchial aspergillosis as a cause of hemoptysis in sarcoidosis. Am J Med Sci, 1998, 315: 56–58.

Gerson SL, Talbot GH, Hurwitz S, Strom BL, Lusk EJ, Cassileth PA. Prolonged granulocytopenia: the major risk factor for invasive pulmonary aspergillosis in patients with acute leukemia. Ann Intern Med, 1984, 100: 345–351.

Glimp RA, Bayer AS. Pulmonary aspergilloma. Diagnostic and therapeutic considerations. Arch Intern Med, 1983, 143: 303–308.

Herbrecht R, Letscher-Bru V, Oprea C, et al. Aspergillus galactomannan detection in the diagnosis of invasive aspergillosis in cancer patients. J Clin Oncol, 2002, 20: 1898–1906.

Hohl TM, Feldmesser M. *Aspergillus fumigatus*: principles of pathogenesis and host defense. Eukaryot Cell, 2007, 6: 1953–1963.

Maertens J, Van Eldere J, Verhaegen J, Verbeken E, Verschakelen J, Boogaerts M. Use of circulating galactomannan screening for early diagnosis of invasive aspergillosis in allogeneic stem cell transplant recipients. J Infect Dis, 2002, 186: 1297–1306.

Maertens J, Verhaegen J, Lagrou K, Van Eldere J, Boogaerts M. Screening for circulating galactomannan as a noninvasive diagnostic tool for invasive aspergillosis in prolonged neutropenic patients and stem cell transplantation recipients: a prospective validation. Blood, 2001, 97: 1604–1610.

Morrison VA, Haake RJ, Weisdorf DJ. The spectrum of non-Candida fungal infections following bone marrow transplantation. Medicine (Baltimore), 1993, 72: 78–89.

Moss RB. Pathophysiology and immunology of allergic bronchopulmonary aspergillosis. Med Mycol, 2005, 43: S203–S206.

Ponikau JU, Sherris DA, Kern EB, et al. The diagnosis and incidence of allergic fungal sinusitis. Mayo Clin Proc, 1999, 74: 877–884.

Salonen J, Lehtonen OP, Terasjarvi MR, Nikoskelainen J. Aspergillus antigen in serum, urine and bronchoalveolar lavage specimens of neutropenic patients in relation to clinical outcome. Scand J Infect Dis, 2000, 32: 485–490.

Segal BH. Role of macrophages in host defence against aspergillosis and strategies for immune augmentation. Oncologist, 2007, 12: 7–13.

Severo LC, Geyer GR, Porto NS. Pulmonary *Aspergillus* intracavitary colonization (PAIC). Mycopathologia, 1990, 112: 93–104.

Shoham S, Levitz SM. The immune response to fungal infections. Br J Haematol, 2005, 129: 569–582.

Stevens DA, Kan VL, Judson MA, et al. Practice guidelines for diseases caused by *Aspergillus*. Clin Infect Dis, 2000, 30: 696–709.

Stevens DA, Schwartz HJ, Lee JY, et al. A randomized trial of itraconazole in allergic bronchopulmonary aspergillosis. N Engl J Med, 2000, 342: 756–762.

Swanink CMA, Meis JFGM, Rijs AJMM, Donnelly JP, Verweij PE. Specificity of a sandwich enzyme-linked immunosorbent assay for detecting Aspergillus galactomannan. J Clin Microbiol, 1997, 35: 257–260.

Tekaia F, Latge J-P. *Aspergillus fumigatus*: saprophyte or pathogen? Curr Opin Microbiol, 2005, 8: 385–392.

Verweij PE, Latge J-P, Rijs AJMM, et al. Comparison of antigen detection and PCR assay using bronchoalveolar lavage fluid for diagnosing invasive pulmonary aspergillosis in patients receiving treatment for hematological malignancies. J Clin Microbiol, 1995, 33: 3150–3153.

Viscoli C, Machetti M, Gazzola P, et al. Aspergillus galactomannan antigen in the cerebrospinal fluid of bone marrow transplant recipients with probable cerebral aspergillosis. J Clin Microbiol, 2002, 40: 1496–1499.

Walsh TJ, Anaissie EJ, Denning DW, et al. Treatment of aspergillosis: clinical practice guidelines of the Infectious Diseases Society of America. Clin Infect Dis, 2008, 46: 327–360.

WEB SITES

The Aspergillus Website, Fungal Research Trust, Copyright © 2007, The Fungal Research Trust. All rights reserved: http://www.aspergillus.org.uk

Utah State University Intermountain Herbarium: http://herbarium.usu.edu/

MedLine Plus Medical Encyclopedia, entry for Aspergillosis:

http://www.nlm.nih.gov/medlineplus/ency/imagepages/2330.htm

Mold-Help.org – non-profit organization specializing in the study of molds and their effect on human health and environments © Mold-Help.org 2003. All rights reserved: http://www.mold-help.org/index.php

MULTIPLE CHOICE QUESTIONS

The questions should be answered either by selecting True (T) or False (F) for each answer statement, or by selecting the answer statements which best answer the question. Answers can be found in the back of the book.

1. **Which of the following are characteristics of *A. fumigatus*?**

    A. *A. fumigatus* can only grow at low temperatures.

    B. Its natural habitat is soil.

    C. *A. fumigatus* produces large spores.

    D. *Aspergillus* is a filamentous fungus with branching hyphae.

    E. Wounds are the main port of entry of *A. fumigatus*.

2. **Which of the following are risk factors for the development of IA?**

    A. Aggressive chemotherapy of leukemia and lymphoma patients.

    B. Intensive and prolonged treatment with steroids.

    C. Bone marrow transplantation.

    D. Alzheimer's disease.

    E. Graft-versus-host reaction.

3. **Which of the following are the most frequent clinical presentations of ABPA?**

    A. Severe joint pain.

    B. Wheezing, cough, fever, malaise, weight loss.

    C. Skin rash.

    D. High levels of serum IgE.

    E. Recurrent pneumonia.

4. **What are the normal immune responses to *A. fumigatus* infection?**

    A. Innate immunity plays a leading role in the disposal of this fungus.

    B. Neutrophils are more efficient in killing of conidial forms while resident macrophages dispose of the hyphal stage of *A. fumigatus*.

    C. Killing of the fungal cells by CD8+ cytotoxic T cells is essential.

    D. Specific antibodies play an important protective role against *A. fumigatus*.

    E. Type I pro-inflammatory cytokines help to kill this pathogen.

5. **Which of the following are true of aspergilloma?**

    A. It can be reliably diagnosed only by CT scan or radiography.

    B. Aspergilloma pathogenesis is based on acute immune inflammation.

    C. Aspergilloma mostly develops in pre-existing lung cavities.

    D. There could be multiple aspergillomas.

    E. It requires intensive antifungal therapy.

6. **Which of the following are true of ABPA?**

    A. ABPA patients have specific symptoms and the condition is easy to diagnose.

    B. Secondary ABPA is particularly frequent in patients with Type I diabetes and SLE.

    C. ABPA pathogenesis is based on type II hypersensitivity reactions.

    D. It is associated with blood eosinophilia.

    E. It presents as bronchial asthma.

7. **Which of the following tests are used for the diagnosis and monitoring of invasive aspergillosis (IA)?**

    A. Microscopy of sputum.

    B. Sputum or bronchoalveolar lavage cell cultures.

    C. Neutrophil cell counts.

    D. CT scan.

    E. ELISA.

8. **Which of the following drugs are used in the treatment of invasive aspergillosis (IA)?**

    A. Amphotericin B.

    B. Erythromycin.

    C. Itraconazole.

    D. Voriconazole.

    E. Tetracycline.

Case 2

Borrelia burgdorferi and related species

A 45-year-old woman was on vacation in Cape Cod, Massachusetts and decided to attend an outdoor music festival. In order to get there she walked 3 miles each way at night, through a dark, wooded, grassy area. Shortly thereafter she developed nonspecific symptoms that included fever, headache, muscle aches, mild neck stiffness, and joint pain. She also noticed an oval 'bull's eye' **rash** on her right arm, which got larger and cleared in the center. Over the ensuing month she felt increasingly fatigued and developed facial paralysis (**Bell's palsy**) (Figure 1), which precipitated a visit to her family physician. She couldn't remember being bitten by a tick but based on her description of the rash and her other symptoms her doctor suspected Lyme disease and took a blood sample for **serology. Enzyme immunoassay** and **Western blot** confirmed the presence of *Borrelia burgdorferi*-reactive antibodies. After confirming that the patient was not pregnant she was prescribed doxycycline 100 mg twice daily for 30 days.

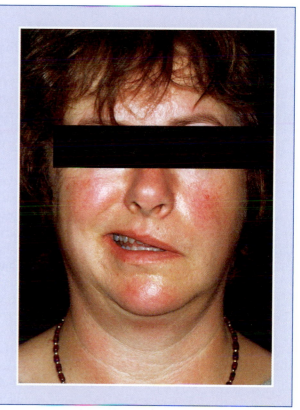

Figure 1. Bell's palsy: this is demonstrated by drooping at the left corner of the mouth, loss of the left naso-labial fold, and inability to completely close the left eye (not shown in image). Reprint permission kindly granted by Dr. Charles Goldberg, MD, and Regents of the University of California.

1. What is the causative agent, how does it enter the body and how does it spread a) within the body and b) from person to person?

Causative agent

The patient has Lyme disease. The causative agent of Lyme disease is *Borrelia burgdorferi sensu lato* (meaning in the broad sense). There are at least 11 genospecies within the *Borrelia burgdorferi* complex (*B. burgdorferi sensu lato*) worldwide. The three main pathogenic genospecies comprising this group are *B. burgdorferi sensu stricto*, *B. garinii*, and *B. afzelii*. Strains found in North America belong to *B. burgdorferi sensu stricto* whereas all three species are found in Europe and Asia. Borreliae are microaerophilic **spirochetes** (Figure 2) that are extremely difficult to culture because of their complex nutrient requirements. Thus, they are usually detected by the immune response that they induce in blood of the infected person (see above and Section 4). The bacteria have a gram-negative wall structure

Figure 2. *Borrelia burgdorferi* **is a spirochete: it is 0.2–0.3 micrometers (μm) wide and its length may exceed 15–20 μm.**

and have a spiral mode of motility produced by axial filaments termed endoflagella. In contrast to the usual type of flagella exhibited by gram-negative bacteria that are anchored in the cytoplasmic membrane and extend through the cell wall into the external environment of the cell, the endoflagella of borreliae are found within the periplasmic space contained between a semi-rigid peptidoglycan layer and a multi-layer, flexible outer membrane sheath. Rotation of the endoflagella within the periplasmic space causes the borreliae to move in a cork-screw fashion (Figure 2). In addition, *Borrelia* species, instead of having circular chromosomes, have linear chromosomes and contain circular and linear plasmids, with some species containing more than 20 different plasmids.

Ticks – the vectors of B. burgdorferi

The bacteria are maintained in an enzootic cycle involving hard-bodied ticks belonging to the *Ixodes ricinus* species complex and a wide range of reservoir vertebrate hosts. The global distribution of *Ixodes* species is shown in Figure 3. In the eastern United States the vector is primarily *Ixodes scapularis* and in the western US it is *I. pacificus*. *I. ricinus* and *I. persulcatus* are the vectors in Europe and Eurasia, respectively. These ticks have a 2-year life cycle (Figure 4). Adult ticks feed and mate on large animals, especially white-tailed deer, in the autumn and early spring. However, white-tailed deer are not considered reservoirs of *B. burgdorferi* because they do not support a sufficiently high level of spirochetes in their blood to infect ticks. Nevertheless, deer are important in tick reproduction and serve to increase tick numbers in an area and spread ticks into new areas. Female ticks then drop off the animals and lay eggs on the ground. By summer, the eggs hatch into larvae, which feed on mice and other small mammals and birds through to early autumn; then they become inactive until the following spring when they molt into nymphs. Nymphs feed on small rodents and other small mammals and birds during the late spring and summer and molt into adults in the autumn, completing the 2-year life cycle. Larvae and nymphs typically become infected with borreliae when they feed on infected small animals, particularly the white-footed mouse. The tick remains infected with the borreliae as it matures from larva to nymph or from nymph to adult. Infected nymphs and adult ticks then bite and transmit the bacteria to other small rodents, other animals, or humans in the course of their normal feeding behavior. The ticks are slow feeders,

Figure 3. The global distribution of *Ixodes* spp. ticks able to transmit the agent of Lyme disease, *Borrelia burgdorferi*. Modified from http://geo.arc.nasa.gov/sge/health/sensor/disease/lyme.html

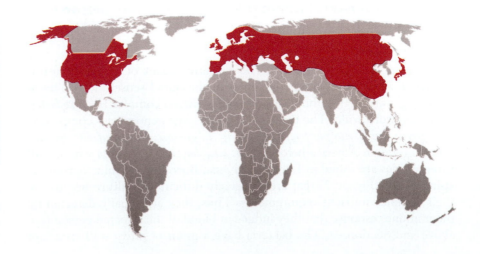

Figure 4. Tick life cycle.

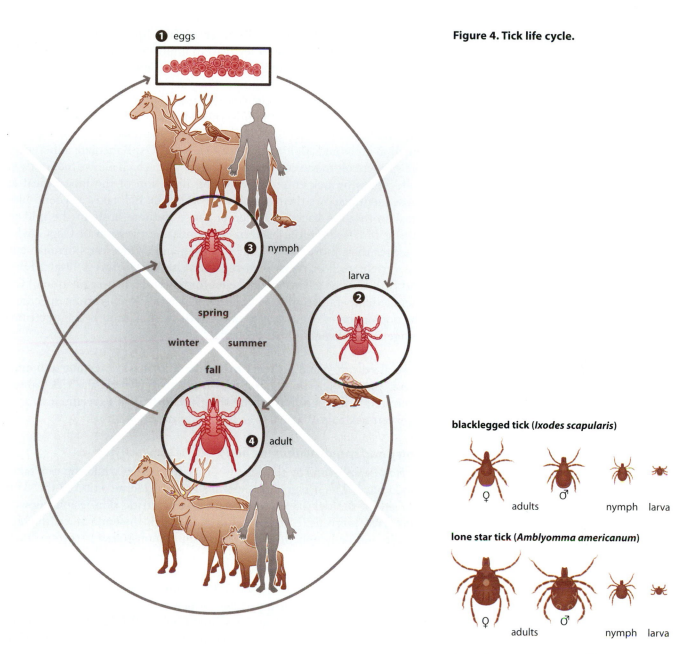

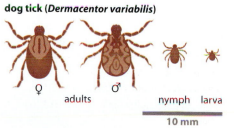

blacklegged tick (*Ixodes scapularis*)

lone star tick (*Amblyomma americanum*)

dog tick (*Dermacentor variabilis*)

10 mm

Figure 5. Appearance and relative sizes of adult male and female, nymph, and larval ticks including deer ticks (*Ixodes scapularis*), lone star ticks (*Amblyomma americanum*), and dog ticks (*Dermacentor variabilis*). Of those pictured, only the *I. scapularis* ticks are known to transmit Lyme disease.

requiring several days to complete a blood meal. Transfer of the borreliae from the infected tick to a vertebrate host probably does not occur unless the tick has been attached to the body for 36 hours or so.

Ticks transmit Lyme disease to humans generally during the nymph stage, probably because nymphs are more likely to feed on a person and are rarely noticed because of their small size (< 2 mm) (Figure 5). Although tick larvae are smaller than nymphs they rarely carry borreliae at the time of feeding and are probably not important in the transmission of Lyme disease to humans. While adult ticks can transmit borreliae they are less likely to do so than nymphs. This is because their larger size means that they are more likely to be noticed and removed from a person's body within a few hours and they are most active during the cooler months of the year, when outdoor activity is limited. It should be noted that dogs, horses, cattle, deer, and other animals are also susceptible to Lyme disease.

Borreliae express a number of outer surface lipoproteins termed Osps (outer surface proteins) and they play an important role in the life cycle of the spirochete by interacting with intercellular and cellular components of its arthropod and vertebrate hosts. They are also important in the evasion of the host immune system (see Section 3). *B. burgdorferi* selectively expresses specific Osps during distinct phases of its life cycle and in specific tissue locations. OspA and OspB are expressed on spirochetes in unfed nymphs and adult ticks. Both OspA and OspB mediate adherence of the spirochetes to the cells of the tick mid-gut, which allows them to avoid endocytosis by tick enterocytes during digestion of the blood meal and subsequently allows their detachment when the tick takes a second blood meal so that the bacteria can enter the vertebrate host. In the mid-gut, during tick feeding, the bacteria up-regulate expression of OspC, which presages their move toward the salivary glands. Once the borreliae enter the vertebrate host they down-regulate OspA and express OspC, DbpA, and BBK32. The environmental cues for up- and down-regulation of Osps include temperature and pH. The sequence of OspC is strain-specific, so the population of *B. burgdorferi* injected by the tick expresses a spectrum of antigenically distinct OspC proteins.

Several salivary gland proteins are induced during tick feeding and one of these, Salp15, has immunosuppressive activity where the tick saliva is deposited in the skin. It has been shown that there is a specific interaction between tick Salp15 and OspC, both *in vitro* and *in vivo*.

Entry and spread within the body

B. burgdorferi sensu lato spirochetes enter the tissues while the tick takes a blood meal. The bacteria may establish a localized infection in the skin at the site of the tick bite, producing a characteristic skin lesion known as **erythema migrans** (EM). In addition, they may disseminate via the blood stream and/or lymphatics. The organism demonstrates a tropism for the central nervous system, heart, joints, and eyes, all of which may become chronically infected, giving rise to neurological disease, **carditis**, **arthritis**, and **conjunctivitis** (see Section 3). Even in the absence of systemic symptoms it appears that as many as half of persons with EM have evidence of borreliae in the blood or cerebrospinal fluid (CSF). Furthermore, borreliae can also persist in skin and perhaps the central nervous system (CNS) for years without causing symptoms.

Person to person spread

Lyme disease is not spread from person to person. It is possible in a woman who contracts Lyme disease during pregnancy for the borreliae to cross the placenta leading to infection of the fetus, but this occurs rarely. The consequence of fetal infection remains unclear. For this reason the Centers for Disease Control and Prevention (CDC) maintains a registry of pregnant women with Lyme disease to accrue data on the effects of Lyme disease on the developing fetus.

Epidemiology

Lyme disease is the most common tick-borne disease in North America and Europe. Although Lyme disease has now been reported in 49 of 50 states in the US, almost all reported cases are confined to New England (Connecticut, Maine, Massachusetts, New Hampshire), the Mid-Atlantic region (Delaware, Maryland, New Jersey, Pennsylvania), the East-North Central region (Wisconsin), and the West North-Central region (Minnesota). In 2005 the overall incidence of Lyme disease in the United States was 7.9 cases per 100 000 persons. However, in the 10 states where Lyme disease is most common (see above), the incidence was 31.6 cases per 100 000 persons. Although Lyme disease is common in the United States and Scandinavia and has been reported in other countries in Western and Eastern Europe, Japan, China, and Australia, it is not a common disease in the UK, with less than 200 cases per year being reported in England and Wales in recent years. The incidence of Lyme disease in Europe is shown in Table 1.

Table 1. Reported cases or estimated cases and incidence by European country

Year Country	2001 Incidence [cases]	2002 Incidence [cases]	2003 Incidence [cases]	2004 Incidence [cases]	2005 Incidence [cases]
Slovenia	163 [3232]	169 [3359]	177 [3524]	193 [3849]	206 [4123]
Austria (estimate***)	– [–]	– [–]	– [–]	– [–]	135 [–]
Netherlands (estimate)*	74 [12000]	– [–]	– [–]	– [–]	103 [17000]
Czech Republic	35 [3547]	36 [3658]	36 [3677]	32 [3243]	36 [3640]
Lithuania	33 [1153]	26 [894]	106 [3688]	50 [1740]	34 [1161]
Länder of Former East Germany	– [–]	18 [3029]	24 [3991]	26 [4497]	– [–]
Finland	13 [691]	17 [884]	14 [753]	22 [1135]	24 [1236]
Latvia	16 [379]	14 [328]	31 [714]	31 [710]	21 [493]
Estonia	25 [342]	23 [319]	42 [562]	36 [480]	21 [281]
Slovakia	13 [675]	11 [568]	14 [726]	13 [677]	16 [843]
Belgium	9.7 [997]	12 [1269]	11 [1118]	16 [1607]	16 [1644]
Bulgaria	4.5 [364]	6.5 [514]	7 [550]	12 [949]	13 [979]
Poland	6.4 [2473]	5.3 [2034]	9.4 [3575]	10 [3822]	12 [4406]
Norway	2.7 [125]	2.4 [111]	3.2 [144]	5.5 [251]	6 [280]
Hungary	13 [1258]	12 [1238]	12 [1208]	12 [1208]	– [–]
Britain : • England and Wales** • Scotland	0.5 [268] 0.6 [28]	0.6 [340] 1.7 [85]	0.6 [335] 1.6 [81]	0.9 [500] 1.7 [86]	1.1 [595] 1.9 [96]
Italy	0.02 [14]	0.05 [29]	0 [0]	0.02 [10]	0.001 [4]
Portugal	0.03 [3]	0.02 [2]	0.01 [1]	0.01 [1]	0.04 [4]

Source Eurosurveillance Editorial Advisors and others.

Methods used to acquire data vary in different European countries. Incidence is the number of new cases per 100 000 population per year

*estimated number of erythema migrans case-patients

**voluntary reporting

***estimate based on physician survey

Taken from http://www.eurosurveillance.org/ew/2006/060622.asp

2. What is the host response to the infection and what is the disease pathogenesis?

Because the spirochete is delivered to the host via the bite of a tick it bypasses the physical barrier of the intact skin and the antimicrobial factors in sweat.

Innate immunity

Components of the innate immune system are brought into play to combat the spirochetes at the site of their delivery by the interaction of **pathogen-associated molecular patterns** (**PAMPs**) on the surface of the borreliae with **Toll-like receptors** (**TLRs**) on epithelium and leukocytes. *B. burgdorferi* activates both the innate and classical pathways of the complement cascade but is resistant to complement-mediated lysis because OspE and other proteins on its surface bind the complement control glycoprotein, factor H, which inactivates complement factor C3b. The **membrane attack complex** (**MAC**) can also be inactivated in a similar manner. Resident macrophages in the area of the inoculum are able to bind, phagocytose, and kill borreliae without the need for **opsonization** by complement or antibody. Binding may be mediated by the mannose-binding receptor and/or the Mac-1 receptor. Polymorphonuclear leukocytes (PMNs) can also kill borreliae without opsonization.

Adaptive immunity

There is little doubt that **IgM** and **IgG** antibodies play the principal role in the clearance of *B. burgdorferi*. Furthermore, these antibodies do not need to be complement fixing. Murine IgG and IgM **monoclonal antibodies** have been developed that are bactericidal for borreliae in the absence of complement. The finding that mice deficient in α/β T cells or deficient in α/β- and γ/δ T cells can clear spirochetemia indicates that T cells are not required for spirochete clearance. Lyme arthritis appears to result from a constellation of factors that includes production of pro-inflammatory **cytokines** and **immune complexes**, and arthritis is linked to HLA-DR4 and HLA-DR2.

How do the borreliae evade these host defense mechanisms? First of all as mentioned earlier, Salp15 salivary gland protein induced during tick feeding has immunosuppressive activity where the tick saliva containing the borreliae is deposited in the skin. Salp15 inhibits the IgG antibody response by blocking **CD4+** T-cell activation and so may protect *B. burgdorferi* by suppressing production of neutralizing antibodies. This mechanism may be particularly important in Lyme-endemic areas where infected ticks frequently feed on their primary hosts that may possess pre-existing antibodies against *B. burgdorferi*.

Furthermore, borreliae can stimulate **interleukin**-10 (IL-10) production by macrophages, mast cells, and **Th2** CD4+ T cells. IL-10 inhibits synthesis of pro-inflammatory cytokines and suppresses antigen presentation to CD4+ helper T cells by antigen-presenting cells.

B. burgdorferi undergoes **antigenic variation** and modulates the expression of Osps on the cell surface during infection. VlsE is an example of an Osp that undergoes antigenic variation. The vls locus of *B. burgdorferi* is located on a 28-kb linear plasmid and consists of an expression site (vlsE) and 15 silent vls cassettes. Another candidate may be OspE, since the gene for this Osp has two hypervariable domains and repeat regions that allow recombination with other genes, which may result in the formation of new antigens.

Pathogenesis

The tissue injury in Lyme disease is mediated by inflammation induced by *B. burgdorferi*. The manner in which the bacterium induces inflammation in the host is not fully understood. Spirochetemia results in the invasion of tissues such as the heart and joints and the host reponds with a vigorous inflammatory response. The role of *B. burgdorferi* PAMPs–TLR interactions in the induction of pro-inflammatory cytokines has been questioned by results of knockout experiments, which indicate that TLR-2-deficient mice or mice deficient in TLR signaling molecules develop arthritis and have a much larger burden of *B. burgdorferi*. It has been suggested that **chemokines** produced at the site of infection may be more important in the influx of inflammatory cells to the site of infection

3. What is the typical clinical presentation and what complications can occur?

Whether the disease develops following *B. burgdorferi* infection depends on the balance between the pathogen and the host's immune response. There are three potential outcomes of the borrelia–host interaction. The spirochete may be cleared without any manifestations of disease, the only indicator of infection being that the individual is **seropositive**. Alternatively, the spirochete establishes in the skin and after a variable incubation period ranging from a few days to a month produces a characteristic spreading rash termed erythema migrans. The rash begins as a small **macule** (a visible change in the color of the skin that cannot be felt) or **papule** (a small, solid and usually conical elevation of the skin), which then expands, ranging in diameter to between 5 and 50 cm (Figure 6). The rash has a flat border and central clearing so that it resembles a 'bull's-eye.' Erythema migrans is probably caused by the inflammatory response to the spirochete infection. From the initial focus of infection in the skin the spirochete spreads throughout the body. Systemic spread of the spirochete results in malaise, headaches, chills, joint pain, **myalgia**, **lymphadenopathy**, and severe fatigue. This phase may last for up to a month. Unless treated, over two-thirds of infected individuals manifest neurological and cardiac symptoms. These manifestations may occur as early as a month or as late as 2 years or more post-infection. Neurological sequelae include **meningitis**, **encephalitis**, and **peripheral nerve neuropathy**, particularly seventh cranial nerve palsy (Bell's palsy). Cardiac sequelae include heart block, **myopericarditis**, and congestive heart failure. Neurological and cardiac sequelae may be followed by **arthralgia** and arthritis. About two-thirds of patients with untreated infection will experience intermittent bouts of arthritis, with severe joint pain and swelling. Large joints are most often affected, particularly the knees. These manifestations may last for months to years with little evidence of bacterial invasion. The manifestations of Lyme disease are related to the particular genospecies of *Borrelia* involved. In Europe *B. garinii* is associated with neurologic disease, while *B. afzelii* is associated with a dermatologic manifestation known as **acrodermatitis chronica atrophicans**, a progressive fibrosing skin process.

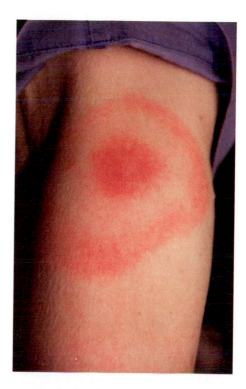

Figure 6. This 2007 photograph depicts the pathognomonic erythematous rash in the pattern of a 'bull's-eye,' which manifested at the site of a tick bite on this Maryland woman's posterior right upper arm. She subsequently contracted Lyme disease. Lyme disease patients who are diagnosed early and receive proper antibiotic treatment usually recover rapidly and completely. A key component of early diagnosis is recognition of the characteristic Lyme disease rash called *erythema migrans*. This rash often manifests itself in a 'bull's-eye' appearance, and is observed in about 80% of Lyme disease patients.

The existence of an entity termed 'chronic Lyme disease' is controversial and is the subject of a recent critical appraisal by the Ad Hoc International Lyme Disease Group comprising clinicians and microbiologists from North America and Europe (see References section). The term chronic Lyme disease is used in North America and in Europe as a diagnosis for patients with persistent pain, neurocognitive symptoms, fatigue, either separately or together, with or without clinical or serologic evidence of previous early or late Lyme disease. The conclusions of the International Lyme Disease Group are that '... the assumption that chronic, subjective symptoms are caused by persistent infection with *B. burgdorferi* is not supported by carefully conducted laboratory studies or by controlled treatment trials. Chronic Lyme disease, which is equated with chronic *B. burgdorferi* infection, is a misnomer, and the use of prolonged, dangerous, and expensive antibiotic treatments for it is not warranted.'

However, persistent joint swelling lasting as long as several years is seen in about 10% of adult patients with Lyme arthritis following appropriate antibiotic therapy. The inability to detect borreliae in joint aspirates or tissues using **polymerase chain reaction (PCR)** has led to a proposed autoimmune etiology.

Other diseases transmitted by hard-bodied ticks

Hard-bodied ticks belonging to the *I. ricinus* species complex are also vectors for the infectious agents *Ehrlichia phagocytophila*, *Babesia microti*, and tick-borne encephalitis (TBE) virus. *E. phagocytophila* is a small intracellular gram-negative coccobacillus that parasitizes neutrophils (human granulocytic ehrlichiosis – HGE). Once taken up into phagosomes the bacteria prevent fusion with lysosomes and replicate forming membrane-enclosed masses termed morulae. Disease presents as a flu-like illness with **leukopenia** and **thrombocytopenia**. Most infected individuals require hospitalization and severe complications are not infrequent. *B. microti* is an intracellular sporozoan parasite that causes babesiosis. The infectious stage, termed pyriform (pear-shaped) bodies, enter the bloodstream and infect erythrocytes. Within the erythrocyte trophozoites replicate by binary fission forming tetrads. The erythrocytes lyse releasing merozoites, which may infect new red blood cells. Disease presents as a flu-like illness leading to **hemolytic anemia** and renal failure. **Hepatomegaly** and **splenomegaly** may be observed in advanced disease. TBE virus is a member of the Flaviviridae. The disease spectrum ranges from a mild **febrile** illness to meningitis, encephalitis or **meningoencephalitis**. Chronic or permanent neuropsychiatric sequelae are observed in as many as 20% of infected patients.

4. How is the disease diagnosed, and what is the differential diagnosis?

In patients with signs and symptoms consistent with Lyme disease a diagnosis is confirmed by antibody detection tests. The current recommendation from the US Centers for Disease Control and Prevention (CDC) (http://www.cdc.gov/mmwr/preview/ mmwrhtml/00038469.htm) is for a two-test approach consisting of a sensitive enzyme immunoassay (EIA) or **immunofluorescence assay (IFA)** followed by a Western immunoblot. Similar tests are used worldwide. All specimens positive or equivocal by the EIA or IFA should be tested by a standardized Western immunoblot. Specimens negative by the EIA or IFA do not require further testing unless clinically indicated. The EIA or IFA can be performed either as a total Lyme titer or as separate IgG and IgM titers. If a Western immunoblot is performed during the first 4 weeks of disease onset both IgM and IgG immunoblots should be performed. A positive IgM test result alone is not recommended for use in determining active disease in persons with illness of greater than 1 month's duration because the likelihood of a false-positive test result for a current infection is high for these persons. If a patient with suspected early Lyme disease has a negative serology, serologic evidence of infection is best obtained by testing paired acute- and convalescent-phase serum samples. Serum samples from persons with disseminated or late-stage Lyme disease almost always have a strong IgG response to *B. burgdorferi* antigens. For an IgM immunoblot to be considered positive two of the following three bands must be present: 24 or 21 kDa (OspC) (the apparent molecular mass of OspC is dependent on the strain of *B. burgdorferi* being tested), 39 kDa (BmpA), 41 kDa (Fla). For an IgG immunoblot to be considered positive five of the following 10 bands must be present: 18 kDa, 21 or 24 kDa (OspC, see above), 28 kDa, 30 kDa, 39 kDa (BmpA), 41 kDa (Fla), 45 kDa, 58 kDa (not GroEL), 66 kDa, and 93 kDa. The different genospecies of *B. burgdorferi sensu lato* in Europe make the serological diagnosis of Lyme disease more complex and the standardization of tests, particularly the Western blot, more difficult.

Differential diagnosis

The following conditions should be considered in the differential diagnosis: calcium pyrophosphate deposition disease, **fibromyalgia**, gonococcal arthritis, **gout**, meningitis, **psoriatic arthritis**, **rheumatoid arthritis**, **syncope**, **systemic lupus erythematosus (SLE)**, and **urticaria**.

5. How is the disease managed and prevented?

Management

Routine use of antibiotics or serological testing after a tick bite is not recommended. Persons who remove attached ticks (see below) should be monitored for up to 1 month for signs and symptoms of Lyme and other tick-borne diseases. Persons developing erythema migrans or other illness should seek medical attention.

The recommended antimicrobial regimens and therapy for patients with Lyme disease are shown in Tables 2 and 3. More information on antibiotic therapy can be obtained from the Infectious Diseases Society of America guidelines for treatment of Lyme Disease (see References section).

Prevention

The best method of preventing Lyme disease is to avoid tick-infested areas. If this is not feasible then the following are recommended:

- wear light-colored clothing so that ticks can be more easily spotted;
- tuck trouser cuffs into socks or boots and tuck shirts into trousers so that ticks cannot crawl under clothing;
- wear a long-sleeved shirt and hat;
- use DEET (*meta*-N,N-diethyl toluamide) tick repellent on skin;

- spray clothing and boots with permethrin;
- check the entire body for ticks every day;
- remove any attached ticks as soon as possible since it takes about 36 hours of attachment before borreliae are transmitted

Table 2. Recommended antimicrobial regimens for treatment of patients with Lyme disease

Drug	Dosage for adults	Dosage for children
Preferred Oral Regimens:		
Amoxicillin	500 mg three times per day	50 mg kg^{-1} per day in three divided doses (maximum, 500 mg per dose)
Doxycycline	100 mg twice per day	Not recommended for children aged < 8 years
		For children aged ≥ 8 years, 4 mg kg^{-1} per day in two divided doses (maximum, 100 mg per dose)
Cefuroxime axetil	500 mg twice per day	30 mg kg^{-1} per day in two divided doses (maximum, 500 mg per dose)
Alternative oral regimens:		
Selected macrolides for patients intolerant of tetracyclines, penicillins, and cephalosporins	azithromycin, 500 mg per day	azithromycin, 10 mg kg^{-1} per day (max. of 500 mg per day)
	clarithromycin, 500 mg twice per day	clarithromycin, 7.5 mg kg^{-1} twice per day (max. of 500 mg per dose)
	erythromycin, 500 mg four times per day	erythromycin, 12.5 mg kg^{-1} four times per day (max. of 500 mg per dose)
Preferred parenteral regimen:		
Ceftriaxone	2 g intravenously once per day	50–75 mg kg^{-1} intravenously per day in a single dose (maximum, 2g)
Alternative parenteral regimens		
Cefotaxime	2 g intravenously every 8 h	150–200 mg kg^{-1} per day intravenously in three or four 4 divided doses (maximum, 6 g per day)
Penicillin G	18 – 24 million U per day i. v. divided every 4 h	200 000–400 000 U kg^{-1} per day divided every 4 h (not to exceed 18–24 million U per day)

Vaccines

Currently, no vaccine is available for the prevention of Lyme disease. However, in December 1998 the US Food and Drug Administration licensed the LYMErix™ vaccine against Lyme disease for human use. In February 2002 the vaccine was withdrawn from the market, reportedly because of poor sales. LYMErix™ contained lipidated recombinant OspA from *B. burgdorferi sensu stricto*. The vaccine was targeted for use in persons aged 15–70 years at high risk of exposure to infected ticks. Interestingly, OspA is not expressed by spirochetes in infected humans, but the vaccine worked because anti-OspA IgG antibodies in human blood were taken up by the infected tick during feeding and killed the borreliae in the tick hind-gut, preventing transmission.

Table 3. Recommended therapy for patients with Lyme disease

Indication	Treatment	Duration, days (range)
Tick bite in the United States	Doxycycline, 200 mg in a single dose; (4 mg kg^{-1} in children ≥ 8 years of age) and/or observation	...
Erythema migrans	Oral regimen	14 (14–21)
Early neurologic disease		
Meningitis or radiculopathy	Parenteral regimen	14 (10–28)
Cranial nerve palsy	Oral regimen	14 (14–21)
Cardiac disease	Oral regimen or parenteral regimen	14 (14–21)
Borrelia lymphocytoma	Oral regimen	14 (14–21)
Late disease		
Arthritis without neurologic disease	Oral regimen	28
Recurrent arthritis after oral regimen	Oral regimen or parenteral regimen	28 for oral regimen & 14 (14–28) for parenteral regimen
Antibiotic-refractor arthritis	Symptomatic therapy	......
Central or peripheral nervous system disease	Parenteral regimen	14 (14–28)
Acrodematitis chronica atrophicans	Oral regimen	21 (14–28)
Post-Lyme disease syndrome	Consider and evaluate other potential causes of symptoms; if none is found, then administer symptomatic therapy	...

SUMMARY

1. What is the causative agent, how does it enter the body and how does it spread a) within the body and b) from person to person?

- Lyme disease is the most common tick-borne disease in North America and Europe.

- There are three genospecies within the *Borrelia burgdorferi* complex (*B. burgdorferi sensu lato*), *B. burgdorferi* (*sensu stricto*), *B. garinii*, and *B. afzelii*.

- Strains found in North America belong to *B. burgdorferi sensu stricto*, whereas the other two species are found in Europe and Asia.

- The bacteria are maintained in an enzootic cycle involving hard-bodied ticks belonging to the *Ixodes ricinus* species complex and a wide range of reservoir vertebrate hosts.

- Because the ticks are slow feeders transfer of the borreliae from the infected tick to a vertebrate host probably does not occur unless the tick has been attached to the body for about 36 hours.

- Ticks transmit Lyme disease to humans generally during the nymph stage.

- Borreliae are microaerophilic spirochetes that are extremely difficult to culture because of their complex nutrient requirements.

- *B. burgdorferi sensu lato* are detected by the immune response that they induce in blood of the infected person.

- The bacteria have a gram-negative wall structure and have endoflagella within the periplasmic space.

- *Borrelia* species have linear chromosomes and contain circular and linear plasmids with some species containing more than 20 different plasmids.

- Borreliae are adept at evading host immunity by varying the lipoproteins on their outer surface.

- The Osps play an important role in the life cycle of the spirochete by interacting with intercellular and cellular components of its arthropod and vertebrate hosts.

- *B. burgdorferi* selectively expresses specific Osps during distinct phases of its life cycle and in specific tissue locations.

- *B. burgdorferi sensu lato* establishes a localized infection in the skin at the site of the tick bite, producing a characteristic skin lesion known as erythema migrans (EM).

- In addition, it may disseminate via the bloodstream and/or lymphatics to the central nervous system, heart, joints, and eyes, all of which may become chronically infected.

- Lyme disease is not spread person to person.

- Rarely borreliae may cross the placenta leading to infection of the fetus.

2. What is the host response to the infection and what is the disease pathogenesis?

- Because the spirochete is delivered to the host via the bite of a tick it by-passes the physical barrier of the intact skin and the antimicrobial factors in sweat.

- The innate pathways of the complement cascade, macrophages, and **dendritic cells** are the first line of defense in the skin.

- *B. burgdorferi* activates both the innate and classical pathways of the complement cascade but is resistant to complement-mediated lysis.

- Macrophages and polymorphonuclear leukocytes can phagocytose and kill borreliae without the need for opsonization by complement or antibody.

- IgM and IgG antibodies play the principal role in the clearance of *B. burgdorferi*.

- *B. burgdorferi* undergoes antigenic variation and modulates the expression of Osps on the cell surface during infection.

- The pathogenesis of Lyme disease is not well understood.

- Tissue injury results from the inflammatory response mounted against the borreliae.

- Chemokines may be the principal mediators of inflammation in Lyme disease.

3. What is the typical clinical presentation and what complications can occur?

- Lyme disease can be divided into three stages: (1) erythema migrans and some associated symptoms; (2) intermittent arthritis, cranial nerve palsies and nerve pain, atrioventricular node block, and severe malaise and fatigue; (3) prolonged arthritis, chronic encephalitis, myelitis, and parapareses (partial paralysis of the lower limbs).

- The manifestations of Lyme disease are related to the particular genospecies of *Borrelia* involved.

4. How is the disease diagnosed, and what is the differential diagnosis?

- In patients with signs and symptoms consistent with Lyme disease a diagnosis is confirmed by antibody detection tests comprising an enzyme immunoassay (EIA) or immunofluorescence assay (IFA) followed by a Western immunoblot.

- All specimens positive or equivocal by the EIA or IFA should be tested by a standardized Western immunoblot. Specimens negative by the EIA or IFA do not require further testing.

- The following conditions should be considered in the differential diagnosis: calcium pyrophosphate deposition disease, fibromyalgia, gonococcal arthritis, gout, meningitis, psoriatic arthritis, rheumatoid arthritis, syncope, systemic lupus erythematosus, and urticaria.

5. How is the disease managed and prevented?

- The preferred oral regimens for erythema migrans and early disease are amoxicillin 500 mg three times daily, doxycycline 100 mg twice daily or cefuroxime 500 mg twice daily for 14 days.

- Meningitis or radiculopathy require parenteral ceftriaxone 2 g i.v. for 14 days.

- Late disease can be treated for 28 days with the oral regimen, except for central or peripheral nervous system disease, which should be treated using the parenteral regimen.

FURTHER READING

Murray PR, Rosenthal KS, Kobayashi GS, Pfaller MA. Medical Microbiology, 5th edition, Mosby, St Louis, MO, 2005: Chapter 43, 433–438.

Murphy K, Travers P, and Walport M. Janeway's Immunobiology, 7th edition, Garland Science, New York, 2008.

REFERENCES

Feder HM Jr, Johnson BJB, O'Connell S, Shapiro ED, Steere AC, Wormser GP. A critical appraisal of "chronic Lyme disease." N Engl J Med, 2007, 357: 1422–1430.

Guerau-de-Arellano M, Huber BT. Chemokines and Toll-like receptors in Lyme disease pathogenesis. Trends Mol Med, 2005, 11: 114–120.

Kurtenbach K, Hanincova K, Tsao JI, Margos G, Fish D, Ogden NH. Fundamental processes in the evolutionary ecology of Lyme borreliosis. Nat Rev Microbiol, 2006, 4: 660–669.

Nigrovic LE, Thompson KM. The Lyme vaccine: a cautionary tale. Epidemiol Infect, 2007, 135: 1–8.

Steere AC. Lyme borreliosis in 2005, 30 years after initial observations in Lyme Connecticut. Wien Klin Wochenschr, 2006, 118: 625–633. (This issue is dedicated to Lyme and other borrelial infections.)

Wilske B, Fingerle V, Schulte-Spechtel U. Microbiological and serological diagnosis of Lyme borreliosis. FEMS Immunol Med Microbiol, 2007, 49: 13–21.

Wormser GP. Clinical practice. Early Lyme disease. N Engl J Med, 2006, 354: 2794–2801.

Wormser GP, Dattwyler RJ, Shapiro ED, et al. The clinical assessment, treatment, and prevention of Lyme disease, human granulocytic anaplasmosis, and babesiosis: clinical practice guidelines by the Infectious Diseases Society of America. Clin Infect Dis, 2006, 43: 1089–1134.

WEB SITES

eMedicine, Inc., 2008, provider of clinical information and services to physicians and healthcare professionals. Lyme disease by John O. Myerhoff, MD. http://www.emedicine.com/med/topic1346.htm

Centers for Disease Control and Prevention (CDC), Department of Vector-Borne Infectious Diseases, 2008, United States Public Health Agency. http://www.cdc.gov/ncidod/dvbid/lyme/index.htm

European Concerted Action on Lyme Borreliosis, 2008, Editor: Jeremy Gray, University College Dublin. http://meduni09.edis.at/eucalb/cms/index.php?lang=en

MULTIPLE CHOICE QUESTIONS

The questions should be answered either by selecting True (T) or False (F) for each answer statement, or by selecting the answer statements which best answer the question. Answers can be found in the back of the book.

1. **A 23-year-old woman presents to her doctor on a Monday morning after having spent the weekend camping. She asks to have a tick removed that became affixed to her lower leg during a long hike the day before. Which one of the following is the most appropriate treatment after removing the tick?**

   A. Cefuroxime axetil.

   B. Doxycycline.

   C. Amoxicillin.

   D. Any of the above antibiotics would be effective.

   E. Antibiotics are not indicated.

2. **Which one of the following is the least likely clinical manifestation of Lyme disease?**

   A. Skin rash.

   B. Arthritis.

   C. Nausea, vomiting, diarrhea.

   D. Myocarditis.

   E. Malaise and fatigue.

3. **A 5-year-old girl travels with her family to Maryland in June for vacation. A week later, her father finds a 2 mm black spot behind her ear that he thought was a scab. Four days later, a red enlarging circular lesion appears around the same ear and fades within a week. They return to Minnesota and visit their pediatrician who notes that the child's smile is not quite symmetrical. Which one of the following is the most likely diagnosis?**

   A. Tinea capitis.

   B. Hypersensitivity to a mosquito bite.

   C. Lyme disease.

   D. Erythema infectiosum.

   E. Enteroviral exanthema.

4. **During June in a suburb of Trenton, NJ, a patient presents to her doctor concerned about her risk for Lyme disease. She lives near a wooded area, and 3 days previously, she pulled off a small tick from behind her knee. Currently, she is asymptomatic and her physical exam is normal. Which one of the following is the correct course of action?**

   A. Obtain a Lyme serologic test, begin empiric doxycycline therapy, and repeat the serologic testing in 6 weeks.

   B. Begin empiric doxycycline therapy and obtain a Lyme serologic test in 6 weeks.

   C. Do nothing unless clinical signs of early Lyme disease develop over the next weeks.

   D. Obtain a Lyme serologic test. If positive, begin therapy with doxycycline. If negative, give no therapy.

   E. Explain to the patient that Lyme disease has never been reported in New Jersey, and she has nothing to worry about.

5. **A 17-year-old white male goes to his doctor worried about Lyme disease. He recently returned from a camp in the Upper Peninsula of Michigan where he went hiking in the woods. He recalls no tick bite, but about 1 week after returning home he developed a low-grade fever, myalgia, and fatigue. He had had these symptoms for 1 week when he presented to his doctor. A physical examination was completely normal. What is the single most important diagnostic clue in establishing the diagnosis of Lyme disease?**

   A. New-onset bundle-branch block.

   B. Nuchal rigidity compatible with meningitis.

   C. High, spiking fevers.

   D. Erythema migrans.

   E. Acute arthritis of a large joint.

6. **All of the following are features of *B. burgdorferi sensu lato*, EXCEPT?**

   A. Microaerophilic.

   B. Difficult to culture because of their exacting nutrient requirements.

   C. Usually detected in the blood of infected patients by microscopy.

   D. They have endoflagella.

   E. They have a linear chromosome and contain circular and linear plasmids.

7. **All of the following are features of the pathogenesis of *B. burgdorferi* in Lyme disease, EXCEPT?**

   A. *B. burgdorferi* is resistant to complement-mediated lysis.

   B. IgM and IgG antibodies play the principal role in the clearance of *B. burgdorferi*.

   C. Arthritis is linked to HLA-DR4 and HLA-DR2.

   D. CD8 T cells are important in combating intracellular borreliae.

   E. *B. burgdorferi* displays antigenic variation.

Case 3

Campylobacter jejuni

A 35-year-old man had been feeling unwell for a few days with nonspecific aches and pains in his joints and a slight headache. He put this down to a barbecue he had attended a few days previously, where he had also drunk a considerable amount of alcohol.

The following day he felt considerably worse with severe colicky abdominal pain and he developed bloody diarrhea, going to the lavatory 10 times during the day. This persisted overnight and he attended his local hospital's Emergency Department.

On examination he was noted to be dehydrated and rather pale. He was admitted to hospital for intravenous rehydration and blood and feces samples were sent for culture. He was started on antibiotics and over the subsequent few days he improved with lessening of the symptoms and was discharged home.

Some weeks later he began to develop weakness in his feet, which gradually affected his legs. He contacted his primary care physician who admitted him to hospital once again. Over the subsequent few days the paralysis affected his upper leg muscles, and gradually over the ensuing weeks slowly resolved with treatment.

1. What is the causative agent, how does it enter the body and how does it spread a) within the body and b) from person to person?

Causative agent

Campylobacter jejuni is a slender, motile nonspore-forming curved gram-negative bacterium (Figure 1) measuring 0.2–0.5 µm wide by 0.5–5.0 µm long, with a single unsheathed polar flagellum. Its genome has been sequenced. Several species of *Campylobacter* exist, two of which cause the majority of human disease: *C. jejuni* and *C. coli* (Table 1). Recent taxonomic studies on the genus have reassigned some campylobacters to two new genera: *Arcobacter* and *Helicobacter*.

C. jejuni has a typical gram-negative cell wall structure with a hydrophobic outer membrane consisting of **lipopolysaccharide** (**LPS**). The major fatty acids are tetradecanoic, hexadecanoic, hexadecenoic, octadecanoic, and 19 cyclopropane. *C. jejuni* is subdivided into Penner serogroups based on the **antigenic variation** of the LPS and further subdivided by **phage types**. Although serogrouping is not used routinely in epidemiological studies it is important as certain Penner groups can cause immune-mediated diseases (see below). *Campylobacter* can be typed by molecular methods such as **restriction fragment length polymorphism** (**RFLP**) and **multi-locus enzyme electrophoresis** (**MLEE**).

The organism is microaerophilic, requiring an atmosphere of 5–10% oxygen and 10% CO_2.

The two main human pathogens are thermotolerant and can grow at 42°C, which is used as a selective condition for clinical isolates.

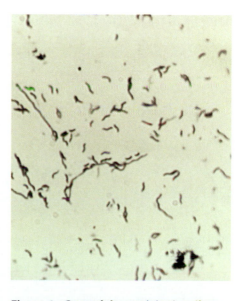

Figure 1. *Campylobacter jejuni*: **a silver stain showing the 'seagull' morphology of *Campylobacter* sp.**

Table 1. The different species of *Campylobacter* and the three genera in the Campylobacteraceae	
Campylobacter	*C. fetus*
	C. coli
	C. concisus
	C. curvus
	C. fetus subsp. *fetus*
	C. fetus subsp. *veneralis*
	C. gracilis
	C. helveticus
	C. hominis
	C. hyoilei
	C. hyointestinalis subsp. *hyointestinalis*
	C. hyointestinalis subsp. *lawsonii*
	C. insulaenigrae
	C. jejuni subsp. *doylei*
	C. jejuni subsp. *jejuni*
	C. lanienae
	C. lari
	C. mucosalis
	C. rectus
	C. showae
	C. sputorum subsp. *bubulus*
	C. sputorum subsp. *mucosalis*
	C. sputorum subsp. *sputorum*
	C. upsaliensis
Arcobacter	*A. butzleri*
	A. cryaerophilus
	A. nitrofigilis
Helicobacter	See case study 12

C. jejuni is found widely in the animal kingdom and is part of the normal flora of many food-source animals, for example poultry, lamb, pig. The commonest meat source is poultry. Campylobacteriosis is thus a **zoonosis**, with infection being acquired principally from contact with animals or eating poorly cooked meats, for example barbeques, or from sandwiches containing contaminated meat. Infection can also be acquired from raw milk contaminated at source, contaminated water sources, or less frequently from birds breaking the top of milk bottles on the doorstep and thereby contaminating the milk. Infection can also be acquired from close contact with animals such as in children's zoos or with infected dogs.

Entry into the body

Entry of the organism is via the mucosal surfaces of the intestine where it mostly remains. Most cases are caused by **serovar** O19, although other serovars have also been implicated, for example O1 and O5. The infectious dose is about 500–1000 organisms and colonization affects the small and large intestine, with the terminal ileum and colon affected most often. Mesenteric **adenitis** regularly occurs. The shape and motility of the organisms enables them to penetrate the mucus layer where they adhere to the enterocytes and are internalized utilizing microfilament/microtubule-dependent mechanisms. Important in the internalization are the

Campylobacter invasion antigens (Cia), which are secreted by a flagellar secretion apparatus (the evolutionary forerunner of the **type III secretion apparatus**).

Spread within the body

C. jejuni rarely causes a **bacteremia**, and remains in the gastrointestinal tract, causing a **terminal ileitis** and **colitis**. Systemic infections are more commonly associated with *Campylobacter fetus* subsp. *fetus*, which principally causes diseases of the reproductive system in sheep or cattle and infections in immunocompromised humans. *C. fetus* subsp. *fetus* possesses an S-layer (protein microcapsule), which is antiphagocytic and may explain its tendency to cause **septicemia**.

Person to person spread

Direct person to person spread is uncommon.

Epidemiology

It is difficult to give figures on a worldwide basis of the incidence of disease due to the differences in reporting such cases. However, in general, the incidence of disease is about 78 per 100 000 population, with a peak at 0–4 years of age and a second peak at 15–24 years, although all ages are susceptible. In the UK the number of reported cases has increased from 24 809 in 1986 to 46 236 in 2006. There is a seasonality to infection, with most cases in different countries occurring in the spring–summer and declining in the winter months. In the UK the highest rate of laboratory-confirmed cases occurs in June. In the UK it is the principal cause of gastroenteritis and in the USA it is second to *Salmonella* infection. Confirmed disease is notifiable in Austria, Denmark, Finland, Germany, Italy, Sweden, and Norway. In the UK, a case of *C. jejuni* infection would be notifiable as a 'food poisoning' case without giving the identity of the organism.

2. What is the host response to the infection and what is the disease pathogenesis?

Innate immunity

Following invasion of the enterocytes, an acute inflammatory reaction occurs with infiltration of the **lamina propria** by granulocytes.

Adaptive immunity

Soon after infection the host mounts an antibody response, which peaks at about 2 weeks and declines over the following weeks. The main antigens are flagella, outer membrane proteins (OMPs), and LPS. Since antibodies to OMP and LPS cross-react with myelin components this can lead to the development of **Guillain-Barre** or the **Miller Fisher syndromes** (see below).

Little is known about the role of the cell-mediated response. However, CD4+ T-helper cells are involved in the production of **IgG** and **IgA** antibodies to the bacterial antigens.

Pathogenesis

Ulceration of the epithelium with crypt **abscesses** occurs following infection.

Cytotoxins have been identified from some strains and apparently non-toxigenic strains can still cause disease. **Cytolethal distending toxin** is present in most strains of *C. jejuni* and induces cell cycle arrest leading to **apoptosis**. It is coded for by three genes – *cdtA*, *cdtB*, and *cdtC* – and is internalized by **endocytosis**. The toxin prevents dephosphorylation (and therefore activation) of CDC2, the catalytic subunit of the cyclin-dependent kinase, which is necessary for cell cycle events and thus causes G_2 arrest. One of the genes (*cdtB*) codes for a deoxyribonuclease and it may be related to the induction of cell cycle arrest. However, its precise role in disease is unclear. Neurological complications (e.g. Guillain-Barre syndrome) can occur, caused by antibodies cross-reacting with gangliosides of the myelin sheath (Figure 2).

3. What is the typical clinical presentation and what complications can occur?

The incubation period is commonly 1–3 days but can be as long as 8 days. Infection may be symptomless in a proportion of individuals. A typical clinical presentation is one of generalized systemic upset with fever and **myalgia** accompanied by abdominal pain and diarrhea. Diarrhea may range from a few loose motions to profuse and watery or grossly bloody. The symptoms may last from a day or two to over a week in about 20% of patients and are generally self-limiting, although relapses may occur in about 10% of patients.

In severe cases of **enterocolitis**, **toxic megacolon** may occur and rare cases of **cholecystitis** or **pancreatitis** have been reported. Three major extraintestinal complications are recognized. The commonest complication is reactive **arthritis**, which occurs in about 1% of cases and occurs several weeks after the initial infection.. It is more common in subjects who are HLA B27. The **hemolytic uremic syndrome** may also follow

Figure 2. Putative mechanism for development of Guillain-Barre syndrome. Antibodies raised against the lipopolysaccharide (LPS) component of the cell wall of some *Campylobacter* strains cross-react with GM1 gangliosides in the myelin sheath of the nerve, due to a structural similarity, leading to damage, loss of nerve conduction, and paralysis.

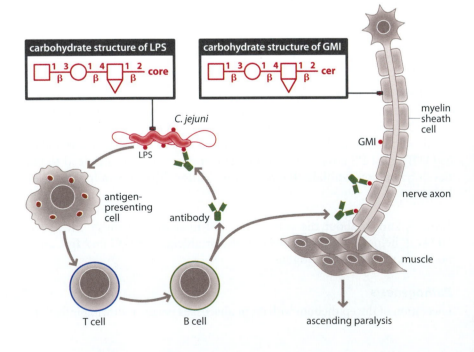

infection and usually comes on within a few days. Infection may be followed by the Guillain-Barre or the Miller Fisher syndrome variant. Although a rare complication, because campylobacteriosis is so common, an infection with *C. jejuni* is the cause of these immune pathologies in about 50% of cases. Other extraintestinal complications that have occasionally been reported are **IgA nephropathy** and **interstitial nephritis**.

4. How is this disease diagnosed and what is the differential diagnosis?

Campylobacter enteritis is confirmed by the isolation of the organism from the feces using a charcoal-based selective medium and culturing under microaerobic conditions at 42°C. In resource-limited countries a rapid means of provisional diagnosis is to look at a wet preparation of feces under the microscope, where the rapid motion of the *Campylobacter* can be seen – in contrast to *Shigella* dysentery where there is no obvious microbial movement as *Shigella* are nonmotile. **Pus** cells are also prominent. *C. jejuni* may also be detected using the **polymerase chain reaction (PCR)**.

Differential diagnosis

It is not possible to make a microbiological diagnosis from the clinical presentation, so all the other causes of gastroenteritis are in the differential diagnosis such as *Shigella, Salmonella, Escherichia coli*, and so forth. Additionally, as the presentation may be severe with bloody diarrhea, **inflammatory bowel disease** may also be confused with campylobacteriosis.

5. How is the disease managed and prevented?

Management

The disease is self-limiting and in most cases no therapy is required. In severe cases, particularly if it lasts for several days, the patient may have to be admitted to hospital for rehydration. Also antibiotics should be given in severe cases. Erythromycin or ciprofloxacin are the two most frequently used, although for both agents, antibiotic-resistant isolates are increasing in prevalence due to overuse of antibiotics in both human and veterinary practice. In the presence of macrolide- or quinolone-resistant isolates, alternatives agents such as tetracycline (except in children) or clindamycin may be used. *Campylobacter* is not susceptible to β-lactams, with the exception of co-amoxiclav. If systemic spread occurs gentamicin can be used but is not useful for the enteric infection.

Prevention

Preventive measures rely mainly on adequate food hygiene and establishing disease-free animal stocks.

A vaccine does not yet exist but is being actively developed.

SUMMARY

1. What is the causative agent, how does it enter the body and how does it spread a) within the body and b) from person to person?

- *Campylobacter* is microaerophilic.

- The cell wall has a typical gram-negative structure.

- The antigenic variation of the LPS is the basis for the Penner groups.

- It is one of the commonest causes of gastroenteritis globally.

- Infection occurs at any age, with peaks in children and young adults.

- Infection is a zoonosis.

- The commonest food source is poultry.

- The organism is located in the small and large intestine.

2. What is the host response to the infection and what is the disease pathogenesis?

- Motility and the organism's spiral shape are important in colonization.

- The infectious dose is about 500–1000 organisms.

- Cell invasion involves *Campylobacter* invasion antigens.

- Invasion involves the microtubules.

- *Campylobacter* produces a cytolethal distending toxin, which causes cell cycle arrest and apoptosis.

- Histologically there is epithelial ulceration and infiltration by granulocytes.

- Antibodies produced in response to the organism can cross-react with myelin.

3. What is the typical clinical presentation and what complications can occur?

- The incubation period is 1–8 days, usually 1–3 days.

- Presents with constitutional symptoms.

- Gastrointestinal symptoms are colicky abdominal pain and diarrhea, which may contain blood.

- Complications include megacolon, reactive arthritis, hemolytic uremic syndrome, and Guillain-Barre syndrome.

4. How is this disease diagnosed, and what is the differential diagnosis?

- Diagnosis is by culture on selective medium incubated microaerobically.

- Differential diagnosis is other bacterial causes of diarrhea and dysentery including *Shigella*, *Salmonella*, *E. coli*, and inflammatory bowel disease.

5. How is the disease managed and prevented?

- Most infections require no treatment.

- In severe cases rehydration and antibiotics may be required.

- Prevention is largely by adequate food processing and hygiene.

FURTHER READING

Ketley JM, Konkel ME. *Campylobacter*: Molecular & Cellular Biology. Garland Science, Abingdon, UK, 2005.

Nachamkin I, Blaser MJ. *Campylobacter*. Blackwell Publishing, Oxford, 2000.

REFERENCES

Desvaux M, Hebraud M, Henderson IR, Pallen MJ. Type III secretion: what's in a name? Trends Microbiol, 2006, 14: 157–160.

Dingle KE, Colles FM, Wareing DRA, et al. Multilocus sequence typing for *Campylobacter jejuni*. J Clin Microbiol, 2001, 39: 14–23.

Karenlampi R, Rautelin H, Hanninen ML. Evaluation of genetic markers and molecular typing methods for prediction of sources of *Campylobacter jejuni* and *C. coli* infections. Appl Environ Microbiol, 2007, 73: 1683–1685.

Konkel ME, Monteville MR, Rivera-Amill V, Joens LA. The pathogenesis of *Campylobacter jejun*i-mediated enteritis. Curr Issues Intest Microbiol, 2001, 2: 55–71.

Nylen G, Dunstan F, Palmer SR, et al. The seasonal distribution of campylobacter infection in nine European countries and New Zealand. Epidemiol Infect, 2002, 128: 383–390

Parkhill J, Wren BM, Mungall E, et al. The genome sequence of the food borne pathogen *Campylobacter jejuni* reveals hypervariable sequences. Nature, 2000, 403: 665–668.

Payot S, Bolla JM, Corcoran D, Fanning S, Megraud F, Zhang Q. Mechanisms of fluoroquinolone and macrolide resistance in *Campylobacter* spp. Microb Infect, 2006, 8: 1967–1971.

Prokhorova TA, Nielsen PN, Petersen J, et al. Novel surface polypeptides of *Campylobacter jejuni* as travellers diarrhoea vaccine candidates discovered by proteomics. Vaccine, 2006, 24: 6446–6455.

Solomon T, Willison H. Infectious causes of acute flaccid paralysis. Curr Opin Infect Dis, 2005, 16: 375–381.

WEB SITES

Centers for Disease Control and Prevention (CDC), Division of Foodborne, Bacterial and Mycotic Diseases, 2008, United States Public Health Agency. http://www.cdc.gov/ncidod/dbmd/diseaseinfo/campylobacter_g.htm

Centers for Disease Control and Prevention (CDC), Morbidity and Mortality Weekly Report, April 13, 2007, 56(14): 336–339. http://www.cdc.gov/mmwr/preview/mmwrhtml/mm5614a4.htm

Centre for Infections, Health Protection Agency, HPA Copyright, 2008. http://www.hpa.org.uk/infections/topics_az/campy/data_ew_month.htm

Centre for Infections, Health Protection Agency, HPA Copyright, 2008. http://www.hpa.org.uk/infections/topics_az/campy/menu.htm

United States Food and Drug Administration, Center for Food Safety and Applied Nutrition, Foodborne Pathogenic Microorganisms and Natural Toxins Handbook, Copyright M. Walderhaug, January 1992 with periodic updates. http://www.cfsan.fda.gov/~mow/chap4.html

MULTIPLE CHOICE QUESTIONS

The questions should be answered either by selecting True (T) or False (F) for each answer statement, or by selecting the answer statements which best answer the question. Answers can be found in the back of the book.

1. **How can *C. jejuni* be typed?**
    A. Based on the lipopolysaccharide.
    B. By the Widal reaction.
    C. By phage typing.
    D. By the isoprenoid content.
    E. By MLEE.

2. **Which of the following are true of *C. jejuni*?**
    A. The organism grows in 48 hours on agar.
    B. It is a strict anaerobe.
    C. It is acquired by the oral route.
    D. Infection occurs in childhood.
    E. It is a rare infection.

3. **Which of the following are correct statements concerning the host response to *C. jejuni*?**
    A. Few granulocytes are recruited to the area.
    B. There is a strong IgA response.
    C. *C. jejuni* is resistant to complement.
    D. Epithelial ulceration occurs.
    E. Crypt abscesses occur.

4. **Which of the following are virulence factors of *C. jejuni* involved in pathogenesis of disease?**
    A. Type IV secretion apparatus.
    B. Cytolethal distending toxin.
    C. Arabinogalactan.
    D. Lipopolysaccharide.
    E. VacA.

5. **Which of the following may be due to infection with *C. jejuni*?**
    A. Megacolon.
    B. Miller Fisher syndrome.
    C. Enterocolitis.
    D. Keratitis.
    E. Septicemia.

6. **Which of the following are typical signs and symtoms of *C. jejuni* infection?**
    A. Colicky lower abdominal pain.
    B. Fever.
    C. Bloody diarrhea.
    D. Skin rash.
    E. Splenomegaly.

7. **Which of the following are important in the routine diagnosis of *C. jejuni* infection?**
    A. Culture.
    B. Urea breath test.
    C. Serology.
    D. PCR.
    E. String test.

8. **Which of the following antibiotics are used to treat gastroenteritis caused by *C. jejuni*?**
    A. Erythromycin.
    B. Metronidazole.
    C. Spiromycin.
    D. Gentamicin.
    E. Ciprofloxacin.

9. **Which one of the following complications of infection affects the nervous system?**
    A. Guillain-Barré syndrome.
    B. Hemolytic uremic syndrome.
    C. Reiter syndrome.
    D. Miller Fisher syndrome.
    E. Anti-Tourette syndrome.

Case 4

Chlamydia trachomatis

A 19-year-old woman was seen by her doctor for a routine gynecological examination and complained about some mid-cycle bleeding. She had been with her current boyfriend for a year, had a pregnancy termination 2 years previously, and was taking birth control pills.

Internal examination revealed a **mucopurulent** discharge at the external cervical os (Figure 1). The cervix was friable and bled easily. The doctor suspected chlamydial infection and collected an endocervical swab specimen for a *Chlamydia* test. The woman returned for the results and was told that the test for *Chlamydia* was positive.

The doctor prescribed a course of doxycycline for 1 week and explained to the patient that this treatment is sufficient and effective in more than 95% of cases, and no repeat testing to prove the eradication of this infection is necessary. However, due to the high risk of re-infection in sexually active young adults, the doctor recommended her to return for a follow-up visit in 6 months.

He further stated that a timely cleared chlamydial infection does not normally lead to infertility, although around 10% probability of ectopic pregnancy remains. The woman was given a leaflet on chlamydial infection and on other sexually transmitted infections (STIs).

The patient was counseled regarding safe-sex practices. The doctor also advised her to contact the local genitourinary clinic to be tested for other STIs including

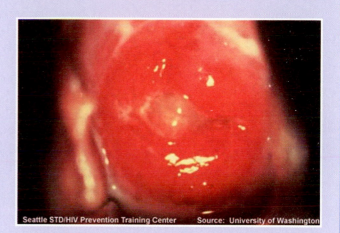

Seattle STD/HIV Prevention Training Center   Source: University of Washington

Figure 1. Mucopurulent cervical discharge caused by chlamydial infection, showing ectopy and edema. Reprint permission kindly granted by the Seattle STD/HIV Prevention Training Center, University of Washington, Seattle.

HIV, and made all necessary arrangements according to the national guidelines.

As a part of the disease management, the doctor carried out all the appropriate actions recommended in the national guidelines to notify the patient's partner and to advise him to visit a genitourinary clinic or to see his doctor for *Chlamydia* and other STI tests. The couple were advised to abstain from sex until both were cleared of the infection.

1. What is the causative agent, how does it enter the body and how does it spread a) within the body and b) from person to person?

Causative agent

This patient was infected with *Chlamydia trachomatis*, which belongs to the Family Chlamydiaceae of the Order Chlamydiales. Until recently Chlamydiaceae was believed to consist of one genus *Chlamydia*, but now another genus, *Chlamydophila*, has been identified within this Family. Bacteria of this Family are obligate intracellular human and animal pathogens. The term 'Chlamydia' is derived from the word 'chlamys,' which means cloak (Khlamus) in Greek, an appropriate name reflecting the cloak-like chlamydial inclusion around the host cell nucleus (see below).

Chlamydiaceae are some of the most widespread bacterial pathogens in the world and there are several species that infect a variety of hosts based on a wide range of tissue tropism. Two species, *Chlamydia trachomatis* and *Chlamydophila pneumoniae*, are human pathogens and are responsible for various diseases that represent a significant economic burden. *Chlamydophila psittaci* and *C. pecorum* are mainly bird/animal pathogens, although **zoonotic** transmission of the former to humans can occur resulting in the disease psittacosis.

Chlamydiaceae species share some important structural features. Like other gram-negative bacteria they have inner and outer membranes, but have a specific **lipopolysaccharide (LPS)** that differs from that of other bacteria. The extracellular osmotic stability of Chlamydiaceae is provided by several complex disulfide cross-linked membrane proteins, the main ones being a 40 kDa major outer-membrane protein (MOMP, a product of the *ompA* gene); a hydrophilic cysteine-rich 60 kDa protein (OmcA); and a low molecular weight cysteine-rich lipoprotein (OmcB). The Chlamydiaceae are thought to have little or no muramic acid, the hallmark constituent of peptidoglycan (PG). It is still unclear whether Chlamydiaceae are at all able to express surface PGs, which are important receptor molecules facilitating entry of intracellular bacterial parasites into host cells (see the *Mycobacterium leprae* and *Mycobacterium tuberculosis* cases). No transfer RNAs were identified in chlamydial cells, thus confirming the parasitism of this bacterium.

There are two human biological variants (biovars) of *C. trachomatis*: **trachoma** and lymphogranuloma venereum (LGV), and one biovar *C. pneumoniae* infecting mice, and causing mouse pneumonitis (which will not be discussed here).

Fifteen serological variants (**serovars**) have been identified in the trachoma biovar (A–K, Ba, Da, Ia, and Ja). These mostly infect columnar and squamo-columnar epithelial cells of mucous membranes (see below). Serovars D–K, Da, Ia, and Ja typically infect genitourinary tissues, but were also found in the mucous membranes of the eye conjunctiva and epithelial tissues in the neonatal lung. Serovars A, B, Ba, and C generally infect the conjunctiva and cause trachoma. The LGV biovar consists of four serovars (L1, L2, L2a, and L3), which predominantly infect monocytes and macrophages passing through the epithelial surface to regional lymphoid tissue. *C. pneumoniae* has only one serovar and infects epithelial cells.

Proteomic analysis of the pathogen is very difficult since Chlamydiaceae species are all obligate intracellular parasites and cannot grow in a cell-free system. As a result, most of the data obtained on the structural and functional proteins and biochemical pathways utilized by Chlamydiaceae are derived from gene sequencing and indirect evidence. The genome of *Chlamydia trachomatis* (serovars A, D, and L2) has been fully sequenced.

Chlamydiaceae species have a more complicated biphasic developmental life cycle than other bacteria in that they have two different forms, a metabolically inert infectious elementary body (EB) and a larger noninfectious reticulate body (RB). Interestingly the term 'elementary body' belongs to the virology world and is derived from the time when Chlamydiaceae were initially considered to be viruses. The EB form of the bacterium survives outside the host cell whereas the RB form lives and replicates in a specialized vacuole of the host cell called an inclusion (another virology-derived term). The biphasic developmental life cycle of *Chlamydia* is shown in Figure 2.

Entry into the body

The infectious EB form of the majority of *Chlamydia* strains is typically 0.2–0.3 µm in diameter. MOMP makes up 60% of its cell wall. Due to their rigid outer membrane, EBs are able to survive outside the eukaryotic host cells. *C. trachomatis* infecting genital tissue usually does so through small abrasions in the mucosal surfaces. The EBs infect nonciliated columnar, cuboidal or transitional epithelial cells, but can also infect macrophages. Bacterial receptors for host cell attachment are unknown, but there is some suggestion that mannose and other sugars associated with MOMP play a role at least for some serovars. EBs probably bind to the host cell membrane through irreversible interactions with cholesterol- and glycosphingolipid-rich microdomains (rafts) followed by restructuring of the actin cytoskeleton within the host cell. This facilitates chlamydial entry through **phagocytosis**, receptor-mediated **endocytosis**, and **pinocytosis**.

Early intracellular phase (0–2 hours after infection). Once inside the cell, EB-containing vacuoles move towards the perinuclear region. The vacuole membrane phospholipids promote homotypic fusion of EB vacuoles with each other, but not with **lysosomes** – an important feature that helps the

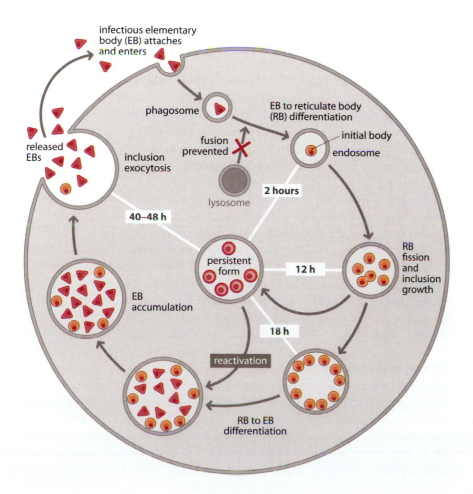

Figure 2. Biphasic developmental cycle of *C. trachomatis*.

pathogen to avoid intracellular destruction. The homotypic fusion is specific to *C. trachomatis* only and not to other *Chlamydia* species. This results in the formation of a single fusion vacuole (a nascent inclusion) containing several EBs. Some of these vacuoles may contain different serovars such as F and E, leading to the possibility for genetic exchange to occur.

The inclusions then move towards the microtubule organization center where they are supplied with nutrients via the host cell Golgi apparatus. Bacterial proteins are directly secreted into the host cell cytosol.

Inclusion development (2–40 hours after infection). The EB forms now undergo a lengthy and complex development using host cell ATP and nutrients as a source of energy. *Chlamydia* species do not seem to be able to produce their own ATP, but there is some evidence that *C. trachomatis* produces its own energy through a unique pyruvate kinase that can be activated by a host cell fructose-2,6 biphosphate.

Still remaining in the inclusion the EB forms now transform into RB forms, which are typically 0.8–1.0 μm in diameter. RBs multiply by binary fission so that the resulting inclusions may contain 500–1000 progeny RBs and occupy up to 90% of the cell cytoplasm. After several rounds of replication RB forms revert to the infectious EB forms. In tissue culture the productive infectious cycle of *Chlamydia* lasts about 48–72 hours depending on the serovar. Eventually the EBs are released to infect other adjacent cells. The mechanisms of the release differ in chlamydial species. In *C. trachomatis* and *C. pneumoniae* infections the inclusions are extruded by reverse endocytosis or the host cells apoptose, while in the case of *C. psittaci* the host cell and the inclusion membranes lyse.

C. trachomatis is able to regulate host cell **apoptosis** throughout the early and productive growth stages of the developmental cycle leading to successful bacterial multiplication. Nonreplicating forms of *Chlamydia* inhibit apoptosis. However, late in the life cycle the pathogen produces a caspase-independent pro-apoptotic Bax protein, which facilitates apoptosis of the host cell thus freeing the secondary EB forms.

Under unfavorable for a pathogen conditions such as the presence of **interferon-γ (IFN-γ)**, lack of nutrients or drug treatment (see Section 5) *Chlamydia* may enter a nonreplicating mode called persistence (Figure 2). During persistent infection the developmental cycle is lengthened or aborted and RB forms are produced that do not divide or differentiate back into the EB forms. Persistent infection with *C. trachomatis* may lead to serious clinical conditions that are difficult to treat. However, dormant forms can revert to metabolically active forms if the unfavorable conditions are removed.

Spread within the body

EBs of *C. trachomatis* infect cervical columnar epithelial cells, but the bacteria can spread by ascending into the endometrium and the fallopian tubes, causing **pelvic inflammatory disease** (PID), ectopic pregnancy, and infertility (see Section 3). Sexually transmitted *C. trachomatis* serovars D–K can also lead to conjunctivitis through autoinoculation or ocular–genital contact.

Spread from person to person

Genital tract infections

C. trachomatis (serovars D–K) is sexually transmitted in vaginal fluid or semen containing the EB form, through vaginal intercourse but occasionally via oral and anal sex. These serovars can also be vertically transmitted from mother to child during birth through an infected birth canal, causing conjunctivitis (**ophthalmia neonatorum**) or chlamydial **pneumonia**.

C. trachomatis (biovar LGV) is also sexually transmitted. In some parts of Africa, Asia, South America, and the Caribbean it is largely found in heterosexuals. In recent outbreaks in industrialized countries the cases are mostly confined to male homosexuals with multiple sexual partners.

Ocular infections

C. trachomatis (serovars A–C) is found predominantly in areas of poverty and overcrowding. Infection can be transmitted from eye to eye by fingers, shared cloths or towels, by eye-seeking flies, and by droplets (coughing or sneezing). The latter route is possible because *C. trachomatis* can exist in the nasopharynx and external nasal exudates of children with trachoma.

Importantly the undiagnosed and untreated children can contribute to a so-called 'age-reservoir effect' responsible for the continuous transmission within the community. Children younger than 5–10 years of age, even those below 12 months, have the highest ocular chlamydial load, and represent a significant reservoir of infection, particularly for their mothers and other 'nanny' figures. These must be identified by the screening programs and treated to prevent further spread of the pathogen in communities.

Epidemiology

C. trachomatis (serovars D–K) is the leading bacterial cause of STIs, with over 50 million new cases occurring yearly worldwide and 4 million new cases each year in the United States. The highest infection rates are detected in African Americans, American Indian/Alaska Natives, and Hispanics. The prevalence is lowest in the north-east and highest in southern USA, with a peak incidence in the late teens/early twenties, particularly those changing sexual partners. In general practice around 1 in 20 sexually active women aged less than 25 years may be infected.

Every year *C. trachomatis* (serovars A–C) is a major cause of 500 000 cases of trachoma worldwide (Figure 3). Active trachoma affects some 85 million people, more than 10 million have **trichiasis** (turned-in eyelashes that touch the eye globe, Figure 4), and about 6 million people suffer visual loss and blindness. Active disease is most commonly seen in children, and in adults the prevalence of trichiasis is about three times higher in women than in men. Trachoma is **endemic** in large areas of Africa and the Middle East, and focal areas of disease are found in India, South-West Asia, Latin America, and Aboriginal communities in Australia. The disease is generally found in clusters in certain communities or even households, indicating the existence of local risk factors in addition to the generally accepted poverty and lack of water and sanitation.

C. trachomatis (biovar LGV – serotypes L1–L3) causes sexually transmitted disease that is prevalent in parts of Africa, Asia, South America, the

Figure 3. World distribution of trachoma according to WHO. *C. trachomatis* (biovar trachoma) is endemic in large areas of Africa and the Middle East, and focal areas of disease are found in India, South-East Asia, and Latin America. Trachoma affects some 150 million people worldwide, more than 10 million have trichiasis (corneal scarring), and about 6 million people suffer visual loss and blindness.

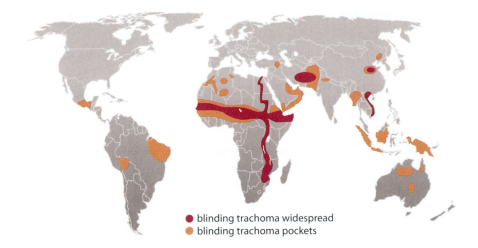

● blinding trachoma widespread
● blinding trachoma pockets

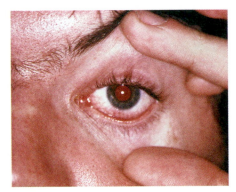

Figure 4. Trachomatous trichiasis: at least one eyelash rubs on the eyeball.

Caribbean, and increasingly in Europe and the USA. Humans are the only natural host, with male homosexuals being the major reservoir of the disease. In the USA the incidence is 300–500 cases per year.

2. What is the host response to the infection and what is the disease pathogenesis?

Host immune responses

Both innate and adaptive immune responses are induced during *C. trachomatis* infection. However, these responses are often insufficient, providing only partial clearance of the pathogen from the body. This can lead to protracted chronic infections and chronic inflammation contributing to pathogenesis.

Chlamydial infection initially induces influx to the infection site of polymorphonuclear cells and macrophages as a part of acute inflammation. This is facilitated by the release of **cytokines** and **chemokines** by the infected epithelial cells, particularly **interleukin**-8 (**IL**-8), which is a powerful neutrophil attractant. Infiltration with neutrophils and macrophages is followed by accumulation of B cells, T cells, and **dendritic cells** in submucosal areas launching T-cell responses and antibody production.

Antibody responses. A humoral response is invoked resulting in production of mucosal secretory **IgA** and circulating **IgM** and **IgG** antibodies. These antibodies are mostly specific for MOMP, the immunodominant antigen, and Hsp60 (a stress protein). Anti-chlamydial IgG antibodies are also found in ocular secretions in trachoma patients. However, the effectiveness of these antibodies in damaging or blocking further entry of EBs to the cells adjacent to the infected ones is unclear.

Cellular responses. It is likely that at least some professional antigen-presenting cells (APCs) at the site of infection engulf EBs, process and present *Chlamydia* peptides via the major histocompatibility complex (MHC) class II-mediated pathway. This leads to activation of **CD4+** T cells and hence effector immune responses involving activation of B cells, **CD8+** cytotoxic T lymphocytes (CTLs), other inflammatory cells and production of cytokines. T-cell-derived cytokines (especially IFN-γ) are believed to exert some control over the infection.

However, once *C. trachomatis* enters epithelial cells, which typically do not express MHC class II molecules, recognition by CD4+ T cells is unlikely. In epithelial cells, recognition of *Chlamydia* peptides may occur via the MHC class I presentation pathway, which activates CD8+ T cells. These can recognize proteins present in the host cell cytosol or cytosolic domains of membrane proteins. MOMP is one of the potential antigenic targets for CD8+ CTLs.

That both *Chlamydia*-specific CD4+ and CD8+ T cells are involved in controlling *C. trachomatis* infection is indicated by their expansion in the site of infection. CD4+ T cells are required for optimal stimulation of *C. trachomatis*-specific CD8+ T cells. Both CD4 and CD8+ T-cell subsets respond by secreting IFN-γ and CD8+ T cells by directly killing infected cells. However, no evidence has been found so far of CD8+ T-cell-mediated killing of the infected cells that would disrupt pathogen replication and intracellular survival.

It appears that one of the most powerful anti-chlamydial T-cell-mediated immune mechanisms is the production of the **Th1** cytokine IFN-γ. IFN-γ inhibits the growth of *Chlamydia* in cell culture, and in experimental models disruption of its production enhances host susceptibility to *Chlamydia* infection. It is thought that IFN-γ can potentially limit *C. trachomatis* infection by the following mechanisms:

- activation of macrophages and their phagocytic potential;
- up-regulation of the expression of MHC molecules by professional and nonprofessional APCs leading to optimization of presentation of microbial antigens;
- expression of **indoleamine 2,3-dioxygenase** (IDO), a host enzyme that degrades intracellular tryptophan essential for *C. trachomatis* growth;
- up-regulation of **inducible nitric oxide synthase** (iNOS), which catalyzes the production of nitric oxide (NO) and other reactive nitrogen intermediates, that can enhance damage to intracellular pathogens;
- down-regulation of transferrin receptor on infected cells, resulting in an intracellular iron deficiency that may limit *C. trachomatis* replication.

Even though IFN-γ plays a role in protective immunity to *C. trachomatis* infection its overall effect is limited. Infection does not stimulate long-lasting immunity and repeated re-infections are common, which results in a prolonged inflammatory response and subsequent tissue damage (see Section 3).

One of the reasons for poor anti-chlamydial immunity is the ability of the pathogen to evade immune responses.

How does C. trachomatis evade the host immune responses?

This bacterium employs several strategies to evade the host immune response.

- Its intracellular location protects it from antibodies and complement.
- It down-regulates the expression of MHC class I molecules on the surface of infected cells, blocking recognition and MHC class I-restricted

CD8+ T-cell-mediated cytotoxicity. The pathogen uses a protease-like activity factor (CPAF) able to degrade host transcription factors required for MHC gene activation.

- Fusion of the pathogen-containing **phagosome** with host cell **lysosomes** is prevented.
- *Chlamydia*-infected macrophages induce apoptosis of CD4+ and CD8+ T cells, which are essential in chlamydial clearance, by paracrine effects and **tumor necrosis factor-α (TNF-α)**.

Pathogenesis

Tissue damage is the result of the host inflammatory response to the persistent infection as well as direct damage to infected cells by the bacteria. Various species of *Chlamydia* produce cytotoxins that can deliver immediate cytotoxicity of host cells if infected with large doses of the pathogen.

With unsuccessful initial elimination by the innate and adaptive immune systems leading to persistence of the pathogen, the site of infection becomes infiltrated with macrophages, plasma cells, and eosinophils. The continuous production of cytokines and chemokines results in development of lymphoid follicles and tissue scarring due to **fibrosis**. It is believed that IFN-γ production by T cells at the site of infection is reduced as the bacterial load decreases. This in turn supports replication of *C. trachomatis* and the inflammatory process resumes, entering into a vicious circle.

Chronic inflammation leads to many of the clinical symptoms seen with *C. trachomatis*. For example, the bouts of **salpingitis** or conjunctivitis resulting in infertility or blindness, respectively.

Trachoma is characterized by the conjunctival lymphoid follicle formation, which contain germinal centers consisting predominantly of B lymphocytes, with CD8+ T lymphocytes in the parafollicular region. The inflammatory infiltrate contains plasma cells, dendritic cells, macrophages, and polymorphonuclear leucocytes. Inflammatory infiltrates taken from patients with scar tissue are characterized by the expansion of CD4+ T lymphocytes.

In the Gambia, an association has been found between trachoma scarring and HLA class II alleles and polymorphisms in the promoter region of the gene for TNF-α. These findings suggest genetic differences in the immune response to *C. trachomatis* as well as a possible genetic predisposition.

3. What is the typical clinical presentation and what complications can occur?

Chlamydia is known as the 'silent epidemic,' since it may not cause any symptoms sometimes for months or years before being discovered. There are several conditions caused by various biovars and serovars of *C. trachomatis* (Table 1).

Urogenital infections

Most people with *C. trachomatis* infection do not have any symptoms and are unaware of the infection. However, when symptoms develop treatment is urgently required to prevent complications.

Table 1. *Chlamydia trachomatis* serovars and associated human diseases

Serovars	Human disease	Method of spread	Pathology
A, B, Ba, and C	Ocular trachoma	Hand to eye, fomites, and eye-seeking flies	Conjunctivitis and conjunctival and corneal scarring
D, Da, E, F, G, H, I, Ia, J, Ja, and K	Oculogenital disease	Sexual and perinatal	Cervicitis, urethritis, endometritis, pelvic inflammatory disease, tubal infertility, ectopic pregnancy, neonatal conjunctivitis, and infant pneumonia
L1, L2, and L3	Lymphogranuloma venereum	Sexual	Submucosa and lymph node invasion, with necrotizing granuloma and fibrosis

Reproduced with permission from Brunham RC and Rey-Ladino J. Immunology of *Chlamydia* infection: implication for *Chlamydia trachomatis* vaccine. (2005). Nature Reviews Immunology, 5: 149–161.

Males. When the symptoms develop the patient may suffer from nongono-coccal **urethritis** (NGU), which can result in discharge from the penis or pain and burning sensation when urinating. If not treated, it can lead to inflammation near the testicles with considerable pain. Spread to the testicles, may cause **epididymitis** and, rarely, sterility. *Chlamydia* causes more than 250 000 cases of epididymitis in the USA each year. Post-gonococcal urethri-tis may occur in men infected with both *Neisseria gonorrhoeae* and *C. trachoma-tis* who receive antibiotic treatment effective solely for gonorrhoea.

Females. The incubation period is usually 1–3 weeks after which the symp-toms of urethritis and **cervicitis** may develop: **dysuria** and **pyuria**, cervical discharge or vaginal spotting, and lower abdominal pain. Physical exami-nation reveals yellow or cloudy mucoid discharge from the os (see Figure 1). If untreated, *Chlamydia* may spread through the uterus to the fal-lopian tubes, causing salpingitis. In the USA chlamydial infection is the leading cause of first trimester pregnancy-related deaths. Women infected with *Chlamydia* have a three- to five-fold increased probability of acquiring HIV due to the increased behavioral risk.

More than 4 billion US dollars are spent annually on the treatment of the most common and severe complication of the sexually transmitted *C. tra-chomatis* – pelvic inflammatory disease (PID). Over 95% of women with uncomplicated and effectively treated chlamydial infection will not develop tubal infertility. Even in the case of PID more than 85% of women remain fertile. However, an approximate 10% risk of ectopic pregnancy – a potentially life-threatening condition – remains after clinical PID. Women with PID often suffer later from abdominal pain and may require a hysterectomy. It is difficult to give a precise prognosis of infertility until a patient tries to have a child.

In both sexes an asymptomatic infection may be present in either the throat or the rectum if the patient has had oral and/or anal intercourse.

Inclusion conjunctivitis

This condition is caused by *C. trachomatis* (serovars D–K) being associated with genital infections. It is often transferred from the genital tract to the eye by contaminated hands. The main symptom is a sensation of a foreign body in the eye, redness, irritation. Other symptoms include mucosal discharge later replaced by purulent discharge, large lymphoid follicles, and papillary hyperplasia of conjuctiva, corneal infiltrates, and vascularization. Corneal scarring is rare and happens mostly in the chronic stage followed by epithelial **keratitis**. Ear infection and **rhinitis** can accompany the ocular disease.

Infant pneumonia

Infants vertically infected with *C. trachomatis* (serovars D–K) from their mother at birth can develop pneumonia presented by staccato cough and **tachypnea** and often preceded by conjunctivitis.

Lymphogranuloma venereum (LGV)

The causative agent is *C. trachomatis* biovar LGV. The first symptom of the infection is the development of a primary lesion – a small painless **papule** or ulcer at the site of infection, often the penis or vagina. Several weeks after the primary lesion patients develop painful inguinal and/or femoral **lymphadenopathy**. In the case of extragenital infection, the lymphadenopathy can occur in the cervix. Patients develop a fever, headache, and **myalgia** followed by inflammation of the draining lymph nodes. As a result the lymph nodes become enlarged and painful and may eventually rupture. **Elephantiasis** of the genitalia, more often in women, can develop due to obstruction of the lymphatics. In females, lymphatic drainage occurs usually in perianal sites and can involve **proctitis** and **recto-vaginal fistulae**. In males, proctitis develops from anal intercourse or from lymphatic spread from the urethra.

Ocular infections

Trachoma is the most serious of the eye infections caused by *C. trachomatis*. The word 'trachoma' in Greek means rough (trakhus) and reflects the roughened appearance of the conjuctiva. Repeated re-infection with the ocular serovars A, B, Ba, and C results in chronic **keratoconjunctivitis**. Following infection there is an incubation period of 5–12 days, after which the symptoms start to appear and include a mild conjunctivitis and eye discharge. Initial infection is often self-limiting and heals spontaneously. A repeated infection, however, leads to the development of chronic inflammation (see Section 2), characterized by swollen eyelids and swelling of lymph nodes in front of the ears. Years of re-infection and chronic inflammation may result in fibrosis and in scarring in the upper subtarsal conjunctiva. The progress of scarring over many years causes distortion of the lid margin and the lashes turn inwards and rub against the cornea. This is called trichiasis (Figure 4). If untreated, persistent trauma can result in ulceration of the cornea, corneal opacity, and blindness.

Inflammation and scarring in the eye may block the natural flow of tears, which represent an important first line of defense against bacteria. This can facilitate secondary re-infection with *C. trachomatis* as well as infection with other bacteria or fungi.

Ocular lymphogranuloma venereum

Ocular infection with *C. trachomatis* biovar LGV can lead to conjunctivitis and **preauricular lymphadenopathy**.

Reiter's syndrome

Reiter's syndrome is associated with infection by a variety of bacteria. Approximately 50–65% of patients have *C. trachomatis* infection, although shigellosis and infection with *Yersinia enterocolitica* can also be present. Reiter's syndrome is a reactive **arthritis**, accompanied by urethritis, cervicitis, conjunctivitis, and painless mucocutaneous lesions.

4. How is the disease diagnosed and what is the differential diagnosis?

Clinical diagnosis of genital infections

The following clinical indicators are used for the diagnosis and screening of chlamydial infection in women:

- less than 25 years of age, sexually active;
- more than one sexual partner;
- mucopurulent vaginal discharge (Figure 1);
- burning sensation when passing urine;
- friable cervix or bleeding after sex or between menstrual periods;
- lower abdominal pain, or pain during sexual intercourse.

Because of the possibility of multiple STIs, all patients with any STI should be evaluated for chlamydial infection and offered an HIV test.

Clinical diagnosis of ocular infections

Examination of an eye for the clinical signs of trachoma involves careful inspection of the lashes and cornea, then aversion of the upper lid and inspection of the upper tarsal conjunctiva using binocular loupes. In 1981 the World Health Organization (WHO) designed a grading system for the assessment of the prevalence and severity of disease in communities, followed by a simplified version in 1987. However, this system is not applicable for diagnosis of individual cases. Clinical diagnosis is best made based on investigation of the history of living in a trachoma-endemic environment, in combination with clinical signs. The WHO grading system can be found at http://www.who.int/blindness/causes/priority/en/index2.html

Clinical diagnosis of lymphogranuloma venereum

The clinical symptoms may initially be unclear since they overlap with those of other STIs. Some men may have had treatment for a range of conditions including inflammatory bowel disease, **Crohn's disease**, and so forth. Diagnosis is largely based on the history of the disease, physical examination, and laboratory tests. The clinical course of LGV is divided into three stages.

1. Primary lesion, which develops after incubation of 3–30 days. This is painless. A papule or ulcer can be found on the genitalia (glans of the penis, vaginal wall, labia, or cervix) or in some cases in the oral cavity.

2. Secondary lesion/**lymphadenitis**. It is a regional dissemination causing inguinal and femoral lymphadenopathy and possibly bubo formation that ulcerates and discharges pus. Lymphadenopathy is usually unilateral involving the retroperitoneal lymph nodes in women and the inguinal lymph nodes in men.

3. Tertiary stage/genito-ano-rectal syndrome. The majority of patients will recover from the second stage without sequelae. However, a few may develop proctitis and fibrosis that may result in chronic genital ulcers or fistulas, rectal strictures, and genital elephantiasis. Early symptoms of LGV **proctocolitis** include anal **pruritus** and discharge, fever, rectal pain, and **tenesmus**.

Genital ulcers and inguinal symptoms are rare.

Laboratory diagnosis

Sample collection. For genital infections, swabs are collected from the cervix or vagina of women or the urethra of men. Swabs should be taken from the throat or rectum if there is a possibility of infection there.

With the use of nucleic acid-based tests (see below) a noninvasive urine test can be used for screening for *Chlamydia* with sufficient sensitivity instead of swabs. It is particularly useful in asymptomatic cases where genital examination and sampling may not be justified.

For suspected LGV infections in the primary stage, a swab of the lesion can be taken. In the secondary stage, bubo pus, saline aspirates of the bubo, swabs of the rectum, vagina, urethra, urine, serum, or biopsy specimens of the lower gastrointestinal tract are used.

For conjunctival specimens, epithelial cells are collected by rubbing a dry swab over the everted palpebral conjunctiva.

The following tests can be made on the samples collected.

Cytology is used to detect the inclusion bodies in stained cell scrapings (Figure 5), but this method lacks sensitivity and is time-consuming.

Cell culture. For many years cell culture has been the gold standard for the diagnosis of *C. trachomatis* and is very specific for this pathogen, but nucleic acid amplification tests are more sensitive and represent a new gold standard (see below). Cultures of susceptible McCoy or HeLa cells (continuously cultured carcinoma cell lines used for tissue culture) are infected with the collected sample. After incubation for 40–72 hours (depending on the biovar) the infected cells are stained with Giemsa or iodine, or by fluorescent antibodies and examined for the presence of iodine-staining inclusion bodies. Iodine stains glycogen, which is only found in the inclusions of *C. trachomatis*, not in other *Chlamydia* species.

Nucleic acid amplification tests (NAATs) are now replacing culture techniques. They include first of all the **polymerase chain reaction (PCR)**, but also transcription-mediated amplification (TMA). NAATs are characterized by high sensitivity (> 89%) and specificity (95–99.5%) for endocervical, urethral, and conjunctival samples. NAATs have abolished a requirement for

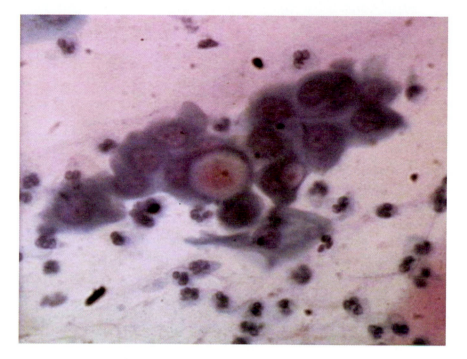

Figure 5. Chlamydial inclusions, which may contain 100–500 RB progeny.

the viability of chlamydial sample and allowed the usage of urine for non-invasive testing. NAATs are the only testing techniques currently recommended by the English National Chlamydia Screening Programme. However, urine, particularly female, contains some compounds such as nitrites, crystals or hormones that may inhibit nucleic acid-based tests, and special test conditions must be observed to avoid mistakes. NAATs can be used for detection of the bacterium as well as for the quantification of the bacterial load. The wider use of NAATs worldwide in screening strategies is, however, limited by the costs and complexity of the equipment required.

Direct hybridization probe tests have been used in the past, but now are largely replaced by NAATs. They include Gen-Probe PACE 2 test, which uses a synthetic single-stranded DNA probe complementary to the chlamydial rRNA region, and Digene HC-II-CT-II probe based on **enzyme immunoassay (EIA)** for the detection of the RNA probe binding to the single-stranded bacterial DNA.

Antibody-based laboratory techniques include direct fluorescent antibody (DFA) detection and **enzyme-linked immunosorbent assay (ELISA)**. The latter can be used for determination of the serovars. Usually diagnostic antibodies target group-specific LPS or serovar-specific outer-membrane proteins (OMPs). These methods are not as good as cell culture, particularly when samples contain few bacteria in asymptomatic patients. However, ELISA-based technology is widely used in screening programs and is of great help in reducing the overall chlamydial infection and incidence of PID and nongonococcal urethritis.

Serological tests are of some value but mainly in neonatal infection because antibody titers can persist for prolonged periods in adult urogenital infections. A high titer of IgM antibodies against *Chlamydia* is indicative of a recent infection, while anti-chlamydial IgG is mostly associated with the past infection. Elevation of *Chlamydia*-specific IgA in infected mucosa has been found, but its diagnostic value is unclear, apart from chronic chlamydial **prostatitis**.

Complement fixation tests are used for patients with LGV who have high complement-fixing antibody titers. Fifteen percent of men with urethritis and 45% of women with endocervical infection have antibody titers of 1:16 or greater.

Differential diagnosis

Differential diagnosis of genital chlamydial infection includes gonorrhea, *Ureaplasma urealyticum*, *Mycoplasma genitalium*, *M. hominis*, and *Trichomonas vaginalis*, as well as urinary tract infections and bacterial vaginosis. Some other conditions such as periurethral **abscess**, **endometriosis**, urethral/vaginal foreign body, other causes of PID, prostatitis, and **epididymo-orchitis** must be ruled out too.

In the case of trachoma, a differential diagnosis can be made with the following conditions: adult inclusion conjunctivitis; other bacterial infections, especially with *Moraxella* species and *Streptococcus pneumoniae*; viral infections including adenovirus, herpes simplex virus, and molluscum contagiosum; *Pediculosis palpebrarum*; toxic conjunctivitis, which is secondary to topical drugs or eye cosmetics; Axenfield's follicular conjunctivitis; Parinaud's oculoglandular syndrome; and vernal conjunctivitis.

For LGV and inguinal **adenopathy**, differential diagnosis of the genital ulcer includes chancroid, herpes, syphilis, and donovanosis. For proctitis the differential diagnosis includes inflammatory bowel disease.

The differential diagnosis is mostly based on the laboratory findings, particularly PCR analysis, supplemented by clinical observations.

5. How is the disease managed and prevented?

Management

Urogenital infections

For urogenital infections the antibiotic doxycycline (Doryx®, Vibramycin®) is the drug of choice and 100 mg is taken twice daily for 7 days. This is effective in 95% of the cases but can interfere with the contraceptive pill and can cause photosensitivity and stomach irritation.

If the infected woman is pregnant or lactating, 500 mg of erythromycin is given four times a day for 7 days or twice daily for 14 days. When erythromycin is used all patients, especially pregnant women, are recommended to have a test of cure 3 weeks after completing therapy, since it is less effective than doxycycline. Alternatively a single-dose therapy with 1 g of azithromycin can be used, which is usually effective in clearing *C. trachomatis*. Azithromycin is a derivative of erythromycin and is characterized by improved bioavailability and ability to maintain high tissue concentrations, particularly at sites of inflammation. Azithromycin has been also proven to be neonatally safe.

Other treatments may include ofloxacin 400 mg twice daily for a week, levofloxacin 500 mg once a day for a week, and in pregnant women amoxicillin 500 mg three times a day for 7–10 days.

Since genital chlamydial infections often have no symptoms, particularly in women, the sexual partner could have become infected months ago. In case of multiple sexual partners, all of them should be tested for chlamydial infection and treated if positive. The patients should also be tested for other STIs. Having sex is not recommended during treatment and for at least a week after the completion of the treatment, particularly if both partners are infected (see the case).

Trachoma

For the treatment of active trachoma two antibiotic regimens are currently recommended: tetracycline ointment applied twice daily for 6 weeks or one 20 mg kg^{-1} dose of azithromycin can be used instead. It must be noted that the application of the ointment is inconvenient in children. Azithromycin is also effective for treating extraocular reservoirs of chlamydial infection, although antibiotic resistance could eventually develop.

Currently, annual mass treatment for 3 years is recommended by the WHO in districts and communities where the prevalence of follicular trachoma in children aged 1–9 years is equal to or greater than 10%. In patients with trichiasis surgery may also be necessary to fix eyelid deformities.

Prevention

Prevention of genital chlamydial infections involves safe sexual practices, using a condom during sexual intercourse, and prompt treatment of infected patients and their sexual partners. It is recommended to have a *Chlamydia* test under the following conditions:

- after sex with a new or casual partner;
- immediately if symptoms occur (see Section 3);
- if a sexual partner has *Chlamydia* or symptoms of *Chlamydia*.

The best way to avoid becoming infected with *Chlamydia* and other STIs is not to have sexual contact or to be in a long-term, mutually monogamous relationship with a partner who is not infected.

To address the problem of eliminating trachoma as a public health disease a strategy has been developed by the WHO known by the acronym SAFE (Surgery, Antibiotics, Facial cleanliness, Environment improvement). Its goals are:

1. identification of endemic trachoma in a community;
2. lowering the immediate risk of visual loss in patients with trichiasis;
3. antibiotic treatment aimed at the reduction of the severity of inflammation and suppression of transmission of ocular *C. trachomatis*;
4. removing the risk factors for transmission of infection.

These risk factors include flies, which can act as physical vectors for transmission of *C. trachomatis*, and are able to carry *Chlamydia* that may transmit ocular infection, particularly in children in areas with poor sanitation. The presence of cattle pens has been associated with trachoma in some African countries. Crowded living conditions in the family unit constitute another factor increasing the risk of trachoma.

Interestingly it is accepted that trachoma was eliminated in Europe and North America as a result of improved living conditions, not treatment. Therefore special socio-economic measures have been recommended to reduce the risk of transmission of ocular *C. trachomatis* infection. They include increasing access to water, fly control interventions, education, and improved living conditions.

Vaccine

Due to its high incidence, relapses, prolonged bouts of infection, and significant morbidity *C. trachomatis* is an important target for vaccine development. A safe vaccine administered prior to adolescence and effective through child-bearing age would have a significant impact on the spread of the disease. The ideal vaccine should protect against multiple serovars, limit the duration of infection, and prevent chronic sequelae, including blindness, PID, and infertility. However, the attempts made so far have not been successful, with the immunity being short-lived and with some individuals experiencing more severe disease as compared with nonvaccinated individuals when contracting chlamydial infection. Current attempts are focused on the generation of antibodies as well as the induction of cellular memory response to *C. trachomatis*.

The major target for the vaccine development has been MOMP, which is an immunodominant antigen and determines serotype-specific immunity. **Monoclonal antibodies** to MOMP neutralize *C. trachomatis* infection *in vitro* and *in vivo*. However, MOMP-based vaccines have not been shown to protect against ocular *C. trachomatis* infection. Not surprisingly, identification of potential vaccine antigens is an active area of research that is greatly helped by the availability of the complete *C. trachomatis* genome sequence. This allows the identification and testing of candidate proteins based on their similarity to proteins important in protective immunity against other bacterial pathogens. Attempts have been made to create a multi-subunit vaccine and a DNA vaccine. No human trials are currently in progress.

SUMMARY

1. What is the causative agent, how does it enter the body and how does it spread a) within the body and b) from person to person?

- *Chlamydia trachomatis* belongs to the family Chlamydiaceae, which contains two genuses: *Chlamydia* and *Chlamydophila*. Bacteria of this family are obligatory intracellular human and animal pathogens.

- *Chlamydia trachomatis* and *Chlamydophila pneumoniae* are human pathogens. *C. trachomatis* can cause sexually transmitted urogenital infections, neonatal conjunctivitis and pneumonia, ocular trachoma, and urogenital infection aasociated with lymphadenopathy and lymphadenitis (LGV). *C. pneumoniae* can cause bronchitis, sinusitis, and pneumonia, and accelerates atherosclerosis.

- All Chlamydiaceae species have specific lipopolysaccharides and envelope proteins: 40 kDa major outer-membrane protein (MOMP), a hydrophilic cysteine-rich 60 kDa protein, and a low molecular weight cysteine-rich lipoprotein.

- *C. trachomatis* contains two human biological variants (biovars): trachoma and LGV, and one biovar infecting mice. Fifteen serological variants (serovars) have been identified in trachoma biovar. MOMP confers serovar specificity.

- *Chlamydia* infects nonciliated columnar epithelial cells, and LGV biovars can also infect macrophages. It attaches and enters the host cell through phagocytosis, receptor-mediated endocytosis, and pinocytosis.

- Chlamydiaceae species are characterized by a biphasic developmental cycle. The infectious EB form is typically 0.2–0.6 μm in diameter and resistant to unfavorable environmental conditions outside of their eukaryotic host cells.

- EB entry is followed by translocation of EB-containing endosomes to the perinuclear region and their homotypic fusion with each other forming a fusion vacuole – a nascent inclusion. EBs accumulate in an inclusion where they use a supply of nutrients via Golgi apparatus and mature into the RB form.

- The RB form (up to 1.5 μm in diameter) multiplies by binary fission. After several rounds of replication during 48–72 hours the RB form reverts to the infectious EB form. Eventually the EBs are released through the extrusion of the inclusions by reverse endocytosis or through apoptosis, or are released after cell lysis.

- *C. trachomatis* (serovars D–K) is sexually transmitted in the EB form, usually through vaginal intercourse but occasionally can be transmitted by oral and anal sex. Spread to the fallopian tubes in females it can lead to pelvic inflammatory disease (PID). It can also lead to conjunctivitis through autoinoculation.

- Genital infection of *C. trachomatis* (serovars D–K) is the leading bacterial cause of sexually transmitted infection with over 50 million new cases occurring yearly worldwide.

- Ocular infection of *C. trachomatis* (serovars A–C) is spread in areas of poverty and overcrowding. Every year *C. trachomatis* is a major cause of 500000 cases of trachoma worldwide, approximately 85 million people worldwide have active trachoma, more than 10 million have trichiasis (inturned eyelashes that touch the globe), and about 6 million people suffer visual loss and blindness.

- Infection can be transmitted from eye to eye by fingers, shared cloths or towels, by eye-seeking flies, and by droplets (coughing or sneezing).

- *C. trachomatis* (biovar LGV) causes sexually transmitted disease that is prevalent in Africa, Asia, and South America. Humans are the only natural host.

- Vertically *C. trachomatis* can be passed from mother to child during birth, causing conjunctivitis or pneumonia.

2. What is the host response to the infection and what is the disease pathogenesis?

- Both innate and adaptive immune responses are induced during *C. trachomatis* infection, but they are usually inefficient at controlling the infection. This may lead to the partial clearance of the pathogen from the body resulting in bacterial persistence, chronic inflammation, tissue damage, and severe clinical symptoms.

- Infection with *Chlamydia* invokes production of secretory IgA and circulatory IgM and IgG antibodies mostly directed to MOMP and Hsp60. Their effectiveness is unclear.

- One of the most powerful anti-chlamydial T-cell-mediated immune mechanisms is production of the Th1 cytokine IFN-γ. IFN-γ can limit *C. trachomatis* infection by activating macrophages, up-regulating expression of MHC, suppressing tryptophan production in the host cells, enhancing production of nitric oxide, and down-regulating transferrin receptor on the host cells.

- Infection does not stimulate long-lasting immunity and repeated episodes of infection are common. Re-infection results in an inflammatory response and subsequent tissue damage.

- One of the reasons for poor anti-chlamydial immunity is the ability of the pathogen to evade or block host immune responses by intracellular location, down-regulation of MHC class I molecules, prevention of the formation of phagolysosome, and induction of apoptosis of cytotoxic T cells.

- Pathogenesis is a result of a direct killing of infected cells and of chronic inflammation. The chronic inflammation induced by *C. trachomatis* infection leads to episodes of PID or conjunctivitis resulting in tubal infertility or blindness, respectively.

3. What is the typical clinical presentation and what complications can occur?

- Urogenital infection in men causes urethritis, which can produce a discharge from the penis or pain and burning sensation when urinating. If not treated, it can lead to epididymitis and, rarely, sterility.

- Urogenital infection in women leads to urethral discharge, dysuria and pyuria, urethritis, and cervical discharge and friability. If untreated, the chlamydial infection may spread through the uterus to the fallopian tubes, causing salpingitis, infertility, or ectopic pregnancy.

- A severe complication of the sexually transmitted *C. trachomatis* is PID.

- Inclusion conjunctivitis is associated with genital infections transferred from the genital tract to the eye by contaminated hands. The main symptom is a sensation of a foreign body in the eye, redness, irritation.

- Infants infected with *C. trachomatis* vertically from their mother at birth can develop pneumonia, which presents as a tachypnea, staccato cough and is often preceded by conjunctivitis.

- The first symptom of LGV is the development of a primary lesion, often in the penis or vagina. Several weeks after the primary lesion patients develop painful inguinal and/or femoral lymphadenopathy, fever, headache, and myalgia followed by inflammation of the draining lymph nodes.

4. How is the disease diagnosed and what is the differential diagnosis?

- Clinical diagnosis of genital chlamydial infection in women is based on age, sexual activity, more than one sexual partner, mucopurulent vaginal discharge, burning when passing urine, friable cervix or bleeding after sex or between menstrual periods, lower abdominal pain, or pain during sexual intercourse.

- Clinical diagnosis of LGV is based on the clinical presentation and the course of the disease assisted by differential diagnostic and laboratory tests.

- Eye examination for the clinical signs of trachoma involves careful inspection of the lashes and cornea, then reversion of the upper lid and inspection of the upper tarsal conjunctiva. A grading system is used by WHO for the assessment of the prevalence and severity of trachoma.

- For genital infections, the most common types of sample for diagnosis include cervical and vaginal swabs for women, and urethal swabs and first-void urine for men.

- For suspected LGV infections, bubo pus, saline aspirates of the bubo, swabs of the rectum, or biopsy specimens of the lower gastrointestinal tract are used.

- For conjunctival specimens, epithelial cells are collected by rubbing a dry swab over the everted palpebral conjunctiva.

- Cytological techniques are used to detect the inclusion bodies in stained cell scrapings, but this method lacks sensitivity.

- Cell culture has traditionally been the gold standard for the diagnosis of *C. trachomatis* and is specific for this pathogen. However, it is labor-intensive and has been gradually replaced by NAATs.

- Antibody-based laboratory techniques – immunofluorescence and enzyme-linked immunosorbent assay (ELISA) – are widely used for determining the serovars in screening programs.

- Serological tests are of some value but only in neonatal infection. A high titer of IgM antibodies is indicative of a recent infection, while anti-chlamydial IgG is mostly associated with past infection.

- The complement fixation test has been used for patients with LGV who have high complement-fixing antibody titers, but is now being replaced by NAATs.

- Nucleic acid amplification tests (NAATs), including polymerase chain reaction (PCR), are now increasingly used for diagnosis and differential diagnosis of chlamydial infections, although their wider use in screening strategies worldwide is limited by the costs and complexity of the equipment required.

5. How is the disease managed and prevented?

- For urogenital infections, the antibiotic doxycycline (Doryx®, Vibramycin®) is given 100 mg twice daily for 7 days. If the infected woman is pregnant or lactating 500 mg of erythromycin is given four times a day for 7 days or twice daily for 14 days. Alternatively a single-dose therapy with 1 g of erythromycin-derivative azithromycin can be used, which is usually effective for clearance of *C. trachomatis*.

- Since genital chlamydial infection often has no symptoms, all sexual partners should be tested for chlamydial infection and treated if positive. Patients should also be tested for other sexually transmitted infections.

- The recommended treatment for active trachoma is topical tetracycline, twice daily for 6 weeks orally. One 20 mg kg^{-1} dose of azithromycin can be used instead.

- Communities or districts with endemic trachoma should be mass treated with antibiotics to avoid re-infection. A strategy has been developed by WHO aimed at eliminating trachoma and known by the acronym SAFE.

- Environmental improvement includes removing the risk factors for transmission of infection such as flies, the presence of cattle pens, and crowded living conditions.

- Safe sexual practices, using a condom during sexual intercourse, and prompt treatment of infected patients and their sexual partners can prevent genital infections with *C. trachomatis*.

- The ideal vaccine against *C. trachomatis* should protect against multiple serovars, limit the duration of infection, and prevent chronic sequelae, including blindness, PID, and infertility.

- The major target for vaccine development is the immunodominant protein MOMP. The current knowledge of the complete genome sequence should allow the identification and testing of candidate proteins able to induce Th1 immune responses that are effective in the elimination of this pathogen.

FURTHER READING

Everett KDE. Chlamydiae. Encyclopedia of Life Sciences. John Wiley & Sons, 2002: www.els.net.

Mims C, Dockrell HM, Goering RV, Roitt I, Wakelin D, Zuckerman M. Medical Microbiology, 3rd edition. Mosby, Edinburgh, 2004.

Murphy K, Travers P, and Walport M. Janeway's Immunobiology, 7th edition, Garland Science, New York, 2008.

Murray PR, Rosenthal KS, Pfaller MA. Medical Microbiology, 5th edition. Elsevier Mosby, Philadelphia, PA, 2005.

REFERENCES

Brunham RC, Rey-Ladino J. Immunology of *Chlamydia* infection: implication for *Chlamydia trachomatis* vaccine. Nat Rev Immunol, 2005, 5: 149–161.

Burton MJ, Holland MJ, Faal N, et al. Which members of a community need antibiotics to control trachoma? Conjunctival *Chlamydia trachomatis* infection load in Gambian village. Invest Ophthalmol Vis Sci, 2003, 44: 4215–4222.

Corsaro D, Greub G. Pathogenic potential of novel *Chlamydiae* and diagnostic approaches to infections due to these obligate intracellular bacteria. Clin Microbiol Rev, 2006, 19: 283–297.

Dautry-Varsat A, Subtil A, Hackstadt T. Recent insights into the mechanisms of *Chlamydia* entry. Cell Microbiol, 2005, 7: 1714–1722.

Eko FO, He Q, Brown T, et al. A novel recombinant multisubunit vaccine against *Chlamydia*. J Immunol, 2004, 173: 3375–3382.

Gambhir M, Basáñez M-G, Turner F, Kumaresan J, Grassly NC. Trachoma: transmission, infection, and control. Lancet Infect Dis, 2007, 7: 420–427.

Hotez PJ, Ferris MT. The antipoverty vaccines. Vaccine, 2006, 24: 5787–5799.

Kelly KA. Cellular immunity and *Chlamydia* genital infection: induction, recruitment and effector mechanisms. Int Rev Immunol, 2003, 22: 3–41.

Loomis WP, Starnbach MN. T cell responses to *Chlamydia trachomatis*. Curr Opin Microbiol, 2002, 5: 87–91.

Mabey DCW, Solomon AW, Foster A. Trachoma seminar. Lancet, 2003, 362: 223–229.

McClarty G, Caldwell HD, Nelson DE. Chlamydial interferon gamma immune evasion influences infection tropism. Curr Opin Microbiol, 2007, 10: 47–51.

Miyari I, Byrne GI. *Chlamydia* and programmed cell death. Curr Opin Microbiol, 2006, 9: 102–108.

Pal S, Peterson EM, De La Maza LM. Vaccination with the *Chlamydia trachomatis* major outer membrane protein can elicit an immune response as protective as that resulting from inoculation with live bacteria. Infect Immun, 2005, 73: 8153–8160.

Reddy BC, Rastogi S, Das B, Salhan S, Verma S, Mittal A. Cytokine expression pattern in the genital tract of *Chlamydia trachomatis* positive infertile women – implication for T-cell responses. Clin Exp Immunol, 2004, 137: 552–558.

Solomon AW, Holland MJ, Burton MJ, et al. Strategies for control of trachoma: observational study with qualitative PCR. Lancet, 2003, 362: 198–204.

Stephens RS, Kalman S, Lammel C, et al. Genome sequence of an obligate intracellular pathogen of humans: *Chlamydia trachomatis*. Science, 1998, 282: 754–759.

Subtil A, Dautry-Varsat A. *Chlamydia*: five years A.G. (after genome). Curr Opin Microbiol, 2004, 7: 85–92.

West SK. Trachoma: new assault on an ancient disease. Prog Retin Eye Res, 2004, 23: 381–401.

Wright HR, Taylor HR. Clinical examination and laboratory tests for estimation of trachoma prevalence in a remote setting: what are they really telling us? Lancet Infect Dis, 2005, 5: 313–320.

Wyrick PB. Intracellular survival by *Chlamydia*. Cell Microbiol, 2000, 2: 275–282.

WEB SITES

Centers for Disease Control and Prevention, Atlanta, GA, USA: www.cdc.gov

Encyclopaedia of Life Sciences: www.els.net

Health Protection Agency, UK. © Health Protection Agency: www.hpa.org.uk

International Trachoma Initiative, Copyright © 2008 The International Trachoma Initiative. All rights reserved: http://www.trachoma.org

Microbiology and Immunology Online, School of Medicine, University of South Carolina: http://pathmicro.med.sc.edu/book/virol-sta.htm

National Institutes of Health, Department of Health and Human Services, USA: www.nih.com

Sexually Transmitted Diseases Diagnostics Initiative, World Health Organization, © WHO/OMS 2001: www.who.int/std_diagnostics

WEB SITES

Sexually Transmitted Diseases Services Clinic, Royal Adelaide Hospital, Adelaide, Australia. Copyright © Department of Health 2005: www.stdservices.on.net/ std/chlamydia

Student.bmj.com is a website owned by the BMJ Publishing Group Limited: www.studentbmj.com

Wikipedia – The Free Encyclopedia: http://en.wikipedia.org/wiki/Main_Page

World Health Organization, © WHO 2008: www.who.int

MULTIPLE CHOICE QUESTIONS

The questions should be answered either by selecting True (T) or False (F) for each answer statement, or by selecting the answer statements which best answer the question. Answers can be found in the back of the book.

1. **Which of the following are true?**
   A. Chlamydiaceae are obligate intracellular pathogens.
   B. *Chlamydia trachomatis* is the only known human pathogen in the Chlamydiaceae Family.
   C. *Chlamydia* species are characterized by tissue tropism.
   D. Their principle membrane protein is 40 kDa major outer-membrane protein (MOMP).
   E. *Chlamydia* species express a high density of surface proteoglycans (PGs).

2. **What is *Chlamydia trachomatis* characterized by?**
   A. A biphasic developmental cycle.
   B. Three human biological variants (biovars): trachoma, genital, and lymphogranuloma venereum (LGV).
   C. One serological variant (serovar).
   D. High spread worldwide.
   E. Production of cytotoxins.

3. **Is the following true for the life cycle of *C. trachomatis*?**
   A. The infectious EB form of *C. trachomatis* infects host epithelial cells and phagocytes.
   B. The entry of the EB form occurs through phagocytosis, endocytosis, and pinocytosis forming endosomes called inclusions.
   C. EB-containing endosomes fuse with host cell lysosomes.
   D. EB forms replicate in the inclusion.
   E. Progeny EB forms are released from the host cell by lysis, reverse endocytosis, or apoptosis.

4. **How is *C. trachomatis* spread?**
   A. By blood transfusion (ocular infection).
   B. Coughing and sneezing (ocular infection).
   C. By fingers, shared cloths or towels (ocular infection).
   D. Through vaginal intercourse (genital infection).
   E. From mother to child during birth.

5. **Which of the following is true about the host response to *C. trachomatis* infection?**
   A. Immune response is insufficient and evokes chronic inflammation.
   B. The most powerful T-cell-mediated immune response is production of IL-4.
   C. Specific IgG and IgM antibodies are directed towards bacterial lipopolysaccharide.
   D. *C. trachomatis* blocks expression of MHC molecules on host cells.
   E. Interferon-γ enhances phagocytosis of *C. trachomatis*.

6. **Is the following true of urogenital infection caused by *C. trachomatis* in women?**
   A. The symptoms develop immediately after infection.
   B. There is a cervical discharge in women.
   C. Dysuria and pyuria occur.
   D. Cervicitis and urethritis can occur.
   E. Even treated infection always leads to infertility.

7. **Which of the following laboratory tests are used for the diagnosis of chlamydial infection?**
   A. Nucleic acid amplification tests (NAATs).
   B. Blood cell count.
   C. Microscopy of biopsies.
   D. Susceptible cell cultures.
   E. Polymerase chain reaction (PCR).

8. **Which of the following drugs are used for the treatment of chlamydial infections?**
   A. Erythromycin and azithromycin.
   B. Tetracyclines.
   C. Amoxicillin.
   D. Rifampicin.
   E. Doxycycline

Case 5

Clostridium difficile

An elderly lady, of no fixed abode, arrived at the hospital's emergency department after having fallen. She was admitted to hospital for fixation of a fracture of the hip. Shortly after admission she developed signs of a chest infection and was started on a cephalosporin, which she remained on for a week. Subsequently she developed profuse watery diarrhea and abdominal pain. A feces sample was sent to the laboratory to test for the toxins of *Clostridium difficile*, which proved positive and she was commenced on oral vancomycin. Despite treatment the diarrhea persisted and it also failed to respond to a course of metronidazole. The condition of the patient worsened and a **sigmoidoscopy** was performed, revealing that she had **pseudomembranous colitis** (Figure 1). Her clinical condition deteriorated and she developed **toxic megacolon** and an emergency **colectomy** was performed. The patient died shortly after the operation.

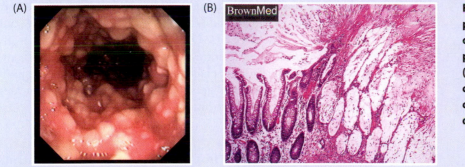

(A)   (B) BrownMed

Figure 1. Endoscopic view (A) of a patient with pseudomembranous colitis showing plaques of pseudomembranes. Histologically (B) the pseudomembranes are composed of mushroom-shaped collections of neutrophils, cellular debris, and fibrin.

1. What is the causative agent, how does it enter the body and how does it spread a) within the body and b) from person to person?

Causative agent

Clostridium difficile is an anerobic spore-forming motile gram-positive rod measuring $0.5–1.0 \times 3.0–16.0$ μm (Figure 2). It produces irregular white colonies on blood agar (Figure 3). The species was named 'difficile' because initially it was hard to culture. It is saccharolytic and produces three **exotoxins**. Toxin A (TcdA) and B (TcdB) induce glucosylation of G-proteins (Rho, Rac, and Cdc42) and ultimately affect actin polymerization leading to loss of tissue integrity. These two toxins, particularly toxin B, are associated with disease in humans. The toxin A-negative strains are mentioned in Section 2 below. The third toxin is a binary toxin that has ADP-ribosylating activity. The clinical correlates of this toxin are currently unclear, but it may be associated with increased **virulence**. The structure of the toxin gene is shown in Figure 4. Toxin A and B production occurs maximally in stationary phase and during nutrient limitation. Carbohydrate limitation results in a switch to amino acid fermentation, which in turn results in a shortage of amino acids and toxin production.

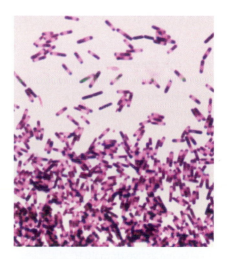

Figure 2. Gram stain of *Clostridium difficile* showing that it is a gram-positive rod. The spore can be clearly seen as a subterminal clear area in most of the bacilli.

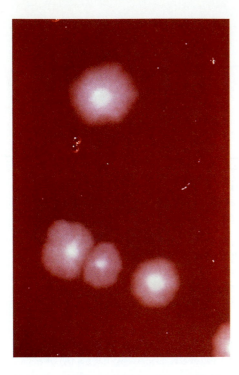

Figure 3. Colonies of *Clostridium difficile* showing the typical appearance of white colonies with crenated edges. Under UV light the colonies fluoresce a greenish-yellow color.

Low levels of biotin also increase toxin production. During exponential growth, TcdC is high and it is believed to be an anti-sigma factor negatively regulating toxin production. TcdC mutants have been described. These strains produce increased amounts of toxins A and B *in vitro*, particularly in stationary phase. These mutant strains appear to be particularly virulent. TcdD is not an exotoxin *per se*, but a positive regulator of transcription of toxin genes and is also responsible for the temperature-dependent regulation of toxin production. The sigma factor TcdR is also a positive regulator, although it is itself under some form of regulation. Additionally, *C. difficile* has a luxS-type quorum-sensing signaling system, although evidence suggests that addition of the inducer (AI-2) has no effect on toxin production.

Several methods of typing *C. difficile* are available, the most frequently used being **pulse field gel electrophoresis**, and **ribotyping** (Figure 5), although **multilocus enzyme electrophoresis** (**MLEE**) and analysis of variable number of tandem repeats (VNTR) may also be used. In the UK the common ribotypes are 001, 106, and more recently 027, these three accounting for 70% of all isolates sent to the reference laboratory.

Phylogenomic studies of *C. difficile* using micro-arrays demonstrate four clades: a hypervirulent **clade,** a TcdA-negative/TcdB-positive clade, and two other human and animal clades. Differences between the clades are related to virulence, ecology, antibiotic resistance, motility, adhesion, and metabolism. The hypervirulent clade is characterized by a single nucleotide deletion in a regulatory gene (TcdC) resulting in a truncated protein and increased toxin production.

Entry into the body

Following the ingestion of spores from contaminated feces or contaminated environment, the organism colonizes mainly the large intestine of the gastrointestinal tract. The dynamics of the complex ecosystem that is the microflora of the gastrointestinal tract is little understood. Many studies have demonstrated that the microflora in an individual is rather stable, although variations have been shown in relation to diet and stress.

Figure 4. This figure shows the arrangement of the operon for toxins A and B of *Clostridium difficile*. The expression of the toxin genes (*TcdA* and *TcdB*) is controlled by positive (*TcdD*) and negative (*TcdC*) regulators, which respond to changes in the environment. The operon also carries a gene, *TcdE*, which produces a protein that is thought to assist in the release of the toxins across the bacterial cell wall. Toxin A is a 308 kDa protein and toxin B is a 270 kDa protein. Each protein has an enzyme active domain E (glucosyltransferase); a domain responsible for binding B, which consists of multiple oligopeptide repeat units; and a domain T that is involved in translocation of the toxin across the host cell membrane.

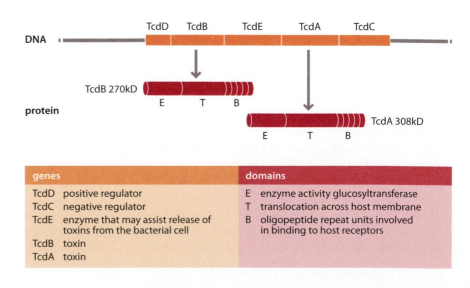

genes		domains	
TcdD	positive regulator	E	enzyme activity glucosyltransferase
TcdC	negative regulator	T	translocation across host membrane
TcdE	enzyme that may assist release of toxins from the bacterial cell	B	oligopeptide repeat units involved in binding to host receptors
TcdB	toxin		
TcdA	toxin		

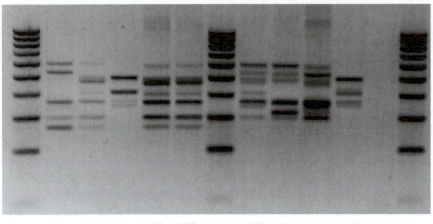

Figure 5. Ribotyping of *Clostridium difficile* – photograph of a gel showing some different common ribotypes of *C. difficile*. The pattern of bands produced by the restriction enzyme is characteristic for the different ribotypes and the patterns can be used to determine if there is cross-infection or a common source outbreak where the bands would be identical. Because some ribotypes are currently very common, e.g. 027, a more discriminatory typing system, e.g. variable number of tandem repeats (VNTR) may be used to subdivide the common ribotypes.

C. difficile is present in the intestines of 2–3% of asymptomatic adults, but may increase to 30% in hospital, and in between 60 and 70% of neonates. *C. difficile* is a **commensal** and normally does not cause any disease. In neonates it is believed, based mostly on animal data, that they do not have the necessary adhesion to allow the toxin to bind. In adults who develop the illness the usual history is of an elderly patient coming into hospital, either already carrying *C. difficile* or acquiring it in hospital, who is given antibiotics and develops diarrhea associated with the *C. difficile*. The toxin is then produced causing cell damage and leading to disease. Current thinking is that first antibiotics are given, which disturb the normal flora and appear to create an ecological niche where *C. difficile* can proliferate. Those who arrive in the hospital already carrying the organism appear relatively protected from disease. Thus, arriving as a carrier may represent prior exposure with subsequent development of some type of immunity, as yet poorly defined. The environmental factors that stimulate the toxin genes to be switched on and cause disease are unknown. In experimental systems it is known that consumption of some amino acids or the production of volatile fatty acids by the indigenous microflora inhibits the proliferation of *C. difficile*. It is also known that viable intestinal microflora is important in inhibition of outgrowth of *C. difficile*, thus supporting some metabolic function. Whether these mechanisms occur in humans is unknown but it is possible. It is certainly known that patients can be symptomatic with *C. difficile* isolated from the feces many days before the detection of toxin in the feces. Why some patients get mild disease and some get severe disease is not known. Why disease affects the elderly more commonly than younger patients is also not known, although the disease is occurring increasingly in the younger age group, which might be associated with more virulent strains (see Section 3).

Spread within the body

The organism remains within the gastrointestinal tract and only rarely is extra-intestinal disease reported.

Spread from person to person

Patients with diarrhea, especially if it is severe or accompanied by incontinence, may unintentionally spread the infection to other patients in hospital. In addition, the ability of *C. difficile* to form spores enables it to survive for long periods in the environment, for example on floors and around toilets, and protects it against heat and chemical disinfectants.

Epidemiology

C. difficile is a well-recognized hospital 'superbug' that has received considerable media attention and generated much public concern. It is the commonest cause of hospital-acquired diarrhea in the USA, with 250 000 cases annually at a cost to the health-care system of $1 billion. In the UK, cases are subject to mandatory reporting and in 2004 the total number of cases was 44 488, which was an increase of 23% on 2003 figures. In 2007 the number of cases recorded in patients over 65 years was 49 785 and since mandatory reporting of cases less than 65 years old was introduced in April 2007 the number of cases reported in this age group to the end of 2007 was 7597. In the UK the hypervirulent strain ribotype 027 in particular has been associated with hospital outbreaks such as in Stoke Mandeville Hospital and several other hospitals in the South East of England. In Canada there is a similar epidemiology with the newly recognized hypervirulent strain (NAP1, ribotype 027). This strain produces large amounts of toxin A and B and in some strains a binary toxin. This strain has been isolated across Europe, North America, and globally, and has been associated with severe disease that is refractory to treatment, both in hospitals and in community-acquired cases – in the latter apparently in the absence of prior antibiotic exposure. However, severe disease can also occur with other ribotypes.

2. What is the host response to the infection and what is the disease pathogenesis?

Immune responses

Although some information has accumulated on the development of immune responses during infection with *C. difficile*, their role is unclear in both controlling the tissue damage and protection against relapses, commonly seen in adult patients following initial treatment with antibiotics.

Both TcdA and B toxins induce an inflammatory response involving a number of factors including **interleukin** (IL)-8, IL-6, **tumor necrosis factor-α** (**TNF-α**), leptin, and **substance P**. The precise pathways are not clear. TcdA stimulates the vanilloid receptor, VR1, which leads to the release of substance P from sensory neurons (and increases the expression of its receptor – neurokinin 1-R), which in turn leads to the release of **neurotensin** and mast cell degranulation. TcdA also increases the release of endocannabinoids, which increase the release of more substance P via stimulation of VR1. Both toxins also induce an inflammatory response with infiltration of polymorphonuclear leukocytes (PMNs) into the **lamina propria** by up-regulation of IL-8 secretion from sensory neurons and colonocytes. The polymorphism within the IL-8 gene may be related to the presentation of more severe disease.

Surface layer proteins from the organism induce a mixed pro-/anti-inflammatory **cytokine** response with the release of IL-1β, IL-6, IL-10, and IL-12, but the significance of this is unclear.

Systemic and mucosal antibodies (**IgG** and **IgA**) are produced by the host to the A and B toxins in some patients following *C. difficile* diarrhea and antibodies to surface layer proteins of the organism are also found. Some of these also appear to be neutralizing antibodies. Low levels of antibodies with low avidity for A and B toxins, typically acquired early in life, have been found in the serum of up to 70% of asymptomatic adults, which are probably not effective in preventing the initial infection.

Understanding the immune response to this commensal is clearly of importance in protection against the severe consequences of infection.

Pathogenesis

TcdA binds to carbohydrate Gal-β 1,4N-Acetyl Glucosamine by repeat units at the C-terminal end of the toxin. Both TcdA and B are taken up into the cytosol of host colonocytes and are glucosyltransferases that inactivate Rho, Rac, and Cdc42, thereby affecting actin polymerization. This leads to impairment of tight junctions, cell rounding up, and cell death (Figure 6). TcdB activates Ca influx, which is necessary for the disintegration of the actin cytoskeleton. TcdA inhibits stimulated mucin exocytosis. Both TcdA and B induce fluid loss.

3. What is the typical clinical presentation and what complications can occur?

C. difficile is associated with antibiotic-associated diarrhoea (CDAD) at one extreme to pseudomembranous colitis (PMC) at the other. The latter has a significant mortality. *C. difficile* is largely linked to hospital admission and has been associated with hospital outbreaks. Infection has usually occurred in those aged over 65 years and usually in association with antibiotics, bowel surgery or chemotherapeutic agents. The illness may follow antibiotics taken briefly or even several weeks previously. Recently community

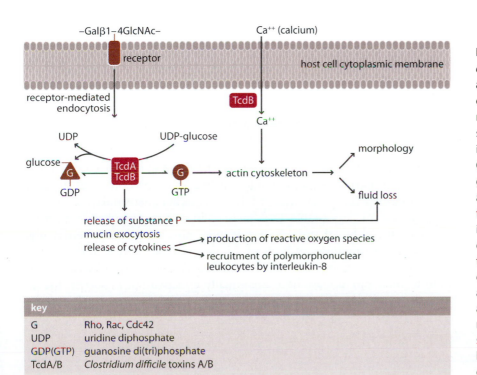

Figure 6. Mode of action of *Clostridium difficile* toxins A/B. This figure shows the action of the toxins A and B on the host cell. The toxins are taken up by receptor-mediated endocytosis after binding to specific glycoprotein receptors. The toxins inhibit the normal functioning of G proteins Rho, Rac, and Cdc42 by glucosylating them. This inactivation affects the actin cytoskeleton, which in turn affects cell morphology and the integrity of the tight junctions between the cells, thereby increasing fluid loss. The toxins also stimulate the release of cytokines, e.g. IL-8, which recruits PMNs and establishes an inflammatory response and the production of reactive oxygen radicals. The toxins increase the release of substance P, a peptide that enhances fluid loss from the intestines and decreases the exocytosis of the protective mucus barrier.

cases of *C. difficile*-associated disease have been reported in a younger age group who have not necessarily been taking antibiotics. These relatively rare cases have been caused by hypervirulent strains such as ribotype 027.

Patients present with colicky abdominal pain, bloating, and watery diarrhea with an almost characteristic smell for *C. difficile*. In severe cases (PMC) there may be blood in the feces and the colon may perforate. Toxic mega-colon may occur, which can paradoxically cause diarrhea to stop or in some cases not to appear at all, and colectomy may be required. Thus *C. difficile* with little or no diarrhea, and pain predominating may be an ominous sign.

Severe damage to the large bowel can result in rupture or perforation.

The rare patients with extra-intestinal disease almost always have long-term carriage of the organism or severe gastrointestinal disease. Diseases reported include **bacteremia**, **abscess**, and prosthetic hip infection.

4. How is this disease diagnosed and what is the differential diagnosis?

The routine diagnosis of *C. difficile*-associated disease is detection of the toxins in the feces by **enzyme immunoassays (ELISA)**. An additional assay that is likely to be useful is the detection of glutamate dehydrogenase in the feces as a surrogate marker for the organism. Culture is also available and is important for the investigation of suspected outbreaks where the isolates can be ribotyped. The organism may be grown on cycloserine/cefoxitin/egg yolk agar and a variety of other more sensitive media, always anaerobically producing 2–3 mm gray-white colonies. **Polymerase chain reaction (PCR)** for toxin-related genes is also being used in some laboratories. Pseudomembranous colitis is diagnosed by colonoscopy where raised yellowish plaques may be seen on the mucosa. If severe disease is suspected limited **sigmoidoscopy** may be used as colonoscopy may result in perforation.

The differential diagnosis includes other infective causes of diarrhea, inflammatory bowel disease, and **diverticulitis**.

5. How is the disease managed and prevented?

Management

For patients who are on antibiotics and who develop diarrhea, one should stop the antibiotics if at all possible. Such a decision is on a case by case basis. The first-line treatment of CDAD is either metronidazole or vancomycin for 14 days (Table 1). Vancomycin and metronidazole must be given orally, although metronidazole may also be given **parenterally**. Pulsed or tapering doses of both antibiotics have been given at the end of the course of treatment and vancomycin has also been given by enema. In mild to moderate disease both are equally effective but in severe disease vancomycin was more effective in a controlled trial. *In vitro* studies, on the

Table 1. Treatment of *C. difficile*-associated disease	
First-line	Vancomycin 125–500 mg four times daily ODS 14 days
	OR
	Metronidazole 250–500 mg four times daily QDS 14 days
Alternatives	Cholestyramine 4 g TDS 14 days
	Bacitracin-fuscidic acid/rifaximin
	Probiotic/fecal enema
	Intravenous immunoglobulin

other hand, indicate that metronidazole has a more rapid bactericidal effect compared with vancomycin. Other antibiotics that have been used are bacitracin, fusidic acid, nitazoxanide, and rifaximin. Tolevamer and difimicin also show some promise in the treatment. Rifampin, which had been recommended, has not been effective in a controlled trial. Alternative treatments are also available including cholestyramine (which binds the toxin but also binds vancomycin, so the two should not be given simultaneously); **monoclonal** anti-toxin **antibodies**; **intravenous immune globulin** (although there are no controlled trials and reported results are contradictory); probiotics such as *Saccharomyces boulardiae* (which is really a strain of *S. cerevesiae*) have been used in conjunction with antibiotics; and fecal enema (or fecal administration by oral–duodenal tube) from a close relative, to re-establish a 'normal' flora.

Prevention

In hospitals and nursing homes, prevention is by strict control-of-infection procedures. Contact precautions for any patients in whom CDAD is suspected are essential. The patient should be isolated in a single room with dedicated toilet facilities and the use of personal protective equipment (PPE), that is gloves and apron. If more than one patient has the illness the patients should be nursed in a dedicated ward. A decision whether to close the ward/unit must be made on an individual basis but if there is evidence of widespread cross-infection then the unit should be closed. Horizontal and 'high touch' surfaces and other items within the immediate environment, particularly toilet facilities, should be regularly cleaned with a hypochlorite disinfectant. All items and rooms after discharge of the patient(s) should be cleaned in the same manner. Attention to hand-washing practices is important and should be undertaken with soap and water rather than alcohol, as the latter has no effect on the spores of *C. difficile*. In fact, the spores of *C. difficile* are highly resistant to many of the generic disinfectants.

There is currently no vaccine although research is ongoing with at least two candidate vaccines.

SUMMARY

1. What is the causative agent, how does it enter the body and how does it spread a) within the body and b) from person to person?

- *Clostridium difficile* is a gram-positive sporulating anaerobe.

- Several typing methods are available including ribotyping and PFGE.

- The spore is acquired by **feco-oral** spread or from the environment and infects colonocytes of the host.

- Up to 70% of neonates may be colonized without disease.

- Asymptomatic colonization may occur in up to 30% of patients in hospital.

- A hypervirulent strain (ribotype 027) first isolated in Canada produces large amounts of toxin and has spread across Europe, North America, and globally.

- *C. difficile* produces three toxins: toxin A and B and sometimes a binary toxin.

- Important risk factors for disease are increased age (> 60 years) and taking antibiotics.

- The organism only rarely causes extra-intestinal disease.

2. What is the host response to the infection and what is the disease pathogenesis?

- Antibodies to toxins A and B are produced following *C. difficile* diarrhea, some of which are neutralizing.

- The toxins bind to host receptors on colonocytes and are taken up into the cell.

- Toxins A/B induce disaggregation of the actin cytoskeleton by inactivating GTPase proteins.

- Impairment of tight junctions and cell death occurs leading to fluid loss and a localized inflammatory response.

- Both the toxins and the inflammatory response itself are thought to be involved in tissue damage.

3. What is the typical clinical presentation and what complications can occur?

- The patient presents with abdominal pain and diarrhea.

- Frequently the patient is aged > 60 years, in hospital, and is on antibiotics.

- Community-acquired disease in younger patients who have not been on antibiotics is relatively rare and associated with hypervirulent strains.

- The disease may be mild to moderate CDAD or severe PMC.

- Complications include perforation or toxic megacolon.

4. How is this disease diagnosed and what is the differential diagnosis?

- The disease is routinely diagnosed by detection of the toxins in the feces using enzyme immunoassays.

- Culture is important for typing the organism in identifying and managing outbreaks.

- PMC is diagnosed by colonoscopy or sigmoidoscopy.

- Differential diagnosis includes other causes of gastroenteritis, diverticulitis and inflammatory bowel disease, and antibiotic-associated diarrhea not due to *C. difficile*.

5. How is the disease managed and prevented?

- The first-line treatment is oral metronidazole or vancomycin for 14 days.

- Alternative treatments include agents to bind the toxin, probiotics, or fecal enema.

- In hospital, prevention is by standard control of infection methods.

FURTHER READING

Cimolai N. Laboratory Diagnosis of Bacterial Infections. Marcel Dekker, NY, 2001: 731–734.

Mandell GL, Bennet JE, Dolin R. Principles & Practice of Infectious Diseases, 6th edition, Vol 3. Elsevier/Churchill Livingstone, 2005: 1249–1259.

Mayhall GC. Hospital Epidemiology & Infection Control, 3rd edition. Philadelphia, Lippincott Williams & Wilkins, 2004: 623–631.

Murphy K, Travers P, and Walport M. Janeway's Immunobiology, 7th edition, Garland Science, New York, 2008.

REFERENCES

Ausiello CM, Cerquetti M, Fedele G, et al. Surface layer proteins from *Clostridium difficile* induce inflammatory and regulatory cytokines in human monocytes and dendritic cells. Microbes Infect, 2006, 8: 2640–2646.

Durai R. Epidemiology, pathogenesis and management of *Clostridium difficile* infection. Dig Dis Sci, 2007, 52: 2958–2962.

Matamouros S, England P, Dupuy B. *Clostridium difficile* toxin expression is inhibited by the novel regulator TvdC. Mol Microbiol, 2007, 64: 1274–1288.

McMaster-Baxter NL, Musher DM. *Clostridium difficile*: recent epidemiological findings and advances in therapy. Pharmacotherapy, 2007, 27: 1029–1039.

Owens RC. *Clostridium difficile* associated disease: changing epidemiology and implications for management. Drugs, 2007, 67: 487–502.

Peled N, Pitlik S, Samra Z, Kazakov A, Bloch Y, Bishara J. Predicting *Clostridium difficile* toxin in hospitalized patients with antibiotic associated diarrhoea. Infect Control Hosp Epidemiol, 2007, 28: 377–381.

Smith JA, Cooke DL, Hyde S, Borriello SP, Long RG. *Clostridium difficile* toxin binding to human intestinal epithelial cells. J Med Microbiol, 1997, 46: 953–958.

Voth DE, Ballard JD. *Clostridium difficile* toxins: mechanism of action and role in disease. Clin Microbiol Rev, 2005, 18: 247–263.

WEB SITES

Centers for Disease Control and Prevention (CDC), Division of Healthcare Quality Promotion, 2008, United States Public Health Agency: http://www.cdc.gov/ncidod/dhqp/id_Cdiff.html

Centre for Infections, Health Protection Agency, HPA Copyright, 2008: http://www.hpa.org.uk/infections/topics_az/clostridium_difficile/C_diff_faqs.htm

Department of Health, National Health Service, Crown Copyright 2008: http://www.dh.gov.uk/en/Policyandguidance/Healthandsocialcaretopics/Healthcareacquiredinfection/Healthcareacquiredgeneralinformation/DH_4115800

Infection Control Services, Ltd, 2007, consultant microbiologists at University College London Hospitals providing advice about infection control to hospitals in the private sector: http://www.infectioncontrolservices.co.uk/

Institute of Biomedical Sciences, 2005, the professional body for biomedical scientists in the United Kingdom: http://www.ibms.org/pdf/bs_articles_2005/clostridium_difficile.pdf

MULTIPLE CHOICE QUESTIONS

The questions should be answered either by selecting True (T) or False (F) for each answer statement, or by selecting the answer statements which best answer the question. Answers can be found in the back of the book.

1. **Which of the following are true of *Clostridium difficile*?**
   A. It is an anaerobic bacterium.
   B. It produces spores.
   C. It produces a zinc metalloprotease as the only toxin.
   D. Toxin production is controlled by a sigma factor.
   E. A **phage typing system** is most useful in differentiating strains.

2. **Which of the following are true concerning the epidemiology of *Clostridium difficile*?**
   A. Asymptomatic carriage in hospital is uncommon.
   B. Commonly associated with outbreaks in hospital.
   C. Rarely causes extra-intestinal disease.
   D. A recognized hypervirulent strain is ribotype 027.
   E. Neonates may be colonized but do not acquire disease.

3. **Which of the following are known to be important in the pathogenesis of *Clostridium difficile*-associated disease?**
   A. Attaching and effacing lesions.
   B. Production of glucosyltransferases by the organism.
   C. Vacuolating cytotoxin.
   D. Taking antibiotics.
   E. Disruption of the actin cytoskeleton.

4. **Which of the following are important in the differential diagnosis of *Clostridium difficile*-associated disease?**
   A. Leptospirosis.
   B. Toxic megacolon.
   C. *Giardia* gastroenteritis.
   D. Inflammatory bowel disease.
   E. Duodenal ulcer.

5. **Which of the following are true statements concerning the diagnosis of *Clostridium difficile*?**
   A. The Gram stain of biopsy material is useful.
   B. Detection of toxins in the blood is the main way of diagnosis.
   C. Culture of the organism is useful for epidemiological investigation.
   D. Colonoscopy is important.
   E. Ribotyping is important.

6. **Which of the following are used in the treatment of *Clostridium difficile*-associated disease?**
   A. Ampicillin.
   B. Cephalosporins.
   C. Metronidazole.
   D. Vancomycin.
   E. Clindamycin.

Case 6

Coxiella burnetii

In 1998 a 35-year-old woman underwent surgery for **aortic coarctation** with the interposition of a **Gortex tube**. The operation was successful and there were no complications. In summer 2006 the woman had spent a month living in a village in Sri Lanka where she had been helping with charity work and was in close contact with newborn cows. In January 2007, the patient was admitted for investigation because of high fever (39.8°C). It turned out that during the previous year she had had several episodes of fever, which resolved without medical help. On physical examination her pulse rate was 120 beats/minute and an arterial pressure 110/46 mmHg. Femoral pulses were present with a **bruit** on the left side, and distal pulses were perceptible. There were no neurological abnormalities and the chest X-ray was normal. However, laboratory tests showed a white blood cell count of 8500 mm^{-3} and creatinine of 1.9 mg dl^{-1}.

Transthoracic echocardiography (TTE) ruled out the diagnosis of cardiac **endocarditis**, but detected large vegetations in the prosthetic tube (Figure 1A). Although blood cultures were sterile, serological tests revealed elevated levels of **IgG** (titer 6400) and **IgA** (titer 6400) antibodies to phase I *Coxiella burnetii* consistent with the chronic Q fever diagnosis. **IgM** titers were low confirming the chronic type of infection.

Antibiotics were immediately prescribed: doxycycline 200 mg per day and ofloxacin 400 mg per day. After 1 week, the fever decreased, but TEE still showed large vegetations on the graft. The patient was referred for surgery where the prosthetic graft was resected and several large vegetations revealed including those in the lumen of the tube. Due to the infection the prosthetic graft was replaced by a homograft. *C. burnetii* was later detected in the aortic specimen (Figure 1B) by **polymerase chain reaction (PCR)** and was isolated in Vero cell culture.

After surgery, a different combination of antibiotics was prescribed: doxycycline and hydroxychloroquine. The patient was discharged after 6 weeks and continued antibiotic treatment with serologic monitoring. After 10 months the patient was well, although antibiotic therapy was continuing. There was no recurrent infection of the homograft tube.

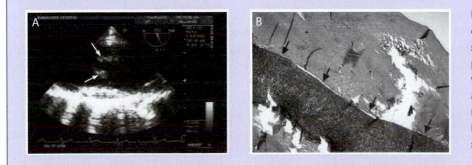

Figure 1. (A) Transthoracic echocardiography (TEE) with arrows showing large vegetations in the prosthetic graft. **(B)** Vascular tissue with foreign material indicated by arrows, and showing fibrosis and nonspecific chronic inflammatory infiltrate (hematoxylin and eosin stain, ×200).

1. What is the causative agent, how does it enter the body and how does it spread a) within the body and b) from person to person?

The patient is infected with *Coxiella burnetii* and is suffering from chronic Q fever. *C. burnetii* is an obligate intracellular bacterium with a complex life cycle and related morphological heterogeneity. It is a pleomorphic coccobacillus 0.3–1.0 μm in size (Figure 2). *C. burnetii* has a cell wall built of approximately 6.5 nm thick outer and inner membranes, which are

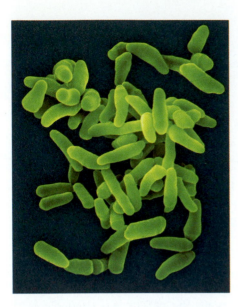

Figure 2. *C. burnetii*, microscopic image (×2200).

separated by a peptidoglycan layer (Figure 3). A spore-like stage does not have dipicolinic acid or a spore coat with cysteine characteristic for other gram-positive bacterial spores. Due to these features *C. burnetii* is considered to be a gram-negative bacterium, although it is almost impossible to stain *C. burnetii* by the Gram technique. The Gimenez staining method is usually used instead.

C. burnetii has been traditionally classified with the Rickettsiales order, the Rickettsiaceae family, and the Rickettsiae tribe, which it shared with the genera *Rickettsia* and *Rochalimaea*. However, subsequent rRNA sequence analysis has demonstrated that the *Coxiella* genus belongs to the gamma subdivision of Proteobacteria and is closely related to the genera *Legionella* and *Rickettsiella*. *C. burnetii* genomic analysis revealed more genes regulating metabolic processes than in other obligate intracellular bacteria such as *Chlamydia* (see Case 4) and *Rickettsia*.

Q fever is a **zoonosis** and *C. burnetii* is found in almost all animal species, particularly in wild and domestic mammals, birds, and arthropods such as ticks. Infected animals most often shed *C. burnetii* into the environment during parturition. As many as 10^9 bacteria can be found in 1 g of placenta at birth. *C. burnetii* can also contaminate the environment via animal urine, feces, and milk. Coxiellosis is more frequent in dairy cows and goats than in sheep. More rarely *C. burnetii* infection has been identified in horses, camels, buffalos, swine, rabbits, rats, and mice, and has been isolated from chickens, ducks, geese, turkeys, and pigeons.

Humans usually become infected from cattle, goats, sheep, domestic ruminants, and pets or contaminated dung and bedding. *C. burnetii* may be transmitted to humans also from consumption of raw eggs, raw milk, and cheese or inhalation of infected **fomites**. Cats and dogs can become infected through tick bites, eating contaminated placentas or milk, and by the aerosol route. There are over 40 tick species naturally infected with *C. burnetii*. *C. burnetii* multiplies in the gut cells of infected ticks and is shed with their feces onto the skin of the animal host. Although *C. burnetii* in ticks are in highly infectious phase I stage (see below), the tick-mediated route of infection is not considered essential in domestic animals compared with their close contact. However, ticks may contribute to the infection of wild animals such as rodents and wild birds. The role of other arthropods in coxiellosis transmission – mosquitoes, lice, mites, flies, and fleas – is controversial. Unlike rickettsial diseases, *C. burnetii* is not believed to be transmitted to humans via tick bites.

Genetic variability

There are a variety of *C. burnetii* strains (genovars) that are closely associated with their **virulence**. Genomic groups I, II, and III belong to so-called

Figure 3. Diagrammatic illustration of *C. burnetii* cell structure.

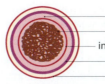

periplasmic space
core
intracellular membrane
peptidoglycan
outer membrane

acute strains, which infect animals, ticks, and humans and cause acute Q fever in humans. Genomic groups IV and V belong to so-called chronic strains and cause human Q fever endocarditis (see Section 3). Genomic group VI *C. burnetii* was isolated from feral rodents in Dugway (Utah, USA) and has unknown pathogenicity. Acute strains all have QpH1 plasmid, and chronic strains have QpRS plasmid or integrated QpRS sequences. Although genomic strain variations are associated with the geographic distribution of isolates it appears that host susceptibility to *C. burnetii* is more important for infection.

There are two variations of *C. burnetti*, based on their **lipopolysaccharide (LPS)** structure, as has been shown *in vitro*. The virulent natural form (called phase I variant or simply phase I) has 'smooth' LPS while an avirulent laboratory strain (called phase II variant or phase II) has 'rough' LPS. The LPS from the latter variation is truncated due to large chromosomal deletions. Detection of antibodies to the antigens expressed by these two forms is used in diagnosis (see Section 4).

Entry into the body

C. burnetii has a complex life cycle. It exists as two distinct forms called 'small-cell variant' (SCV) and 'large-cell variant' (LCV). SCV is a form of the bacterium that can exist extracellularly and resist environmental conditions such as low or high pH, treatment with ammonium chloride, disinfectants, and UV radiation. Only exposure to high (≤ to 5%) concentrations of formalin for no less than 24 hours may kill *C. burnetii* SCV and this can be used for sanitation. SCVs are small (204 × 450 nm in size), rod-shaped, and spore-like. This form is metabolically inactive and resistant to osmotic pressure. Its cell wall is rich with proteins and peptidoglycan that may render the bacterium high resistance to harsh environmental conditions.

C. burnetii is inhaled directly from aerosols of infected animals or their contaminated body fluids, newborns, placenta, and wool. The SCV form infects **alveolar macrophages**. It is believed that the organisms enter the macrophages by attaching to integrin-associated proteins and **Toll-like receptor 4 (TLR4)** and complement receptor 3 (CR3) molecules. Which of these is involved in attachment determines successful internalization and survival in macrophages and monocytes (see Section 2).

After phagolysosomal fusion, the SCVs are activated, multiply, and may transform into LCVs, which is the metabolically active intracellular form of *C. burnetii* (Figure 4). LCVs are up to 2 μm long, more pleomorphic, rounded, granular, sometimes with fibrillar cytoplasm and with dispersed nucleoid filaments. Six strain types of LCV have been described: Hamilton, Bacca, Rasche, Biothere, Corazon, and Dod.

LCVs also divide by binary fission and can undergo sporogenic differentiation leading to the formation of spore-like endogenous forms of bacteria. These further develop into the metabolically inactive SCVs, which are then released from the infected host cell either by cell **lysis** or possibly by **exocytosis** and can spread via the bloodstream to other body organs (see below). At this stage multiplication of *C. burnetti* in the alveolar macrophages may lead to **pneumonia** (see Section 3).

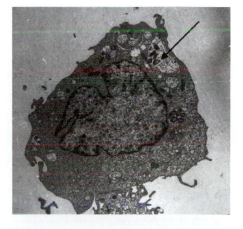

Figure 4. Electron micrograph of large cell variant (LCV, indicated by an arrow) of *C. burnetii* multiplying in a human macrophage.

Spread within the body

Following growth in the alveolar macrophages, the organism is disseminated to other organs, including the liver where they infect and reside in the **Kupffer cells** (liver macrophages). The dissemination route is mostly via the blood monocytes and, rarely, through the gastrointestinal route by drinking raw milk or eating raw cheese or raw eggs.

Spread from person to person

Person to person transmission of the bacteria is extremely rare. There are some cases described when human Q fever infection has occurred in an obstetrician who performed an abortion on a pregnant parturient woman, or during autopsies, intradermal inoculation, and blood transfusion. Congenital infection can very rarely result from transplacental transmission.

Sexual transmission of *C. burnetii*, although demonstrated in experimental mice, is controversial in humans and more information is required.

Risk factors

Q fever is considered to be an occupational hazard for farmers, veterinarians, abattoir workers, and laboratory personnel. People in charge of pet dogs and cats are also at risk, particularly if present at their birth.

Infected areas and soil may remain contaminated for weeks, since *C. burnetii* is resistant to killing in the natural environment, hence the bacteria may be spread by the wind and infect patients who are not even in direct contact with animals.

Age appears to be a risk factor for Q fever: symptomatic Q fever is more likely to occur in those 15 years old or above than in those under 15 years of age. Children in general are less likely to be diagnosed with Q fever. There is some indication that although men and women seem to be equally susceptible to infection by *C. burnetii*, symptomatic Q fever is more severe in men. The reasons for age and sex bias remain unknown, although this may be explained by the hormonal influences.

Epidemiology

Q fever has been reported in almost every country, except for New Zealand. However, since this disease does not reach epidemic proportions its detection and identification are not very accurate.

In the United States fewer than 30 cases of Q fever are reported annually. The disease is **endemic** in California; a cat-associated outbreak was reported in 1989 in Goldsboro, Maine. In Canada, Q fever was diagnosed in Nova Scotia in 1981; most of the recent cases are associated with exposure to infected cats, dogs, wild hares, and deer. The incidence is equal to 0.73 per 1000000 per year.

In Europe, the majority of acute Q fever cases are diagnosed in spring and early summer, outside the lambing period, which correlates with increased contamination of the environment with *C. burnetii*. The main lambing season in October is not associated with coxiellosis. Interestingly, detection of Q fever is very high in areas contaminated with rickettsia, although this can be due to a better organized epidemiological service.

In France, Q fever is quite common, particularly in the south near Marseille, which may reflect the activity of a nearby National Reference Center (NRC) for rickettsial diseases. Incidence of acute Q fever is approximately 50 per 100 000 inhabitants per year, while Q fever-induced endocarditis is estimated to be 5% of all national endocarditis cases. Due to the current decline in rural population, Q fever is increasingly diagnosed in the urban population, after occasional exposure to infected animals or contaminated raw milk with a peak in spring or early summer.

The average annual incidence of Q fever in the UK has been estimated as 0.15–0.35 cases per 100 000 population per year. It accounts for about 3% of all endocarditis cases in England and Wales. The outbreaks in the UK and in Northern Ireland were related to exposure to cattle and domestic animals.

In Germany, between 27 and 100 cases are reported annually. Several outbreaks were associated with living near flocks of sheep, and here the prevalent mode of contamination was considered to be air-borne.

In Spain, Q fever was diagnosed predominantly in the cattle-raising Basque country and Navarra, and in Madrid (quality of the reference centre?). The majority of cases show seasonal correlation with the peak in the lambing between March and July. The main symptoms vary from pneumonia in the Basque country to hepatitis in Andalusia in southern Spain.

A large outbreak of Q fever was reported in Switzerland in the Val de Bagnes (Valais) in the autumn of 1983 through infected sheep. Ten years later, in 1993 an outbreak of Q fever happened near Vicenza in north-eastern Italy caused by the crossing of populated areas by flocks of sheep. Otherwise both these countries are characterized by a low frequency of the disease.

An average of 291 cases of Q fever per year in 37 administrative territories has been diagnosed in Russia, although this may be an underestimate. The main regions involved were Povolzhje, West Siberia, and central Chernozemje, mostly in Astrakhan, Novosibirsk, and Voronezh.

In Greece, particularly in Crete, infection is mostly associated with contact with animals or the consumption of fresh milk and cheese. The risk of infection peaked in age groups 20–39 years and 80–89 years old and the majority of cases were diagnosed between January and June.

The national incidence of Q fever in Israel, where this disease is endemic, is approximately 0.75 cases per 1 000 000 per year, through possible exposure to cattle or sheep.

Q fever is also endemic in animals in Japan, with some acute and chronic cases reported and complications with pneumonia and endocarditis. The incidence is due to contact with infected animals and contaminated animal products.

In Australia Q fever is a notifiable disease, with an incidence estimated to be 3.11–4.99 per 100 000 population, with no seasonal prevalence, although this could be an underestimate. Most cases were diagnosed in Queensland and New South Wales, but none in northern Tasmania. In Australia the disease has been associated with livestock and the meat industry.

As mentioned before, New Zealand seems to be free from coxiellosis and thus from human Q fever, although no explanation of this phenomenon has been offered.

2. What is the host response to the infection and what is the disease pathogenesis?

Innate immunity

In order to enter human macrophages, virulent phase I *C. burnetii* interact with phagocytic αvβ3 integrin, integrin-associated protein (IAP), and TLR4, but do not engage CR3. This pathway does not lead to efficient internalization of the pathogen but if internalized, natural phase I *C. burnetii* can survive successfully within macrophages and monocytes (Figure 5).

Unlike the virulent phase I organisms, but similar to *Legionella* and *Mycobacterium* species (see Cases 22 and 23), avirulent *C. burnetii* phase II bacteria mostly engage CR3 receptors together with αvβ3 integrin and IAP, but do not use the TLR4 pathway. CR3-mediated internalization is effective, and so is phase II bacteria intracellular killing in phagolysosomes. Inability to bind to TLR4 is due to the truncated structure of the phase II bacterial LPS, which is a major ligand for TLR4.

It appears that the type of receptor engaged in binding defines the intracellular fate of *C. burnetii* in phagocytic cells (Figure 5).

On internalization into macrophages, both the virulent phase I and avirulent phase II *C. burnetii* are present in **phagosomes** that mature into acidic

Figure 5. Entry of *C. burnetii* into macrophages and monocytes. Virulent phase I *C. burnetii* interacts with αvβ3 integrin and integrin-associated protein (IAP) and Toll-like receptor 4 (TLR4), but not with complement receptor 3 (CR3) on the surface of macrophages. αvβ3 integrin-mediated internalization induces reorganization of cytoskeleton, activation of tyrosine kinase, and phosphorylation of endogenous substrates via TLR4 pathway. This leads to the intracellular survival of the pathogen. Avirulent phase II *C. burnetii* interacts with αvβ3 integrin and IAP and CR3, but not with TLR4 due to the truncated bacterial LPS. The subsequent lack of cytoskeleton reorganization and tyrosine kinase activation leads to the intracellular killing of the pathogen via CR3.

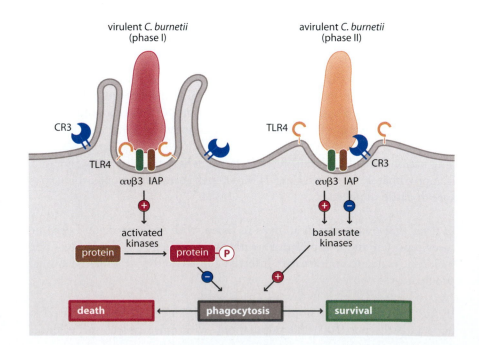

phagolysosomes. *C. burnetii* is an **acidophilic** bacterium that has become adapted to the phagolysosomal pH of eukaryotic cells and can multiply in acidic environments (pH = 4.7–5.2).

In most cases, since the majority of *C. burnetii* infections are asymptomatic, the killing of the bacteria by the macrophage/monocyte appears to be effective (with the help of T cells – see below). However, macrophage effector mechanisms can clearly be overcome and *C. burnetii* has a number of strategies to achieve this.

1. It is able to inhibit phagosome maturation that is necessary for intracellular killing.

2. It can release enzymes, for example superoxide dismutase and **catalase**, that are capable of degrading **reactive oxygen intermediates** (**ROIs**) produced by the macrophages and monocytes as part of their microbicidal action. In fact, infection with *C. burnetii* induces little release of ROI by the infected cells.

3. Monocytes synthesize **tumor necrosis factor-α** (**TNF-α**) in response to the phase I *C. burnetii*. Although this **cytokine** has bactericidal properties, it nevertheless further stimulates internalization of phase I *C. burnetii* by monocytes and macrophages by up-regulating expression of adhesion receptors (Figure 5) that leads to the infection of adjacent host cells.

4. Activation of the host immune responses through human **dendritic cells** is probably avoided through masking of adhesion/activation ligands for the organisms. Phase I *C. burnetii* LPS, lipoproteins, and lipopeptides do not activate host macrophages via TLR4 and TLR2, respectively, by possibly masking TLR ligands. This leads not only to decreased killing by phagocytes, but also to the lack of surveillance by dendritic cells resulting in the persistence of the bacteria.

Adaptive immunity

T-cell immunity

T-cell immunity is the most effective form of immunity to control *C. burnetii* infection and often to eradicate it, but in some cases a persistent intracellular infection may be established. Chronic infection is often attributable to the immunocompromised status of the patients, particularly in the monocyte/phagocytic and T-cell compartments. T cells assist intracellular killing by monocytes and macrophages through the production of **interferon-γ** (**IFN-γ**), which enhances the release of ROIs and induces **apoptosis** of *C. burnetii*-infected macrophages. IFN-γ also restores phagosome maturation allowing intracellular killing of *C. burnetii*. It induces vacuole alkalinization and controls ion metabolism in macrophages, thus inhibiting the intracellular multiplication of *C. burnetii*. Formation of **granulomas** is T-cell-mediated (see pathogenesis below).

Impairment of IFN-γ and IL-2 production results in increased susceptibility to *C. burnetii* infection. Development of a serious complication of chronic coxiellosis – Q fever-mediated endocarditis (see Section 3) – is associated with decreased numbers of CD4+ T lymphocytes producing these cytokines and their impaired proliferative activity in response to the *C. burnetii* antigens. In patients with Q fever endocarditis, T-cell-mediated immunity and IFN-γ production may be suppressed by **prostaglandin** E_2

synthesized by mononuclear cells. During chronic Q fever the immune response is not only ineffective, but may even be harmful, leading to **vasculitis** and **glomerulonephritis**.

In contrast to the immunity-enhancing **Th1** cytokines (e.g. IFN-**γ**), production of **Th2** cytokines such as **transforming growth factor-β (TGF-β)** and IL-10 is associated with the development of complications (endocarditis) and Q fever relapses. IL-10 has been shown to inhibit Th1 immune responses to *C. burnetii* and thus the ability of the host to control infection.

In conclusion, the balance between Th1 and Th2 cytokines determines the outcome of infection with *C. burnetii*. Prevalence of IFN-**γ** in an immunocompetent host represents an acute immune response leading to the elimination of the infection. Predominant production of Th2 cytokines and/or the immunocompromised status of the patient will favor establishment of a persistent chronic infection and the development of clinical complications (see pathogenesis).

Antibodies

Chronic Q fever is accompanied by the production of IgG and IgA (particularly in case of Q fever endocarditis) antibodies to phase I *C. burnetii* antigens. Although not considered to be of primary importance for immune defense against *C. burnetii*, they are believed to facilitate bacterial entry into host cells and thus speed up development of immune responses. However, *C. burnetii* continues to multiply intracellularly despite the high concentrations of IgG, IgM, and IgA.

Even though they do not appear to be protective in humans, antibodies to *C. burnetii* can be used for diagnostic purposes (see Section 4) together with nonspecific clinical signs and evidence of possible exposure. High levels of specific antibodies may provide evidence of recent infection and lead to timely treatment.

Pathogenesis

Autoantibodies

Patients with **hepatitis** in acute and chronic Q fever frequently have autoantibodies to smooth muscle, cardiolipin, phospholipids, circulating anticoagulant, and antinuclear antibodies, which might participate in the pathogenesis of the disease. Such autoantibodies could be involved in the vasculitis and/or glomerulonephritis seen in some patients with Q fever.

Granulomas

More severe cases of acute Q fever have liver cell **necrosis** with granulomas. *C. burnetii* infection often results in the formation of granulomas in infected organs (see Section 3). The granulomas are formed by monocytes migrating through the vascular endothelium. It has been documented that expression of TLR2, and possibly TLR4, by monocytes is essential for the formation of granulomas.

Q fever granulomatous hepatitis is accompanied by portal triaditis, Kupffer cell hyperplasia, and moderate fatty change. Kupffer cells may contain *C. burnetii* bacteria, which induce local inflammation and the formation of granulomas. Granulomas specific for Q fever hepatitis have a particular shape: a central clear space and a fibrin ring, and are called doughnut granulomas. They are characteristic of acute, but not chronic Q fever hepatitis, the latter being due to T-cell deficiency. Bone marrow lesions also contain doughnut granulomas (Figure 6).

Endocarditis

Endocarditis in Q fever usually involves inflammation of the aortic and mitral valves and prosthetic valves (see the case on page 73), leading to their destruction, perforation of valvular cusps and the Valsalva sinus, and the development of **aneurysms**. *C. burnetii*-infected monocytes in immunocompromised patients may attach to the damaged endothelium of cardiac valves and cause their inflammation (endocarditis) and colonization (Figure 7A). *C. burnetii* is found in foamy macrophages and in **histiocytes** as a single intracytoplasmic cluster (Figure 7B). Infected cardiac valves may contain thrombi with fibrin and platelets, necrosis, calcification or collagen. Bacterial vegetation may cause infarcts in the spleen, kidneys, and brain. Emerging immune complexes can also contribute to the valvular damage. Their deposition in kidneys may lead to glomerulonephritis.

3. What is the typical clinical presentation and what complications can occur?

C. burnetii has the ability to induce asymptomatic persistent infections both in humans and animals with bacteria shed in feces and urine. It is mostly associated with immunodeficiency or pregnancy, where persistent infection may lead to the contamination of the placenta with *C. burnetii*, followed by abortion or low birth weight.

In the majority of immunocompetent hosts (approximately 60% of all infected individuals), infection with *C. burnetii* remains asymptomatic and often goes undiagnosed. Thirty-eight percent of the remaining 40% of infected patients are symptomatic, but develop a mild nonspecific flu-like disease and a transient **bacteremia**. In only 2% of cases do patients require hospitalization with severe acute or chronic Q fever manifestations. In some cases, chronic Q fever can become evident many years after initial exposure.

Symptomatic acute Q fever manifests itself primarily as a self-limited **febrile** illness, atypical pneumonia, or a granulomatous hepatitis, whereas endocarditis is the more common presentation of chronic Q fever. However, because the clinical manifestations of Q fever are often nonspecific, the disease should be considered in febrile patients with recent contact with parturient animals.

It is believed that the same *C. burnetii* strains induce both acute and chronic Q fever.

Acute Q fever

The incubation period ranges from 14 to 26 days. The onset of the acute Q fever is usually abrupt and presents as a flu-like febrile illness with headache, **myalgia**, high-grade fever with chills, sweating, nausea, fatigue, and **photophobia**. The disease may be accompanied by a significant 6–12 kg (1–2 stones) weight loss. The flu-like illness is self-limiting and

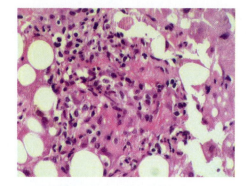

Figure 6. Granulomatous inflammation of bone marrow in a patient with Q fever.

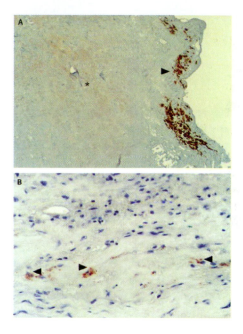

Figure 7. Histology of the heart valve of a patient with chronic Q fever endocarditis. (A) Focal and small inflammatory infiltrates with macrophages (arrowhead). The valve stroma is reorganized and fibrotic (*). (Immunoperoxidase staining with an anti-CD68 monoclonal antibody (mAb) ×100). (B) Immunohistochemical detection of *C. burnetii* in a resected cardiac valve using a mAb and hematoxylin counterstain. Note the intracellular location of the bacteria in the macrophages (arrowheads) ×400.

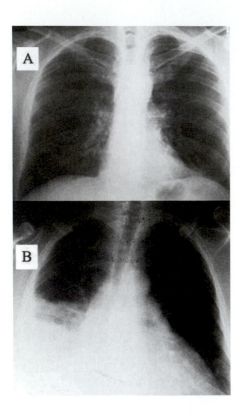

Figure 8. (A) Normal chest radiograph and (B) chest radiograph demonstrating consolidation of the right lower lobe due to the acute pneumonia caused by infection with *C. burnetii*.

may last for 1–3 weeks, but most frequently symptoms disappear in 10 days. In some cases, a feeling of general malaise might persist for months.

Complications

Further development of an atypical pneumonia or hepatitis is also common. Usually pneumonia is clinically highly symptomatic or mild and develops as a result of *C. burnetii* penetration into the lungs (Figure 8). It presents with nonproductive cough, fever, **pleuritic** and chest pain, and insignificant auscultatory changes. Acute and severe respiratory distress and a fatal outcome are very rare.

Dissemination of *C. burnetii* can cause hepatitis, clinically asymptomatic or accompanied by **hepatomegaly**, although **jaundice** is rare. As described above more severe cases have liver cell necrosis with granulomas.

Acute Q fever may rarely present with several atypical conditions such as **meningoencephalitis, Guillain-Barré** syndrome, **cerebellar ataxia**, inflammation of the optic or other cranial nerves, **myopericarditis, hemolytic anemia, thyroiditis, pancreatitis, lymphadenopathy, erythema nodosum**, glomerulonephritis, **orchitis**, skin **rash**, and others. These symptoms should be taken into account for the differential diagnosis of Q fever (see Section 4).

As mentioned in Section 1, Q fever in children is very rare, but it may be underdiagnosed and may present with subclinical manifestations such as febrile pneumonia. However, in some cases characteristic adult symptoms of Q fever can also be present.

The range of the symptoms, course, and outcome of the disease are dependent on a number of factors including the following.

- The route of *C. burnetii* infection. Aerosol route of infection will lead to the development of pneumonia, while infection via the gastrointestinal route (although rare) will predominantly lead to granulomatous hepatitis.
- The inoculating dose of *C. burnetii*. It has been suggested that only a large bacterial dose would lead to **myocarditis**.
- The virulence potential of the infecting strain of *C. burnetii*, related to geographic and genetic variation.
- Immunocompetence of the host. Patients with acquired immunosuppression through the treatment of cancer or with HIV infection, and pregnant women have a worse prognosis.
- In immunocompromised individuals, chronic Q fever is often accompanied by serious complications such as endocarditis, chronic interstitial lung disease, **osteomyelitis**, chronic hepatitis, septic infection of aneurysm and vascular grafts.

Chronic Q fever

Chronic Q fever is defined by a disease course of more than 6 months and the presence of IgG and IgA antibodies to phase I *C. burnetii* antigens (see Section 4). More than 7% of patients with acute Q fever further develop chronic endocarditis, which accounts for 60–70% of all chronic Q fever cases and causes damage to heart valves. A history of cardiac valve defects is found in over 90% of Q fever endocarditis patients. The predisposing heart disease may be congenital, rheumatic, syphilitic, or degenerative. Immunosuppression, immunodeficiency (including AIDS), and pregnancy (see below) also count as predisposing conditions. Prognosis of chronic Q fever is serious. If untreated it can be fatal, but appropriate antibiotic therapy (see Section 5) prevents the mortality from Q fever endocarditis in more than 90% of cases. However, relapse is frequent (50% of all cases) after withdrawal of the therapy. Timely diagnosis of Q fever endocarditis has an enormous prognostic value.

Clinical symptoms are related to cardiac involvement such as heart failure or cardiac valve dysfunction. Peripheral manifestations such as **purpuric** rash and clubbing have also been described. Nonspecific features include a low-grade intermittent fever with chills (particularly at the beginning of the disease), night sweats, weakness, fatigue, weight loss. A chest X-ray may reveal **cardiomegaly** while electrocardiography may detect arrhythmia and ventricular hypertrophy. Hepatomegaly and **splenomegaly** may also be present.

As well as endocarditis in chronic Q fever, other multiorgan involvement can be also found including chronic hepatitis, chronic infections of vascular aneurysms or prostheses (see the case), chronic osteomyelitis and osteoarthritis, lung tumors, and pneumonic fibrosis. **Immune complex** glomerulonephritis may evolve to renal insufficiency. Embolic complications of brain, arm, and leg vessels may develop too.

Immunocompromised adults and children may present with *C. burnetii* infections of bones including osteomyelitis, osteoarthritis, and aortic graft infection with contiguous spinal osteomyelitis. Bone infections may be under-reported and should be suspected in cases with tuberculoid-type bone lesions but no mycobacterial infection.

An additional complication is that in convalescing patients chronic fatigue syndrome may develop similar to that occasionally observed in patients with chronic typhoid fever or chronic brucellosis. This condition is often under-diagnosed and it is believed that up to 20% of acute Q fever patients develop chronic fatigue syndrome. The syndrome is accompanied by non-specific manifestations such as fatigue, myalgia, **arthralgia**, night sweats, mood swings, and abnormal sleep patterns.

Q fever during pregnancy

Q fever during pregnancy has been reported in 18–39-year-old women in the United States, Canada, France, the UK, Italy, former Czechoslovakia, and Israel. Q fever in pregnant women may be asymptomatic or may present with flu-like symptoms, accompanied by fever, severe **thrombocytopenia**, and atypical pneumonia. Since pregnancy is a serious predisposition for the development of chronic Q fever, pregnant women should avoid contact with domestic animals, especially cats, to decrease the risk of Q fever and toxoplasmosis.

Infection during the first trimester results in miscarriage, and in the second trimester it mostly results in prematurity. Death *in utero* has been reported in two cases, one with co-infection with *Chlamydia psittaci* infection (see

Chlamydia case). Only five healthy infants have been born to mothers with *C. burnetti* infection. However, the mechanisms leading to a fatal outcome remain unclear since infection with *C. burnetii* only causes **placentitis**. Immune complexes have been implicated, which may cause vasculitis or vascular **thrombosis** of the placenta with resulting placental insufficiency, although direct fetal injury cannot be excluded.

4. How is the disease diagnosed and what is the differential diagnosis?

Clinical diagnosis

There is a combination of three main symptoms that is highly characteristic of Q fever: prolonged fever, pneumonia, and hepatitis (Figure 9).

All day fever with severe headaches, which may increase up to 39–40°C and reach a plateau within 2–4 days, is a characteristic symptom. After 5–14 days and with treatment, the temperature returns to normal. In untreated patients, fever may last from 5 to 57 days, particularly in older patients. Thus, acute Q fever can be a cause of prolonged fever of unknown etiology. In the majority of cases the fever is persistent, but around 25% of patients experience a biphasic fever. Severe headaches are often retro-orbital.

Duke's criteria are used worldwide to calculate a diagnostic score for infective endocarditis. They include two major criteria (isolation of a typical microorganism from two separate blood cultures and detection of cardiac vegetation on **echocardiogram**) and several minor criteria (fever, previous valvulopathy, etc.). The combination of two major criteria or one major and three minor criteria is considered diagnostic for infective endocarditis. Recently Duke's criteria were modified to include the presence of an antiphase I IgG titer of ≥ 800 (see below).

Laboratory diagnosis

Collection and handling of clinical specimens

Since *C. burnetii* is a highly infectious agent, there is a significant risk for laboratory-acquired Q fever. Collection of clinical material and handling of *C. burnetii*-infected cell cultures or infected animals should be restricted

Figure 9. Pathophysiology of Q fever (from Rault et al., Natural history and pathophysiology of Q fever).

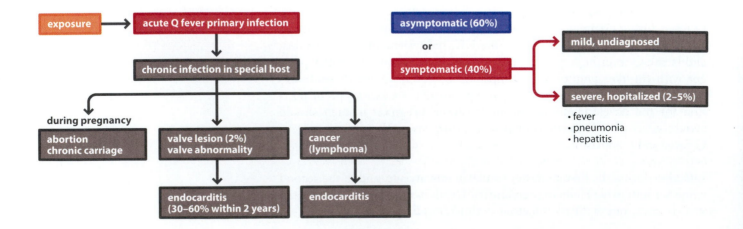

to experienced laboratory personnel wearing protective clothing and masks. They must work exclusively in biosafety level III laboratories. It is preferable that laboratory personnel are vaccinated against *C. burnetii* (see Section 5). A high risk of cross-contamination between infected and uninfected laboratory animals should also be taken into account.

Cultures

Human embryonic lung fibroblasts (HEL cells) are often used for *in vitro* cultures because of their high susceptibility to *C. burnetii* infection and simple growth conditions. Other cell lines used are monkey kidney cells or Vero cells (see the case). *C. burnetii* can also be cultured in embryonated hen yolk sacs or in laboratory guinea pigs and mice.

Human tissue specimens such as blood, bone marrow, cardiac valve, cerebrospinal fluid, vascular aneurysm, bone or liver biopsies, milk, placenta, and aborted fetal material are also used for *C. burnetii* culture. For microscopic observation *C. burnetii* is stained using the Gimenez method or immunofluorescence.

Blood examination

The white blood cell count is usually normal in acute Q fever, but can be raised to $14–21 \times 10^9 \ l^{-1}$ in about a quarter of the patients. Platelet counts decrease in the acute phase, but thrombocytosis is more common during recovery. The **erythrocyte sedimentation rate** (**ESR**) and levels of **C-reactive protein** (**CRP**) are mildly raised. The levels of liver enzymes and creatinine kinase are usually moderately increased.

A high ESR is more often seen in chronic Q fever, and is accompanied by **anemia** and thrombocytopenia. Liver enzyme levels are high as well as creatinine concentrations. Cryoglobulinemia, polyclonal hypergamma-globulinemia, and **hematuria** are quite common.

However, it must be stated that all these findings are not specific to *C. burnetii* infection, but have complementary value.

Pathology and immunohistology

Fresh or formalin-fixed and paraffin-embedded tissue biopsy specimens are used for **immunohistochemical** staining with specific antibodies. Immunohistochemistry of removed cardiac valves is usually used in combination with *C. burnetii* culture and only for the confirmation of Q fever endocarditis (Figure 7). Infected mononuclear cells in cardiac valves usually contain a large intracytoplasmic mass of *C. burnetii*, which cannot be found extracellularly. This may explain why *C. burnetii* vegetations in cardiac valves in Q fever endocarditis are often undetected by echocardiography.

Serology

Serology remains the primary diagnostic tool for Q fever because of the low sensitivity of cell culture and molecular biology methods (see below). Antibody levels against killed phase I and phase II cellular antigens provide solid evidence in support of a clinical diagnosis of acute Q fever. Relative titers of antibodies to phase I and phase II *C. burnetii* help to differentiate acute from chronic Q fever. Antibodies to phase II antigen prevail in acute Q fever. A rise in IgM antibody to phase II antigen is followed by IgG antibody, although IgM might persist for many months after the primary infection. Chronic Q fever is characterized by excess or equal titers of IgA and IgM anti-phase I antibodies as compared with anti-phase II antibodies. Antibody concentrations can also be used to monitor the course of the disease. Rising or persistent levels of antibodies to phase I *C. burnetii* may indicate chronic infection. It must be noted, however, that antibodies can only be detected from 2 to 3 weeks following the disease onset.

Immunofluorescence assays (IFAs), enzyme-linked immunosorbent fluorescence assays (ELIFA), microagglutination (and, these days rarely complement fixation) are the most frequently used serological methods for the diagnosis of Q fever.

Indirect IFA remains the reference method, and allows detection of IgG, IgM, and IgA classes of antibodies. The source of phase I and phase II *C. burnetii* antigens are Nine Mile strains. Phase I antigen is isolated from spleens of infected mice, while phase II antigen is obtained from *C. burnetii* growing in cell culture. An anti-phase II IgG titer of ≥ 200 and an anti-phase II IgM titer of ≥ 50 represent significant evidence of acute Q fever. An anti-phase I IgG titer of ≥ 800 is an indication of Q fever endocarditis. Phase I IgA is no longer considered in the primary diagnosis of chronic Q fever and is only used for confirmation and follow-up (Table 1). However, this method is not optimal for wide-scale screening in the case of an act of bioterrorism (see Section 5), and in this situation **ELISA** is more appropriate.

ELISA is a more sensitive test than IFA and can be used for the detection of both anti-phase I and anti-phase II antibodies. The following cut-off values are commonly used: in case of acute Q fever ≥ 1024 for anti-phase II IgG and ≥ 512 for anti-phase II IgM, and in case of chronic Q fever ≥ 128 for anti-phase I IgG and IgM antibodies. ELISA is used more for seroepidemiological studies rather than as a routine test because the results are more difficult to interpret by inexperienced personnel in clinical laboratories.

Table 1. Diagnostic titers of human serum antibodies in acute and chronic Q fever using indirect immunofluorescence

Antibodies		*Acute Q fever*	*Chronic Q fever*
Antibodies to phase I *C. burnetii*	IgM	Positive, but lower, than antibodies to phase II	Positive or negative
	IgG	Positive, but lower, than antibodies to phase II	> 1:800
	IgA	Positive, but lower, than antibodies to phase II	> 1:100
Antibodies to phase II *C. burnetii*	IgM	≥ 1:50	Positive, but lower than or equal to phase I antibodies or negative.
	IgG	≥ 1:200	Positive, but lower than or equal to phase I antibodies
	IgA	Positive or negative	Positive, but lower than or equal to phase I antibodies

Microagglutination is a simple and sensitive test but requires significantly larger amounts of antigen compared with IFA or ELISA. A highly specific and sensitive modification of this technique has been developed recently using high-density particle agglutination.

The complement fixation test is inferior in sensitivity to IFA or ELISA and is more time-consuming. It is characterized by a high frequency of false-negative results due to cross-reactivity with hen egg antigens and a prozone effect.

Western blotting, dot immunoblotting, indirect hemolysis test, and radioimmunoassay are not routinely used for diagnosis since they require special conditions and are costly and time-consuming.

Molecular biology methods

Radiolabeled DNA probes and PCR-based techniques for 16S rDNA, *sodB*, and *gltA* genes can be used for identification of *C. burnetii* strains in clinical specimens or cultures. Application of PCR-based methods to blood samples did not prove very successful due to the high number of false-positive results. In clinical practice, PCR is currently used only to test tissue samples, especially cardiac valve specimens, or to confirm Q fever diagnosis in cases when cell cultures are negative. However, PCR is also less hazardous and can be used simultaneously on many samples. This is particularly valuable when the detection of *C. burnetii* and other pathogens is required in biological warfare (see Section 5). PCR can be used for a rapid simultaneous detection of *C. burnetii*, *Yersinia pestis*, *Bacillus anthracis*, and *Brucella melitensis*.

Differential diagnosis

Serological cross-reactivity has been described between *C. burnetii* and *Legionella pneumophila*, *L. micdadei*, *Bartonella quintana*, and *B. henselae*. This cross-reactivity must be taken into account in the case of atypical pneumonia. A differential diagnosis can be established by quantitative serological methods when antibody titers against anti-phase I and anti-phase II *C. burnetii* antigens are high in true Q fever.

5. How is the disease managed and prevented?

Management

The majority of patients with Q fever require no treatment and symptoms will disappear on their own. However, treatment may be required if the symptoms are severe, if they continue for several weeks, or if they continue to return after the initial infection.

Antibiotic therapy

Doxycycline (as used in the case on page 73) and other tetracyclines are normally used to treat patients with acute Q fever. Fluoroquinolones, rifampin, co-trimoxazole, and macrolides (especially clarithromycin) are also used (see below). *C. burnetii* is resistant to β-lactam compounds and aminoglycosides, so their use is not recommended for the treatment of Q fever. Various strains of *C. burnetii* have demonstrated heterogeneity in susceptibility to different antibiotics. The mechanisms of antibiotic resistance of

C. burnetii are not well understood and in some cases may be due to individual amino acid substitutions in some bacterial enzymes.

Since it is very difficult to isolate the bacteria in each outbreak case, the use of an optimal antibiotic therapy may not always be achieved. A suboptimal treatment regimen may lead to difficulty in eradicating *C. burnetii* in chronically infected patients.

The antibiotic susceptibility of *C. burnetii* can be evaluated using an embryonated egg model, an animal model, or cell culture.

Acute Q fever

In immunocompetent infected individuals, symptomatic acute Q fever most often has a mild course and resolves spontaneously within 2 weeks.

In severely ill patients empirical antibiotic therapy is recommended because delay in the treatment may be dangerous. Patients with suspected Q fever pneumonia must begin antibiotic treatment during the first 3 days of the illness. Until recently, mostly tetracycline has been used at 500 mg four times a day (q.i.d.), which can reduce the duration of fever by 50%. Currently, doxycycline at 100 mg twice a day (b.i.d.) for 14 days is recommended since it has better pharmacokinetic properties. However, neither tetracycline nor doxycycline should be prescribed to patients with serious gastric intolerance, children younger than 8 years old, or pregnant women.

Adult patients with gastric side effects to tetracyclines should be treated for no less than 14–21 days with fluoroquinolones such as ofloxacin (200 mg three times a day, t.i.d.) and pefloxacin (400 mg b.i.d.), but these antibiotics are not recommended for children and pregnant women, where macrolides such as erythromycin may be used instead. Erythromycin (500 mg q.i.d.) is currently recommended for the treatment of atypical pneumonia, but it is not very effective in severe cases of Q fever pneumonia since *C. burnetii* strains vary in their susceptibility to erythromycin. Other macrolides – clarithromycin and roxithromycin – may prove more effective but require **clinical trials**.

Co-trimoxazole is recommended for pregnant women since it prevents fetal death and miscarriage, but it does not prevent the disease progressing to the chronic infection since it is not bactericidal for *Coxiella*. Therefore following delivery, conventional treatment with doxycycline plus hydroxychloroquine for 1 year must be resumed to eliminate the infection.

Additional treatment with prednisone may be recommended for patients with Q fever hepatitis.

Chronic Q fever

Q fever endocarditis is fatal in almost all cases if untreated, although the course of the disease can be slow and may last for years. However, even treatment with antibiotics, as a monotherapy or as a combination, does not cure chronic Q fever in the majority of cases. This is apparently due to the decreased activity of antibiotics in the acidic environment of the *C. burnetii* phagolysosome.

Monotherapy with tetracyclines is usually recommended and is effective, but patients often relapse when antibiotic therapy is withdrawn. Alternative antibiotic monotherapies such as co-trimoxazole, rifampin, and fluoroquinolones are used to treat endocarditis patients with gastric intolerance to tetracyclines, but relapse usually occurs when the drugs are stopped. As a result combined antibiotic therapy should be administered to treat chronic Q fever.

The doxycycline-fluoroquinolone (ofloxacin, pefloxacin) combination was effective in treatment of Q fever endocarditis with the reduction of the mortality rates down to 6%. However, at least 3 years of antibiotic therapy is required to avoid relapses, which otherwise happen in 50% of cases after the withdrawal of antibiotics. More recently the combination of doxycycline and an alkalinizing agent such as chloroquine for the duration of 18 months has been used to reduce the acidity of phagolysosomes. With this latter regimen the relapse rate was significantly lower (14.3%) and patients improved rapidly. However, patients receiving chloroquine treatment have an increased risk of retinopathy and will require regular ophthalmologic checks.

The efficiency of the clinical application of antibiotics is usually measured by a reduction in the titers of serum anti-phase I IgG and IgA antibodies to 1:200 or less.

Blood cell count (particularly CD4/CD8 ratio), evaluation of hepatic enzymes, and creatinine levels in serum are also used for the monitoring. In Q fever endocarditis, an echocardiogram should be performed every 3 months.

In some severe cases of Q fever endocarditis with hemodynamic failure, cardiac surgery for valve replacement can be performed together with administration of antibiotics. However, increased cardiac surgery may lead to prosthetic valve endocarditis caused by *Staphylococcus* species (see the *Staphylococcus aureus* case).

Prevention

Animal vaccines

One approach would be to develop a vaccine for wild and domestic animals to control *C. burnetii* infection in animal hosts. Vaccines based on phase I bacteria have been shown to be more protective than those prepared from phase II *C. burnetii*. Importantly, Q fever vaccination has protected cattle against abortion, low fetal weight, and infertility, and therefore had a significant economic effect. A combined vaccine to *C. burnetii* and *Chlamydia psittaci* has been offered in Europe to protect cattle and goats, although it has not proved very efficient in protection against Q fever. Q fever vaccination of domestic animals is currently not routinely used and is not considered to be cost-effective.

Human vaccines

Human vaccines are made from whole cells of killed, purified phase I *C. burnetii*, with LPS and protein antigens. However, its protective capacity may differ in various geographic areas since it is based on a single strain of the bacterium (monovalent vaccine). The alternative would be to create a polyvalent vaccine designed from phase I *C. burnetii* antigens pooled from different strains.

Side effects have been a particular problem in the development of a successful vaccine. The most successful vaccine today, Qvax®, based on formalin-killed whole cell *C. burnetii* from the Henzerling strain (Qvax®, CSL Limited, Parkville, Victoria, Australia) has been produced and licensed in Australia, and reported to provide 100% protection against Q fever for over 5 years. It has been suggested that the outstanding efficacy of this vaccine is due to its cross-protection from *C. burnetii* variants. However, this vaccine is not yet commercially available worldwide.

Other vaccine candidates have been produced using genetic recombination such as a fusion protein, consisting of *C. burnetii* outer-membrane protein 1 and heat-shock protein B (HspB) as well as a mixture of eight recombinant *C. burnetii* proteins expressed in *E. coli*. Vaccination with single recombinant *C. burnetii* proteins Rcom, rP1, rCbMip, or rP28 also demonstrated a certain level of protection. A cocktail of recombinant subcellular components administered with an optimal adjuvant is a focus of modern Q fever vaccine design.

C. burnetii infection is an occupational hazard, therefore professionals who could be exposed should be considered for vaccination. These include livestock and animal product handlers, dairy workers, veterinarians, and laboratory personnel handling *C. burnetii*-infected animals. Vaccination should be also considered for immunocompromised patients with cardiac valve defects or artificial **prostheses** or vascular aneurysms. To prevent local and systemic reactions in patients with chronic Q fever a single-dose vaccine has been proposed. Vaccination of patients post-infection is not recommended in order to prevent hypersensitivity reactions.

However, since epidemiological observations show that approximately one-seventh of the population have been exposed to *C. burnetii* it is unclear what scale of vaccination should be adopted.

Bioterrorism

C. burnetii is also significant in biological warfare. *C. burnetii*-induced Q fever is a danger of bioterrorism due to its high infectivity at low doses, aerosol exposure route, environmental stability of the infectious spore-like forms, rapid onset of the disease, and significant Q fever morbidity. Soldiers or civilians who may be exposed as the result of a bioterrorism or biowarfare attack should be considered for vaccination. Animals could also be the target of bioterrorism. *C. burnetii* is classified as a Category B biothreat agent by the Centers for Disease Control and Prevention in the USA.

SUMMARY

1. What is the causative agent, how does it enter the body and how does it spread a) within the body and b) from person to person?

- *C. burnetii* is an obligate intracellular bacterium with a complex life cycle. It is a pleomorphic coccobacillus 0.3–1.0 µm in size, with a cell wall built of approximately 6.5 nm thick outer and inner membranes separated by a peptidoglycan layer. Although it is considered to be a gram-negative bacterium, *C. burnetii* cannot be stained by the Gram technique and the Gimenez staining method is usually used instead.

- *Coxiella* genus belongs to the gamma subdivision of Proteobacteria and is closely related to the genera *Legionella* and *Rickettsiella*.

- Q fever is a zoonosis and *C. burnetii* is found in almost all animal species, particularly in wild and domestic mammals, birds, and arthropods such as ticks. Infected animals most often shed *C. burnetii* into the environment during parturition. *C. burnetii* can also contaminate the environment via animal urine, feces, and milk.

- Humans usually become infected from cattle, goats, sheep, domestic ruminants, and pets or contaminated dung and bedding. *C. burnetii* may also be transmitted to humans from consumption of raw eggs or cheese, from drinking raw milk or inhalation of infected fomites.

- There are over 40 tick species naturally infected with *C. burnetii*. *C. burnetii* multiplies in the gut cells of infected ticks and is shed with their feces onto the skin of the animal host, particularly in wild animals such as rodents and wild birds.

- There are a variety of *C. burnetii* strains (genovars) associated with their virulence. Genomic groups I, II, and III belong to so-called acute strains, which infect animals, ticks, and humans and cause acute Q fever. Genomic groups IV and V belong to so-called chronic strains and cause human Q fever endocarditis. Genomic group VI *C. burnetii* has unknown pathogenicity.

- There are two variations of *C. burnetti*, based on their lipopolysaccharide (LPS) structure. The virulent natural form (called phase I) has 'smooth' LPS while an avirulent laboratory strain (called phase II) has 'rough' LPS.

- *C. burnetii* has a complex life cycle. It exists as two distinct forms called small-cell variant (SCV) and large-cell variant (LCV).

- SCV is a small (204 × 450 nm in size) rod-shaped, spore-like form of the bacterium, which can exist extracellularly and resists environmental conditions. It is inhaled directly from aerosols of infected animals, placenta, and wool and infects alveolar macrophages by attaching to integrin-associated proteins, Toll-like receptor (TLR4), and complement receptor 3.

- After phagolysosomal fusion, the SCVs are activated, multiply, and may transform into LCVs – the metabolically active, up to 2 µm long intracellular form of *C. burnetii*.

- LCVs also divide by binary fission and can undergo differentiation leading to the formation of spore-like endogenous SCVs, which are then released from the infected host cell either by cell lysis or by exocytosis and can spread via the bloodstream to other body organs, including the liver where they infect and reside in the Kupffer cells.

- Person to person transmission of the bacteria is extremely rare and is confined to obstetricians who perform abortions or during autopsies, intradermal inoculation, and blood transfusion. Congenital infection can very rarely result from transplacental transmission.

- Q fever is considered to be an occupational hazard for farmers, veterinarians, abattoir workers, and laboratory personnel. People in charge of pet animals are also at risk. Infected areas and soil may remain contaminated for weeks.

- Q fever has been reported in almost every country, except for New Zealand, although the reasons for this exception are not yet understood.

2. What is the host response to the infection and what is the disease pathogenesis?

- Virulent phase I *C. burnetii* interacts with phagocytic αvβ3 integrin, integrin-associated protein (IAP), and TLR4, but does not engage complement receptor 3 (CR3). This pathway does not lead to efficient internalization but if internalized natural phase I *C. burnetii* can survive successfully within macrophages and monocytes.

- Avirulent *C. burnetii* phase II bacteria mostly engage CR3 receptor together with αvβ3 integrin and IAP, but do not use the TLR4 pathway. This internalization is effective, and so is phase II bacteria intracellular killing in phagolysosomes.

- Upon internalization in macrophages *C. burnetii* are included in phagosomes that mature into acidic phagolysosomes. *C. burnetii* is an acidophilic bacterium, which has adapted to the phagolysosomal pH of eukaryotic cells and can multiply in an acidic environment (pH = 4.7–5.2).

- Innate immune responses can be overcome by *C. burnetii* through inhibiting the maturation of the phagosomes, release of enzymes that degrade reactive oxygen intermediates (ROIs), inducing the synthesis of tumor necrosis factor-α (TNF-α) and through masking its adhesion and activation ligands leading to the lack of surveillance by dendritic cells.

- T-cell immunity is the most effective to control *C. burnetii* infection and often to eradicate it through the production of interferon-γ (IFN-γ), which enhances maturation of the phagosomes, release of ROIs, and intracellular killing by monocytes and macrophages. IFN-γ also induces vacuole alkalinization and controls ion metabolism in macrophages, thus inhibiting the intracellular multiplication of *C. burnetii*.

- In contrast to the immunity-enhancing Th1 cytokines, production of Th2 cytokines such as transforming growth factor-β (TGF-β) and IL-10 is associated with the development of complications (endocarditis) and Q fever relapses.

- The balance between Th1 and Th2 cytokines determines the outcome of the infection with *C. burnetii*. Prevalence of IFN-γ in an immunocompetent host represents an acute immune response leading to the elimination of the infection. Predominant production of Th2 cytokines and/or immunocompromised status of the patient will favor establishment of a persistent chronic infection and the development of clinical complications.

- Chronic Q fever is accompanied by the production of high titers of IgG and IgA (particularly in case of Q fever endocarditis) antibodies to phase I *C. burnetii* antigens, although this does not prevent intracellular multiplication of *C. burnetii*.

- Patients with hepatitis in acute and chronic Q fever frequently have autoantibodies to smooth muscle, cardiolipin, phospholipids, circulating anticoagulant, and antinuclear antibodies, which might participate in the pathogenesis of the disease.

- *C. burnetii* infection often results in the formation of granulomas in infected organs. The granulomas are formed by monocytes migrating through the vascular endothelium. Granulomas specific for Q fever hepatitis have a particular shape, a central clear space and a fibrin ring, and are called doughnut granulomas. They are characteristic for acute but not chronic Q fever hepatitis.

- *C. burnetii*-infected monocytes in immunocompromised patients may attach to the damaged endothelium of cardiac valves and cause their inflammation (endocarditis) and colonization. Endocarditis in Q fever usually involves inflammation of the aortic and mitral valves, prosthetic valves. Infected cardiac valves may contain thrombi with fibrin and platelets, necrosis, calcification or collagen. Emerging immune complexes can also contribute to the valvular damage and to glomerulonephritis.

3. What is the typical clinical presentation and what complications can occur?

- In approximately 60% of all infected immunocompetent individuals infection with *C. burnetii* remains asymptomatic. Thirty eight percent of the remaining 40% of infected patients are symptomatic, but develop a mild nonspecific flu-like disease and a transient bacteremia. In only 2% of cases do patients require hospitalization with severe acute or chronic Q fever manifestations.

- The incubation period of acute Q fever ranges from 14 to 26 days. The onset is usually abrupt and presents as a flu-like febrile illness with headache, myalgia, high-grade fever with chills, sweating, nausea, fatigue, photophobia, and significant weight loss. The flu-like illness is self-limiting and may last for 1–3 weeks, but most frequently symptoms disappear in 10 days.

- Atypical pneumonia develops as a result of *C. burnetii* penetration in the lungs and presents with nonproductive cough, fever, pleuritic and chest pain.

- Dissemination of *C. burnetii* can cause hepatitis, clinically asymptomatic or accompanied by hepatomegaly, although jaundice is rare. More severe cases have liver cell necrosis with granulomas.

- In immunocompromised individuals, chronic Q fever is often accompanied by serious complications such as endocarditis, chronic interstitial lung disease, osteomyelitis, chronic hepatitis, septic infection of aneurysm and vascular grafts.

- Chronic Q fever is defined by a disease course of more than 6 months and the presence of IgG and IgA antibodies to phase I *C. burnetii* antigens. More than 7% of patients with acute Q fever further develop chronic endocarditis, which accounts for 60–70% of all chronic Q fever cases and causes damage to heart valves.

- A history of congenital, rheumatic, syphilitic or degenerative cardiac valve defects is found in over 90% of Q fever endocarditis patients. Immunosuppression, immunodeficiency, and pregnancy also count as predisposing conditions. If untreated Q fever endocarditis can be fatal. Appropriate antibiotic therapy prevents the mortality in more than 90% cases, but relapse happens in 50% of all cases after withdrawal of the therapy.

- Clinical symptoms include heart failure, cardiac valve dysfunction, and peripheral manifestations such as purpuric rash and clubbing. Cardiomegaly, arrhythmia, and ventricular hypertrophy are detected. Hepatomegaly and splenomegaly may also be present.

- In up to 20% of convalescing patients chronic fatigue syndrome may develop.

- Q fever cases during pregnancy may be asymptomatic or present with flu-like symptoms, accompanied by fever, severe thrombocytopenia, and atypical pneumonia, or may lead to the development of chronic Q fever. Pregnant women should avoid contact with domestic animals, especially cats, to decrease the risk of Q fever and toxoplasmosis.

- Infection during the first trimester results in miscarriage, and in the second trimester it mostly results in prematurity.

4. How is the disease diagnosed and what is the differential diagnosis?

- There is a combination of three main symptoms that is highly characteristic of Q fever: prolonged fever, pneumonia, and hepatitis.

- All day fever (with severe headaches), which may increase up to 39–40°C, is a characteristic symptom. In untreated patients, fever may last for 5–57 days, especially in older patients. Fever is usually persistent, but around 25% of patients experience a biphasic fever.

- To calculate a diagnostic score for infective endocarditis, Duke's criteria are used worldwide.

- For *in vitro* *C. burnetii* cultures human embryonic lung fibroblasts (HEL cells), monkey kidney cell or Vero cell lines, embryonated hen yolk sacs or guinea pigs and mice, as well as human tissue specimens are used.

- In acute Q fever the white blood cell count is usually normal, but can be raised to 14–21×10^9 l^{-1} in about a quarter of the patients. Platelet counts decrease in the acute phase, but during recovery thrombocytosis is more common. Erythrocyte sedimentation rate (ESR) and the levels of CRP, liver enzymes, and creatinine kinase are usually moderately increased.

- In chronic Q fever, a high ESR is more often seen and is accompanied by anemia and thrombocytopenia. Liver enzyme and creatinine levels are high. Cryoglobulinemia, polyclonal hypergammaglobulinemia, and hematuria are quite common.

- Serological methods remain the leading diagnostic tools. Antibodies to phase II antigen prevail in acute Q fever. A rise in IgM antibody to phase II antigen is followed by IgG antibody, although IgM might persist for many months after the primary infection. Chronic Q fever is characterized by excess or equal titers of IgA and IgM anti-phase I antibodies as compared with anti-phase II antibodies. Antibody concentrations can also be used to monitor the course of the disease.

- Indirect IFA remains the reference method, and allows detection of the IgG, IgM, and IgA classes of antibodies. However, in case of an act of bioterrorism, this method is not optimal for a wide-scale screening. Microagglutination is also used for diagnosis.

- Radiolabeled DNA probes and PCR-based techniques can be used for identification of *C. burnetii* strains in clinical specimens or cultures. PCR can be used for a rapid simultaneous detection of *C. burnetii*, *Yersinia pestis*, *Bacillus anthracis*, and *Brucella melitensis*.

- Serological cross-reactivity between *C. burnetii* and *Legionella pneumophila*, *L. micdadei*, *Bartonella quintana*, and *B. henselae* must be taken into account in case of atypical pneumonia.

5. How is the disease managed and prevented?

- In immunocompetent infected individuals symptomatic acute Q fever most often has a mild course and resolves spontaneously within 2 weeks. Treatment is required if the symptoms are severe, if they continue for several weeks, or if they continue to return after the initial infection.

- Patients with suspected Q fever pneumonia must start antibiotic treatment during the first 3 days of the illness. In the past mostly tetracycline has been used, which can reduce the duration of fever by 50%. More recently, however, doxycycline at 100 mg twice a day (b.i.d.) for 14 days is recommended since it has better pharmacokinetic properties.

- Adult patients with gastric side effects to tetracyclines are treated with fluoroquinolones such as ofloxacin and pefloxacin, but these antibiotics are not recommended for children and pregnant women, where macrolides such as erythromycin, clarithromycin, and roxithromycin – may be used instead.

- Co-trimoxazole is recommended for pregnant women, but it does not prevent the disease progressing to chronic infection since it is not bactericidal for coxiella. When pregnancy is over, conventional treatment with doxycycline plus hydroxychloroquine must be resumed to eliminate the infection.

- For the treatment of Q fever endocarditis monotherapy with tetracyclines is usually recommended and is effective, but patients often relapse when antibiotic therapy is withdrawn. Alternative antibiotic monotherapies such as co-trimoxazole, rifampin, and fluoroquinolones are used in the cases of gastric intolerance to tetracyclines.

- The doxycycline-fluoroquinolone (ofloxacin, pefloxacin) combination was effective in treatment of Q fever endocarditis, with the reduction of the mortality rates down to 6%. More recently the combination of doxycycline and chloroquine for the duration of 18 months reduced the relapse rate down to 14.3%.

- Decrease in the titers of serum anti-phase I IgG and IgA antibodies down to 1:200 and CD4/CD8 cell ratio, serum levels of the hepatic enzymes, and creatinine are used for the monitoring of the treatment as well as echocardiogram for Q fever endocarditis patients.

- In some severe cases of Q fever endocarditis with hemodynamic failure, cardiac surgery for valve replacement can be performed together with administration of antibiotics.

- Vaccines for wild and domestic animals in order to control *C. burnetii* infection in animal hosts are based on phase I bacteria. A combined vaccine to *C. burnetii* and *Chlamydia psittaci* has been offered, but Q fever vaccination of domestic animals is currently not routinely used and is not considered to be cost-effective.

- Human vaccines are made from the whole cells of killed, purified phase I *C. burnetii* with LPS and protein antigens, or antigens pooled from different strains. An effective vaccine (Qvax®) based on formalin-killed whole cell *C. burnetii* has been produced and licensed in Australia, but it is not yet commercially available worldwide.

- *C. burnetii* is classified as a Category B biothreat agent by the Centers for Disease Control and Prevention in the USA. *C. burnetii*-induced Q fever is a danger of bioterrorism due to its high infectivity at low doses, aerosol exposure route, environmental stability of the infectious spore-like forms, rapid onset of the disease, and significant Q fever morbidity. Animals could also be the target of bioterrorism.

FURTHER READING

Cutler SJ, Bouzid M, Cutler RR. Q fever. J Infect, 2007, 54: 313–318.

Raoult D, Mege JL, Marrie TJ. Q fever: queries remaining after decades of research. In: Scheld M, Craig WA, Hugues JM, editors. Emerging Infections 5. ASM Press, Washington, DC, 2001: 29–56.

Woldehiwet Z. Q fever (coxiellosis): epidemiology and pathogenesis. Res Vet Sci, 2004, 77: 93–100.

REFERENCES

Baca OG, Paretsky D. Q fever and *Coxiella burnetii*: a model for host-parasite interactions. Microbiol Rev, 1983, 47: 127–149.

Cutler S, Paiba G, Howells J, Morgan K. Q fever – a forgotten disease? Lancet Infect Dis, 2002, 2: 717–718.

Dellacasagrande J, Ghigo E, Capo C, Raoult D, Mege JL. *Coxiella burnetii* survives in monocytes from patients with Q fever endocarditis: involvement of tumor necrosis factor. Infect Immun, 2000, 68: 160–164.

Fenollar F, Fournier PE, Carrieri MP, Habib G, Messana T, Raoult D. Chronic endocarditis following acute Q fever. Clin Infect Dis, 2000, 33: 312–316.

Fournier PE, Casalta JP, Habib G, Messana T, Raoult D. Modification of the diagnostic criteria proposed by the Duke Endocarditis Service to permit improved diagnosis of Q fever endocarditis. Am J Med, 1996, 100: 629–633.

Fournier PE, Marrie TJ, Raoult D. Diagnosis of Q fever. J Clin Microbiol, 1998, 36: 1823–1834.

Georghiou GP, Hirsch R, Vidne RA, Raanani E. *Coxiella burnetii* infection of an aortic graft: surgical view and a word of caution. Interact Cardiovasc Thorac Surg, 2004, 3: 333–335

Ghigo E, Honstettre A, Capo C, Gorvel JP, Raoult D, Mege JL. Link between impaired maturation of phagosomes and defective *Coxiella burnetii* killing in patients with chronic Q fever. J Infect Dis, 2004, 190: 1767–1772.

Heinzen RA, Hackstadt T, Samuel JE. Developmental biology of *Coxiella burnetii*. Trends Microbiol, 1999, 7: 149–154.

Honstettre A, Ghigo E, Moynault A, et al. Lipopolysaccharide from *Coxiella burnetii* is involved in bacterial phagocytosis, filamentous actin reorganization, and inflammatory responses through Toll-like receptor 4. J Immunol, 2004, 172: 3695–2703.

Honstettre A, Imbert G, Ghigo E, et al. Dysregulation of cytokines in acute Q fever: role of interleukin-10 and tumor necrosis factor in chronic evolution of Q fever. J Infect Dis, 2003, 187: 956–162.

Houpikian P, Habib G, Mesana T, Raoult D. Changing clinical presentation of Q fever endocarditis. Clin Infect Dis, 2002, 34: E28–E31.

Karakousis PC, Trucksis M, Dumler JS. Chronic Q fever in the United States. J Clin Microbiol, 2006, 44: 2283–2287.

Kobbe R, Kramme S, Gocht A, et al. Travel-associated *Coxiella burnetii* infections: three cases of Q fever with different clinical manifestation. Travel Med Infect Dis, 2007, 5: 374–379.

Madariaga MG, Rezai K, Trenholme GM, Weinstein RA. Q fever: a biological weapon in your backyard. Lancet Infect Dis, 2003, 3: 709–721.

Maltezou HC, Raoult D. Q fever in children. Lancet Infect Dis, 2002, 2: 686–691.

Marrie TJ. *Coxiella burnetii* pneumonia. Eur Respir J, 2003, 21: 713–719.

Maurin M, Raoult D. Q fever. Clin Microbiol Rev, 1999, 12: 518–553.

Raoult D, Fenollar F, Stein A. Q fever during pregnancy – diagnosis, treatment, and follow-up. Arch Intern Med, 2002, 162: 701–704.

Raoult D, Marrie T, Mege J. Natural history and pathophysiology of Q fever. Lancet Infect Dis, 2005, 5: 219–226.

Rodolakis A. Q fever, state of art: epidemiology, diagnosis and prophylaxis. Small Rum Res, 2006, 62: 121–124.

Seshadri R, Paulsen IT, Eisen JA, et al. Complete genome sequence of the Q-fever pathogen *Coxiella burnetii*. Proc Natl Acad Sci, 2003, 100: 5455–5460.

Thomas DR, Salmon RL, Coleman TJ, et al. Occupational exposure to animals and risk of zoonotic illness in a cohort of farmers, farmworkers, and their families in England. J Agric Safety Hyg, 1999, 5: 373–382.

Waag D. *Coxiella burnetii*: host and bacterial responses to infection. Vaccine, 2007, 25: 7288–7295.

Zhang G, Samuel JE. Vaccines against *Coxiella* infection. Expert Rev Vaccines, 2004, 3: 577–584.

WEB SITES

Centers for Disease Control and Prevention, National Center for Infectious Diseases, Atlanta, GA, USA: http://www.cdc.gov/ncidod/dvrd/qfever/

MedLine Plus Medical Encyclopedia, Entry for Q-Fever: http://www.nlm.nih.gov/medlineplus/ency/article/001337.htm

Nation Master website, this website was created by Rapid Intelligence, a web technology company based in Sydney, Australia: http://www.nationmaster.com/encyclopedia/ Q-fever

Patient UK, a joint venture between PiP and EMIS (Egton Medical Information Systems, Copyright © 2008 EMIS & PIP: http://www.patient.co.uk/showdoc/40000453/

Wikipedia – The Free Encyclopedia: http://en.wikipedia.org/wiki/Q_fever

MULTIPLE CHOICE QUESTIONS

The questions should be answered either by selecting True (T) or False (F) for each answer statement, or by selecting the answer statements which best answer the question. Answers can be found in the back of the book.

1. **Which of the following are true of *C. burnetii*?**
   A. *C. burnetii* is an obligate intracellular pathogen.
   B. It only infects humans.
   C. *C. burnetii* species has limited genetic variability and only consists of one genovar.
   D. It is characterized by two so-called phase variations.
   E. *C. burnetii* has a simple life cycle.

2. **Is *C. burnetii* characterized by?**
   A. Wide geographic distribution.
   B. Aerosol route of transmission.
   C. Frequent sexual transmission in humans.
   D. Ability to live and multiply in flies.
   E. Dissemination through blood to other organs.

3. **Is the following true for the life cycle of *C. burnetii*?**
   A. The SCV form of *C. burnetii* infects alveolar macrophages and monocytes.
   B. It survives in phagosomes and phagolysosomes with an alkaline pH.
   C. SCVs divide by binary fission inside phagolysosomes and transform into LCVs.
   D. LCVs do not divide and transform back into SCVs.
   E. SCVs are released from the host cell exclusively by macrophage apoptosis.

4. **Which of the following is true about the host response to *C. burnetii* and the ways the bacterium evades immune responses?**
   A. In most cases the immune response is insufficient and evokes chronic inflammation.
   B. Specific IgG, IgM, and IgA represent a powerful immune response to the infection.
   C. *C. burnetii* can release enzymes degrading reactive oxygen intermediates (ROIs).
   D. *C. burnetii* inhibits maturation of phagosomes.
   E. Interferon-γ produced by T cells enhances phagocytosis of *C. burnetii*.

5. **Which of the following are symptoms of acute Q fever?**
   A. Incubation period is about 14–26 days.
   B. It presents as a flu-like febrile illness.
   C. Q fever presents with typical pneumonia.
   D. Q fever hepatitis is a common symptom.
   E. Meningoencephalitis is a common symptom.

6. **Which of the following are the symptoms or predisposing factors of chronic Q fever?**
   A. The disease course is about 3 months.
   B. Chronic endocarditis is the main symptom.
   C. Underlying cardiac valve defects are predisposing factors.
   D. Immunosuppression and pregnancy are predisposing factors.
   E. Doughnut-type granulomas develop in the liver and in the bone marrow.

MULTIPLE CHOICE QUESTIONS (continued)

7. **Which of the following laboratory tests are used for the diagnosis of infection with *C. burnetii*?**

    A. Blood cell count changes are highly characteristic of Q fever.

    B. Duke's criteria are used for the clinical diagnosis of Q fever endocarditis.

    C. High levels of IgG and IgM antibodies to phase I *C. burnetii* indicate acute Q fever.

    D. Excess or equal titers of IgA and IgM anti-phase I antibodies as compared to anti-phase II antibodies are highly indicative of chronic Q fever.

    E. For microscopic observations tissue samples are stained with the Gram technique.

8. **Which of the following drugs are used for the treatment of Q fever?**

    A. Doxycycline.

    B. Tetracycline.

    C. Erythromycin and azithromycin.

    D. Chloroquine.

    E. Penicillin.

Case 7

Coxsackie B virus

A 22-year-old medical student walked into an Emergency Department complaining of a headache that had begun 3 hours before and seemed to be getting worse. He felt generally unwell, but without any other specific symptoms. On examination, he was febrile (38°C), and was noted to have neck stiffness. He complained when a bright light was shone into his eyes. The casualty officer made a diagnosis of acute meningitis, and admitted him to the medical ward.

A lumbar puncture was performed and 2 hours later the following results were received:

- cerebrospinal fluid (CSF) – clear and colorless;
- white cell count 120×10^6/l, 75% mononuclear cells, no red cells;
- CSF protein 4.2 g/l (normal 1.5–4.0 g/l);
- CSF glucose 4.3 mmol/l, blood glucose 5.7 mmol/l;
- Gram stain – no organisms seen.

A throat swab and a sample of feces were sent to the laboratory for viral culture.

Three days later, the patient was feeling much better and was ready to be discharged home. The ward received a telephone call from the microbiology laboratory informing them that the CSF sample was positive for enterovirus RNA. A further result 2 days later confirmed that the virus present in the CSF was coxsackie B4.

1. What is the causative agent, how does it enter the body and how does it spread a) within the body and b) from person to person?

Causative agent

Coxsackie viruses belong to the *Enterovirus* genus within the family of viruses known as the *Picornaviridae*. 'Pico' means small, so this translates as the small RNA viruses. The *Picornaviridae* carry a genome of positive sense single-stranded RNA in an unenveloped viral particle. The other important human pathogens within this family are the rhinoviruses, the commonest cause of the common cold, and hepatitis A virus.

More than 70 enteroviruses infect humans (Table 1), and their nomenclature is somewhat eclectic. Historically, the most important are the three serotypes of poliovirus. Coxsackie is the name of the town where a new enterovirus was isolated from patients with a polio-like illness, and subsequently the numerous coxsackie viruses were split into two groups, A and B, on the basis of their growth properties in mice and also *in vitro*, in tissue culture. The next group of enteroviruses to be described were the echoviruses, where echo stands for enterocytopathic human orphan, describing the fact that these viruses produce an enterovirus-like **cytopathic** effect in routine cell culture, that they are isolated from humans, and that initially no particular disease manifestation was associated with them. The more recently discovered enteroviruses are now given the initials EV (for enterovirus) followed by a number, for example EV70, EV71. All these viruses are serologically distinct, that is infection with any one

Table 1. The human enteroviruses
Polioviruses (serotypes 1, 2, and 3)
Coxsackie A viruses (23 serotypes, numbered A1–A24, as coxsackie A23 is the same virus as echovirus 9)
Coxsackie B viruses (serotypes B1–B6)
Echoviruses (31 serotypes, numbered 1–33, as echoviruses 10 and 28 have been reclassified into other virus families)
Enterovirus serotypes (EV 68–71)

enterovirus induces antibody production that does not cross-neutralize any of the other enteroviruses.

Entry and spread within the body

The term enterovirus reflects the fact that these viruses enter and leave the human host via the enteric tract (otherwise known as the gastrointestinal tract). Thus, they are swallowed in contaminated food or water, being able to survive the highly acidic pH of the stomach (nonenveloped viruses are in general more hardy than enveloped ones). Initial replication in the small bowel wall results in viral excretion in the feces, which may persist for several days. At some stage, virus passes through the small bowel wall and into the host bloodstream. This **viremic** stage allows access of the virus to many different cells and tissues of the body, resulting in a number of different clinical manifestations (see below).

Spread from person to person

Spread between individuals therefore may arise through contamination of food and water, for example through inadequate hand hygiene after defecation. Enterovirus contamination of water supplies through inadequate disposal of sewage can give rise to outbreaks of infection.

Epidemiology

Enteroviral meningitis may arise at any age, but is most common in infants and young children. There is a seasonality to enterovirus infections, most arising in the summer months in temperate climates. Enterovirus infections are common worldwide.

2. What is the host response to the infection and what is the disease pathogenesis?

Immune responses

Initial innate immune responses serve to slow down spread of infection. **Natural killer (NK)** cells appear to be important in the early response to coxsackie B viruses, not through cytotoxicity but rather through their production of interferon (IFN)-γ. In fact, mice lacking the IFN receptor die quickly after challenge with coxsackie B viruses.

Antibody responses against coxsackie B viruses appear 7–10 days after infection.

T-cell responses are also made but the humoral immune response appears to be more important for this group of viruses than the cell-mediated response, as evidenced by the following.

1. In patients with antibody deficiency, enterovirus infection may become chronic, for example persistence of live poliovirus vaccines given to patients with **hypogammaglobulinemia**, with a particular risk of serious central nervous system manifestations, for example chronic enteroviral **encephalitis**.

2. Enteroviruses are not recognized as particularly serious pathogens in the context of patients with cell-mediated immune deficiencies, for example those with HIV infection, or transplant recipients.

IgA might play a role in protection by preventing entry via the intestinal epithelium.

Coxsackie B viruses and other enteroviruses have a number of strategies to prevent immune responses. These include: down-regulation of surface MHC class I molecules to thwart **CD8+** T-cell responses and inhibition of **dendritic cell** function (seen with echoviruses but not coxsackie B viruses).

Pathogenesis

Evidence has been accumulating that coxsackie B viruses and other enteroviruses may play a role in autoimmune disease. Enteroviruses have been implicated in the etiology of type I diabetes through 'molecular mimicry.' There is some evidence that antibodies to coxsackie viruses cross-react with human islet cell antigens and it has been reported that some epitopes of coxsackie B viruses cross-react at the T-cell level with glutamic acid decarboxylase (GAD).

Coxsackie B viruses are thought to be the main etiologic agents of viral **myocarditis**, which is a common cause of idiopathic dilated cardiomyopathy, a severe pathological condition that often requires heart transplantation.

Myocarditis in experimental animal models infected with coxsackie B virus leads to autoimmunity.

3. What is the typical clinical presentation and what complications can occur?

As mentioned above, the clinical manifestations of enteroviral infection are **protean**, although, despite the name enterovirus, these viruses do not cause gastroenteritis. The most common outcome is probably no disease at all, that is asymptomatic infection. Febrile illnesses with nonspecific **rashes** are common in children, and some at least are due to enterovirus infections. Coxsackie A viruses may give rise in children to **herpangina**, or hand, foot, and mouth disease (**vesicles** in the mouth and on the hands and feet). **Conjunctivitis** is particularly associated with EV70.

The disease poliomyelitis has been known for centuries, and arises from infection with one of the three serotypes of poliovirus. Infection of nerve cells supplying muscle tissue (collectively referred to as lower motor neurons, whose cell bodies lie within the anterior horn of the spinal cord) results in death of those cells and flaccid paralysis of the relevant muscle

Figure 1. This image shows muscle wasting (right lower limb) arising from poliomyelitis.

(Figure 1). This is potentially life-threatening, especially if muscles involved with respiration are affected. Very occasionally, poliomyelitis-like illness can arise from infections with other enteroviruses, for example coxsackie viruses, or even with completely unrelated viruses, for example Japanese encephalitis or West Nile viruses.

Meningitis

As in our specific case, meningitis typically presents with a fever, irritability and/or lethargy, especially in young children, neck stiffness (although the absence of this symptom/sign does not exclude a diagnosis of meningitis, especially in young children), and an aversion to bright light – **photophobia**.

Complications of enteroviral meningitis are unusual in an immunocompetent child or adult. Spread of virus from the **meninges** to the brain, resulting in **meningoencephalitis**, may occur rarely. This may be heralded by an abrupt deterioration in mental state, or the onset of **seizures**. In patients with immunodeficiencies, particularly those associated with impaired antibody production, meningoencephalitis is much more common, and may become chronic.

Muscle infections

Coxsackie viruses can also infect muscle tissue itself, giving rise to **Bornholm's disease** or **myocarditis**. Enteroviruses are also the commonest cause of viral **pericarditis**.

Respiratory infections

Within the respiratory system, enteroviruses occasionally cause upper respiratory tract manifestions such as the common cold or, especially in young babies, can involve the lower respiratory tract, presenting with **bronchiolitis** or even **pneumonia**.

Neonatal infections

Enteroviruses are particularly feared pathogens in neonates, in whom they may cause devastating disseminated infections, resulting in multisystem life-threatening disease including myocarditis, **hepatitis**, and encephalitis. Furthermore, patient to patient spread within a neonatal ward has been reported on a number of occasions.

4. How is this disease diagnosed and what is the differential diagnosis?

Diagnosis and differential diagnosis

Meningitis is a medical emergency, and appropriate diagnostic tests must be initiated as soon as possible to expedite effective therapy (if available). Initial investigations should include blood cultures and a lumbar puncture (although this is contraindicated if there is evidence of raised intracranial pressure) to obtain a sample of CSF. Laboratory examination of the CSF usually allows distinction between bacterial and viral meningitis. In the former, CSF protein is grossly raised, CSF glucose is reduced to less than half of the blood sugar level, and the predominant cellular infiltrate is with polymorphonuclear leucocytes (PMNs) (Figure 2A). Gram staining of centrifuged CSF, that is the pellet, will often reveal the presence of bacteria. In viral meningitis, CSF protein is only marginally raised, CSF glucose is usually normal, and the predominant cellular infiltrate is with mononuclear cells, that is lymphocytes and monocytes (Figure 2B) (although an early predominance of PMNs can sometimes be seen very early in the course of viral meningitis). A Gram stain will be negative.

Mumps virus used to be the commonest cause of viral meningitis, but with the introduction of effective mumps vaccines (given universally precisely because of this manifestation of infection) enteroviruses now account for the majority of cases of viral meningitis. Herpes simplex virus can also cause meningitis, usually in young women with extensive genital herpes infection. Confirmation that viral meningitis is due to an enterovirus, as opposed to other viral causes, can be made in a number of ways. The most sensitive approach is to detect the presence of enteroviral RNA in the CSF sample by a genome amplification technique such as the **polymerase chain reaction (PCR)**. By sequencing the PCR product it is possible to determine which particular enterovirus is the culprit. Many (but not all) enteroviruses may also be isolated from CSF in routine tissue culture. If CSF is not available, then alternative samples are a throat swab or a sample of feces. Isolation of enterovirus from feces may be possible up to 2 weeks after initial clinical presentation.

Serological assays are available for detection of an antibody response to an enterovirus. **IgM** anti-enterovirus antibodies are indicative of a recent infection, but this test is usually only available within a reference laboratory. An alternative approach is to demonstrate a rise in **IgG** anti-enterovirus antibody titers in a sample taken a week or two after the initial presentation (known as a convalescent sample) compared with those in a sample taken at the time of acute illness. Thus, this is not a technique that allows the diagnosis to be made when the patient first presents acutely ill.

Other clinical manifestations may be evident on examination of the patient, depending on the particular causative agent, for example meningococcal meningitis is often associated with a spreading, nonblanching **purpuric rash** (see Case 25).

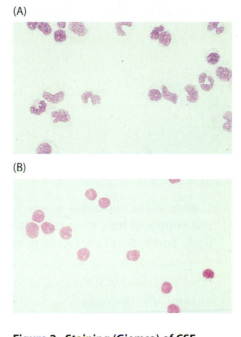

(A)

(B)

Figure 2. Staining (Giemsa) of CSF deposit showing:
(A) polymorphonuclear cells typical of bacterial meningitis; (B) mononuclear cells typical of viral meningitis.

5. How is the disease managed and prevented?

Management

Enteroviral meningitis will usually resolve spontaneously, and the vast majority of patients will make a complete recovery. There are no licensed antiviral drugs with activity against enteroviruses, so management is supportive only.

Prevention

There are two types of vaccines available for the prevention of poliovirus infection (but neither offers any protection against infection with other enteroviruses). The Sabin vaccine consists of a mixture of the three poliovirus serotypes in a live attenuated form. It is administered orally (i.e. mimicking the natural route of infection), and therefore induces immunity within the gut mucosa, as well as within the bloodstream. The Salk vaccine is an inactivated vaccine, again containing all three poliovirus serotypes. The World Health Organization (WHO) is currently co-ordinating a campaign for the elimination of poliovirus infection. Although the ultimate goal has not yet been reached, at the time of writing there are only a few countries in the world that continue to suffer wild-type poliovirus infections (for example Nigeria, India, Pakistan, Afganistan).

SUMMARY

1. What is the causative agent, how does it enter the body and how does it spread a) within the body and b) from person to person

- The enteroviruses are nonenveloped positive single-stranded RNA viruses belonging to the family *Picornaviridae*.
- They comprise poliovirus types 1–3, coxsackie A viruses (23 types), coxsackie B viruses (6 types), echoviruses (31 types), and enteroviruses 68–71.
- Virus entry is via the gastrointestinal tract. Initial replication occurs within the small bowel, associated with excretion of virus in the feces.
- This is followed by viremic spread to various target organs.
- Human to human spread is via the **fecal–oral route**.

2. What is the host response to the infection and what is the disease pathogenesis?

- Antibody responses are essential to control enterovirus infections.
- In antibody-deficient individuals, infection becomes chronic and can be life-threatening.

3. What is the typical clinical presentation and what complications can occur?

- Enteroviruses cause a wide range of diseases, including the following:
- Febrile illness with or without skin rashes.
- Respiratory tract infections.
- Herpangina.
- Hand, foot, and mouth disease.
- Conjunctivitis.
- Meningitis.
- Encephalitis (in antibody-deficient patients).
- Myocarditis and pericarditis.
- Bornholm's disease (epidemic pleurodynia).
- Disseminated multisystem disease in the newborn.
- Poliomyelitis.

4. How is this disease diagnosed and what is the differential diagnosis?

- Enteroviral meningitis is diagnosed on the basis of typical CSF findings (mononuclear cell infiltrate, protein raised only marginally, glucose normal, Gram stain negative) plus any of the following:
- Isolation of the virus in tissue culture from CSF, throat swab or feces.
- Genome amplification (for example PCR) within CSF.
- IgM antibodies, or rising titers of IgG antibodies.
- Other causes of viral meningitis include mumps virus, herpes simplex virus, some arboviruses.

5. How is the disease managed and prevented?

- Enterovirus infections, including meningitis, are almost always self-limiting in immunocompetent hosts and resolve spontaneously.
- There are no licensed antiviral drugs with proven activity against enteroviruses (although pleconaril, an investigational receptor-blocking drug, has undergone clinical trials with some success).
- Poliovirus infection can be prevented by use of either live attenuated (Sabin) or inactivated (Salk) vaccines.

FURTHER READING

Humphreys H, Irving WL. Problem-orientated clinical microbiology and infection, 2nd edition. Oxford University Press, Oxford, 2004.

Murphy K, Travers P, and Walport M. Janeway's Immunobiology, 7th edition, Garland Science, New York, 2008.

Richman DD, Whitley RJ, Hayden FG. Clinical Virology, 2nd edition. ASAM Press, Washington, DC, 2002.

Zuckerman AJ, Banatvala JE, Pattison JR, Griffiths PD, Shaub BD. Principles and Practice of Clinical Virology, 5th edition. Wiley, Chichester, 2004.

REFERENCES

Chadwick DR. Viral meningitis. Br Med Bull, 2006, 75-76: 1–14. (Review.)

Lee BE, Davies HD. Aseptic meningitis. Curr Opin Infect Dis, 2007, 20: 272–277. (Review.)

Logan SAE, MacMahon E. Clinical review: viral meningitis. BMJ, 2008, 336: 36–40.

Palacios G, Oberste MS. Enteroviruses as agents of emerging infectious diseases. J Neurovirol, 2005, 11: 424–33. (Review.)

Pallansch MA, Sandhu HS. The eradication of polio – progress and challenges. N Engl J Med, 2006, 355: 2508–2511.

WEB SITES

All the Virology on the WWW Website, developed and maintained by Dr David Sander, Tulane University: http://www.virology.net/garryfavweb13.html#entero

Centers for Disease Control and Prevention, National Center for Immunization and Respiratory Diseases, Atlanta, GA, USA: http://www.cdc.gov/ncidod/dvrd/revb/ enterovirus/non-polio_entero.htm

Centre for Infections, Health Protection Agency, HPA Copyright, 2008: http://www.hpa.org.uk/infections/topics_az/enterovirus/menu.htm

Website of Derek Wong, a medical virologist working in Hong Kong: http://virology-online.com/viruses/ Enteroviruses.htm

MULTIPLE CHOICE QUESTIONS

The questions should be answered either by selecting True (T) or False (F) for each answer statement, or by selecting the answer statements which best answer the question. Answers can be found in the back of the book.

1. **Which of the following viruses are classified as enteroviruses?**
    A. Coxsackie B5 virus.
    B. Echovirus 33.
    C. Norwalk virus.
    D. Poliovirus type 2.
    E. Rotavirus.

2. **Which of the following clinical syndromes may arise in an immunocompetent 5-year-old child as a result of enterovirus infection?**
    A. Conjunctivitis.
    B. Herpangina.
    C. Foot and mouth disease.
    D. Gastroenteritis.
    E. Lower motor neuron paralysis.

3. **Which of the following features in a cerebrospinal fluid sample are suggestive of a viral cause of meningitis?**
    A. Cloudy fluid.
    B. A mononuclear cell infiltrate.
    C. CSF sugar less than half the value of the blood sugar.
    D. The presence of intracellular gram-negative diplococci.
    E. CSF protein level at the upper limit of normal.

4. **Which of the following viruses are well-recognized causes of meningitis?**
    A. Coxsackie A9.
    B. Echovirus 33.
    C. Herpes simplex virus.
    D. Measles virus.
    E. Mumps virus.

5. **Which of the following statements regarding enteroviral meningitis are true?**
    A. Enteroviruses account for up to 30% of cases of viral meningitis.
    B. Poor hygiene is a significant factor in the spread of illness.
    C. Most cases occur in children under the age of 14 years.
    D. Infection most often occurs in the winter months in temperate climates.
    E. Long-term sequelae including nerve deafness are common.

6. **Which of the following laboratory assays can be used in the diagnosis of enteroviral meningitis?**
    A. Demonstration of a rise in enterovirus-specific antibodies in paired acute and convalescent blood samples.
    B. Electron microscopy of a cerebrospinal fluid sample.
    C. Gram staining of a cerebrospinal fluid sample.
    D. Reverse transcriptase polymerase chain reaction assay on a sample of cerebrospinal fluid.
    E. Virus isolation in cell culture from a fecal sample taken 10 days after the onset of illness.

7. **Which of the following antiviral drugs have activity against enteroviruses?**
    A. Aciclovir.
    B. Amantadine.
    C. Oseltamivir.
    D. Ritonavir.
    E. Zidovudine.

8. **Poliovirus type 2 is isolated from a stool sample from a 20-year-old male nurse. The nurse recalls being vaccinated against poliovirus infection as a child with Sabin (live attenuated vaccine), and receiving a booster dose 4 years ago before travelling to India. Investigation of this person's immune system is most likely to reveal which of the following?**
    A. No abnormality.
    B. Antibody deficiency.
    C. T-cell (cell-mediated) deficiency.
    D. Abnormalities of phagocytosis.
    E. Complement deficiency.

Case 8

Echinococcus spp.

A 28-year-old Kurdish refugee complained of upper abdominal pain for 4 months. He had recently noticed some pain on the right side of his chest. He had not noticed any fever and his weight was stable. His doctor requested a chest X-ray. This showed a mass lesion in the right lung. He was referred to the Thoracic Surgical Department at the local hospital. A **CT scan** of his chest and abdomen was requested. This demonstrated a cystic structure in his liver, with another **cyst** in his chest (Figure 1). The surgeons excised the lung lesion. The histopathologists reported that it was a **hydatid cyst**. The patient was referred to the Infectious Diseases Department. A serological test was positive for *Echinococcus*. He was given treatment with albendazole and his remaining cyst was monitored serially on scans.

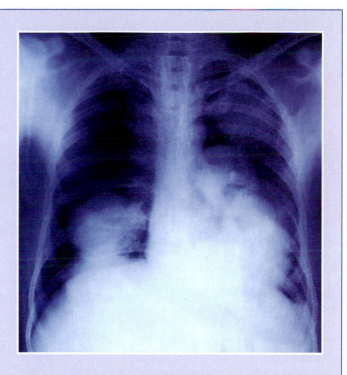

Figure 1. Pulmonary hydatid cysts. Chest X-ray showing hydatid cysts (opaque areas at center right and center left) in a patient's lungs. The cysts are caused by the parasitic tapeworm *Echinococcus* sp. Humans are not a natural host for the parasite, but may become infected from ingesting eggs shed in the feces of an infected dog or other canids. Cysts may be formed in the liver, lungs or any other organ in the body. Large pulmonary cysts may cause symptoms, including shortness of breath, chest discomfort and sometimes wheeze. Treatment involves giving antiparasitic drugs and careful surgical resection when possible.

1. What is the causative agent, how does it enter the body and how does it spread a) within the body and b) from person to person?

Causative agent

Echinococcus species are tapeworms, also known as cestodes. Adults are only 3–8.5 mm in length. Two species are responsible for the majority of human infections. *Echinococcus granulosus* arises from dogs and other canids and has been described worldwide. It causes cystic echinococcosis. *Echinococcus multilocularis* arises from arctic or red foxes and is restricted to the Northern Hemisphere. It causes alveolar echinococcosis. Infection with the latter seems to be spreading as red fox populations have been growing. Adult tapeworms live in the intestines of the dog or wild canid hosts. They attach to the mucosa with suckers and hooks. There are different parts to the structure of the tapeworm, and these are shown in Figure 2. The adult tapeworms lay eggs, which pass out in the feces and contaminate the soil.

sucker   uterus with eggs

Echinococcus granulosus

Figure 2. An adult tapeworm of
Echinococcus granulosus.

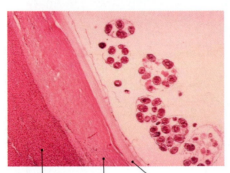

inflammatory cells   germinal layer
fibrosis

acellular layer

**Figure 3. Histopathology of the
structure of a hydatid cyst**. Round
protoscoleces are seen on the right. They
arise from a thin germinal layer, which is
surrounded by an acellular layer; the host
inflammatory cells can be seen to the left
of this.

These may be swallowed by an intermediate host such as sheep for *E. granulosus* and small mammals for *E. multilocularis*.

Entry and spread within the body

Humans are infected if they accidentally swallow eggs. The eggs have a tough shell that breaks down and releases a larval form called the oncosphere. The oncosphere burrows into the intestinal wall and then spreads to other parts of the body along blood vessels or lymphatics. The liver is the commonest site of infection as most of the intestinal drainage of blood is along the portal vein to the liver. However, infection could arise anywhere in the body.

When the oncosphere lodges in an organ it develops into a vesicle. This progressively grows into a cyst. The cyst wall includes an inner germinal layer. From this buds a stage called protoscolex within brood capsules. Around the germinal layer is an acellular, laminated layer. For *E. granulosus* this cyst is unilocular and known as a hydatid cyst. For *E. multilocularis* the cyst is multilocular, or referred to as alveolar. Occasionally cysts can rupture giving rise to secondary, 'daughter' cysts. The structure of a hydatid cyst due to *E. granulosus* is shown in Figure 3.

Person to person spread

Humans do not spread *Echinococcus* spp. from person to person. Humans are an end host, with occasional exceptions.

Intermediate hosts

Animals such as sheep or cattle may ingest *E. granulosus* eggs shed in dog feces. They can act as intermediate hosts. When sheep or cattle are killed, dogs may eat their meat or offal. When the dogs swallow cysts within infected tissues the protoscoleces are released. They attach to the dogs' intestinal mucosa to mature into the adult tapeworms. Thus the life cycle is completed (Figure 4). Dogs are the definitive hosts. This life cycle for *E. granulosus* occurs on farms but for *E. multilocularis* there is a rural, or **sylvatic**, life cycle with small mammals serving as the intermediate host and arctic or red foxes as the definitive host.

Epidemiology

Given the life cycle, hydatid disease is most common where dogs and animals such as sheep and cattle mix. In the United Kingdom hydatid disease has been mainly found in sheep farming areas such as North Wales. Worldwide the Turkana district of Kenya has one of the highest incidences of human cystic echinococcosis. Dogs are a key part of Turkana communities that herd cattle. The Turkana do not always bury their dead. Dogs may then scavenge human remains and continue the life cycle.

2. What is the host response to the infection and what is the disease pathogenesis?

Potentially the host may mount a response to the oncospheres, after they have entered the intestine, or against the cysts, which become surrounded by the acellular, laminated layer. There may also be a host response to fluid that leaks from a ruptured cyst.

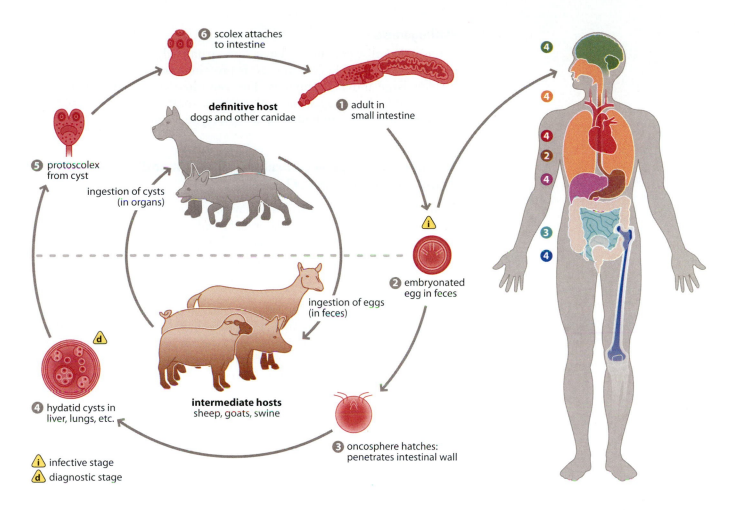

⑥ scolex attaches
to intestine

definitive host
dogs and other canidae

① adult in
small intestine

⑤ protoscolex
from cyst

ingestion of cysts
(in organs)

ⓘ

② embryonated
egg in feces

ingestion of eggs
(in feces)

ⓓ

④ hydatid cysts in
liver, lungs, etc.

intermediate hosts
sheep, goats, swine

③ oncosphere hatches:
penetrates intestinal wall

ⓘ infective stage
ⓓ diagnostic stage

④④④②④③④

Little is known of the host response to the oncospheres. A single or small number of oncospheres may not pose a sufficient antigenic stimulus to trigger an adaptive immune response. The innate immune response, utilizing phagocytic cells or complement, may attack them after invasion of the intestinal mucosa. Once a hydatid cyst has developed an antibody response occurs that is initially **IgM** followed by an **IgG** and **IgE** response. If these antibodies also recognize the oncospheres they could play a part in preventing further ingested oncospheres from causing additional infection.

An inflammatory response occurs around the maturing cyst. This is triggered by complement activation. There is an infiltration of macrophages and lymphocytes and these include both **CD4+** and **CD8+** T lymphocytes. Both **Th1-** and **Th2**-type **cytokines** are found, such as **interferon (IFN)-γ, interleukin** (IL)-4, and interleukin-10. Eventually a host-derived, fibrous layer forms around the cyst. Complement activation and inflammation subside as the acellular layer accumulates. It is thought that there is local immunomodulation with impaired macrophage responses, impaired **dendritic cell** differentiation, and T-regulatory cells. Cyst-derived antigens, including an antigen B, affect the maturation and function of dendritic cells, such that when they present antigen to lymphocytes the immune response skews to a Th2 pattern. The latter is characterized by IL-4 production and increased IgG and IgE levels.

Figure 4. Life cycle of *Echinococcus granulosus*. Adult worms living in the dog intestine shed eggs in dog feces. These contaminate the environment and may be ingested by intermediate hosts such as sheep. In the intestine oncospheres hatch from the eggs and penetrate the intestinal wall. The oncosphere spreads to other organs and eventually matures into a cyst. Within the cyst protoscolices develop. When a cyst is ingested by a dog the protoscolices are released, attach to the intestinal mucosa as a scolex, and mature into adult worms.

Pathogenesis

Large cysts cause pressure and may cause pain or dysfunction of the organ affected. If a hydatid cyst ruptures its proteinaceous contents leak, with a sudden, large antigenic stimulus. This can cause an anaphylactic reaction if IgE antibodies trigger the activation of mast cells and the systemic release of histamine and other mediators (**type 1 hypersensitivity**).

3. What is the typical clinical presentation and what complications can occur?

Hydatid cysts grow gradually. They may not cause any physical problems and may remain asymptomatic. Just over half of infected individuals are asymptomatic. When the cyst is large pressure effects may cause pain or dysfunction of the organ affected. If a cyst ruptures, daughter cysts may cause additional problems and cystic fluid can trigger life-threatening **anaphylaxis**.

Hydatid cysts may be found in any part of the body but about 90% occur in the liver or lung or both. Large cysts in the liver can cause upper abdominal pain. The pain may localize more to the right side of the abdomen, as the right lobe of the liver is most frequently affected. The liver can enlarge and become palpable. The cyst may become secondarily infected with bacteria. If the cyst presses on the biliary tree obstructive **jaundice** can develop. Rupture into the biliary tree may lead to daughter cysts in the common bile duct. Again these can cause an obstructive jaundice. In the lung large cysts may cause pain. This may be felt as a central chest ache. When cysts abut the chest wall they cause localized pain at these sites. Very large cysts may compromise breathing and make individuals short of breath.

Hydatid disease due to *E. granulosus* has a case fatality rate of about 2%. Infection due to *E. multilocularis* has a higher mortality rate. It can cause any of the clinical features associated with hydatid disease, but can have a progressive course with growing cysts that enlarge the liver and compromise its function.

If cysts rupture an abrupt antigenic challenge may trigger anaphylaxis. A classical anaphylactic reaction comprises swelling of lips, tongue, vocal chords, and bronchial airway mucosa; urticarial reactions in the skin; and a fall in blood pressure. **Eosinophilia** may be observed if there is cyst leakage but is not uniformly present with intact cysts.

4. How is the disease diagnosed and what is the differential diagnosis?

The case history illustrates that sometimes the diagnosis is made after surgical removal of a lesion and histopathological examination. However, it is preferable to make a diagnosis before surgery as accidental incision of a cyst and leakage of its contents may trigger anaphylaxis.

Either ultrasound of the liver or a CT scan of thorax or abdomen may reveal a cystic structure. Some features are highly suggestive of a hydatid cyst. There is a thick cyst wall. Septa may be seen within the cyst. If the germinal layer detaches from the cyst wall the inner septa may float in the fluid like a 'water lily.' Mature, degenerating cysts may become calcified.

Serological tests are available for *Echinococcus* antibodies. Methods used include **enzyme-linked immunosorbent assay (ELISA)**, indirect hemagglutination assays, and latex agglutination assays. The antigens used in assays are derived from hydatid cyst fluid. There may be cross-reactions with other tapeworm infections affecting specificity. Sensitivity may be as low as 50% for infection in the lung, but more than 80% for liver.

Differential diagnosis

The differential diagnosis is from other causes of cystic structures. The range of possibilities is larger for the lung (see Ryu and Swensen, 2003) than for the liver. In the liver benign cysts with thin walls can occur. Liver tumors may have the same shape, but are more solid structures. Serological tests will also help to eliminate other pathologies.

5. How is the disease managed and prevented?

Management

Treatment by surgery and chemotherapy may not be required for small asymptomatic cysts. Surgical removal may be required for troublesome cysts if technically feasible.

Prior treatment with albendazole has been shown to benefit the management of liver hydatid cysts (see Gil-Grande et al., 1993). Albendazole given continuously for 1–3 months reduces cyst viability and can lead to shrinkage, making surgery safer. For small, troublesome cysts albendazole treatment alone may be sufficient. Combination treatment with praziquantel may be more effective.

Another approach to the management of some cases of cystic echinococcosis has been aspiration of accessible cysts with a needle and injection of an agent such as 95% ethanol to kill the protoscoleces inside the cyst. The ethanol is then aspirated back after 20 minutes. Microscopy of the aspirate shows free protoscoleces, which are sometimes referred to as 'hydatid sand.' This method is referred to as PAIR, which stands for **P**uncture, **A**spiration, **I**njection, **R**easpiration. The key risk is anaphylaxis if cyst fluid leaks out. Experienced units report success with PAIR but its place in management requires further investigation.

Prevention

Humans need to maintain high standards of hand washing to avoid swallowing eggs. Prevention of *Echinococcus* infection requires breaking the natural life cycle. Dogs have been treated with arecoline or praziquantel. Praziquantel baiting has been used to treat red foxes. On farms dogs are kept away from offal.

Vaccines have been trialed in sheep and dogs. A recombinant DNA vaccine containing the gene for an oncosphere antigen, and designated EG95, has shown >90% protective efficacy in sheep.

Despite efforts, *Echinococcus* infection continues to be a problem in many countries.

SUMMARY

1. What is the causative agent, how does it enter the body and how does it spread a) within the body and b) from person to person?

- The causative agent is *Echinococcus granulosus* or *Echinococcus multilocularis*. These are tapeworms, or cestodes.

- Adult *Echinococcus* tapeworms live in the intestine of dogs or foxes and shed eggs in the feces. Dogs and foxes are the definitive hosts.

- The eggs are accidentally swallowed by other species. Once swallowed oncospheres emerge from the eggs, penetrate the intestinal wall, and spread to other parts of the body where they form cysts.

- Dogs are infected when they eat meat or offal containing cysts. This happens for instance on farms, where they might eat offal from sheep or cattle.

- Species such as sheep or cows are therefore referred to as intermediate hosts. Humans are usually an end host.

2. What is the host response to the infection and what is the disease pathogenesis?

- Little is known about the host response to the oncospheres when they penetrate the intestinal wall.

- Once the oncospheres develop into cysts in tissues there is a surrounding inflammatory response triggered by complement activation, involving macrophages and then a CD4+ and CD8+ T-lymphocyte response.

- Eventually an acellular and fibrous layer develops around the cyst with down-regulation of the immune and inflammatory response.

- The pathogenesis of disease depends on the size of cysts. Large cysts can pose a physical problem in the infected organ.

- If cysts rupture and leak an IgE-mediated response can be triggered.

3. What is the typical clinical presentation and what complications can occur?

- Small cysts may remain asymptomatic.

- Hydatid cysts due to *E. granulosus* are mostly found in the liver or lung or both. The cysts of *E. multilocularis* are primarily found in the liver.

- Large hydatid cysts in the liver can cause pain, may obstruct the flow of bile, and may rupture into the biliary tree.

- Disease due to *E. multilocularis* can be steadily progressive.

- Cyst rupture and leakage of proteinaceous contents can trigger an anaphylactic reaction.

4. How is the disease diagnosed and what is the differential diagnosis?

- Diagnosis is based on imaging such as ultrasound or CT scan. Typical radiological features may be present to diagnose hydatid disease.

- Additional tests include **serology**.

- Cross-reactions with other tapeworm infections affect the specificity of serology. Sensitivity ranges from 50 to 80%.

- The differential diagnosis is from other causes of cystic lesions.

5. How is the disease managed and prevented?

- Uncomplicated, asymptomatic cysts may be left alone.

- Symptomatic cysts may be treated with albendazole to reduce cyst viability.

- Surgical resection may then be performed if necessary and if feasible.

- Some experienced centers also undertake a procedure called PAIR, which involves percutaneous puncture of cysts, aspiration of contents, injection of etanol, and then reaspiration.

- Prevention requires breaking the life cycle by treating dogs and preventing the consumption of offal by dogs. To avoid accidental infection humans must maintain high standards of hand hygiene.

FURTHER READING

Gottstein B, Reichen J. Echinococcosis/hydatidosis. In: Cook GC, Zumla A, editors. Manson's Tropical Diseases, 21st edition. Saunders/Elsevier, London/Philadelphia, 2003: 1561–1582.

REFERENCES

Craig PS, McManus DP, Lightowlers MW, et al. Prevention and control of cystic echinococcosis. Lancet Infect Dis, 2007, 7: 385–394.

Gil-Grande LA, Rodriguez-Caabeiro F, Prieto JG, et al. Randomised controlled trial of efficacy of albendazole in intra-abdominal hydatid disease. Lancet, 1993, 342: 1269–1272.

McManus DP, Zhang W, Li J, Bartley PB. Echinococcosis. Lancet, 2003, 362: 1295–1304.

Ryu JH, Swensen SJ. Cystic and cavitary lung diseases: focal and diffuse. Mayo Clin Proc, 2003, 78: 744–752.

WEB SITES

Centers for Disease Control and Prevention (CDC), 2008, United States Public Health Agency: www.cdc.gov/

Centre for Infections, Health Protection Agency, HPA Copyright, 2008: www.hpa.org.uk

World Health Organization, Initiative for Vaccine Research, Copyright WHO 2008: www.who.int

MULTIPLE CHOICE QUESTIONS

The questions should be answered either by selecting True (T) or False (F) for each answer statement, or by selecting the answer statements which best answer the question. Answers can be found in the back of the book.

1. **Which of the following are true about *Echinococcus*?**

   A. *Echinococcus* is a cestode tapeworm.

   B. *Echinococcus granulosus* is restricted in its geographical distribution to the Northern Hemisphere.

   C. Adult worms live in the intestine of humans, sheep or cattle.

   D. The life cycle for *Echinococcus granulosus* involves dogs.

   E. *Echinococcus multilocularis* does not cause human infections.

2. **Which of the following are true about the life cycle of *Echinococcus* infection?**

   A. Humans are infected by swallowing eggs.

   B. The larval form that arises from the eggs is called an oncosphere.

   C. Cysts usually develop from the oncosphere in the intestinal wall.

   D. Humans spread infection when they pass eggs in their feces.

   E. For *E. granulosus* dogs continue the life cycle when they eat sheep or cattle offal.

3. **Which of the following are true about the epidemiology of *Echinococcus* infection?**

   A. Infection is principally an urban disease.

   B. The high incidence of infection observed in the Turkana district of Kenya is due to local fox populations.

   C. Infection due to *E. multilocularis* can occur in arctic regions.

   D. Red fox populations are diminishing with a corresponding decline in *E. multilocularis*.

   E. *E. granulosus* infection is disappearing worldwide.

MULTIPLE CHOICE QUESTIONS (continued)

4. **Which of the following components are included in the host response to *Echinococcus*?**

   A. A Th2-type response has been clearly established against the invading oncospheres.

   B. Complement activation occurs as the oncosphere matures into cysts.

   C. Both CD4+ and CD8+ T lymphocytes are found around the cysts.

   D. An inflammatory response occurs that continues unabated around the cysts.

   E. Anaphylactic reactions occur in response to cysts that rupture and leak.

5. **Which of the following statements are true about hydatid cysts?**

   A. Cysts are always symptomatic.

   B. The spleen is the commonest location for cysts.

   C. Cysts are usually limited in size and do not enlarge to make organs palpable.

   D. More than one cyst can develop.

   E. Cysts can cause pain.

6. **What are the recognized complications of infection?**

   A. Rupture of cysts leading to secondary daughter cysts.

   B. Secondary infection with bacteria.

   C. Obstructive jaundice when the biliary tree is compressed by a cyst.

   D. Increased blood pressure during an anaphylactic reaction to cyst rupture.

   E. Death from progressive infection.

7. **What is the diagnosis of *Echinococcus* infection based on?**

   A. A very specific serological test.

   B. A highly sensitive serological test.

   C. Recognition of characteristic cyst features on ultrasound or CT scanning.

   D. The presence of eosinophilia in the blood.

   E. A diagnostic aspiration of cysts to look for protoscoleces.

8. **Which of the following should be considered in the differential diagnosis of *Echinococcus* infection?**

   A. Benign cysts of the liver.

   B. Developmental cysts in the lung.

   C. Liver tumors.

   D. Cirrhosis of the liver.

   E. Amebic liver abscess.

9. **Which of the following are relevant for the treatment of *Echinococcus* infection?**

   A. Surgical resection is always indicated.

   B. Antiparasitic drug treatment before surgery is beneficial.

   C. For drug treatment albendazole is always combined with praziquantel.

   D. Drug treatment is used for 10 days.

   E. Cysts can also be killed by the injection of 95% ethanol.

10. **Which of the following are true for the control of *Echinococcus* infection?**

    A. It is helped by regular drug treatment of dogs to clear infection.

    B. It requires the avoidance of feeding farm dogs with offal.

    C. It requires the routine use of vaccination in animals.

    D. It is helped by vaccination of humans.

    E. It is helped by good hand hygiene measures in farm workers.

Case 9

Epstein-Barr virus

A 20-year-old woman went to her doctor complaining of a sore throat. It started 4 days previously with associated episodes of fever and chills. On examination her doctor noticed that her tonsils and **uvula** were red and slightly swollen and that she had cervical **lymphadenopathy**. Suspecting a bacterial infection her doctor prescribed a course of ampicillin and suggested that she took a couple of days off work. The patient returned a week later with worsening symptoms. She was now very lethargic with a temperature of 38.5°C and a widespread maculopapular rash as shown in Figure 1.

On the second visit to her doctor she had patches of white exudate on the tonsils (Figure 2), **petechial** hemorrhages on the soft palate, and generalized lymphadenopathy. The doctor could also palpate an enlarged spleen and noticed that she was slightly tender over the right **hypochondrium**. He suspected that she might have infectious mononucleosis and sent her for hematological and antibody tests.

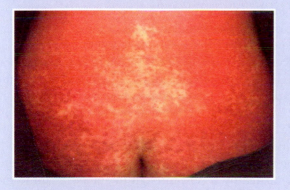

Figure 1. Infectious mononucleosis: ampicillin-induced rash. Note the resemblance to measles.

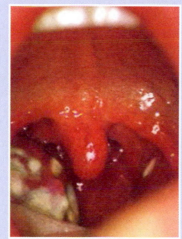

Figure 2. Infectious mononucleosis: follicular exudate on the tonsil, which is very swollen; the uvula is red and edematous.

1. What is the causative agent, how does it enter the body and how does it spread a) within the body and b) from person to person?

Causative agent

The patient has the clinical symptoms of primary Epstein-Barr virus (EBV) infection – infectious mononucleosis or glandular fever as it is commonly called because of the concomitant swollen lymph nodes. EBV, first discovered in 1964, is one of the most common human viruses. It belongs to the *Herpesviridae* family, of which there are eight identified to date as being human pathogens (Table 1). They include herpes simplex virus (HSV-1 and HSV-2) cytomegalovirus (CMV), varicella-zoster virus (VZV), and others. EBV is a gamma herpesvirus and is also called human herpesvirus *4* (HHV4). All herpesviruses are enveloped with a linear dsDNA genome of about 175 kbp. The icosahedral **capsid** is approximately 100–110 nm in

Table 1. Human herpesviruses
Herpes simplex virus type 1
Herpes simplex virus type 2
Varicella-zoster virus
Epstein-Barr virus
Cytomegalovirus
Human herpesvirus type 6
Human herpesvirus type 7
Human herpesvirus type 8 (also known as Kaposi's sarcoma-associated herpesvirus)

diameter, containing 162 tubular (i.e. with a hole running down the long axis) capsomeres. The **virus envelope** has glycoprotein spikes on its surface – these envelope proteins include gp42, gH, gL, and gp350.

Entry and spread within the body

Primary infection. The virus is transmitted to an uninfected from an infected person via saliva. Although the precise cellular route of infection has not yet been elucidated, it appears that either tonsillar B cells or the overlying squamous epithelial cells are probably the first cells to be infected. Most evidence points to the tonsil as the primary site of infection. Here virus replication involves a lytic cycle producing many more **virions**, which then pass via the afferent lymphatics into the cervical lymph nodes where further B cells are infected. For entry into B cells, gp350 on the viral envelope binds to a complement receptor (**CR2**, also called CD21), on the surface of B cells. gp42 binds MHC class II, which serves as co-receptor on the cell surface and is responsible for the fusion between the virus envelope and the host cell membrane. The virus is also carried by B cells through the bloodstream to the spleen and other lymphoid tissues where EBV causes B-lymphocyte proliferation (proliferating EBV-infected lymphocytes are sometimes referred to as transformed B cells – see later) resulting in lymphadenopathy at both cervical and other sites of the body and **splenomegaly**. The incubation time of this infection (i.e. time from initial infection to development of symptoms – if they occur; see later), is usually 4–8 weeks. Even after resolution of symptoms the virus is not cleared from the body and remains **latent** throughout life in some of the memory B lymphocytes leading to a permanent carrier state. These memory B cells contain the latent EBV genome as an **episome** that expresses only part of its genetic information and serves upon reactivation of the virus from its latent state (see later) as a seed for further lytic infection at mucosal sites. This results in shedding of EBV in saliva but it is unclear whether the source of EBV in saliva is infected B or epithelial cells. Whatever the exact mechanism, most, if not all long-term carriers continue to shed low levels of virus throughout their life.

Natural killer (NK) cells and T cells may occasionally be the targets of EBV during primary infection but the molecules used for entry are at present unknown. Infection might be responsible for the rare development of NK-cell and possibly T-cell **lymphomas** (see complications of EBV infection).

Person to person spread

As described above, EBV is present in saliva throughout life. Spread of the infection is usually through intimate contact between an infected and an uninfected person – often through kissing. Hence this disease is known as the 'kissing disease.' EBV has also been shown to be present in both vaginal and male genital secretions. This, together with recent evidence that penetrative sex increases the frequency of **seroconversion** suggests that EBV is also sexually transmitted. Spread of the virus in children is probably mainly through fingers contaminated with saliva and other close contact.

Epidemiology of EBV infection and infectious mononucleosis

Globally the majority of adults are infected with EBV (95–98%). In developing countries children become infected with the virus early in life and the infection is usually asymptomatic or so mild that it is undiagnosed. In developed countries, infection with EBV occurs relatively late and infectious mononucleosis is most common in young adults such as college students; recent studies in Europe show that at least 75% of school leavers are EBV **seropositive** but only about 30% develop the symptoms of infectious mononucleosis. The incidence of infectious mononucleosis shows no consistent seasonal peak and the disease is rare in older adults.

2. What is the host response to the infection and what is the disease pathogenesis?

Following primary infection with EBV several virus-encoded molecules are produced, which induce an antibody response. These include the following.

- *Early antigens*   EA complex
- *Late antigens*   virus capsid antigens (VCA complex) – membrane especially gp 350
- *Latency antigens*   some EB nuclear antigens (EBNAs)
  three latent membrane proteins (LMPs).

In the infected cell, EA complex is produced early in the EBV replication cycle, whereas the VCA complex, a virus structural protein, is produced later. EBNA-1 is an important **latency** factor (a latent DNA replication factor), EBNA-2 is a viral oncogene and EBNAs 3A and C are transcriptional regulators that regulate the function of EBNA 2. These are not expressed during the establishment of latency. LMP1 is a viral oncogene that initiates B-cell activation and proliferation and also plays a role in the establishment of latency by inhibiting B-cell **apoptosis**, whereas LMP2 plays a major role in the maintenance of latency.

Both antibody and T-cell responses are induced against EBV antigens.

Antibody responses

Figure 3 shows the typical sequence of appearance of antibodies made to some of the EBV-encoded antigens in a primary response. **IgM** to VCA is the first to appear followed by **IgG**. Antibodies of the IgM class to VCA are indicative of recent EBV infection, whereas IgG alone suggests infection some time previously. IgG VCA antibodies are present throughout

Figure 3. Basic pattern of serum antibodies to Epstein-Barr virus-associated antigens before, during, and after primary infection. HA, heterophil antibody; EA(D), diffuse form of early antigen; VCA, virus capsid antigen; EBNA, Epstein-Barr nuclear antigen.

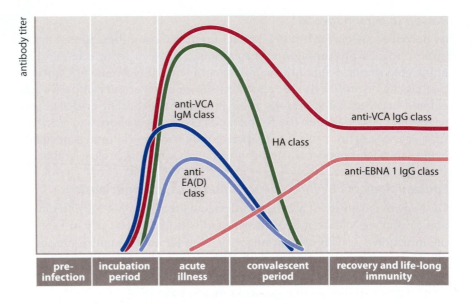

the life of individuals infected with EBV and are used to determine the status of an individual regarding prior infection with the virus. Antibodies to the viral membrane glycoprotein gp350 are neutralizing since they block binding of the virus to CD21 on the B-cell surface and therefore have the potential to protect an individual from primary infection. Antibodies to EA are highest during the acute phase of infection but are often detectable at varying levels thereafter and are thus not a reliable indicator of acute infection. Antibody responses to EBNAs arise approximately 4 weeks after onset of symptoms, and persist for life.

NK-cell and T-cell responses

The role of NK cells in primary infection is unclear but there is evidence that these cells are 'activated' by virus-infected cells in the oropharynx. Cytotoxic **CD8+** T cells are the main cells that deal with primary EBV infection and their strong antiviral response determines the clinical features of the disease.

Figure 4 shows the virus–cell interactions and the virus-induced CD8+ T-cell response in primary EBV infection as seen in infectious mononucleosis patients and healthy virus carriers. T cells initially respond to the EBV-infected cells by proliferation, firstly **CD4+** T cells and then CD8+ T cells and outnumber the B cells by a factor of about 50 to 1. These T cells appear in the circulation as 'atypical lymphocytes' and are used in diagnosis (see Section 4). During the early response to the virus, up to 50% of the CD8+ T cells in the circulation can be specific for EBV antigens. Infectious mononucleosis is an immunopathological disease with the immune response itself being responsible for the clinical symptoms and pathogenesis. The symptoms of infectious mononucleosis result from the intense immunological activity of CD8+ T cells specifically recognizing EBV-encoded antigens expressed by infected cells. This results in a strong inflammatory response. It has been suggested that the almost complete absence of this disease in young children (although not proven) is due to less mature T-cell responses, thus resulting in a weaker but still effective

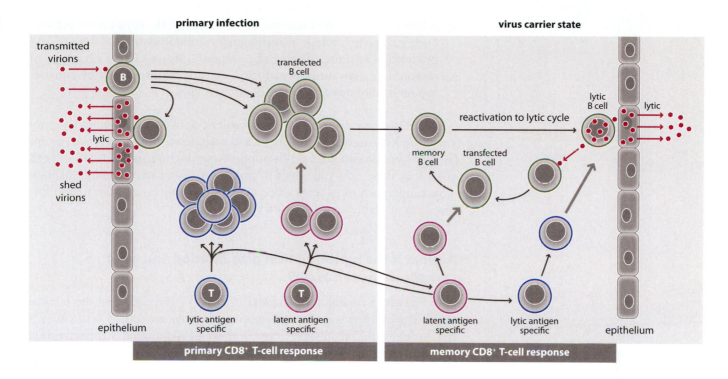

immune response. Exposure to lower viral doses in this age group might also play a role.

In infectious mononucleosis the cytolytic CD8+ T cells are able to control the infection by killing most of the virus-infected cells (transformed B cells) but note that the high-level shedding of virus in the throat of infectious mononucleosis patients continues. This might be because of the poor recruitment of the cytolytic CD8+ T cells to the tonsillar tissue. A few memory B cells containing the latent virus evade the immune response by down-regulation of EBV antigens. These B cells enter a resting state as members of a long-lived memory cell pool, with about 1–50 cells per million circulating B cells latently infected with EBV, some 1000-fold below that typically seen in acute infectious mononucleosis. Following recovery from the primary infection, cytotoxic memory CD8+ T cells persist in the blood at a frequency of 0.2–2% of cytotoxic CD8+ T cells, which increases with age. Some individuals aged over 60 years can have up to 40% of their CD8+ T cells specific for EBV antigens.

Virus evasion mechanisms

The virus has several strategies that help it to evade the immune system and in particular to avoid attack by CD8+ T cells. These include the following.

1. EBV is able to 'switch off' expression of viral antigens by all B cells constituting the latent reservoir of EBV and thus removes the targets for CD8+ T cells.

2. The envelope protein gp42 is actively shed as a soluble **truncated** molecule during lytic infection and binds directly to MHC class II/peptide complexes and inhibits activation of CD4+ T cells that are required for the generation of CD8+ cytolytic T cells.

Figure 4. Diagrammatic representation of virus–cell interactions and of virus-induced CD8+ T-cell responses in primary EBV infection, as seen in infectious mononucleosis patients, and in healthy virus carriers. Red arrows denote transmission of virus; black arrows denote movement of cells (where the arrow is dotted, this reflects uncertainty as to the level of cell movement via this route); broad shaded arrows denote effector T-cell function. Adapted from the article "Cellular Responses to Viral Infection in Humans: Lessons from Epstein-Barr Virus", by Andrew D. Hislop, Graham S. Taylor, Delphine Sauce and Alan B. Rickinson. Reprint permission kindly given by the *Annual Review of Immunology*, Volume 25 © by Annual Reviews www.annualreviews.org. Additional reprint permission given by the lead author, Alan B. Rickinson.

3. A late antigen, BCRF1 (the viral homologue of **IL**-10) secreted actively during the lytic cycle is also thought to inhibit cytolytic CD8+ T-cell generation by either driving the response into a CD4+ **Th2** type of response, or activating regulatory T cells. It might also act by enhancing survival of newly infected B lymphocytes.

Nevertheless the presence of the cytolytic CD8+ T-cell memory pool keeps the EBV-infected lymphocytes in control and it is only if the patient becomes immunosuppressed (leading to a reduction in the CD8+ T-cell pool) that these cells can begin to proliferate in an uncontrolled manner. This can happen in patients who are immunosuppressed prior to and following transplantation (see below).

3. What is the typical clinical presentation and what complications can occur?

The majority of individuals infected with EBV do not show the clinical symptoms of infectious mononucleosis. However, around 98% of patients presenting with the disease have a triad of symptoms – **pharyngitis**, **pyrexia**, and enlargement of cervical lymph nodes and tonsils. Fatigue is common and may persist for months after the initial presentation. **Myalgia** is also common. Splenomegaly is frequent and some studies show an incidence of up to 100% in patients with infectious mononucleosis but usually resolving by 4 weeks after onset of symptoms. Splenic rupture is a rare but a potentially life-threatening complication. Other complications that are less common for these patients include **encephalitis**, upper airway obstruction (due to massive lymphoid hyperplasia of the tonsil) – **tracheostomy** may be required to relieve this, abdominal pain, and liver involvement including **hepatomegaly** and **jaundice** – although an increase in hepatocellular enzymes is relatively common. Older patients are less likely to have sore throat and lymphadenopathy but more likely to have liver involvement. Platelet abnormalities are relatively common with patients having mild **thrombocytopenia**. Autoantibodies induced by EBV are believed to be the cause of the thrombocytopenia and in some cases this can develop into the more **severe idiopathic thrombocytic purpura** with hemorrhagic consequences. **Aplastic anemia** (**pancytopenia**) has also been described following EBV infection with a similar autoimmune etiology proposed.

Rashes occur in about 5% of patients and may be macular, petechial, scarlatiniform, **urticarial**, or **erythema multiforme**. Also, 90–100% of patients with infectious mononucleosis who receive ampicillin develop a **pruritic** maculopapular rash usually 7–10 days after administration. This was seen in the patient described here (see Figure 1). It is not caused by an allergy to penicillin but rather a transient hypersensitivity reaction, since ampicillin-binding antibodies have been detected in patients with infectious mononucleosis.

Clinical consequences of EBV infection

As already indicated, recovery is the norm after infectious mononucleosis. However, in addition to the complications already mentioned above, there are a number of serious clinical consequences that can result from EBV infection.

X-linked lymphoproliferative disease (XLPD) is a rare familial condition affecting young boys and characterized by extreme susceptibility to EBV infection. Primary infection results in a particularly severe form of infectious mononucleosis with a large expansion of infected B cells and T cells infiltrating the organs of the body, especially the liver. Patients with this condition have several abnormalities of the immune system including defects in NK cell activity, overactivity of **Th1** cells, and poor Th2 responses leading to overexpansion of CD8+ T cells and overproduction of Th1 inflammatory **cytokines** and hence macrophage activation and **hemophagocytosis**. The primary defect has been mapped to a small cytoplasmic protein, SAP, involved in NK and T lymphocyte signaling. In 60% of cases, the disease is fatal. Patients with XLPD generally die of acute liver **necrosis** or multi-organ failure. Survivors are also at high risk of developing B-cell **lymphoma**.

Cancers

Cancer is the uncontrolled proliferation of cells in the body. When EBV infects B cells they are induced to proliferate presumably to increase the amount of virus during primary infection but also increase the number of B cells that can harbor the latent virus. The process that results in B-cell 'transformation' is mediated by some EBV proteins interfering with the control of the cell cycle. Persistent lack of cell cycle control and the result of additional unknown factors are believed to lead to 'malignant transformation' resulting in a B-cell tumor.

Burkitt's lymphoma (BL) is a malignant tumor associated with EBV. This is based on finding the EBV genome in tumor cells and the frequent association of this tumor with an elevated antibody titer against EBV VCA. It is **endemic** to central parts of Africa and New Guinea and has an annual incidence of 6–7 cases per 100 000 and a peak incidence at 6 or 7 years of age. BL is most frequently found in the 'lymphoma belt,' a region extending from West to East Africa The region is characterized by high temperature and humidity, ideal for the malaria parasite, and is probably the reason why malaria was initially suspected to be associated with BL. In Uganda, the association of BL with EBV is very strong (97%), but is weaker outside the 'lymphoma belt;' (85% in Algeria; only 10–15% in France and the USA).

Other B-cell malignancies. EBV has also been associated with 50% of Hodgkin's lymphomas (HL) and causes 90% of lymphoproliferative disease in immunosuppressed transplant patients – post transplant lymphoproliferative disease (PTLD) – and sometimes in XLPD patients. EBV-associated B-cell lymphomas also occur in patients with AIDS.

Non-B-cell malignancies

Nasopharyngeal cancer (NPC), a malignancy of epithelial cells, is relatively rare (less than 1 per 100 000 in most populations) except in southern China, where there is an annual incidence of more than 20 cases per 100 000. There is also a high incidence in isolated northern populations such as Inuit and Greenlanders, while the incidence is moderate in North

Africa, Israel, Kuwait, the Sudan, and parts of Kenya and Uganda. The rate of incidence generally increases from ages 20 to around 50 and men are twice as likely to develop NPC as women. In the USA, Chinese-Americans comprise the majority of NPC patients, together with workers exposed to fumes, smoke, and chemicals, implying a role for chemical carcinogenesis. There appears to be an association between eating highly salted foods and the incidence of NPC. Genetic studies have indicated an HLA association, which may account for the higher disease incidence in Southern China.

EBV is also thought to cause some T- and NK-cell lymphomas and other epithelial tumors such as gastric cancers (10%).

4. How is this disease diagnosed and what is the differential diagnosis?

Diagnosis of infectious mononucleosis is made on presentation with pharyngitis, pyrexia, and cervical lymphadenopathy and is confirmed by laboratory tests.

Hematological tests

The white blood cell count (WBC) is moderately raised by the second week and a blood film will contain 'atypical lymphocytes' (Figure 5A). Figure 5B shows a normal blood film for comparison. The large atypical cells make up to 10–20% of the blood mononuclear cells (hence mononucleosis), are characterized by the presence of bean-shaped or lobulated nuclei, and may persist in the blood for several months. They are cytotoxic CD8+ T cells, produced to control the EBV-induced B-cell proliferation. Hoagland's criteria are most widely used for the diagnosis of infectious mononucleosis. These are: at least 50% of the WBC being lymphocytes; at least 10% of them being atypical lymphocytes; the presence of fever, pharyngitis, and lymphadenopathy; and confirmed by a positive antibody test (see below). These criteria while quite specific are not extremely sensitive and only about half of the patients with infectious mononucleosis meet all of Hoagland's criteria.

Antibody tests

Tests for **heterophil antibodies** are often used for diagnosis. In primary infection, some of the EBV-activated B cells are induced to differentiate into plasma cells. As a result of this polyclonal activation they secrete a variety of unrelated antibodies, including the so-called heterophil antibodies and these form the basis of the Paul-Bunnell test. Heterophil antibodies (produced in 90% of patients with infectious mononucleosis) are able to agglutinate sheep, horse, or ox erythrocytes. The classic Monospot test uses horse erythrocytes and newer tests use heterophil antigens attached to latex beads. *Heterophil antibodies are often absent in children, especially under age 4.*

EBV-specific antibody tests may also be employed for the diagnosis of infectious mononucleosis (see Figure 3). The most useful of these antibody assays is that which detects IgM against the Epstein-Barr VCA. Anti-VCA of the IgM class develops early in the illness and declines rapidly over the

(A)

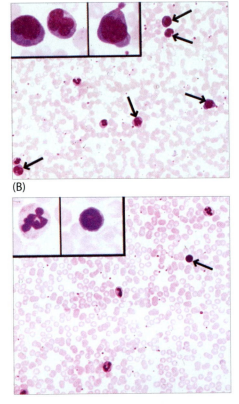

(B)

Figure 5. Photomicrographs of peripheral blood films stained by the Giemsa method (×40; insets ×100).
(A) From a patient with infectious mononucleosis; numerous large atypical mononuclear cells are present (arrows and insets) in addition to the normal cells.
(B) From a normal individual showing four neutrophils (detail in left inset) and one lymphocyte (arrow and right inset) among a mass of red cells.

following 3 months. **Enzyme immunoassays (ELISAs)** can be used to quantitate IgM anti-VCA but detection is often by indirect **immunofluorescence assay** using EBV-infected lymphoblastoid cell lines as the antigenic substrate. Patients also seroconvert to IgG anti-VCA antibodies in primary infection and these serum antibodies persist for life. Antibodies against EBNA are absent early in typical infectious mononucleosis but persist indefinitely thereafter. They may therefore be helpful in differentiating EBV reactivation from primary infection.

Differential diagnosis

Patients with streptococcal pharyngitis or one of several other viruses can present with sore throat, fatigue, and **adenopathy**. In the case presented here the doctor attempted treatment with ampicillin on the supposition of a streptococcal infection. Unfortunately the patient responded adversely to this antibiotic with a rash (see Section 3). This is an idiosyncratic response peculiar to infectious mononucleosis.

Acute CMV infection, toxoplasmosis, and acute HIV infection share many clinical and laboratory features of infectious mononucleosis, including splenomegaly, hepatomegaly, **lymphocytosis**, atypical lymphocytosis, and even (but rarely) false-positive results in the heterophil antibody test. Distinguishing clinically between infectious mononucleosis caused by EBV infection and an infectious mononucleosis-like syndrome caused by toxoplasmosis or CMV may not be possible, or perhaps not useful, since the management of these syndromes is the same. However, diagnostic testing is needed in pregnant women because toxoplasmosis and CMV infections may cause congenital infection resulting in severe damage to the unborn child. If acute HIV infection is suspected, tests for antibody and viral nucleic acid should be performed. The differential diagnosis for suspected infectious mononucleosis is summarized in Table 2.

Table 2. Infectious mononucleosis: differential diagnosis	
Diagnosis	**Key distinguishing features**
Streptococcal pharyngitis	Absence of splenomegaly or hepatomegaly; fatigue less prominent.
Other viral pharyngitis (eg. adenoviruses, patients withenteroviruses)	Patients are less likely to have lymphadenopathy, tonsillar exudates, fever of absence of cough than streptococcal pharyngitis or infectious mononucleosis.
Cytomegalovirus (CMV)	Sore throat less severe, lymphadenopathy may be minimal or absent. Confirm by specific antibody tests.
Toxoplasmosis	Sore throat less severe; confirm by specific antibody tests.
Acute HIV infection	Mucocutaneous lesions, rash, nausea, vomiting, diarrhoea and weight loss.

The differential diagnosis also includes **leukemia**/lymphoma although these would be less likely.

5. How is the disease managed and prevented?

Management

Usually there is no treatment given for infectious mononucleosis other than good supportive care, including adequate hydration and throat lozenges or sprays, or gargling with a 2% lidocaine (Xylocaine®) solution to relieve pharyngeal discomfort. **Nonsteroidal anti-inflammatory drugs** (**NSAIDs**) or acetaminophen are given for fever and myalgias; aspirin should not be given to children because it may cause **Reye's syndrome**.

Patients are recommended to rest and return to usual activities based on their energy levels. Although the exact period of increased risk is unknown, patients might also be advised to avoid contact sports or other activities that may lead to splenic rupture for a period of several weeks.

Corticosteroids may be considered in patients with significant pharyngeal **edema** that might cause or threaten respiratory compromise.

Prevention

There is currently no approved vaccine for EBV but one is eagerly awaited. It would be useful to protect adolescents and young adults against infectious mononucleosis. A vaccine could also be used to prevent primary infection in all age groups and hence presumably reduce the burden of EBV-associated cancers. It could be used in EBV seronegative individuals with XLPD, seronegative individuals receiving a seropositive transplant, those children at risk of BL and individuals at risk of developing NPC, especially in southern China.

The principal target of EBV neutralizing antibodies is the major virus surface glycoprotein gp350 and several vaccine candidates based on this molecule have been developed. Live recombinant vaccinia virus vectors have been used to express the gp350 antigen and were found to confer protection in primates and induce antibodies in EBV seronegative Chinese infants.

Recent results (2007) from two double-blind randomized (phase I and phase I/II) trials using a recombinant subunit gp350 trial carried out on 148 healthy adult volunteers in Belgium showed that the vaccine formulations had a good safety profile, were immunogenic, and induced gp350-specific antibody responses including neutralizing antibodies.

Clinical trials of an EBNA-3A peptide are being conducted in Australia.

SUMMARY

1. What is the causative agent, how does it enter the body and how does it spread a) within the body and b) from person to person?

- Epstein-Barr virus (also called human herpesvirus 4) is a gamma herpesvirus with linear dsDNA, an icosahedral capsid containing 162 capsomeres, which is approximately 100–110 nm in diameter.

- During primary infection the virus probably infects pharyngeal squamous epithelial cells and undergoes a lytic cycle producing many more virions that directly infect B cells in transit through the tonsil.

- For entry into B cells, gp350 on the viral envelope binds to a complement receptor CD21 (CR2) on the surface of B cells. Another envelope glycoprotein, gp42, is responsible for the fusion between the virus envelope and the host cell membrane.

- EBV causes proliferation of the B cells (B-cell transformation) resulting in lymphadenopathy at both cervical and other sites of the body and splenomegaly.

- Memory B cells maintain the latent EBV genome as an episome that expresses only part of its genetic information.

- EBV virions are shed into the saliva at high level for several months after primary infection and spread of the infection is usually through intimate contact with an uninfected person. Spread of the virus in children is probably mainly through fingers and close contact.

- The majority of adults globally are infected with EBV (95–98%). In developing countries more children may become infected with the virus early in life. In developed countries around 50–70% of adolescents and young adults undergo primary infection with EBV but only about 30% develop the symptoms of infectious mononucleosis.

2. What is the host response to the infection and what is the disease pathogenesis?

- In primary infection EBV antibodies are produced at different times as they are recognized by the immune system of the host. These include: early antigens, EA; late antigens such as virus capsid antigens (VCA); and latency antigens including EBNAs and LMPs.

- IgM antibody to VCA is the first to appear followed by IgG. Antibodies to the latency antigens, for example EBNA, are seen later.

- Cytotoxic CD8+ T cells are the main cells that control EBV infections. During the early response to the virus, up to 50% of the CD8+ T cells in the circulation can be specific for EBV antigens.

- The symptoms of infectious mononucleosis are attributable to the activation of B cells through polyclonal activation and the intense immunological response to them mounted by CD8+ T cells.

- The cytotoxic T cells are able to control the primary infection but a few resistant B cells containing the latent EBV enter a resting state as members of a long-lived memory cell pool.

- Strategies that help EBV to evade the immune system include: 'switching off' the expression of viral antigens during latency; shedding of gp42, which binds to MHC class II/peptide complexes and inhibits CD4+ T cell activity; late antigen BCRF1 inhibits CD8 T-cell activity.

3. What is the typical clinical presentation and what complications can occur?

- The majority of individuals infected with EBV do not show clinical symptoms of infectious mononucleosis. Nearly all patients with the disease have a triad of symptoms – pharyngitis, pyrexia, and cervical lymph node and tonsillar enlargement.

- Older adults are less likely to have sore throat and adenopathy but more likely to have hepatitis and jaundice (9% of cases). Hepatocellular enzymes are mildly increased in 90% of infectious mononucleosis cases. Encephalitis occurs rarely.

- Patients with X-linked lymphoproliferative disease (XLPD), a rare familial condition affecting young boys, have extreme sensitivity to EBV. In 60% of cases, the disease is fatal resulting from macrophage activation, hemophagocytosis, and destruction of all of the lymphoid tissue.

- Burkitt's lymphoma (BL), a malignant tumor associated with EBV, is endemic to central parts of Africa and New Guinea, with an annual incidence of 6–7 cases per 100 000 and a peak incidence at 6 or 7 years of age. In African countries such as Uganda, in the lymphoma belt, the association of BL with EBV is very strong (97%), but is weaker elsewhere (85% in Algeria; only 10–15% in France and the USA).

- Nasopharyngeal cancer (NPC) is relatively rare (less than 1 per 100 000 in most populations) except in populations in southern China, where there is an annual incidence of more than 20 cases per 100 000. The rate of incidence generally increases from age 20 to around 50 and men are twice as likely to develop NPC as women.

- EBV causes PTLD and has been shown to be associated with Hodgkin's lymphoma (HL) as well as with some T- and NK-cell lymphomas and other epithelial tumors such as gastric cancers.

4. How is this disease diagnosed and what is the differential diagnosis?

- Diagnosis of infectious mononucleosis includes persistent pharyngitis with pyrexia and cervical lymphadenopathy and confirmed by laboratory tests.

- The white blood count is moderately raised and a blood film contains 'atypical lymphocytes' that can make up to 10–20% of the blood mononuclear cells. These are the cytotoxic CD8+ T cells.

- Tests for 'heterophil antibodies' such as the Paul-Bunnell test or Monospot tests are positive.

- EBV-specific IgM anti-VCA can be detected by indirect immunofluorescence or ELISA.

- Differential diagnosis includes: streptococcal pharyngitis, several viruses inducing pharyngitis, acute CMV infection and toxoplasmosis, acute HIV infection.

5. How is the disease managed and prevented?

- Usually no treatment is given other than good supportive care.

- Adequate hydration is required; throat lozenges or sprays, or gargling with a 2% lidocaine (Xylocaine®) solution to relieve pharyngeal discomfort.

- Nonsteroidal anti-inflammatory drugs are given for fever and myalgias.

- Patients are recommended to rest and return to usual activities based on their energy levels.

- Corticosteroids may be considered in patients with significant pharyngeal edema that might cause or threaten respiratory compromise.

- Vaccines targeting the membrane gp350 have already been trialed (phase I and phase I/II) and shown to be safe and induce an antibody response including neutralizing antibodies. Vaccines against EBNA-3A are also being developed.

FURTHER READING

Humphreys H, Irving WL. Problem-orientated Clinical Microbiology and Infection, 2nd edition. Oxford University Press, Oxford, 2004.

Knipe DM, Howley PM, Griffin DE, et al. Fields Virology, 5th edition. Lippincott Williams and Wilkins, Philadelphia, 2007.

Mandell GL, Bennett JE, Dolin R. Mandell's Principles and Practice of Infectious Diseases, 6th edition. Elsevier Health Sciences, Philadelphia, USA, 2005.

Mims C, Dockrell HM, Goering RV, Roitt I, Wakelin D,

Zuckerman M. Medical Microbiology, 3rd edition. Mosby, Edinburgh, 2004.

Murphy K, Travers P, and Walport M. Janeway's Immunobiology, 7th edition, Garland Science, New York, 2008.

Richman DD, Whitley RJ, Hayden FG. Clinical Virology, 2nd edition. ASM Press, Washington, DC, 2002.

Zuckerman AJ, Banatvala JE, Pattison JR, Griffiths PD, Shaub BD. Principles and Practice of Clinical Virology, 5th edition. Wiley, 2004.

REFERENCES

Cohen JI. Medical progress: Epstein-Barr virus. N Engl J Med, 2000, 343: 481–492.

Hislop AD, Taylor GS, Sauce D, Rickinson AB. Cellular responses to viral infection in humans: lessons from Epstein-Barr virus. Ann Rev Immunol, 2007, 25: 587–617.

Moutschen M, Leonard P, Sokai EM, et al. Phase I/II studies to evaluate safety and immunogenicity of a recombinant gp350

Epstein-Barr virus vaccine in healthy adults. Vaccine, 2007, 25: 4697–4705.

Thorley-Lawson DA, Gross A. Persistence of the Epstein-Barr virus and the origins of associated lymphomas. N Engl J Med, 2004, 350: 1328–1337.

Williams H, Crawford DH. Epstein-Barr virus: the impact of scientific advances on clinical practice. Blood, 2006, 107: 862–869.

WEB SITES

American Academy of Family Physicians, 'American Family Physician', October 1, 2004. Written by MH Ebell, Athens, Georgia: http://www.aafp.org/afp/20041001/ 1279.html

Australian Academy of Science, 'Nova – Science in the News,' published November 1997: http://www.science.org.au/nova/ 026/026key.htm

Centers for Disease Control and Prevention (CDC), National Center of Infectious Diseases, 2008, United States Public Health Agency: http://www.cdc.gov/ncidod/diseases/ebv.htm

World Health Organization, Initiative for Vaccine Research, Copyright WHO 2008: http://www.who.int/vaccine_research/ diseases/viral_cancers/en/index1.html

MULTIPLE CHOICE QUESTIONS

The questions should be answered either by selecting True (T) or False (F) for each answer statement, or by selecting the answer statements which best answer the question. Answers can be found in the back of the book.

1. **Which of the following are true of Epstein-Barr virus?**

   A. It is an RNA virus.

   B. It has a helical capsid.

   C. It mainly infects T cells.

   D. It is a zoonosis.

   E. It always causes infectious mononucleosis.

2. **Which of the following diseases are associated with EBV?**

   A. X-linked lymphoproliferative disease.

   B. Graves' disease.

   C. Burkitt's lymphoma.

   D. Cervical cancer.

   E. Nasopharyngeal cancer.

3. **Which of the following laboratory tests would be helpful in the diagnosis of infectious mononucleosis?**

   A. Paul-Bunnell test.

   B. Monospot test.

   C. Electron microscopy of pharyngeal cell scraping.

   D. Culture of the virus on fibroblasts.

   E. Presence of serum IgM VCA antibodies.

4. **What are the typical symptoms of a patient presenting with infectious mononucleosis?**

   A. Pharyngitis, pyrexia, and cervical lymphadenopathy with tonsillar enlargement.

   B. Pharyngeal inflammation and transient palatal petechiae.

   C. Encephalitis.

   D. Osteomyelitis.

   E. Hepatitis and jaundice in young patients.

5. **How would you treat a patient with infectious mononucleosis?**

   A. Just for symptoms.

   B. With acyclovir.

   C. Aspirin.

   D. Steroids.

   E. Ampicillin.

Case 10

Escherichia coli

A 50-year-old worker at a school cafeteria went to her primary care physician complaining of tiredness, shaking chills, a pain in her loin, and a burning sensation on passing urine, which she was doing more frequently than normal.

On examination, her doctor noted that she seemed a bit pale and that she had some suprapubic tenderness. He tested her urine with a dipstick and found she had a positive result for nitrite, **pus** cells, and protein.

The doctor took a blood and urine sample for confirmation and sent them to the local hospital laboratory. A diagnosis of **pyelonephritis** was made and she was started on antibiotics. The following day the laboratory results were available. The full blood count showed a **neutrophilia** and a pure culture of *Escherichia coli* was grown from the urine (Figure 1).

Figure 1. *E. coli* cultured from the urine. This shows the growth of lactose-fermenting (yellow colonies) *E. coli* on a medium containing lactose and a pH indicator. In routine practice urine is cultured on cystine lactose electrolyte-deficient (CLED) medium. Cystine is a growth factor for some organisms; lactose is incorporated to differentiate utilization of lactose from nonutilization (e.g. *Pseudomonas*); and the electrolyte deficiency inhibits the swarming of *Proteus*, which may obscure bacterial colonies.

1. What is the causative agent, how does it enter the body and how does it spread a) within the body and b) from person to person?

Causative agent

Uropathogenic *E. coli* (UPEC) is the major cause of urinary tract infections (UTIs) in anatomically normal, unobstructed urinary tracts. *Escherichia coli* is a motile nonsporing gram-negative bacillus approximately 0.6–1 μm wide and 2–3 μm long (Figure 2). It has a typical gram-negative cell wall with an outer hydrophobic membrane containing **lipopolysaccharide** (LPS). It has peritrichous flagella and possesses **fimbriae**, which are important in adhesion (Figure 3). Eighty percent of all urinary tract infections are caused by *E. coli* in the community and about 60% in hospital.

The LPS is composed of lipid A, an inner core of polysaccharide linked to the lipid A by ketodeoxyoctonate (KDO) and an outer variable polysaccharide (Figure 4). The general arrangement of LPS is similar in most

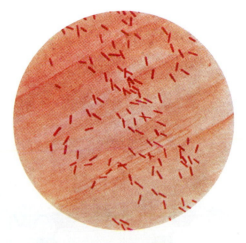

Figure 2. *E. coli* organisms. This is a Gram stain of *E. coli* showing the red (gram-negative) appearance.

**Figure 3. The hair-like fimbriae on
E. coli.** This electron-micrograph shows the
hair-like fimbriae on *E. coli*, which are
important for adhesion. Fimbriae were first
demonstrated by Prof. Duguid at the
University of Dundee in the 1960s.

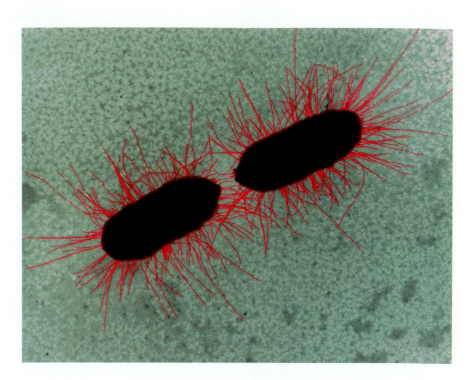

**Figure 4. The structure of *E. coli*
lipopolysaccharide (LPS).** This illustrates
the structure of the LPS found in the outer
membrane of gram-negative bacteria. It
consists of a lipid A component consisting
of β-hydroxy fatty acids attached to
carbohydrate, a core carbohydrate
component consisting of monomers of
different carbohydrates, e.g. heptoses and
ketodeoxyoctonic acid. This inner region of
the LPS is virtually identical in all gram-
negative bacteria, although the fatty acids
may vary in different genera of bacteria.
The outer part of the LPS consists of
repeating units consisting of
carbohydrates, e.g. mannose, glucose,
galactose, rhamnose, etc., as illustrated by
the different shading in the illustration. This
part is highly variable even within the
same species of bacteria and is called the
variable region or the 'O' antigen. Variation
between different species also exists in the
degree of phosphorylation of LPS. The
differences in the fatty acids and level of
phosphorylation affect the endotoxin
activity of the LPS.

gram-negative bacteria, with the highest variability appearing in the outer
'O' antigen component. The inner lipid A component comprises different
fatty acids (such as hexanoic (6:0), dodecanoic (12:0), tetradecanoic (14:0),
hexadecanoic (16:0), and octadecanoic (18:0), with their monoenoic equiv-
alents) linked by ester amide bonds to KDO via *N*-acetylglucosamine.
This part of LPS is the **endotoxin** – which is phosphorylated – and can
have dramatic effects on the clotting and **kallikrein** cascades, complement
activation, and **cytokine** production. The core region comprises 6- and 7-
carbon sugars (glucose, galactose, heptose) linked to the outer variable
region comprising 6-carbon sugars (e.g. glucose, mannose, tyvelose, rhamnose). The polysaccharide sequences of the variable region may be
branched and repeated to give long chains projecting from the surface of
the bacterium.

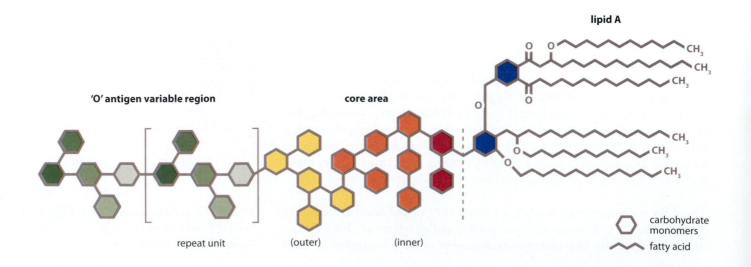

Not all *E. coli* strains cause UTI and, in fact, *E. coli* (mainly of the K12 strain) is part of the normal flora of the intestine (i.e. a **commensal**) that normally does not cause any problems. However, different strains (pathovars) carry different **virulence** characteristics and are associated with different clinical outcomes (Table 1).

Other agents that cause UTI are enterococci; other members of the Enterobacteriaceae such as *Enterobacter, Klebsiella, Staphylococcus saprophyticus* in young women, *Proteus* spp., and *Pseudomonas* spp.

Entry into the body

The gastrointestinal tract in man is colonized by *E. coli* within hours to a few days after birth. The organism is ingested in food or water or obtained directly from other individuals handling the infant. It attaches through specific **adhesins** and, like many commensals, it helps to prevent attachment of pathogenic microorganisms. *E. coli* is present in large amounts in feces.

Spread within the body

In females, if the *E. coli* organisms from the feces have the appropriate virulence factors (see below), they can attach to the epithelium of the vagina in the **introitus**, which then becomes colonized. The organisms can enter the urinary tract via the ureter and may eventually be introduced into the bladder where **cystitis** can develop. Once in the bladder and in the presence of vesicourethral reflux, the organism may spread up the ureter to the renal pelvis and **pyelonephritis** may ensue as it did in this patient.

Some cases of pyelonephritis are hematogenous, the organism localizing in the kidneys from the blood, for example *Staphylococcus aureus*, although this route is uncommon with *E. coli*, as the ascending route of infection predominates.

Spread from person to person

Person to person spread is not relevant for *E. coli* causing urinary tract infection, although enteropathogenic *E. coli* that causes **gastroenteritis** can be transmitted by **feco–oral** spread.

Table 1. The different pathovars of *E. coli*

Strain	Designation	Disease
E. coli K12	K12	Commensal
Enteropathogenic *E. coli*	EPEC	Gastroenteritis
Enterotoxigenic *E. coli*	ETEC	Gastroenteritis
Enteroinvasive *E. coli*	EIEC	Gastroenteritis
Enterohemorrhagic *E. coli*	EHEC (O157)	Gastroenteritis, hemolytic uremic syndrome
Enteroaggregative *E. coli*	EAEC	Gastroenteritis
Diffusely adherent *E. coli*	DAEC	Gastroenteritis
E. coli K1	K1	Neonatal meningitis
Uropathogenic *E. coli*	UPEC	Pyelonephritis

Epidemiology of UTI

Because of the anatomic differences in the length of the urethra between the male and female, the female generally is far more likely to have a urinary tract infection compared with the male, except in neonates where it occurs more commonly in boys. In neonates the prevalence is about 1–2%. In schoolchildren the prevalence is 4.5% in girls and 0.5% in boys and where it occurs in the latter it frequently signifies renal tract congenital abnormalities. In adulthood, up to 40% of females will have a urinary tract infection at some time in their lives. Sexual intercourse and pregnancy are risk factors for developing a UTI. In men the prevalence is less than 0.1%. A risk factor in males is an enlarged prostate and the prevalence of UTI in elderly men is up to 10%. In post-menopausal women the lack of estrogen is a risk factor, as the microbial flora changes from one mainly of lactobacilli to one with gram-negative bacteria. In both males and females a significant risk factor in hospitals is catheterization. In fact, it is known that catheter-associated infection is the most frequent hospital-acquired infection.

2. What is the host response to the infection and what is the disease pathogenesis?

Some virulence characteristics of *E. coli* are shown in Table 2 and are often carried on plasmids or on **pathogenicity islands** (PAIs). These PAIs are DNA sequences that are recognized as having a GC content that differs from the average GC content of the organism and occur due to horizontal transfer from other organisms.

UPEC have specific virulence characteristics that allow them to cause infection in the renal tract. These virulence characteristics include (a) specific fimbriae (e.g. P-fimbriae) that bind to the P-blood group antigen present on uroepithelial cells; (b) specific capsular antigens, for example K1 associated with serum resistance; (c) the production of **hemolysin**; (d) cytotoxic necrotizing factor; and (e) secreted autotransported toxin (SAT).

(a) The P-fimbriae are composed of the main fimbrial protein (PapA) and the adhesin (PapG). PapG adhesin has several allelic variants, with allele II variant causing most cases of pyelonephritis and allele III variant found mainly in children and women. P-fimbriae binding is not inhibited by mannose, but UPEC can bind to mannose receptors on the epithelial cells of the urinary tract by type I fimbriae, which are present in most *E. coli* strains and are composed of the Fim-A structural protein and the Fim-H adhesin, which binds mannoproteins. S-fimbriae bind to sialyllactose residues found on the renal tubules and glomeruli. Expression of Dr fimbriae, which bind to **decay accelerating factor** (a natural complement inhibitory protein found on epithelial cells) is associated with cell invasion and related to cystitis in children and pyelonephritis in pregnant women. Many other types of adhesins with different binding specificities have also been identified and expression of certain of these adhesins is thought to determine whether the *E. coli* moves further up the tract to infect the kidney. Individuals who are of the nonsecretor genotype (do not secrete blood group antigens into, e.g. saliva, urine, mucus) are more prone to binding of *E. coli* and thus to UTI.

Table 2. Virulence characteristics of *E. coli*

Virulence characteristic	Strain	Binds	Effect/disease
Adhesins			
Type I fimbriae	Most strains	Mannose	Pyelonephritis/gastroenteritis
P-fimbriae	UPEC	Galα1-4gal	Pyelonephritis
S-fimbriae	UPEC	Sialyl2-3lactose	Pyelonephritis
Dr-fimbriae	UPEC	DAF	Cystitis/pyelonephritis
Nonfimbrial adhesins	Most strains	Various ligands	Pyelonephritis/gastroenteritis
CFA I/II	ETEC		Gastroenteritis
AAF type III	EAggEC		Gastroenteritis
K-antigen K12	Commensals		
K1	UPEC		Antiphagocytic and serum resistance Pyelonephritis/neonatal meningitis
Toxins			
Hemolysin	Most	–	Cytotoxic/antiphagocytic
Siderophores	Most	–	Sequester iron
LT (heat-labile toxin)	ETEC	–	Fluid loss/traveler's diarrhea
ST (heat-stable toxin)	ETEC	–	Fluid loss/traveler's diarrhea
Verotoxin, EspP	EHEC	–	Dysentry/hemolytic uremic syndrome
SAT	UPEC	–	Cytotoxic/pyelonephritis
PET/PIC	EAggEC	–	Gastroenteritis
EspC	EPEC	–	Gastroenteritis

DAF, decay accelerating factor; CFA, colonization factor; SAT, secreted autotransported toxin; PET, plasmid-encoded toxin; PIC, protein involved in intestinal colonization; Esp, *E. coli* secreted protein.

(b) UPEC protect themselves from the host defenses with the K-capsule, which is antiphagocytic and also provides resistance to complement (serum resistance).

(c) UPEC also produces a number of toxins, which are also secreted by other *E. coli* pathovars. The hemolysin is a pore-forming toxin and is cytotoxic.

(d) Cytotoxic necrotizing factor affects intracellular signaling by modifying the Rho GTP binding proteins, actin polymerization and is cytotoxic.

(e) UPEC also secrete an autotransporter toxin (SAT): a member of the SPATE (serine protease autotransporters of Enterobacteriaceae) group of toxins, which is also cytotoxic.

Binding of *E. coli* to the uroepithelial cells in the bladder induces widening of the junctions between the squamous epithelium, exposing the underlying basal cells to which the organism can readily bind, and infection also increases desquamation of the superfical uroepithelium. Adhesion of bacteria also induces an acute inflammatory response stimulating the synthesis of **Th1** cytokines by the uroepithelium, **interleukin (IL)**-1 and IL-6 with IL-8, which recruits granulocytes and then macrophages and the patient develops a temperature. Production of these cytokines is probably related to the binding of the LPS to **Toll-like receptors** (especially TLR4) on the uroepithelium. In the renal medulla, the high ammonium concentration, osmolarity, and low pH all contribute to **immune-paresis** favoring the organism. The host secretes **defensins** from the uroepithelium and **Tamm Horsfall protein**, which binds *E. coli*, aiding its removal. Binding of the organism also results in exfoliation contributing to the *E. coli* that are found in the urine. The host develops a normal acquired immune response to the presence of the organism in the upper urinary tract, with the production of immunoglobulins of all isotypes. Infection of the bladder has a much reduced antibody response. Cell-mediated immunity to infection appears to be of little importance in the urinary tract with the exception of cytokine secretion.

The high numbers of granulocytes that accumulate in some parts of the renal tract cause damage to the host in a bystander effect through release of **oxygen free radicals** and proteolytic enzymes. The *E. coli* invade the uroepithelium forming intracellular bacterial colonies and inducing **necrosis** of the cells leading to further inflammation.

The role of the adaptive immune response in pyelonephritis is poorly understood, although **sIgA** antibodies are produced and may inhibit the binding of the organism.

3. What is the typical clinical presentation and what complications can occur?

The clinical presentation of UTI depends on the age of the person. In neonates the symptoms are nonspecific with vomiting, fever, and a 'floppy' infant. In older children and adults, localized symptoms occur. Cystitis presents with frequency, dysuria, suprapubic pain, and fever. The urine may contain blood. In cases of pyelonephritis the patient will present with fever, rigors, loin pain, frequency, and dysuria and **hematuria**. Elderly patients may present with a typical picture or with fever, incontinence, dementia or signs suggestive of a chest infection. A 'colicky' pain radiating from the loin to the groin is suggestive of renal stones – which may occur in the absence of infection, although renal calculi are a risk for infection and often associated with *Proteus*, which has an enzyme (urease) that can hydrolyze urea.

Complications include renal scarring, **septicemia**, papillary necrosis, which if bilateral can lead to renal failure, parenchymal **abscess**, and perinephric abscess. There is no clear relationship between pyelonephritis and the sequential development of **chronic interstitial nephritis** and **hypertension**.

4. How is this disease diagnosed and what is the differential diagnosis?

The organism can be isolated by culturing a mid-stream specimen of urine (see Figure 1) and in 30% of cases the organism will also be found in the blood culture. Work by Kass showed that if the number of organisms in the urine was greater than 10^5 bacteria per ml then this correlated well with clinical disease and this figure is considered as a 'significant' **bacteriuria**. However, this study was in asymptomatic healthy women and it is recognized that urinary infections can occur with fewer organisms in the urine. The presence of a single species of bacteria in the urine is also taken as evidence of significance but again urinary infections can occur when more than one species is present.

A number of automated methods for assessing bacteriuria are available, such as turbidometric, bioluminescence, electrical impedance, flow cytometry, and radiometric tests, although none are in general routine use in the UK.

Microscopy is generally not useful in the diagnosis of infection, although the presence of **white cell casts** is suggestive but not diagnostic of pyelonephritis.

The differential diagnosis of pyelonephritis depends on the context and age of the patients. Dermatological conditions such as **shingles** (before the appearance of the rash) and musculoskeletal injury may give rise to loin pain but without urinary symptoms or a fever in the case of musculoskeletal injury. Renal vein **thrombosis** can give rise to severe pain and fever; renal abscess, papillary necrosis, and **urolithiasis** will give rise to pain and fever. The presence of a 'sterile' **pyuria** can indicate renal tuberculosis, urolithiasis or neoplasm.

5. How is the disease managed and prevented?

There is no strong reason for treatment of asymptomatic bacteriuria in either nonpregnant women or the elderly. On the other hand in pregnant women and children, particularly in the latter if there are renal congenital abnormalities allowing **vesico-ureteric reflux**, then treatment is required even in the absence of symptoms. If not treated, renal scarring or even renal failure may occur.

The choice of antimicrobial agent to treat UTI is often empirical but is usually based on the recognition that the commonest infectious agent is *E. coli* and a knowledge of the local antibiotic resistance patterns. Cystitis can be treated with 3 days of appropriate antibiotics. Trimethoprim, nitrofurantoin or amoxycillin are often used.

In cases of pyelonephritis, as there is parenchymal disease, a 2-week course of antibiotics is required, for example cefuroxime or a third-generation cephalosporin, an aminoglycoside or a fluoroquinolone depending on the resistance pattern.

Because of the spread of extended-spectrum β-lactamases (ESBLs), particularly in bacteria such as *Enterobacter*, no β-lactam antibiotic is effective (except the carbapenems) and a different antibiotic class should be used, for example a fluoroquinolone.

Frequent relapses of UTIs may require prophylaxis, usually with nitrofurantoin or trimethoprim taken at night, so high concentrations of the antibiotic are present in the bladder overnight.

SUMMARY

1. What is the causative agent, how does it enter the body and how does it spread a) within the body and b) from person to person?

- The commonest infecting agent for urinary tract infections is *E. coli*.

- Other agents causing UTI are enterococci, *Proteus*, *Staph. saprophyticus*, *Pseudomonas*, and other enteric bacteria.

- *E. coli* has a typical gram-negative cell wall structure with an outer lipopolysaccharide (LPS) membrane.

- LPS consists of an inner lipid A component, a common core region of polysaccharides and an outer variable region of polysaccharides.

- The infection is endogenous, the source of the bacteria being the fecal flora.

- Organisms gain entry to the bladder from the perineum and in the presence of vesico-ureteric reflux may infect the renal parenchyma.

- UTI is more common in females because of the short urethra.

2. What is the host response to the infection and what is the disease pathogenesis?

- UPEC have several adhesins that bind to uroepithelial cells.

- UPEC produce a number of toxins that increase desquamation of uroepithelial cells and are also cytotoxic.

- UPEC are protected from the innate host defense system by the K1 capsular antigen, which is anti-phagocytic.

- In the renal tract the immune defenses are relatively inactive due to the osmolarity and the concentration of ammonium ions.

- Binding of UPEC to uroepithelial cells stimulates the production of IL-8 and the recruitment of granulocytes to the renal tract.

- The role of the adaptive immune system in protection is uncertain although IgA and **IgG** are produced.

3. What is the typical clinical presentation and what complications can occur?

- Infection may be asymptomatic.

- In pregnant women and children asymptomatic infection can have serious consequences.

- Cystitis presents with frequency of micturition, dysuria, and suprapubic pain.

- Pyelonephritis presents with loin pain, fever, rigors, frequency of micturition, and dysuria.

- Neonates have a nonspecific presentation with fever, vomiting, and failure to thrive.

- Elderly patients may have fever, incontinence, and dementia.

- Complications are septicemia, renal scarring, renal abscess, and renal failure.

4. How is this disease diagnosed and what is the differential diagnosis?

- Microscopy is generally unhelpful although a sterile pyuria may indicate renal tuberculosis.

- Laboratory diagnosis is by culture.

- A significant bacteriuria is a single species in numbers greater than 10^5 organisms per ml of urine.

- Differential diagnosis includes musculoskeletal injury, renal vein thrombosis, urolithiasis.

5. How is the disease managed and prevented?

- Asymptomatic bacteriuria in pregnant women or children should be treated.

- Antibiotic use depends upon the local resistance pattern.

- Three days treatment with an appropriate antibiotic is usually sufficient for uncomplicated cystitis.

- Two weeks treatment with the same antibiotic is required for pyelonephritis.

FURTHER READING

Murphy K, Travers P, and Walport M. Janeway's Immunobiology, 7th edition, Garland Science, New York, 2008.

Zaslau S. Bluprints Urology. Blackwell Publishing, UK, 2004: Chapter 6.

REFERENCES

Browne P, Ki M, Foxman B. Acute pyelonephritis among adults: cost of illness and considerations for the economic evaluation of therapy. Pharmacoeconomics, 2005, 23: 1123–1142.

Katchman EA, Milo G, Paul M, Christiaens T, Baerheim A, Leibovici L. Three day versus longer duration of antibiotic treatment for cystitis in women: systematic review and meta-analysis. Am J Med, 2005, 118: 1196–1207.

Kau AL, Hunstad DA, Hultgren SJ. Interaction of uropathogenic *Escherichia coli* with host uroepithelium. Curr Opin Microbiol, 2005, 8: 54–59.

Larcombe J. Urinary tract infection in children. Clin Evid, 2005, 14: 429–440.

MacDonald RA, Levitin H, Mallory GK, Kass EH. Relation between pyelonephritis and bacterial counts in the urine. N Engl J Med, 1957, 256: 915–922.

Neumann I, FernandaRojas M, Moore P. Pyelonephritis in non-pregnant women. Clin Evid, 2005, 14: 2352–2357.

Sheffield JS, Cunningham FG. Urinary tract infection in women. Obstet Gynecol, 2005, 106: 1085–1092.

WEB SITES

American Urological Association: http://www.auanet.org/

eMedicine, Inc., 2007, provider of clinical information and services to physicians and healthcare professionals: http://www.emedicine.com/med/topic2841.htm

European Association of Urology: http://www.uroweb.org/

Internet Pathology Laboratory for Medical Education, Mercer University School of Medicine, Savannah, GA, USA. This website and all contents and design, including images, are protected under US Copyright © 1994–2008 by Edward C. Klatt MD: http://www-medlib.med.utah.edu/WebPath/RENAHTML/RENALIDX.html

Todar's Online Textbook of Bacteriology, written and edited by Kenneth Todar, University of Wisconsin-Madison, Department of Bacteriology. All rights reserved. © 2008 Kenneth Todar, University of Wisconsin-Madison, Department of Bacteriology: http://textbookofbacteriology.net/ e.coli.html

MULTIPLE CHOICE QUESTIONS

The questions should be answered either by selecting True (T) or False (F) for each answer statement, or by selecting the answer statements which best answer the question. Answers can be found in the back of the book.

1. **Which of the following is true of *E. coli*?**

    A. It is a significant cause of urinary tract infections.

    B. It is part of the normal flora of the intestine.

    C. It lacks endotoxin in the cell wall.

    D. It is a significant cause of gastroenteritis.

    E. It is gram-positive.

2. **Which of the following are correct statements concerning urinary tract infection?**

    A. It is an exogenous infection.

    B. In the neonate, it is more common in boys than in girls.

    C. *Staphylococcus saprophyticus* is a common cause in women.

    D. Asymptomatic bacteriuria may have severe consequences in some individuals.

    E. It is not very common in hospitals.

3. **Which of the following are virulence factors of uropathogenic *E. coli* (UPEC)?**

    A. Cytolethal distending toxin.

    B. Type I fimbriae.

    C. P-fimbriae.

    D. S-layer.

    E. Serum resistance.

4. **The host responds in which of the following ways in a case of *E. coli* pyelonephritis?**

    A. Poor antibody response.

    B. Recruitmet of granulocytes.

    C. Development of a temperature.

    D. Inefficient immune defences in the renal medulla.

    E. Secretion of IL-6 and IL-8.

5. **Which of the following are typical of pyelonephritis?**

    A. Frequency of micturition.

    B. Dysuria.

    C. Rigors.

    D. Loin pain.

    E. Rash.

6. **Which of the following are important in the routine diagnosis of urinary tract infection?**

    A. Microscopy.

    B. Culture.

    C. Significant bacteriuria.

    D. Automated methods of assessing bacteria may be used.

    E. Antibiotic sensitivity testing.

7. **Which of the following are likely to be correct to treat an extended-spectrum β-lactamase (ESBL)-producing *Enterobacter* causing pyelonephritis?**

    A. Three days of cefuroxime.

    B. A carbapenem.

    C. Two weeks of nitrofurantoin.

    D. Gentamicin.

    E. A fluoroquinolone.

Case 11

Giardia lamblia

A 24-year-old man went on a 3-month backpacking trip across India. He drank bottled water and reportedly ate well-cooked food in hotels and restaurants. While in India his stools were looser than normal. In the week before his return he developed frequent watery, nonbloody diarrhea. This settled enough for him to fly home. He immediately went to his doctor and a stool culture grew *Campylobacter*. His bowels improved over 10 days without treatment but 2 weeks after his return he developed more diarrhea, with loss of appetite, bloating, and flatulence. For the first time his stools failed to flush away completely in the toilet and were particularly offensive in smell. He began to lose weight. His doctor requested three stool specimens for culture and also microscopy for ova, cysts, and parasites. One out of three specimens contained *Giardia* cysts. He was treated with a course of metronidazole and his symptoms improved.

1. What is the causative agent, how does it enter the body and how does it spread a) within the body and b) from person to person?

Causative agent

Giardia was discovered in the 17th century by Anton van Leeuwenhoek examining his own stools by microscopy. This protozoan has two stages to its life cycle (Figure 1): (a) a trophozoite (feeding and pathology-causing stage) that is flagellated (with four pairs of flagellae), pear-shaped, with two nuclei, a ventral 'sucking' disk, and median bodies. It also has a rigid cytoskeleton composed of microtubules and microribbons. It measures 9–21 μm long by 5–15 μm wide; (b) a cyst, with a highly resistant wall that enables it to remain viable outside the body of the host for long periods. The cyst is smooth-walled and oval in shape, measuring 8–12 μm long by 7–10 μm wide. *Giardia* has been isolated from mammals, amphibians, and birds. The species that infects humans is called *G. duodenalis*, *lamblia* or *intestinalis*.

The *Giardia* genome has recently been sequenced. There are two main genotypes of human infective *Giardia* isolates, A and B, with other genotypes restricted to animals other than humans.

Entry into the body

Cysts are ingested from contaminated water or food and having passed through the stomach begin to open up at one end (excystation), releasing very short-lived excyzoites, which then divide to become trophozoites (Figure 1). They settle in the small intestine (predominantly in the mid-jejunum). A minimum of 10–25 cysts are necessary to produce an infection. The trophozoites attach to the intestinal wall through their ventral 'sucking' disk and feed on nutrients (Figure 2). They increase in number by binary fission and colonize large areas of epithelial surface causing diarrhea and damage to the epithelium (see Section 2). At regular intervals and

Figure 1. The life cycle of *G. lamblia*.
Both cysts and trophozoites are found in feces (1). The cysts are hardy and can survive 2–3 months or more in cold water. Cysts in contaminated water, food, or by the fecal–oral route (hands or fomites) cause infection (2). In the small intestine, the cysts give rise to trophozoites (each cyst producing two trophozoites) (3). The trophozoites multiply by binary fission and remain in the lumen of the small bowel where they can be free in the mucus or attached to the epithelial cells by their ventral sucking disk (4). Trophozoites encyst on transit towards the colon. The cyst is the stage found most commonly in nondiarrheal feces (5). The cysts are infectious when passed in the stool or shortly afterwards and if ingested by another person the cycle begins again.

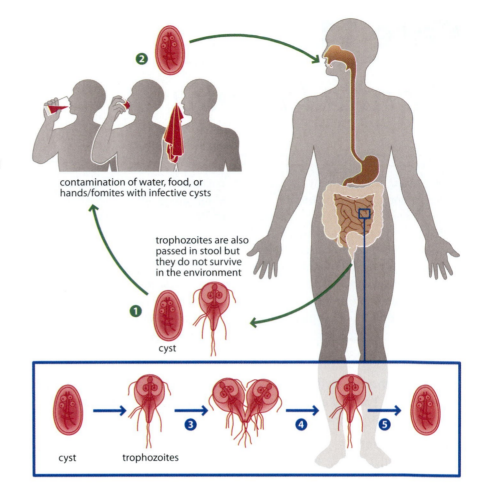

contamination of water, food, or hands/fomites with infective cysts

trophozoites are also passed in stool but they do not survive in the environment

cyst

cyst   trophozoites

Figure 2. *G. lamblia* attached to microvilli in the small intestine. This colored TEM (transmission electron micrograph) shows a *G. lamblia* trophozoite attached by means of its ventral sucking discs to microvilli in the human small intestine.

following detachment and movement down towards the colon (and probably exposure to biliary secretions), some of the trophozoites become encysted (encystation), with each trophozoite forming a single cyst. Both trophozoites and cysts pass out of the body in the feces.

Spread within the body
Penetration of the epithelial surface by the trophozoites is very rare, as is migration of the trophozoites to systemic sites. Invasion of the gallbladder, pancreas, and urinary tract have been reported but the trophozoites normally remain in the intestine/colon.

Spread from person to person
Cysts can remain dormant for up to 3 months in cold water. Spread is through ingestion of contaminated food and also via the **fecal–oral** route (hands and **fomites**), although water is probably the main source. Contamination of public drinking supplies has led to giardiasis epidemics. When children become infected, up to 25% of their family members also become infected. Individuals can shed cysts in their feces and remain symptom-free but are an important source of person to person transmission (see Section 3). Sexual transmission of *Giardia* has been described in homosexual males. *Giardia* is found in a wide variety of different animal species and has been regarded as a **zoonosis**, although there is little evidence for animals being a significant source of human giardiasis.

Epidemiology

Giardiasis is one of the commonest causes of diarrhea worldwide. Infection is linked to poor hygiene and sanitation and is more prevalent in warm climates. It is more commonly found in children where, in developing countries, it is estimated that up to 20% are infected.

Giardia is widespread in the United States, with carrier rates as high as 30–60% among children in day-care centers, institutions, and on Native American reservations. Infection is most common between July and October in children younger than 5 years and adults aged 25–39 years. Water-borne outbreaks appear to be the most common source of infection. In Canada, an infection rate of 19.6 per 100 000 population per year has been described, with a similar seasonal variation to that seen in the USA. In the Western world, giardiasis is more likely to be diagnosed as a cause of diarrhea that occurs or persists after travel to a developing country. This is due to its relatively long incubation period and persistent symptoms. Thus the organism is a cause of 'traveler's diarrhea,' also called 'backpackers diarrhea' and 'beavers' fever' (since it was originally believed to be transmitted from beavers to man).

It is not unusual for outbreaks of *Giardia* to occur on cruise ships.

2. What is the host response to the infection and what is the disease pathogenesis?

In a study of prison 'volunteers' in the 1950s given the same infectious dose, 50% of subjects developed asymptomatic infections, 35% self-limiting symptomatic infections, and 15% troublesome persistent diarrhea. Given that they received the same dose and strain this illustrates the impact of host susceptibility and resistance. However, host defenses against *G. lamblia* are not well characterized but they are believed to involve both nonimmunological mucosal processes and immune mechanisms.

Host defenses

Intestinal epithelial cells are shed and replaced every 3–5 days and therefore the trophozoites need to constantly detach and reattach to new epithelial surfaces. The mucus produced by the goblet cells also has some protective effect on attachment of the trophozoites, at least in preventing immediate access. **Commensal** organisms may also play a role in preventing attachment/inhibiting proliferation.

Immune responses

There appears to be little or no mucosal inflammation in human *Giardia* infection, which indicates that local defense must be occurring without systemic recruitment. Most of the data on immune responses to *Giardia* come from experimental animal models.

Innate immunity

Antimicrobial peptides such as **defensins** and lactoferrin secreted by intestinal epithelial cells have anti-giardial activity *in vitro* and may have activity *in vivo*. Nitric oxide (NO) inhibits growth, encystation, and excystation *in vitro*, but has no effect on viability. However, any NO produced by epithelial cells is probably inactivated by *Giardia*, as are **reactive oxygen species** also produced by epithelial cells. Although monocytes/macrophages can kill trophozoites *in vitro* by oxidative mechanisms, very few are found in the human intestinal lumen during infection.

Adaptive immunity

T cells appear to play a role in *Giardia* infections in mice in that specific depletion of **CD4+** T cells results in the development of chronic giardiasis. That specific T cells are activated by *Giardia* in humans has been demonstrated by their **interferon-γ (IFN-γ)** production in response to *Giardia* lysates. However, any protective role of these cells is unclear.

Antibodies to *Giardia* are found in both mucosal secretions and serum. Patients with **hypogammaglobulinemia** tend to acquire more chronic giardiasis, suggesting that antibody responses to human *Giardia* do have some protective role. Murine models of giardiasis show that **IgA** antibodies play more of an important role in late infection whilst B-cell-independent mechanisms are important in early stages of infection. Specific IgA antibodies are found in human saliva and breast milk and can protect children against infection in early life. Antibodies to the variant surface proteins (VSP: see below), cytoskeletal proteins unique to *Giardia* (α- and β-tubulin, α- and β-giardin), and enzymes (e.g. arginine deaminase, orthinine carbamoyl transferase, and enolase) are found. These enzymes have been found to be induced following contact of the parasite with the intestinal epithelial cells. **IgG** antibodies to *Giardia* have been shown to kill *Giardia* trophozoites *in vitro* through complement. Although it is unlikely that this mechanism could occur in the intestinal lumen it might be one explanation as to why *Giardia* does not invade.

Giardia-specific antigens and antigenic variation

Genomic analysis of *Giardia* has identified around 150 different genes encoding the immunodominant VSP antigens. Each trophozoite expresses only one VSP, which enables the trophozoites to switch their surface coats. The exact mechanism of this variation is unknown but does not appear to be through DNA rearrangement but rather through gene expression differences at the mRNA level. It is likely that the variation is driven by antibody. This **antigenic variation** (at least in mice) is thought to be a mechanism whereby the trophozoites can avoid the immune system. An alternative, but not mutually exclusive, biological explanation for antigenic variation is their adaptation to different intestinal environments. There is some evidence for this possibility in that different VSPs have different susceptibility to proteases.

Interestingly, the repertoire of VSP antigens is much smaller than that seen in trypanosomes (see Case 9) and the mechanisms leading to their 'switching' are different.

Pathogenesis

The mechanism by which giardiasis causes diarrhea and malabsorption is unclear. There is little evidence of any exotoxins producing epithelial damage. Initially it was thought that the attachment of the trophozoites acted as a mechanical barrier to absorption but this is unlikely since the size of the absorptive epithelial area in the intestine is enormous. Another postulated mechanism is the damage to the brush border, which provides a large surface area for absorption. Interestingly, although more common in animal models of giardiasis, biopsies from only 3% of patients with infection showed villus shortening and there was little inflammation. In experimental infection of 10 human volunteers with *Giardia* type B genotype, only 5 individuals developed symptoms and only 2 of these showed any change to the brush border. Thus microvillus shortening and inflammation are not directly correlated with the symptoms and indeed clearance of the organism from the intestinal tract. From *in vitro* studies there is some evidence for *Giardia* inducing a change in the cytoskeleton of human duodenal cells with increased **apoptosis** and disruption of tight junctions in monolayers of intestinal cells. Although disruption of tight junctions has not been confirmed by clinical observation, there is evidence for a correlation of infection with impairment of both absorption and digestive functions.

Giardia infections can produce symptoms that persist long after infection, although again the mechanisms for this are unclear.

3. What is the typical clinical presentation and what complications can occur?

The clinical effects of *Giardia* infection range from asymptomatic carrier status to severe malabsorption (see Section 2). Factors contributing to the variations in presentation include the **virulence** of particular *Giardia* strains (see Section 2), their genotype (A or B), the numbers of cysts ingested, the age of the host, and the state of the immune system (see later). Carriers often have a large number of cysts in their stools.

If symptoms are present, they occur about 1–3 weeks after ingestion of the parasite. These include:

- watery diarrhea with abdominal cramps;
- severe flatulence;
- nausea with or without vomiting;
- fatigue;
- and possibly fever.

A slower onset may occur with development of yellowish loose, soft and foul-smelling stools – often floating due to the high lipid content. Stools may be watery or even constipation can occur. Initial symptoms usually last 3–4 days or can become chronic leading to recurrent symptoms, severe malabsorption and debilitation may occur. Other symptoms include anorexia, malaise, and weight loss.

Children with malabsorption syndrome often show failure to thrive and **protein-losing enteropathy** can be a complication leading to stunted growth of children, commonly seen in Africa. Reduced uptake of lipids across the gut epithelium causes deficiency in lipid-soluble vitamins, which is an additional problem for children.

Poor nutrition can also contribute to an increased risk of a person having symptoms with the infection. More serious infections, which can lead to

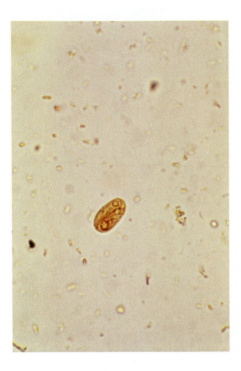

Figure 3. *G. lamblia* **cyst in a wet mount stained with iodine.**

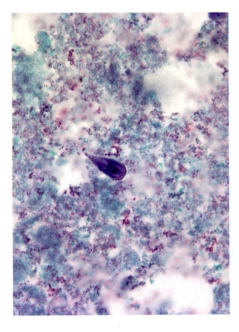

Figure 4. *G. lamblia* **trophozoite stained with trichrome.**

death, are seen in people with a weakened immune system, such as patients with HIV/AIDS, cancer, transplant patients, and the elderly. *Helicobacter pylori* may predispose to *Giardia* due to **hypochlorhydria.**

Patients may develop lactose intolerance.

4. How is the disease diagnosed and what is the differential diagnosis?

Clinical diagnosis is often difficult because the same symptoms can occur with a number of intestinal parasites. Giardiasis is therefore diagnosed by the identification of cysts (Figure 3) or trophozoites (Figure 4) in the feces. Usually three stool samples are taken to determine the presence of the parasite. Three samples are taken as shedding of cysts from infected individuals is highly variable. Direct mounts as well as concentration procedures may be used. Samples can be stained with iodine or trichrome.

Alternate methods for detection of the parasite in the stool samples include antigen detection tests by **ELISA** and by a direct fluorescence assay (DFA) (Figure 5). Commercial kits for both of these are available.

Polymerase chain reaction (PCR) testing for *Giardia* DNA can also be used if available.

A 'string' test (entero test) can also be performed. This involves swallowing a weighted gelatine capsule on a piece of string. After the gelatine dissolves in the stomach the weight carries the string into the duodenum. The string is left for 4–6 hours or overnight while the patient is fasting and then examined for bilious staining. This indicates successful passage into the duodenum and mucus from the string can be examined for trophozoites after fixation and staining.

A duodenal biopsy can be taken and this may be the most sensitive test. This is often taken in cases of unexplained diarrhea.

Differential diagnosis

Other causes of gastroenteritis need to be considered including amebiasis, bacterial overgrowth syndromes, Crohn ileitis, *Cryptosporidium* enteritis, irritable bowel syndrome, celiac sprue, and tropical sprue.

5. How is the disease managed and prevented?

Management

There are several drugs that can be used in the treatment of giardiasis. They include three classes: nitroimidazoles (metronidazole (Flagyl®), tinidazole, ornidazole, and nimorazole); nitrofuran derivatives (furazolidone); acridine compounds (mepacrine and quinacrine).

Metronidazole is the most common antibiotic treatment for giardiasis. An adult dose is 200–250 mg, three times daily for 5 days; children – 5 mg/kg, three times daily for 10 days.

Furazolidone (Furoxone) treatment comprises 100 mg, four times daily for 7 days for adults and 25–50 mg four times daily for 7 days for children.

One kind of drug alone has not proven to be effective in all cases, but in situations of resistant infections or recurrent infections, combination drug therapy or single medication given long term can be used. Quinacrine, although used less often than metronidazole because of side effects, has a success rate of about 95%.

Different countries may have a preference for different drugs. For example, in Europe mepacrine and furazolidine are not used compared with the Americas. In addition, from a worldwide perspective, albendazole (a benz-imidazole compound) is used, which has a much broader range of action than metronidazole and the other agents listed. It kills *Giardia* very well, but also *Entamoeba*, *Ascaris*, *Enterobius*, hookworms, and so forth and can do this in a single dose. In developing countries a single dose albendazole is being given to schoolchildren and has been associated with improved school attendance and educational attainment. They feel better for being cleared of protozoa and helminths.

Pregnant patients

Treatment of pregnant patients with *Giardia* is difficult because of the potential adverse effects of anti-*Giardia* agents on the fetus. If possible, drug treatment should be avoided during the first trimester. Mildly symptomatic women should have their treatment delayed until after delivery. If left untreated, however, adequate nutrition and hydration maintenance is important.

Prevention

- Good hygiene is very important.
- Contaminated water should be avoided: untreated water should not be consumed. Outbreaks of giardiasis in developed countries are often traced back to breakdown in filtration systems of drinking water supplies.
- Individuals traveling to warm climates where *Giardia* is found should take extra care with drinking water and consumption of raw food – boil drinking water, and so forth.
- There is no known chemoprophylaxis for humans.
- There is no vaccine for humans although there is an effective killed vaccine for dogs (Giardiavax®).

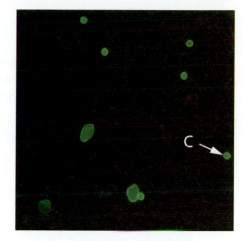

Figure 5. Identification of cysts of *G. intestinalis* by fluorescent-labeled *Giardia* antibodies. A formalin-fixed preparation stained with commercially available fluorescent antibodies to *Giardia* and visualized under a fluorescence microscope. Cysts of *Giardia* are seen as large green ovoid objects (labeled C). Oocysts of *Corynebacterium parvum* are also seen in this preparation.

SUMMARY

1. What is the causative agent, how does it enter the body and how does it spread a) within the body and b) from person to person?

- *Giardia* is a protozoan flagellate. It has two stages – a trophozoite and a cyst with a highly resistant wall.

- The species that infects humans are variously referred to as *G. duodenalis*, *lamblia* or *intestinalis*. There are two main genotypes of human *Giardia* isolates, A and B, with other genotypes in different mammals.

- The main infectious stage is the cyst; cysts are ingested in contaminated water or food. They lose their cell wall in the duodenum and emerge as trophozoites, which attach to the intestinal wall through their ventral 'sucking' disk and feed. They colonize large areas of epithelial surface. They rarely invade the epithelium and spread. They become encysted again and both trophozoites and cysts pass out of the body in stools.

- Contamination of public drinking supplies has led to giardiasis epidemics. When children become infected, up to 25% of their family members also become infected. Individuals can shed cysts in their feces and remain symptom-free. Sexual transmission of *Giardia* has been described in homosexual males.

- *Giardia* is found in a wide variety of different animal species and has been regarded as a zoonosis, although there is little evidence for animals being a significant source of human giardiasis.

- Giardiasis is one of the most commont causes of diarrhea worldwide. It is more commonly found in children where, in developing countries, it is estimated that up to 20% are infected.

- *Giardia* is widespread in the United States, with carrier rates as high as 30–60% among children in day-care centers, institutions, and on Native American reservations. Water-borne outbreaks appear to be the most common source of infection.

- In the Western world, giardiasis is often the cause of diarrhea that occurs or persists after travel to a developing country – 'traveler's diarrhea,' 'backpacker's diarrhea,' and 'beavers' fever.'

2. What is the host response to the infection and what is the disease pathogenesis?

- The host defenses are clearly effective since individuals infected with *Giardia* are often asymptomatic and some are able to clear the organism without treatment.

- Host defenses are not well characterized but include both nonimmunological and immunological mechanisms.

- Mucus prevents immediate access of trophozoites. Commensals may also play a role in preventing attachment/inhibiting proliferation.

- Antimicrobial peptides such as defensins and lactoferrin secreted by intestinal epithelial cells have anti-giardial activity *in vitro* and may have activity *in vivo*. Monocytes/macrophages can kill trophozoites *in vitro* by oxidative mechanisms; very few are found in the human intestinal lumen during infection.

- Most information on immune responses to *Giardia* infections comes from animal models. CD4+ T cells appear to play a role in *Giardia* infections in mice but their protective role in humans is unclear.

- Antibody responses to human *Giardia* do have some protective role. Specific IgA antibodies in human saliva and breast milk can protect children against infection in early life. Immunodominant antigens include VSPs, cytoskeletal structures, giardin, and enzymes.

- *Giardia* shows antigenic variation, with around 150 different genes encoding the VSP antigens. Each trophozoite has one VSP expressed and these switch providing escape from the immune system.

- The mechanism by which giardiasis causes diarrhea and malabsorption is unclear. Damage to the brush border has been suggested but villus shortening and inflammation is not frequently found in human disease. There is, however, evidence for a correlation of infection with impairment of both absorption and digestive functions. *Giardia* can induce changes in the cytoskeleton of human duodenal cells but the significance of this is unclear.

3. What is the typical clinical presentation and what complications can occur?

- The clinical effects of *Giardia* infection range from asymptomatic carrier status to severe malabsorption.

- Factors contributing to the variations in presentation include the virulence of particular *Giardia* strains, their genotype (A or B), the numbers of cysts ingested, the age of the host, and the state of the immune system. Carriers often have a large number of cysts in their stools.

- If symptoms are present, they occur about 1–3 weeks after ingestion of the parasite. These include: watery diarrhea with abdominal cramps, severe flatulence, nausea with or without vomiting, fatigue, and possibly fever.

- Infection can become chronic leading to recurrent symptoms, severe malabsorption, and debilitation. Other symptoms include anorexia, malaise, and weight loss. Children with malabsorption syndrome often show failure to thrive. Patients may develop lactose intolerance.

- Poor nutrition can also contribute to an increased risk of a person having symptoms with the infection.

- Patients with a weakened immune system such as patients with HIV/AIDS, cancer, transplant patients or the elderly can develop more severe infections.

4. How is the disease diagnosed and what is the differential diagnosis?

- Giardiasis is diagnosed by the identification of cysts or trophozoites in the feces. Usually three stool samples are taken to determine the presence of the parasite. Direct mounts as well as concentration procedures may be used. Samples can be stained with iodine or trichrome.

- Commercial kits are available for antigen detection tests by ELISA and by immunofluorescence. PCR testing for *Giardia* DNA can also be used.

- A 'string' test (entero test) can also be performed.

- Differential diagnosis for other causes of gastroenteritis includes amebiasis, bacterial overgrowth syndromes, Crohn ileitis, *Cryptosporidium* enteritis, irritable bowel syndrome, celiac sprue, and tropical sprue.

5. How is the disease managed and prevented?

- Three classes of drugs are used: nitroimidazoles (e.g. metronidazole (Flagyl®), tinidazole, ornidazole, and nimorazole); nitrofuran derivatives (e.g. furazolidone), and acridine compounds (e.g. mepacrine and quinacrine).

- Metronidazole is the most common antibiotic treatment for giardiasis.

- Treatment of pregnant patients with these drugs is difficult because of the potential adverse effects of anti-*Giardia* agents on the fetus.

- Prevention should include: good hygiene, avoidance of contaminated food and water, extra care during traveling to warm climates, boiling water, etc.

- No known chemoprophylaxis and no human vaccine as yet. There is an effective vaccine for dogs.

FURTHER READING

Goering RV, Dockrell HM, Zuckerman M, et al. Medical Microbiology, 4th edition. Mosby, Edinburgh, 2008.

Murphy K, Travers P, and Walport M. Janeway's Immunobiology, 7th edition, Garland Science, New York, 2008.

Murray PR, Rosenthal KS, Pfaller MA. Medical Microbiology, 5th edition, Elsevier Mosby, Philadelphia, PA, 2005.

REFERENCES

Cacciò SM, Beck R, Lalle M, Marinculic A, Pozio E. Multilocus genotyping of *Giardia duodenalis* reveals striking differences between assemblages A and B. Int J Parasitol, 2008 May 15 (Epub ahead of print).

Eckmann L. Mucosal defences against *Giardia*. Parasite Immunol, 2003, 25: 259–270.

Morrison HG, McArthur AG, Gillin FD, et al. Genomic minimalism in the early diverging intestinal parasite *Giardia lamblia*. Science, 2007, 317: 1921–1926.

Ringqvist E, Palm JE, Skarin H, et al. Release of metabolic enzymes by *Giardia* in response to interaction with intestinal epithelial cells. Mol Biochem Parasitol, 2008, 159: 85–91.

Roxstrom-Lidquist K, Palm D, Reiner D, Ringquist E, Svärd SG. *Giardia* immunity – an update. Trends Parasitol, 2006, 22: 26–31.

WEB SITES

Centers for Disease Control and Prevention, Traveler's Yellow Book 2008, Atlanta, GA, USA: http://wwwn.cdc.gov/travel/yellowBookCh4-Giardiasis.aspx

World Health Organization, Water Sanitation and Health Program, © WHO 2008: http://www.who.int/water_sanitation_health/dwq/en/admicrob5.pdf

Karolinska Institutet, University Library, Stockholm, Sweden © 2008: http://www.mic.ki.se/Diseases/C03.html

MULTIPLE CHOICE QUESTIONS

The questions should be answered either by selecting True (T) or False (F) for each answer statement, or by selecting the answer statements which best answer the question. Answers can be found in the back of the book.

1. Which of the following statements are true of *Giardia*?
   A. The feco–oral route is a common route of infection.
   B. Infection is through ingestion of eggs.
   C. The upper part of the small intestine is the site of attachment of trophozoites.
   D. Trophozoites invade the epithelium and spread throughout the body.
   E. Encystation takes place in the duodenum.

2. What is the usual clinical presentation of giardiasis?
   A. Watery diarrhea.
   B. Pulmonary infection.
   C. Malabsorption.
   D. Trachoma.
   E. Blood in the stools of infected individuals.

3. Which of the following statements are true of *Giardia*?
   A. Most infections produce serious symptoms.
   B. It is the most common cause of traveler's diarrhea.
   C. It is only found in humans.
   D. It has an incubation period of 8 weeks.
   E. It shows antigenic variation.

4. Which of the following are used for the treatment of giardiasis?
   A. Amoxicillin.
   B. Prednisone.
   C. Metronidazole (nitroimidazoles).
   D. Mepacrine.
   E. Furazolidone.

5. How is *Giardia* infection diagnosed?
   A. Identification of cysts in the stools.
   B. Identification of trophozoites in urine.
   C. With fluorescent antibodies to the cysts.
   D. The 'rope' test.
   E. The breath test.

6. Which of the following statements are true of *Giardia*?
   A. Infection is more serious in immunocompromised individuals.
   B. Trophozoites possess four pairs of flagellae.
   C. Cysts can be boiled and still remain viable.
   D. An effective human vaccine is available.
   E. A prophylactic drug is available for travelers to warm climates.

7. Which of the following complications are associated with giardiasis?
   A. Malabsorption.
   B. Blind loop syndrome.
   C. Failure to thrive.
   D. Lactose intolerance.
   E. Intussusception.

Case 12

Helicobacter pylori

A 50-year-old advertising executive consulted his primary health-care provider because of tiredness, lethargy, and an abdominal pain centered around the lower end of his sternum, which woke him in the early hours of the morning. The pain was relieved by food and antacids. His uncle had died of stomach cancer and he was worried that he had the same illness.

On examination his doctor noted that he seemed a bit pale and that he had a **tachycardia**. His blood pressure was low. He was slightly tender in his upper abdomen but there was no guarding or rebound tenderness.

The doctor took blood and feces samples and organized for an upper gastrointestinal **endoscopy**. The full blood count showed a hypochromic normocytic **anemia** with a hemoglobin of 8.9 consistent with iron-deficiency anemia. The gastroscopy showed a 3 cm ulcer in the prepyloric region of the stomach (Figure 1). The fecal antigen test for *Helicobacter pylori* was positive. The patient was started on routine treatment for a duodenal ulcer.

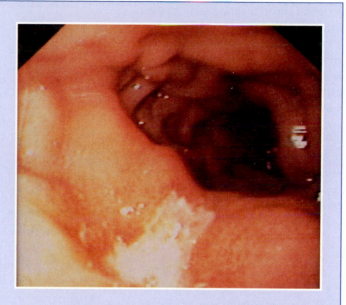

Figure 1. Gastroscopy showing a duodenal ulcer in the prepyloric region of the stomach.

1. What is the causative agent, how does it enter the body and how does it spread a) within the body and b) from person to person?

Causative agent

Helicobacter pylori is a nonspore-forming, motile gram-negative bacterium with a helical shape measuring $2.5–4.5 \times 0.5–1.0$ µm. It has one to five unipolar sheathed flagellae. In addition to its helical shape, curved forms occur and the bacillus also converts to a coccoid morphology when under environmental stress. The organism has one of the highest rates of polymorphism and during long-term colonization undergoes evolutionary changes in the genome that may relate to the ability of the organism to become a chronic infection. The genome has at least five regions that it may have acquired from other organisms by lateral transfer of DNA. These are called 'pathogenicity islands' (PAIs) and since they carry **virulence** genes are found in many bacteria. In the case of *H. pylori* the cag-PAI (Figure 2) comprises 30 genes that are responsible for the production of a **type IV secretion apparatus** (Figure 3), which is used to transfer the CagA protein into host cells (see host response). Although the organism has a typical gram-negative cell wall, the **lipopolysaccharide** (**LPS**) is much less of an **endotoxin** compared with *Escherichia coli* due to the reduced phosphorylation sites and different fatty acids. *H. pylori* grows in an atmosphere of

Figure 2. The cytotoxin-associated gene (CagA) pathogenicity island (cag-PAI).
This shows the general structure of the cag-PAI. Strains with a cag-PAI are called Type I strains and those lacking a cag-PAI Type II strains. Type I strains are more likely to be linked to severe disease than Type II strains. The cag-PAI comprises 30 genes involved in the synthesis of the Type IV secretion apparatus and a gene for the Cag A protein. The PAI may be complete, it may be separated into two sections (CagI and CagII) by an insertion element as indicated in the upper part of the diagram, part of the PAI may be missing or the strain may lack a PAI. Variation occurs in the 3' end of the PAI and so far 4 types (A-D) have been identified. The sequence in this region of the PAI also varies into one found principally in strains isolated from Western countries (WSS) and one from Asian countries (EASS).

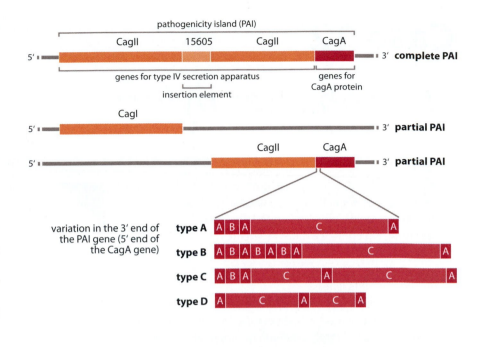

5–15% oxygen, 5–12% carbon dioxide, and 70–90% nitrogen (i.e. it is **micro-aerobic**) taking up to 5 days to grow on primary isolation and producing small (1–2 mm diameter) colonies on horse blood agar (5% horse blood in Columbia agar base). The colonies are domed, glistening, entire, gray or water-clear, and are sufficiently characteristic to suggest the presence of the organism (Figure 4). It can metabolize glucose, but its main carbon and energy source is from catabolism of amino acids. The organism has

Figure 3. Type IV secretion apparatus.
The Type IV secretion apparatus is a complex structure that acts as a micro-syringe and is used to transfer material such as the CagA protein and part of the peptidoglycan of the cell wall of *H. pylori* into the host cell. The CagA protein affects cellular signalling events. Proteins from the Type IV secretion apparatus are responsible for the release of NFκB, transfer to the nucleus of the host cell and up-regulation and synthesis of IL-8 and other pro-inflammatory cytokines.

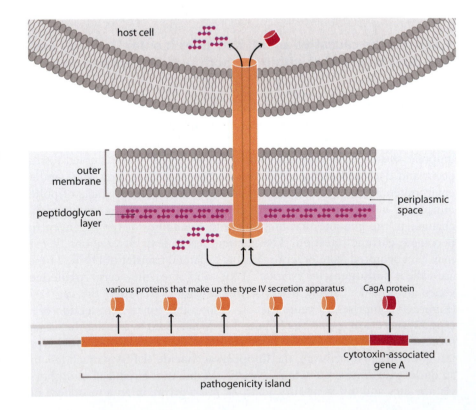

a urease, which is found both on the surface of the bacterium and in the cytoplasm and which is important for regulating the periplasmic pH. *H. pylori* can survive an acid environment for a short time but is not an **acidophile**. Since its isolation in 1983 there have been many other *Helicobacter* species isolated from a wide range of animals. Some of these *Helicobacter* species can cause **gastroenteritis** in humans (e.g. *Helicobacter cinedae*) and some cause stomach ulcers in the animal host. Many are associated with the lower intestinal tract and particularly the hepato-biliary system where, in animals, the organisms are the cause of **hepatoma**.

Entry into the body

H. pylori enters via the mouth. Once in the stomach it can only survive for a short time before it is killed by the acid. However, the presence of the enzyme urease on its surface (by hydrolyzing urea and producing ammonium ion) protects it for sufficient time for it to penetrate the mucus layer. The spiral shape, motility, and the production of phospholipases and the ammonium ion (which affects the tertiary structure of the mucus making it thin and watery) allow the *Helicobacter* to penetrate the mucus layer very quickly. *H. pylori* can only adhere to gastric epithelial tissue, which is found in the stomach and in the first part of the duodenum as islands of gastric **metaplasia**. Many **adhesins** on the surface of the organism have been identified, but the principle adhesin is BabA, which binds to the Lewis b blood group antigen expressed on gastric tissue. The organism is found mainly in the antrum of the stomach but can also be found in all parts of the stomach and duodenum.

Spread within the body

The organism only grows within the stomach either only in the antrum or throughout the whole of the stomach lining. It does not normally penetrate the gastric epithelium and enter the submucosal layers and therefore remains restricted to the stomach and duodenum.

Spread from person to person

Spread is via the **feco–oral** route or **oro–oral** route.

Epidemiology

The global incidence of *H. pylori* is shown in Figure 5. Sero-epidemiological studies have identified several risk factors. Thus, infection usually occurs in childhood and infection is higher in Social Class IV and V compared with I and II. Infection in industrialized countries (West Europe, North America) is about 5–10% in the first decade rising to 60% in the sixth decade. In nonindustrialized countries (Africa, South America, Middle and Far East) infection in the first decade is about 60–70% with little increase with age. Overcrowding and a poor public hygiene infrastructure are factors in the spread of *Helicobacter*. Spread within families has been documented by molecular typing techniques. In industrialized countries the low rate in childhood and the high rate in the elderly can be explained by improving social standards of housing (with less overcrowding), potable water supplies, and sewage disposal. The high rate in the elderly can be explained as a cohort effect, reflecting the housing and public health standards when they were children. The route of infection is not clear but is either feco–oral or oro–oral. In some countries (e.g. Peru), there is some evidence that infection may be acquired from sewage contamination of

Figure 4. Colonies of *H. pylori*.

Figure 5. Global incidence of *H. pylori* infection. The prevalence of *H. pylori* infection correlates with socio-economic status rather than race. In the United States, probability of being infected is greater for older persons (> 50 years = > than 50%), minorities (African Americans 40–50%) and immigrants from developing countries (Latino > 60%, Eastern Europeans > 50%). Infection is less common in more affluent Caucasians (< 40 years = 20%).

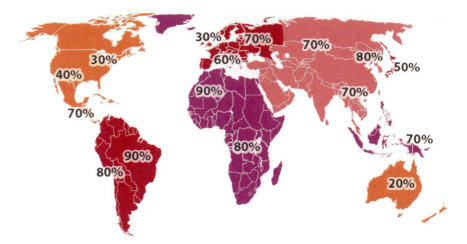

water supplies. Although some other species (see Causative agent, above) can infect humans, there is no evidence that *H. pylori* can be acquired from animals (i.e. is a **zoonosis**).

2. What is the host response to the infection and what is the disease pathogenesis?

The overall sequence of infection with *H. pylori* is shown in Figure 6, illustrating its macroscopic and microscopic location and the location of ulcer formation. On initial infection of the gastric tissue the host responds strongly to the presence of *H. pylori* by both the innate and acquired immune systems.

Immune responses

Even though the organism does not normally penetrate the gastric epithelium, damage caused by factors shown below result in an acute inflammatory reaction at first, followed by a more chronic inflammatory reaction. Initially there is a strong recruitment of granulocytes to the area but **phagocytosis** of *H. pylori* by both granulocytes and macrophages is inhibited in some way, not yet clearly understood. The acquired immune

Figure 6. General scheme of infection with *H. pylori* leading to development of a peptic ulcer. (A) a cartoon of *H. pylori* (where the unipolar flagellae can be seen); (B) the location of *H. pylori* in the antrum of the stomach; (C) a cartoon of the microscopic location of the organisms in the mucus layer and on the surface epithelium and (D) the location of a duodenal and gastric ulcer. The former can be in the first part of the duodenum (as shown here) or the pre-pyloric region, and the latter is usually on the lesser curvature of the corpus of the stomach.

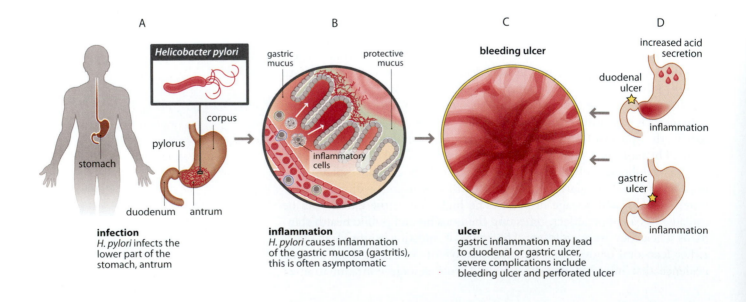

A

infection
H. pylori infects the lower part of the stomach, antrum

B

inflammation
H. pylori causes inflammation of the gastric mucosa (gastritis), this is often asymptomatic

C

ulcer
gastric inflammation may lead to duodenal or gastric ulcer, severe complications include bleeding ulcer and perforated ulcer

D

system is also activated with the production of both local (stomach) and systemic antibodies of **IgM**, **IgG**, and **IgA** class. It is supposed that IgA secreted into the stomach is unable to penetrate the mucus and presumably fails to block successfully the adhesins expressed by the organisms. *H. pylori* is susceptible to complement **lysis** *in vitro*, yet *in vivo* it is protected by natural complement inhibitors. These include coating of the organism with anti-complementary proteins such as **CD59**, which prevent lysis of the organism. Antibodies of the IgM and IgG class suggest that at least parts of the organism are taken up by **dendritic cells** leading to a systemic immune response. Thus, although a good antibody and cellular immune response is mounted, the organism is able to evade the immune defenses. An initial intense acute inflammatory response is invoked with infiltration of the **lamina propria** by granulocytes and as the inflammation becomes chronic a mononuclear cell infiltrate occurs and a chronic infection of the stomach ensues (Figure 7).

Despite the identification of several virulence characteristics such as motility, urease, vacuolating cytotoxin, CagA (cytotoxin-associated gene A), and NapA (neutrophil activating protein A) proteins, the precise route to the development of peptic ulcer disease or gastric cancer is unclear. The final clinical outcome may also depend upon the host polymorphism as individuals with certain IL-1β **polymorphisms** are more at risk of developing gastric cancer. Colonization by *H. pylori* leads to an inflammatory response that lasts a life-time.

Pathogenesis

There are five main ways in which tissue damage can occur (Figure 8). These are: (A) local damage caused by a vacuolating cytotoxin (Figure 9), the ammonium ion as a result of the urease activity, and the production of phospholipase, which contribute to the formation of a poor quality mucus barrier; (B) alteration of gastric physiology with enhanced acid production – gastric cell dynamics are affected by interference with normal cell signaling events caused by introduction of the CagA protein and peptidoglycan of *Helicobacter*; (C) bystander damage is caused by release of free radicals

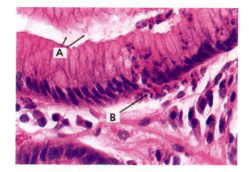

Figure 7. *Helicobacter pylori* – mediated inflammation in the stomach. Haematoxylin and eosin (H&E) stain of the gastric mucosa showing in the mucus and on the epithelial surface (A). *Helicobacter pylori* stains poorly with H&E and is better seen with a silver or Giemsa stain. Infiltrating granulocytes and mononuclear cells can be seen in the epithelial layer (lamina propria) indicating that the infection is becoming chronic (B).

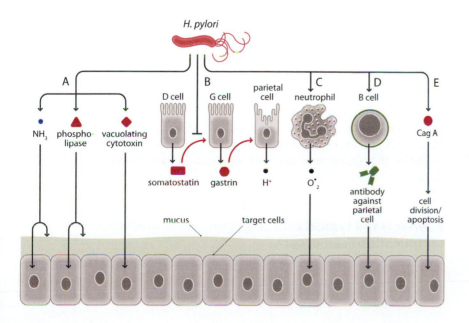

Figure 8. Mechanisms of pathogenesis of *H. pylori*. There are five main ways in which tissue damage can occur. These are: (A) local damage caused by a vacuolating cytotoxin (VacA), the ammonium ion as a result of the urease activity and the production of phospholipase that contribute to the formation of poor quality mucus barrier; (B) alteration of gastric physiology with enhanced acid production; (C) bystander damage caused by activation of granulocytes; (D) autoimmunity; (E) alteration of the balance of cell division and apoptosis.

Figure 9. The vacuolating cytotoxin. The gene for the vacuolating cytotoxin (VacA) has a different leader sequence depending on the strain of *H. pylori*. The toxin is activated by the acid of the stomach, and the monomers of the toxin oligomerize in the host cytoplasmic membrane, penetrate and affect the normal endocytic cycle, producing large acidified vacuoles that lead to cell death. Various polymorphisms are found in the gene. The VacA gene is not carried within the PAI and its leader sequence varies with the strain: the S1 leader sequence is found in type I strains and the S2 in type II strains, which produces less cytotoxin. Polymorphisms within the S1 leader sequence S1a, S1b, S1c are found in different geographical regions of the world, are associated with different racial groupings, and can be used as a surrogate marker for human migration patterns. S1a is found mainly in North America and Europe; S1b is found mainly in Iberia and South America, and S1c mainly in the Far East.

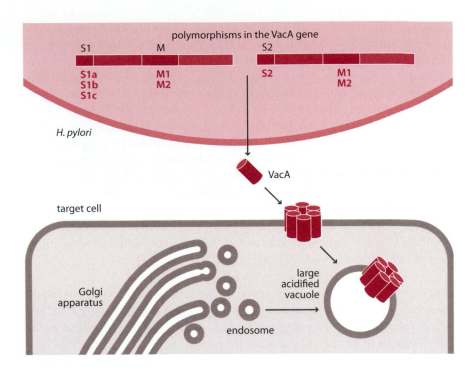

from the granulocytes; (D) autoimmunity – autoantibodies are induced by *Helicobacter* that kill the acid-secreting parietal cells; (E) alteration of the balance of cell division and apoptosis.

Colonization by *H. pylori* leads to excess acid in the stomach and both a **hypergastrinemia** and **hyperpepsinogenemia**, unless gastric **atrophy** occurs.

The route to duodenal ulcer (high acid, antral **gastritis**, low cancer risk) or gastric ulcer (low acid, pan gastritis, high cancer risk) is not obvious and may depend on the interaction between host and organism polymorphisms. The development of atrophic gastritis (atrophy of the mucosal epithelium) is a risk factor for the development of gastric cancer and the evolution to cancer proceeds via stages of metaplasia, dysplasia, and eventual cancer. The ability of *H. pylori* to alter the rate of cell division/apoptosis may be instrumental in the development of cancer and along with the low acid may allow colonization of the stomach by a wide range of other bacteria, perhaps allowing the local production of carcinogens. Other factors are also important in the eventual development of cancer such as the amount of dietary antioxidants and salt consumed (Figure 10). That *H. pylori* is still involved at the later stages of cancer development is suggested by the observation that with eradication of the organism, metaplastic changes can reverse.

3. What is the typical clinical presentation and what complications can occur?

Most persons colonized by *H. pylori* will remain symptom-free. About 20% will go on to develop peptic ulcer disease and about 1% gastric cancer. *H. pylori* is the principal cause of peptic ulcer disease (gastric or duodenal ulcer).

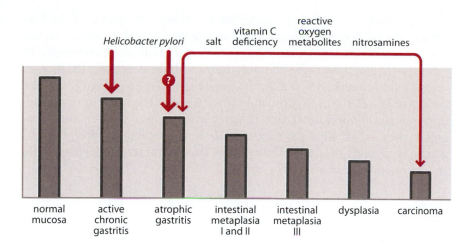

Figure 10. Model of the sequential steps in the development of gastric cancer mediated by *H. pylori*. This shows the multiple stages to development of gastric cancer with *H. pylori* being the initial insult but other factors such as salt intake, and a lack of vitamin C, also playing a role. Some evidence suggests that in the early stages of dysplasia, eradication of *H. pylori* can lead to reversal of the histological changes although the ultimate endpoint of whether this will prevent cancer is still an open question.

H. pylori is a good example of a 'slow' infection, as infection occurs in childhood but related diseases occur in adulthood. Ulcer disease presents with epigastric pain, heartburn or **dyspepsia** or may be totally asymptomatic. Anemia (due to blood loss from the ulcer) and weight loss may also occur and signs and symptoms of perforation (acute abdominal pain, abdominal rigidity and **guarding**, rebound tenderness and shock).

The role of *H. pylori* in nonulcer dyspepsia (NUD) is controversial but it may be that a subset of individuals who have NUD do benefit from eradication of the organism. Its role in **inflammatory bowel disease** is similarly controversial. In addition to gastrointestinal diseases it has been suggested that *H. pylori* may be related to a wide range of extra-gastrointestinal disease such as coronary heart disease, **stroke**, migraine, **idiopathic thombocytopenic purpura** (ITP), **rosaceae**, and gallbladder disease. Some evidence exists for most of these but again there is contrary evidence.

The most serious consequence of *H. pylori* infection is the development of cancer.

H. pylori is a Class I carcinogen and the cause of the majority of cases of gastric **adenocarcinoma** (except that of the cardia of the stomach) and mucosal associated lymphoid tissue (MALT) **lymphoma**. Clinical presentation of carcinoma is very nonspecific and is usually associated with gastric ulcer rather than duodenal ulcer. It presents with abdominal pain or a mass with so-called 'alarm symptoms' – weight loss and anemia. Gastric cancer originating at the **cardia** (gastro-esophageal junction) does not appear to be related to colonization by *H. pylori*.

4. How is this disease diagnosed and what is the differential diagnosis?

There are numerous ways in which *H. pylori* can be diagnosed. The first general method involves an **endoscopy** and biopsy. The biopsy can then be cultured under micro-aerobic conditions for *H. pylori*, which takes 5 days; histology can demonstrate the characteristically shaped organism on the surface of the epithelial cells (Giemsa or Genta stain) and the inflammatory cell type can be identified (hematoxylin and eosin); the urease of the bacterium can be detected using a rapid urease test (which involves putting one of the biopsies into a urea solution with a pH indicator); and finally *H. pylori* can be detected using the **polymerase chain reaction (PCR)** with appropriate primers: 16S rRNA, the urease gene (*ure* A, B or C), the flagella gene (*flaA*), the *cagA* gene, and the vacA gene. These methods of diagnosis are usually performed in a research setting or at a gastroenterology unit as they are invasive and expensive. It is mandatory, however, that if a patient is 55 years or over or if they show alarm symptoms, they must have an endoscopy to exclude gastric cancer.

Routinely, in the primary care setting, diagnosis is by noninvasive tests. These are:

1. **serology** – the detection of IgG antibodies to *H. pylori*;
2. the stool antigen test – these are immunoassay tests that use either polyclonal or **monoclonal antibodies** to detect the *Helicobacter* antigen in the feces;
3. the urea breath test – this test is performed by giving the patient a drink containing labeled urea (^{13}C or ^{14}C) and 20 minutes later collecting the breath and measuring the amount of labeled CO_2.

The basis of the breath test is that the *Helicobacter* urease hydrolyzes the labeled urea to labeled CO_2. *H. pylori* urease has such a high affinity for urea that any hydrolysis of the urea is caused by *Helicobacter* rather than other urease-containing bacteria. The latter two tests indicate if the person is currently infected with *Helicobacter*, whereas the serology test only indicates exposure and not that the person has a current infection. Either of these latter two tests is recommended for use in the Maastriche Guidelines (a consensus document from gastroenterologists in Europe) in the cost-effective 'Test & Treat' policy – a person complaining of upper gastrointestinal symptoms will be tested for *Helicobacter* and if positive will be treated.

Additional information on the degree of atrophy of the stomach can be obtained by combining a test for serum antibodies to *H. pylori* with assays for pepsinogen I, pepsinogen II, and gastrin 17 (G-17). Low levels of G-17 indicate antral atrophy and low levels of pepsinogen I with a low I/II ratio indicate corpus atrophy.

The other major cause of ulcer disease is **nonsteroidal anti-inflammatory drug (NSAID)** usage. Other causes are **Crohn's disease** and hypersecretory states such as gastrinoma (**Zollinger Ellison syndrome**), antral G cell hyperplasia, **mastocytosis**, and multiple endocrine neoplasms (MEN-1). Acute stress ulcers are caused by excess alcohol use, burns, trauma to the central nervous system, **cirrhosis**, chronic pulmonary disease, renal failure, radiation, and chemotherapy.

5. How is the disease managed and prevented?

Management

The standard first-line therapy for *H. pylori* eradication is a proton pump inhibitor (PPI) combined with two antibiotics: omeprazole (20 mg BD) or lansoprazole (30 mg bd) with clarithromycin (500 mg bd) and amoxicillin

(1000 mg bd) or metronidazole (500 mg bd) and amoxicillin (1000 mg bd), all given for 7–10 days. The regimen containing clarithromycin would be the first choice, as resistance to this antibiotic ranges from 5 to 25%, whereas resistance to metronidazole in many parts of the world is over 50%. Ranitidine bismuth citrate can be substituted (400 mg bd) for the PPI. These regimens can deliver a 90% eradication rate although more frequently the eradication rate is in the 70s due to resistance. Rescue regimens should be guided by sensitivity of the isolate but some useful ones are: PPI plus bismuth citrate (240 mg bd) plus tetracycline (500 mg bd) plus furazolidone (200 mg bd) for 14 days or PPI plus levofloxacin (250 mg bd) plus amoxicillin (1000 mg bd) for 10–14 days.

Prevention

Improvement of public health standards in developing countries may help to decrease the incidence of transmission.

Although a successful vaccine has been produced in animal models, a vaccine does not yet exist for *Helicobacter* in humans. This might be due to the fact that appropriate 'protective' antigens have not yet been defined for human disease.

SUMMARY

1. What is the causative agent, how does it enter the body and how does it spread a) within the body and b) from person to person?

- The cell wall is a typical gram-negative structure.

- The lipopolysaccharide has considerably less endotoxin activity compared with other gram-negative bacteria.

- *H. pylori* occurs in over 50% of the global population.

- Colonization occurs in childhood.

- Transmission is feco–oral or oro–oral. In some locations transmission may be from water supplies.

- Colonization is related to local social conditions and the public health infrastructure.

- The organism colonizes the gastric tissue either in the stomach or the duodenum.

2. What is the host response to the infection and what is the pathogenesis of disease?

- There is a strong innate immune response with infiltration by granulocytes (acute inflammation).

- *Helicobacter* avoids the innate immune response by inhibiting phagocytosis.

- There is a strong acquired immune response with antibody production but this is generally ineffective.

- *Helicobacter* avoids complement lysis by coating itself with anti-complementary proteins such as CD59.

- Certain virulence markers, e.g. CagA, VacA, are associated with more severe disease.

- Direct damage is brought about by the secretion of enzymes that destroy the mucus barrier and vacuolating cytotoxin that kills the surface epithelial cells.

- Gastric regulation of acid production is disturbed by the inhibition of somatostatin caused by the LPS of *Helicobacter*.

- Autoantibodies are induced by *Helicobacter* that kill the acid-secreting parietal cells.

- Gastric cell dynamics are affected by interference with normal cell signaling events caused by introduction of the CagA protein and peptidoglycan of *Helicobacter*.

- Bystander damage is caused by release of free radicals from the granulocytes.

3. What is the typical clinical presentation and what complications can occur?

- *Helicobacter* is a 'slow' infection, with colonization occurring in childhood and disease occurring years later.

- The vast majority of persons colonized by *H. pylori* remain asymptomatic.

- *H. pylori* is the main cause of peptic ulcer disease and gastric cancers.

- *H. pylori* may be associated with some extra-gastrointestinal diseases.

4. How is this disease diagnosed, and what is the differential diagnosis?

- Diagnosis is by invasive (endoscopy) or noninvasive tests.

- Invasive tests are culture, histology, rapid urease test, PCR.

- Noninvasive tests are serology, antigen detection, and the urea breath test.

- A cost-effective strategy is 'Test & Treat.'

- Recommended tests are the urea breath test and the fecal antigen tests.

- Anyone over 55 years or showing alarm symptoms must have an endoscopy.

5. How is the disease managed and prevented?

- The first-line treatment is a PPI plus clarithromycin and amoxycillin or metronidazole and amoxycillin for 7–10 days.

- Resistance to metronidazole is high.

- Rescue regimens should be guided by the sensitivity of the isolate to antibiotics.

FURTHER READING

Lydyard P, Lakhani S, Dogan A, et al. Pathology Integrated: An A–Z of Disease and its Pathogenesis. Edward Arnold, London, 2000: 254–256.

Mims C, Dockrell HM, Goering RV, Roitt I, Waklin D, Zuckerman M. Medical Microbiology, 3rd edition. Mosby, London, 2004: 232–235.

REFERENCES

Eslick GD. *Helicobacter* infection causes gastric cancer? A review of the epidemiological, meta-analytic and experimental evidence. World J Gastroenterol, 2006, 12: 2991–2999.

Ford AC, Delaney BC, Forman D, Moayyedi P. Eradication therapy for peptic ulcer disease in *Helicobacter pylori* positive patients. Cochrane Database Systematic Review, 2006, 2: CD003840.

Kusters JG, Van Vliet AH, Kuipers EJ. Pathogenesis of *Helicobacter pylori* infection. Clin Microbiol Rev, 2006, 19: 449–490.

O'Morain C. Role of *Helicobacter pylori* in functional dyspepsia. World J Gastroenterol, 2006, 12: 2677–2680.

WEB SITES

European Helicobacter Study Group: www.helicobacter.org

European Society for Primary Care Gastroenterology ©2008: www.espcg.org

Helicobacter Foundation © Helicobacter Foundation 2006: www.helico.com

United European Gastroenterology Federation: www.uegf.org

MULTIPLE CHOICE QUESTIONS

The questions should be answered either by selecting True (T) or False (F) for each answer statement, or by selecting the answer statements which best answer the question. Answers can be found in the back of the book.

1. **Which of the following statements are true of the cell wall of *Helicobacter*?**
   A. It increases somatostatin levels.
   B. It has high endotoxin activity.
   C. The lipopolysaccharide has low levels of phosphorylation.
   D. It does not contain lipopolysaccharide.
   E. It is gram-positive.

2. **Which of the following are true of *H. pylori*?**
   A. The organism grows in 24 hours on agar.
   B. It is a strict anaerobe.
   C. It is acquired by the oral route.
   D. Infection occurs in childhood.
   E. Over 50% of the global population are affected.

3. **Which of the following statements are correct concerning the host response to *Helicobacter*?**
   A. Few granulocytes are recruited to the area.
   B. There is a strong IgG response.
   C. *Helicobacter* is sensitive to complement *in vitro*.
   D. There is a poor IgA response.
   E. *Helicobacter* resists phagocytosis.

4. **Which of the the following virulence factors of *Helicobacter* are involved in pathogenesis of disease?**
   A. A type IV secretion apparatus.
   B. Cytolethal distending toxin.
   C. CagA protein.
   D. Lipopolysaccharide.
   E. Teichoic acid.

5. **Which of the following may be due to infection with *Helicobacter pylori*?**
   A. Gastric ulcer
   B. Idiopathic thrombocytopenic purpura.
   C. Gastric adenocarcinoma at the cardia.
   D. Food poisoning.
   E. Septicemia.

6. **Which of the following are typical signs or symptoms of duodenal ulcer?**
   A. Lower abdominal pain.
   B. Fever.
   C. Pain relieved by food.
   D. Pain occurs in the early hours of the morning.
   E. Diarrhea.

7. **Which of the following are important in the routine diagnosis of *Helicobacter pylori*?**
   A. Culture.
   B. Urea breath test.
   C. Serology.
   D. Pepsinogen I/II ratio.
   E. Fecal antigen test.

8. **Which of the following antibiotics are used to eradicate *Helicobacter pylori*?**
   A. Erythromycin.
   B. Metronidazole.
   C. Clarithromycin.
   D. Gentamicin.
   E. Amoxicillin.

Case 13

Hepatitis B virus

A 28-year-old stockbroker has been feeling generally unwell for the last 4 days, and off his food. Although usually a smoker (10–15 cigarettes per day), he hasn't been able to face lighting up since his illness began. Since yesterday, he has complained of a vague ache below his ribs on the right side. He noticed that his urine had become very dark, and his friends told him today that his eyes looked yellow (Figure 1). He has no relevant past medical history.

On examination by his local primary care physician, he was **pyrexial** (38.5°C) and clinically jaundiced. The only other sign of note was some right upper quadrant abdominal tenderness, but no **guarding**.

A clotted blood sample was sent to the laboratory, and 3 hours later, the laboratory called to convey the preliminary results, suggesting that the patient was suffering from acute hepatitis B virus infection.

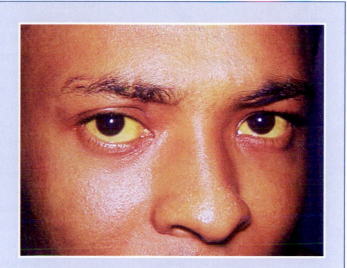

Figure 1. Jaundice as demonstrated by a yellow discoloration of the sclera of the eye.

1. What is the causative agent, how does it enter the body and how does it spread a) within the body and b) from person to person?

Causative agent

Hepatitis B virus (HBV) belongs to the virus family *Hepadnaviridae* – that is hepa (referring to the liver), DNA (referring to the nature of the viral genome), viruses – each member of which infects the liver of its particular host species. HBV is the member of the family that infects humans.

The *Hepadnaviridae* have an unusual partially double-stranded circular DNA genome (Figure 2). The genome is small – around 3200 bases – and therefore encodes only a small number of viral proteins, or antigens. There are four open reading frames (ORFs) identifiable within the genome, as follows.

- ORF C encodes the core antigen (HBcAg), which forms a protective **nucleocapsid** around the genome.
- ORF S encodes the surface antigen (HBsAg). This is composed of the large, middle, and small surface proteins, the different sizes arising from use of different start codons within the gene (large contains the pre-S1, pre-S2, and S regions; middle contains the pre-S2 and S regions; small contains just the S region – see Figure 2A), which are embedded in a lipid bilayer derived from internal membranes of the infected hepatocyte, surrounding the nucleocapsid.

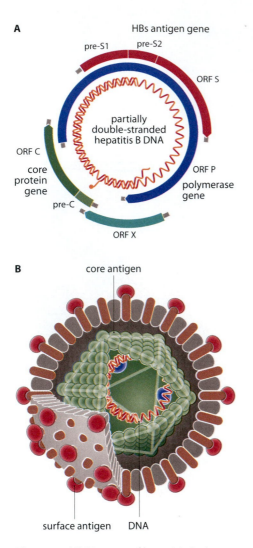

A

HBs antigen gene

pre-S1 pre-S2

ORF S

ORF C

core
protein
gene

pre-C

ORF P
polymerase
gene

partially
double-stranded
hepatitis B DNA

ORF X

B

core antigen

surface antigen   DNA

Figure 2. **(A)** Diagram of hepatitis B virus (HBV) genome showing the partially double-stranded DNA genome, and the four open reading frames (S, encodes hepatitis B surface antigen; C, encodes hepatitis B core antigen; P, encodes DNA polymerase enzyme; X, encodes X antigen) from which mRNA is synthesized. **(B)** Schematic diagram of the structure of an HBV particle. The partially double-stranded DNA genome is enclosed within the core antigen/protein, which is in turn surrounded by an envelope consisting of a lipid bilayer derived from internal membranes of the hepatocyte into which are embedded the large, middle, and small surface proteins, which together constitute the surface antigen.

- ORF P encodes the polymerase enzyme, which has both DNA- and RNA-dependent polymerase activities.
- ORF X encodes the X antigen, which is known to act as a transactivator of transcription, and is likely involved in carcinogenesis.

The hepatitis B viral structure is shown schematically in Figure 2B.

The replication cycle of HBV is unique among human viral pathogens. After entry into a susceptible cell (the cellular receptor(s) for HBV have yet to be definitively identified) and uncoating of the viral genome, the first step is synthesis of the missing part of the positive DNA strand, to form a complete double-stranded (ds) DNA molecule. A number of species of RNA are transcribed from the DNA, encoding the various viral proteins referred to above, but importantly, there is also a 3.5 kb RNA copy of the whole genome (pregenomic RNA). This RNA is packaged within newly synthesized core protein, together with the viral polymerase, to form immature new virus particles. The final step in viral maturation is then the reverse transcription of this pregenomic RNA into a DNA copy, followed by synthesis of an incomplete complementary DNA strand to yield partial dsDNA.

Entry and spread within the body

HBV enters a new host via the genital tract or following direct inoculation of virus into the bloodstream (see below). Once within the blood, virus travels to the liver, where it infects hepatocytes. Once within a liver cell, the virus replicates, and new virus particles are released directly into the bloodstream. From here, they gain access to every bodily compartment, including the genital tract.

Spread from person to person

HBV infection is spread by three routes.

1. *Mother to baby, or vertical transmission.* Babies acquire infection at the time of birth, through exposure to infected maternal blood and/or genital tract secretions. On a global scale, this is by far the most important route of infection. Over 90% of babies of carrier mothers become infected.

2. *Sexual transmission.* In an infected individual, both seminal fluid (male) and the female genital tract will contain virus. Unprotected sexual intercourse may therefore result in transmission of infection from one partner to another. This is especially the case with male homosexuals, where the act of intercourse may also involve exposure to blood through mucosal tears.

3. *Direct blood to blood transfer of virus (parenteral transmission).* This is a somewhat artificial, but nevertheless important, route of transmission, as humans are not normally exposed to each others' blood (except during childbirth). The most obvious way in which this can arise is via blood transfusion – if the blood donor is infected with HBV, then the blood will transmit infection to the recipient. More subtle ways of achieving this include the sharing of contaminated needles/syringes when injecting recreational drugs; exposure to contaminated needles, for example via tattooing, body piercing or acupuncture; needlestick injuries as suffered by health-care workers, i.e. accidental stabbing of a

needle derived from an infected patient into the health-care worker's own finger; or by sharing contaminated razors or toothbrushes.

Epidemiology

The World Health Organization (WHO) estimates that there are more than 300 million carriers (see below) of HBV. Carriage rates vary across the globe (Figure 3). WHO classifies countries into three groups according to their rates of HBV carriage – 'high' means that over 8% of a country's population are carriers, 'intermediate' equates to 2–8%, while 'low' is <2%. High prevalence countries include China, Japan, SE Asia, and much of sub-Saharan Africa and South America – carriage rates in these countries may exceed 20%. Northern Europe and the United States (except for parts of Alaska) are low prevalence areas – the carriage rate in the UK is 0.1%.

Armed with the above information on the routes of transmission and epidemiology of HBV infection, it is possible to draw up a list of individuals who have a higher risk of being infected (Table 1). An understanding of risk factors is helpful when constructing policies both for screening for HBV infection and for selective vaccination (see Section 5 below).

2. What is the host response to the infection and what is the disease pathogenesis?

Innate immunity

In the early stages of infection within the liver, innate immune responses, particularly **interferon** (**IFN**) induction, and **natural killer** (**NK**) cell activity are thought to be important in determining the eventual outcome of infection. Locally released IFN binds to surface receptors on neighboring cells, triggering the activation of various IFN response genes within those cells, the net effect of which will be to render the cells relatively resistant to virus infection.

Adaptive immunity

The virus encodes a number of distinct antigens. In addition to HBsAg and HBcAg, there is a third important antigen, namely the 'e' antigen, or

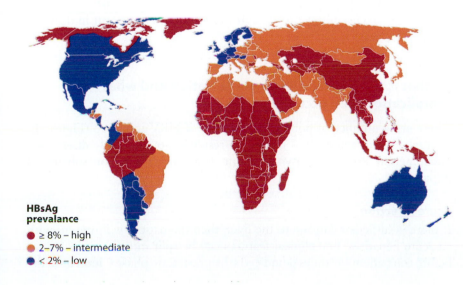

HBsAg prevalance
- ≥ 8% – high
- 2–7% – intermediate
- < 2% – low

Figure 3. Global map of hepatitis B virus (HBV) carriage rates.

Table 1. Individuals at high risk of being HBV infected
Babies, children of HBV-infected mothers
Sexual partners of HBV-infected individuals
Household contacts of HBV-infected individuals
Members of ethnic groups with high rates of HBV carriage
People with multiple heterosexual or homosexual partners, including sex workers
Injecting drug users who share needles/paraphernalia
Health-care workers (exposed to blood from infected patients)
Prisoners (high rates of injecting drug use)
Patients who receive regular transfusion of blood or blood products (e.g. hemophiliacs)
Patients with chronic renal failure (because of risks from hemodialysis)
Patients with chronic liver disease (where additional HBV infection may be life-threatening)

HBeAg. This is a soluble protein released from infected cells. It is, in fact, derived from the same gene as the core antigen. In an immunocompetent individual, the adaptive immune response should lead to the generation of antibodies to all viral antigens – those routinely measured in the laboratory are to the surface, core, and e antigens, i.e. anti-HBs, anti-HBc, and anti-HBe, respectively. The production of anti-HBs antibodies is particularly important in enabling the host to overcome the infection and eliminate the virus, as these antibodies are potentially neutralizing, and can therefore prevent newly released virus particles from infecting susceptible hepatocytes. Cellular immune responses are also generated, which lead to virus-specific cytotoxic **CD8+** T lymphocytes (CTLs) within the liver being able to kill infected hepatocytes through recognition of viral antigens present on the infected cell surface in association with HLA class I molecules.

Pathogenesis

Most damage to hepatocytes in HBV infection is thought to arise through the host CTL response killing infected cells in this way – HBV replication within hepatocytes of itself is not cytolytic and does not result in death of the infected cell.

3. What is the typical clinical presentation and what complications can occur?

There are a number of potential outcomes of HBV infection (Figure 4). Around 55% of acute HBV infections result in no detectable disease – a phenomenon known as asymptomatic **seroconversion**, or subclinical **hepatitis**.

Acute infection

If there is sufficient damage to the liver, then the patient will present with an acute hepatitis. The clinical features can be split into those that occur before the patient becomes jaundiced (the pre-icteric phase – **icterus** is the

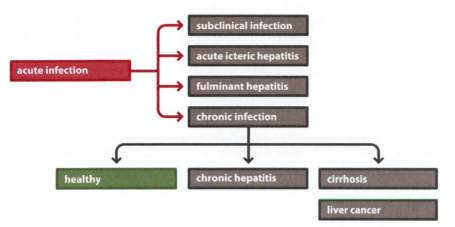

Figure 4. Outcomes following hepatitis B virus (HBV) infection of an adult. Around 55% of acute HBV infections are asymptomatic; 1% of infections result in fulminant hepatitis, i.e. acute liver failure; 5–10% of adults fail to clear infection and become chronic carriers. The potential outcomes of such chronic infection are also shown.

same as **jaundice**), and those that arise once the patient is jaundiced (icteric phase). The pre-icteric phase is fairly nonspecific – the patient may complain of lethargy, loss of appetite, (anorexia), nausea, alcohol and cigarette intolerance, and fever. The only clue to the fact the disease process is happening in the liver is that the patient may also mention right upper quadrant abdominal pain, which arises as the inflamed liver swells and stretches its innervated capsule. One of the many functions of the liver is to process pigments derived from hemoglobin in the blood, resulting in excretion of these into the bile, and thence in the feces. As the liver becomes damaged, this process may start to fail, resulting in these pigments accumulating in the bloodstream. This gives rise to the clinical sign of jaundice (due to an excess of circulating **bilirubin**) – a yellow discoloration of the skin, most easily seen by looking at the sclera of the eyes. The circulating pigments are filtered in the kidney and excreted into the urine, which becomes very dark, while the absence of their excretion into the bowel means that the feces become very pale.

Although acute hepatitis is not a trivial illness, and patients may take many weeks to recover full health, most patients will survive. However, there may be such overwhelming liver damage that the patient goes into acute liver failure – a condition known as fulminant hepatitis. This occurs in about 1% of all acute HBV infections, and carries a high mortality (i.e. >70%), the only really effective form of therapy being a liver transplant.

Chronic infection

A proportion of patients acutely infected with HBV will fail to eliminate virus from their liver, and therefore become chronically infected. In chronic infection, virus continues to replicate in infected hepatocytes, and is continually released from these cells into the bloodstream, but without interfering with the normal lifespan of the cells. Such chronic carriers serve as the source of infection for other individuals, via the routes of transmission discussed above. The chances of an acute HBV infection becoming chronic are dependent on a number of factors, most importantly the age and immune status of the patient. Babies infected from their carrier mothers almost always become chronic carriers themselves, reflecting the immaturity of the immune system at birth. About 10% of infected children and 5% of infected adults become carriers – in these individuals, it is believed that carriage results from a failure of the IFN response at the time of acute infection to facilitate viral clearance. Immunodeficient patients (for example those with HIV infection) are also highly likely to become chronic carriers once infected, as the absence of an effective CTL arm of the immune response mitigates against clearance of HBV-infected hepatocytes.

Chronic carriers may not be symptomatic – the liver has a large functional reserve, and liver function may be preserved even though there are virally infected hepatocytes present. However, such individuals are at risk of long-term chronic inflammatory hepatitis, as the infected cells are killed by the host CTL response. The usual response to death of host cells within the liver is the laying down of fibrous tissue. Over time, the continual death of liver cells, and their replacement by fibrous tissue results in complete loss of normal liver architecture and function – a condition known as **cirrhosis** of the liver, which has a number of potentially life-threatening complications. The average time from acute infection to cirrhosis is around 20 years. It is estimated that about 20% of chronic HBV carriers will develop cirrhosis and its complications within their lifetimes.

One further serious complication of chronic HBV infection is the development of **hepatocellular carcinoma** (HCC), that is, a malignant proliferation of hepatocytes. There are a number of molecular mechanisms underlying this, including a possible role for the X protein of HBV, which may interfere with normal cell division regulatory mechanisms. HBV carriers are about 300 times more likely to develop HCC than individuals who are not carriers. Thus, in parts of the world where HBV carriage is common, primary liver cell tumors are one of the commonest forms of malignant disease – globally, this is the fourth commonest tumor. On average, HCC appears about 5 years later than cirrhosis.

Note that the propensity for HBV infection of babies to become chronic, plus the fact that the life-threatening complications of chronic carriage do not usually become manifest for at least 20 years, provides a rational explanation of the epidemiology of HBV infection. Infected baby girls become carriers, but reach child-bearing age without yet having suffered the serious consequences of disease. Thus, they in turn infect their babies. Passage of virus from generation to generation in this way results in very high rates of carriage within a population.

4. How is this disease diagnosed and what is the differential diagnosis?

Laboratory diagnosis of HBV infection depends on the detection of various markers of infection (Table 2). Potentially, the laboratory can detect three viral antigens (HBsAg, HBcAg, and HBeAg) and antibodies to these three antigens (anti-HBs, anti-HBc, and anti-HBe, respectively), and the availability of sensitive genome detection techniques also means that HBV DNA levels can be measured.

The first test to be performed is to detect HBsAg in a blood sample. This protein is excreted in vast quantities by infected liver cells, and is easy to detect. The presence of HBsAg in a blood sample means only one thing – on the day the blood sample was taken, the patient was infected with HBV.

Table 2. Laboratory markers and their interpretation

Status	HBsAg	IgM anti-HBc	IgG anti-HBc	HBeAg	Anti-HBe	Anti-HBs
Acute HBV infection	+	+	+	+	–	–
Cleared HBV infection	–	–	+	–	+	+
Chronic HBV infection, high risk	+	–	+	+	–	–
Chronic HBV infection, low risk	+	–	+	–	+	–
Responder to HBV vaccine	–	–	–	–	–	+

Of itself, this marker cannot distinguish between acute or chronic infection, as it will be present in both.

All patients infected with HBV will make an antibody response to the core antigen (anti-HBc). This marker can be used to distinguish between acute (recent) and chronic infection, as **IgM** anti-HBc will be present in the former, IgM antibodies being a marker of recent infection.

Patients diagnosed with acute infection (HBsAg-positive, IgM anti-HBc-positive) must be followed to see if they eliminate virus over the following 6 months. This will be manifest by the disappearance of HBsAg from the peripheral blood, followed shortly by the appearance of anti-HBs antibodies.

If HBsAg persists for longer than 6 months, then the patient is, by definition, a chronic carrier. However, not all chronic carriers are equally infectious or equally at risk of chronic liver disease. The HBeAg/anti-HBe system distinguishes between two types of carrier. HBeAg-positive carriers are extremely infectious, and at increased risk of long-term disease (eAg is derived from the core gene, and is essentially a surrogate marker of virus replication). Over time, some HBeAg-positive carriers will lose HBeAg from their blood, as virus replication slows down, and shortly afterwards, it is possible to detect anti-HBe. An anti-HBe-positive carrier is much less infectious, and at much lower risk of chronic liver disease. Note that there are mutants of HBV that do not obey the above generalizations, particularly those that have stop mutations within the coding region of e antigen and are therefore incapable of synthesizing this antigen. These viruses may nevertheless be fully replication competent, and a patient infected with such a mutant may have high levels of viral DNA in the absence of HBeAg. Further discussion of these viral mutants is beyond the scope of this text.

Anti-HBs arises not only in patients who clear infection, but also in individuals who are vaccinated (see below) – measurement of anti-HBs responses in vaccines is an important marker of protection.

HBV DNA is a true marker of viral replication and infectivity. HBV DNA monitoring has become an essential part of the management of patients undergoing antiviral therapy.

Differential diagnosis

There are many causes of acute hepatitis, and therefore the differential diagnosis of an acutely jaundiced patient is a long one. Acute alcoholic hepatitis can present in an identical fashion. There are also a number of other infectious agents that can damage the liver. In particular, there are a series of viruses that have a tropism for the liver, which are known by letters of the alphabet. Note that these are quite distinct viruses and not related to each other – the only thing they have in common is that they infect the liver.

Other hepatitis viruses

The important features of these viral pathogens are as follows.

Hepatitis A virus (HAV) – this is a **picornavirus** and therefore has a positive single-stranded RNA genome. An infected individual excretes virus in the feces, and therefore the transmission route is fecal–oral. The possible outcomes of HAV infection include asymptomatic seroconversion (the majority), acute hepatitis (clinically indistinguishable from acute HBV infection), and fulminant hepatitis (less common than with HBV). Note that there are no chronic sequelae of HAV infection – as patients recover from infection, virus is eliminated from the body and the patient is then immune from further infection. Diagnosis of acute infection is by detection of IgM antibodies against HAV.

Hepatitis E virus – also has an RNA genome (with unusual genomic organization – HEV has not yet been classified into a known virus family), and bears many clinical and epidemiological similarities to HAV. Spread is via the **fecal–oral route**, and huge outbreaks involving thousands of individuals have been reported from fecal contamination of water supplies in various countries. Mortality from acute HEV is greater than with acute HAV, especially in pregnant women (10%). Chronic carriage of HEV does not arise in immunocompetent hosts. Diagnosis of acute infection is by detection of IgM antibodies against HEV, although available IgM assays are not as reliable as their counterparts for detection of IgM against HAV.

Hepatitis C virus – this is a ***flavivirus***, with a positive single-strand RNA genome. This virus bears many clinical and epidemiological similarities to HBV. It is a blood-borne virus, with risk groups being injecting drug users and recipients of contaminated blood or blood products. In contrast to HBV, mother to baby transmission only occurs in about 3% of cases, and there is very little evidence that HCV is transmitted sexually. HCV infection can become chronic – indeed it does so with greater frequency in adults than HBV, with only around 25% of acute infections being spontaneously cleared. Chronic infection predisposes the patient to the development of cirrhosis and also an increased risk of hepatocellular carcinoma. Diagnosis is first by detection of antibodies to HCV, which indicate that an individual has been infected with the virus. A subsequent test for HCV RNA (i.e. a genome detection test, usually the **polymerase chain reaction (PCR)** assay) distinguishes those individuals who have cleared the infection (HCV RNA-negative) from those who are chronic carriers (HCV RNA-positive).

Hepatitis D virus – this is an incomplete virus. It has a small RNA genome, encoding its capsid protein, referred to as the delta antigen. However, it requires an outer protein coat and it uses HBV as a helper virus to provide it with HBsAg to enable cellular egress and entry. Thus, HDV infection can only occur in patients already infected with HBV (superinfection), or at the same time as acute HBV infection (co-infection). The risk groups for HDV infection are therefore essentially the same as for HBV infection. Co-infection increases the risk of acute fulminant hepatitis, while patients who become chronic carriers of both HBV and HDV are at increased risk of the development of serious liver disease. Diagnosis is by the detection of delta antigen and antibodies to delta antigen.

5. How is the disease managed and prevented?

Management

There is no specific treatment for acute hepatitis B infection – advice is mainly for bed rest and avoidance of alcohol (a potent liver toxin). In cases of fulminant hepatitis, liver transplantation may be the only resort.

Carriers of HBV should be advised that they are potentially infectious to others, particularly their sexual partners and close household contacts. Under no circumstances should they donate blood, or share razors or toothbrushes. Blood spillages (for example from cuts) should be appropriately dealt with.

The management of patients with chronic hepatitis B has changed dramatically over the last 5 years, as the convoluted replication cycle of the virus has been elucidated. In particular, the recognition that there is a reverse transcription (RT) step in this process has led to the use of RT inhibitors in chronic hepatitis B, the relevant drugs being developed initially as RT inhibitors for the treatment of patients with HIV infection. Options for therapy (see Table 3) therefore now include the following.

1. Interferon-alpha (IFN-α) – this acts as an immunomodulatory drug in this context, not as an antiviral. It induces the expression of class I HLA molecules on the hepatocyte surface, and therefore facilitates recognition of virally infected cells by circulating virus-specific CTLs, which kill the infected hepatocytes and thereby reduce the burden of infected cells within the liver. This mechanism will only be effective in patients with an intact immune system whose T cells are capable of cytotoxic

---

Table 3. List of agents available for treatment of patients with chronic HBV infection

(a) Interferons
Standard interferon-alpha
Pegylated interferon-alpha

(b) Reverse transcriptase inhibitors
Lamivudine (Epivir®)
Adefovir (Hepsera®)
Entecavir (Baraclude®)
Telbivudine (Tyzeka®)
Tenofovir (Viread®)

activity against HBV-infected hepatocytes – thus it is unlikely to work in immunodeficient patients (for example HIV co-infected ones), or in patients infected with HBV at birth (the vast majority of HBV carriers globally), in whom immunological tolerance is induced such that their immune system does not recognize the virus as foreign. IFN-α is given as a 6-month course of thrice-weekly injections. More recently, preparations of IFN-α linked to polyethylene glycol have become available (pegylated- or PEG-IFN). This has a longer half-life than standard IFN, and therefore only requires once-weekly injection. The longer maintenance of plasma levels over time with PEG-IFN is also associated with a better therapeutic activity. IFNs may induce a plethora of unwanted side effects. Muscle aches, fatigue, and headache are common, but most patients become tolerant to these effects. Bone marrow suppression (low white cell count) and psychiatric manifestations such as depression and even suicide are more worrying.

2. HBV DNA polymerase inhibitors – currently licensed ones are lamivudine (3TC, 3-thia-cytidine), adefovir dipivoxil, and entecavir. These are effective against the polymerase activity of HBV, and therefore directly suppress viral replication. Note that their use does not eliminate virally infected hepatocytes, and therefore there is a possibility that virus may re-emerge if therapy is stopped. As with HIV, there is also a risk that mutations may arise within the viral polymerase (*Pol*) gene, which confer resistance to these drugs. Around 75% of patients treated with lamivudine for more than 3 years will have such drug-resistant mutants, but resistance rates to adefovir dipovoxil and entecavir are substantially lower. Combination therapy with more than one antiviral agent should decrease the chances of emergence of viral resistance.

Management of patients with chronic HBV infection is complex, and should be performed by physicians with appropriate training and expertise (hepatologists). Difficulties include when and in whom to initiate therapy, and with what therapeutic modality (IFN- or nucleoside analog-based).

Prevention

HBV infection can be prevented by means of vaccination, using the HBsAg as the immunizing antigen. Original vaccines consisted of HBsAg purified from human plasma, and plasma-derived vaccines are still used in some parts of the world. However, in Europe and the USA the current vaccines are subunit vaccines prepared by recombinant DNA technology. The gene encoding the HBsAg has been cut out of the virus and inserted into a yeast. As the yeast replicates in culture, large amounts of HBsAg are synthesized within the cytoplasm of the cell. Cell lysis followed by purification yields HBsAg, which is administered by intramuscular injection (usually as a course of three injections given at 0, 1, and 6 months). As the only part of the virus present in the vaccine is HBsAg, the response to the vaccine can be easily measured by quantifying the anti-HBs in the vaccinee 6–8 weeks after the last dose. Protection against HBV infection is proportional to the amount of anti-HBs produced. Around 10% of adults do not mount an anti-HBs response to this vaccine and therefore remain susceptible to infection. A further 10% will make a rather weak response, which may offer only limited protection.

The generation of an effective vaccine against HBV has undoubtedly been a major medical advance. However, there are controversies surrounding the most effective way in which to use it. The WHO recommends that the vaccine be included as one of the routine immunizations in childhood, and most countries, including the United States, have adopted this universal policy. However, in the UK, there is a selective vaccination policy, that is the vaccine is targeted at those subgroups of the population at particular risk of infection. Such subgroups are relatively easy to list – babies of carrier mothers, sexual partners of carriers, sexually promiscuous individuals (particularly homosexual males and sex workers), injecting drug users, and health-care workers (occupationally at risk) – but in practice, many of these individuals are hard to identify and may not access health-care services at which they can be vaccinated.

Vaccines for other hepatitis viruses.
There is no effective antiviral agent against HAV, but there is a vaccine – heat-killed whole virus. This is currently offered to individuals at risk of infection, for example travelers to countries where HAV infection is **endemic**.

There is currently no effective antiviral agent or licensed vaccine against HEV, although there are experimental vaccines undergoing clinical trials that have shown initial promise.

While there is no vaccine to protect against HCV infection, antiviral therapy of infected individuals does achieve some success. The current optimal treatment regimen consists of combination therapy with PEG-IFN-α injections and ribavirin (given orally). This has to be given for at least 6 months, confers significant adverse side effects, and overall, only about 50% of patients with chronic infection will clear the virus. The success rate is much greater (>95%) if patients are treated at the acute stage of infection, but this is not easy to do as most patients who acquire HCV infection are asymptomatic and therefore are not aware that they have become infected.

There is currently no antiviral treatment for HDV, but successful vaccination against HBV also prevents infection with HDV.

SUMMARY

1. What is the causative agent, how does it enter the body and how does it spread a) within the body and b) from person to person

- Hepatitis B virus, a hepadnavirus.

- Entry is via the genital tract, or through direct inoculation of the virus into the bloodstream.

- Once virus has entered hepatocytes, viral replication results in release of new virus particles directly into the bloodstream, and from there into every body compartment.

- The three routes of transmission are mother to baby (infected birth canal, exposure to infected maternal blood); sexual; and parenteral, for example via contaminated needles.

2. What is the host response to the infection and what is the disease pathogenesis?

- IFN production within the liver renders neighboring noninfected cells relatively resistant to infection.

- Exposure to viral antigens leads to the development of both humoral and cellular responses.

- Cytotoxic CD8+ T lymphocytes are crucial in eliminating virally infected hepatocytes.

- Antibody against the surface protein (anti-HBs) is vital to enable neutralization of released virus particles.

- An inadequate immune response results in failure to eliminate the virus from the liver. The patient then becomes chronically infected.

- HBV is not a cytolytic virus. Chronic liver damage arises from the host inflammatory response, particularly CD8+ T-cell destruction of hepatocytes.

3. What is the typical clinical presentation and what complications can occur?

- Acute infection may be asymptomatic, that is there is insufficient damage to the liver to cause disease.

- Patients may present with acute hepatitis – initially (pre-icteric) the features are nonspecific – fever, malaise, lethargy, anorexia, right upper quadrant abdominal pain.

- Liver damage results in failure to excrete pigments derived from hemoglobin, resulting in dark urine, pale stools, and clinical jaundice.

- Liver damage may be so overwhelming that the patient goes into liver failure – fulminant hepatitis.

- Recovery from acute hepatitis does not always equate with clearance of infection: 90% of neonates, 10% of children, and 5% of adults will become chronically infected (carriers).

- HBV carriers may be healthy, or may suffer varying degrees of chronic inflammatory liver disease, leading eventually (i.e. after 20+ years) to cirrhosis of the liver.

- Carriers are also at a 300-fold increased risk of developing hepatocellular carcinoma.

4. How is this disease diagnosed and what is the differential diagnosis?

- There are a number of laboratory markers of HBV infection.

- The presence of HBsAg in a serum sample indicates that HBV infection is present, but does not distinguish between acute and chronic infection.

- The presence of IgM anti-HBc indicates recent infection.

- Chronic infection is defined as the persistence of HBsAg for more than 6 months.

- Chronic carriers can be split into those who are highly infectious and at increased risk of liver disease, or those much less infectious and at lower risk of liver disease on the basis of the HBeAg/anti-HBe system.

- There are mutants of HBV in which HBeAg testing is misleading, with high levels of viral DNA in the absence of HBeAg.

- Measurement of HBV DNA levels is useful in monitoring the effectiveness of antiviral therapy.

- There are many causes of acute hepatitis. Of particular relevance, a number of distinct viruses have a hepatic tropism.

- HAV is a picornavirus, is spread by the fecal–oral route, causes asymptomatic, acute, and fulminant hepatitis, but does not result in chronic disease.

- HEV is an unclassified RNA virus, spread via the fecal–oral route, and has a higher mortality than HAV in the acute stage. Chronic infection does not arise.

- HCV is a flavivirus, spread via blood-borne routes, and usually results in chronic infection, with consequent risks of cirrhosis and hepatocellular carcinoma.

- HDV is an incomplete virus that requires the presence of HBV in order to replicate. This virus often results in chronic infection with an accelerated progression of liver disease.

5. How is the disease managed and prevented?

- Management of acute infection is symptomatic only. Fulminant hepatitis may necessitate liver transplantation.

- Carriers of HBV should be educated about their condition and advised as to how to prevent transmission.

- Chronic HBV infection may be treated with IFN-α, which acts as an immunomodulatory agent, or with HBV polymerase inhibitors such as lamivudine, adefovir, and entecavir.

- Long-term therapy may result in the emergence of resistant viral mutants.

- Prevention is by means of a subunit vaccine consisting of purified HBsAg, given as a course of three intramuscular injections.

- WHO recommends universal vaccination in childhood. Some countries (UK included) still have selective vaccination policies.

- Vaccine-induced protection is related to the titer of anti-HBs produced. Not all vaccinees generate adequate anti-HBs responses.

- HAV can be prevented by vaccination using a killed whole virus vaccine, currently offered only to high risk groups.

- Treatment of HCV infection consists of PEG-IFN-α plus ribavirin combination therapy for at least 6 months. This results in viral clearance in about 50% of patients with chronic infection, and >95% of patients with recent acute infection.

FURTHER READING

Humphreys H, Irving WL. Problem-orientated Clinical Microbiology and Infection, 2nd edition. Oxford University Press, Oxford, 2004.

Murphy K, Travers P, and Walport M. Janeway's Immunobiology, 7th edition. Garland Science, New York, 2008.

Richman DD, Whitley RJ, Hayden FG. Clinical Virology, 2nd edition. ASM Press, Washington, DC, 2002.

Zuckerman AJ, Banatvala JE, Pattison JR, Griffiths PD, Shaub BD. Principles and Practice of Clinical Virology, 5th edition. Wiley, Chichester, 2004.

REFERENCES

Craig AS, Schaffner W. Prevention of hepatitis A with the hepatitis A vaccine. N Engl J Med, 2004, 350: 476–481.

Dusheiko G, Antonakopoulos N. Current treatment of hepatitis B. Gut, 2008, 57: 105–124.

Ganem D, Prince AM. Hepatitis B virus infection – natural history and clinical consequences. N Engl J Med, 2004, 350: 1118–1129.

Lavanchy D. Hepatitis B virus epidemiology, disease burden, treatment, and current and emerging prevention and control measures. J Viral Hepat, 2004, 11: 97–107.

Lok AS. The maze of treatments for hepatitis B. N Engl J Med, 2005, 352: 2743–2746.

Patel K, Muir AJ, McHutchison JG. Diagnosis and treatment of chronic hepatitis C infection. BMJ, 2006, 332: 1013–1017.

WEB SITES

All the Virology on the WWW Website, developed and maintained by Dr David Sander, Tulane University: http://www.virology.net/garryfavweb12.html#Hepad

Centre for Infections, Health Protection Agency, HPA Copyright, 2008: http://www.hpa.org.uk/infections/topics_az/hepatitis_b/menu.htm

Centers for Disease Control and Prevention (CDC), 2008, United States Public Health Agency: http://www.cdc.gov/ncidod/diseases/hepatitis/

Website of Derek Wong, a medical virologist working in Hong Kong: http://virology-online.com/viruses/HepatitisB.htm

MULTIPLE CHOICE QUESTIONS

The questions should be answered either by selecting True (T) or False (F) for each answer statement, or by selecting the answer statements which best answer the question. Answers can be found in the back of the book.

1. Which of the following viruses contains DNA?
    A. Hepatitis A virus.
    B. Hepatitis B virus.
    C. Hepatitis C virus.
    D. Hepatitis D virus.
    E. Hepatitis E virus.

2. Which of the following are recognized routes of transmission of hepatitis B virus?
    A. Fecal–oral.
    B. Via contaminated needles.
    C. Via blood or blood products.
    D. From mother to baby.
    E. Sexual.

3. Which of the following viruses are usually transmitted by contaminated food and water?
    A. Hepatitis A virus.
    B. Hepatitis B virus.
    C. Hepatitis C virus.
    D. Hepatitis D virus.
    E. Hepatitis E virus.

4. Infection with which of the following viruses carries an increased risk of development of hepatocellular carcinoma?
    A. Epstein-Barr virus.
    B. Hepatitis A virus.
    C. Hepatitis C virus.
    D. Hepatitis E virus.
    E. Yellow fever virus.

5. Which of the following statements relating to hepatitis viruses is/are true?
    A. The risk of transmission of hepatitis B virus from mother to baby is greater than that of hepatitis C virus.
    B. The chances of an adult becoming a chronic carrier are greater following infection with hepatitis B virus than with hepatitis C virus.
    C. The risk of acute liver failure (fulminant hepatitis) is greater for infections with hepatitis B virus than it is for hepatitis A virus.
    D. The mortality associated with acute hepatitis E virus infection is greater than that with acute hepatitis A virus infection.
    E. Hepatitis D virus can only replicate in cells already infected with hepatitis C virus.

6. In a chronic carrier of hepatitis B virus, which of the following markers may be present in the patient's serum?
    A. Antibody to hepatitis B core antigen.
    B. Antibody to hepatitis B e antigen.
    C. Antibody to hepatitis B surface antigen.
    D. Hepatitis B surface antigen.
    E. HBV DNA.

7. Which of these statements, concerning hepatitis B e antigen (HBeAg) and antibody to HBeAg (anti-HBe), is/are true?
    A. The presence of HBeAg in serum means the patient is at increased risk of serious chronic liver disease.
    B. The presence of anti-HBe in serum means that the patient is of low infectivity.
    C. The e antigen is a breakdown product of the surface antigen.
    D. There are mutants of HBV that do not synthesize e antigen.
    E. The presence of HBeAg in a serum sample indicates

MULTIPLE CHOICE QUESTIONS (continued)

that infection with hepatitis B virus has taken place within the last 3 months.

8. **In relation to the treatment of patients with chronic HBV infection, which of the following statements are correct?**

   A. IFN-α therapy works primarily by rendering uninfected cells resistant to virus replication.

   B. IFN therapy of chronic HBV infection is less likely to succeed in patients who are co-infected with HIV.

   C. The main mode of action of lamivudine is to prevent assembly of mature virus particles.

   D. Polymerase mutations leading to lamivudine resistance arise in the majority of patients treated for more than 3 years.

   E. Treatment with HBV polymerase inhibitors usually leads to elimination of HBV from an infected liver.

9. **Which ONE of the following statements about hepatitis B vaccine is true?**

   A. Current vaccines are derived from human plasma.

   B. It is a live attenuated vaccine.

   C. Multiple doses have to be given to ensure an adequate immune response.

   D. Response can be monitored by measuring antibody to hepatitis B core antigen in the recipient's serum.

   E. It will also provide protection against infection by hepatitis E virus.

10. **Which of the following groups is at increased risk of HBV infection, and should therefore be targeted in a selective HBV vaccination policy?**

    A. Health-care workers.

    B. Sex workers.

    C. Frequent travelers.

    D. Sexual partners of known hepatitis B carriers.

    E. Babies of HBV mothers.

Case 14

Herpes simplex virus 1 (HSV-1)

A 13-year-old boy was taken to the hospital's emergency department by ambulance. His mother had called out the emergency services as he had had a generalized **fit** that morning. Since he stopped fitting, he had been very drowsy. His mother reported that he was well until 24 hours previously, when he started acting strangely – including wandering around the house not knowing where he was. He vomited once the previous evening, but otherwise has not complained of any specific problems.

On examination, he was drowsy but responsive. He was unable to give a coherent history, and had difficulty in understanding where he was. He was **febrile** (38.5°C), but had no other abnormal physical signs.

An intravenous line was set up, and he was started on empirical antibiotic therapy, together with intravenous aciclovir. A **magnetic resonance imaging** (**MRI**) **scan** of his brain was organized for later in the day, which was reported by the duty radiologist as showing 'an area of low attenuation in the right temporal lobe extending into the frontal lobe gray matter. There was mass effect with displacement of the right middle cerebral artery, appearances most compatible with herpes simplex **encephalitis**.'

A lumbar puncture was performed that evening. The cerebrospinal fluid (CSF) was noted to be slightly blood-stained. The microbiology technician reported the presence of 500×10^6 red blood cells l^{-1}, 57×10^6 white cells l^{-1}, predominantly lymphocytes, a normal CSF sugar, and a CSF protein level of 4.8 g l^{-1}, just above the upper limit of normal. No organisms were seen on a gram-stained film.

A provisional diagnosis of herpes simplex encephalitis was made. This was confirmed 2 days later when a report was phoned through from the virology reference laboratory indicating that the CSF was positive for the presence of HSV DNA as tested by **polymerase chain reaction** (**PCR**).

1. What is the causative agent, how does it enter the body and how does it spread a) within the body and b) from person to person

Causative agent

Herpes simplex virus type 1 is a herpesvirus. The *Herpesviridae* family of viruses is characterized by having a double-stranded DNA genome of 125–240 kb in size (i.e. enough genetic material to encode for 150–200 viral proteins), surrounded by an icosahedral **capsid**, and a **lipid envelope**. The space between the capsid and the envelope is referred to as the **tegument.** This contains a number of virally encoded proteins, some of which are thought to play a role in viral transport within nerves (see later). Embedded within the envelope are several virally encoded glycoproteins, which are important for binding of the virus to target cells and subsequent cell entry. Thus far, eight herpesviruses have been identified as pathogens of humans – see Table 1 in the Epstein-Barr virus case (Case 9).

The genome sequences of HSV-1 and HSV-2 share considerable homology and the biological properties of these two viruses are indeed similar. Disease arising from infection with each virus is clinically indistinguishable. However, HSV-1 infections tend to occur 'above the waist,' that is

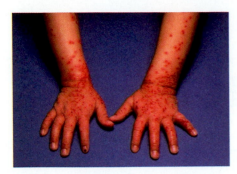

Figure 1. Eczema herpeticum. Virus is able to spread across nonintact skin in patients with eczema, giving rise to eczema herpeticum.

oral, while HSV-2 infections are classically 'below the waist,' that is genital. Herpes simplex encephalitis (HSE) is almost always due to HSV-1 infection.

Entry and spread within the body

HSV infects at mucosal surfaces (oropharynx and nasopharynx, conjunctivae, genital tract). Intact skin is impervious to HSV, the difference being that skin is covered by a thick layer of keratin, while mucosa has nonkeratinized epithelium. If, however, skin is not intact, for example in patients with a chronic dermatitis such as **eczema**, virus may spread across the skin and cause very extensive lesions – so-called eczema herpeticum (Figure 1). This is potentially life-threatening, as virus may gain access to the bloodstream through the skin abrasions, and thereby seed internal organs.

The interaction of HSV with epithelial cells is **cytolytic**, with cell death occurring 24–48 hours after infection. Virus is released from infected cells by budding, and one infected cell may give rise to thousands of progeny virus particles. These new virus particles infect neighboring cells, leading potentially to extensive local spread at the site of infection.

Latency of HSV

At some stage during this process, virus enters nerve terminals and travels in a retrograde direction up the nerve axon to reach the nerve cell body, the site of **latency** of HSV (Figure 2). Following HSV infection in the mouth, the trigeminal ganglia are typically infected, with occasional extension to cervical ganglia; genital infection, in contrast, results in infection of the sacral nerve root ganglia. Within a latently infected cell, viral genome is present as an **episome** within the nucleus, but there is no detectable production of virus proteins, and no production of new virus particles. Molecular mechanisms underlying entry into, maintenance of, and subsequent reactivation from latency are poorly understood, although well-recognized triggers for reactivation include local trauma, immunosuppression, ultraviolet light, and 'stress.' While virus is in a latent state, there is no evident damage to the infected cell. Reactivation from latency occurs when the viral replication cycle resumes, resulting in new virus particles that travel down the axon to reach the periphery, where they may give rise to clinical disease. The first exposure of an individual to HSV infection is referred to as the *primary* infection, while a reactivated infection is

A Primary infection

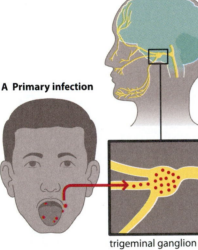

trigeminal ganglion

B Latent phase

C Recurrence

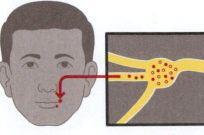

Figure 2. Schematic diagram illustrating herpes simplex virus latency.
(A) During primary infection of the mouth, there is active viral replication within the oropharyngeal mucosa (red dots). Some of this virus enters nerve terminals and travels up the nerve axon towards the nerve cell body – in this case the trigeminal ganglion – which is the site of latency for this virus. (B) After recovery from the acute infection, the patient enters the latent phase. There is no viral replication within the mouth or trigeminal ganglion (absence of red dots), but the viral genome lies latent within the latter (open red circles). (C) Following reactivation, latent virus (open red circles) within the trigeminal ganglion may start to replicate (red dots), and newly formed viral particles travel back down the nerve axon to reach the periphery (i.e. the mouth), where further virus replication takes place, although this is limited by the effects of the host immune system. The clinical manifestation of this recurrence, i.e. a cold sore, is therefore much more localized than in primary infection.

referred to as a *recurrent*, or *secondary*, infection. The clinical manifestations of these two forms of herpesvirus infection are often quite distinct, as in the former case the infection is occurring in an immunologically naïve individual, while in the latter case virus is replicating in a host whose immune system has seen the virus before.

In patients with oropharyngeal (or genital) HSV infection, an additional risk is spread to other mucosal sites via direct self-inoculation of virus, for example on contaminated fingers. The propensity for HSV to spread across nonintact skin has been mentioned above. Virus may (rarely) gain access to the bloodstream, especially in patients with eczema herpeticum – such **viremia** is potentially life-threatening, as infection of the internal organs may result.

Spread of virus into the brain, an event which necessarily precedes the onset of herpes simplex encephalitis, is discussed as part of the disease pathogenesis raised under Question 2 below.

Spread from person to person

Herpetic lesions present externally, for example in the **oropharynx** or genital tract, contain very large amounts of virus. In addition to the development of symptomatic recurrent disease, reactivation of virus may occur in the absence of clinical disease. Such asymptomatic recurrences result in transient (for example 24 hours) shedding of virus from the mouth or genital tract. Infection can then be transmitted to others through direct contact with infected mucosal sites, for example through kissing or unprotected sexual intercourse.

The important (but fortunately rare) route of mother to baby spread of HSV infection is discussed in more detail in the case of HSV-2, Case 15.

Epidemiology

Oral and genital HSV infections occur worldwide. Oral infection is usually with HSV-1, and genital infection with HSV-2, but these boundaries have become blurred with increasing safe-sex practices leading to oro-genital transmission of HSV-1 (and vice versa, i.e. genital–oral transmission of HSV-2). In the UK, up to half of all new genital herpes infections may be due to HSV-1.

2. What is the host response to the infection and what is the disease pathogenesis?

Herpes simplex virus infection is common, and may give rise to disease at a number of distinct sites within the body – see Table 1. Herpes simplex encephalitis (HSE) is fortunately rare – estimated incidence is around 1–3 per million population per year.

HSV-1 is the cause of encephalitis in more than 90% of cases. More than two-thirds of cases of HSV-1 encephalitis (HSE) are the result of reactivation of latent virus in previously infected individuals, with the remainder due to primary infection in children and young adults. Neonatal HSE is distinct in that HSV-2 is the more usual pathogen because of maternal genital infection resulting in intrapartum transmission.

Table 1. Diseases caused by herpes simplex infection

Orolabial infection

    Primary - gingivostomatitis

    Recurrent - cold sore

Genital infection

Keratitis (infection of the cornea)

Herpetic whitlow - infection of the nail bed

Eczema herpeticum

Herpes encephalitis

Herpes gladiatorum - scratching of virus into the skin, e.g. in wrestlers

Patients suffering from HSE undoubtedly have virus present within the brain substance itself, but it is unclear when and how virus reaches this site. The precise anatomic pathways whereby virus enters the brain substance are not known with certainty. Virus may travel via either or both the trigeminal and olfactory nerve tracts. In patients with reactivated disease, there is also controversy concerning the site of the latent virus that reactivates. Virus may reactivate peripherally (within the trigeminal ganglion or olfactory bulb) and enter the central nervous system by retrograde transport, or may reactivate centrally within latently infected brain tissue. HSV DNA has been demonstrated by sensitive genome amplification assays within post-mortem brain tissue in up to a third of asymptomatic HSV **seropositive** adults. However, the possibility of reactivation of virus from this site remains a speculative hypothesis at the moment.

Despite the above uncertainties, what is not in doubt is that virus is indeed present and replicating within the brain substance in a patient with HSE. HSV infection in this situation is cytolytic. The presence of replicating virus generates an acute inflammatory host response, with an infiltration of mononuclear cells, the characteristic immune response to a virus infection. Cell death arises from both virus infection and the host immune response, resulting in hemorrhagic **necrosis**, the pathological hallmark of HSE (see Figure 3). Involved necrotic brain tissue becomes liquefied, and the disease spreads outwards as new virus particles are released. Microscopic examination of brain tissue reveals damage extending much further than that visible macroscopically.

3. What is the typical clinical presentation and what complications can occur?

HSE can arise in any part of the brain, and the neurological manifestations are dependent on the precise anatomic site. Large case series reported from the National Institutes of Health in the USA show that around 75% of patients demonstrate altered behavior or personality change – reflecting the commonest underlying anatomic site of disease in the frontotemporal regions. Also, 80% of patients complain of headache, 90% have fever, and almost all have evidence of decreased level of consciousness. Other presenting manifestations include **seizures** (67%), vomiting (46%), **hemiparesis** (33%), and memory loss (25%).

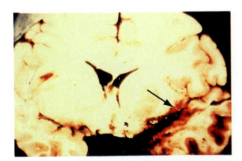

Figure 3. Herpes simplex encephalitis: gross pathology. The arrow indicates a large area of hemorrhagic necrosis in the left frontotemporal region of the brain.

The pathological process underlying the disease (described above) is such that the only relevant 'complication' of HSE is continuing and worsening brain damage, leading to death. Untreated, the case mortality of HSE is around 70%, and only a small fraction (10%) of survivors are able to return to normal function after recovery from the acute illness. These figures have changed with the advent of antiviral therapy, although mortality may still be as high as 20–30%.

4. How is this disease diagnosed and what is the differential diagnosis?

A diagnosis of HSE should be considered in any patient presenting with altered consciousness, fever, and/or focal neurological signs, including seizures. Confirmation of the diagnosis may require both virological and neuroimaging investigations, and may not be straightforward.

Examination of cerebrospinal fluid (CSF) obtained via a lumbar puncture is a routine investigation when assessing a patient with possible encephalitis. However, the findings in HSE are not specific. The CSF typically demonstrates elevated protein and **pleocytosis**, with mononuclear cells of between 10 and 500 cells/mm^3 and red cells predominating, but even a completely normal CSF does not exclude the possibility of HSE. The most sensitive and specific test on CSF is detection of HSV DNA using a genome amplification technique such as PCR, but this may be negative, especially if the sample is taken less than 72 hours from onset of symptoms. This approach has more or less replaced the need for brain biopsy, which used to be regarded as the definitive test, and may still be necessary in cases of diagnostic difficulty. Note that despite the presence of HSV DNA in CSF, successful isolation of HSV in cell culture is very unusual.

Testing of serum samples for the presence and titer (amount) of anti-HSV antibodies, whether taken in the acute stage, at a later stage, or both, is not particularly helpful in the diagnosis of HSE. Most patients with HSE already have serum antibodies at the time of onset of illness, as the disease represents a reactivation of virus rather than a primary infection. Even a rise in antibody titer taken a few days apart does not confirm HSE, as reactivation of virus may be a consequence of fever or general illness, and can result in an increase in viral antibody titers. However, measurement of antiviral antibody synthesis within the CSF may be helpful, provided that serial samples demonstrate a rise in this antibody titer, although such a rise may take several days if not weeks to occur.

The cardinal feature of HSE revealed by neuroimaging is the focal nature of the disease process and its predilection for frontotemporal lobe involvement – as opposed to other causes of encephalitis, which may affect the brain more globally (Figure 4). Thus, whichever imaging technique is used, the demonstration of a focal lesion or lesions (disease may spread bilaterally, especially if therapy is delayed) should be interpreted as indicating HSE until proven otherwise. Techniques include (proceeding from the more historical to the most modern) **electroencephalogram (EEG)**, **radionuclide brain scanning**, **CT scan** or **magnetic resonance imaging (MRI)**. MRI scans can often show evidence of HSE before the damage is evident on a CT scan.

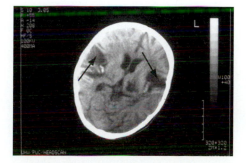

Figure 4. Neuroimaging in herpes simplex encephalitis: CT scan of a patient with herpes simplex encephalitis. The arrows point to bilateral areas of low attenuation in the frontotemporal regions.

Differential diagnosis

The differential diagnosis of HSE is that of any encephalitic illness, and there is therefore a potentially long and complex list of diseases that can mimic HSE. The more common of these are listed in Table 2.

5. How is the disease managed and prevented?

Management

The treatment of HSV diseases is with aciclovir (acycloguanosine, see Figure 5) and its derivatives. Thymidine kinase, a virally encoded enzyme, is able to mono-phosphorylate aciclovir, which is then di- and tri-phosphorylated by host cell enzymes. Aciclovir triphosphate then competes with GTP for incorporation into the growing viral DNA chain. Once aciclovir is incorporated, the absence of a complete deoxyribose ring means that there are no hydroxyl groups available to participate in formation of phosphodiester linkages with any incoming bases, and therefore further synthesis of the DNA chain is terminated. The antiviral selectivity of aciclovir arises because:

- the drug can only be phosphorylated in virally infected cells as host cells do not possess the necessary thymidine kinase enzyme;

- not only is the drug only activated in infected cells, but because phosphorylation of the molecule effectively reduces the concentration of free aciclovir within the cell, more free drug diffuses into the cell from the extracellular space, resulting in concentration of the drug specifically in virally infected cells;

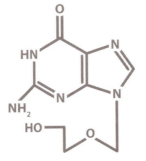

Figure 5. Structure of aciclovir. Aciclovir is acicloguanosine - note the absence of a complete deoxyribose ring.

Table 2. Differential diagnosis of herpes simplex encephalitis	
Other viral infections that can cause encephalitis	Cytomegalovirus
	Epstein-Barr virus
	Influenza A virus
	Enteroviruses
	West Nile virus
	Mumps virus
	Nipah virus
Space-occupying lesions	Abscess/subdural empyema
	Bacterial infections (e.g. *Listeria, Mycobacterium tuberculosis, Mycoplasma*)
	Other infectious agents (e.g. *Rickettsia, Toxoplasma gondii, Cryptococcus neoformans*)
	Intracerebral tumor
	Subdural hematoma
Systemic diseases	Vascular disease
	Systemic lupus erythematosus
	Toxic encephalopathy

• in addition to acting as a DNA chain terminator, aciclovir triphosphate binds to, and strongly inhibits, viral DNA polymerase with a much greater affinity than it does cellular DNA polymerases.

Valaciclovir is a valine ester of aciclovir that is hydrolyzed after absorption into valine and aciclovir. However, it is much better absorbed when administered orally than is aciclovir. Penciclovir is a similar molecule to aciclovir, and works in a similar fashion, that is it requires initial phosphorylation which can only be achieved by viral thymidine kinase, and the triphosphate derivative acts as a potent viral DNA polymerase inhibitor. It is marketed as famciclovir, a complex ester of penciclovir that is better absorbed orally.

Untreated HSE has a high mortality and survivors are left with major brain damage. In this setting, aciclovir therapy is the mainstay of HSE treatment, and can be life-saving. It should be administered intravenously in high dosage, 10 mg kg^{-1} every 8 hours, and continued for 2–3 weeks. The longer the delay in starting antiviral therapy, the more residual damage the patient will be left with, so it is vital that therapy be given as soon as possible. Thus, if a diagnosis of HSE is thought to be a clinical possibility in any given patient, aciclovir therapy must be started immediately, long before the necessary diagnostic tests have been conducted to confirm or refute the diagnosis. In clinical practice, this means that many more patients receive intravenous aciclovir therapy than turn out actually to have HSE, but given the safety profile of the drug, and the catastrophic consequences of not treating the disease promptly, this is acceptable.

Disease outcome varies from death to full recovery. Key poor prognostic indicators include older age, lower **Glasgow coma scale score** at presentation (i.e. decreasing levels of consciousness), and increasing length of time between onset of illness and start of antiviral therapy.

Prevention

There is currently no prophylactic vaccine available that protects against infection with herpes simplex virus.

SUMMARY

1. What is the causative agent, how does it enter the body and how does it spread a) within the body and b) from person to person?

- Herpes simplex virus is a herpesvirus.

- Herpesviruses are characterized as having a double-stranded DNA genome in an icosahedral capsid surrounded by a lipid envelope.

- HSV enters at mucosal surfaces (oropharynx and nasopharynx, conjunctivae, genital tract) and replicates within epithelial cells.

- Spread is local, through release of new virus particles from dying cells. Intact skin is an effective barrier to HSV infection.

- Virus is also able to enter nerve cells and travel retrograde to the nerve cell body, the site of latency of HSV.

- Virus may be reactivated from latency, leading to the presence of virus at the same mucosal sites. Such shedding of virus most commonly is asymptomatic.

- Person to person spread arises through direct contact with virus shed from mucosal surfaces gaining access to recipient mucosal surfaces.

2. What is the host response to the infection and what is the disease pathogenesis?

- Routes of entry into the brain substance are not clear, but include via the trigeminal nerve and the olfactory nerve.

- Timing of viral entry into the brain is also controversial – this may occur at the time of primary infection, or at the time of peripheral reactivation of virus in the trigeminal ganglion or olfactory bulb.

- Virus infects brain cells in a cytolytic fashion, resulting in cell death.

- Virus particles released from dying cells spread to adjacent brain tissue.

- The host acute inflammatory response is characterized by an influx of mononuclear cells.

- The pathologic process is one of hemorrhagic necrosis with liquefaction of affected brain tissue.

3. What is the typical clinical presentation and what complications can occur?

- Clinical presentation is nonspecific, common symptoms and signs including altered behavior, headache, fever, seizures, and altered consciousness.

- Less common manifestations include vomiting, paralysis (hemiparesis), and memory loss.

- Untreated, the disease is progressive, with a high mortality (70%). Survivors may be left with considerable brain damage.

4. How is this disease diagnosed and what is the differential diagnosis?

- HSE should be considered in any patient with altered levels of consciousness, fever, and/or acute onset of focal neurology, including seizures.

- CSF findings are not specific. The commonest CSF picture is an increase in red cells, around 100 white cells μl^{-1}, mostly mononuclear, increased protein, and normal sugar. However, CSF may be completely normal.

- The detection of HSV DNA in CSF by PCR is the most sensitive and specific diagnostic test, but this may be negative, especially early in the course of the disease.

- Neuroimaging, whether by EEG, or radionuclide, CT, or MRI scan, is helpful, the cardinal feature being the presence of focal, rather than diffuse, disease.

- There is a long list of differential diagnoses, including other viral (many), bacterial, fungal or parasitic infections; intracerebral space-occupying lesions such as tumor; systemic disease, for example vasculitis.

5. How is the disease managed and prevented?

- High dose aciclovir therapy given intravenously for a minimum of 2 weeks is essential.

- Aciclovir should be administered as soon as the diagnosis is considered.

- There is no prophylactic vaccine against HSV infection.

FURTHER READING

Humphreys H, Irving WL. Problem-orientated Clinical Microbiology and Infection, 2nd edition. Oxford University Press, Oxford, 2004.

Murphy K, Travers P, and Walport M. Janeway's Immunobiology, 7th edition. Garland Science, New York, 2008.

Richman DD, Whitley RJ, Hayden FG. Clinical Virology, 2nd edition. ASM Press, Washington, DC, 2002.

Zuckerman AJ, Banatvala JE, Pattison JR, Griffiths PD, Shaub BD. Principles and Practice of Clinical Virology, 5th edtion. Wiley, Chichester, 2004.

REFERENCES

Elbers JM, Bitnun A, Richardson SE, et al. A 12-year prospective study of childhood herpes simplex encephalitis: is there a broader spectrum of disease? Pediatrics, 2007, 119: e399–e407.

Whitley RJ, Kimberlin DW. Herpes simplex encephalitis: children and adolescents. Semin Pediatr Infect Dis, 2005, 16: 17–23.

Whitley RJ, Roizman B. Herpes simplex virus infections. Lancet, 2001, 357: 1513–1518.

WEB SITES

All the Virology on the WWW Website, developed and maintained by Dr David Sander, Tulane University: http://www.virology.net/garryfavweb12.html#Herpe

Neuroland.com, this site is maintained and supported by Charles Tuen, MD, Neurologist at Methodist Medical Center in Dallas, Texas: http://neuroland.com/id/herpes_ence.htm

Patient UK, a joint venture between PiP and EMIS (Egton Medical Information Systems), Copyright © 2008 EMIS & PIP: http://www.patient.co.uk/showdoc/40000500/

Website of Derek Wong, a medical virologist working in Hong Kong: http://virology-online.com/viruses/HSV.htm

MULTIPLE CHOICE QUESTIONS

The questions should be answered either by selecting True (T) or False (F) for each answer statement, or by selecting the answer statements which best answer the question. Answers can be found in the back of the book.

1. **Which of the following are features of herpesviruses?**
   A. They carry their genome in the form of single-stranded DNA.
   B. The genome size is of the order of 150–200 kb.
   C. They are nonenveloped viruses.
   D. They have a helical capsid.
   E. There is a reverse transcriptase step in their replication cycle that renders them sensitive to reverse transcriptase inhibitors.

2. **Which of the following may be outcomes of herpes simplex virus infection of a cell?**
   A. Acute cytolytic infection.
   B. Chronic infection.
   C. Integration of the viral genome into a host cell chromosome.
   D. Latent infection.
   E. Transformation.

3. **Which of the following statements regarding the pathogenesis of herpes simplex encephalitis are true?**
   A. About one-third of cases of HSE arise as a result of primary HSV infection.
   B. Brain cell loss occurs through both virus-induced cell death and the action of host cytotoxic T cells.
   C. Hemorrhage is a typical pathologic feature.
   D. The predominant cellular infiltrate in affected brain is composed of polymorphonuclear leukocytes.
   E. Visually normal brain tissue adjacent to affected areas of brain is also microscopically normal.

4. **Which of the following are common clinical manifestations evident in a patient with herpes simplex encephalitis (i.e. occur in >50% of patients)?**
   A. Bilateral muscle weakness.
   B. Diarrhea.
   C. Drowsiness.
   D. Headache.
   E. Personality change.

5. **Which of the following statements are true of the CSF findings in patients with herpes simplex encephalitis?**
   A. A normal CSF excludes the diagnosis of HSE.
   B. Red cells are commonly found in the CSF.
   C. The typical white cell response is of mononuclear cells.
   D. The CSF protein is usually below the lower limit of normal.
   E. The CSF sugar is usually less than half that of the blood sugar.

6. **Which of the following statements regarding the use of neuroimaging techniques in the diagnosis of herpes simplex encephalitis are correct?**
   A. A focal lesion in the temporofrontal region is the most common finding.
   B. Bilateral lesions may be seen.
   C. CT scanning is the most sensitive neuroimaging technique for the diagnosis of HSE.
   D. EEG is not usually abnormal in a patient with HSE.
   E. The detection of diffuse cerebral inflammation on an MRI scan confirms the diagnosis of HSE.

7. **Aciclovir works through which of the following mechanisms?**
   A. DNA polymerase inhibition.
   B. Neuraminidase inhibition.
   C. Protease inhibition.
   D. Reverse transcriptase inhibition.
   E. Viral entry inhibition.

8. **Aciclovir is a selective antiviral agent for which of the following reasons (more than one may apply)?**
   A. It becomes concentrated specifically in virally infected cells.
   B. It binds to a target enzyme which is not present in host cells.
   C. It binds to a viral enzyme with much greater affinity than it does to the corresponding cellular enzyme.
   D. It requires activation, the first step of which cannot be performed by the host cell and therefore requires the presence of virus within the cell.
   E. It is only able to diffuse into cells that are infected with virus.

9. **Which of the following drugs will inhibit the replication of herpes simplex virus?**
   A. Azidothymidine.
   B. Foscarnet.
   C. Ganciclovir.
   D. Penciclovir.
   E. Valaciclovir.

Case 15

Herpes simplex virus 2 (HSV-2)

A 17-year-old female went to see her doctor as she was feeling generally unwell, feverish, and had noticed a painful blistering **rash** on her external genital area (vulva), and a whitish vaginal discharge. She also complained of painful swellings in both her groins, and that passing urine earlier in the day gave her a severe stinging pain.

When asked specific questions, she reported that she had had unprotected intercourse with a new sexual partner 4 days ago. She has had no serious illnesses in the past, and was not on any medications.

On examination, there was a widespread ulcerative rash on the labia and vaginal walls (Figure 1). On passing a speculum, blistering lesions were seen on the cervix, together with a profuse **mucopurulent** discharge (Figure 2). A clinical diagnosis of primary genital herpes simplex infection was made and a swab was taken of one of the **vesicles** and sent to the laboratory for confirmation. Two days later, the laboratory reported detection of herpes simplex type 2 (HSV-2) DNA by **polymerase chain reaction (PCR)** assay.

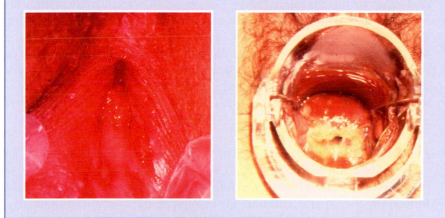

Figure 1 (left). Genital ulceration due to HSV infection. Bilateral ulcerative lesions seen in a patient with symptomatic primary genital herpes simplex virus infection.

Figure 2 (right). Primary genital HSV infection with profuse mucopurulent discharge. Passage of a speculum in a patient with primary genital herpes simplex virus infection reveals vesicular lesions on the cervix and a profuse whitish mucopurulent cervical discharge.

1. What is the causative agent, how does it enter the body and how does it spread a) within the body and b) from person to person

Causative agent

Herpes simplex virus type 2 is a herpesvirus, see HSV-1 case (Case 14). This extensive virus family is characterized by having a double-stranded DNA genome of 125–240 kb in size. The HSV-2 genome is 152 kb and encodes around 100 viral proteins. The genome is surrounded by an icosahedral **capsid**, and a **lipid envelope**. The region between the capsid and the envelope is referred to as the **tegument**, and the virus encodes a number of tegument proteins, some of which are thought to be involved in transport of the virus within nerves. The envelope is studded with several virally encoded glycoproteins, which are important in cell attachment and entry processes. Thus far, eight herpesviruses have been identified as pathogens of humans – see Table 1 in the Epstein-Barr virus case (case 9).

Entry and spread within the body

Intact skin is impervious to HSV infection (and most other viruses). HSV therefore infects by direct contact at mucosal surfaces, that is the **oropharynx**, the genital area, and the conjunctivae. In this case, the patient has acquired infection through unprotected intercourse with a partner who would have had virus present somewhere in his genital area. Glycoprotein B in the envelope appears to be one of the molecules involved in attachment of HSV-2 to the mucosal epithelial cells. Infection is **cytolytic**, that is the virus replicates within the cells, and over a period of about 24–48 hours, this results in acute cell death with release of thousands of new virus particles, which are then able to infect neighboring cells. Direct spread of virus leads to the development of extensive vesicular lesions (which may subsequently enlarge into blisters), which may affect the entire genital (up to and including the cervix), perineal, and perianal mucocutaneous area.

As with all herpesviruses, HSV-2 infection results in the establishment of **latency**. At some stage during the acute infection, virus enters the sensory nerve cell terminals, and travels up the axoplasm to reach the nerve cell bodies within the lumbosacral spinal ganglia, the site of latency of genital HSV. Once the initial infection is brought under control, either by the host immune system, or by appropriate antiviral therapy, no virus will be detectable within the genital tract, but virus will remain latent within the lumbosacral ganglia. In this latent state, the viral genome is present, but there is no virus replication, and no damage to the nerve cell. However, at some future date, in response to a number of stimuli, latent virus may be reactivated, replicate, travel back down the axon to reach the periphery, and may give rise to symptomatic disease. The molecular mechanisms whereby herpesviruses become latent and are maintained in latency are poorly understood, as are the triggers that lead to reactivation.

Further spread of virus may occur by accidental direct self-inoculation, for example infection of the finger nail bed (known as a herpetic whitlow, or **paronychia**); virally contaminated fingers (from the genital area or from the mouth) can then spread infection, for example to the eyes, with conjunctival infection and the risk of **keratitis**. Virus may be scratched into the skin, resulting in a local crop of vesicles. If skin is not intact, or example in patients with a chronic dermatitis such as **eczema**, virus may spread across the skin and cause very extensive lesions – so-called eczema herpeticum (this is usually associated with oral, and HSV-1, infection, see HSV-1 case (Case 14)–encephalitis). This is potentially life-threatening, as virus may gain access to the bloodstream through the skin abrasions, and thereby seed internal organs.

Spread from person to person

Virus is present at high titer within the vesicular fluid. Spread of genital herpes occurs by direct contact during unprotected intercourse. Oral herpes infection can also be transmitted in this way to the genital area via orogenital sex.

One very important route of person to person spread of genital HSV is from mother to baby. This may arise when a baby is born through an infected birth canal due to maternal primary or even reactivated infection, and may have devastating consequences for the baby (see clinical features below).

Epidemiology of herpes genital infections

Oral and genital HSV infections occur worldwide. Oral infection is usually with HSV-1, and genital infection with HSV-2, but these boundaries have become blurred with increasing safe-sex practices leading to oro-genital transmission of HSV-1 (and vice versa, i.e. genital–oral transmission of HSV-2). In the UK, as many as half of all new genital herpes infections may be due to HSV-1. However, as genital HSV-2 infection is much more likely to recur than is genital HSV-1, HSV-2 remains the principal cause of recurrent genital herpes.

Large-scale seroprevalence surveys looking for antibodies specific to HSV-2, mostly conducted within the USA at a time before genital HSV-1 infection was common, suggest that up to 20% of the adult population is HSV-2 **seropositive**. Seroprevalence increases with age and number of lifetime sexual partners, and is higher in black as compared with white populations. However, only a small proportion of individuals with antibodies to HSV-2 give a history of primary or recurrent genital herpes, demonstrating the largely asymptomatic nature of the infection.

2. What is the host response to the infection and what is the disease pathogenesis?

Immune response

Innate immunity is key to prevention or control of a primary infection. The importance of intact skin has been referred to above. Initial infection at mucosal sites will lead to **interferon (IFN)** production, which in turn will help to protect uninfected neighboring epithelial cells. In fact, there is a strong correlation in mice for the production of IFN-β and innate immunity to HSV-2. The IFN-β is produced by epithelial cells but also by **dendritic cells** and macrophages in the submucosa (probably through interaction with cell surface Toll-like receptor 9 (TLR9) in the case of the latter two cell types.

Carriage of the virus by immature dendritic cells to draining lymph nodes leads to the development of an adaptive immune response. Both **CD4+** T cells and cytotoxic **CD8+** T cells are produced in response to infection and play a role in elimination of the virus. Memory T lymphocytes appear to be effective in response to local episodes of recurrent infection and home efficiently to sites of infection.

In experimental mouse models of genital herpes infections, **IgG** antibodies appear to be protective but **IgA** is less so.

Once virus has become latent in nerve cell bodies, the adaptive immune response, particularly the T-cell arm, is vital in preventing, or at least limiting, disease caused by reactivation. Thus, patients with cellular immunodeficiency, such as HIV-infected individuals, or transplant recipients, are much more prone to get symptomatic recurrent disease.

Infection with HSV-1 does not induce protective immunity against HSV-2 infection, and vice versa.

3. What is the typical clinical presentation and what complications can occur?

As explained above, most primary genital infections with HSV do not give rise to clinical disease. The assumption must be that expression of disease is a result of the race between replication and spread of the virus on the one hand, and the ability of the innate immune response to control this on the other.

In the minority of patients where disease does become clinically evident, primary genital herpes is not a trivial disease. Lesions are extensive and bilateral, extending from the labia to the cervix. There may be spread onto adjacent skin. There will be painful inguinal **adenopathy**, as in this patient, and a systemic response (e.g. fever) reflecting the release of **cytokines** from the intense inflammatory reaction taking place. In females, passing urine may be exquisitely painful if there are lesions near the urethral meatus. It may take 2–3 weeks before the patient fully recovers.

In marked contrast, recurrent disease, even when symptomatic, is usually mild, as the reactivation is taking place in an individual whose immune system has seen the virus before. Thus, the lesions will be much more localized, mostly unilateral, with no spread onto the adjacent skin; no, or only unilateral, inguinal adenopathy; and no, or only mild, systemic symptoms. The natural history of recurrent lesions is resolution within 5–7 days. Many recurrences are asymptomatic – and it must be individuals who are not aware of asymptomatic reactivation of disease who are the source of spread to their sexual partners.

There are a number of possible complications. In primary disease, there may be meningeal irritation, such that the patient presents with a clinical diagnosis of **meningitis**. This may rarely happen even with recurrent disease – so-called **Mollaret's meningitis**. HSV meningitis is more common in women. Irritation of the sacral nerve roots (**radiculomyelopathy**) may present with aching pain in the sacral **dermatomes** associated with **parasthesiae** or **dysasthesiae** in the lower limbs. Accidental inoculation elsewhere in the body has been mentioned above, for example giving rise to herpetic keratitis (infection of the cornea). One disastrous complication of genital herpes in a female is spread to her baby, resulting in neonatal herpes (Figure 3). This usually arises in women suffering a primary attack of genital herpes in late pregnancy, of which she may be unaware. However, the birth canal is rich in virus, and there has not yet been time for the mother to generate and pass protective antibodies transplacentally to the fetus. The majority of neonates infected in this way acquire internally disseminated infection, including herpes **encephalitis**, which has a high mortality and morbidity even with appropriate therapy. Only about half of these neonates have herpetic lesions evident on their skin or mucous membranes, making the diagnosis very difficult. It is only the babies whose infection is limited to the skin and mucous membrane who make a complete recovery – only about 10–15% of all neonatally infected babies. Fortunately, neonatal HSV infection is not common – about 1 in 50 000 births in the UK, but perhaps 1 in 5000 in the USA.

In immunocompromised patients, for example with HIV infection, or those on cytotoxic therapy, HSV recurrences are not only more frequent,

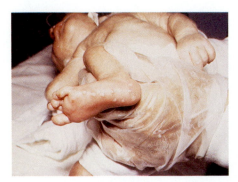

Figure 3. Neonatal herpes.

but may also be somewhat atypical, with extensive local lesions, and the potential for systemic and life-threatening spread.

HSV-2 infection has also been shown to increase HIV spread in patients and infected dendritic cells also promote HIV infection.

4. How is this disease diagnosed and what is the differential diagnosis?

The clinical picture of a patient such as described in this chapter, with extensive ulceration, is characteristic enough to make a diagnosis, although it is still imperative to confirm this by sending an appropriate sample to the laboratory, as a diagnosis of genital herpes has important implications both for the individual patient and for his/her sexual partners. In addition, typing of the virus has prognostic significance, as type 2 genital infections are more likely to recur than are type 1 infections. Note that in any patient with one sexually transmitted infection (STI), it is also appropriate to test for the presence of other STIs.

Recurrent disease, where there may only be one or two ulcers, is much more difficult, and laboratory confirmation should always be sought, as the differential diagnosis of genital ulceration is not straightforward.

Samples sent to the laboratory should be taken by abrading the base of an ulcer or vesicle, and breaking the swab off into viral transport medium (isotonic fluid containing antibiotic to prevent bacterial overgrowth). There are many ways in which the laboratory can make the diagnosis. The most sensitive and widely used nowadays is to demonstrate the presence of HSV DNA by a genome amplification assay such as PCR. By use of appropriate primers it is possible to distinguish between HSV-1 and HSV-2 using this approach. Alternatively, immunofluorescence of the abraded cells with **monoclonal antibodies** against HSV-1 and HSV-2 can be used, or virus can be isolated in cell culture. There is even sufficient virus within vesicle fluid to be visualized by electron microscopy, although very few laboratories will perform that these days.

Differential diagnosis

The differential diagnosis is anything that can cause genital ulceration. In addition to genital herpes, this can be due to physical trauma including excoriation of marked acute candidal vulvitis, chemical burns from disinfectants, **pyogenic infection**, fixed drug eruption, systemic disease (e.g. **Behcet's syndrome, erythema multiforme**), other infections, for example the **chancre** of primary syphilis, **chancroid, granuloma** inguinale, lymphogranuloma venereum (see *Chlamydia trachomatis*, Case 4), and dermatological conditions (e.g. **lichen sclerosis** et **atrophicus**).

5. How is the disease managed and prevented?

Management

As for HSV-1, aciclovir (Figure 4) and its derivatives are used for treatment (see page 182 for detailed description of the mechanism of action of aciclovir). Briefly, aciclovir requires triphosphorylation, the first step of which

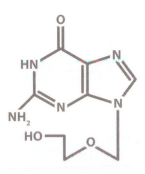

Figure 4. Structure of aciclovir. Aciclovir is acicloguanosine - note the absence of a complete deoxyribose ring

is mediated by a virally encoded enzyme, thymidine kinase. Aciclovir triphosphate competes with guanosine triphosphate (GTP) for incorporation into the growing viral DNA chain. The acyclic nature of the molecule results in chain termination. In addition, aciclovir triphosphate has a high affinity for viral DNA polymerase (much greater than for cellular DNA polymerases), and acts as a strong viral DNA polymerase inhibitor.

In patients with symptomatic primary genital HSV infection, antiviral therapy should be initiated as soon as possible. This results in rapid cessation of viral replication and resolution of lesions several days faster than in the absence of therapy. Management of recurrent genital herpes is, however, more controversial. This is, as explained above, a much less severe clinical condition, with a natural history of evolution of lesions to clearance of the order of 6–7 days. Oral therapy with aciclovir (or derivatives), even if initiated by the patient as soon as he/she is aware that a recurrence is imminent, results in shortening of this disease period by around 24 hours. Thus, in general, there is not a great deal of benefit to be gained by routine therapy of recurrent genital HSV infection. However, there are exceptions to this. Some unfortunate individuals suffer from rather atypical genital herpes, with frequent attacks (e.g. once a month), which may last for up to 2 weeks. The immunological reasons for this are poorly understood, but such disease can be managed by use of continuous prophylactic aciclovir. This may continue for several months or even years.

Viral resistance to aciclovir may arise through a number of mechanisms. Viral variants lacking a thymidine kinase enzyme (TK⁻) exist in nature and are selected for when aciclovir (or a similar agent) is used. However, the TK gene is an important **virulence** factor for HSV, and TK⁻ variants are not pathogenic. Mutations may arise in the TK gene (TK mutants) such that the enzyme no longer phosphorylates aciclovir, but again, such mutants have decreased virulence and are not a clinical problem. Mutations may also arise in the viral DNA polymerase gene such that the enzyme no longer binds aciclovir triphosphate. Such DNA pol mutants are fully virulent and are a clinical problem, but have thus far only been described in heavily immunocompromised individuals who take prolonged courses of aciclovir because of frequent and severe recurrences.

Prevention

Genital ulceration of any cause, and certainly including that due to HSV infection, increases the risk of spread of HIV infection, most likely through multiple mechanisms, for example damage to the skin barrier, attraction of susceptible T cells and macrophages to the site, and up-regulation of HIV replication. HSV-2 infection is associated with a threefold increased risk of sexually acquired HIV infection. Recent data suggest that one way to reduce the spread of HIV infection may be to treat individuals with anti-herpes drugs to prevent the occurrence of genital ulceration.

There have been several attempts to produce a vaccine against genital herpes, but thus far, no vaccine has demonstrated efficacy in appropriate **clinical trials**.

SUMMARY

1. What is the causative agent, how does it enter the body and how does it spread a) within the body and b) from person to person?

- Herpes simplex virus type 2, a herpesvirus.
- Sexually transmitted through direct contact with infected genital tract secretions.
- Direct spread from the site of inoculation through release of large numbers of newly formed virus particles from lysed cells.
- Virus replicating in the genital mucosa gains access to sensory nerve terminals and travels up the nerve axon to the nerve cell body in the lumbosacral sensory ganglia.
- Nerve cell body is site of virus latency.
- Reactivation from latency results in virus particles traveling down to the nerve terminals, giving rise to recurrent genital herpes, which acts as a source of infection for sexual partners.

2. What is the host response to the infection and what is the disease pathogenesis?

- An antibody response develops, which is first detectable about 7 days after onset of the disease, initially IgM only, followed by IgG.
- T-cell responses are key to preventing symptomatic recurrent disease, as patients with impaired T-cell-mediated immunity are at risk of frequent and severe recurrences.

3. What is the typical clinical presentation and what complications can occur?

- The majority of primary genital HSV infections are asymptomatic.
- When disease occurs, it is often extensive, bilateral, spread onto adjacent skin, with a systemic response evidenced by regional **lymphadenopathy** and a **febrile** response, lasting for 2–3 weeks.
- Recurrent disease is milder, with fewer crops of lesions, unilateral, no spread onto adjacent skin; no, or only mild, systemic response; and lasting 5–6 days.
- Complications include meningitis (usually women), lumbosacral meningeal irritation presenting as radiculomyelopathy, self-inoculation to other parts of the body, for example herpetic paronychia, conjunctival infection.
- Genital HSV in pregnancy (usually primary) may give rise to neonatal herpes, which has a high morbidity and mortality.

4. How is this disease diagnosed and what is the differential diagnosis?

- Demonstration of HSV DNA in vesicle fluid or in a swab of an ulcer base by genome amplification (usually PCR).
- Immunofluorescent antigen detection using monoclonal anti-HSV antibodies to stain cells abraded from an ulcer base.
- Isolation of virus in tissue culture.
- Differential diagnosis includes physical trauma, chemical burns, pyogenic or other infections, fixed drug eruptions, systemic blistering diseases, dermatological conditions.

5. How is the disease managed and prevented?

- HSV is sensitive to aciclovir (and derivatives).
- Treatment is recommended for symptomatic primary disease.
- Prophylactic therapy may be offered to patients with frequent, atypically aggressive recurrent disease.
- There is no vaccine licensed for prevention of genital HSV infection.

FURTHER READING

Humphreys H, Irving WL. Problem-orientated Clinical Microbiology and Infection, 2nd edition. Oxford University Press, Oxford, 2004.

Minson AC. Alphaherpesviruses: herpes simplex and varicella zoster. In: Mahy BWJ, ter Meulen V, editors. Topley and Wilson's Microbiology and Microbial Infections. Hodder Arnold, London, 2005: Chapter 26.

Murphy K, Travers P, and Walport M. Janeway's Immunobiology, 7th edition. Garland Science, New York, 2008.

Richman DD, Whitley RJ, Hayden FG. Clinical Virology, 2nd edition. ASM Press, Washington, DC, 2002.

Zuckerman AJ, Banatvala JE, Pattison JR, Griffiths PD, Shaub BD. Principles and Practice of Clinical Virology, 5th edition. Wiley, Chichester, 2004.

REFERENCES

Gupta R, Warren T, Wald A. Genital herpes. Lancet, 2007, 370: 2127–2137.

Kimberlin DW, Rouse DJ. Genital herpes. N Engl J Med, 2004, 350: 1970–1977.

Sen P, Barton SE. Genital herpes and its management. BMJ, 2007, 334: 1048–1052.

WEB SITES

All the Virology on the WWW Website, developed and maintained by Dr David Sander, Tulane University: http://www.virology.net/garryfavweb12.html#Herpe

Centre for Infections, Health Protection Agency, HPA Copyright, 2008: http://www.hpa.org.uk

Centers for Disease Control and Prevention, Sexually Transmitted Diseases, Atlanta GA, USA: http://www.cdc.gov/std/Herpes/default.htm

Website of Derek Wong, a medical virologist working in Hong Kong: http://virology-online.com/viruses/HSV.htm

MULTIPLE CHOICE QUESTIONS

The questions should be answered either by selecting True (T) or False (F) for each answer statement, or by selecting the answer statements which best answer the question. Answers can be found in the back of the book.

1. **Which of the following viruses belong to the herpesvirus family?**
   A. Coxsackie B virus.
   B. Cytomegalovirus.
   C. Epstein-Barr virus.
   D. Varicella-zoster virus.
   E. Variola major virus.

2. **All herpesviruses undergo the phenomenon of latency. Which of the following statements regarding latency are correct?**
   A. Viral nucleic acid is found in the cell nucleus.
   B. Viral capsid proteins are found on the plasma membrane of the cell.
   C. There is continuous release of new viral particles from the cell.
   D. Viral particles are located in the cell cytoplasm but are not released from the cell.
   E. The presence of latent virus inhibits the specialized functioning of the cell.

3. **Which of the following are recognized routes of spread of herpes simplex virus?**
   A. Air-borne.
   B. Direct contact.
   C. Mother to baby.
   D. Sexual.
   E. Water-borne.

4. **Viruses that undergo latency may cause both primary and secondary (or reactivated) infections. Which of the following statements regarding the clinical manifestations of herpes simplex virus infection are true?**
   A. Inguinal lymphadenopathy is a common feature of recurrent genital herpes.
   B. Infection in neonates who acquire infection from mothers undergoing a primary genital infection late in pregnancy is usually confined to the skin and mucous membranes.
   C. Person to person spread of genital herpes most likely arises from a source patient with asymptomatic genital shedding of virus.
   D. Primary infection may give rise to systemic, as well as local, symptomatology.
   E. The majority of primary infections in the genital area result in severe symptomatic disease.

5. **Recognized complications of genital HSV infection include which of the following?**
   A. Eczema.
   B. Increased risk of acquisition of HIV infection.
   C. Cholecystitis.
   D. Meningitis.
   E. Urinary retention.

6. **Which of the following components of the immune system is the most important in controlling the outcome of reactivated genital HSV infection?**
   A. B cells.
   B. Complement.
   C. Macrophages.
   D. NK cells.
   E. T cells.

7. **Regarding the diagnosis of genital HSV infection, which of the following statements are correct?**
   A. Genome detection assays demonstrating the presence of HSV DNA in a vesicle swab are the most sensitive assays available.
   B. The clinical features of recurrent genital herpes are so distinctive that laboratory confirmation is not usually required.
   C. The diagnosis can be made by demonstrating the presence of antibodies specific for HSV-2 in a serum sample.
   D. The diagnosis can be made by immunofluorescent antigen detection using cells scraped from the base of a genital ulcer.
   E. The virus can be isolated in tissue culture from vesicle fluid.

8. **Which of the following drugs are routinely used to treat genital HSV infection?**
   A. Amantadine.
   B. Azidothymidine.
   C. Famciclovir.
   D. Ganciclovir.
   E. Valaciclovir.

9. **Resistance to aciclovir can arise in HSV infection through which of the following mechanisms?**
   A. Acquisition of an aciclovir efflux pump.
   B. Deletion of the viral thymidine kinase gene.
   C. Mutations in the UL97 gene.
   D. Mutations in the viral DNA polymerase gene.

Case 16

Histoplasma capsulatum

A 60-year-old resident of Louisville (Ohio) had suffered from **rheumatoid arthritis** for 9 years and was currently being treated with 10 mg of methotrexate weekly and 8 mg of methylprednisolone daily followed by monthly injections of 3 mg kg^{-1} infliximab **monoclonal antibody**. Ten weeks after the start of infliximab, he felt severely ill and was hospitalized with the symptoms of **dyspnea** and cough, quickly followed by respiratory failure, requiring mechanical ventilation. A chest radiograph revealed bilateral nodular infiltrates (Figure 1). **Bronchoalveolar lavage** fluid contained yeast forms resembling *Histoplasma capsulatum*. Laboratory tests showed normal blood cell counts, but positive *Histoplasma* urine antigen (10.3 U, normal levels <1 U). The findings were confirmed by yeast cell culture and complement fixation titers 1:2048 to the mycelial M antigen and 1:256 to the yeast Y antigen (normal levels < 1:8). The diagnosis of histoplasmosis was further confirmed by immunodiffusion and the patient was given antifungal drugs amphotericin B lipid complex 5 mg kg^{-1} per day for 11 days, followed by itraconazole 200 mg per day for 2 months. Therapy resulted in improvement of the respiratory function, although the patient required ventilatory support throughout the treatment.

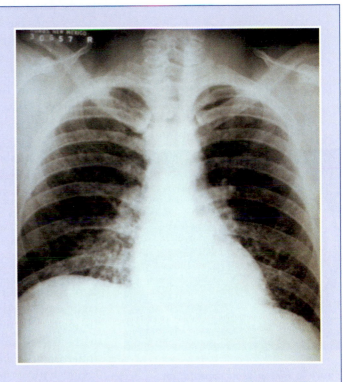

Figure 1. Chest radiograph of the patient infected with *H. capsulatum*, revealing bilateral nodular infiltrates.

1. What is the causative agent, how does it enter the body and how does it spread a) within the body and b) from person to person?

Causative agent

Histoplasma capsulatum causes a systemic endemic **mycosis** called histoplasmosis (sometimes called Darling's disease). The genus *Histoplasma* (Ajellomyces) from the family Onygenales contains one species, *Histoplasma capsulatum*. There are three varieties: *H. capsulatum* var. *capsulatum*, which causes the common histoplasmosis, *H. capsulatum* var. *duboisii*, a cause of African histoplasmosis (histoplasmosis duboisii), and *H. capsulatum* var. *farciminosum*, which causes **lymphangitis** in horses. Some *Histoplasma* isolates may resemble species of *Sepedonium* and *Chrysosporium*.

H. capsulatum is a thermally dimorphic ascomycete, which means that it can survive at two different temperatures. At ambient temperatures below

30°C *H. capsulatum* remains in a saprophytic mycelial mold form, but at mammalian body temperature, (37°C) it grows as a parasitic yeast. The two varieties of *H. capsulatum* (*capsulatum* and *duboisii*) that infect humans are similar in **saprophytic** mold form but differ in their parasitic tissue morphology (see later).

The saprophytic mycelial growth of *H. capsulatum* requires an acidic damp soil environment with high organic content. This is provided by bird droppings, particularly those of chickens and starlings, or excrement of bats. The fungus has been found in poultry house litter, caves, areas harboring bats, and in bird roosts. Birds cannot be infected by *Histoplasma* and do not transmit the disease, but their excretions enrich the soil and support the growth of the fungal mycelium. In contrast, bats can become infected, and they transmit the fungus through droppings. Contaminated soil is the common natural habitat for *Histoplasma* and it remains potentially infectious for years.

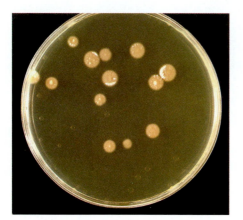

Figure 2. *Histoplasma capsulatum* colonies growing at 25°C.

At 25°C on Sabouraud dextrose agar (SDA) or brain heart infusion agar (BHIA) supplemented with 5–10% sheep blood *H. capsulatum* grows slowly into granular suede-like to cottony colonies, brown with a pale yellow-brown or yellow-orange reverse (Figure 2). Initially the colonies are white, but then become buff-brown with age. They can also be glabrous or verrucose, sometimes with a red pigmented strain. The colonies are not sensitive to cycloheximide in the culture media. When observed under the microscope hyphae are septate and hyaline. *H. capsulatum* produces hyphae-like short, hyaline, undifferentiated conidiophores, which arise at right angles to the parent hyphae. Macroconidia appear as large (8–14 μm in diameter), thick-walled, round, unicellular, hyaline, and tuberculate with finger-like projections on the surface (Figure 3A). Microconidia (microaleurioconidia) are small (2–4 μm in diameter), unicellular, hyaline, round with a smooth or rough wall, and borne on short branches or directly on the sides of the hyphae.

However, at 37°C the fungal morphology of *H. capsulatum* is totally different: numerous small round to oval budding yeast-like cells, 2–4 μm in size can be observed under the microscope (Figure 3B). Colonies are creamy, smooth, moist, white, and yeast-like (Figure 4). This change in morphology under temperature-controlled regulation is used as a diagnostic test for *Histoplasma* (see Section 4).

Figure 3. Two forms of *Histoplasma capsulatum* **demonstrating features of a thermal dimorph.** In nature at about 25°C it grows as a mycelial filamentous form with macroconidia and smaller microconidia (A). At body temperature of 37°C it grows as a yeast (B).

Entry into the body

Conidia and mycelial fragments of *H. capsulatum* from contaminated disrupted soil can be air-borne, are found in aerosols, and can be inhaled. Once inhaled they settle in bronchioles and alveolar spaces where they encounter phagocytic macrophages. Binding of the microconidia is thought to be through their surface expression of heat-shock protein 60 (HSP60), which presumably binds to **Toll-like receptor** 4 (TLR4) on the macrophage. Other surface molecules important for the binding and internalization of microconidia include the integrins **CD11/CD18**. **Opsonization** via antibodies and complement is also important for uptake of the fungus. Conversion from the mycelial to the pathogenic yeast phase is critical for infectivity of *H. capsulatum* and occurs intracellularly inside the macrophages. The stimuli that drive the conversion are not completely understood, although the temperature change appears to be a major factor. Upon conversion the intracellular budding yeasts enlarge and reach approximately 3 μm in diameter.

Figure 4. Culture of *H. capsulatum* on BHIA with 10% sheep blood, incubated at 37°C.

It must be stated that cultures of *H. capsulatum* represent a severe biohazard to laboratory staff and must be handled with extreme caution and under appropriate conditions in a safety cabinet.

Spread within the body

Following the initial infection *H. capsulatum* may spread – carried via pulmonary macrophages in draining lymphatics and the bloodstream – to many organs that contain mononuclear phagocytes. These include the liver and spleen and regional lymph nodes.

Infection with *H. capsulatum* var. *duboisii* rarely involves the lungs but mostly causes cutaneous histoplasmosis. It can also infect the liver, lymphatic system, and subcutaneous and bony tissues.

H. capsulatum var. *duboisii* usually grows as a large yeast (7–15 μm in diameter) within the cytoplasm of **histiocytes** and multinucleate giant cells, but sometimes can appear as small yeast cells similar to those of var. *capsulatum*. The yeasts may form rudimentary pseudohyphae consisting of four or five cells and aggregates, which can be observed within giant cells and extracellularly following **necrosis** of the host tissue. The infection presents as nodular and ulcerative cutaneous and osteolytic bone lesions, disseminated or localized.

Spread from person to person

This organism is generally not spread from person to person.

Epidemiology

H. capsulatum is found in temperate climates, predominantly in river valleys between latitudes 45° north and 30° south in North and Central America. It is **endemic** to the Ohio, Missouri, and Mississippi River valleys in the United States (Figure 5). In these regions of the USA, approximately 90% of residents have been exposed to the fungus, and quite likely on a continuous basis. Africa, Australia, and parts of East Asia, in particular India and Malaysia, are also endemic regions. However, it is believed that the fungus can grow anywhere under appropriate soil conditions. *H. capsulatum* var. *duboisii* is endemic in Central and West Africa and in the island of Madagascar.

Figure 5. Distribution of histoplasmosis in the United States, with dark red areas indicating the highest incidence of the disease.

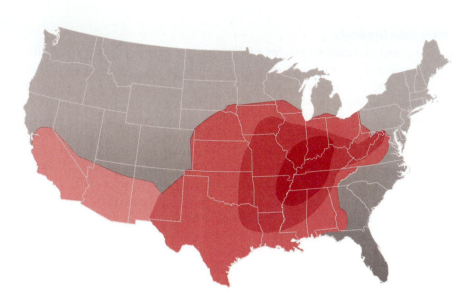

Approximately 250 000 individuals are infected annually with *H. capsulatum* in the USA and 500 000 worldwide. Of these, 50 000–200 000 develop symptoms of histoplasmosis and 1500–4000 require hospitalization. Infection is more prevalent in immunocompromised individuals (see Section 2) and therefore *Histoplasma* is regarded as an opportunist.

2. What is the host response to the infection and what is the disease pathogenesis?

Innate immunity

As well as being the site of the infection, phagocytes, especially macrophages are the main cells responsible for removing the fungus (Figure 6). After ingestion, *H. capsulatum* yeasts reside in phagocytic **phagosomes** and replicate there approximately every 15–18 hours. One mechanism used by the fungus to reduce intracellular killing is by inhibiting phagolysosomal fusion and thus preventing exposure to the lysosomal hydrolytic enzymes. The fungus also increases the phagosomal pH to 6.5, which potentially enhances the availability of iron it needs for growth. Decrease in free iron via constitutive and inducible iron sequestration represents an important host antimicrobial defense.

Human macrophages respond to ingestion of *H. capsulatum* by a vigorous oxidative burst mediated by NADPH oxidase, and the release of nitric oxide (NO) and other nitrogen intermediates, yet the fungus is still able to survive this response and even replicate in the presence of **reactive oxygen species** (**ROS**). It eventually lyses the alveolar macrophage and the released yeasts are ingested by other resident macrophages and polymorphonuclear neutrophils (PMNs) newly recruited to the site of infection. In immunocompromised individuals, the infection/lytic cycle may be repeated several times.

Figure 6. Two macrophages with engulfed small yeast-like fungus cells of *Histoplasma capsulatum* var. *capsulatum* (hematoxylin & eosin stain).

Intracellular growth of the fungus is inhibited by **interleukin** (**IL**)-3, granulocyte-macrophage colony stimulating factor (**GM-CSF**) and macrophage-CSF (M-CSF), although the mechanism of their action remains unknown.

PMNs from immunocompetent hosts can inhibit the growth of *H. capsulatum* through release of **defensins**. With the development of immunity mediated by neutrophils and macrophages, yeast growth ceases within 1–2 weeks after exposure.

Antibody-mediated immunity

Humoral immunity has little or no role in host defense against the fungus. Although exposure to *H. capsulatum* does induce an antibody response, and the **IgG** fraction of the sera from infected individuals contains complement-fixing and precipitating antibodies, these paradoxically are associated with progressive disease. Passive transfer of immune serum has not mediated protection from the fungus and B-lymphocyte-deficient mice were not susceptible to infection. No correlation was found between the titers of antibodies specific to *H. capsulatum* and immunity to the infection. However, recent studies have shown that a 17 kDa cell wall histone H2B-like protein, apparently restricted to *H. capsulatum*, represents a good target for the generation of protective antibodies.

Cell-mediated immunity

Of all the human mycoses, histoplasmosis illustrates best the importance of the T-mediated immune system in limiting the extent of *H. capsulatum* infection. Susceptibility to dissemination of *H. capsulatum* is associated with T-cellular immunodeficiency. In immunocompetent individuals delayed-type hypersensitivity to histoplasma is observed 3–6 weeks after exposure to the fungus and in up to 85–90% of cases a positive response to skin antigen test for *Histoplasma* can be detected (see Section 4).

It is thought that *H. capsulatum*, like other foreign organisms, is taken up by immature **dendritic cells** and processed with antigens presented to T cells.

Specific T cells contribute to protective immunity to *H. capsulatum* mainly via production of **cytokines** that activate phagocytes, with **interferon-γ (IFN-γ)** and **tumor necrosis factor-α (TNF-α)** being particularly important. Deficiency in IFN-γ and TNF-α causes a decrease in NO production by phagocytic cells, which is essential for controlling *H. capsulatum* infection. IL-12 regulates the induction of IFN-γ, which is a critical **Th1** cytokine in primary host resistance. During the acute phase of the infection IL-12, TNF-α, and IFN-γ are released and enhance the influx of the myeloid cells into the lungs followed by T and B cells. GM-CSF is also important for the production of TNF-α, IFN-γ, and NO, and is responsible for down-regulation of **Th2** cytokines IL-4 and IL-10. High levels of Th2 cytokines weaken the efficiency of protective immunity by inducing inflammatory responses and chronic infection.

For secondary immune responses, TNF-α is critical in both pulmonary and disseminated infections. Its deficiency leads to the dramatic elevation of IL-4 and IL-10 in the lungs. GM-CSF also affects secondary immunity in the pulmonary infection, but not substantially. It seems that GM-CSF is required rather for optimization of the protective immune response to *H. capsulatum*.

In most immunocompetent individuals, activation of a protective T-lymphocyte-mediated immunity results in containment of the infection.

However, even in normal individuals, but particularly in immunodeficient hosts, the organism may not be completely eradicated and may persist as a latent infection. This can erupt as active disease at a later time when and if the host–pathogen balance is disrupted or host immune responses decline with age or through viral infection.

Pathogenesis

The chronic infection will lead to the inflammatory responses that can last over weeks to months and result in the development in the affected organs of calcified fibrinous granulomatous lesions with areas of caseous necrosis. *H. capsulatum* may remain latent in healed **granulomas** and recur, resulting in impairment of cell-mediated immunity.

3. What is the typical clinical presentation and what complications can occur?

In about 80–90% of infected immunocompetent individuals histoplasmosis is asymptomatic, subclinical (showing self-limiting influenza-like symptoms) or benign. Clinical symptoms of histoplasmosis develop mostly in immunocompromised individuals.

Acute pulmonary histoplasmosis

The onset of the disease in symptomatic cases develops 3–14 days after exposure to the fungus. Common symptoms include fever, headache, malaise, **myalgia**, abdominal pain, and chills. Individuals exposed to a large inoculum of fungus may develop severe dyspnea due to diffuse pulmonary involvement. Diffuse or localized **pneumonitis** may be severe enough to require ventilatory support (see the case). Weight loss, night sweats, and fatigue may persist for weeks after the acute symptoms resolve. A small number of patients may show rheumatological manifestations such as **erythema multiforme**, **arthritis**, and **erythema nodosum**, which can be of a diagnostic value. Sometimes pulmonary **auscultation** may detect **rales** or wheezes. Severe **hypoxemia** and associated acute respiratory distress syndrome only develop in patients inhaling a high inoculum. In a small group of patients (about 5%), **pericarditis** may be present. Approximately 10% of patients have asymptomatic pleural **effusions**. Hepatosplenomegaly is rare.

Lymphadenopathy reflected by enlarged hilar and mediastinal lymph nodes is present in 5–10% of patients. In some cases it can cause local obstructions such as superior vena cava (SVC) syndrome, which results from the compression of the SVC. Obstruction of venous drainage may lead to cerebral symptoms: headache, visual distortion, **tinnitus**, and altered consciousness. Compression of the pulmonary airway and circulation, or of trachea or bronchi presents as cough, hemoptysis, dyspnea, and/or chest pain. Compression of the esophagus with subsequent **dysphagia** is rare.

Occasionally in immunocompetent hosts infection with *H. capsulatum* may manifest as histoplasmoma with a single pulmonary parenchymal nodule appearing as a coin-like lesion on chest radiographs. They are formed as a result of **fibrosis** developing in response to the yeast antigens, which create fibrotic capsule thus enlarging the lesions. Histoplasmoma is usually asymptomatic.

Chronic pulmonary histoplasmosis

This form occurs mostly in patients with underlying pulmonary chronic lung disease such as emphysema and may mimic tuberculosis. Histoplasmosis may actually coexist with tuberculosis, actinomycosis, other mycoses, and **sarcoidosis**. In individuals with underlying pulmonary disease chronic cavitary histoplasmosis is the condition most commonly observed, with a frequency of 1 per 100 000 persons per year in endemic areas. Approximately 90% of patients develop cavities that if enlarged may lead to necrosis. The main symptoms include cough, weight loss, fever, and malaise. In case of **cavitations**, additional symptoms of hemoptysis, sputum production, and increasing dyspnea develop. Untreated cases may result in progressive pulmonary fibrosis and subsequently in respiratory and cardiac failure and recurrent infections.

Chronic fungal infection leads to the development of well-organized solitary pulmonary nodules with a circumferential rim of calcification, which allows their identification on chest X-ray (see Figure 1). Fungi within these nodules are usually dead. Older lesions show well-developed granuloma and have a central area of **caseation** resembling tuberculosis. These centers are usually occupied by the fungi.

Progressive disseminated histoplasmosis

Progressive disseminated histoplasmosis occurs in 1:2000 cases of infected immunocompetent adults and in 4–27% of infected immunosuppressed individuals with impaired cellular immunity, children, or the elderly. Symptoms vary depending on duration of illness. There is an acute and subacute form of progressive disseminated histoplasmosis.

The acute form is associated with fever, worsening cough, weight loss, malaise, and dyspnea. Dissemination sites include the central nervous system (CNS; 5–20% of patients) and hematopoietic systems, liver, spleen, and eye. In the bone marrow **anemia**, **leukopenia**, and **thrombocytopenia** are detected. Adrenal glands undergo enlargement, and **Addison's disease** may develop as well. Genitourinary tract symptoms include **hydronephrosis**, bladder ulcers, penile ulcers, and **prostatitis**. Oral ulcers and small bowel micro and macro ulcers develop in the gastrointestinal tract. Skin symptoms include **papular** to nodular **rash**, while ocular problems include **uveitis** and choroiditis. Among CNS-related conditions chronic **meningitis** and **cerebritis** should be noted. The acute form, if untreated, results in death within weeks. About 10% of individuals develop fatal hyperacute syndrome.

The subacute form of disseminated histoplasmosis presents with a wide spectrum of symptoms based on dissemination in the affected organs. Gastrointestinal involvement leads to diarrhea and abdominal pain, while cardiac dissemination results in valvular disease, cardiac insufficiency, dyspnea, peripheral edema, angina, and fever. Pericarditis may develop, often with pleural effusions. CNS involvement is reflected by headaches, visual and gait disturbances, confusion, **seizures**, altered consciousness, and neck stiffness or pain. If untreated the subacute form is fatal within 2–24 months. Approximately 5–10% of patients, treated or not, develop adrenal insufficiency.

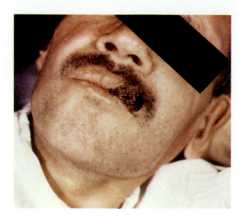

Figure 7. Mouth lesions due to *Histoplasma capsulatum* **dissemination into oral cavity.**

In 50–60% of patients with the chronic form of disseminated histoplasmosis constitutional symptoms such as mouth and gum pain are present due to mucosal ulcers. Extensive ulceration including the oropharyngeal area, buccal mucosa, tongue, gingiva, and larynx is characteristic of chronic progressive disseminated histoplasmosis (Figure 7). Rare isolated lesions can be found in individuals who are immunocompetent.

The patients susceptible to the development of disseminated and potentially fatal histoplasmosis are mostly immunocompromised individuals, particularly HIV-infected patients in the endemic areas (see Section 1), children less than 2 years old, elderly persons, and people exposed to a very large inoculum.

Presumed ocular histoplasmosis syndrome

This syndrome is only found in 1–10% of individuals who live in endemic areas. Ocular histoplasmosis damages the retina of the eyes and leaves scar tissue leading to retinal leakage and a loss of vision similar to macular degeneration. Atrophic scars with lymphocytic cell infiltration (histo spots) may be seen posterior to the equator of the eye. The condition is bilateral in approximately 10% of patients. In the case of macular scarring, retinal hemorrhage, detachment or edema may be present.

Clinical manifestations of histoplasmosis are more frequent in males than in females, with a ratio of 4:1.

4. How is the disease diagnosed and what is the differential diagnosis?

A social and occupational history is of great help for the initial evaluation. This includes travel to or residence in an endemic area, interaction with birds or bats, either recent or past. Any possible immunodeficiency in patients due to a medical condition or drug treatment must be taken into account (see case).

Imaging

A chest X-ray may be taken to help in the diagnosis of the kind of histoplasmosis. It usually does not show any irregularities in *acute pulmonary histoplasmosis*, although sometimes enlarged hilar and mediastinal nodes, patchy infiltrates in the lower lung fields may be detected. However, diffuse pulmonary involvement caused by exposure to a large inoculum may present with a reticular nodular or **miliary** pattern. Pleural effusions are found in fewer than 10% of uncomplicated cases. Cavitations are very rare. Histoplasmomas can be detected as lesions and residual nodules, 1–4 cm in diameter.

In *chronic pulmonary histoplasmosis* hilar lymphadenopathy is rare, although some calcified nodules from prior healed infections may be discovered. In contrast to acute pulmonary histoplasmosis, here the cavitations are found in almost 90% of patients, and are located in the upper lobes of the lungs. Often **emphysematous** changes are present as well as fibrotic scarring, particularly in long-standing cases.

Around 50% of patients with *acute progressive disseminated histoplasmosis* have hilar lymphadenopathy with diffuse nodular infiltrates. At the onset of the disease no chest radiography changes are detected in 33% of patients, but as the disease progresses pulmonary involvement becomes gradually more apparent.

In contrast, *chronic progressive disseminated histoplasmosis* does not present with chest radiography changes.

Head **CT scanning** helps to detect cerebral histoplasmosis before performing a lumbar puncture. Abdominal CT can confirm *subacute progressive disseminated histoplasmosis*, which leads to adrenal infection in 80% of patients and presents with bilateral adrenal enlargement on the scan.

Respiratory functional tests

Pulmonary functional tests help to characterize the scale of pulmonary involvement. They include evaluation of the restrictive defect, detection of a small airway obstruction, diffusion impairment, and hypoxemia. The functional tests are also used for monitoring the progression of pulmonary disease in patients with *chronic pulmonary histoplasmosis*.

Laboratory tests

Specimens obtained for analysis include: blood, sputum, bronchial lavage, and cerebrospinal fluid (CSF), by lumbar puncture, if CNS involvement is suspected. Tissue biopsies may be taken of pulmonary lesions and lymph nodes by **bronchoscopy** or thoracoscopy. Biopsies can be also taken from oropharyngeal ulcers.

Serology

In immunocompetent patients antibodies to *H. capsulatum* can be detected by immunodiffusion and/or complement fixation tests, but in immunosuppressed patients antibody measurement is unreliable since 20–50% of patients usually test negative. False-negative results can also be detected iduring the first 2 months after exposure to the fungus.

The titers of complement-fixing (CF) antibodies to yeast and mycelial-phase antigens (Y and M, see the case on page 197) are considered positive at dilutions greater than 1:8. A titer of 1:32 or more suggests active histoplasmosis infection. CF antibody to Y antigen is detectable in primary infection, while CF antibody to M antigen appears later. Cross-reactivity with *Blastomyces dermatitidis* and *Coccidioides immitis* must be ruled out for differential diagnosis. Positive results are usually found in 5–15% of cases of acute pulmonary infection 3 weeks after exposure and in 75–95% of cases at 6 weeks. The test usually turns negative in the course of months with resolution of infection. However, in chronic infection the CF test may remain positive for a long time in 70–90% of patients with chronic pulmonary histoplasmosis or chronic progressive disseminated histoplasmosis.

An immunoprecipitation test is aimed at the detection of antibodies to two fungal glycoproteins by precipitation bands called H and M. H band detects antibody to β-glucosidase, whilse M band detects antibody to a **catalase**, both enzymes secreted by the pathogen. The M band is usually detected in 50–80%, while the H band is found in only 10–20% of patients. The level of anti-M antibody remains high for years, and anti-H antibody becomes undetectable within 6 months of the resolution of infection. However, a positive test for anti-H antibody is more specific for active histoplasmosis.

Antigen detection in urine and other fluids

Antigen detection by **ELISA** is useful in individuals who are immunocompromised when antibody production may be impaired. This method is more sensitive for urine samples compared with serum, plasma, CSF, and bronchoalveolar lavage fluid.

Detection of the capsular antigen of *H. capsulatum* in urine or serum is a very powerful tool for the diagnosis of the disseminated forms of histoplasmosis. In cases of acute progressive disseminated histoplasmosis detection rates are 50% with serum assay and 90% with urine assay. Lower detection rates are observed in acute or chronic pulmonary histoplasmosis.

Cross-reactivity with *Blastomyces* and *Coccidioides* species causes false-positive results.

Some patients with acute histoplasmosis may have high serum levels of angiotensin-converting enzyme. This may cause a diagnostic confusion with sarcoidosis, particularly if the patient with histoplasmosis also has hilar **adenopathy**.

Histology and cytology

In rare cases, a tissue biopsy may reveal the presence of large yeast cells with a false capsule. Special stains may reveal budding yeast in areas of necrosis from histoplasmomas and calcified lymph nodes, but the organisms can be mistaken for *Pneumocystis carinii* and other fungal organisms.

Yeast is almost undetectable in circulating neutrophils and monocytes using Wright-Giemsa staining. Therefore direct microscopic evidence of histoplasma infection showing characteristic yeast-like cells from any specimen taken from the patient is considered significant diagnostic proof.

Cell culture

Isolation of *H. capsulatum* in culture remains the gold standard method for diagnosis of histoplasmosis. However, the fungus takes up to 4 weeks to grow *in vitro*. Specimens are inoculated onto SDA or BHIA supplemented with 5–10% sheep blood and observed for morphology of the growing colonies (see Section 1). A positive culture of diagnostic value has to show conversion of the mold form to the yeast phase by growth at 37°C. Sputum culture results are usually positive in 60% of patients with chronic pulmonary histoplasmosis and in approximately 10–15% of patients with acute pulmonary histoplasmosis. Blood cultures are positive for histoplasma in 50–90% of patients with acute progressive disseminated histoplasmosis. Cultures of CSF are positive in 30–60% of patients with histoplasmal meningitis.

Skin tests

Positive histoplasmin skin tests were demonstrated by almost 80% of the people living in areas endemic for *H. capsulatum* and therefore their diagnostic value is very low. For histoplasmin skin test both H and M antigens are usually used to induce complement-fixing antibodies or precipitins (see above).

PCR

A diagnostic **polymerase chain reaction (PCR)** assay is not routinely used in parasitology and mycology laboratories, since mostly PCR protocols have not yet provided satisfactory sensitivity and specificity for fungal identification. However, PCR is used to detect *H. capsulatum* var. *capsulatum* in contaminated soil. No PCR has been developed for *H. capsulatum* var. *duboisii*. Development of a reliable PCR assay for histoplasmosis in various specimens is under way and will be particularly useful in areas where *H. capsulatum* is endemic and where detection of antigens is impossible such as French Guiana.

Differential diagnosis

A differential diagnosis for *H. capsulatum* with *Chrysosporium* and *Sepedonium* species that also inhabit soil and plant material should be carried out.

The conidia of *Chrysosporium parvum* morphologically resemble those of *H. capsulatum*. *Chrysosporium* species are also dimorphic, but this is a false dimorphism since they do not convert from mold to yeast at 37°C. *Chrysosporium* species have the antigens secreted, and found in fungal extracts, cross-reacting with those used for the diagnostic kits of *H. capsulatum*, which results in nonspecific precipitaion bands (see above).

Macroconidia of *Sepedonium* species also resemble those of *Histoplasma*.

In the case of *H. capsulatum* var. *duboisii* differential diagnosis must be made from *B. dermatitidis* since they both occur in Africa.

The most reliable methods for differential diagnosis are: (a) only *H. capsulatum* shows conversion to a yeast phase at 37°C; (b) detection with molecular biology methods such as DNA hybridization probe; (c) morphological observation – *Chrysosporium* does not produce tuberculate macroconidia and *Sepedonium* is not dimorphic and does not produce microconidia.

5. How is the disease managed and prevented?

As discussed in Section 3, in immunocompetent individuals infection with *Histoplasma* species in most cases is self-limiting and does not require therapy. Therapeutic intervention is recommended in cases of prolonged infection, systemic infection or if patients are immunocompromised.

Management

The main approaches for the treatment of different forms of histoplasmosis are outlined below.

Acute pulmonary histoplasmosis: Mild symptoms in immunocompetent individuals need monitoring only. In patients with prolonged symptoms or extensive pulmonary involvement, therapy is recommended.

Chronic pulmonary histoplasmosis: No treatment is required for asymptomatic immunocompetent individuals without serious underlying disease. Mild interstitial pneumonitis and/or thin-walled (<3 mm) cavities with serial chest radiographs require monitoring for 2–4 months. Medical treatment is recommended if the lesions are persistent or if patients develop cavities with thick walls (>3 mm). Treatment is indicated for all immunocompromised patients.

Progressive disseminated histoplasmosis: All patients with the symptoms of meningitis require medical therapy. In the case of severe CNS infection, intravenous antifungal therapy should be supplemented with **intrathecal** or intraventricular injections.

If the presence of pleural fluid is causing respiratory and hemodynamic distress, immediate procedural interventions are required such as thoracentesis or pericardiocentesis with severe pleural effusions and pericardial tamponade. In some cases pericardial window placement is recommended.

Cutaneous and rheumatologic lesions are usually self-limiting and medical intervention is recommended only for prolonged course of the disease episodes or in immunocompromised patients.

Maculopathy in ocular histoplasmosis is treated with steroids.

Surgery may be required in some cases for resection of pulmonary cavitation lesions, repair of infected heart valves, and **aneurysms**.

Antifungal drugs

Severe cases of acute histoplasmosis and all cases of symptomatic chronic and disseminated disease require treatmnent with antifungal drugs. The accepted scheme includes treatment with amphotericin B, followed by oral itraconazole, although in mild cases itraconazole only can be used. Second choice drugs include fluconazole, a less active compound, and ketoconazole. Imidazole, a broad-spectrum antifungal agent, can also be prescribed.

Amphotericin B (AmB) is used to treat acute pulmonary histoplasmosis, chronic pulmonary histoplasmosis, all forms of progressive disseminated histoplasmosis, meningitis, and endovascular histoplasmosis and also as a maintenance therapy for acute progressive disseminated histoplasmosis. In adults as well as in children $0.7–1$ mg kg^{-1} per day of AmB or up to a total dose of 35 mg kg^{-1} is used intravenously. A minimum total dose should be no less than 2 g. In case of underlying severe immunodeficiency (such as AIDS) when a life-long antifungal maintenance therapy is required to treat acute progressive disseminated histoplasmosis, the induction dose of $0.7–1$ mg kg^{-1} per day AmB to a total of $20–25$ mg kg^{-1} is used, followed by a maintenance therapy of 50 mg once per week.

Itraconazole is a fungistatic drug that suppresses fungal cell growth. Indications include mildly symptomatic or prolonged acute pulmonary histoplasmosis with cutaneous or rheumatologic manifestations in immunocompromised patients or those with a prolonged courses of illness. It can be used as an alternative to AmB in the treatment of chronic pulmonary histoplasmosis or chronic and subacute progressive disseminated histoplasmosis, as well as for alternate maintenance therapy after induction with AmB in acute progressive disseminated histoplasmosis.

Ketoconazole has fungistatic activity while imidazole is fungicidal. Indications for the use and the doses are similar to itraconazole. They can be used as an alternative to AmB in the treatment of chronic pulmonary histoplasmosis or chronic and subacute progressive disseminated histoplasmosis. However, since these drugs do not cross the blood–brain barrier, they are not effective in the treatment of fungal meningitis.

Anti-inflammatory drugs

Apart from the fungicidal and fungistatic drugs, nonsteroidal anti-inflammatory drugs (NSAIDs) are used with analgesic, anti-inflammatory, and antipyretic action. They are particularly recommended for patients with pericarditis. Among the most used is ibuprofen given orally at 400 mg every 4–6 hours, 600 mg every 6 hours, or 800 mg every 8 hours, with a maximal dose of 3.2 g per day while symptoms persist. The pediatric dose is 20–70 mg kg^{-1} per day with a low induction dose and maximal dose of 2.4 g per day. Usual NSAIDs contraindications apply such as peptic ulcer disease, recent or possible gastrointestinal bleeding or perforation, and renal insufficiency.

Among other anti-inflammatory drugs corticosteroids, mostly prednisone, can be used to decrease hypersensitivity to *Histoplasma*. High-dose steroids are used in patients with extensive maculopathy. Usually they are administered orally at 60–80 mg per day in adults and 4–5 mg m^{-2} per day in children or, alternatively, 0.05–2 mg kg^{-1} divided b.i.d./q.i.d. Administration must be tapered over 2 weeks as symptoms resolve since abrupt discontinuation of glucocorticoids may cause adrenal crisis. Usual contraindications include viral infection, peptic ulcer disease, hepatic dysfunction, connective tissue infections, and fungal or tubercular skin infections.

Since IFN-γ augments antifungal activity of PMNs and macrophages it is considered for immunotherapy of *H. capsulatum*.

Prognosis

Prognosis for a complete recovery is good for acute pulmonary histoplasmosis. The relapse rate in chronic pulmonary histoplasmosis is 20%, while in treated acute progressive disseminated histoplasmosis it is 50%. If lifelong antifungal maintenance is administered the relapse rate drops to 10–20%. The course of chronic progressive disseminated histoplasmosis can last for years with long asymptomatic periods.

Fatal outcome of acute progressive disseminated histoplasmosis and subacute progressive disseminated histoplasmosis is imminent without treatment.

Full recovery from histoplasmal meningitis with therapy is 50%.

Prevention

Prevention includes chemical disinfection and respiratory barrier protection in high-risk areas or during high-risk activities. Since the outbreaks of histoplasmosis are associated with disruption of the soil in endemic areas, decontamination of the infected soil with a 3% formalin solution is highly recommended.

The NIOSH/NCID document 'Histoplasmosis: Protecting Workers at Risk' contains information on safety procedures and personal protective equipment aimed at reducing the risk of infection.

Individuals residing in or traveling to endemic areas, particularly those with a history of immunodeficiency (AIDS, lymphoma, immunosuppressive treatment), must be educated/briefed about exposure risks. The elderly and children are also at risk.

In some areas special precautions are taken by placing warning signs around particularly contaminated soil.

Vaccines

Although it has been shown that there is no immunity to *Histoplasma* by prior infection and no effective vaccine is currently available, recent studies have indicated that vaccination with highly immunogenic HSP60 protein protects mice from lethal and sublethal histoplasmosis by stimulating **CD4+** T lymphocytes. More perspectives for the development of a vaccine are associated with targeting the histone H2B-like protein on the surface of *H. capsulatum* (see Section 2).

SUMMARY

1. What is the causative agent, how does it enter the body and how does it spread a) within the body and b) from person to person?

- The genus *Histoplasma* contains one species, *Histoplasma capsulatum*, with three varieties: *H. capsulatum* var. *capsulatum*, which causes the common histoplasmosis, *H. capsulatum* var. *duboisii*, which causes African histoplasmosis, and *H. capsulatum* var. *farciminosum*, which causes lymphangitis in horses.

- *H. capsulatum* is thermally dimorphic and can survive at two different temperatures. Below 30°C *H. capsulatum* remains in a saprophytic mycelial mold form, but at an average mammalian body temperature 37°C it grows as a parasitic yeast.

- For the saprophytic mycelial growth *H. capsulatum* requires an acidic damp soil with high organic content provided mostly by bird droppings or excrement of bats. Contaminated soil remains potentially infectious for years. Bats can become infected, and they transmit the fungus through droppings.

- In a saprophytic form at 25°C *H. capsulatum* colonies grow slowly into granular suede-like to cottony colonies brown with a pale yellow-brown or yellow-orange reverse. Macroconidia are large and tuberculate, microconidia are small and round. At 37°C *H. capsulatum* grows with a different morphology: the colonies are creamy, smooth, white and yeast-like.

- In the lungs inhaled mycelial fragments and microconidia of *H. capsulatum* are ingested by resident phagocytes. Conversion from the mycelial to the pathogenic yeast phase occurs inside the macrophages, mostly due to the temperature change.

- *H. capsulatum* can bind directly to the CD11/CD18 integrins on macrophages supposedly by fungal surface HSP60. The internalization is facilitated by opsonization with antibody and/or complement.

- The initial pulmonary infection disseminates to many organs that contain mononuclear phagocytes and produces extrapulmonary manifestations in the liver and spleen, or in regional lymph nodes.

- Infection with *H. capsulatum* var. *duboisii* mostly involves cutaneous, subcutaneous, liver, lung, lymphatic, and bony tissues. It usually grows as a large yeast within multinuclear giant cells.

- *H. capsulatum* is endemic to the Ohio, Missouri, and Mississippi River valleys in the United States, where approximately 90% of residents are repeatedly exposed to the fungus; and also Africa, Australia, India, and Malaysia. Approximately 250 000 individuals are infected annually with *H. capsulatum* in the USA and 500 000 worldwide. Of these, 50 000–200 000 develop symptoms of histoplasmosis and 1500–4000 require hospitalization.

2. What is the host response to the infection and what is the disease pathogenesis?

- After they are ingested *H. capsulatum* yeasts reside in phagosomes and replicate approximately every 15–18 hours. The yeasts inhibit phagolysosomal fusion, prevent exposure to the lysosomal hydrolytic enzymes, block accumulation of vacuolar ATPase, and increase phagosomal pH to 6.5.

- In response to ingestion of *H. capsulatum* macrophages enhance oxidative burst and the release of nitrogen intermediates. The yeasts are ingested by newly recruited PMNs, which release fungistatic defensins.

- T cells contribute to protective immunity via production of cytokines that activate phagocytes, particularly IFN-γ and TNF-α. GM-CSF is important for the production of TNF-α, IFN-γ, and NO, and down-regulation of Th2 cytokines IL-4 and IL-10, which are involved in pathogenesis.

- In most immunocompetent individuals, protective T-mediated immunity results in containment of the infection. However, even in normal individuals, and particularly in immunodeficient hosts, the organism may persist as a chronic infection and may lead to inflammatory responses and fibrinous granulomatous lesions with necrosis.

- Minimal cellular reaction to *H. capsulatum* var. *duboisii* was detected apart from large numbers of giant cells and macrophages. Neutrophils were present during the necrotic stage.

- Humoral immunity has little role in host defense against H. capsulatum, although exposure to it induces antibody response.

3. What is the typical clinical presentation and what complications can occur?

- In about 80–90% of infected immunocompetent individuals histoplasmosis is asymptomatic and subclinical with self-limiting influenza-like symptoms or benign.

- The symptoms of severe acute pulmonary syndrome are nonspecific: fever, chills, myalgias, cough, and chest pain. The syndrome can be mild (lasting 1–5 days) or severe (lasting 10–21 days).

- In a small group of the infected individuals (5–10%) histoplasmosis presents as chronic progressive lung disease, chronic cutaneous or systemic disease or an acute fatal systemic disease. The patients are mostly immunocompromised, children less than 2 years old, the elderly, and people exposed to a very large inoculum.

- Chronic pulmonary histoplasmosis occurs in patients with underlying pulmonary chronic lung disease (emphysema) and may mimic tuberculosis. The symptoms include cough, weight loss, fever, and malaise. In case of cavitations, hemoptysis, sputum production, and increasing dyspnea develop and lead to necrosis.

- The acute form of progressive disseminated histoplasmosis is associated with fever, worsening cough, weight loss, malaise, and dyspnea. The symptoms related to the organ of dissemination include: anemia, leukopenia and thrombocytopenia (bone marrow); hydronephrosis, bladder and penile ulcers, prostatitis (genitourinary tract); oral ulcers, small bowel ulcers (gastrointestinal tract); papular to nodular rash (skin); uveitis and choroiditis (eyes); chronic meningitis and cerebritis (CNS). The acute form, if untreated, results in death within weeks.

- The subacute form of progressive disseminated histoplasmosis presents with diarrhea and abdominal pain, valvular disease, cardiac insufficiency, dyspnea, peripheral edema, angina, fever, pericarditis, CNS involvement, adrenal insufficiency. If untreated the subacute form is fatal within 2–24 months.

- In 50–60% of patients with the chronic form of progressive disseminated histoplasmosis painful granulomatous mucosal ulcers appear as nodular ulcerative or vegetative lesions localized on the oral mucosa, tongue, palate, or lips.

- Presumed ocular histoplasmosis syndrome is only found in 1–10% of individuals who live in endemic areas. It damages the retina of the eyes and leaves scar tissue, leading to the retinal leakage and a loss of vision similar to macular degeneration.

4. How is the disease diagnosed and what is the differential diagnosis?

- A social and occupational history is important for the initial evaluation, such as travel to or residence in an endemic area, interaction with birds or bats, both recent or past, and possible immunodeficiency of the patient.

- Chronic pulmonary histoplasmosis can be detected through pulmonary auscultation as rales, wheezes with a history of underlying pneumonitis, consolidation, or cavitation.

- In acute progressive disseminated histoplasmosis the leading symptoms are hepatosplenomegaly and lymphadenopathy. CNS-related symptoms include a mass lesion, encephalopathy, and meningitis. Cutaneous lesions, ulcerations, or purpura may be present.

- Subacute progressive disseminated histoplasmosis presents as abdominal mass or intestinal ulcers and lesions. CNS dissemination may lead to mass lesions or meningismus, muscle weakness, ataxia, altered consciousness, or focal deficits. Endocarditis, murmurs, peripheral edema, petechiae indicate cardiac dissemination.

- Extensive ulceration in the oropharyngeal area, buccal mucosa, tongue, gingiva, and larynx is characteristic of chronic progressive disseminated histoplasmosis.

- In presumed ocular histoplasmosis syndrome atrophic scars with lymphocytic cell infiltration (histo spots) may be seen posterior to the equator of the eye. The condition is bilateral in approximately 10% of patients.

- Around 50% of patients with acute progressive disseminated histoplasmosis have hilar lymphadenopathy with diffuse nodular infiltrates. As the disease progresses pulmonary involvement becomes more apparent on radiography.

- Abdominal CT can confirm subacute progressive disseminated histoplasmosis in 80% of patients, presenting with bilateral adrenal enlargement on the scan.

- Direct microscopy evidence demonstrating characteristic yeast-like cells from any specimen taken from the patient is considered significant diagnostic proof.

- Using sputum and blood samples diagnosis is made on the basis of conversion of the mold form to the yeast phase by growth at 37°C. Sputum culture results are usually positive in 60% of patients with chronic pulmonary histoplasmosis and in approximately 10–15% of patients with acute pulmonary histoplasmosis. Blood cultures have positive fungal yields in 50–90% of patients with acute progressive disseminated histoplasmosis. Cultures of the lumbar punctures are positive in 30–60% of patients with histoplasmal meningitis.

- The titer of complement-fixing antibody is considered positive at dilutions greater than 1:8. Positive results are usually found in 5–15% of cases of acute pulmonary infection 3 weeks after exposure and in 75–95% of cases at 6 weeks. In chronic infection the test may remain positive for a long time in 70–90% of patients.

- The immunoprecipitation test detects two fungal glycoproteins, β-glucosidase (H) and catalase (M), secreted by the pathogen. Anti-M antibody is detected in 50–80% and anti-H antibody in only 10–20% of patients. The level of anti-M antibody remains high for years, and more specific anti-H antibody becomes undetectable within 6 months of the resolution of infection.

- Differential diagnosis needs to be carried out of *H. capsulatum* with *Chrysosporium* and *Sepedonium* species. In case of *H. capsulatum* var. *duboisii* differential diagnosis must be made with *B. dermatitidis*, since they both occur in Africa.

5. How is the disease managed and prevented?

- In most cases infection with *Histoplasma* species of immunocompetent individuals is self-limiting and does not require therapy, which is recommended in cases of prolonged infection, systemic infection or immunocompromised patients.

- All patients with progressive disseminated histoplasmosis and the symptoms of meningitis require medical therapy.

- Surgical resection of pulmonary cavitary lesions is required when repeated relapses or progressive disease occurs despite repeated intensive medical therapy.

- Severe cases of acute histoplasmosis and all cases of chronic and disseminated disease require treatment with antifungal drugs. The accepted scheme includes treatment with amphotericin B, followed by itraconazole. Second choice drugs are fluconazole and ketoconazole or a broad-spectrum antifungal drug, imidazole.

- Amphotericin B (AmB) in both adults and children is prescribed at 0.7–1 mg kg^{-1} per day or up to a total dose of 35 mg kg^{-1} is used intravenously. In case of severe immunodeficiency (such as AIDS) a life-long antifungal maintenance therapy is required of 50 mg once per week.

- Itraconazole is a fungistatic drug. The doses used are 200 mg perorally (maximum 400 mg per day) for 3–6 weeks in acute pulmonary histoplasmosis or for 6–12 months for all other manifestations of histoplasmosis. A dose of 200 mg of itraconazole is used for life-long maintenance therapy.

- Ketoconazole has fungistatic activity while imidazole is fungicidal. Indications for use and doses are similar to itraconazole. However, since these drugs do not cross the blood–brain barrier, they are not effective in the treatment of fungal meningitis.

- Nonsteroidal anti-inflammatory drugs (NSAIDs) such as ibuprofen are used with analgesic, anti-inflammatory, and antipyretic action, particularly for patients with pericarditis.

- Corticosteroids, mostly prednisone, can be used to decrease hypersensitivity to *Histoplasma*. High-dose steroids are used in patients with extensive maculopathy.

- Since IFN-γ augments antifungal activity of PMNs and macrophages it is considered for immunotherapy of *H. capsulatum*.

- Prognosis for a complete recovery is good for

acute pulmonary histoplasmosis. The relapse rate is 20% in chronic pulmonary histoplasmosis and 50% in treated acute progressive disseminated histoplasmosis, which drops to 10–20% in case of a life-long antifungal maintenance.

- Prevention includes chemical disinfection and respiratory barrier protection in high-risk areas or during high-risk activities. Individuals residing in or traveling to endemic areas, particularly those with a history of immunodeficiency, the elderly, and children, must be educated/briefed about exposure risks.

- Attempts to develop vaccines are currently under way targeting HSP60 protein and the histone H2B-like protein expressed on the surface of *H. capsulatum*.

FURTHER READING

Ajello L, Hay RJ. Medical Mycology, Vol 4, Topley & Wilson's Microbiology and Infectious Infections, 9th edition. Arnold, London, 1997.

Chandler FW, Kaplan W, Ajello L. A Colour Atlas and Textbook of the Histopathology of Mycotic Diseases. Wolfe Medical Publications, London, 1980.

Mandell GL, Bennett JE, Dolin R. Histoplasmosis. In: Principles and Practice of Infectious Diseases, 6th edition. Oxford, Churchill Livingstone, 2004.

Murray JF, Nadel JA. Histoplasmosis. In: Textbook of Respiratory Medicine, 4th edition (online). Saunders/Elsevier, 2001: 1045–1055.

Sutton DA, Fothergill AW, Rinaldi MG, editors. Guide to Clinically Significant Fungi, 1st edition. Williams & Wilkins, Baltimore, 1998.

REFERENCES

Antachopoulos C, Walsh TJ. New agents for invasive mycoses in children. Curr Opin Pediatr, 2005, 17: 78–87.

Bennish M, Radkowski MA, Ripon JW. Cavitation in acute histoplasmosis. Chest, 1983, 84: 496–497.

Buckley HR, Richardson MD, Evans EG, Wheat LJ. Immunodiagnosis of invasive fungal infection. J Med Vet Mycol, 1992, 30(Suppl 1): 249–260.

Darling ST. The morphology of the parasite (*Histoplasma capsulatum*) and the lesions of histoplasmosis, a fatal disease of tropical America. J Exp Med, 1909, 11: 515–530.

Deepe GS Jr. Immune response to early and late *Histoplasma capsulatum* infections. Curr Opin Microbiol, 2000, 3: 359–362.

Goldman M, Johnson PC, Sarosi GA. Fungal pneumonias. The endemic mycoses. Clin Chest Med, 1999, 20: 507–519.

Goodwin RA, Shapiro JL, Thurman GH, Thurman SS, Des Prez RM. Disseminated histoplasmosis: clinical and pathologic correlations. Medicine, 1980, 59: 1–33.

Guimaraes AJ, Pizzini CV, De Matos Guedes HL, et al. ELISA for early diagnosis of histoplasmosis. J Med Microbiol, 2004, 53: 509–514.

Kauffman CA. Histoplasmosis: a clinical and laboratory update. Clin Microbiol Rev, 2007, 20: 115–132.

Kurowski R, Ostapchuk M. Overview of histoplasmosis. Am Fam Physician, 2002, 66: 2247–2252.

Maubon D, Simon S, Aznar C. Histoplasmosis diagnosis using a polymerase chain reaction method. Application on human samples in French Guiana, South America. Diagn Microb Infect Dis, 2007, 58: 441–444.

Nosanchuk JD, Steenbergen JN, Shi L, Deepe GS, Casadevall A. Antibodies to a cell surface histone-like protein protect against *Histoplasma capsulatum*. J Clin Invest, 2003, 112: 1164–1175.

Picardi JL, Kauffman CA, Schwarz J, et al. Pericarditis caused by *Histoplasma capsulatum*. Am J Cardiol, 1976, 37: 82–88.

REFERENCES

Powell GM, Toledo TM. Pulmonary tuberculosis complicated by pulmonary histoplasmosis. Tex Med, 1970, 66: 46–51.

Salzman SH, Smith RL, Aranda CP. Histoplasmosis in patients at risk for the acquired immunodeficiency syndrome in a nonendemic setting. Chest, 1988, 93: 916–921.

Segal BH, Kwon-Chung J, Walsh TJ, et al. Immunotherapy for fungal infections. Clin Infect Dis, 2006, 42: 507–515.

Wheat JL. Current diagnosis of histoplasmosis. Trends Microbiol, 2003, 11: 488–494.

Wheat LJ. Antigen detection, serology, and molecular diagnosis of invasive mycoses in the immunocompromised host. Transpl Infect Dis, 2006, 8: 128–139.

Wheat LJ, Conces D, Allen SD, et al. Pulmonary histoplasmosis syndromes: recognition, diagnosis, and management. Semin Respir Crit Care Med, 2004, 25: 129–144.

Wheat LJ, Musial CE, Jenny-Avital E. Diagnosis and management of central nervous system histoplasmosis. Clin Infect Dis, 2005, 40: 844–852.

Woods JP. Knocking on the right door and making a comfortable home: *Histoplasma capsulatum* intracellular pathogenesis. Curr Opin Microbiol, 2003, 6: 327–331.

WEB SITES

American Academy of Pediatrics, Red Book Online, The Report of the Committee of Infectious Diseases, Copyright © 2008 American Academy of Pediatrics: http://aapredbook.aappublications.org/

Doctor Fungus Corporation Website: www.doctorfungus.org

eMedicine, Inc., 2007, provider of clinical information and services to physicians and health-care professionals: www.emedicine.com

Microbiology and Immunology Online, School of Medicine, University of South Carolina: http://pathmicro.med.sc.edu/mycology/opportunistic.htm

Website maintained by ESun Technologies, LLC: www.antimicrobe.org

Wikipedia – The Free Encyclopedia: www.wikipedia.org

MULTIPLE CHOICE QUESTIONS

The questions should be answered either by selecting True (T) or False (F) for each answer statement, or by selecting the answer statements which best answer the question. Answers can be found in the back of the book.

1. **Which of the following are characteristic of *H. capsulatum*?**

   A. Exclusive growth at 37°C temperature.

   B. Ability to infect both birds and bats.

   C. Only one variety: *H. capsulatum* var. *capsulatum*.

   D. Ability to produce microconidia as well as macroconidia.

   E. Being endemic to the Ohio, Missouri, and Mississippi River valleys.

2. **What are the main characteristics of the interaction between *H. capsulatum* and the host immune system?**

   A. *H. capsulatum* undergoes transformation from mold to yeast form inside alveolar macrophages.

   B. In many cases *H. capsulatum* prevents formation of phagolysosome.

   C. T-cell immunity plays a leading role in eradication of *H. capsulatum*.

   D. Specific antibodies play an important protective role against *H. capsulatum*.

   E. IL-4 and IL-10 are crucial for anti-histoplasma immune responses.

3. **Which of the following is true for infection with *H. capsulatum*?**

   A. Microconidia and mycelial fragments of *H. capsulatum* are inhaled.

   B. It is mostly asymptomatic or self-limiting in immunocompetent hosts.

   C. Progressive disseminated histoplasmosis is a mild disease.

   D. Chronic pulmonary histoplasmosis develops in patients with underlying pulmonary diseases.

   E. Histoplasmosis never disseminates to the central nervous system.

4. **Which are the most frequent symptoms of acute progressive disseminated histoplasmosis?**

   A. Oral and bowel ulcers.

   B. Shrinking of the adrenal glands.

   C. Anemia, leukopenia, and thrombocytopenia.

   D. Lung cavities.

   E. Meningitis and cerebritis.

5. **Which of the following is true about *H. capsulatum* var. *duboisii*?**

   A. Its cells are smaller than those of *H. capsulatum* var. *capsulatum*.

   B. It causes African histoplasmosis.

   C. Strong T-cell responses have been detected.

   D. Host response involves giant cells and macrophages.

   E. It often involves the lungs.

6. **Which of the forms of histoplasmosis can be detected by chest radiography?**

   A. Acute pulmonary histoplasmosis.

   B. Chronic pulmonary histoplasmosis.

   C. Acute progressive disseminating histoplasmosis.

   D. Chronic progressive disseminating histoplasmosis.

   E. Presumed ocular histoplasmosis syndrome.

7. **Which of the following laboratory tests are the most informative for the diagnosis of histoplasmosis?**

   A. Cell culture.

   B. Histology.

   C. Lymphocyte cell counts.

   D. CT scan.

   E. Serology.

8. **Which of the following drugs are used for the treatment of progressive disseminating histoplasmosis?**

   A. Clarithromycin.

   B. Itraconazole.

   C. Penicillin.

   D. Amphotericin B.

   E. Tetracycline.

Case 17

Human immunodeficiency virus (HIV)

A 37-year-old accountant began to feel extremely unwell on returning from a 2-week holiday in South America. He had been feeling 'under the weather' for 8 weeks before his holiday and had had several infections during that time. He was homosexual and often had several sexual partners during the same week. He did not believe in the use of condoms. Over the past 4 days he had developed a dry cough, had noticed increasing shortness of breath, and had begun to feel feverish. A painless lump had also developed in his mouth, and on his upper gum.

Because he felt so unwell, he went to the emergency department at his local hospital where they admitted him to one of the wards. On examination of his chest there were few physical signs but a chest X-ray showed widespread shadowing (Figure 1). Sputum swabs were taken for bacteriology but were found to be negative. A bronchial lavage was therefore performed to determine the cause of the shadowing. The stained cell pellet showed the presence of the organism *Pneumocystis jiroveci* (Figure 2). This is an example of opportunistic infection that is only observed in people with immunodeficiency.

On examination of his mouth, a purplish nodular swelling was visible (Figure 3) and this was biopsied during his hospital stay. This lesion histologically showed disordered angiogenesis, characteristic for the so-called **Kaposi sarcoma**, suggesting a severe immunodeficiency, most probably due to HIV infection leading to AIDS. This was confirmed by HIV **seropositivity**. The patient then was immediately treated with co-trimoxazole and **highly active antiretroviral therapy** (**HAART**).

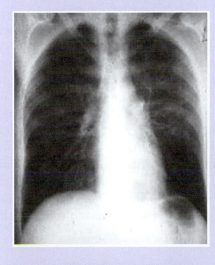

Figure 1. Chest X-ray showing bilateral shadowing from both hila to the periphery.

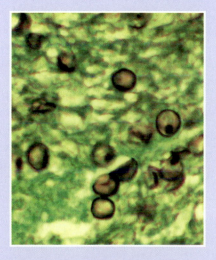

Figure 2. Cysts of *Pneumocystis jiroveci*. Histopathology of lung showing characteristic cysts with cup forms and dot-like cyst wall thickenings. These cysts were detected in the bronchial lavage of the patient. (Methenamine silver stain.)

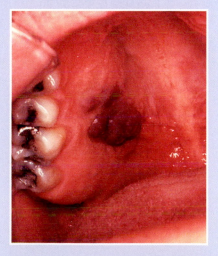

Figure 3. Kaposi's sarcoma seen as a purple-red growth in the patient's oral cavity. (Courtesy of Sol Silverman Jr, DDS, University of California, San Francisco: CDC Public Health Image Library # 6070.)

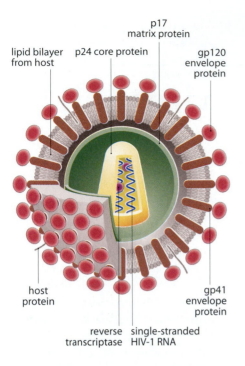

p17
matrix protein

lipid bilayer from host

p24 core protein

gp120 envelope protein

host protein

gp41 envelope protein

reverse transcriptase

single-stranded HIV-1 RNA

Figure 4. The structure of HIV. The HIV-1 genome is composed of approximately 9 kb of RNA, which consists of 9 different genes coding for 15 proteins. Some of these proteins are seen in the figure.

1. What is the causative agent, how does it enter the body and how does it spread a) within the body and b) from person to person?

Causative agent

The causative agent of AIDS is the human immunodeficiency virus (HIV). HIV, first discovered in 1983/84 by Montagnier and Gallo, is a **retrovirus** of about 100–120 nm in diameter belonging to the lentivirus group. It contains two single strands (ss) of positive sense RNA surrounded by a protein coat, which is in turn surrounded by a **lipid envelope**. Two types of HIV are recognized, HIV-1 (frequent) and HIV-2 (rare). The structure of HIV is shown in Figure 4. The HIV-1 genome is composed of approximately 9 kb of RNA, which consists of 9 different genes that result in 15 proteins. Three of the genes code for polyproteins that are cleaved to produce nine different proteins. The main proteins are: p24 (**capsid** protein), gp120 and gp41 (envelope glycoproteins that are involved in viral attachment and entry into cells), reverse transcriptase, integrase, and protease. HIV-2 is similar to HIV-1 in many of its properties including life cycle and spread but is different in its global distribution and the resulting disease is less aggressive.

Entry into the body

The virus enters the body mainly via mucosal surfaces or is injected directly into the bloodstream via contaminated needles.

The principal way in which HIV is transmitted is through sexual intercourse. Virus and virus-infected cells are present in vaginal fluid and in semen. Globally, heterosexual spread of HIV dominates but in Europe and America homosexual transmission is more important. Plasma and blood products that contain the HIV strain are infectious. Transfer of blood, either deliberately by transfusion or as a consequence of using contaminated needles and syringes, can also transmit HIV. Since the virus may be present at high level in plasma, in the past blood products such as clotting factor concentrates have infected many hemophiliacs. Today these products are carefully screened, virally inactivated, and do not pose a risk for contracting HIV. In those areas of the world where there is an HIV epidemic via heterosexual transmission, transmission of the virus from mother to infant (vertical transmission) contributes significantly to the burden of infection; infection can occur *in utero*, at birth, and from breast-feeding.

HIV is not particularly infectious since HIV is also a relatively labile virus. And many more **virions** are necessary for infection than with other viral pathogens such as hepatitis B. It is thought that the mucosal surfaces have some protection through innate immunity and that small lesions in the mucosal cell wall are required for successful infection.

Cell entry

On entering the body, the virus binds to and infects **CD4+** helper T lymphocytes where it replicates and produces more viruses by budding (Figure 5). At least two surface receptors are required for attachment, fusion of virus envelope with cell membrane, and entry of HIV into the cell. The virus attachment protein gp120 first binds to the cellular receptor CD4. Subsequent interactions between the virus and the **chemokine**

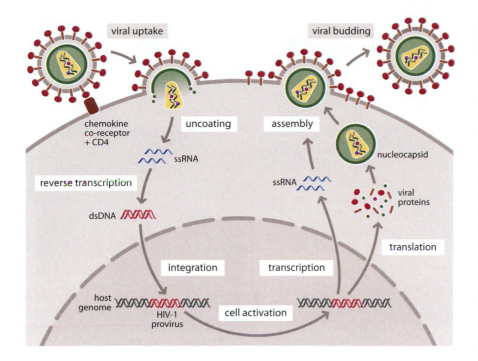

Figure 5. The life cycle of HIV. HIV-1 attaches to the cell surface through gp120 binding to CD4 and a chemokine receptor. CD4 is predominantly found on T-helper lymphocytes but also on macrophages and related cells; the chemokine receptors CXCR4 and CCR5 are also found on these cells. Fusion of the envelope with the membrane causes release of the viral capsid into the cell, which uncoats and through the enzyme reverse transcriptase (RT) the ssRNA is converted into double-stranded (ds) DNA. This becomes integrated into the host DNA as a provirus by the viral enzyme integrase. Transcription of the proviral DNA only occurs following activation of the cell , and produces new ssRNA strands and mRNA for a number of viral proteins including: reverse transcriptase (RT); the integrase involved in integrating the provirus into the host genome; the protease that hydrolyzes viral polyproteins into functional protein products that are essential for viral assembly and subsequent activity; other proteins (regulatory proteins) including Vif, Vpr and nef that are important in the infectivity of HIV. The ssRNA strands and viral proteins are packaged together to form a mature HIV virion that buds from the surface of the cells to infect more cells.

co-receptors (CCR5 or CXCR4) trigger irreversible conformational changes in the envelope glycoproteins, allowing fusion of the viral envelope with the cell membrane, and release of the virus core into the cytoplasm. The absolute requirement for a co-receptor in addition to the CD4 receptor for entry of the virus has been documented through the observation that male homosexuals – known to have been exposed to the virus for many years – have not shown the manifestations of viral infection. These individuals were subsequently shown to have homozygous mutations in their CCR5 gene, although these data are contentious.

Because of their high expression of CCR5, memory CD4+ T lymphocytes are the primary target of HIV early in infection. **Dendritic cells** are also important in the infection process. The normal skin as well as the prepuce of the penis is enriched in dendritic cells (that in this location are referred to as Langerhans cells; Figure 6). These dendritic cells also express CD4 and a surface HIV receptor called DC-Sign that binds to the virus. Dendritic cells at mucosal surfaces migrate to local lymph nodes where they present processed HIV peptides to CD4+ T cells and in so doing the whole virus comes into contact with CD4 and co-receptor CCR5 on the helper T cell leading to their infection (Figure 7). The mechanism of viral infection of CD4+ T cells via binding to DC-Sign on dendritic cells is independent of specificity of antigen to the CD4+ T cell.

Spread within the body

Following initial infection, the virus passes into draining lymph nodes and with increasing viral load in the bloodstream, to primary lymphoid organs (i.e. thymus and bone marrow) and other mucosal surfaces, especially the gut-associated lymphoid tissue, other secondary lymphoid tissues, including other lymph nodes, and even the lungs and the brain. This is presumably through infected CD4+ T cells or blood-borne virus infecting **alveolar macrophages** and **microglial cells**, respectively.

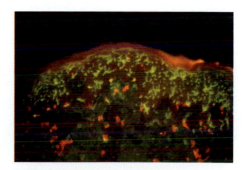

Figure 6. The prepuce of the penis showing the distribution of CD4+ dendritic cells in the epidermal/dermal layer (stained green) with the CD4+ T cells labeled orange (green and yellow) in the subepidermal tissues. (Image kindly provided by Prof. Len Poulter.)

Figure 7. Infection of CD4+ T cells through dendritic cells in mucosal surfaces: model of infection through DC-Sign. Dendritic cells (DCs) in the lamina propria bind HIV via their surface DC-Sign. The DCs then migrate to the draining lymph nodes and 'present' intact HIV to CD4+ T cells via their CD4 and co-receptors. Infection of CD4+ T cells through DCs is also a mechanism for other routes of entry of HIV.

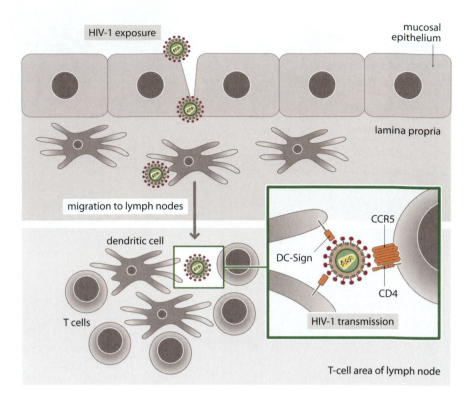

Infection of large numbers of both resting and activated memory CD4+ T cells in the gut may fuel **viremia**. In this early stage of infection more than half of all memory CD4+ T cells are lost, especially those in the gut. The body is left with a very restricted repertoire of CD4+ T cells with which to fight infection; some of these T cells include a stable reservoir of HIV-infected resting memory cells.

Epidemiology of HIV

Since the discovery of AIDS in 1981, over 25 million children and adults have died of the disease, with the major number of deaths having occurred in sub-Saharan Africa. It is estimated that a total of 39.5 million (34.1–47.1 million) individuals are currently infected with HIV-1 (Figure 8). In 2006, ~11 000 new infections were estimated to occur daily worldwide. An estimated 2.9 million (2.5–3.5 million) died of AIDS in 2006.

Figure 8. The numbers of children and adults estimated to be living with HIV infection in 2006 (from UNAIDS report 2006). Adapted with kind permission of UNAIDS, www.unaids.org.

HIV-2 is endemic in West African countries such as Senegal, Nigeria, Ghana, and the Ivory Coast, and as yet is not showing significant spread from these countries.

2. What is the host response to the infection and what is the disease pathogenesis?

Much of the data on the development of a host immune response and the mechanisms of pathogenesis have been derived from animal studies modeled on simian immunodeficiency virus in nonhuman primates. The interaction between HIV and the immune system is complex and not fully understood but there is an immune response to the virus that unfortunately rapidly attenuates since the viral target is the immune system itself. There are three phases to the infection and the immune response plays a role in both the acute and chronic phases of infection and in the developing third phase: AIDS.

Acute phase

Following primary HIV infection via the vaginal, anal or (less likely) oral route, dendritic cells in the **lamina propria** carry the virus to draining lymph nodes where production of virus by CD4+ T cells leads to a detectable viremia at about 7–14 days after infection (Figure 9). The higher the level of viremia reached, the greater the spread of virus to other sites of the body (see spread within the body in Section 1) and the greater the chance the individual has of passing on HIV. There is evidence at this stage of an increase in HIV-specific **CD8+** cytotoxic T-cell responses, which are thought to reduce the viral load 10–100-fold over a period of 2–6 months. Antibodies to HIV-1 are initially against isolated viral proteins (e.g. antibodies to p24 are raised early in disease) but rapidly evolve to target other viral proteins (e.g. gp120 and gp41,) and then persist at high level for the duration of the infection. In spite of the increasing level of HIV-binding antibodies, neutralizing antibodies do not appear until the viremia has significantly subsided, questioning their role in decreasing the viral load. In fact, the neutralizing antibodies are thought to merely drive the generation of escape mutants in the gp120 hypervariable region, thus making it even harder for the immune system to control it (see treatment in Section 5). During this phase the levels of circulating CD4+ T cells decrease significantly but increase slightly as the immune response begins

Figure 9. Acute and chronic phases of HIV infection. Following primary infection CD4+ T cells begin to decrease after about 6 weeks and then increase slightly during the chronic phase of infection (blue line). During the acute infection, virus levels increase in the circulation and disseminate the virus to lymphoid organs and tissues (red line). The immune system finally manages to produce an immune response and destroy many virus-containing cells through cytotoxic T cells, leading to a reduction of viral load in the circulation. This level of virus in the circulation after the initial immune response is called the 'set point' and appears to determine the rate of disease progression, which might take up to 15 years. A pool of virus-infected cells remains during the chronic phase during which there is clinical latency. There is a gradual increase in viremia and a gradual decrease in CD4+ T cells to a level that the patient is not able to cope with infections. Patients are also prone to develop tumors at this stage.

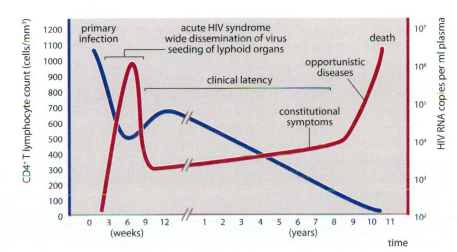

to exert some control over the infection. However, the immune system is unable to clear the virus completely and a reservoir of latently infected CD4+ T cells and dendritic cells 'harboring' the virus remains.

Chronic phase

After the acute phase, the level of viremia settles to a particular point called the 'set point' or point of equilibrium, which varies from individual to individual, and presumably represents the balance between production of new virus and control by the immune system. During the chronic phase: (a) the CD4+ T cell levels gradually decrease further and (b) the viremia gradually increases. The length of this phase varies from person to person and during this period there is a so-called clinical **latency** period, that is no symptoms (Figure 9), although the immune impairment progresses, sometimes associated with a variety of nonspecific symptoms and general malaise.

AIDS

Eventually, when the CD4 count drops to around 200 cells/mm^3 or less, the patient may present with an AIDS-defining illness, usually of infective origin. In other words, the late disease known as acquired immune deficiency syndrome, AIDS, is a very carefully defined condition. Progression of the disease to full-blown AIDS (see Section 3 – Clinical presentation) can occur from a few months up to and beyond 10 years (patients in the latter group are called long-term nonprogressors).

The pathogenesis of the disease continues to be elucidated, nevertheless, the key event leading to immunodeficiency is the death of CD4+ T cells.

Several mechanisms have been proposed for the death of CD4+ T cells, the most important of which are the following.

- Direct death of infected CD4+ T cells as the result of massive infection of the cell.

- **Apoptosis** of CD4+ T cells mediated by cytotoxic CD8+ T cells recognizing the infected CD4+ T cell presenting with HIV peptides via HLA class 1 molecules.

The CD4+ T cell is the 'conductor' of the immunological orchestra, which together with dendritic cells controls the development and function of most immunological cell types. Thus, ablation of the majority of CD4+ T cells results in major deficiencies in all arms of the adaptive immune response and also in **natural killer (NK) cell** function, which plays a role in innate responses against viruses (Figure 10).

As described above, the dendritic cell is an important cell involved in the infection of CD4+ T cells with HIV. Whether it becomes grossly infected itself and produces significantly large amounts of virus is presently unclear. It is, however, certain that the antigen presentation function of dendritic cells is severely handicapped as a result of HIV infection. Infected dendritic cells could also affect the early differentiation of Th0 cells into either **Th1** or **Th2** or indeed other subsets early in an immune response, thus disrupting all immune responses.

A number of pathological changes also occur in primary and secondary lymphoid organs and tissues that contribute to immunodeficiency.

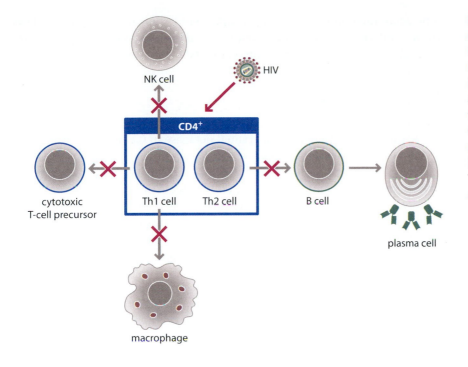

Figure 10. HIV compromises the pivotal role of CD4 T cells in immune responses. Both Th1 and Th2 cells can be infected with HIV. Th1 cells help to induce natural killer (NK) cell function in the innate response against viruses. Th1 cells are important in the development of cytotoxic effector CD8+ T cells from their cytotoxic T-cell precursors and helping macrophages to kill intracellular parasites. Finally, Th2 and also Th1 cells help B cells to differentiate into antibody-producing plasma cells.

- *Changes to the primary lymphoid organs.* Thymus (in young children): thymocytes become infected and there is a decrease in the output of CD4+ T cells. Late effects include changes to stromal cell function and loss of tissue architecture. The bone marrow depression seen late in the disease is probably due to effects on stromal cells but the mechanism remains unclear.

- *Changes to lymph nodes* include polyclonal activation of CD8+ T cells and B cells, as well as a reduction in the number of CD4+ T cells. The possible mechanisms are **cytokine**-induced activation and differentiation mediated by HIV products and other microbial products thought to be present due to the breakdown of the barrier that normally prevents entry of gut lumen microorganisms. Recent data have indicated that patients infected with HIV have increased plasma levels of **lipopolysaccharides** (**LPS**) and microbial components. One model explaining the continuous ongoing 'inflammation-like' response in lymph nodes is from T-cell activation through microbial products trapped within lymph nodes interacting with **Toll-like receptors** (**TLRs**) on the immune cells. The inflammation leads to **fibrosis** and later in the disease breakdown in the structure of the lymph nodes including germinal centers.

3. What is the typical clinical presentation and what complications can occur?

Acute HIV infection may be asymptomatic, but from 25 to 65% of patients have some symptoms that usually begin about 2–8 weeks after exposure – ranging from an infectious mononucleosis-like illness with fever, **maculopapular rash**, sore throat, night sweats, malaise, **lymphadenopathy**, diarrhea, to mouth and genital ulcers to neurological disease (aseptic **meningitis, neuropathy, myelopathy, encephalitis**). The symptoms are usually mild and subside spontaneously.

The acute infection is followed by a period of clinical latency, which lasts for several years, although about one-third of patients develop *persistent generalized lymphadenopathy* – involving two or more sites outside the inguinal and other areas.

Progression to AIDS

As the infection progresses (see Section 2 and Figure 9), viral load in the blood increases, paralleled by a gradual decrease in CD4+ count leading to a decline in immune function.

When the CD4+ count falls below 200×10^6 cells/mm^3, and HIV-infected patients present with **opportunistic infections** (OIs) and/or tumors, full-blown AIDS is diagnosed. These OIs include a variety of bacterial, viral, and fungal infections, as well as tumors such as Kaposi's sarcoma and nonHodgkin's **lymphomas**. These complications of HIV disease are summarized in Table 1. It is important to emphasize that HIV-positive individuals are most likely to die of opportunistic infections rather than HIV directly.

Kaposi's sarcoma

The patient in this case study had Kaposi's sarcoma, which is the most common neoplasm affecting HIV-infected individuals. It more frequently presents on the skin of the patient but other common sites are the gums of the mouth. Lesions are usually violaceous in colour. Human herpesvirus type 8 (HHV8), also known as Kaposi's sarcoma-associated herpesvirus (KSHV), is invariably found in tumor tissue; this is an opportunistic infection – the result of immune compromise.

The World Health Organization (WHO) have categorized the clinical stages of HIV disease into four stages (Table 2) – www.avert.org/HIVstages.

Table 1. AIDS-defining illnesses

Nonviral infections
- Recurrent pneumonia
- *Pneumocystis* pneumonia
- Candida of esophagus or trachea/bronchi/lungs
- Disseminated or extrapulmonary *Mycobacterium* infections
- Extrapulmonary cryptococcosis
- Coccidioidomycosis and histoplasmosis – disseminated
- Chronic intestinal cryptosporidiosis, isosporiasis
- Cerebral toxoplasmosis

Viral
- Herpes simplex chronic ulcers, bronchitis or esophagitis
- Cytomegalovirus disease
- Progressive multifocal leukoencephalopathy (PML – JC virus)
- Wasting syndrome due to HIV
- HIV-associated encephalopathy (dementia)

Tumors
- Kaposi's sarcoma and other human herpesvirus type 8 (HHV8)-related tumors
- Lymphomas – Burkitt's or immunoblastic
- Primary central nervous system (CNS) lymphoma
- Invasive cervical cancer

Table 2. WHO categorization of the clinical stages of HIV disease

Clinical stage I
- Asymptomatic
- Persistent generalized lymphadenopathy

Clinical stage II
- Moderate unexplained weight loss (under 10% of presumed or measured body weight)
- Recurrent respiratory tract infections (sinusitis, tonsillitis, otitis media, pharyngitis)
- Herpes zoster
- Angular chelitis
- Recurrent oral ulceration
- Papular pruritic eruptions
- Seborrheic dermatitis
- Fungal nail infections

Clinical stage III
- Unexplained severe weight loss (over 10% of presumed or measured body weight)
- Unexplained chronic diarrhea for longer than 1 month
- Unexplained persistent fever (intermittent or constant for longer than 1 month)
- Persistent oral candidiasis
- Oral hairy leukoplakia
- Pulmonary tuberculosis
- Severe bacterial infections (e.g. pneumonia, empyema, pyomyositis, bone or joint infection, meningitis, bacteremia)
- Acute necrotizing ulcerative stomatitis, gingivitis or periodontitis
- Unexplained anemia (below 8 g dl^{-1}), neutropenia (below 0.5 billion l^{-1}) and/or chronic thrombocytopenia (below 50 billion l^{-1})

Clinical stage IV
- HIV wasting syndrome
- *Pneumocystis* pneumonia
- Recurrent severe bacterial pneumonia
- Chronic herpes simplex infection (orolabial, genital or anorectal of more than 1 month's duration or visceral at any site)
- Esophageal candidiasis (or candidiasis of trachea, bronchi or lungs)
- Extrapulmonary tuberculosis
- Kaposi sarcoma
- Cytomegalovirus infection (retinitis or infection of other organs)
- Central nervous system toxoplasmosis
- HIV encephalopathy
- Extrapulmonary cryptococcosis including meningitis
- Disseminated nontuberculous mycobacteria infection
- Progressive multifocal leukoencephalopathy
- Chronic cryptosporidiosis
- Chronic isosporiasis
- Disseminated mycosis (extrapulmonary histoplasmosis, coccidiomycosis)
- Recurrent septicemia (including nontyphoidal *Salmonella)*
- Lymphoma (cerebral or B-cell nonHodgkin)
- Invasive cervical carcinoma
- Atypical disseminated leishmaniasis
- Symptomatic HIV-associated nephropathy or HIV-associated cardiomyopathy
- Some additional specific conditions can also be included in regional classifications (such as the reactivation of American trypanosomiasis (meningoencephalitis and/or myocarditis) in the WHO Region of the Americas and penicilliosis in Asia)

4. How is this disease diagnosed and what is the differential diagnosis?

The diagnosis of HIV-1 infection is based on detecting specific antibodies, antigens, or both, and many commercial kits are available. Rapid testing kits for HIV-1 antibodies can now provide results in 20 minutes and are used to test plasma, serum, whole blood or saliva. Antibodies are the diagnostic tool from which comes the vernacular phrase 'testing HIV-positive.' Current antibody tests are highly specific but false positives occasionally occur and the diagnosis must always be confirmed. Some laboratories use **Western blotting** analysis, while others use combinations of anti-HIV assays of other formats (e.g. particle agglutination and **ELISA**) for this purpose. Real-time **polymerase chain reaction** (**PCR**) can also be used for direct detection of viral RNA. It is also essential to obtain a second blood sample from the same patient, just to confirm that the laboratory result belongs to the correct patient.

Antibody tests are not immediately useful during the earliest phase of infection when antibodies are undetectable – the so-called 'diagnostic window.' Because of this problem combined tests are used to screen for HIV that detect both antibody and p24 antigen. In newborn infants – due to maternal antibody passing across the placenta from an HIV-positive mother and persisting for up to 15 months – direct virus detection is necessary; tests for viral nucleic acid, usually proviral DNA, are required for early diagnosis of HIV in the newborn. The relative decrease of CD4+ T cells in relation to the other T-cell population, the CD8+ T cells, can also be diagnostic of viral infection in babies.

Levels of CD4+ T cells and viral load are measured for staging purposes. Usually flow cytometry is used for measuring CD4+ T-cell counts and several kits are available to determine plasma HIV-1 RNA copies. Viral load determines the rate of destruction of the immune system and the levels of CD4+ T cells indicate the degree of immunodeficiency and therefore the likelihood of developing opportunistic infections.

Advanced laboratory infrastructure and resources are needed for quantifying CD4+ T cells and viral loads and especially for short turn-around times on sample results. The availability and standardization of tests for diagnosis and monitoring remain a problem in countries where the national health budget is limited. New equipment designed to cost a fraction of that of commercial flow cytometers is available and is being tested. The Gates Foundation is also currently funding diagnosis and treatment of HIV disease in resource-limited countries. Information on clinical and laboratory strategies to monitor HIV-infected individuals in developing countries can be obtained from the WHO website.

Differential diagnosis

The case under discussion presented with **pneumonia**. The differential diagnosis of pneumonia in an HIV-infected person includes *Pneumocystis* pneumonia, **miliary** tuberculosis, nonHodgkins lymphoma, **pyogenic** bacterial pneumonia, and visceral Kaposi's sarcoma.

5. How is the disease managed and prevented?

Management

Currently patients with HIV infections cannot be 'cured' and the aim of therapy is to reduce the replication of a persistently expressed chronic virus infection to an undetectable level. In this way the morbidity and mortality associated with HIV infection are reduced by decreasing viral damage to the immune system and allowing some recovery of immune function. However, this does not lead to total clearance of the HIV from the body, since some HIV is integrated as a provirus into human DNA and the majority of 'successfully' treated individuals apparently remain infected for life.

There are five points in the HIV replication cycle for which antiretroviral (ARV) drugs are currently in use. The drugs comprise the reverse transcriptase inhibitors (RTIs), protease inhibitors (PIs), fusion inhibitors, CCR5 inhibitors, and HIV integrase inhibitors (Table 3) – the list given in this table is not exhaustive and there are about 20 different anti-HIV drugs available.

Reverse transcriptase inhibitors

Zidovudine, also known as azidothymidine (AZT), was the first RTI to be developed for use against HIV. The RTIs inhibit viral replication after cell entry but before integration of the provirus into the host nucleus. They prevent the conversion of viral ssRNA into dsDNA.

Table 3. Examples of anti-HIV drugs	
Class of antiretroviral drug	**Example**
Nucleoside and nucleotide RT inhibitors (NRTIs)	
Nucleosides	Abacavir
	Didanosine
	Lamivudine
	Stavudine
	Zidovudine
Nucleotide	Tenofovir
Nonnucleoside RT inhibitors (NNRTIs)	Efavirenz
	Nevirapine
Protease inhibitors (PIs)	Amprenavir
	Indinavir
	Lopinavir 'boosted' with ritonavir
	Nelfinavir
	Saquinavir
Fusion inhibitor	Enfuvirtide
Chemokine co-receptor antagonist (CCR5 inhibitor)	Maraviroc
Integrase inhibitor	Raltegravir

RT, reverse transcriptase.

There are two classes of RTIs:

- the nucleoside and nucleotide analogs, together referred to as NRTIs, bind competitively to the reverse transcriptase;
- the nonnucleoside RTIs (NNRTIs) have a different mode of action and bind noncompetitively to reverse transcriptase.

Protease inhibitors (PIs)

These are synthetic analogs of the natural substrate designed to bind with the active site of the viral protease, inhibiting its function and thereby preventing post-translational cleavage of polyproteins and virus particle maturation. Therefore new virions are not produced. There are currently 10 PIs approved by the FDA (US Food and Drug Administration).

Fusion inhibitors (FIs).

Enfuvirtide binds to the gp41 subunit of the viral envelope glycoprotein and prevents the conformational change required for viral fusion and entry into cells.

Chemokine co-receptor antagonist (CCR5 inhibitor).

Maraviroc blocks the binding of HIV to its co-receptor, CCR5, which is essential early in HIV infection for entering CD4+ cells. Maraviroc is not effective in the late stages of HIV infection when viruses with CXCR4 tropism predominate.

Integrase inhibitors.

Raltegravir blocks the HIV integrase and prevents HIV DNA from entering the chromosomal DNA of CD4+ cells.

Combination antiretroviral treatment

The high rate of viral replication (at least 10^{10} new virions produced per day), a high rate of mutation (about one mutation per virion), the ability to recombine, and the fact that RT RNA-dependent DNA polymerase is not of high fidelity and has no proof-reading ability, all contribute to the huge diversity of HIV-1 strains in chronically infected individuals. Thus emergence of drug-resistant strains within individuals poses a major problem in the therapy of HIV/AIDS. When HIV monotherapy was first introduced with the nucleoside RT inhibitor zidovudine (AZT), incomplete viral suppression allowed continued replication in the face of drug selective pressure. Thus a combination of three or more active drugs is required to effectively treat patients with HIV. For example, two or more NRTIs together with an NNRTI or a PI agent. This combined therapy is known as highly active antiretroviral treatment (HAART) and is designed to suppress viral replication to undetectable levels (viral load < 50 copies per ml) so as to prevent emergence of drug-resistant HIV.

Drug interactions and side effects also frequently limit the usefulness of combination therapy. Nevertheless, HAART has been shown to slow down the progression of HIV to AIDS and has led to dramatic improvements in the quality of life of many people with HIV infection.

Resistance testing

Indirect evidence of the emergence of *in vivo* drug resistance is the presence of falling CD4 counts and climbing HIV RNA levels in the face of apparently adequate antiviral therapy. In the case of our patient a test for drug resistance would be performed before embarking on a particular HAART regimen. Resistance testing in the clinic is now based on gene amplification by PCR and analysis by direct sequencing; it is now possible to do this routinely for RT and protease genes, allowing, in principle, 'tailor-made' drug therapy for each patient, albeit at a considerable cost.

Unfortunately, there are only about 20–25% of patients with HIV disease who are treated with HAART globally. The main reason is financial cost. It is perceived that the global fight against HIV is a losing battle, since for every patient treated successfully with HAART, six new cases of HIV infection are diagnosed.

Prevention

Perhaps the most important aspect of control is increasing awareness about HIV and the way the infection is transmitted. National awareness at the governmental level of a country's burden of infection is essential in encouraging acceptance of the problem by that country's people. Intervention can be sought in the following ways.

1. *Sexual health*: education, practice of safer sex (condoms), HIV testing and counseling.
2. *Drug abuse*: education, needle and syringe exchange, outreach and support.
3. *Blood plasma products:* HIV screening, donor selection, viral inactivation procedures.
4. *Antenatal screening:* identification of HIV-infected mothers during pregnancy, allowing antiviral therapy, intervention at the time of delivery, and replacing breast milk with formula milk.

All of these interventions require access to accurate, timely, cost-effective, and confidential HIV testing.

Microbicides

A number of microbicides have been developed for use as topical products to prevent infection. Although proof of concept for their use has been elusive because of the absence of surrogate markers of protection, a number of products incorporating ARV drugs including entry and NRTIs are in phase III **clinical trials**.

Vaccines

The major problem in producing HIV vaccines is the continuous **antigenic drift** due to the high mutation rates (see above). Thus although there is a great deal of research taking place, no successful vaccines against HIV have yet emerged. Much of the initial effort focused on producing vaccines to generate neutralizing antibodies; however, those that have reached the stage of phase 3 clinical trials have all failed to prevent infection.

Recently a number of candidate vaccines against HIV have been developed that induce primarily T-cell responses and are in phase I and phase II clinical trials (Johnson and Fauci, 2007). Development of a vaccine that would prevent the initial establishment of HIV infection by producing an immune response capable of totally clearing the virus before it infects cells and integrates into the host chromosome and forming a latent viral reservoir, remains the ultimate goal.

In the meantime, although CD8+ T cells do not seem to target the pool of latently infected CD4+ T cells containing the '**provirus**', they clearly are important in reducing the viremia during early infection, and the 'set point' level of viremia predicts the progression of the disease. Therefore a vaccine that enhances the early HIV-specific CD8+ T-cell response will be important. The idea of these 'nonperfect' vaccines would be to reduce the initial burst of viremia and the frequency of latently infected cells instead of preventing infection. The aim is to increase the time to progression and to reduce transmission of the virus, which is most likely to occur during high viral load in either early or late infection.

Treatment of Kaposi's sarcoma is usually to 'watch and wait,' as it may regress with HAART regimens.

Treatment of *Pneumocystis jiroveci* infection is with co-trimoxazole.

SUMMARY

1. What is the causative agent, how does it enter the body and how does it spread a) within the body and b) from person to person?

- AIDS is caused by the human immunodeficiency virus (HIV), a retrovirus of about 100–120 nm in diameter belonging to the lentivirus group. The virus contains two single strands (ss) of positive sense RNA surrounded by an envelope. There are two viruses, HIV-1 and HIV-2.

- The HIV-1 genome is composed of approximately 9 kb of RNA, which consists of 9 different genes coding for 15 proteins. The main proteins are: p24 (capsid protein), gp120 and gp41 (envelope proteins that are involved in viral attachment and entry into cells), and the enzymes reverse transcriptase, integrase, and protease.

- The principal ways in which HIV is transmitted are through sexual intercourse (heterosexual or homosexual), transfer of blood (transfusions or contaminated needle sharing) and contaminated blood products (e.g. factor VIII), and maternal–infant transmission (infection occurs either *in utero*, at birth or from breast-feeding).

- HIV binds to and infects CD4+ helper T cells where it replicates and gives rise to new virions by budding. CD4 and a chemokine co-receptor (usually CCR5 or CXCR4) are both required for binding and entry of HIV.

- Memory CD4+ T cells express mainly CCR5 and most of the CD4+ T cells in the lamina propria at mucosal surfaces are memory cells and are targeted by the virus. Dendritic cells are also important in the infection process and bind HIV via surface DC-Sign. When dendritic cells present HIV or other antigenic peptides to CD4+ T cells, the CD4+ T cells become infected with the virus.

- Following acute infection, the virus passes via the bloodstream to; primary lymphoid organs, that is thymus and bone marrow; other mucosal surfaces – especially the gut-associated lymphoid tissue; other secondary lymphoid tissues – including other lymph nodes; and the lungs and the brain.

- It is estimated that a total of 39.5 million (34.1–47.1 million) individuals are currently infected with HIV-1, with an estimated 11000 new infections occurring daily worldwide. An estimated 2.5–3.5 million people died from complications associated with AIDS in 2006.

2. What is the host response to the infection and what is the disease pathogenesis?

- Much of the data on the development of a host immune response and the mechanisms of pathogenesis have been derived from animal studies modeled on the simian immunodeficiency virus in nonhuman primates.

- There are three phases to the infection and the immune response plays a role in all three – acute and chronic phases and AIDS.

- Following primary HIV infection, for example via the vaginal route, the virus is carried by dendritic cells in the lamina propria to draining lymph nodes where production of the virus by CD4+ T cells leads to a viremia detected at about 7–14 days after infection. There is evidence at this stage showing an increase in HIV-specific CD8+ cytotoxic T-cell responses, which reduce the viral load 10–100-fold over a period of 2–6 months.

- The level of viremia settles to a particular point called the 'set point' or point of equilibrium, which varies in level from individual to individual. During this phase the levels of circulating CD4+ T cells decrease significantly but increase slightly as the immune response begins to have some control over the infection.

- The immune system is unable to clear the virus completely and a reservoir of latently infected CD4+ T cells and dendritic cells 'harboring' the virus remains. During this phase, which is generally clinically asymptomatic: (a) the CD4+ T-cell levels gradually decrease further; (b) the viremia gradually increases. The length of this phase varies from person to person and during this period there is so-called clinical latency, that is asymptomatic, although the immune impairment progresses.

- At some stage when the CD4+ count drops towards 200 cells/mm^3 the patient may develop an illness that heralds the onset of progressive disease called acquired immune deficiency syndrome, AIDS.

- The key event leading to the immunodeficiency is the death of CD4+ T cells. The mechanisms for this are thought to include direct apoptosis mediated by the virus and via CD8+ T-cytotoxic cells recognizing the HIV peptide with MHC class I on the surface of infected CD4+ T cells.

- Infection of thymus cells (in young children) results in a decrease in the output of CD4+ T cells. Late effects include changes to stromal cell function and loss of tissue architecture. The bone marrow depression observed late in the disease is probably due to effects on stromal cells but the mechanism remains unclear.

- Lymph nodes become sites of activated CD8+ T cells and B cells, which eventually results in breakdown of follicular structure and fibrosis.

3. What is the typical clinical presentation and what complications can occur?

- Primary HIV infection may be asymptomatic, but from 25 to 65% of patients may have some symptoms that usually begin about 2–8 weeks after exposure – ranging from an infectious mononucleosis-like illness with fever, maculopapular rash, sore throat, night sweats, malaise, lymphadenopathy, diarrhea, to mouth and genital ulcers to neurological disease (aseptic meningitis, neuropathy, myelopathy, encephalitis). The symptoms are usually mild and resolve spontaneously.

- The acute infection is followed by a period of clinical latency, which lasts for several years, although about one-third of patients develop *persistent generalized lymphadenopathy*.

- As the infection progresses, viral load in the blood increases, paralleled by a decrease in CD4 counts and a decline in immune function.

- When the CD4 count falls below 200×10^6 cells/mm^3, patients present with opportunistic infections and tumors and they are diagnosed with AIDS. These include a variety of bacterial, viral, and fungal infections and tumors such as Kaposi's sarcoma and nonHodgkin's lymphomas.

- Kaposi's sarcoma (seen in this patient) is a neoplasm affecting HIV-infected individuals. It is associated with human herpesvirus type 8 (HHV8), which is thought to act as an opportunistic infection as the result of immune compromise.

- The World Health Organization (WHO) have categorized the clinical stages of HIV disease into four stages.

4. How is this disease diagnosed and what is the differential diagnosis?

- The diagnosis of HIV-1 infection is based on detecting specific antibodies, antigens, or both. Rapid testing kits for HIV-1 antibodies can now provide results in 20 minutes and are used to test plasma, serum, whole blood or saliva.

- Because of occasional false positives detected in antibody assays the diagnosis must always be confirmed, depending on the quality of reagents initially used. Some laboratories use Western blot analysis and others use combinations of anti-HIV assays of other formats (e.g. particle agglutination) for this purpose.

- Antibody tests are not useful during the earliest phase of infection when antibodies are undetectable or in newborn infants – due to maternal antibody passing across the placenta. Therefore direct virus detection is necessary. This includes quantification of viral RNA or p24 antigen in heat-denatured serum.

- For staging, levels of CD4+ T cells and viral load are measured. Usually flow cytometry is used for CD4+ T-cell counts and several kits are available to determine plasma HIV-1 RNA copies per ml.

- Advanced laboratory infrastructure and resources are required for quantifying CD4+ T cells and viral loads and especially for short turn-around times on samples.

- The availability and standardization of tests for diagnosis and monitoring remain a problem in countries where the national health budget is limited.

5. How is the disease managed and prevented?

- Currently patients with HIV infections cannot be 'cured' and therapy is carried out to reduce the replication of a persistently expressed chronic virus infection to undetectable levels (viral load <50 copies per ml). HIV is integrated as a provirus into chromosomal DNA and 'successfully' treated individuals remain infected for life.

- The five kinds of antiretroviral (ARV) drugs are reverse transcriptase inhibitors (NRTI and NNRTIs), protease inhibitors (PIs), fusion inhibitors, chemokine co-receptor antagonists (CCR5 inhibitor), and integrase inhibitors.

- Some examples of RT inhibitors are the following: nucleoside analog, for example zidovudine; nucleotide analog, for example tenofovir; and nonnucleoside, for example nevirapine. These drugs inhibit viral replication after cell entry but before integration of the provirus into the host nucleus.

- Protease inhibitors (PIs) prevent post-translational cleavage of polyproteins and virus particle maturation and therefore new virions are not produced.

- Fusion and CCR5 inhibitors block the entry of the virus into the target cell, for example enfuvirtide, maraviroc.

- Integrase inhibitors prevent the formation of HIV provirus.

- Combination antiretroviral treatment is given because of the huge diversity of HIV-1 in chronically infected individuals. Thus emergence of drug-resistant viral strains within individuals poses a real problem in the therapy of AIDS.

- A combination of three or more active drugs is needed to effectively treat patients with HIV. For example, two or more NRTIs together with an NNRTI or PI. This highly active antiretroviral therapy (HAART) is designed to suppress viral replication to undetectable levels so as to prevent the emergence of drug-resistant HIV-1.

- Resistance testing is required to determine appropriate HAART regimens and to identify possible viral resistance.

- Transmission of HIV can be reduced through the following: sexual health and drug abuse education; screening of blood and plasma products for HIV; antenatal screening; and replacing breast milk with formula milk.

- A number of microbicides have been developed incorporating ARVs including entry inhibitors and nucleotide and nucleoside RT inhibitors. These are in phase III clinical trials.

- The major problem in producing HIV vaccines is continuous antigenic drift due to point mutations and high mutation rates. Recent efforts have focused on development of vaccines enhancing the early HIV-specific CD8+ T-cell response to reduce the initial burst of viremia and the frequency of latently infected cells, aimed at increasing the time to progression and reducing transmission of the virus.

- Treatment of Kaposi's sarcoma is usually 'watch and wait' and it may regress with HAART. Treatment of *Pneumocystis jiroveci* infection is with co-trimoxazole.

FURTHER READING

Mims C, Dockrell HM, Goering RV, Roitt I, Wakelin D, Zuckerman M. Medical Microbiology, 3rd edition. Mosby, Edinburgh, 2004.

Murphy K, Travers P, and Walport M. Janeway's Immunobiology, 7th edition. Garland Science, New York, 2008.

Roitt I, Brostoff J, Roth D, Male D. Immunology, 7th edition. Mosby, Elsevier, 2006.

REFERENCES

Boggiano C, Littman DR. HIV's vagina travelogue. Immunity, 2007, 26: 145–147.

Deeks SG. Antiretroviral treatment of HIV infected adults. BMJ, 2006, 332: 1489–1493.

Gotch F, Gilmour J. Science, medicine and research in the developing world: a perspective. Nat Immunol, 2007, 8: 1273–1276.

Johnson MI, Fauci AS. An HIV vacine – evolving concepts. N Engl J Med 2007, 356: 2073–2081.

Simon V, Ho DD, Karim QA. HIV/AIDS epidemiology, pathogenesis, prevention, and treatment. Lancet, 2006, 368: 489–504.

Wu L., KewalRamani VN. Dendritic-cell interactions with HIV: infection and viral dissemination. Nat Immunol, 2006, 6: 859–868.

WEB SITES

Averting HIV and AIDS – International AIDS charity: www.avert.org

Centers for Disease Control and Prevention (CDC), 2008, United States Public Health Agency: http://www.cdc.gov/hiv/

Department of Health, National Health Service © Crown copyright 2008: http://www.dh.gov.uk/en/Policyandguidance/Healthandsocialcaretopics/HIV/index.htm

HIV InSite Gateway to HIV and Aids Knowledge, University of California San Francisco: http://ivinsite.ucsf.edu

Mayo Clinic, Mayo Foundation for Medical Education and Research, Copyright ©2001-2008 www.mayoclinic.org/kaposis-sarcoma

World Health Organization: http://www.who.int/hiv/en/

MULTIPLE CHOICE QUESTIONS

The questions should be answered either by selecting True (T) or False (F) for each answer statement, or by selecting the answer statements which best answer the question. Answers can be found in the back of the book.

1. Which of the following statements regarding the human immunodeficiency virus are correct?

    A. It belongs to the flavivirus family.

    B. It carries a double-stranded RNA genome.

    C. It is an enveloped virus.

    D. The genome encodes both a reverse transcriptase and a protease.

    E. Infection of humans has arisen through transfer of simian immunodeficiency viruses into the human population.

2. Which of the following are recognized routes of infection of human immunodeficiency virus?

    A. Fecal–oral.

    B. Droplet spread.

    C. Percutaneous.

    D. Sexual.

    E. Transplacental.

MULTIPLE CHOICE QUESTIONS (continued)

3. **Which of the following statements regarding the entry of HIV into target cells are correct?**

   A. The initial binding event occurs between viral gp120 and cellular CD8 molecules.

   B. CCR5 is a co-receptor for HIV binding and entry.

   C. Infection of CD4+ T cells can arise through interaction with dendritic cells.

   D. Viral entry proceeds through fusion of the viral envelope with the cell membrane.

   E. Viral entry can be inhibited by anti-gp120 antibodies.

4. **Which of the following statements relating to the HIV replication cycle are correct?**

   A. Reverse transcription results in production of a double-stranded DNA copy of the virus.

   B. The RNA genome of HIV can exist in a latent state inside infected cells.

   C. The viral genes are **polycistronic**, resulting in the translation of polyproteins.

   D. The virus requires the availability of cellular enzymes to integrate its genome into host cell chromosomes.

   E. Maturation of viral particles requires the presence of a viral protease enzyme.

5. **Which of the following are true about HIV disease?**

   A. The virus causes immunodeficiency by mainly killing cytotoxic T cells.

   B. Killing of helper T cells can be through apoptosis.

   C. HIV causes mortality of the patient directly.

   D. Lymphadenopathy results from increased proliferation of B cells and cytotoxic T cells.

   E. Macrophages and dendritic cells act as a reservoir of HIV.

6. **Which of the following are true for laboratory diagnosis of HIV?**

   A. A negative anti-HIV result excludes the possibility that the patient is infected with HIV.

   B. It is necessary to confirm a positive anti-HIV test result in a number of different assays.

   C. Cell culture of HIV is routinely performed to assess drug sensitivities.

   D. Detection of anti-HIV antibodies in a 3-month-old baby indicates that mother to baby transmission of infection has occurred.

   E. In late stage disease, anti-HIV antibodies may disappear, leading to a false-negative test result.

7. **Which of the following are AIDS-defining illnesses?**

   A. Herpes zoster.

   B. Recurrent cold sores.

   C. Esophageal candidiasis.

   D. Cytomegalovirus disease.

   E. Kaposi's sarcoma.

8. **Which of the following are recognized manifestations of AIDS?**

   A. Cerebral toxoplasmosis.

   B. Oral hairy leukoplakia.

   C. Nasopharyngeal carcinoma.

   D. Progressive multifocal leukoencephalopathy.

   E. *Pneumocystis jiroveci* infection.

9. **Which of the following are true of seroconversion illness?**

   A. It occurs in only 10% of infected individuals.

   B. It has many features of a glandular fever-like illness.

   C. Anti-HIV antibodies are always detectable at this stage of infection.

   D. Patients are not infectious at this stage of infection.

   E. It is often followed by a long period when the patient is asymptomatic.

10. **Which of the following statements regarding therapy are correct?**

    A. In the early stages of infection, monotherapy with a nucleoside analog reverse transcriptase inhibitor is recommended.

    B. Drug resistance occurs through the accumulation of point mutations in the genes encoding the target enzyme.

    C. HAART should be given until such time as the viral load becomes undetectable, and then stopped, until the viral load exceeds 10 000 copies per ml.

    D. Monitoring of therapy is by CD4 count measurement.

    E. Successful therapy results in restoration of immune function.

Case 18

Influenza virus

A 59-year-old woman went to see her doctor, as she had been unwell for the past 3 days. She initially noticed a nonproductive cough, and then she became abruptly worse with a marked fever, headache, and shivering. Since then she had developed muscle aches all over her body, especially in the legs, and her eyes had become watery and painful to move. She was a nonsmoker, previously fit and well, and on no regular medication. On examination, she was **febrile** (38.2°C), and had difficulty in breathing through her nose, but there were no other abnormal physical signs.

A throat swab was taken, broken off into viral transport medium, and sent to the laboratory. Immunofluorescent staining with a **monoclonal antibody** against influenza A virus was positive, confirming a diagnosis of acute influenza virus infection.

1. What is the causative agent, how does it enter the body and how does it spread a) within the body and b) from person to person?

Causative agent

Influenza A virus belongs to the *Orthomyxoviridae* virus family (myxo = affinity for mucin). The viral genome consists of 8 segments of negative single-strand RNA (i.e. RNA that cannot be translated directly on the ribosome, but has to be first copied into its complementary, positive, strand), which collectively encode 10 (or possibly 11) viral proteins (Figure 1). Each RNA segment is closely associated with the nucleoprotein, to form a helical

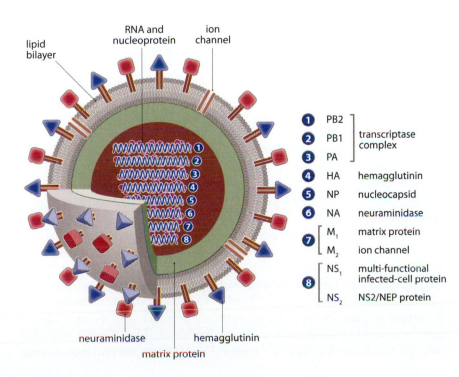

❶	PB2	transcriptase complex
❷	PB1	
❸	PA	
❹	HA	hemagglutinin
❺	NP	nucleocapsid
❻	NA	neuraminidase
❼	M_1	matrix protein
	M_2	ion channel
❽	NS_1	multi-functional infected-cell protein
	NS_2	NS2/NEP protein

Figure 1. Schematic diagram of an influenza virus. The eight segments of RNA are enclosed within a nucleocapsid, which is in turn surrounded by a lipid envelope into which are inserted two surface glycoproteins, the hemagglutinin and neuraminidase. The helical nucleocapsid contains eight segments of ssRNA each coated with nucleoprotein. This is surrounded by a layer of M1 (membrane or matrix) protein, which in turn is surrounded by a lipid envelope into which are inserted two viral glycoproteins (hemagglutinin and neuraminidase) and a small amount of the M2 ion channel protein.

ribonucleoprotein (RNP), or **nucleocapsid**. The RNPs are in turn surrounded by a matrix protein and then a **lipid envelope**, which contains two viral glycoproteins, **hemagglutinin** (H or HA) and neuraminidase (N or NA), and also small amounts of the nonglycosylated M2 ion channel protein. Influenza viruses are grouped into types on the basis of the nature of the NP, which occurs in one of three antigenic forms, hence types A, B, or C influenza viruses. Influenza type A viruses are widespread in nature, infecting many avian species, but also humans, pigs, horses, and occasionally other species such as cats. Influenza B virus is an exclusively human pathogen, while influenza C viruses are not serious pathogens in humans. Influenza type A viruses are further subdivided into subtypes depending on the nature of their two external glycoproteins. Thus far, 16 distinct hemagglutinins and nine different neuraminidases have been identified, where each HA or NA molecule differs by at least 20% of its amino acid sequence from all other HA and NA molecules. When referring to an influenza A virus isolate, it is therefore necessary to specify precisely which subtype it is, for example influenza A/H1N1 or influenza A/H7N7.

Entry and spread within the body

Influenza virus enters via the nasal or oral mucosa. In humans and other mammalian species, the virus is **pneumotropic** (in avian species, the virus infects a variety of tissues and is primarily spread through the **fecal–oral** route), that is it preferentially binds to, and infects, respiratory epithelial cells, all the way from the **oropharynx** and nasopharynx right down to the alveolar walls. Influenza virus attaches to target cells via an interaction between the viral ligand, hemagglutinin, and a cellular receptor, comprising sialic acid residues, a component of the carbohydrate within glycoproteins, on the surface of respiratory epithelial cells. This can occur throughout the length of the respiratory tract. The virus enters the host cell via **vesicles** and uncoats. The virus then replicates and new virions are released by the infected cells by budding at the plasma membrane of the host cell. With infections of the lower respiratory tract, direct infection of **pneumocytes** and macrophages can occur. Given the systemic nature of the illness caused by influenza virus infection (see below), it is perhaps surprising that the virus itself does not usually spread beyond the respiratory tract.

Spread from person to person

Transmission of influenza viruses from person to person is believed to be via large droplets (≥ 5 μm diameter), which are generated from an infected respiratory tract during coughing, sneezing, or even talking. The droplets are deposited on the nasal or oral mucosa of a new susceptible host leading to infection.

Epidemiology

The epidemiology of influenza has several unusual characteristics (Figure 2). Annual outbreaks of infection are highly seasonal, arising each winter in temperate climates, with a considerable percentage (e.g. 10%) of the population acquiring infection, with concomitant increases in hospital admissions and influenza-related deaths. The size of these outbreaks varies from year to year. In the UK, an outbreak is referred to as an epidemic only when the consultation rate for 'influenza-like illness' recorded through the Royal College of General Practitioners surveillance scheme

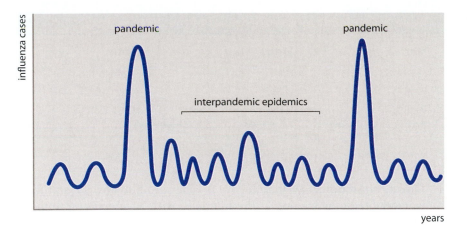

Figure 2. Epidemiology of influenza. This diagram shows the number of cases of influenza occurring over time. Each peak corresponds to a winter season, illustrating the annual epidemics. Superimposed on that, at irregular intervals averaging about once every 30–40 years, there is a massive peak corresponding to an influenza pandemic.

exceeds 400 per 100 000 population. However, superimposed on this regular annual cyclical pattern, unpredictable global epidemics occur, on a scale much greater than the annual outbreaks, sweeping across the world with huge numbers of infections, and considerable morbidity and mortality. These latter phenomena are referred to as pandemics, and experience in the 20th century plus careful reading of historical records suggest that these have occurred about every 30 years or so. Influenza epidemics and pandemics arise from the processes of antigenic drift and antigenic shift, respectively.

Antigenic drift results in the emergence of new strains each year. It arises from random spontaneous mutation occurring within the influenza virus genome as it replicates. Virus causing an outbreak in a particular year will have up to 1% genome sequence difference from virus that caused the previous year's outbreak. Although this will occur across the whole viral genome, it is the mutations within the genes encoding the HA and NA surface glycoproteins that are important in this context. The HA protein contains five highly immunogenic regions (Figure 3) to which the antibody response to infection is directed. Mutations within these **epitopes** may therefore allow virus to escape the inhibitory effects of antibodies that would otherwise bind to these regions and prevent virus–cell interactions. The important amino acid differences that accumulate year-on-year within this protein are clustered precisely within these five epitopes. Thus, antigenic drift is an excellent example of Darwinian evolution – mutations occur randomly within the genome, but only those that confer a selective advantage to the virus emerge in the epidemic strain, the selective pressure being the population immune response generated by the previous year's epidemic. Drift occurs in both influenza A and B viruses.

Antigenic shift, which generates the new pandemic strains, is an altogether different process. The viruses causing the influenza pandemics of the 20th century are shown in Table 1. Each pandemic arose from the emergence of a new influenza A subtype into the human population. As the new pandemic strain appeared, so the old circulating strain disappeared – thus, in 1956–57, H2N2 completely replaced H1N1, only to be replaced itself by H3N2 virus in 1968 (an unusual exception to this, arising in 1976, is discussed later). There are two possible underlying mechanisms that can give rise to new pandemic strains, as described below.

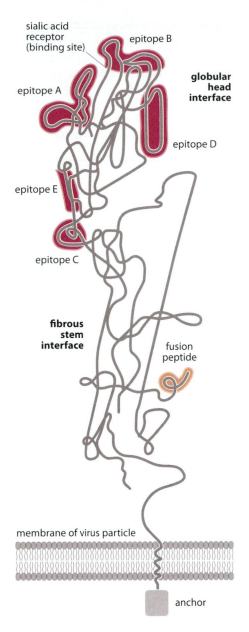

sialic acid receptor (binding site)

epitope B

epitope A

globular head interface

epitope D

epitope E

epitope C

fibrous stem interface

fusion peptide

membrane of virus particle

anchor

Figure 3. Hemagglutinin (HA) structure showing five key epitopes. This shows the protein chain of a single subunit of influenza A hemagglutinin trimer. Epitopes A, B, C, D, and E are positions where antibody molecules have been shown to bind to HA. The three dimensional structure was determined by X-ray diffraction of the crystalline protein.

Table 1. Pandemic influenza viruses of the 20th century		
Year	*Virus*	*Name*
1918–19	H1N1	Spanish flu
1956–57	H2N2	Asian flu
1968	H3N2	Hong Kong flu

1. Direct transfer of an avian influenza A virus into humans. This process is undoubtedly happening at the moment, with an increasing number of human infections with the avian H5N1 virus (responsible for large avian epidemics, particularly among chickens) being reported worldwide. However, virus that crosses a species barrier in this way is often not well adjusted for replication in its new host. Currently, avian H5N1 virus does not replicate to high titer within infected humans. Person to person spread is therefore very inefficient, as infected individuals are not releasing large amounts of virus in their respiratory secretions. H5N1 virus has thus not emerged (yet) as a new human pandemic virus. There is a worry, however, that as it replicates within human cells, this virus may acquire mutations that could result in adaptation to efficient replication within human cells, at which point person to person spread will become more likely, and a true pandemic might eventuate. There is some evidence that the H1N1 virus that caused the 1918–19 pandemic was entirely avian in origin, and that it had been causing sporadic infections within humans for several years before its emergence as a pandemic virus in 1918. The presumption is that during those preceding years the virus acquired the necessary mutations to allow adaptation to increased replication within human cells.

2. Genetic reassortment of human and avian viruses within a co-infected host (Figure 4). Influenza viruses have a segmented genome. Thus, if a cell is infected with two different influenza viruses, it is possible that reassortment (mixing) of these gene segments can occur, such that progeny virus can contain gene segments derived from either one of the 'parent' viruses. In Figure 4, the emergent virus has six RNA segments derived from the human parent virus, plus segments 4 and 6 from the avian parent. Segments 4 and 6 encode the HA and NA proteins, respectively. Thus, the progeny virus will be one that is well adapted for growth in human cells (all its internal proteins are derived from the human parent), but has two entirely new proteins on its surface (each HA and NA protein differing by at least 20% amino acid sequence from all other HA and NA proteins). Such a virus would cause devastating infection across the whole human population, as no-one would have any immunity against these new surface proteins. The H2N2 and H3N2 pandemic viruses from 1956 and 1968 do indeed contain genes derived from both human and avian viruses. It is believed that the reassortment process that generated these viruses took place within pigs (hence referred to as the 'mixing vessel'), which are uniquely susceptible to infection with both human and avian viruses. However, it has become at least a theoretical possibility that humans themselves could act as the mixing vessel, for example if a human was co-infected simultaneously with an avian A/H5N1 and a human A/H1N1 or A/H3N2 virus. The

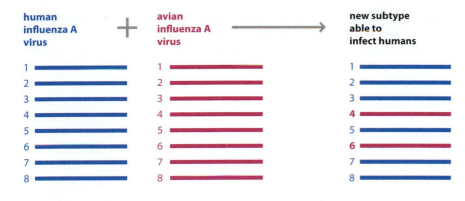

Figure 4. Genetic reassortment.
Each RNA segment (numbered 1–8) is represented by a horizontal line. The human virus is blue, the avian virus is red. When co-infecting the same cell, emergent viruses may possess RNA segments from either 'parent' virus.

chances of the latter happening will clearly be increased the more humans become infected with avian A/H5N1 virus.

It is worth emphasizing the differences between antigenic drift and shift. The former occurs in both influenza A and B viruses, and is a result of random genetic mutation followed by Darwinian selection, resulting in up to 1% differences in amino acid sequences focused within key epitopes within the surface HA and NA molecules. The latter only occurs in influenza A viruses (presumably because influenza B viruses do not have an animal reservoir), and is a result of either direct cross-species transfer or of genetic reassortment, resulting in the generation of viruses with surface HA and NA proteins that differ by over 20% in amino acid sequence from those in previously circulating strains.

In 1976, Russian flu emerged, caused by an influenza A/H1N1 virus. This is not presented in Table 1 as a pandemic, because the mechanism of emergence of this virus is not believed to have been a natural occurrence. Instead, there is evidence that this virus emerged from a laboratory – its gene sequences were remarkably similar to those of the last previous isolates of H1N1 virus in 1957. Thus its reappearance in the human population was most likely due to human error. Interestingly, this virus did not displace the A/H3N2 virus and ever since 1976, both A/H1N1 and A/H3N2 influenza viruses have co-circulated among humans, together with influenza B viruses. Any one of the three circulating viruses can predominate in a particular year.

2. What is the host response to the infection and what is the disease pathogenesis?

Damage to the respiratory epithelial surface occurs due to the **cytolytic** interaction of the virus and the host cell, that is the infected host cells undergo acute cell death. In effect, the virus strips off the inner lining of the respiratory tract, and in so doing, removes two important innate immune defence mechanisms – mucus-secreting cells, and the **muco-ciliary escalator**. The production of mucus by cells within the respiratory epithelium allows entrapment of inhaled particulate matter (e.g. bacteria). The muco-ciliary escalator then transports any inhaled particulate matter towards the pharynx, to be coughed out in sputum or swallowed. Removal of these defenses results in potential exposure of the lower respiratory tract to inhaled particulate matter, such as bacteria.

The above processes result in impairment of lung function to a greater or lesser extent, so patients present with **rhinorrhea**, sore throat, cough, and shortness of breath. However, there is also an important systemic element to the disease influenza (see below). This arises because influenza viruses are potent inducers of **cytokines** such as **interferon**-α (**IFN**-α) and **interleukin** (**IL**)-6, and it is these cytokines, not the virus, that circulate in the bloodstream and give rise to the systemic manifestations of fever, headache, muscle aches and pains, and severe malaise. Administration of IFN, for example as treatment for chronic hepatitis C virus infection, reproduces this symptomatology. Note that the above applies strictly to infection with the currently circulating human influenza viruses – there is emerging evidence that infection of humans with avian influenza A/H5N1 may result in **viremia** and spread to other organs beyond the lungs.

In addition to this innate immune response to infection, adaptive humoral and cellular immune responses are also stimulated. Antibodies to the surface proteins, particularly hemagglutinin, may be neutralizing, that is they can prevent the interaction of the HA protein with cellular sialic acid residues and thereby prevent infection. However, antigenic drift results in the generation of strains of virus that can escape this protective immunity. T-cell responses to influenza virus are mostly directed against antigens derived from the internal viral proteins, for example the nucleoprotein. These proteins are much more conserved within influenza types than the surface proteins, so T-cell immunity may offer some protection each year to emerging drifted viruses.

3. What is the typical clinical presentation and what complications can occur?

There are two distinct components to the illness that arises following infection with influenza virus – a respiratory tract component, plus a marked systemic illness characterized by fever, headache, and **myalgia**. Infection does not necessarily result in clinical disease – this will be dependent on the pre-existing state of the patient's lung function, the infecting dose of virus, the presence of pre-existing immunity and the extent to which that immunity is able to cross-react with a new viral strain. However, symptomatic influenza virus infection is not a trivial illness. There is considerable morbidity, and it may take several days before patients are well enough to return to their normal daily activities.

The commonest life-threatening complication of influenza virus infection is **pneumonia**, of which there are two pathological types.

Primary influenzal pneumonia. The virus itself infects right down to the alveoli. There is a mononuclear cell infiltrate into the alveolar walls, and the airspaces become filled with fibrinous inflammatory exudates. This can occur in previously healthy individuals of any age.

Secondary bacterial pneumonia. In recent years, this has been considerably more common than viral pneumonia. Bacteria gain access to the lower respiratory tract for reasons explained above. There is a polymorphonuclear cell infiltrate into the alveoli. This complication is more common in the elderly and in those with pre-existing lung disease, of whatever etiology, for example chronic bronchitis.

There is some evidence that influenza infection can also result in a **myocarditis** – certainly patients with pre-existing cardiovascular disease are at increased risk of mortality should they acquire infection. An **encephalitis** (inflammation of the brain substance) is also well recognized. This is not due to the virus itself gaining access to the brain – as explained above, virus is restricted to the lungs. Thus, this is believed to be an immune-mediated phenomenon, or so-called post-infectious encephalitis. In certain individuals, the immune response generated to the influenza virus infection can cross-react with antigens present within the brain, resulting in an encephalitis.

Human infection with avian influenza A H5N1 carries a very high mortality (>50%), and yet, as outlined above, this virus does not replicate efficiently within human cells. This apparent paradox is explained by the fact that this virus induces an explosive acute inflammatory reaction within the lungs – referred to as a **cytokine storm**. The high mortality thus arises from the inappropriate hyperactivity of the host immune response to infection, as opposed to the cytolytic properties of the virus itself.

4. How is this disease diagnosed and what is the differential diagnosis?

Infections with a number of different agents (mostly viruses) can result in presentation with an 'influenza-like illness.' Infection with respiratory syncytial virus, especially in the elderly, is the most common mimic of influenza virus infection. Other possibilities include human metapneumovirus, adenoviruses, and *Mycoplasma pneumoniae*. Clinical 'end-of-the-bed' diagnosis is therefore neither sensitive nor specific enough for practical purposes – with the advent of antiviral drugs that are absolutely specific for influenza viruses, accurate diagnosis is necessary to ensure that these drugs are used appropriately and effectively. As the efficacy of these drugs is dependent on initiation of their use as soon as possible after infection, there is a need for rapidity as well as accuracy.

There are two main approaches to the rapid diagnosis of influenza virus infection. Historically, most laboratories relied on immunofluorescent (IF) antigen detection using monoclonal anti-influenza antibodies (Figure 5). This technique relies on the fact that cells in which a virus is replicating will express viral antigens somewhere within the cell. Thus, cells from the patient (e.g. from a throat swab, or ideally, a nasopharyngeal aspirate) are spotted down onto a glass slide, and a fluorescently tagged monoclonal antibody is added. The antibody will bind to cells infected with virus, but not to uninfected cells. After an incubation period, any unbound antibody is washed off, and the cells are examined using a fluorescence microscope. The presence of brightly fluorescent cells is a positive result indicating that the patient was infected with the virus to which the monoclonal antibody was raised. The whole process takes about 2 hours. More recently, some laboratories have adopted a genome detection technique such as the **polymerase chain reaction (PCR)** assay (with prior reverse transcription, as influenza virus carries an RNA genome, but PCR only amplifies DNA). PCR-based assays have a major advantage in terms of their incredible sensitivity, as theoretically they result in several logs of amplification of the targeted nucleic acid sequences. Real-time PCR assays are also rapid. In

Figure 5. Detection of influenza virus by immunofluorescence. A throat swab (or ideally, a nasopharyngeal aspirate) is spotted onto a glass slide, and a fluorescently tagged monoclonal antibody is added. The antibody binds to cells infected with virus, but not to uninfected cells. After an incubation period, any unbound antibody is washed off, and the cells are examined under ultraviolet light. The presence of brightly fluorescent cells is a positive result indicating that the patient is infected with the virus to which the monoclonal antibody was raised. The whole process takes about 2 hours.

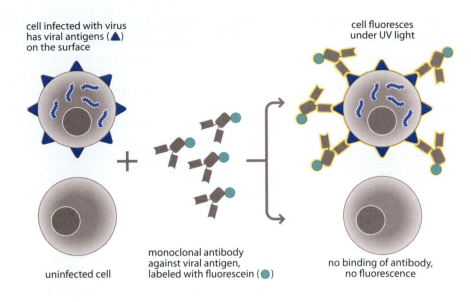

cell infected with virus has viral antigens (▲) on the surface

cell fluoresces under UV light

uninfected cell

monoclonal antibody against viral antigen, labeled with fluorescein (●)

no binding of antibody, no fluorescence

the context of anxieties about the spread of avian influenza H5N1 into humans, and the possible emergence of a new pandemic strain, most countries have adopted pandemic plans that rely on diagnosis of the first case (or cases) of infection as soon as possible, in order to institute emergency infection control procedures. Real-time PCR assays are recognized as the gold standard in this context.

If rapidity of diagnosis is not an issue (e.g. in epidemiological monitoring of infection), then alternative diagnostic approaches include isolation of virus in cell culture, and demonstration of a fourfold or greater rise in anti-influenza virus antibody titers in peripheral blood samples taken 7 days or more apart. Influenza viruses grow well in cell culture, but may take several days before their presence is revealed by the development of a **cytopathic** effect. **Serological** diagnosis (i.e. a rise in antibody titer) is also slow, since it requires, by definition, a second sample taken some days after the patient presents with acute illness.

5. How is the disease managed and prevented?

Management

Many cases of influenza virus infection require symptomatic relief only, for example with anti-pyretics and analgesics such as paracetamol, or other proprietary over-the-counter or home-grown remedies.

For the more seriously ill patients there are two classes of anti-influenzal agents shown to be effective. The first of these include amantadine and rimantadine. These drugs work by preventing the uncoating of influenza virions that have entered a target cell. They do this by binding to the viral matrix M2 protein and thereby blocking ion channels, whose function is essential for the pH-mediated dissolution of the viral capsid, that is uncoating (see Section 1). While appropriately conducted **clinical trials** of amantadine have provided clear evidence of efficacy, there are a number of drawbacks to the use of the agents. Firstly, they only work against influenza A viruses and therefore are of no benefit in a patient infected with influenza B virus. Secondly, these drugs are also dopamine agonists and therefore have marked central nervous system stimulatory activity – in fact, amantadine was originally developed for the treatment of Parkinson's disease. Thus, it is very poorly tolerated in the elderly, the precise group of patients who are most likely to require antiviral therapy, giving rise to hallucinations, insomnia, and agitation. Thirdly, resistance to amantadine emerges within a few days of onset of therapy, due to point mutations in the M2 protein. Finally, many of the avian influenza viruses, including H5N1 strains, are inherently resistant to amantadine. For all of the above reasons, this class of drugs is therefore not widely recommended.

The second class of anti-influenzal drugs comprise the neuraminidase inhibitors (Figure 6). The development of these inhibitors was a purposeful effort to design effective antiviral drugs through a logical process – these molecules can therefore be regarded as 'designer drugs.' The influenza neuraminidase was purified, crystallized, and its three-dimensional structure was elucidated. Small molecules were then designed to bind to the active site of the enzyme. One advantage of this class of drugs is that they have activity against all known influenza neuraminidase subtypes. Both of the currently licensed members of this family, zanamavir and oseltamivir, are effective in inhibiting production of infectious viral particles, and are effective in randomized clinical trials. Currently, their use in the UK is reserved for the treatment of seriously ill patients admitted to hospital, although in the USA they may be prescribed by primary care physicians. The importance of these drugs is illustrated by the decision of several governments to stockpile millions of doses as part of their influenza pandemic preparedness plans. Unsurprisingly, however, as these drugs are more widely used, there are increasing reports of viral variants

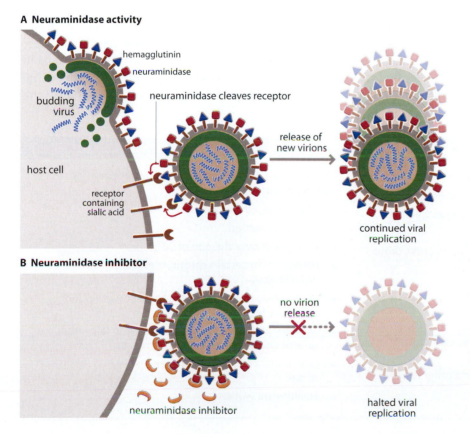

A Neuraminidase activity

hemagglutinin

neuraminidase

budding virus

host cell

neuraminidase cleaves receptor

receptor containing sialic acid

release of new virions

continued viral replication

B Neuraminidase inhibitor

no virion release

neuraminidase inhibitor

halted viral replication

Figure 6. Neuraminidase on the surface of the virus fulfills an essential role in the life cycle of the virus. As newly formed viral particles bud out of an infected cell (A), the hemagglutinin on the viral surface would naturally bind to sialic acid receptors on the surface of the cell. Thus, it would not be possible for these new virus particles to move away from the cell and infect other cells, were it not for the fact that the neuraminidase is there to remove the sialic acid residues and release the viral particles. Thus, inhibition of the viral neuraminidase by small molecule inhibitors (B) prevents virus release from the cell and therefore also prevents any downstream viral infection of and replication within other cells.

Adapted with kind permission from the New England Journal of Medicine Volume 353: 1363 – 1373, Page 1364, Figure 1. © 2005 Massachusetts Medical Society.

emerging with point mutations within the neuraminidase gene that confer drug resistance. Worryingly, oseltamivir-resistant avian H5N1 strains have now been documented to occur in infected humans.

Prevention

Vaccines

Prevention of influenza virus infection is possible through the use of active vaccination. There are two different types of vaccine currently in use. Inactivated vaccines contain either whole virus grown in embryonic hens' eggs and chemically inactivated, or just the surface HA and NA proteins purified from whole virus – the latter are less reactogenic. Live attenuated vaccines are a relatively recent development, containing cold-adapted viruses, and administered by intranasal inoculation. Preliminary data from the use of these novel vaccines suggest that they induce a more robust protection from influenza infection, possibly even from antigenic drift variants. It remains to be seen whether these live vaccines will ultimately replace the use of the inactivated ones. The propensity for influenza viruses to undergo antigenic drift and shift creates major problems for the vaccine manufacturers, as essentially the viruses are moving targets. In practice, the World Health Organization (WHO) monitors circulating influenza viruses through the collaborative activities of a large number of reference laboratories around the world. Each year, the WHO (based on monitoring current strains in its reference laboratories) announces which particular A/H1N1, A/H3N2, and B viral strains should be used for vaccine being manufactured for the following influenza season. The protection offered by these vaccines depends to a large extent on the degree of antigenic match between the vaccine strains and the strains actually circulating during the season – some years this is better than others!

The availability of a vaccine then begs the question of who should be vaccinated. Most countries adopt a selective policy, that is the recommendation is to vaccinate those subgroups within the population who will fare badly should they acquire infection. Table 2 lists those groups in the UK designated in the Department of Health guidelines. There is some controversy

Table 2. Target groups for influenza vaccination	
UK – Department of Health recommendations	**USA – Centers for Disease Control and Prevention recommendations**

• Patients aged 6 months or older with underlying:
    chronic respiratory disease (including asthma)
    chronic heart disease
    diabetes requiring insulin or oral hypoglycemic drugs
    chronic renal disease
    immunosuppression
    chronic liver disease
• Individuals over the age of 65
• Health and social care staff directly involved in patient care
• People who live in nursing homes and other long-term care facilities
	• Children aged 6 months to 5 years
	• Pregnant women
	• Individuals over the age of 50
	• Patients with certain chronic medical conditions (see list above)
	• People who live in nursing homes and other long-term care facilities
	• Household contacts of individuals who fall within the groups above
	• Household contacts of children less than 6 months of age
	• Health-care workers

as to whether children should also be vaccinated. There are also some important contraindications to the use of influenza vaccine. Given the nature of the inactivated vaccine, patients with allergy to egg proteins should not be vaccinated.

Prophylaxis with anti-influenzal drugs is also effective at preventing infection, although clearly the protection mediated by this approach only lasts as long as the drugs are administered. However, this will be important if/when the next pandemic emerges. It will take some months before an effective vaccine against the pandemic strain of virus is developed and manufactured in enough doses to offer realistic protection on a population basis. Thus, initial prevention measures once a new pandemic strain is identified may well include the use of prophylaxis with the neuraminidase inhibitors.

SUMMARY

1. What is the causative agent, how does it enter the body and how does it spread a) within the body and b) from person to person?

- Influenza A, B, and C viruses, carry a segmented ($n = 8$) negative single-stranded RNA genome, are enveloped, and belong to the family *Orthomyxoviridae*.

- Typing into A, B, C is according to the nature of the internal proteins.

- Type A is subtyped according to the nature of the surface H and N proteins.

- Entry is via inhalation, or droplet inoculation onto oropharyngeal mucous membranes.

- Viral tropism is for respiratory epithelial cells, with no evidence of spread beyond the lungs (except perhaps for avian H5N1 infection of humans).

- The viral ligand hemagglutinin binds to cell surface sialic acid receptors.

- On entry the virus uncoats and replication begins.

- Epidemiology is characterized by pandemics arising from antigenic shift, and interpandemic epidemics, arising from antigenic drift.

- Antigenic drift is due to Darwinian selection of variants with mutations in key neutralizing epitopes within the surface glycoproteins in the face of immune selection pressure.

- Antigenic shift is due to the emergence of a new influenza A subtype.

- Possible mechanisms include genetic reassortment of human and avian influenza viruses within a mixing vessel, either pigs or humans, or direct trans-species transfer of avian viruses to humans, with subsequent adaptive mutations.

- The last pandemic, due to A/H3N2 virus, was in 1968.

- Currently, A/H1N1, A/H3N2, and B viruses co-circulate, giving rise to annual interpandemic epidemics.

2. What is the host response to the infection and what is the disease pathogenesis?

- Viral infection results in **lysis** or **apoptosis** of the cell.

- Influenza virus infection therefore effectively strips off the inner lining of the respiratory tract, including mucus-secreting and ciliated epithelial cells.

- This predisposes to inhalation of particulate matter, including bacteria, into the lower respiratory tract.

- Innate immune responses include potent induction of interferon and other cytokines.

- Adaptive immune responses include neutralizing antibody production and T-cell responses.

3. What is the typical clinical presentation and what complications can occur?

- There are two components to the clinical manifestations of influenza virus infection.

- Respiratory tract symptomatology (e.g. rhinorrhea, cough) arises from local cellular damage and inflammation.

- Systemic manifestations (fever, headache, pronounced myalgia) arise from the effects of circulating cytokines.

- Life-threatening complications include primary influenzal pneumonia, secondary bacterial pneumonia, myocarditis, and post-infectious encephalitis.

- Avian H5N1 infection of humans has a high mortality, due to an intense acute inflammatory reaction within the lungs (a cytokine storm).

4. How is this disease diagnosed and what is the differential diagnosis?

- Diagnosis is by demonstration of influenza virus in a respiratory sample – for example nasopharyngeal aspirate, throat swab.

- Immunofluorescent antigen detection with labeled monoclonal anti-influenza antibodies is a rapid, specific, and reasonably sensitive approach.

- Influenza virus RNA detection by genome amplification (e.g. real-time reverse transcriptase polymerase chain reaction assay) is also rapid and much more sensitive. This is the approach used for early diagnosis of human infections with avian A/H5N1 virus.

- Retrospective diagnosis can be made by isolation of virus in cell culture (takes several days), or by demonstration of a rise in specific antibody titers (requires a blood sample taken several days after onset of illness).

- Many other infections may present with an 'influenza-like illness' – the most common is with respiratory syncytial virus.

5. How is the disease managed and prevented?

- The majority of patients infected with influenza virus infection can be managed symptomatically, for example with appropriate analgesia and anti-pyretics.

- Amantadine has activity against some (not all) influenza A viruses. It works by binding to the M2 protein and blocking an ion channel necessary for uncoating of the virus.

- It has central nervous system stimulatory side effects and is not well tolerated, particularly in the elderly.

- Resistance emerges rapidly due to mutations in the M2 protein.

- The neuraminidase inhibitors (zanamavir, oseltamivir) have activity against all known influenza virus neuraminidase enzymes. In the UK, their use is focused on seriously ill hospitalized patients.

- Resistance to the neuraminidase inhibitors has been reported, due to point mutations within the NA gene.

- Vaccination is with either inactivated whole virus, a subunit derivative containing only purified hemagglutinin and neuraminidase, or live attenuated vaccines.

- The vaccines are trivalent (i.e. contain antigens from all three co-circulating viruses).

- Vaccine composition is adjusted annually to take account of antigenic drift.

- Vaccine targeting may differ in different countries. In the UK, recommendations are to vaccinate high-risk subgroups within the general population, that is those with pre-existing respiratory, cardiac, renal, endocrine, or liver disease, immunodeficiency, or those over the age of 65.

- Targeted individuals require annual vaccination.

- An alternative to vaccination is prophylactic use of neuraminidase inhibitors.

FURTHER READING

Humphreys H, Irving WL. Problem-orientated Clinical Microbiology and Infection, 2nd edition. Oxford University Press, Oxford, 2004.

Murphy K, Travers P, Walport M. Janeway's Immunobiology, 7th edition. Garland Science, New York, 2008.

Richman DD, Whitley RJ, Hayden FG. Clinical Virology, 2nd edition. ASM Press, Washington, DC, 2002.

Zuckerman AJ, Banatvala JE, Pattison JR, Griffiths PD, Shaub BD. Principles and Practice of Clinical Virology, 5th edition. Wiley, Chichester, 2004.

REFERENCES

Belshe RB. The origins of pandemic influenza – lessons from the 1918 virus. N Engl J Med, 2005, 353: 2209–2211.

Lim WS, Thomson A, Little P. Preparing for the next flu pandemic. BMJ, 2007, 334: 268–269.

Moscona A. Neuraminidase inhibitors for influenza. N Engl J Med, 2005, 353: 1363–1373.

Osterholm MT. Preparing for the next pandemic. N Engl J Med, 2005, 352: 1839–1842.

Webster RG. H5N1 influenza – continuing evolution and spread. N Engl J Med, 2006, 355: 2174–2177.

Writing Committee of the Second WHO Consultation on Human Influenza A/H5N1. Update on Avian influenza A (H5N1) infection in humans. N Engl J Med, 2008, 358: 261–273.

WEB SITES

All the Virology on the WWW Website, developed and maintained by Dr David Sander, Tulane University: http://www.virology.net/garryfavweb13.html#Ortho

Centers for Disease Control and Prevention, Coordinating Center for Infectious Diseases (CCID) Atlanta, GA, USA: http://www.cdc.gov/flu/

Centre for Infections, Health Protection Agency, HPA Copyright, 2008: http://www.hpa.org.uk/infections/topics_az/influenza/

Website of Derek Wong, a medical virologist working in Hong Kong: http://virology-online.com/viruses/Influenza.htm

MULTIPLE CHOICE QUESTIONS

The questions should be answered either by selecting True (T) or False (F) for each answer statement, or by selecting the answer statements which best answer the question. Answers can be found in the back of the book.

1. Which of the following statements regarding influenza viruses are true?
   A. Their genome consists of eight segments of double-stranded RNA.
   B. They are classified into types A, B, and C on the basis of the nature of their internal proteins, particularly the nucleoprotein.
   C. Type A influenza viruses are further subdivided into subtypes on the basis of the nature of their matrix proteins.
   D. Type B influenza viruses are further subdivided into subtypes on the basis of the nature of their surface proteins.
   E. There is no animal reservoir of type B influenza viruses.

2. The emergence of new pandemic strains of influenza virus may arise from which of the following processes?
   A. Trans-species transfer of an avian influenza virus directly to humans.
   B. Spontaneous mutations in the genes encoding the surface glycoproteins.
   C. Reassortment of avian and human influenza viruses within a single host.
   D. Use of neuraminidase inhibitors resulting in mutations in the neuraminidase gene.
   E. Natural selection of viral variants in an immunized host population.

MULTIPLE CHOICE QUESTIONS (continued)

3. **Which of the following statements regarding the epidemiology of influenza viruses is/are correct?**

   A. Influenza B viruses undergo antigenic shift but not antigenic drift.

   B. Antigenic shift in influenza viruses gives rise to global pandemics of influenza.

   C. Antigenic drift in influenza viruses gives rise to interpandemic epidemics of influenza.

   D. Antigenic shift describes the emergence of new influenza A virus subtypes.

   E. Antigenic drift results in amino acid changes clustered within key epitopes of the viral nucleoprotein.

4. **Which of the following statements regarding disease associated with influenza virus infection is/are true?**

   A. Influenza virus infection of respiratory epithelial cells results in transformation of those cells.

   B. Systemic manifestations of influenza virus infection (e.g. fever, myalgia, headache) arise from the presence of virus circulating in the bloodstream.

   C. Pneumonia arising as a complication of influenza virus infection is usually due to secondary bacterial invasion.

   D. Influenza-related mortality is higher in patients with pre-existing cardiac disease.

   E. Influenza-related encephalitis arises through cross-reactivity of the immune response to infection with the patient's brain tissue.

5. **Which of the following statements regarding the avian H5N1 influenza virus are true?**

   A. Infection of humans results in death in 10–20% of cases.

   B. The pathogenesis of disease arises from explosive release of cytokines within the respiratory tract.

   C. Infection can be prevented by vaccination with vaccines containing antigens derived from A/H1N1 and A/H3N2 viruses.

   D. This virus is always sensitive to oseltamivir.

   E. This virus is a possible candidate for the next influenza pandemic.

6. **Which of the following statements regarding diagnostic tests is/are true?**

   A. Virus isolation in cell culture is a rapid diagnostic technique.

   B. Antigen detection techniques are dependent on the presence of viable virus in the sample sent to the laboratory.

   C. Genome detection techniques are the most sensitive assays for diagnosis of virus infections.

   D. Genome amplification assays cannot be used for RNA viruses.

   E. Demonstration of high antibody titers to the H5 hemagglutinin in serum samples taken from acutely ill

patients will be the mainstay of diagnosis of human infection with avian H5N1 influenza virus.

7. **Which of the following drugs has proven efficacy against influenza A viruses?**

   A. Aciclovir.

   B. Foscarnet.

   C. Indinavir.

   D. Zanamavir.

   E. Zidovudine.

8. **With regard to anti-influenza drugs, which of the following statements are true?**

   A. Amantadine is effective as a prophylactic agent against influenza B virus.

   B. The mode of action of amantadine involves blockage of an ion channel and prevention of viral uncoating.

   C. Neuraminidase inhibitors have no activity against influenza B virus.

   D. Resistance to oseltamivir has not been described in influenza A viruses.

   E. Zanamavir should not be used in patients with a history of egg allergy.

9. **Which of the following statements regarding influenza are true?**

   A. Influenza re-infections occur despite the presence of high levels of serum antibodies.

   B. Influenza pandemics have only occurred since the beginning of the 20th century.

   C. Human influenza viruses only infect humans.

   D. Epidemics of influenza occur in the winter months in the northern hemisphere.

   E. Infections with influenza B virus are less severe than those with influenza C virus.

10. **With regard to influenza vaccines, which of the following statements are true?**

   A. Vaccine-induced immunity is clinically useful for at least 10 years.

   B. Universal vaccination against influenza is currently recommended.

   C. Inactivated influenza vaccines are contraindicated in immunosuppressed individuals.

   D. Influenza vaccines contain antigens derived from A/H1N1, A/H3N2, and B viruses.

   E. Live attenuated influenza vaccines are administered by subcutaneous injection.

Case 19

Leishmania spp.

A 72-year-old gentleman retired to the south of Spain but returned to the UK for the summer months. He began to develop fever, malaise, loss of appetite, and weight loss. He was admitted to hospital and had temperatures reaching 39°C. Both his liver and spleen were palpable. No lymph nodes could be felt. Blood tests showed a **pancytopenia**. Routine investigations for an infection were negative and he did not improve with broad-spectrum antibiotics. His condition deteriorated and the size of the liver and spleen increased (Figure 1). A bone marrow examination did not show any sign of hematological malignancy. No organisms were seen on staining. His history was explored again. Four months before his illness he had been on a camping break in Spain to a coastal area. He recalled seeing many thin dogs in the vicinity. Part of his bone marrow sample was sent to a reference laboratory for *Leishmania* **polymerase chain reaction (PCR)**. This returned positive. He was successfully treated with a course of liposomal amphotericin B and over the ensuing 3 months his liver and spleen became impalpable and his blood tests returned to normal. His diagnosis was visceral leishmaniasis probably due to *Leishmania infantum*.

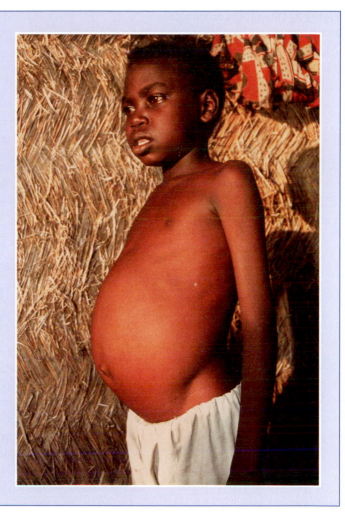

Figure 1. A child with visceral leishmaniasis. As in the patient described in the case history the liver and spleen are enlarged, causing distension of the abdomen.

1. What is the causative agent, how does it enter the body and how does it spread a) within the body and b) from person to person?

Causative agent

Leishmania are protozoan parasites. They have an intracellular form called an amastigote (Figure 2) and an extracellular, flagellated form called a promastigote (Figure 3).

There is variety in the clinical diseases caused, geographical distribution, and animal reservoirs. The genus *Leishmania* is divided into groups, species, and complexes. Classification was classically determined by **isoenzyme typing**; molecular methods (using DNA) are now more common. Table 1 lists species and the diseases they cause.

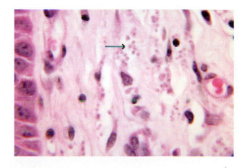

Figure 2. A skin biopsy showing *Leishmania* amastigotes (arrowed).

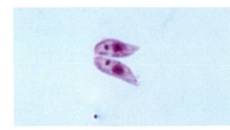

Figure 3. Elongated *Leishmania* promastigotes.

Table 1. Species of *Leishmania* and the diseases they cause

Disease	Species
Cutaneous leishmaniasis	L. aethiopica
	L. killicki
	L. major
	L. tropica
	L. amazonensis
	L. columbiensis
	L. guyanensis
	L. lainsoni
	L. mexicana
	L. naiffi
	L. shawi
	L. venezuelensis
Mucocutaneous leishmaniasis	L. braziliensis
	L. panamensis
Visceral leishmaniasis	L. donovani
	L. infantum
	L. chagasi

Entry and spread within the body

People are infected after the bite of a sandfly laden with *Leishmania* promastigotes. Under the skin the promastigotes are rapidly phagocytosed by macrophages. For cutaneous disease lesions are confined to the locality of the sandfly bite. For *L. braziliensis* and *L. panamensis* cutaneous spread can occur and later this can involve mucous membranes of the mouth or nose. *L. donovani* and *L. infantum* are capable of deeper spread within macrophages to the rest of the **mononuclear phagocytic system**, mainly present in organs such as the liver, spleen, and bone marrow. They are responsible for visceral leishmaniasis. In India, visceral leishmaniasis is called **kala-azar**. Relapse of infection after an interval may be manifest as a widespread cutaneous form of disease, called post kala-azar dermal leishmaniasis (PKDL). This occurs in India and East Africa. The life cycle of *Leishmania* is shown in Figure 4.

Person to person spread

In areas with visceral leishmaniasis sandflies can ingest protozoa when they feed from the skin. Numbers of *Leishmania* in the skin are even higher in PKDL. However, leishmaniasis is largely a zoonosis. Different animal reservoirs occur in different regions. They include rodents, gerbils, hyraxes, sloths, and the domestic dog.

The sandfly vector is a *Phlebotomus* species in the Old World and *Lutzomyia* species in the New World. Sandflies are small, less than 5 mm in size, and bite at dusk or during the night (Figure 5). They are not capable of flying great heights above the ground and usually bite individuals sleeping close to the ground. In the case described above the patient was probably infected through sandflies when he was lying near the ground on his camp bed. He normally lived in a flat. The sandflies will have carried infection from the local dog population.

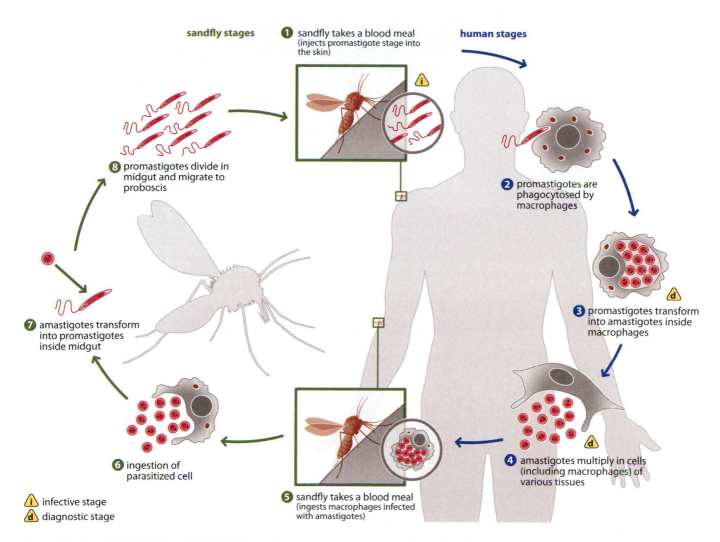

sandfly stages

human stages

① sandfly takes a blood meal (injects promastigote stage into the skin)

⑧ promastigotes divide in midgut and migrate to proboscis

⑦ amastigotes transform into promastigotes inside midgut

⑥ ingestion of parasitized cell

⑤ sandfly takes a blood meal (ingests macrophages infected with amastigotes)

② promastigotes are phagocytosed by macrophages

③ promastigotes transform into amastigotes inside macrophages

④ amastigotes multiply in cells (including macrophages) of various tissues

ⓘ infective stage
ⓓ diagnostic stage

Figure 4. Life cycle of _Leishmania_ spp. _Leishmania_ promastigotes are inoculated by sandflies into human and other animal hosts at the time of taking a blood meal (1). Promastigotes are phagocytosed by macrophages (2). Within macrophages promastigotes transform into amastigotes (3). Amastigotes can multiply in various cell types (4). Macrophages containing amastigotes are ingested by sandflies taking a blood meal (5) and the life cycle continues within the sandfly vector (6–8).

Female sandflies bite and take blood from their target host. Any amastigotes ingested from the skin change into promastigotes. These pass into the sandfly midgut, proliferate, cause damage to the digestive valve system, and are regurgitated to the biting mouthparts and then onto the skin of the next host to be bitten.

Another form of transmission for visceral leishmaniasis has been described among intravenous drug users in Southern Europe. Infection can be passed on with shared needles and equipment. In one study about half of discarded needles in Madrid were positive by PCR for _Leishmania_.

Epidemiology

Notification of cases of leishmaniasis occurs in only 32 of 88 endemic countries. Numbers of people afflicted by the disease are therefore estimates. There are about 0.5 million new cases of visceral leishmaniasis annually, and 90% of these are seen in Bangladesh, Brazil, India, Nepal, and Sudan. Another 1.5 million individuals get cutaneous or mucocutaneous disease.

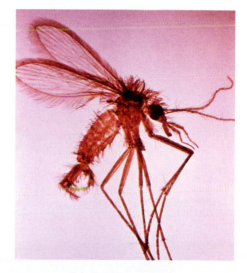

Figure 5. _Phlebotomus_ sandfly.

About 90% of mucocutaneous disease occurs in Bolivia, Brazil, and Peru, while about 90% of cutaneous disease occurs in Afghanistan, Brazil, Iran, Peru, Saudia Arabia, and Syria. In total there is an estimated worldwide prevalence of 12 million cases.

2. What is the host response to the infection and what is the disease pathogenesis?

As promastigotes enter the skin they are phagocytosed by macrophages and neutrophils. They change into amastigotes. Classically any pathogen engulfed by a phagocyte is wrapped within host cell plasma membrane. This forms a phagosome. Various membrane molecules are imported and exported as cytoplasmic vesicles fuse with or erupt from the phagosome. Eventually **lysosomes** fuse with the phagosome and discharge their contents. Lysosomal enzymes lyse susceptible pathogens. On engulfment phagocytes are activated to produce **reactive oxygen species** and reactive nitrogen intermediates. They secrete tumor necrosis factor-α (**TNF-α**), which contributes to their activation state. Macrophages are activated further by T-helper 1 lymphocytes (**Th1**) through interferon-γ (**IFN-γ**). These are stimulated by antigen-presenting cells, most efficiently by **dendritic cells**. They secrete **interleukin**-12 (IL-12). When it takes time to deal with a pathogen the combination of Th1 cells and macrophages organize into **granulomas**.

To survive, *Leishmania* need to subvert the above process. Various effects have been described but the mechanisms by which these occur are not altogether clear. Some leishmanial molecules that have been shown experimentally to play a part are lipophosphoglycan, a surface membrane metalloprotease (gp63), cysteine proteases, and a *Leishmania* homolog of activated C kinase receptor (LACK). The outcome is macrophage activation and the generation of reactive oxygen and nitrogen intermediates is suppressed, *Leishmania* resist lysosomal attack, dendritic function is compromised, and LACK induces a **Th2** response.

Recent *in vitro* work on human cells suggests that *Leishmania* may use neutrophils as a 'Trojan horse.' Neutrophils fail to kill *Leishmania* after **phagocytosis** and undergo **apoptosis**. Apoptotic fragments may contain *Leishmania*. The neutrophils release **MIP-1β**, a **chemokine** that attracts macrophages. When macrophages phagocytose the apoptotic neutrophil fragments, *Leishmania* enter 'silently' and continue to multiply. The macrophages release **transforming growth factor-β (TGF-β)**, which is anti-inflammatory.

In mice experimentally infected with *L. major* there is a clear polarization of Th1 and Th2 responses. Some strains mount a Th1 response and control infection, unlike others (BALB/c), which mount a Th2 response and experience fatal disseminated infection. Humans infected with *L. major* experience localized cutaneous disease. Conversely *L. donovani*, which causes visceral leishmaniasis in humans, can be controlled by BALB/c mice. *L. donovani* does not lead to a polarized Th1 and Th2 response between mouse strains. While there are differences between mice and humans in *Leishmania* infection, the experimental experience with mice indicates the important role of host genetics.

In humans, markers of a Th1 (IFN-γ) and Th2 (interleukin (IL)-4) response are both present at the same time. IL-4 down-regulates Th1 responses and so do other cytokines such as IL-10, IL-13, and TGF-β. These cytokines are more prominent in forms of infection that are not self-limiting like visceral leishmaniasis and PKDL. IL-10 seems to play a greater role in susceptibility to these infections than IL-4. Otherwise an unrestrained Th1 response heals other forms of infection.

3. What is the typical clinical presentation and what complications can occur?

Infection may be asymptomatic. *Leishmania* may reside in the body for years and only cause clinical disease if the host becomes immunocompromised.

Cutaneous leishmaniasis is seen on exposed parts of the body where the sandflies are likely to bite (Figure 6). Thus lesions may be found on the face, arms, and lower legs. Lesions may be single or multiple. They are usually apparent 2–6 weeks after the bite. Initially there is a red **papule**. This gradually enlarges over a few weeks. The lesion may take on a raised ulcerated form or be **papulo-nodular**. Secondary infection is possible and then lesions are more likely to be painful. Without specific treatment lesions will usually self-heal, but over prolonged periods. This may take 6 months to a few years. Over this period lesions may seem to regress and then relapse. All *Leishmania* species are capable of causing cutaneous disease, but the host immune response may alter the clinical picture. A weakened immune response with a high parasite burden causes diffuse cutaneous leishmaniasis with multiple, spreading **papular** lesions. There may also be lymphatic spread with localized nodules along the track of lymphatics. A strong immune response with a low parasite burden causes a condition called leishmania recidivans. The immune response effectively clears the initial site of infection. A series of small papules surround this central clearing and these in turn are cleared.

Mucocutaneous leishmaniasis is associated with *L. braziliensis* and *L. panamensis*. As these species names suggest this form of leishmaniasis is restricted to South America. Cutaneous lesions occur first as in purely cutaneous leishmaniasis. These can self-heal, but the parasite does not disappear from the body. After an interval, sometimes of several years, the parasite re-emerges in the mucous membranes of nose or mouth. Local inflammation results in nasal stuffiness. There is then progressive destruction of the anatomy of the nose or mouth and infection can progress backwards towards the throat and larynx (Figure 7). Eating and drinking become difficult and secondary infections in the upper and lower respiratory tract often occur. These latter effects can prove fatal unless the infection is treated. There can be considerable scarring and disfigurement if treatment is delayed.

Visceral leishmaniasis is associated with *L. donovani* in India and East Africa and with *L. infantum* around the Mediterranean and South America. A cutaneous lesion may not be apparent. After an incubation period of a few months illness is heralded by fevers. These may continue for about a month before abating. The spleen progressively enlarges first and then the liver (Figure 1). Both may become massively enlarged. The enlarged

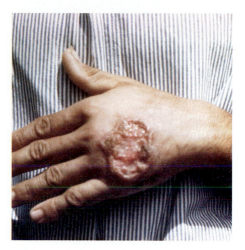

Figure 6. Cutaneous leishmaniasis of an ulcerating form.

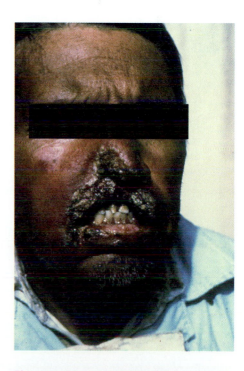

Figure 7. Mucocutaneous leishmaniasis in a patient with progressive destruction of tissues around the lips and nose.

spleen causes **hypersplenism** and consumption of blood cells, but infection within the bone marrow also causes a pancytopenia with **anemia**, **leukopenia**, and **thrombocytopenia**. In dark skins the anemia plus hormonal effects of chronic infection cause an altered appearance. In India the graying of the complexion is called kala-azar. Leukopenia predisposes to secondary infections, which themselves may be life-threatening. Thrombocytopenia can predispose to bleeding and there may be life-threatening hemorrhage. On blood tests there is also a fall in albumin levels. A drop in oncotic pressure can result in **edema**. This may be peripheral in the legs or ascites within the abdomen. There is also a polyclonal stimulation of **IgG** antibodies. The polyclonal stimulation of B lymphocytes can compromise their ability to respond to other infections. Visceral leishmaniasis runs a chronic and progressive course. Patients become wasted. It is invariably fatal unless treated. If treated some parasites may escape killing and return later to cause post kala-azar dermal leishmaniasis (PKDL). In India this interval may be 2–3 years, but shorter intervals have been observed in Sudan. The host now has some immunity from the first spell of visceral leishmaniasis. The parasite is largely confined to the skin, with extensive papulo-nodular lesions starting on the face and peripheries and then spreading to most of the body surface. This may self-cure, only to relapse and remit at a later date.

Leishmaniasis and HIV co-exist in many areas. The immunocompromised nature of HIV has caused more florid manifestations of leishmaniasis. Parasite burdens are higher. Species that may only cause cutaneous disease may become visceral.

4. How is the disease diagnosed and what is the differential diagnosis?

In endemic areas cutaneous and mucocutaneous leishmaniasis may be diagnosed on purely clinical grounds. The clinical picture of fever, **splenomegaly**, and anemia due to visceral leishmaniasis may also be caused by other diseases. Investigations are required to confirm the diagnosis. These could entail direct visualization of the parasite in a tissue sample, detection of antigen, detection of nucleic acid by PCR, or immunodiagnosis. The latter includes **serology** or a leishmanial skin test.

Deep tissue samples may be obtained by a splenic aspirate, bone marrow aspirate, lymph node aspirate, or sometimes liver biopsy. Cutaneous lesions may be squeezed firmly with fingers to exclude blood, superficially incised with a scalpel at their edge, and then tissue fluid expressed and impressed onto a glass slide. On staining of tissue samples intracellular amastigotes are sought. Their appearance is characteristic with a small kinetoplast body adjacent to the nucleus. This is called a **Donovan body** (Figure 8). The sensitivity of tissue sampling varies with the sample – >90% for splenic aspirate, 55–97% for bone marrow, and 60% for lymph nodes. If facilities are available culture can help, but now in resource-rich settings PCR is applied with a sensitivity of >95%. Leishmanial antigen can be detected in urine. A latex agglutination technique called KATEX has shown sensitivities of 68–100% for visceral leishmaniasis. After successful treatment antigen disappears from the urine.

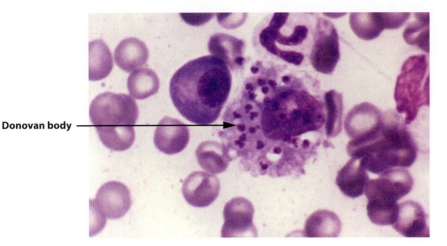

Donovan body

Figure 8. Characteristic *Leishmania* amastigote forms (Donovan bodies, arrowed) on an impression smear. Darkly staining small kinetoplast adjacent to nuclei.

Serological immunological tests are unable to distinguish current from past infection. A commonly used test for anti-leishmanial antibody is the direct agglutination test (DAT). Promastigotes are formalin-fixed onto slides and serum is placed on top. Agglutination is observed after 24 hours. DAT has a sensitivity of about 95% and its specificity is about 86%. A dipstick test has been developed with the K39 antigen impregnated on a reagent strip. Blood is added to the strip and a reaction noted after 20 minutes. The K39 dipstick has a sensitivity of about 94% and specificity of about 90%.

Historically a test similar to the tuberculin skin test for tuberculosis was used for leishmaniasis. This was the Montenegro skin test. Leishmanial antigen was implanted in the forearm and the induration after 48–72 hours was measured. Now standardized antigen preparations are no longer available. When available the skin test had been useful in distinguishing present from past infection.

Differential diagnosis

In endemic areas, cutaneous and mucocutaneous leishmaniasis may have a characteristic appearance. Cutaneous lesions may have to be differentiated from other infected insect bites, tuberculosis, fungal infection, **myiasis**, and skin cancers. Mucosal sites may also be affected by syphilis, histoplasmosis, paracoccidioidomycosis, and leprosy. Visceral leishmaniasis may have to be differentiated from other causes of fever, splenomegaly, and anemia. The differential diagnosis includes malaria, schistosomiasis, typhoid fever, brucellosis, tuberculosis, rickettsial infection, sarcoidosis, **Still's disease**, and hematological malignancy.

5. How is the disease managed and prevented?

Management

Simple cutaneous leishmaniasis may be left to self-heal in geographical areas with *L. major*. Some cutaneous lesions, mucocutaneous disease, and visceral leishmaniasis require treatment. Until recently the mainstay of treatment has been pentavalent antimony compounds. These include sodium stibogluconate and meglumine antimonate. They are administered by intramuscular injection on a daily basis for up to 28 days. The intramuscular injections can be painful and there can be systemic toxicity. Amphotericin B and more recently liposomal amphotericin B represent advances in treatment. However, they require intravenous administration, are also toxic, and are much more expensive, especially liposomal amphotericin B. An oral, tolerable agent is now available for visceral leishmaniasis. Oral miltefosine for 28 days was shown to be equally effective with amphotericin B for visceral leishmaniasis in India. Alternative treatments are very welcome, as about 60% of visceral leishmaniasis infections in Bihar, India are resistant to treatment with pentavalent antimonials.

Prevention

Vaccines have been trialed for leishmaniasis, but have not been encouraging to date. Prevention has therefore focused on sandflies. As they bite at night, sleeping under a bed-net might afford some protection. But the sandflies are small and can get through the mesh of the nets. However, if the nets are impregnated with a pyrethroid insecticide the sandflies are killed. Insecticide-treated bed-nets are also a key component of malaria control programmes. Their distribution and maintenance are a logistic challenge. To date there seems to have been little impact on the incidence of leishmaniasis. Numbers of cases seem to be growing worldwide.

SUMMARY

1. What is the causative agent, how does it enter the body and how does it spread a) within the body and b) from person to person?

- *Leishmania* are protozoan parasites.

- There is a large number of species.

- The extracellular stage is called the promastigote and the intracellular stage is the amastigote.

- Spread is by sandflies either from animal reservoirs or humans with heavy skin loads of parasite. The latter occurs in a condition called post kala-azar dermal leishmaniasis.

- The ability to spread within the body is a function of both the species of *Leishmania* and the host immune response.

2. What is the host response to the infection and what is the disease pathogenesis?

- *Leishmania* are phagocytosed by neutrophils and macrophages.

- T-helper 1 lymphocytes help macrophages to kill *Leishmania*.

- Some species may be more successful at subverting the immune response and causing disseminated infection.

- Subversion of the immune response involves suppression of macrophage activation and diversion towards a T-helper 2 type of response.

- In mice infected with *L. major* there is a clear polarization of T-helper 1 and 2 responses, depending on the genetic background of the mice.

3. What is the typical clinical presentation and what complications can occur?

- Cutaneous leishmaniasis is caused by a large number of species with infection confined to one locality.

- Lesions enlarge gradually over a few weeks, become papulo-nodular or ulcerate.

- Lesions may self-heal.

- Mucocutaneous leishmaniasis is caused by *L. braziliensis* and *L. panamensis*.

- After an initial cutaneous lesion there is a later mucosal lesion, which is progressively destructive of nose or mouth.

- Visceral leishmaniasis is due to *L. donovani* or *L. infantum*.

- Infection spreads through the mononuclear phagocytic system with enlargement of spleen and liver, and bone marrow infiltration.

- Skin complexion changes giving rise to the Indian term, kala-azar.

- After treatment of visceral leishmaniasis relapse may be confined to the skin with post kala-azar dermal leishmaniasis.

4. How is the disease diagnosed and what is the differential diagnosis?

- Diagnosis of cutaneous or mucocutaneous leishmaniasis may be purely clinical.

- Stained tissue samples may show characteristic Donovan bodies.

- Species-specific PCR has a high sensitivity.

- Assays exist for leishmanial antigen or antibody.

5. How is the disease managed and prevented?

- Simple cutaneous lesions may self-heal.

- **Parenteral** pentavalent antimonial drugs have been traditional treatment for all forms of disease.

- Parenteral amphotericin B, either conventional or liposomal, can be used for visceral leishmaniasis.

- Oral miltefosine is a recent advance in treating visceral leishmaniasis.

FURTHER READING

Dedet JP, Pratlong F. Leishmaniasis. In: Cook GC, Zumla AI. Manson's Tropical Diseases, 21st edition. Saunders/Elsevier, London/Philadelphia, 2003: 1339–1364.

Murphy K, Travers P, Walport M. Janeway's Immunobiology, 7th edition. Garland Science, New York, 2008.

REFERENCES

Chappuis F, Rijal S, Soto A, et al. A meta-analysis of the diagnostic performance of the direct agglutination test and rK39 dipstick for visceral leishmaniasis. BMJ, 2006, 333: 723–726.

Herwaldt BL. Leishmaniasis. Lancet, 1999, 354: 1191–1199.

Murray HW, Berman JD, Davies CR, Saravia NG. Advances in leishmaniasis. Lancet, 2005, 366: 1561–1577.

Sundar S, Rai M. Laboratory diagnosis of visceral leishmaniasis. Clin Diagn Lab Immunol, 2002, 9: 951–958.

Sundar S, Jha TK, Thakur CP, et al. Oral miltefosine for Indian visceral leishmaniasis. N Engl J Med, 2002, 347: 1739–1746.

Van Zandbergen G, Klinger M, Mueller A, et al. Neutrophil granulocyte serves as a vector for *Leishmania* entry into macrophages. J Immunol, 2004, 173: 6521–6525.

WEB SITES

Centers for Disease Control and Prevention, Atlanta, Georgia, USA: www.cdc.gov/

Centre for Infections, Health Protection Agency, HPA Copyright, 2008: www.hpa.org.uk

World Health Organization: www.who.int

MULTIPLE CHOICE QUESTIONS

The questions should be answered either by selecting True (T) or False (F) for each answer statement, or by selecting the answer statements which best answer the question. Answers can be found in the back of the book.

1. **Which of the following are true about the causative agent of leishmaniasis?**
    A. The extracellular form of *leishmania* is called the amastigote.
    B. *Leishmania* can be taxonomically subdivided by isoenzyme typing.
    C. The form of clinical disease can depend on the species of *Leishmania*.
    D. There are both animal and human reservoirs of infection.
    E. Spread within the body occurs inside macrophages.

2. **Which of the following are true of the transmission of leishmaniasis?**
    A. The domestic dog can serve as a reservoir of infection.
    B. The insect vector is the mosquito.
    C. The infective form when the vector bites a new host is the amastigote.
    D. Infection can be transmitted between intravenous drug users with shared needles and equipment.
    E. Individuals with post kala-azar dermal leishmaniasis (PKDL) can serve as a human reservoir of infection.

3. **Which of the following are true of the host response to leishmaniasis?**
    A. A Th1 response is associated with susceptibility.
    B. Transforming growth factor-β activates macrophages to increase leishmanicidal functions.

MULTIPLE CHOICE QUESTIONS (continued)

C. In different strains of mice infected with *L. major* there is either a Th1 or Th2 response.

D. Neutrophils fail to kill phagocytosed *Leishmania*.

E. A granulomatous inflammatory response develops in tissues.

4. **Which of the following are true of the pathogenesis of leishmaniasis?**

    A. *Leishmania* may enter macrophages in apoptotic neutrophil fragments.

    B. *Leishmania* are susceptible to lysosomal attack.

    C. *Leishmania* homolog of activated C kinase receptor (LACK) steers the host to a Th1 response.

    D. Dendritic cell antigen presentation is down-regulated by *Leishmania*.

    E. *Leishmania* suppresses the production of reactive oxygen and nitrogen intermediates.

5. **Which of the following are true of cutaneous and mucocutaneous leishmaniasis?**

    A. Cutaneous lesions are usually found in moist areas such as the armpit and groins.

    B. Lesions are manifest 2–6 days after a bite.

    C. Lesions are always painful.

    D. *L. braziliensis* and *L. panamensis* are the key species responsible for mucocutaneous disease.

    E. In mucocutaneous disease the mucosal lesions appear several years after an initial cutaneous lesion.

6. **Which of the following are true of visceral leishmaniasis?**

    A. Massive enlargement of the liver and spleen are possible.

    B. As a reaction to infection there is a rise in the number of platelets.

    C. In the blood there is a hypergammaglobulinemia.

    D. Post kala-azar dermal leishmaniasis occurs as an immediate reaction to treatment for visceral leishmaniasis.

    E. In HIV-infected patients usually cutaneous species of *Leishmania* can cause visceral disease.

7. **Which of the following are true of the diagnosis of leishmaniasis?**

    A. Mucocutaneous disease can be diagnosed clinically in endemic areas.

B. On staining tissue samples the characteristic appearance of the organism is called Donovan bodies.

C. The sensitivity of splenic aspirates is about 50%.

D. PCR has a high sensitivity and specificity.

E. Serological tests can distinguish recent acute infection from past infection.

8. **Which of the following are true of the differential diagnosis of leishmaniasis?**

    A. Cutaneous tuberculosis may cause similar cutaneous lesions.

    B. Mucocutaneous leishmaniasis is the only infection that destroys the nasal septum.

    C. Visceral leishmaniasis may have to be differentiated from malaria.

    D. Clinical features alone are sufficient to differentiate visceral leishmaniasis from other diagnoses.

    E. Noninfectious diseases can cause fever like visceral leishmaniasis.

9. **Which of the following are true of the treatment of leishmaniasis?**

    A. All cutaneous lesions need to be treated to prevent disease progression.

    B. All treatments are parenteral (i.e. need to be given by injection).

    C. Treatment courses can last up to 28 days.

    D. Liposomal amphotericin B can be used to treat visceral leishmaniasis.

    E. Miltefosine is less effective than other treatments for visceral leishmaniasis.

10. **Which of the following are true of the control and prevention of leishmaniasis?**

    A. Mosquito nets are sufficient to prevent sandfly bites.

    B. The distribution of mosquito nets to control malaria has reduced the incidence of leishmaniasis.

    C. Sandflies are intrinsically resistant to pyrethroid insecticides.

    D. There is no effective vaccine to date.

    E. Treatment of individuals with post kala-azar dermal leishmaniasis is sufficient to eliminate the reservoir of infection.

Case 20

Leptospira spp.

A Staff Sergeant was repatriated from an exercise in Belize and admitted to hospital in the UK. He was complaining of fever, headache, and **myalgia**. After an initial improvement he began to deteriorate. He became jaundiced with signs of **pneumonia** and on examination had conjunctival inflammation and hepatosplenomegaly. A chest X-ray indicated bi-basal opacities. His blood count showed a **neutrophilia** with a **thrombocytopenia**. Liver function tests showed an elevated conjugated **bilirubin** with mild elevation of transaminases. He was **oliguric** and **uremic**. A diagnosis of Weil's disease (leptospirosis) was made and he was started on benzylpenicillin and renal dialysis.

1. What is the causative agent, how does it enter the body and how does it spread a) within the body and b) from person to person?

Causative agent

Leptospirosis is caused by a number of pathogenic isolates of *Leptospira*. The current taxonomy of *Leptospira* is based on DNA relatedness with over 17 species. The genome classification of *Leptospira* does not correlate with the original serological classification and some genospecies contain both pathogenic and nonpathogenic isolates. Over 250 pathogenic serovars have been identified. *L. interrogans* contains several pathogenic **serovars** such as Icterohaemorrhagiae, Autumnalis, Copenhageni, Canicola, Hardjo, and Pomona. *L. biflexa* is a nonpathogenic isolate. See Table 1.

Leptospira are motile, very thin, tightly coiled spirochetes measuring 0.1 μm by 10–20 μm with a characteristic curve at either end (Figure 1) and they have a typical gram-negative cell wall structure (see Case 10, *E. coli*). They have two axial flagella that are located between the peptido-glycan and outer membrane layer in the periplasmic space (Figure 2).

Entry into the body

Infection is acquired by contact with water or soil contaminated by one of the pathogenic *Leptospira* serovars. Entry is via skin lesions, or lesions in the mucosae of the respiratory and digestive tracts or conjunctivae. The organism may also be acquired from contaminated aerosols entering the respiratory tract.

A number of animal species act as a reservoir of infection, the common ones being rodents, but dogs (*L. canicola*) and livestock (*L. pomona*) can also act as reservoirs. Infected animals excrete the bacterium in their urine thus contaminating water sources and soil. Human infection can be acquired directly from contact with the animals or from contact with contaminated water. Globally *Leptospira* is the commonest **zoonosis**.

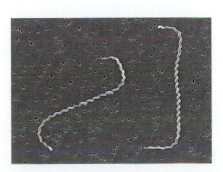

Figure 1. Morphology of *Leptospira*.

Table 1. The named species of *Leptospira* and the associated serovars	
Genospecies	*Serovar**
Leptospira interrogans	Australis
	Autumnalis
	Ballum
	Bataviae
	Canicola
	Celledoni
	Copenhageni
	Cynopteri
	Djasiman
	Grippotyphosa
	Hardjo
	Hebdomadis
	Icterohaemorrhagiae
	Javanica
	Louisiana
	Lyme
	Manhao
	Mini
	Panama
	Pomona
	Pyrogenes
	Ranarum
	Sarmin
	Sejroe
	Shermani
	Tarassovi
Leptospira biflexa	
Leptospira borgpetersenii	
Leptospira broomii	
Leptospira fainei	
Leptospira inadai	
Leptospira noguchii	
Leptospira parva	
Leptospira santarosai	
Leptospira weilii	
Leptospira kirschneri	
Leptospira meyeri	
Leptospira wolbachii	
Leptospira wolffi	
Unnamed genospecies 1–5	

* This is not a complete list of all the named Serovars.

Spread within the body

After entry, the organism is spread around the body by the blood, entering all organs and thus giving rise to a wide spectrum of clinical presentations (see later).

Epidemiology

Leptospirosis is a global disease, although it is primarily a disease of tropical and subtropical regions and is relatively uncommon in temperate climates. It is endemic in many countries but outbreaks are also associated with adverse weather. Examples of this include: an outbreak of leptospirosis in Nicaragua following Hurricane Mitch in 1995; an outbreak in Peru

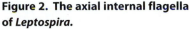

Figure 2. The axial internal flagella of *Leptospira*.

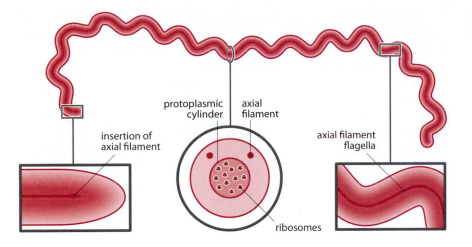

and Ecuador following heavy flooding in 1998; a post-cyclone outbreak in Orissa, India in 1999.

The precise number of human cases worldwide is not known. Incidences range from approximately 0.1–1 per 100 000 per year in temperate climates to 10–100 per 100 000 in the humid tropics; incidence may reach over 100 per 100 000. Reports of cases in the USA and Europe are low. In the UK it varies between 13 and 31 per annum and in continental France is in the order of 300, although in 2005 there were only 212 confirmed cases. In the USA the incidence is between 43 and 93 per annum.

Certain occupations are prone to infection, such as veterinarians, butchers, sewage workers, and farmers. Recreational exposure can occur through adventure holidays such as white water rafting and other water sports.

2. What is the host response to the infection and what is the disease pathogenesis?

Immune response

Infection with *Leptospira* induces antibodies to a number of proteins: p76, p62, p48, p45, p41, p37, and p32. Antibodies are mainly **IgM** to the **lipopolysaccharide** and **IgG** to proteins. The role of T cells in infection is unknown. It is thought that antibodies produced help to clear the organism from the tissues and blood but it is commonly retained in the kidney where it multiplies and is then passed out via the urine to further infect individuals. Complement is thought to be activated by the **alternative pathway** (at least *in vitro*), but the pathological serovars tend to be resistant, probably through binding of the serum inhibitory factor H.

Pathogenesis

In the kidney the organism migrates to the interstitium, renal tubules, and tubular lumen, causing an **interstitial nephritis** and tubular **necrosis**. Liver involvement is seen as centrilobular necrosis with proliferation of **Kupffer cells**. *Leptospira* also invades skeletal muscle, causing **edema**, vacuolization of myofibrils, and focal necrosis. In the lungs *Leptospira* induces intra-alveolar hemorrhages. A consistent pathological finding in all organs is a **vasculitis**. Whether organ damage is due directly to secreted toxins of

Leptospira or is secondary to the vasculitis induced by cell wall components of the organism such as collagenase or bystander damage is unclear. A study has shown that the **peptidoglycan** of *Leptospira* can increase the adhesion of the organism to vascular endothelial cells. The illness is characterized by a bleeding **diathesis** although its pathophysiology is uncertain. Infection causes a thrombocytopenia but it is unclear whether this is due to a direct action of the *Leptospira* on thrombopoiesis or secondary to **disseminated intravascular coagulation** or a specific antiplatelet antibody. Comparative genomics of two *Leptospira* demonstrate many outer membrane proteins and lipoproteins that may be important in relation to pathogenesis and for the development of vaccines.

3. What is the typical clinical presentation and what complications can occur?

The illness was first recognized in sewage workers in 1883. The incubation period is about 10 days. Because the organism spreads to all organs of the body the clinical presentation may vary. The infection may be asymptomatic and exposure only recognized serologically. Symptomatic disease may present as a biphasic illness with an initial nonspecific phase where the patient complains of fever, **pharyngitis**, headache, nausea, diarrhea, and myalgia. There may be conjuctival inflammation and occasionally a pretibial **maculopapular rash**. This phase corresponds to the *leptospiremic phase* and lasts about 7 days. The symptoms abate for a few days then recur with additional symptoms relating to either **meningitis** (headache, stiff neck, **photophobia**); liver failure (**jaundice**); renal failure (**oliguria**); cardiac arrhythmias or pulmonary infection (cough, hemoptysis). The liver and spleen may be enlarged. This phase corresponds to the *immune phase* and may last from 1 to 6 weeks. In some cases the two phases may not be apparent or the illness appears to start with the immune phase. The severity of the disease may vary from a mild illness to more severe disease with complications including cardiac arrhythmias, liver or renal failure, **myelitis**, and the **Guillain-Barré syndrome**. Severe disease (Weil's disease) with liver and renal failure carries a high mortality. A particular serovar may determine clinical manifestations, although any serovar can lead to any of the signs and symptoms, for example jaundice is seen in 83% of patients with L. *interrogans* serovar Icterohaemorrhagiae infection but only in 30% of patients infected with L. *interrogans* serovar Pomona. The pretibial rash is seen in patients with L. *interrogans* serovar Autumnalis infection and gastrointestinal (GI) symptoms predominate in patients infected with L. *interrogans* serovar Grippotyphosa. Aseptic meningitis commonly occurs in those infected with L. *interrogans* serovar Pomona or L. *interrogans* serovar Canicola.

4. How is this disease diagnosed and what is the differential diagnosis?

Leptospira spp. can be detected by dark ground microscopy in blood or cerebrospinal fluid (CSF) during the first week of the illness and in the urine thereafter. The organism can also be cultured from these specimens using Ellinghausen McCullough Johnson Harris, Fletchers (EMJH) or PML5 medium incubated at 30°C for 2–3 weeks. Because of the low sensitivity and specificity of dark ground microscopy and the long time-scales of culture, the routine method of diagnosis is serological. The microscopic agglutination test (MAT), however, is difficult to perform and relies on mixing serum from the patient with live *Leptospira* from the different serogroups and looking for agglutination by dark ground microscopy. This test also has a low sensitivity due to failure to **seroconvert** in a proportion of patients and cross-reactivity with a number of other infections and illnesses including other spirochetal diseases (Lyme disease, relapsing fever, treponemal disease), *Legionella*, HIV, and autoimmune diseases. Although the MAT is serogroup-specific, cross-reactions between serogroups occur and the individual serovars cannot be determined. Higher sensitivity can be obtained using paired sera. IgM ELISA tests are most often used for diagnosis but other formats such as latex agglutination are also commercially available. The PCR is also available for diagnosis but does not give information about the serovar of the strain.

Differential diagnosis

Patients presenting with the initial **febrile** illness may be diagnosed with influenza or **Dengue fever**. If presenting with signs of meningitis then leptospirosis may be misdiagnosed as viral meningitis or if with **petechiae** then meningococcal meningitis. A CSF white cell count would suggest a viral etiology rather than meningococcal, as the cell response in leptospirosis is lymphocytic. If presenting with jaundice the differential diagnosis includes:

- viral hepatitis
- malaria
- schistosomiasis
- relapsing fever
- **tularemia**

and if with jaundice and renal failure then it includes:

- **Legionnaires disease**
- hemolytic uremic syndrome.

If the pulmonary syndrome is prominent it may be confused with Hantavirus.

5. How is the disease managed and prevented?

Management

The illness is treated with benzylpenicillin or doxycycline, which should be given early in the illness during the leptospiremic phase. The **Jarisch-Herxheimer** reaction may occur with penicillin treatment. This reaction is due to the release of **endotoxin** when large numbers of organisms are killed by antibiotics, leading to increase in **immune complexes** and the release of **cytokines**, particularly **tumor necrosis factor** (**TNF**). This also occurs typically with syphilis and Lyme disease. Supportive measures such as hemodialysis may be required in severe disease.

Prevention

As many animal groups are infected with *Leptospira*, prevention depends on identifying and mitigating risks such as immunization of pets, personal protective equipment for specific occupations, and use of chemoprophylaxis.

SUMMARY

1. What is the causative agent, how does it enter the body and how does it spread a) within the body and b) from person to person?

- Two main identified species exist: *L. interrogans* and *L. biflexa*.

- Over 250 serovars of *Leptospira* exist.

- Leptospirosis is a zoonosis, the organism is carried by many animals and excreted in the urine.

- Infection is acquired by contact with contaminated water or soil.

- The organism enters through the conjunctiva, mucosa or skin abrasions.

- *Leptospira* spreads to all body locations by the blood.

2. What is the host response to the infection and what is the disease pathogenesis?

- IgG and IgM antibodies are produced during infection.

- *Leptospira* causes a vasculitis in many organs.

- Liver, lungs, heart, **meninges**, kidneys, and muscles are all affected

- Thrombocytopenia occurs.

3. What is the typical clinical presentation and what complications can occur?

- The incubation period is about 10 days.

- The clinical presentation is classically biphasic.

- The initial illness is nonspecific with fever and myalgia.

- The second phase of the illness may present as meningitis, pneumonia, jaundice or renal failure.

- Complications include cardiac arrhythmia, Guillain-Barre syndrome, renal failure.

4. How is this disease diagnosed and what is the differential diagnosis?

- *Leptospira* can be detected in the blood, CSF or urine by dark ground microscopy.

- The organism can be cultured on a variety of specialist media after incubation at 30°C for 1–2 weeks.

- Serologically the disease can be diagnosed by the macroscopic agglutination test or the IgM ELISA.

- The microagglutination test (MAT) is serogroup-specific, does not give identification of the infecting serovar, and is subject to many cross-reactions with unrelated bacteria.

- The differential diagnosis includes influenza, viral meningitis, viral hepatitis, malaria, and Legionnaires disease.

5. How is the disease managed and prevented?

- Treatment is with benzylpenicillin or doxycycline.

- The Jarisch-Herxheimer reaction may occur with penicillin.

- Prevention is by identifying risks, animal immunization, personal protective equipment, and chemoprophylaxis.

FURTHER READING

Lydyard P, Lakhani S, Dogan A, et al. Pathology Integrated: An A–Z of Disease and its Pathogenesis. Edward Arnold, London, 2000: 254–256.

Mandell GL, Bennet JE, Dolin R. Principles & Practice of Infectious Diseases, 6th edition, Vol 2. Elsevier/Churchill Livingstone, Philadelphia, 2005: 2789–2795.

Mims C, Dockrell HM, Goering RV, Roitt I, Wakwlin D, Zuckerman M. Medical Microbiology, 3rd edition. Mosby, London, 2004: 232–235.

REFERENCES

Bharti AR, Nally JE, Ricaldi JN, et al. Leptospirosis: a zoonotic disease of global importance. Lancet Infect Dis, 2003, 3: 757–771.

De Souza AL. Neuroleptospirosis: unexplored and overlooked. Indian J Med Res, 2006, 124: 125–128.

Guerreiro H, Croda J, Flannery B, et al. Leptospiral proteins recognised during the humoral response to leptospirosis in humans. Infect Immun, 2001, 69: 4958–4968.

Higgins R. A minireview of the pathogenesis of acute leptospirosis. Can Vet J, 1981, 22: 277–278.

Levett PN, Branch SL. Evaluation of two enzyme linked immunosorbent assay methods for detection of immunoglobulin M antibodies in acute leptospirosis. Am J Trop Med Hyg, 2002, 66: 745–748.

McBride AJ, Athanazio DA, Reis MG, Ko AI. Leptospirosis. Curr Opin Infect Dis, 2005, 18: 376–386.

Morey RE, Galloway RL, Bragg SL, et al. Species-specific identification of Leptospiraceae by 16S rRNA gene sequencing. J Clin Microbiol, 2006, 44: 3510–3516.

Nascimemto ALTO, Ko AI, Martins EAL, et al. Comparative genomics of two Leptospira interrogans serovars reveals novel insights into physiology and pathogenesis. J Bacteriol, 2004, 186: 2164–2172.

Wagenaar JFP, Goris MGA, Sakudarno MS, et al. What role do coagulation disorders play in the pathogenesis of leptospirosis. Trop Med Int Health 2007, 12: 111–122.

WEB SITES

Centers for Disease Control and Prevention, Atlanta, GA, USA: http://www.cdc.gov/ncidod/dbmd/diseaseinfo/leptospirosis_g.htm

Centre for Infections, Health Protection Agency, HPA Copyright, 2008: http://www.hpa.org.uk/infections/topics_az/zoonoses/leptospirosis/gen_info.htm

The Leptospirosis Information Center, Content ©2004–2008 all rights reserved: http://www.leptospirosis.org/

MULTIPLE CHOICE QUESTIONS

The questions should be answered either by selecting True (T) or False (F) for each answer statement, or by selecting the answer statements which best answer the question. Answers can be found in the back of the book.

1. **Which of the following statements are true of *Leptospira interrogans*? It:**

   A. Is gram-positive.

   B. Has axial flagella.

   C. Can be detected by dark ground microscopy.

   D. Has only one serovar.

   E. Is nonmotile.

2. **Which of the following are true concerning *Leptospira*?**

   A. They are typically spread from person to person.

   B. They are found in a wide range of animals.

   C. They are spread by contact with animal urine.

   D. All *Leptospira* are pathogenic.

   E. Infection is associated with certain occupations.

3. **Which of the following are true for disease caused by *Leptospira*?**

   A. Vasculitis is a common occurrence.

   B. Pulmonary hemorrhages are frequent.

   C. Acute tubular necrosis may occur.

   D. It is a granulomatous disease.

   E. Thrombocytopenia is common.

4. **Which of the following are true of the clinical presentation of leptospirosis?**

   A. All infection presents in the same fashion.

   B. Any serovar can cause any clinical presentation.

   C. Leptospirosis is a biphasic illness.

   D. Leptospirosis may be confused with Legionnaires disease, viral hepatitis or Hantavirus infection.

   E. Jaundice and renal failure with conjunctival inflammation is a common presentation with *L. interrogans* serovar Icterohaemorrhagiae.

5. **Which of the following are complications of leptospirosis?**

   A. Myocarditis.

   B. Stevens Johnson syndrome.

   C. Weil's disease.

   D. Guillain-Barré syndrome.

   E. Renal failure.

6. **Which of the following are useful in the diagnosis of leptospirosis?**

   A. Gram stain.

   B. Culture.

   C. Microscopic agglutination test.

   D. IgM ELISA.

   E. PCR.

7. **Which of the following are used in the treatment or prevention of leptospirosis?**

   A. Vaccination.

   B. Benzylpenicillin.

   C. Gentamicin.

   D. Personal protective equipment.

   E. Mosquito nets.

Case 21

Listeria monocytogenes

A young woman was admitted to hospital with fever, headache, **myalgia**, and joint pains of 48 hours duration. The previous day she had attended a lunch where she had eaten ham and a soft cheese. She was 26 weeks pregnant. The admitting physician thought the patient may have listeriosis and a blood culture was taken in addition to a full blood count. Gram-positive cocci were reported on the Gram stain of the blood culture and the patient was started empirically on teicoplanin. The following day small hemolytic colonies were present on the blood agar (Figure 1) and a Gram stain revealed gram-positive rods. Further testing showed the organism was motile with tumbling motility and it was biochemically identified as *Listeria monocytogenes*. A diagnosis of listeriosis was made and the patient's treatment was changed to ampicillin and gentamicin. .

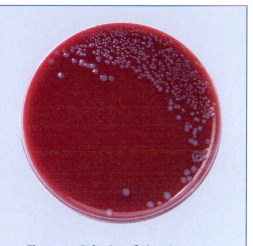

Figure 1. Colonies of *Listeria monocytogenes* on blood agar.

1. What is the causative agent, how does it enter the body and how does it spread a) within the body and b) from person to person?

Causative agent

Listeria monocytogenes is the cause of human disease. It is a gram-positive nonsporing motile bacillus (Figure 2). There are six species of *Listeria*: *L. monocytogenes*, *L. ivanovii*, *L. welshimeri*, *L. innocua*, *L. seeligeri*, and *L. grayi*. The most common cause of human disease is *L. monocytogenes*, although *L. ivanovii* can rarely cause disease. *Listeria* species grow on blood agar (see Figure 1) and *L. monocytogenes* produces a narrow zone of β-hemolysis often only seen beneath the colonies. The organism grows at 37°C but can also grow slowly at 4°C and this can be used as an enrichment technique when examining foodstuff. The organism shows a characteristic tumbling (end over end) movement at 25°C, which is diagnostic. *L. monocytogenes* has several serotypes based on cell wall (O) and flagellar (H) antigens. The majority of disease is caused by serotypes 1/2a, 1/2b, and 4b. Several molecular subtyping techniques (**multilocus enzyme electrophoresis**, **pulse field gel electrophoresis**, and **ribotyping**) have been found useful in epidemiological investigations.

Source of infection

L. monocytogenes has been isolated from a variety of natural sources including soil, water, animals, and vegetables and is widely distributed in the animal kingdom, with over 40 species of wild and food-source animals including birds, crustaceans, and fish being colonized. The organism can

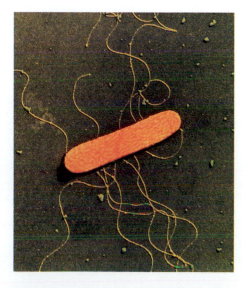

Figure 2. Scanning electron microscopic (EM) image of *Listeria monocytogenes* showing flagella.

also be carried in the intestine of about 5% of the human population without any symptoms of disease. Infection is acquired by consumption of contaminated food (Figure 3) such as fish, salad, pate, soft cheeses, salami, ham, and coleslaw, where contamination rates as high as 70% may occur. Ingestion of *Listeria* probably occurs frequently and the organism can be carried in the intestine of the human population for short periods without any symptoms of disease. Interestingly, disease develops mainly in specific groups of individuals including pregnant women, neonates, and immunocompromised patients (see clinical presentation in Section 3 below).

Entry and spread within the body

The organism enters the epithelial cells of the intestine using specific adhesion molecules (described in Section 2) and then enters the bloodstream. This gives rise to **septicemia** or **meningitis** as the two most common presentations. Localization of the organism in the central nervous system (CNS) is frequently seen, although it is not common in pregnant women for some unidentified reason. An alternative route of infection to the CNS may be intra-axonal spread directly from the peripheral nerves in the gastrointestinal tract. The organism can also cross the placenta to give rise to fetal disease, and subsequent abortion.

Epidemiology

This organism is found worldwide but most countries do not keep accurate records.

In the UK there are about 25 cases per annum. In the USA there were 135 reported cases in 2005, which represents an incidence of 0.3/100000 population and a decrease of 32% from 1996–1998 figures. In France there were 269 cases in 1999, falling to 209 in 2003.

2. What is the host response to the infection and what is the disease pathogenesis?

L. monocytogenes is an intracellular pathogen and a risk factor for infection is lack of cell-mediated immunity. *Listeria* penetrates both nonphagocytic

Figure 3. There are a variety of sources of infection with *Listeria*.

and phagocytic cells, thus avoiding the host immune system. Following entry via the epithelial cells of the gastrointestinal tract *Listeria* enters the **lamina propria** where it enters **dendritic cells** and macrophages. Listeria adhesion protein (LAP) is an important **adhesin** and binds to heat-shock protein 60 expressed on cells. Additionally, *Listeria* has surface proteins, for example Act A and internalin (InlA,B), which are important in cellular penetration. Act A binds to heparin on the cell surface and internalin binds to E-cadherin, which induces actin re-arrangement and **phagocytosis** of the bacterium. *Listeria* is taken up into membrane-bound **vesicles (phagosomes** in phagocytic cells). Uptake of *Listeria* into endosomes by the non-phagocytic epithelial cells is via receptor-mediated **endocytosis**. In order to avoid phagolysosome fusion and thus killing, in professional phagocytic cells, *Listeria* escapes from inside the phagosome by secreting a pore-forming enzyme listeriolysin O and phospholipases that disrupt the phagosome membrane. Escape from the endosomes in nonphagocytic cells occurs in a similar way. Once in the cytoplasm the organism replicates and a protein (Act A) at one pole of the cell precipitates and polymerizes actin. This polymerized actin forms a growing scaffold that pushes the bacterium through the cytoplasm to the plasma membrane where it makes contact with the plasma membrane of an adjacent cell, is promptly ingested, and the whole process is repeated (Figure 4). The infected host cells eventually die by **apoptosis** or **necrosis**.

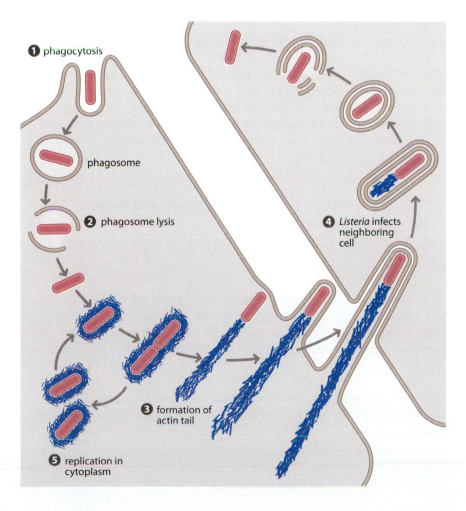

1 phagocytosis

phagosome

2 phagosome lysis

3 formation of actin tail

5 replication in cytoplasm

4 *Listeria* infects neighboring cell

Figure 4. Escape of *Listeria monocytogenes* from the phagosome, actin polymerization, and movement into an adjacent cell. 1, uptake; 2, escape from the phagosome by listeriolysin O and phospholipases; 3, actin polymerization; 4, penetration to an adjacent cell; 5, intracytoplasmic replication.

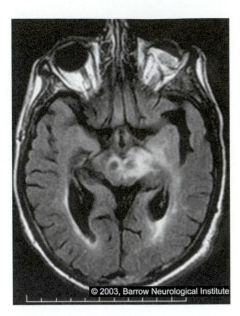

© 2003, Barrow Neurological Institute

Figure 5. *Listeria monocytogenes* meningoencephalitis. MRI scan showing enhancement around the left midbrain and thalamus and cerebral peduncle, indicating meningoencephalitis in a case of *Listeria* infection in an immunocompromised patient.

In the CNS, *Listeria* will predominately penetrate **microglia** and to a lesser extent astrocytes and oligodendrocytes. Direct penetration into neurons is rare, although if co-cultured with macrophages *Listeria* can penetrate neurons following cell to cell contact. *L. monocytogenes* may invade the CNS by several routes: hematogenous spread and direct invasion of endothelial cells of the blood–brain barrier; circulating mononuclear cells carrying *Listeria*; or a direct neural route by intra-axonal spread. In some animals, *Listeria* may initially be taken up by tissue macrophages in the mouth and spread by cell to cell contact with distal nerve axons and thence by intra-axonal spread into the brain. The role of this route in humans is unclear.

Once bacteria arrive in the brain, **encephalitis** is established by direct cell to cell spread. Pathologically in the brain following infection there is necrosis and focal hemorrhages with **meningoencephalitis** (Figure 5), microabscesses, **vasculitis**, and perivascular lymphocytic infiltration. *L. monocytogenes* is present in the necrotic parenchymal lesions and can be seen in the cytoplasm of macrophages and endothelial cells.

Listeria avoids both the innate and humoral immune system by its intracellular localization. However, antibodies to *Listeria* can be detected along with complement. These can enhance **opsonization**, resulting in killing of *Listeria* by **NK (natural killer) cells**, **dendritic cells**, and macrophages. **Interferon-γ (IFN-γ)** is a key cytokine produced by NK cells, **CD4+ Th1** cells, and **CD8+** T cells and is important in the immune response to infection with *Listeria*. Escape from the phagosome in dendritic cells and **interleukin (IL)-10** are both important for the development of protective cytotoxic CD8 memory cells. This is presumably via enhancement of Th1 responses.

Once taken up by activated macrophages many *Listeria* will be killed following phagolysosome fusion and subsequent oxygen-dependent and independent routes. However, bacteria localized in nonactivated macrophages and epithelial cells and those escaping from the phagosome before phagolysosome fusion can survive in **granulomas** produced by overactive CD4+ Th1 cells, as in tuberculosis (TB).

3. What is the typical clinical presentation and what complications can occur?

L. monocytogenes affects mainly pregnant women, neonates, and immunocompromised individuals.

In pregnancy the disease may present as an acute **febrile** flu-like illness and unlike disease in nonpregnant adults, meningitis is an uncommon presentation. Most disease occurs in the third trimester. About 25% of perinatal infections result in fetal death (leading to spontaneous abortion) and of those that survive about 70% will develop neonatal disease.

Infection acquired *in utero* may result in granulomatosis infantiseptica, which has a high mortality and is due to widespread granulomatous **abscesses** in many organs.

Neonatal infection presents either as an early onset **bacteremia** or as a late onset (1–2 weeks post partum) meningitis.

In immunocompromised patients and those over 60 years of age, listeriosis presents most frequently with meningitis (60%) and bacteremia (30%). Brain abscess, abscesses in other organs or **endocarditis** may complicate bacteremia. *Listeria* infection in the CNS may present with a meningitis that is indistinguishable from other causes of meningitis or as a meningoencephalitis with altered consciousness and **seizures**.

Large outbreaks of *Listeria* gastroenteritis in otherwise healthy individuals have occasionally been reported, which presents as diarrhea, fever, and myalgia. This presentation has been linked to consumption of high counts of *Listeria* in food. Rarely otherwise healthy individuals may develop an encephalomyelitis of the brainstem caused by *Listeria*, which presents with a flu-like prodrome and a few days later with an acute onset of cerebellar signs, cranial nerve palsies, and respiratory failure.

4. How is this disease diagnosed and what is the differential diagnosis?

Diagnosis of listeriosis is confirmed by isolation of *L. monocytogenes* from a normally sterile site such as the blood or the cerebrospinal fluid (CSF). Mis-identification of *Listeria* can occur as they may appear like diphtheroids or gram-positive cocci in specimens.

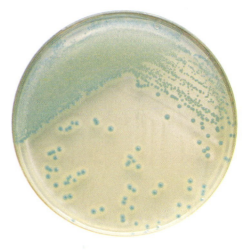

Listeria grows well on routine diagnostic media and on blood agar *L. monocytogenes* (as well as *L. seeligeri* and *L. ivanovii*) are beta-hemolytic. *L. monocytogenes* and *L. seeligeri* have a narrow zone of beta-**hemolysis** whereas *L. ivanovii* has a wide double zone of hemolysis. A number of selective media exist for the isolation of *Listeria* from food/feces (Table 1). A number of chromogenic media have also been developed to differentiate *L. monocytogenes* from other *Listeria* spp. and other bacteria (Figure 6).

Figure 6. BBL CHROMagar showing the characteristic blue-green color of *L. monocytogenes* surrounded by a white opaque halo.

Serology is not useful for diagnosis of acute cases, although listeriolysin O antibodies have been used to identify patients with noninvasive illness.

Table 1. Selective media for the isolation of *Listeria* from food/feces

Selective medium	Constituents
Listeria selective agar	Tryptose, glucose, sodium chloride, thiamine, acriflavine, nalidixic acid
Oxford *Listeria* agar	Peptone, corn starch, sodium chloride, esculin, lithium chloride, ferric ammonium citrate
Al-Zoreky Sadine *Listeria* agar	Acriflavine, ceftazidime, moxalactam
Listeria PALCAM agar	Peptone, starch, sodium chloride, D-mannitol, ammonium ferric citrate, esculin, glucose, lithium chloride, polymyxin, ceftazidime, phenol red
LPM agar	Casein, peptone, beef extract, sodium chloride, lithium chloride, glycine, phenylethanol, moxalactam
HiChrome *Listeria* agar	Meat extract, yeast extract, peptone, rhamnose, sodium chloride, lithium chloride, chromogenic mixture
Listeria selective supplement I	Acriflavine, cyclohexamide, nalidixic acid
Listeria selective supplement II	Acriflavine, cyclohexamide, colistin, cefotetan, fosfomycin

Differential diagnosis

As *Listeria* infection presents as a **pyrexia of unknown origin** (**PUO**), a bacteremia or a meningitis, any of the microorganisms causing these conditions must be considered in the differential diagnosis. There should be a high index of suspicion for *Listeria* infection if PUO, bacteremia or meningitis occur in the at-risk groups. In particular the other main causes of neonatal bacteremia and meningitis (group B *Streptococcus* and *E. coli*) should be considered.

5. How is the disease managed and prevented?

Management

Disease caused by *Listeria* is usually treated with a combination of ampicillin and gentamicin. Co-trimoxazole can be used in patients who are hypersensitive to β-lactams or in place of gentamicin. In patients who are pregnant and who cannot take folic acid inhibitors, a glycopeptide can be used in place of ampicillin.

Bacteremia is usually treated for 2 weeks and meningitis for 3 weeks, although because of the affinity of *Listeria* for the CNS one can argue that bacteremia should also be treated for 3 weeks. Brain abscess or endocarditis should be treated for 6 weeks.

Prevention

At-risk individuals (pregnancy, neonates, immunocompromised) should avoid food in which *Listeria* can grow to high cell numbers (e.g. pate and deli meats, soft cheese, raw seafood, salad, and coleslaw) and correct food-handling precautions will reduce the chance of infection. There are no vaccines for *Listeria* as yet.

SUMMARY

1. What is the causative agent, how does it enter the body and how does it spread a) within the body and b) from person to person?

- There are several species of *Listeria* but *L. monocytogenes* is responsible for the majority of disease.

- *L. monocytogenes* is a motile nonspore-forming gram-positive rod.

- *L. monocytogenes* can grow at 4°C and shows characteristic tumbling motility at 25°C.

- *L. monocytogenes* produces a narrow zone of beta-hemolysis on blood agar.

- *Listeria* spp. are ubiquitous in animals and the environment and therefore found globally.

- Infection is usually acquired from consumption of contaminated food.

- Neonatal infection can be acquired transplacentally.

2. What is the host response to the infection and what is the disease pathogenesis?

- *Listeria* is an intracellular pathogen avoiding the host immune system.

- *Listeria* penetrates both phagocytic and nonphagocytic cells.

- *Listeria* escapes from the phagosome with the aid of listeriolysin O.

- *Listeria* precipitates actin when in the cytoplasm and penetrates adjacent cells.

- *Listeria* infection is limited by CD8 cytotoxic T cells.

3. What is the typical clinical presentation and what complications can occur?

- *L. monocytogenes* typically infects pregnant women, neonates, and immunocompromised individuals.

- Disease in pregnant women presents as a flu-like illness but can cause spontaneous abortion and granulomatosis infantiseptica.

- In neonates early post-partum disease presents as a bacteremia and late onset disease as a meningitis.

- Disease in immunocompromised adults presents as a meningitis or bacteremia.

- Disease caused by *L. monocytogenes* in otherwise healthy adults is infrequent but may present as a brainstem meningoencephalitis or as a febrile gastroenteritis.

- Bacteremia may be complicated by abscesses or endocarditis.

4. How is this disease diagnosed and what is the differential diagnosis?

- Disease caused by *Listeria* is diagnosed by

isolation of the organism from sterile sites such as blood or CSF on routine blood agar.

- Several selective media exist for the isolation of *Listeria* from food or feces.

- Serology is not helpful in diagnosis of acute illness.

- Any of the many causes of bacteremia, endocarditis, and meningitis can mimic the presentation of listeriosis and a high index of suspicion should be held in at-risk groups.

5. How is the disease managed and prevented?

- Treatment is with ampicillin and gentamicin.

- Co-trimoxazole is an alternative treatment in case of penicillin hypersensitivity.

- In penicillin hypersensitivity in pregnancy a glycopeptide should be used.

- At-risk individuals (pregnancy, neonates, immunocompromised) should avoid food in which *Listeria* can grow to high cell numbers.

FURTHER READING

Cimolai N. Laboratory Diagnosis of Bacterial Infections. Marcel Dekker, NY, 2001: 333–375.

Mandell GL, Bennet JE, Dolin R. Principles & Practice of Infectious Diseases, 6th edition, Vol 2. Churchill Livingstone, Philadelphia, 2005: 2478–2484.

Mims C, Dockrell HM, Goering RV, Roitt I, Wakwlin D, Zuckerman M. Medical Microbiology, 3rd edition. Mosby, London, 2004.

Murphy K, Travers P, Walport M. Janeway's Immunobiology, 7th edition. Garland Science, New York, 2008.

Wilson M, McNab R, Henderson B. Bacterial Disease Mechanisms: An Introduction to Cellular Microbiology. Cambridge University Press, London, 2002: Chapter 8.

REFERENCES

Bahjat KS, Liu W, Lemmens EE, et al. Cytosolic entry controls CD8+-T-cell potency during bacterial infection. Infect Immun, 2006, 74: 6387–6397.

Bierne H, Cossart P. *Listeria monocytogenes* surface proteins: from genome predictions to function. Microbiol Mol Biol Rev, 2007, 71: 377–397.

Bierne H, Cossart P. InlB a surface protein of *Listeria monocytogenes* that behaves as an invasin and a growth factor. J Cell Sci, 2002, 115: 3357–3367.

D'Orazio SE, Troese MJ, Starnbach MN. Cytosolic localization of *Listeria monocytogenes* triggers an early IFN gamma response by CD8 T cells that correlates with innate resistance

to infection. J Immunol, 2007, 177: 7146–7154.

Foulds KE, Rotte MJ, Seder RA. IL-10 is required for optimal CD8 T cell memory following *Listeria monocytogenes* infection. J Immunol, 2006, 177: 2565–2574.

Gandhi M, Chikindas ML. *Listeria*: a foodborne pathogen that knows how to survive. Int J Food Microbiol, 2007, 113: 1–15.

Kim KP, Jagadeesan B, Burkholder KM, et al. Adhesion characteristics of *Listeria* adhesion protein (LAP)-expressing *Escherichia coli* to Caco-2 cells and of recombinant PAP to eukaryotic receptor Hsp60 as examined in a surface plasmon resonance sensor. FEMS Microbiol Lett, 2006, 256: 324–332.

REFERENCES

Lecuit M. Understanding how *Listeria monocytogenes* targets and crosses host barriers. Clin Microbiol Infect, 2005, 1: 430–436.

Ramaswamy V, Cresence VM, Rejitha JS, et al. *Listeria* – review of epidemiology and pathogenesis. J Microbiol Immunol Infect, 2007, 40: 4–13.

Shaughnessy LM, Swanson JA. The role of activated macrophages in clearing *Listeria monocytogenes* infection. Front Biosci, 2007, 12: 2683–2692.

WEB SITES

Centers for Disease Control and Prevention, Division of Foodborne, Bacterial and Mycotic Diseases, Atlanta, GA, USA: http://www.cdc.gov/nczved/dfbmd/disease_listing/listeriosis_gi.html

Centre for Infections, Health Protection Agency, HPA Copyright, 2008: http://www.hpa.org.uk/infections/topics_az/listeria/menu.htm

Food Safety and Inspection Service, United States Department of Agriculture: http://www.fsis.usda.gov/Fact_Sheets/Listeriosis_and_Pregnancy_What_is_Your_Risk/index.asp

Food and Drug Administration, United States Department of Health and Human Services: http://www.fda.gov/FDAC/features/2004/104_bac.html

World Health Organization, © 2003 WHO/OMS: http://www.who.int/topics/listeria_infections/en/

MULTIPLE CHOICE QUESTIONS

The questions should be answered either by selecting True (T) or False (F) for each answer statement, or by selecting the answer statements which best answer the question. Answers can be found in the back of the book.

1. **Which of the following are true of *Listeria* species?**
   A. They are gram-positive.
   B. *L. monocytogenes* is nonmotile.
   C. *L. monocytogenes* can grow at 4°C.
   D. All *Listeria* species are pathogenic for humans.
   E. *L. monocytogenes* is alpha-hemolytic.

2. **Which of the following are true concerning *Listeria monocytogenes*?**
   A. It is ubiquitous in the environment.
   B. Infection is a zoonosis.
   C. Soft cheese is a low risk food that is unlikely to be contaminated with large numbers of *L. monocytogenes*.
   D. Infection may be acquired transplacentally.
   E. Infection is associated with certain at-risk groups.

3. **Which of the following are true for *Listeria*?**
   A. The bacteria are rapidly killed by antibodies.
   B. *Listeria* has a tropism for the CNS.
   C. *Listeria* induces uptake into nonphagocytic cells.
   D. *Listeria* hydrolyzes actin in the cytoplasm.
   E. *Listeria* infection is controlled by cytotoxic CD8 cells.

4. **Which of the following are true of the clinical presentation of diseases caused by *L. monocytogenes*?**
   A. Granulomatosis infantiseptica may occur.
   B. Infection in pregnant women may lead to spontaneous abortion.
   C. It is typically associated with necrotizing fasciitis.
   D. Meningitis is common in pregnant women.
   E. Outbreaks are uncommon.

5. **Which of the following are useful in the diagnosis of *Listeria*?**
   A. Gram stain.
   B. Culture.
   C. Serology.
   D. PCR.
   E. Gas-liquid chromatography.

6. **Which of the following are used in the treatment or prevention of *Listeria*?**
   A. Vaccination.
   B. Co-trimoxazole.
   C. Gentamicin.
   D. Education leaflets.
   E. Good kitchen practices.

Case 22

Mycobacterium leprae

A 37-year-old Hispanic woman, a native of Southern Mexico, went to see her primary care physician after complaining of a persistent **rash** throughout her body. The woman had three children, was a nonsmoker, and appeared to be well-nourished.

Her symptoms had started 5 years before with spasms, with needle-like pain, in her arms. She also felt tired and stressed and had been initially diagnosed with depression. Her skin examination indicated **atopic dermatitis** and **urticaria** and she was prescribed ibuprofen, fluoxetine, and hydroxyzine.

Physical examination revealed numerous hypo-pigmented skin lesions (Figure 1), especially those on her arms, nasal bridge area, cheeks, abdomen, and back and her legs. Her eyebrows had started thinning and she had numbness in her forearm.

The patient had a biopsy of skin lesions on the abdomen. Acid-fast bacilli were detected and a diagnosis of lepromatous leprosy was made. The patient was counseled about leprosy and was prescribed a course of treatment. As required by Mexican law, her case was reported to the Public Health Department.

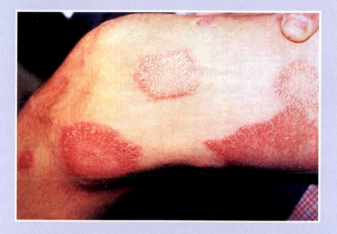

Figure 1. Skin lesions of a patient with lepromatous leprosy. This is usually accompanied by the loss of sensation around the affected area.

1. What is the causative agent, how does it enter the body and how does it spread a) within the body and b) from person to person?

Causative agent

Leprosy (also called Hansen's disease) is a chronic granulomatous disease of the skin and peripheral nervous system caused by *Mycobacterium leprae*. *M. leprae* is a member of the *Mycobacteriaceae* family. It is an obligate intra-cellular gram-positive bacillus that requires the environment of the host macrophage for survival and propagation by binary fission. It shows pref-erential tropism towards macrophages and Schwann cells that surround the axons of nerve cells. The bacilli resist intracellular degradation by macrophages, possibly by escaping from the **phagosome** into the cyto-plasm and preventing fusion of the phagosome with **lysosomes**.

M. leprae is an acid-fast microorganism. This means that it is resistant to decolorization by acids during staining procedures. In common with all pathogens belonging to the genus *Mycobacterium*, *M. leprae* can be stained using the Ziehl-Neelsen method, in which the bacteria are stained bright red and stand out clearly against a blue background (Figure 2). The bacilli

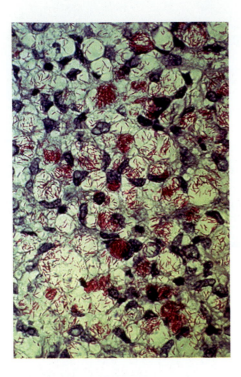

Figure 2. Skin lesion fluid from the leprosy patient stained by the Ziehl-Neelsen acid-fast method. The reagent stains *M. leprae* red against the blue background.

appear as straight or curved rod-shaped organisms, 1–8 μm long and approximately 0.3 μm in diameter.

A Fite modification of this method allows identification of more bacilli using their ability to accumulate large amounts of hyaluronic acid.

M. leprae, unlike *M. tuberculosis*, cannot be cultured in the laboratory. *M. leprae* can, however, be propagated in mouse footpad and in its natural host – the nine-banded armadillo – for experimental purposes.

The genome of *M. leprae* has been sequenced and it includes 1605 genes coding for proteins and 50 genes for stable RNA molecules. The genome of *M. leprae* is smaller than that of *M. tuberculosis* and less than half of the genome contains functional genes but there are many pseudogenes, with intact counterparts in *M. tuberculosis* (Table 1). Gene deletion and decay compared with *M. tuberculosis* have eliminated many important metabolic activities. *M. leprae* is therefore even more dependent on host metabolism and this is why it is characterized by an even slower growth rate than that of *M. tuberculosis*.

The cell wall of *M. leprae* is similar to that of *M. tuberculosis*, but its cell wall includes the species-specific phenolic glycolipid I (PGL-I) widely used for serodiagnosis of lepromatous leprosy and genus-specific lipoarabinomannan among other molecules (see Section 4).

The major reservoir of *M. leprae* is in humans, rarely apes, and some species of monkeys: chimpanzees, nine-banded armadillos, and mangabey monkeys. These, especially armadillos, are characterized by a natural low body temperature that favors the survival of *M. leprae*.

Entry and spread within the body

The major port of entry and exit of *M. leprae* is the respiratory system, particularly the nose, and the microorganism is taken up by **alveolar macrophages** where they live and proliferate very slowly. In early infections subsequent to entry into respiratory mucosa, the bacilli are spread by

Table 1. Comparative genomics of *M. leprae* and *M. tuberculosis* (adapted from Scollard et al., 2006)

Parameter	M. leprae *(strain TN)*	M. tuberculosis *(strain H37Rv)*
Genome size (bp)	3268203	4411532
Protein genes	1614	3993
tRNA genes	45	45
rRNA genes	3	3
Unknown genes	142	606
Pseudogenes	1133	6
Gene density (bases/gene)	2024	1106
% protein coding	49.5	91.2
Single nucleotide polymorphism frequency	1/24000 bp	1/3000 bp

the circulation, and reach neural tissue where they infect Schwann cells. It is still unclear how *M. leprae* travels through blood but it is likely to be via macrophages and **dendritic cells** (**DCs**).

M. leprae binds to Schwann cells via the cell wall PGL-I (an **adhesin**), which attaches to the G-domain of α2-chain of laminin 2 isoform in the basal lamina of Schwann cells and is restricted to peripheral nerves. However, the subsequent uptake of *M. leprae* by the Schwann cells is facilitated by its α-dystroglycan – a receptor for laminin on the cell membrane. The bacilli live in the infected Schwann cells and macrophages and are characterized by slow growth (sometimes for years) that is reflected in the long incubation period of the disease. The average division rate of *M. leprae* is very low for microbes, 10–12 days for one division. Optimal temperature for growth of *M. leprae* is lower than core body temperature (30–35°C), thus it prefers to grow in the cooler parts of the body, that is skin and superficial nerves.

Person to person spread

Leprosy sufferers, particularly those with multibacillary leprosy (predominantly those with lepromatous leprosy (LL) – see Sections 2 and 3) represent the major source of the infection, and transmission is through aerosols to the respiratory tract. Its dissemination through skin lesions seems to be less important. There is increasing evidence that those infected individuals that do not develop the symptoms of the disease may nevertheless transiently excrete the microbe nasally and spread the infection. As a result, close proximal contacts of the infected individuals (household, neighbors, social contacts) have an increased risk of contracting the disease. The risk varies for different forms of the disease (see Section 3): 8–10 times higher for LL and 2–4 times higher for the tuberculoid form (TT). Factors such as probability of a frequent contact, similar genetic and immunological background, and similar environment may all play a role. Contrary to the myth, *M. leprae* does not pass through the skin in either direction and cannot be transmitted through touch.

It is a slowly developing disease with an incubation period from 6 months to up to 40 years or even longer! Again the mean incubation time is different for TT and LL, being 4 years and 10 years, respectively.

Epidemiology

According to the World Health Organization (WHO) in 1997, 2 million people worldwide were infected with *M. leprae*. Since 1998, around 800 000 new cases of leprosy per year have been detected. Although the prevalence of leprosy has diminished in recent years down to approximately 460 000 infections globally, partially due to the effective treatment, the rate of detection of new cases remained constant with a high frequency (17%) of infection in children. Children are the most susceptible and tend to have predominantly the TT form. In adults the LL form is more common in men than women.

Leprosy can affect races all over the world. However, it is most common in warm, wet areas in the tropics and subtropics. Leprosy is endemic in Asia, Africa, the Pacific region, and Latin America. About 75% of the patients have been registered with the disease in South-East Asia, particularly in India where leprosy is epidemic (Figure 3). Significant regional

Figure 3. Global distribution of leprosy in 2003 according to WHO. South-East Asia, Africa, the Pacific region and South America are the regions the most affected.

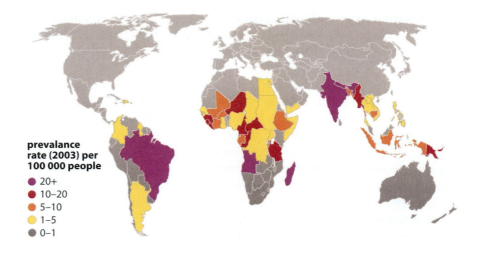

prevalance rate (2003) per 100 000 people
- 20+
- 10–20
- 5–10
- 1–5
- 0–1

clustering has been detected in individual countries and even communities. Interestingly, African blacks have a high incidence of TT, while people with white skin and Chinese people mostly have LL.

In the last decade of the 20th century the case detection rate remained stable (India), decreased (China) or increased (Bangladesh). However, these data rely on the measurement of incidence through the number of patients registered for the treatment that is far from optimal. Often long sufferers from leprosy remain unregistered for various reasons (stigma, costs, lack of health services). On the other hand, double registering, relapses, and self-treatment cases may contribute to inflation of the figures.

Consequently, the information currently available is inadequate and does not anticipate trends in leprosy epidemiology and transmission.

2. What is the host response to the infection and what is the disease pathogenesis?

Host immune responses to *M. leprae* differ for TT and LL types of leprosy, with a prevalence of either T-cell immunity (TT) or antibody production (LL). For the definition of different types of leprosy see Section 3.

Tuberculoid leprosy

Cell-mediated immunity predominates in this form of leprosy. The replication of *M. leprae* in Schwann cells leads to the stimulation of T-cell response and subsequently to chronic inflammation. This is mediated through the bacterial cell wall glycolipid PGL-I. The Schwann cells can express HLA class II molecules and therefore stimulate **CD4+** T cells. **γ-interferon (γ-IFN)**-producing CD4+ T cells are found infiltrating the lesions and forming **granulomas** with epithelioid and giant cells.

In addition, interaction of *M. leprae* with **Toll-like receptors** (TLR1 and TLR2) on Schwann cells in patients with TT is believed to induce **apoptosis** thus leading to nerve damage. A chronic inflammatory reaction results in swelling of perineural areas, **ischemia, fibrosis**, and death of axons. However, very little circulating antibody against *M. leprae* is present in TT.

Surprisingly, though, and differently from *M. tuberculosis*, co-infection with *M. leprae* and HIV does not have any appreciable effect on the clinical symptoms, although HIV infection impairs T-cell-mediated immune responses.

A strong T-cell-mediated immunity combined with a weak antibody production leads to the development of a mild clinical form of TT with few nerves involved and low bacterial load. The form of leprosy with dominating T-cell immune responses is currently called type 1 (reversal) leprosy as opposed to type 2 leprosy. The latter is characterized by high titers of antibodies in combination with nonappreciable cellular immune response and results in widespread skin lesions, extensive nerve involvement, and high bacterial load, mostly contributing to the LL form. The borderline forms are characterized by an increasing reduction of cellular immune responses associated with increasing bacillary load.

Lepromatous leprosy

Antibody-mediated responses with no protective effect are a feature of this form of leprosy. In LL, PGL-I stimulates production of specific **IgM** that is proportional to the bacterial load and this has diagnostic and prognostic importance. These antibodies are not protective and are thought to downregulate cell-mediated immunity. They form **immune complexes** that are deposited in tissues attracting neutrophils to the deposition sites and causing complement activation. This leads to the development of intense and aggravated inflammation and tissue/organ damage. A systemic inflammatory response mediated by the deposition of extravascular immune complexes is called **erythema nodosum** leprosum (ENL). Women with leprosy may develop post-partum nerve damage. Although pregnancy itself does not seem to aggravate the disease or cause a relapse, LL patients experience ENL reactions throughout pregnancy and lactation, possibly associated with earlier loss of nerve function.

As might be expected by the predominant cellular or humoral response, the clinical form of the disease is affected by the balance of **Th1/Th2** cytokines in TT and LL forms of leprosy. Since T-cell immunity is enhanced by the T helper 1 (Th1) **cytokine** pathway (**interleukin (IL)**-2, IL-15, IL-12, IFN-γ), and humoral immunity is supported by Th2 cytokines (IL-4, IL-5 and IL-10), the severity of antibody-dominated LL is dependent upon activation of Th2 cytokines.

Since dendritic cells (DC) are likely to encounter the bacilli in the nasal mucosa or skin abrasions, they play a key role in the subsequent regulation of inflammation and balance of Th1/Th2 cell-mediated immunity – summarized in Figure 4.

Genetic polymorphism was also found to be associated with the pattern of the disease: HLA DR2 and DR3 alleles are associated with TT, and DQ1 with LL forms of leprosy.

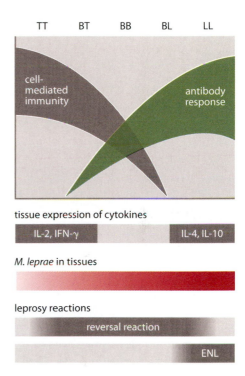

Figure 4. Spectrum of clinical and immunological characteristics of leprosy. Immune responses regulated by Th1 cytokines (IL-2, IFN-γ) are characteristic of the less severe TT form with the prevalence of T-mediated immune responses, whilst Th2 cytokines (IL-4, IL-10) enhance the more aggressive antibody-dependent LL form of leprosy. Intermediate forms (BT, BB, and BL) are characterized by an increasing shift in balance from Th1 towards Th2 cytokine-mediated immune responses accompanied by a concomitant appearance of *M. leprae* in the tissues. Intermediate forms can revert to the less severe form of the disease, but LL is irreversible and presents as erythema nodosum leprosum (ENL).
BT: borderline turberculoid leprosy;
BB: mid-borderline leprosy;
BL: borderline lepromatous leprosy.

3. What is the typical clinical presentation and what complications can occur?

The different forms of leprosy

The forms of leprosy range from the mildest indeterminate form to the most severe lepromatous type. Severe forms arise because of a less effective immune response to the infection (see Section 2). Most of those infected develop an appropriate immune response and never develop signs of leprosy.

The classification of leprosy is currently based on clinical grounds. Some authors identify the least aggressive indeterminate leprosy (IL) as the very first stage of the disease.

According to the accepted Ridley and Jopling (1966) classification there are five more advanced stages of leprosy, given here in increasing severity.

- Tuberculoid leprosy (TT), also called type I leprosy.
- Borderline tuberculoid leprosy (BT).
- Mid-borderline leprosy (BB).
- Borderline lepromatous leprosy (BL).
- Lepromatous leprosy (LL), also called type II leprosy.

Patients with IL, a very early form of leprosy, may either be cured or progress to one of the other forms of leprosy depending on their immune status. Within each type of leprosy, a patient may remain in that stage, improve to a less debilitating form or worsen to a more debilitating form depending on their immune status (see Section 2). Lepromatous leprosy is the only form that never reverts to a less severe form.

Both IL and TT forms are characterized by fewer than five skin lesions, with no bacilli being detected on smear tests and are collectively called paucibacillary (PB) leprosy.

IL is an early form with no loss of sensation and one or several skin hypopigmentation areas and **erythematous** macules. About 75% of these marks heal spontaneously. In some affected individuals, supposedly with weak immune responses, the disease progresses to other forms.

In patients with TT, one or several erythematous plaques are usually found on the face and limbs, but not on intertriginous areas or the scalp. Loss of sensation and hair loss (**alopecia**) are also characteristic. Neural involvement includes thickening of peripheral nerves, their tenderness, and subsequent loss of function. The TT form can either resolve itself spontaneously after several years, or develop to borderline BT, and very rarely can progress to the more aggressive LL form.

LL and, often, BL are characterized by more than five skin lesions and considerable numbers of bacilli in smears and are collectively called multibacillary (MB) leprosy. Patients with LL have small diffuse and symmetric macular lesions. Lepromatous infiltrations are called lepromas or plaques and lead to the development of leonine faces. The loss of sensation and nerve loss progress slowly. Bone and cartilage damage is common as well as alopecia of eyebrows and eyelashes (**madarosis**). Testicular **atrophy** leads to sterility and **gynecomastia**.

Clinical presentation

The initial symptoms of leprosy are not demonstrative and often go unnoticed. Most often patients start to lose sensation to heat and cold years before skin lesions occur (see the case details). The next sensation to be affected is light touch, then pain, and finally hard pressure. The most usual sites for numbness are hands and feet, and the first symptoms that patients report are usually burns and ulcers at the numb sites. However, the disease is often recognized through the development of dermal eruptions. At this stage other symptoms appear such as damage to the anterior chambers of the eyes, chin, earlobes, knees or testes.

Skin

The most common skin presentations are lesions in the form of **macules** or plaques, more rarely **papules** and nodules. The TT form is characterized by few hypopigmented lesions with lost or reduced sensation. In patients with LL there are usually many lesions, sometimes confluent, with not so apparent loss of sensation. Alopecia is common in LL patients and eyebrows and eyelashes are affected. The scalp remains intact.

Nerves

Peripheral nerve damage occurs characteristically near the surface of the skin, particularly in the neck, elbow, wrist, neck of the fibula, and the medial malleolus (posterior tibial nerve). The latter is the most affected, followed by ulnar, median, lateral popliteal, and facial nerves. Involvement of small dermal nerves leads to **hypoesthesia** and **anhidrosis** in TT patients and 'glove and stocking' loss of sensation in the case of LL.

Eyes

Both nerve damage and bacterial invasion can lead to blindness in up to 3% and potential blindness in almost 11% of multibacillary patients (mostly the LL form). Patients suffer from loss of corneal sensation and dryness in the cornea and conjuctiva due to the reduction in blinking. This increases the risk of microtrauma and ulceration of the cornea. **Photophobia** and **glaucoma** can also develop.

Systemic

Systemic damage is mostly associated with the LL form of the disease and is a result of bacillary infiltration. Particularly important is testicular atrophy, followed by **azoospermia** and gynecomastia. Adequate treatment significantly decreases the risk of the development of renal damage and **amyloidosis**. However, **lymphadenopathy** and **hepatomegaly** can develop in LL patients. Joint invasions by the microorgansim may lead to aseptic **necrosis** and **osteomyelitis**.

4. How is the disease diagnosed and what is the differential diagnosis?

Diagnosis of leprosy is mainly clinical, and a panel of clinical criteria has been developed that gives an average reliability of 91–97% for multibacillary and 76–86% for paucibacillary leprosy. The panel consists of three major criteria, as follows.

1. Hypopigmentation or reddish patches of the skin with the loss of sensation. The latter can be tested by touching the skin (tactile test) with a wisp of cotton, while the patient's eyes are closed. Test tubes with hot and cold water may be used to test temperature perception.

2. Detection of acid-fast bacilli in biopsies or in skin smears.

3. Thickening of peripheral nerves. Enlarged nerves, especially in the areas where a superficial nerve may be involved are examined by palpation. Attention should be paid for wincing during nerve palpation, indicating pain.

A skin biopsy for testing the presence of inflamed nerves is currently the standard criterion for diagnosis. In addition, biopsies are examined for the presence of acid-fast bacilli and the morphologic index (MI) is calculated that represents the number of viable bacilli per 100 bacilli. Histological diagnosis is used as the gold standard.

Tissue/skin smear tests using the Ziehl-Neelsen staining method (see Figure 2) or Fite modification allow the visualization of bacilli under the microscope and calculation of the bacterial index (BI). However, this method is relatively insensitive and smears from almost 70% of patients are found to be negative.

Two tests have been introduced to diagnose nerve injury.

1. In the histamine test, a drop of histamine diphosphate is placed on non-affected and affected skin. It normally causes the formation of a wheal on normal skin, but not where the peripheral nerves have been damaged. However, this test is not very clear in dark-skinned patients.

2. In the methacholine sweat test, an intracutaneous injection of methacholine results in sweating in normal skin but no sweating in lepromatous lesions.

Establishment of neural inflammation histologically differentiates leprosy from other granulomatous diseases.

There is still no definitive diagnostic laboratory test for leprosy. However, the tests may be an aid in diagnosis and include the following.

1. **Serological** tests to detect antibodies to *M. leprae*. Usually high titers of antibodies indicate untreated LL. Antibodies to PGL-1 are present in 90% of LL patients, but only in 40–50% of TT cases. They are also detected in 1–5% of healthy individuals.

2. **Polymerase chain reaction (PCR)** assay for *M. leprae* specific genes or repeat sequences has been able to confirm 95% of multibacillary and 55% of paucibacillary cases. So far PCR has not been used as a primary diagnostic tool, only to confirm the diagnosis when acid-fast bacilli are detected, but clinical features are doubtful. However, with the automation of PCR-based and reverse transcription (RT)-PCR assays they are being gradually implemented in many reference laboratories in countries with **endemic** leprosy. Various PCR-based techniques are used for the detection of drug resistance.

The Lepromin test allows the asessment of delayed hypersensitivity to antigens of *M. leprae* or other mycobacteria that cross-react with *M. leprae* by injection of a standardized extract of inactivated bacteria under the skin. A positive test indicates cell-mediated immunity associated with TT. This test is only of value in the classification of leprosy. It should not be used to establish a diagnosis of leprosy.

Early diagnosis is often prevented by difficulties in reaching some affected rural communities in developing countries. Delays in the diagnosis and treatment lead to an increase in leprosy-related disabilities. Overcoming the stigma of leprosy in certain countries and public education about this disease are important. Effective therapy further reduces the stigma attached.

5. How is the disease managed and prevented?

Leprosy control includes early detection of the disease, adequate treatment of patients, effective care for the prevention of disabilities, and patient rehabilitation.

Management

All forms and stages of newly diagnosed leprosy are normally treated with rifampicin, clofazimine, minocycline, ofloxacin, and dapsone as a part of a standard multidrug therapy (MDT) recommended by WHO. MDT was introduced for leprosy control as a result of a resistance to monotherapy with dapsone (diaminodiphenyl sulfone). However, the doses and the combinations of these antibiotics differ for paucibacillary (PB) and multibacillary (MB) leprosy.

1. For the one lesion PB leprosy, a single shot of rifampicin 600 mg, ofloxacin 400 mg, and minocycline 100 mg (also known as ROM) is used, which is both acceptable and cost-effective.

2. For PB leprosy with more than one lesion, a 6-month treatment includes: daily self-administered doses of dapsone 100 mg and a monthly supervised administration of rifampicin 600 mg.

3. For MB leprosy, the treatment is prolonged over 24 months and consists of: daily self-administered combinations of dapsone 100 mg and clofazimine 50 mg, together with monthly supervised administration of rifampicin 600 mg and clofazimine 300 mg.

Other chemotherapeutic agents include levofloxacin (LVFX), sparfloxacin (SPFX), and clarithromycin (CAM), which are also effective against *M. leprae*.

Research is under way to establish a new efficient treatment regimen. It has recently been discovered that thalidomide is effective in treating leprosy. In 1998 thalidomide was approved by the US Food and Drug Administration (FDA) to treat leprosy with several restrictions to avoid birth defects.

Sometimes, the use of immunomodulators such as levamisole is recommended to decrease the persistence of a large pool of dead bacilli via the enhancement of cell-mediated immunity.

Unfortunately, in countries where the disease is most widespread, low treatment completion rates have been reported, 60–90% for PB patients and 40–80% for MB patients. This could be due to a number of social factors including costs of the visits to the health-care units, stigma, and disabilities.

Prevention

Vaccines for leprosy

An effective prophylactic vaccine for leprosy does not yet exist, although it is very desirable. It could help to prevent drug-susceptible as well as drug-resistant disease.

It has been suggested that a previous infection with *M. tuberculosis* or BCG vaccination provides some protection from leprosy with individual degrees of resistance varying from 20% to 80%. BCG administration led to the induction of delayed hypersensitivity to *M. leprae*. It was even more effective when given together with heat-killed *M. leprae*, and caused a shift in immune response from multibacillary to paucibacillary leprosy, which is beneficial for a patient.

Purification of a vaccine material from actual *M. leprae* preparations is limited by poor standards and quality of the bacilli isolation from armadillo tissues.

New approaches to the development of anti-leprosy vaccines include production of a recombinant vaccine. This is based on the current progress of *M. leprae* genome sequencing and bioinformatic tools.

Prevention of disabilities and rehabilitation

Twenty five percent of leprosy patients, particularly with LL and BL forms, have some degree of disability. It increases with age and is more severe in males. The gravity of disability correlates with the duration of the disease. Hands are most frequently affected followed by the feet and eyes.

There are currently around 3 million people with physical deformities and disabilities caused by leprosy.

The socioeconomic impact of leprosy continues to be a burden in endemic countries. Rehabilitation strategies for leprosy should contain physical, psychological, social, and economic aspects. Up to now, most leprosy control programs have been focused on physical rehabilitation only. However, recently socio-economic rehabilitation has been introduced by some local governments, nongovernmental organizations (NGOs), and community-based rehabilitation (CBR) programs focused on the patient's self-awareness and combating the stigma attached to the disease. These are specifically adapted for the local needs and differ among countries and from area to area.

SUMMARY

1. What is the causative agent, how does it enter the body and how does it spread a) within the body and b) from person to person?

- *Mycobacterium leprae* bacilli are the cause of leprosy. They are straight or curved rod-shaped organisms, 1–8 µm long and approximately 0.3 µm in diameter.

- *M. leprae* bacilli live in infected macrophages and Schwann cells, surrounding the axons of nerve cells and are characterized by their slow growth.

- An intracellular parasite, *M. leprae* cannot be cultured by conventional laboratory methods and can be propagated in mouse footpad or in its natural host – the nine-banded armadillo.

- The genome of *M. leprae*, contains less than 50% of the functional genes of *M. tuberculosis*. These lost genes coding for the metabolic pathways have been replaced in *M. leprae* by pseudogenes or inactivated genes.

- The main components of *M. leprae* cell wall are similar to those of *M. tuberculosis*, but include species-specific phenolic glycolipid I (PGL-I) and genus-specific lipoarabinomannan.

- The major port of entry and exit of *M. leprae* is the respiratory system. It is mostly transmitted by aerosol spread or exposure to nasal and oral mucosa.

- *M. leprae* binds to the G-domain of α2-chain of laminin 2 in the basal lamina of Schwann cells. Uptake of *M. leprae* by the Schwann cells is facilitated by α-dystroglycan – a receptor for laminin on cell membrane.

- Optimal temperature for growth for *M. leprae* is low (30–35°C), so it prefers to grow in the cooler parts of the body: skin and superficial nerves. Natural hosts of *M. leprae* – armadillos – are characterized by low body temperature.

- Asymptomatic infected individuals may transiently excrete the microbe nasally and spread the infection among close proximal contacts.

- In 1997 two million people worldwide were infected with *M. leprae*, with around 800 000 new cases of leprosy detected per year.

2. What is the host response to the infection and what is the disease pathogenesis?

- A strong T-cell-mediated immunity combined with a weak antibody production leads to the development of a mild clinical form of tuberculoid leprosy (type I or reversal) with few nerves involved and low bacterial load.

- In the tuberculoid form of leprosy the most damage is inflicted by chronic inflammation caused by immune responses.

- In lepromatous (type II) leprosy high titers of antibodies in combination with nonappreciable cellular immune response result in widespread skin lesions, extensive nerve involvement, and a high bacterial load.

- In LL systemic inflammatory response to the deposition of extravascular immune complexes leads to erythema nodosum leprosum (ENL) resulting in neutrophil infiltration and complement activation.

- The borderline forms of leprosy are characterized by an increasing reduction of cellular immune responses associated with increasing bacillary load.

- The clinical form of the disease is affected by the balance of Th1/Th2 cytokines that favor T-cell- or B-cell- (antibody) mediated immune responses, respectively.

3. What is the typical clinical presentation and what complications can occur?

- The forms of leprosy distinguished by the Ridley-Jopling classification are: indeterminate leprosy (IL), tuberculoid leprosy (TT), borderline forms (BT, BB, and BL) and lepromatous leprosy (LL).

- IL and TT are characterized by fewer than five lesions, no bacilli on smear tests, and are called paucibacillary (PB) disease.

- LL and, often, BL are characterized by more than five lesions with bacilli and are called a multibacillary (MB) disease.

- *M. leprae* is not very pathogenic and most infections do not result in leprosy. Some early symptoms such as skin lesions are self-limiting and can heal spontaneously.

- The initial symptoms of leprosy are the loss of sensation of temperature, followed by the loss of the sensation of a light touch, then pain, and finally hard pressure.

- The disease is mostly recognized through the development of dermal eruptions such as lesions in the form of macules or plaques, more rarely papules and nodules.

- The TT form is characterized by few hypopigmented lesions with lost or low sensation. The LL form is characterized by many lesions, sometimes confluent, with less apparent loss of sensation.

- Peripheral nerve damage appears near the surface of the skin, particularly in the neck, elbow, wrist, neck of the fibula, and medial malleolus.

- Both nerve damage and bacterial invasion can lead to blindness in up to 3% and potential blindness in almost 11% of multibacillary patients.

- Systemic damage in the patients with the LL form is a result of bacillary infiltration and comprises testicular atrophy, azoospermia, and gynecomastia.

4. How is the disease diagnosed and what is the differential diagnosis?

- The three main diagnostic criteria are: hypopigmentation or reddish patches of the skin with the loss of sensation; acid-fast bacilli in biopsies or in skin smears; thickening of peripheral nerves.

- Skin biopsy for testing of the presence of inflamed nerve is currently the standard criterion for diagnosis.

- Tissue/skin smear test using Ziehl-Neelsen staining method or Fite modification allows bacilli to be visualized under the microscope, but

this method is not highly sensitive and smears from almost 70% of all patients are found to be negative.

- Serological tests are used to detect antibodies to *M. leprae*. High titers of antibodies indicate untreated LL.

- To confirm diagnosis, automated PCR- and RT-PCR-based techniques have been implemented in many reference laboratories in countries with endemic leprosy.

- Early diagnosis is often prevented by difficulties in reaching some affected rural communities in the developing countries.

5. How is the disease managed and prevented?

- Leprosy control includes early detection, adequate treatment, comprehensive and effective care for the prevention of disabilities, and rehabilitation.

- Newly diagnosed leprosy is treated with a combination of rifampicin, clofazimine, minocycline, ofloxacin, and dapsone as a part of a standard multidrug therapy (MDT) recommended by WHO.

- For the one lesion PB leprosy a single shot of rifampicin 600 mg, ofloxacin 400 mg, and minocycline 100 mg (also known as ROM) is used.

- For PB leprosy with more than one lesion a 6-month treatment includes: a daily self-administered dose of dapsone 100 mg and a monthly supervised administration of rifampicin 600 mg.

- For MB leprosy the treatment is prolonged to 24 months and consists of: daily self-administered combination of dapsone 100 mg and clofazimine 50 mg together with monthly supervised administration of rifampicin 600 mg and clofazimine 300 mg.

- BCG administration together with killed *M. leprae* leads to a shift in immune response from multibacillary to paucibacillary leprosy, which is beneficial for a patient.

- The development of BCG-based and recombinant vaccines is currently under way.

- Rehabilitation strategies for leprosy should contain physical, psychological, social and economic aspects.

FURTHER READING

Britton WJ. Leprosy. In: Cohen J, Powerly WG, editors. Infectious Diseases, 2nd edition. Mosby, London, 2004: 1507–1513.

Murphy K, Travers P, Walport M. Janeway's Immunobiology, 7th edition. Garland Science, New York, 2008.

REFERENCES

Barker L. *Mycobacterium leprae* interactions with the host cell: recent advances. Indian J Med Res, 2006, 123: 748–759.

Brosch R, Gordon SV, Eiglmeier K, Garnier T, Cole ST. Comparative genomics of the leprosy and tubercle bacilli. Res Microbiol, 2000, 151: 135–142.

Demangel C, Britton WJ. Interaction of dendritic cells with mycobacteria: where the action starts. Immunol Cell Biol, 2000, 78: 318–324.

Eiglmeier K, Simon S, Garnier T, Cole ST. The integrated genome map of *Mycobacterium leprae*. Lepr Rev, 2001, 72: 387–398.

Gebre S, Saunderson P, Messele T, Byass P. The effect of HIV status on the clinical picture of leprosy: a prospective study in Ethiopia. Lepr Rev, 2000, 71: 338–343.

Groenen G, Saha NG, Rashid MA, Hamid MA, Pattyn SR. Classification of leprosy cases under field conditions in Bangladesh: II, reliability of clinical criteria. Lepr Rev, 1995, 66: 134–143.

Haanpaa M, Lockwood DN, Hietaharju A. Neuropathic pain in leprosy. Lepr Rev, 2004, 75: 7–18.

Lockwood DNJ. The management of erythema nodosum leprosum: current and future options. Lepr Rev, 1996, 67: 253–259.

Lockwood DN, Suneetha S. Leprosy: too complex a disease for a simple elimination paradigm. Bull WHO, 2005, 83: 230–235.

Ridley, DS, Jopling, WH. Classification of leprosy according to immunity. A five-group system. Int J Lepr Other Mycobact Dis, 1966, 34: 255–273.

Scollard DM, McCormick G, Allen JL. Localization of *Mycobacterium leprae* to endothelial cells of epineurial and perineurial blood vessels and lymphatics. Am J Pathol, 1999, 154: 1611–1620.

Scollard DM, Adams LB, Gillis TP, Krahenbuhl JL, Truman RW, Williams DL. The continuing challenges of leprosy. Clin Microbiol Rev, 2006, 19: 338–381.

World Health Organization. Leprosy elimination project. Status report 2002–2003. World Health Organization, Geneva, Switzerland.

You EY, Kang TJ, Kim SK, Lee SB, Chae GT. Mutations in genes related to drug resistance in *Mycobacterium leprae* isolates from leprosy patients in Korea. J Infect, 2005, 50: 6–11.

WEB SITES

American Leprosy Missions, Greenville, SC, USA. Copyright © 2008, American Leprosy Missions. All rights reserved: http://www.leprosy.org/LEPinfo.html

Lepra Society: Health in Action, Andhra Pradesh, India © Copyright LEPRA Society – All Rights Reserved: http://www.leprasociety.org

Leprosy Elimination Initiative, World Health Organization, © WHO 2008: http://www.who.int/lep/en

Leprosy Mission International, Middlesex, UK, © The Leprosy Mission International: http://www.leprosymission.org

New Zealand Dermatological Society Incorporated. © 2008 NZDS: http://dermnetnz.org/bacterial/leprosy.html

Skillicorn Family Ministries, Leprosy Sufferers: http://www.webspawner.com/users/LEPROSY

MULTIPLE CHOICE QUESTIONS

The questions should be answered either by selecting True (T) or False (F) for each answer statement, or by selecting the answer statements which best answer the question. Answers can be found in the back of the book.

1. **Which of the following are true of *M. leprae*?**

   A. Its cell wall contains species-specific phenolic glycolipid I (PGL-I).

   B. It binds to the target cells via genus-specific lipoarabinomannan.

   C. The genome of *M. leprae* is decayed compared with *M. tuberculosis*.

   D. Nine-banded armadillos are natural hosts of *M. leprae*.

   E. It is highly contagious.

2. **Which of the following are characteristic of the physiology of *M. leprae*?**

   A. Slow growth.

   B. A requirement of 37°C for optimal growth.

   C. Being disseminated through the digestive route.

   D. Its growth inside macrophages and Schwann cells.

   E. Its preference for growth in the peritoneal cavity.

3. **In which of the following regions of the world is leprosy endemic?**

   A. The Pacific region.

   B. Southern Europe.

   C. Southern America.

   D. East-Southern Asia.

   E. Africa.

4. **Which of the following characterize the tuberculoid (TT) form of leprosy?**

   A. The prevalence of the production of Th1 over Th2 cytokines.

   B. The predominantly T-cell-mediated immune responses.

   C. The development of erythema nodosum leprosum (ENL).

   D. A possibility of a spontaneous resolution.

   E. Belonging to the paucibacillary form of leprosy.

5. **Which of the following characterize the lepromatous form (LL) of leprosy?**

   A. The prevalence of the production of Th1 over Th2 cytokines.

   B. Predominantly antibody-mediated immune responses.

   C. The possibility to revert to the less severe form.

   D. Belonging to the multibacillary form of leprosy.

   E. More than five skin lesions.

6. **Which of the following clinical tests are important for the differential diagnosis of leprosy?**

   A. Hypopigmentation of the skin.

   B. Reddish patches of the skin.

   C. Loss of sensation.

   D. Thickening of peripheral nerves.

   E. Abscesses.

7. **Which of the following laboratory tests are used for the establishment or confirmation of the diagnosis of leprosy?**

   A. Bacillary cell culture.

   B. Microscopy of biopsies.

   C. Serological identification of the specific antibodies.

   D. Polymerase chain reaction.

   E. Lepromin test.

8. **Which of the following are included in the multidrug therapy regime (MDT) of leprosy?**

   A. Clofazimine.

   B. Benzylpenicillin.

   C. Isoniazid.

   D. Rifampicin.

   E. Dapsone.

Case 23

Mycobacterium tuberculosis

A 63-year-old man lived in a hostel for the homeless and sold magazines outside a railway station. He had been finding it difficult to cope with this recently, as he had been feeling weak, had lost weight, and often had a fever at night.

One month ago, he started coughing up blood and feeling breathless, which had really worried him. He was not registered with a primary health-care provider but a friend told him about a walk-in practice for homeless people. Next day he went to the practice and was seen by the physician on duty, who found that the patient had a low-grade fever and detected bronchial breathing when he listened to his chest. The doctor sent him for a chest X-ray and asked him to return for the results. When the X-ray result came back, it showed that he had apical shadowing and large **cavitation** consistent with tuberculosis (TB).

The X-ray is shown in Figure 1. A sputum sample was taken since the doctor suspected that the patient had tuberculosis and the patient was started an antituberculosis therapy.

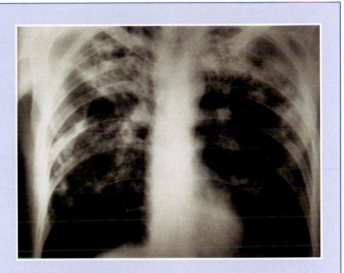

Figure 1. Chest X-ray of patient showing typical apical consolidation with possible cavities. There is also some consolidation adjacent to the mediastinum on the right. The bases are relatively spared. The medial aspect of the right hemi-diaphragm is just visible.

1. What is the causative agent, how does it enter the body and how does it spread a) within the body and b) from person to person?

Causative agent

This patient is infected with *Mycobacterium tuberculosis*. It is a weakly gram-positive mycobacterium classified as an 'acid-fast bacillus' because the dye that is used to stain it is resistant to removal by acid. The stain used to identify mycobacteria is called the Ziehl-Neelsen (ZN) stain and characteristically stains mycobacteria red while all other organisms stain green (Figure 2). *Mycobacterium* is the only genus of medical importance that stains red with ZN stain.

In common with most other bacteria the cell wall contains peptidoglycan. Overlying this is a layer of arabinogalactan, which is covalently linked to the outer layer composed of mycolic acid, long chain fatty acids specific for the mycobacterial genus, with other components such as glycophospholipids and trehalose dimycolate (also called cord factor as on staining the organism it has the appearance of cords). Running vertically through the whole of the cell wall and linked to the cytoplasmic membrane is lipoarabinomannan (Figure 3).

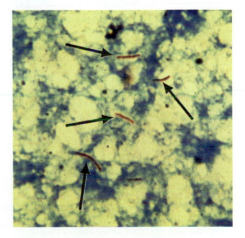

Figure 2. Ziehl-Neelsen stain of sputum: note the red bacilli (arrowed) against a green background stain. Note that this is the only genus of medically important bacteria that stains red with the Ziehl-Neelsen stain.

Figure 3. Model of the structure of the cell wall mycobacteria. Note the mycolatearabinogalactan-peptidoglycan-complex (MAPc). The mycolic acid layer is impervious to many substances necessitating the presence of porin channels to allow entry of hydophilic compounds. This layer plays a major role in the defense of the cell because few antibodies can penetrate it and it is relatively resistant to dessication and some disinfectants.

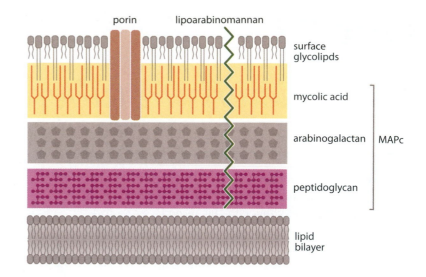

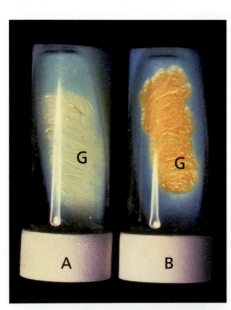

Figure 4. Growth of *Mycobacterium tuberculosis*. (G) on Lowenstein Jensen medium. The growth of *M. tuberculosis* does not produce a pigment (A) but the growth of another mycobacterial species produce a yellow pigment (B).

The genus *Mycobacterium* can be broadly divided into rapid and slow growers and noncultivable species, *in vitro*. *Mycobacterium leprae*, the causative agent of leprosy and a close relative of the tubercle bacillus, cannot be grown on artificial media and can only be propagated in armadillos. Rapid-growing mycobacteria produce colonies within 2–3 days for *M. smegmatis* while slow growers take more than 7 days. *M. tuberculosis*, a slow grower, can take 2–4 weeks to produce colonies. This means that clinical decisions affecting treatment of tuberculosis and leprosy and the diagnosis of these conditions do not rely primarily on culture. However, the gold standard for diagnosis of tuberculosis is culture of *M. tuberculosis* from the patient. The slow growth also raises problems in determining the actual species causing illness (as the treatment may vary depending on the causative agent) and in determining antibiotic resistance. The usual medium for the isolation and growth of mycobacteria is Lowenstein-Jensen, which contains egg yolk and a dye (malachite green) that inhibits the growth of more rapidly growing bacteria (Figure 4).

Entry and spread within the body

In primary infection *M. tuberculosis* enters the body (in aerosol droplets, see below) via the respiratory tract and is deposited in the alveoli of the lungs where it is taken up mainly by **alveolar macrophages**. Entry into the alveolar macrophages is mediated through a variety of surface receptors expressed by these phagocytic cells. These include surface complement receptors, scavenger receptors, and Fc-γ receptors. The macrophages also have **pattern recognition receptors (PRRs)** recognizing **PAMPs (pathogen-associated molecular pattern receptors)**, mannose receptors, and other PRRs such as the **Toll-like-receptors (TLRs)**. The latter receptors also recognize mycobacterial compounds but are not involved in **phagocytosis** but rather in signaling to induce a pro-inflammatory response. Although the organism can multiply extracellularly to some extent within the alveolus, the organism is able to survive and multiply within the macrophages (due to mechanisms that prevent killing within the **phagosome** – see below). Eventually macrophages die by programmed cell death (**apoptosis**) and release mycobacteria. **Dendritic cells** also take up mycobacteria and become activated, which induces their migration to draining lymph nodes where they prime/activate T cells. Activated T cells recognizing mycobacterial antigens migrate into the lung and induce formation of small **granulomas** (see host response).

The site of infection in the lungs tends to be at the base and close to the pleura. After up to 3 weeks and usually before cell-mediated immunity develops to any great extent, the microorganisms are released from the macrophages and spread via the bloodstream to draining regional lymph nodes, for example hilar or mediastinal, as well as to every organ in the body (principally the lung apices, **meninges**, kidneys, and bones). Macrophages can also carry viable microorganisms around the body and how much of the spread is via this mechanism is unclear. Spread via the pulmonary arteries can give rise to **miliary** tuberculosis (TB) of the lungs which is, however, primarily found in immunocompromised patients and small children. Another possibility is that the organisms are swallowed, causing laryngeal TB or intestinal TB.

It should be emphasized that primary infection with *M. tuberculosis* leads to active disease in only a small number of individuals (5%). Thus most individuals are able to control the initial infection, showing either no symptoms or mild clinical manifestations similar to those seen for a common cold. However, most infected individuals carry the organism in a latent state for life under the control of an effective immune system (see below). Some may develop active disease many years after primary infection, often when they become immunosuppressed (reactivation).

Person to person spread

In patients with active tuberculosis, *M. tuberculosis* bacilli from granulomas are released into the bronchi and are spread through coughing. The aerosols produced contain droplet nuclei and survive for quite long periods of time outside the body. It is estimated that each infected person infects on average 20 other individuals. Repeated contact with an infected individual, particularly in a closed environment, produces higher transmission rates than casual contact. Similarly, if the infected person is smear-positive (mycobacteria seen in the sputum by ZN stain), this indicates that bacteria are present in large numbers in the airways; consequently, they are much more contagious, that is 50% of contacts may become infected. Whereas if the index case is smear-negative (i.e. mycobacteria not seen by ZN stain of sputum) but is culture-positive, then only about 5% of contacts will be infected. The number of times an index case coughs is also directly related to the transmission rate.

Epidemiology

Announced as a Global Emergency in 1993 by the World Health Organization (WHO), it is estimated that one-third of the world's population (2 billion) is infected with *M. tuberculosis*, with 8–10 million individuals developing active disease annually and about 2 million dying (Table 1).

It is estimated that someone is newly infected every second and that the mortality from tuberculosis could be as high as 4 million per annum in 2020. Tuberculosis is responsible for 1 in 4 preventable deaths. The main foci of infection are South-East Asia and Africa.

The situation is severely exaggerated by the AIDS epidemic, since (as mentioned above) on average, 5% of individuals in the general population

Table 1. Global distribution of tuberculosis cases

Country	Actual cases (estimate)	Rate/100000 population
India	1932852	168
China	1311184	99
Indonesia	534439	234
Bangladesh	350641	225
Pakistan	291743	181
Nigeria	449558	311
Philippines	247740	287
South Africa	453929	940
Russian Federation	152797	107
Ethiopia	306330	378
Vietnam	148918	173
Congo	14869	403
UK	9358	15
USA	13148	4

WHO Global Surveillance Monitoring Project – 2008 WHO/HTM/TB/2008.393.

infected with *M. tuberculosis* will eventually develop active disease but this increases to 50% if the person is co-infected with HIV. This is due to the immunosuppressive effect of the virus (see HIV case). At-risk groups for development of active disease in the population include prison inmates, the homeless, alcoholics, intravenous drug users (IVDUs), and those who are suffering social deprivation.

2. What is the host response to the infection and what is the disease pathogenesis?

In an infectious process, normally bacteria that are taken up by macrophages are killed within the phagolysosome. However, mycobacteria can live and divide within the macrophages by inhibiting maturation of their phagosomes to prevent fusion with **lysosomes** and phagolysosome formation (through some of their cell wall glycolipids such as the cord factor or trehalose dimycolate) and are thus not exposed to the bactericidal content of the lysosome. The host may also respond by inducing a mechanism called '**autophagy**,' which is not only the recycling system of the host cell but also a mechanism to target intracellular microorganisms to lysosomes.

'Activation' of the macrophages by mycobacterial cell wall components or DNA through TLR2 or RLR9, respectively, leads to **cytokine** production initiating an inflammatory response, which induces the recruitment of further macrophages/monocytes from the circulation. Other important outcomes of macrophage activation include over-riding the *M. tuberculosis*-mediated block of phagosome maturation and up-regulation of numerous antimicrobial effectors (NADPH oxidase, which generates **reactive oxygen species**, iNOS which generates reactive nitrogen species, NRAMP).

Although *M. tuberculosis* inhibits natural killing mechanisms of the macrophage, some intracellular organisms do die and these are broken down by proteolytic enzymes to produce peptides that are then presented via class II human leukocyte antigens (HLA) to **CD4+** T cells, which in the presence of **interleukin** (**IL**)-12/IL-18 produced by macrophages and dendritic cells will differentiate to **Th1** cells. These Th1 cells produce the pro-inflammatory cytokines **interferon-γ** (**IFN-γ**) and **tumor necrosis factor-α** (**TNF-α**) which, together with IFN-γ and IL-1, mainly produced by the macrophages themselves, further activate the bactericidal effectors of macrophages such as **defensins** (such as **cathelicidin**), nitric oxide (NO) production and autophagy induction (Figure 5).

During mycobacterial infection cytotoxic CD8 T cells are induced through peptide antigens presented to them by class I HLA. These T cells produce IFN-γ but are also able to kill cells infected with mycobacteria by **perforin** and **granzymes**. **Granulysin**, released by cytotoxic T cells can also kill mycobacterium. This cellular response can result in some tissue destruction.

Although difficult to prove, in some primary infected individuals the organism is likely to be completely eliminated from the body. However, in most cases the organism will survive and persist within some macrophages (but not proliferate), which leads to continuous activation of both CD4 and **CD8+** T cells. The cytokines released by these T cells and macrophages lead to the development of granulomas (or 'tubercles' – which gave the disease its name), which is a way of limiting the spread of infection and tissue involvement (Figure 6). Not being able to eliminate the organism completely results in the production of a connective tissue layer to 'wall off' the organism from the rest of the body. This is the containment phase (latent infection with dormant mycobacteria) of the disease.

Post infection, 3–5% of individuals get active disease within the first year but this might rise to 15–35% in subsequent years depending on several conditions, especially immunosuppression (e.g. HIV co-infection, etc.). Whether or not someone gets active or quiescent disease depends upon the following factors.

1. Genetic background: alleles of HLA DR (DRB1*1501 increase and 1502 decrease); alleles of NRAMP1; alleles of INF-γ.

2. The infectious dose: high TB dose.

3. Activation state of the immune system: BCG vaccination, HIV co-infection, and so forth.

Additional factors include malnutrition, iron overload, anti-TNF treatment (infliximab) in patients with rheumatoid arthritis (RA), and old age.

Histologically, a granuloma is a collection of activated macrophages called epithelioid cells and a center that frequently shows an area of tissue **necrosis**. In tuberculosis, the necrosis is characteristically 'cheesy' and is called caseous necrosis. Sometimes the macrophages fuse to form giant cells. Lymphocytes, particularly of the CD4 T-cell subset (but also CD8+ T cells), are also present in the granulomas and actively produce cytokines.

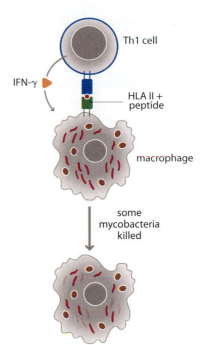

Figure 5. Killing of *M. tuberculosis* in some macrophages mediated by IFN-γ released by Th1 cells. Specific Th1 cells recognize mycobacterial peptides presented by HLA class II molecules on the infected macrophages. IFN-γ 'activates' the killing mechanisms in the macrophages so that some mycobacteria can be killed. Modified from Lydyard, Whelan and Fanger – Instant Notes in Immunology, 2004.

partial activation of macrophage

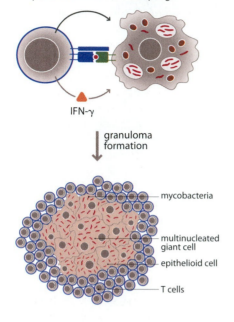

IFN-γ

granuloma formation

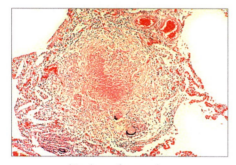

mycobacteria

multinucleated giant cell

epithelioid cell

T cells

histology of a granuloma

Figure 6. Granuloma showing macrophages containing mycobacteria, epithelioid cells, and multinucleate giant cells all surrounded by T cells, which are mainly CD4+.

They are formed by macrophages containing mycobacteria, epithelioid cells, and multinucleate giant cells all surrounded by T cells, which are mainly CD4+.

The persistence of mycobacterial antigens in live and dead mycobacteria means that the T cells are continuously activated and the granulomatous response in tuberculosis is considered to be a **type IV hypersensitivity** response. This is a classic form of chronic inflammation through persistence of the infectious agent. In this form there is an equilibrium set up between the immune system and the organism that keeps live organisms 'in check.'

At anytime during life, the disease can be reactivated, often through the individual becoming immunosuppressed (e.g. co-infection with HIV). This leads to a change in status of these granulomas. Some of the granulomas cavitate (decay into a structureless mass of cellular debris), rupture, and spill thousands of viable, infectious bacilli into the airways (if they are in the lung). This leads to the person becoming infectious, as described above. Reactivation can also take place within granulomas at other sites in the body leading to active disease, for example in the brain causing **meningitis**.

The host response depends on the immune conditions and dose of microorganisms.

1. Strong T-cell immunity and high dosage: there is greater tissue damage and **caseation**.

2. Strong T-cell immunity and low dosage: granulomas are produced.

3. Weak T-cell immunity (e.g. HIV co-infection): there is a poor granulomatous response, and many microorganisms are produced.

Figure 7 shows the possible sequence of events following infection.

3. What is the typical clinical presentation and what complications can occur?

The most common clinical presentation is of a temperature, chronic productive cough that may be streaked with blood (hemoptysis), and weight loss. The release of T-cell and macrophage cytokines, particularly TNF-α, leads to a fever (by its action on the thermoregulatory system of the hypothalamus) and weight loss. Adjacent granulomas in lymph nodes may fuse to produce a sizeable lump, which can be seen in a chest X-ray in the mediastinum or tissue destruction and cavities produced by dead tissue (cavitation, see Figure 1). The tissue destruction in the lungs resulting in cavitation can lead to loss of lung volume and erosion of bronchial arteries (cavitation, see Figure 1). This leads to coughing up blood. Spread of the organism through the body can lead to granulomas developing in other organs such as brain, bone, liver, and so forth; perhaps the most common complication being the 'space-occupying' effects of granulomas, for example in the brain, where it can lead to **seizures**. Tuberculosis can thus present with protean manifestations such as adrenal failure (**Addisons disease**) and fractures if it occurs in bone, for example vertebral collapse (**Potts disease**) (see Further Reading: Lydyard et al, 2000).

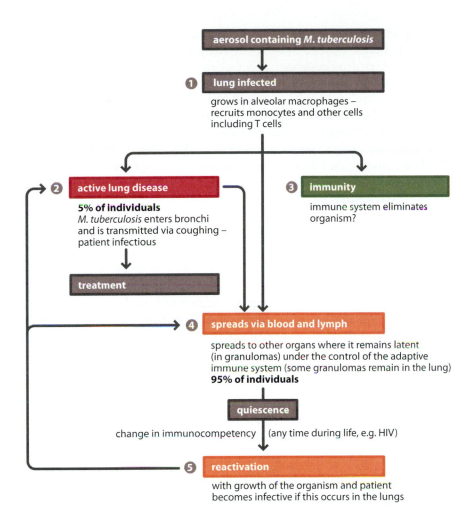

Figure 7. Sequence of possible events following tuberculosis infection.
1) In primary infections, *M. tuberculosis* organisms enter the lungs via aerosols and are taken up by alveolar macrophages and dendritic cells. The infection then moves to the lung parenchyma where monocytes and T cells are recruited through cytokines produced by the infected macrophages. The organism grows slowly within the resident and moncyte derived macrophages and the organisms eventually produce active disease in 5-35% of cases. 2) It is possible, but not proven, in some individuals that the organisms are toally eliminated from the body after primary infection. 3) In active disease, the foci become larger and form granulomas resulting in significant tissue damage, release of the organism into the bronchi and transmission by aerosols (infectious stage). The patient develops symptoms at this stage and can infect other people. 4) During growth of the microorganism in alveolar, foci (Ghon foci) some of them are also spread via the blood stream and lymphatic system to other organs in the body where immune systems keeps the *M. tuberculosis* sequestered in granulomas (inactive). 5) At a later date and probably as the result of a decrease in immunocompetence (for example, HIV co-infection), reactivation of the growth of the *M. tuberculosis* within the granuloma occurs. This results in further spread and growth of the organism. Growth and active disease can develop in several organs other than the lung, for example the brain (meningitis), and in the gut. If reactivation occurs in the lung, the individual becomes infectious.

4. How is this disease diagnosed and what is the differential diagnosis?

The slow growth of *M. tuberculosis* means that clinical decisions affecting treatment of tuberculosis and the diagnosis of this condition do not rely primarily on culture. However, culture remains the gold standard for confirming diagnosis of tuberculosis. The slow growth also raises problems in determining the actual species causing illness (as the treatment may vary depending on the causative agent) and in determining the strain of *M. tuberculosis*, which is important in terms of antibiotic resistance. Therefore diagnosis is principally a clinical decision and treatment is started on that basis. There is no rapid, accurate test for *M. tuberculosis*. The simplest laboratory procedure is the ZN stain on sputum but this is only about 50% sensitive and detects about 5000 organisms per ml. It does not distinguish between the different species of mycobacteria. Specimens (e.g. sputum, bronchoalveolar fluid, early morning urines, gastric aspirates, pus) are cultured on Lowenstein-Jensen medium but *M. tuberculosis* growth can take 2–4 weeks. More rapid growth and detection can be achieved in automated machines by detecting metabolic products of radioactively labeled substrates, but this can still take 7–14 days. Also available are nucleic acid amplification techniques like the **polymerase chain reaction (PCR)**, which amplifies specific sequences of the genome of *M. tuberculosis*, thereby detecting its presence in the specimen and if used with liquid culture can speed up the detection of the organism greatly. PCR is particularly useful in detecting the presence of multidrug-resistant tuberculosis, as drug resistance is correlated with characteristic mutations in specific genes that can be detected by PCR. Serological tests are under development and are available in some laboratories, such as the assay for IFN-γ from lymphocytes activated by mycobacterial antigen (PPD) and an **Elispot test** assaying the T-cell responses to mycobacterial antigens (e.g. early secretory antigen target-6, ESAT 6). The use of transrenal DNA is also being investigated as a source of material for diagnosis.

Differential diagnosis

For the differential diagnosis, it is important that the typical symptoms of weight loss, chronic cough, and fever may also be present with tumors of the lung, for example adenocarcinoma, squamous cell carcinoma, oat cell carcinoma. Because *M. tuberculosis* spreads throughout the body and can present with signs and symptoms referrable to many systems, the infection can mimic many other diseases (e.g. brucellosis or **lymphoma**) and mimic symptoms associated with TB infection can be caused by other diseases (seizures in case of tuberculoma can also be caused by glioma; tuberculous meningitis causes the same symptoms as cryptococcal meningitis).

There are many other causes of granuloma that are noninfectious but are particulate in nature, for example silica, or as in TB, difficult to digest antigens as found in the mycobacterial cell wall. The antigen(s) is unknown in some cases of granulomatous disease, for example **sarcoidosis**.

5. How is the disease managed and prevented?

Management

Infected patients in hospital should be isolated in a negative pressure room (where the air pressure outside the room is greater than that in the room thus any airflow is into the room). Health-care workers should wear close-fitting masks if they are involved in activities likely to induce coughing/expectoration by the patient, for example physiotherapists. Once the patient has been on adequate treatment for 2 weeks they can come out of isolation.

The standard anti-TB regimen in the UK is a combination of rifampicin plus isoniazid plus pyrazinamide for 2 months, when the pyrazinamide is stopped and the rifampicin and isoniazid are continued for a further 4 months, so the total treatment time is 6 months.

Some countries and some physicians start with quadruple therapy including ethambutol as part of the initial regimen. This length of treatment (especially since patients may feel better before the end of treatment) may lead to lack of compliance, a primary factor leading to the increase in drug resistance, as suboptimal treatment can lead to the development of secondary drug resistance. Patients who are thought to have drug-resistant tuberculosis may be started on the above three drugs plus ethambutol until the actual sensitivity of the isolate is determined, when an appropriate combination of drugs can be given. Newer antituberculous agents are under trial, for example levofloxacin as part of the primary regimen, although some authorities recommend that the fluoroquinolones should be reserved for multidrug-resistant tuberculosis (MDR-TB) (see References: Meyers, 2005).

Tuberculosis is a notifiable disease in the UK. The contacts of patients with active tuberculosis should be screened by an X-ray or given prophylaxis if appropriate. Close contacts of children with primary tuberculosis should be screened as there is likely to be a source that should be identified. In many areas the prevalence of tuberculosis is highest in the disadvantaged and homeless (see above). This is also the group that has high levels of drug-resistant tuberculosis (see References: Patel, 1985). Failure of therapy either due to inappropriate treatment or lack of compliance is important in development of drug-resistant strains. In the community this can be controlled by giving directly observed therapy (DOT). This requires a health-care infrastructure and appropriate finances.

Prevention

Vaccination: BCG (Bacillus Calmette-Guerin) vaccine is an attenuated form of *M. bovis* (which causes TB in cattle) grown for many years on artificial medium. This is believed to provide up to 80% protection against *M. tuberculosis* in some areas of Northern Europe, particularly in the UK, and USA, but the efficacy of BCG is highly variable and rather ineffective in parts of the world around the tropics. This has been correlated with high exposure to environmental mycobacteria. It has been suggested that a Th1 response is inhibited in particular by high levels of IL-4 (Th2 response), which inhibit autophagy and the ability of the macrophage to kill mycobacteria.

In most European countries, Canada, and the USA, vaccination was abolished some years ago, due to their low TB transmission status and potential interference of vaccination with skin test diagnosis. In the UK, USA, and some other countries, BCG vaccination is only given to specific high-risk populations, for example nurses and military personnel and children at risk (see the Eurosurveillance website for 2005 European survey of health policies regarding BCG vaccination of children).

A number of new vaccines are currently being tested.

SUMMARY

1. What is the causative agent, how does it enter the body and how does it spread a) within the body and b) from person to person?

- *Mycobacterium tuberculosis* is an acid-fast bacillus, which grows very slowly.

- Because the cell wall is very impervious, few antibiotics can penetrate and it is relatively resistant to desiccation and some disinfectants.

- The route of transmission is aerial, by inhalation of infected droplet nuclei.

- *M. tuberculosis* organisms grow in alveolar macrophages, are released into the bronchi, and are spread through coughing.

- Active disease occurs in only 5% of individuals following primary infection.

- Most infected individuals carry the organism in a latent state for life, under the control of an effective immune system.

- Some infected individuals may develop active disease many years after primary infection if they become immunosuppressed (reactivation).

- One-third of the world's population (2 billion) is infected with *M. tuberculosis*, with 8–10 million individuals developing active disease annually and about 2–3 million dying.

- Tuberculosis is responsible for 1 in 4 preventable deaths worldwide.

- The main foci of infection are South-East Asia and Africa.

2. What is the host response to the infection and what is the disease pathogenesis?

- Mycobacteria taken up by macrophages inhibit the phagolysosome fusion and can thus live within the macrophage.

- Local tissue destruction is caused by the host response to the presence of mycobacterial antigens.

- Immune cells are activated, releasing cytokines, which cause further damage.

- CD4+ T cells are activated to produce cytokines to help elimination of the *M. tuberculosis* organisms from the macrophages. Inability to get rid of all the microorganisms results in cytokine production leading to granuloma formation. This host response limits the spread of mycobacteria. A granuloma is a characteristic histological feature of chronic inflammation, which is a collection of activated and resting macrophages called epithelioid cells surrounding an area of necrosis.

- Reactivation of the microorganisms within granulomas can occur at any time during later life through decreased immunocompetency (e.g. co-infection with HIV).

3. What is the typical clinical presentation and what complications can occur?

- Tuberculosis commonly presents with fever, weight loss, chronic cough, and there may be hemoptysis.

- Many sites other than the respiratory tract can be affected and it may present with signs and symptoms referrable to other organ systems.

- *M. tuberculosis* may cause seizures (CNS), meningitis (CNS), fractures (bones – Pott's disease) or Addison's disease (adrenal gland).

4. How is this disease diagnosed and what is the differential diagnosis?

- Diagnosis of tuberculosis is a clinical decision and treatment is started on that basis.

- The simplest laboratory procedure is the Ziehl-Neelsen (ZN) stain.

- Specimens are cultured on Lowenstein-Jensen medium and culture remains the gold standard for accurate diagnosis of tuberculosis.

- The polymerase chain reaction (PCR) is also used to detect its presence in the specimen and is useful for detecting multidrug-resistant TB.

- Serological tests are under development.

- Tuberculosis may mimic tumors of the lung, brain, bone, intestine, blood, and other systemic granulomatous infections, for example brucellosis.

5. How is the disease managed and prevented?

- The standard anti-TB regimen is a combination of rifampicin plus isoniazid plus pyrazinamide for 2 months, followed by rifampicin and isoniazid for a further 4 months, so the total treatment time is 6 months.

- Some regimens include ethambutol in the initial phase.

- Tuberculosis is a notifiable disease in the UK.

- Effectiveness of BCG vaccination is highly variable between geographic areas. It is thought to be 70–80% effective in the UK but is given only to 'at-risk' groups in the UK, USA, and some other countries.

FURTHER READING

Fitzgerald D, Haas DW. *Mycobacterium tuberculosis*. In: Mandell GL, Bennett JE, Dolin R. Principles & Practice of Infectious Diseases, 6th edition. Elsevier, Philadelphia, 2005: 2852–2885.

Kucers A. Drugs mainly for tuberculosis. In: Kucers A, Crowe S, Grayson ML, Hoy J, editors. The Use of Antibiotics: A Clinical Review of Antibacterial, Antifungal and Antiviral Drugs, 5th edition. Butterworth Heinmann, Oxford, 1997: 1179–1242.

Lydyard P, Lakhani S, Dogan A, et al. Pathology Integrated: An A–Z of Disease and its Pathogenesis. Edward Arnold, 2000: 254–256.

Lydyard PM, Whelan A, Fanger MW. Instant Notes in Immunology, 2nd edition. Bios, Taylor and Francis, 2004.

Mims C, Dockrell HM, Goering RV, Roitt I, Wakwlin D, Zuckerman M. Medical Microbiology, 3rd edition. Mosby, Edinburgh, 2004: 232–235.

Murphy K, Travers P, Walport M. Janeway's Immunobiology, 7th edition. Garland Science, New York, 2008.

Rom WN, Garay SM, Tuberculosis, 2nd edition. Lippincott Williams and Wilkins, Philadelphia, 2004.

REFERENCES

Dye C, Scheele S, Dolin P. Consensus statement: Global burden of tuberculosis: Estimated incidence, prevalence and mortality by country. WHO Global Surveillance and Monitoring Project. JAMA, 1999, 282: 677–687.

Ewer K, Deeks J, Alvarez L. Comparison of T cell based assay with tuberculin skin test for diagnosis of *Mycobacterium tuberculosis* infections in a school tuberculosis outbreak. Lancet, 2003, 361: 1168–1173.

Meyers JP. New recommendations for the treatment of tuberculosis. Curr Opin Infect Dis, 2005, 18: 133–140.

Patel KR. Pulmonary tuberculosis in residents of lodging houses, night shelters and common hostels in Glasgow: a 5 year prospective study. Br J Dis Chest, 1985, 79: 60–66.

Rook AWG, Dheda K, Zumla A. Immune responses to tuberculosis in developing countries: implications for new vaccines. Nat Rev, 2005, 5: 661–667.

Russell DG. Who puts the tubercle in tuberculosis? Nat Rev Microbiol, 2007, 5: 39–47.

Shamputa IC, Rigouts AL, Portaels F. Molecular genetic methods for diagnosis and antibiotic resistance detection of mycobacteria from clinical specimens. APMIS, 2004, 112: 728–752.

Thwaits GE, Tran TH. Tuberculous meningitis: many questions, too few answers. Lancet Neurol, 2005, 4: 160–170.

Underhill DM, Ozinsky A, Smith KD, Aderem A. Toll-like receptor-2 mediates mycobacteria-induced proinflammatory signaling in macrophages. Proc Natl Acad Sci USA, 1999, 7: 14459–14463.

WEB SITES

Could a Tuberculosis Epidemic Occur in London as It Did in New York? Emerging Infectious Diseases Journal Vol. 6, No. 1 Jan–Feb 2000, National Center for Infectious Diseases, Centers for Disease Control and Prevention: http://www.cdc.gov/ncidod/eid/vol6no1/hayward.htm

Eurosurveillance: Europe's Leading Journal on Infectious Disease Epidemiology, Prevention and Control, Volume 11, Number 3, 01 March 2006: http://www.eurosurveillance.org/ViewArticle.aspx?ArticleId=604

Global Tuberculosis Control Report 2008, World Health Organization, © WHO 2008: www.who.int/tb/publications/global_report/

Lesions associated with primary tuberculosis, Atlas of Granulomatous Diseases, Dr Yale Rosen: http://www.granuloma.homestead.com/TB_primary_gross.html

TB in London, UK Coalition of People Living with HIV and Aids, Registered Charity: http://www.ukcoalition.org/tb/london.html

Tuberculosis, Health Topics, University of Iowa Hospital and Clinics: http://www.uihealthcare.com/topics/infectiousdiseases/infe4731.html

Tuberculosis Textbook 2007: Tuberculosis 2007 was made possible by an unrestricted educational grant provided by Bernd Sebastian Kamps and Patricia Bourcillier: http://tuberculosistextbook.com/

Tuberculosis, Treatment of Hospital Infections, British Society for Antimicrobial Chemotherapy, Birmingham, UK: http://www.bsac.org.uk/pyxis/Bone%20and%20joint/Potts%20disease/Potts%20disease.htm

MULTIPLE CHOICE QUESTIONS

The questions should be answered either by selecting True (T) or False (F) for each answer statement, or by selecting the answer statements which best answer the question. Answers can be found in the back of the book.

1. **Which of the following are true of the cell wall of mycobacteria?**
   A. It contains mycolic acid.
   B. It contains peptidoglycan.
   C. It stains green with the Ziehl-Neelsen stain.
   D. It is highly permeable to antibiotics.
   E. It contains lipopolysaccharide.

2. **Which of the following are true of *M. tuberculosis*?**
   A. The organism grows in 3 days.
   B. The organism can only be grown in armadillos.
   C. The organism is disseminated by the aerial route.
   D. Each infected person infects on average 20 other individuals.
   E. About one-third of the world's population is infected.

3. **Which of the following are associated with the host response to *M. tuberculosis*?**
   A. Macrophages.
   B. Granulomas.
   C. Destructive antibody response.
   D. Lipopolysaccharide binding protein.
   E. Cell-mediated immunity.

4. **Which of the following are ways in which *M. tuberculosis* avoids the host defenses?**
   A. Remains a superficial infection.
   B. Adopts an intracellular location.
   C. Prevents phagolysosome fusion.
   D. Resists acidification of the phagolysosome.
   E. Is resistant to TNF-α.

5. **Which of the following may be due to infection with *M. tuberculosis*?**
   A. Osteomyelitis.
   B. Meningitis.
   C. Abscess.
   D. Food poisoning.
   E. Pyelonephritis.

6. **Which of the following are typical signs of pulmonary tuberculosis?**
   A. Chronic cough.
   B. Seizures.
   C. Skin rash.
   D. Weight loss.
   E. Low grade fever.

7. **Which of the following are important in the diagnosis of tuberculosis?**
   A. Culture.
   B. Microscopy.
   C. Serology.
   D. Polymerase chain reaction.
   E. Histopathology.

8. **Which of the following could be confused with extrapulmonary tuberculosis?**
   A. Staphylococcal abscess of the spine.
   B. Addison's disease.
   C. Ringworm.
   D. Glioma.
   E. Cryptococcal meningitis.

9. **Which of the following antibiotics are used to treat tuberculosis?**
   A. Benzylpenicillin.
   B. Isoniazid.
   C. Rifampicin.
   D. Pyrazinamide.
   E. Vancomycin.

10. **Which of the following are true of tuberculosis?**
   A. It is a notifiable disease in the UK.
   B. Contacts should be started on empirical antituberculous therapy.
   C. DOTS means the patient visits a treatment center for their drugs.
   D. Disease prevalence is correlated with social deprivation.
   E. There is no effective vaccine.

Case 24

Neisseria gonorrhoeae

A 15-year-old heterosexual male was brought to the emergency room by his sister. He gave a 24-hour history of **dysuria** and noted some 'pus-like' drainage in his underwear and the tip of his penis (Figure 1). Urine appeared clear and urine culture was negative, although urinalysis was positive for leukocyte esterase and multiple white cells were seen on microscopic examination of urine. He gave a history of being sexually active with five or six partners in the past 6 months. He claimed that he and his partners had not had any sexually transmitted diseases. His physical exam was significant for a yellow urethral discharge and tenderness at the tip of the penis. A Gram stain of the discharge was performed in the emergency room (Figure 2). He was given antimicrobial agents and scheduled for a follow-up visit 1 week later. He was asked to provide the names and addresses of his sexual partners to the Health Department so that they could be examined and treated if necessary. One of the sexual partners, a 15-year-old female who reported having had sexual relations with her boyfriend frequently over the last 6 months, was asymptomatic until 2 days before

coming to the hospital when she developed fever, shaking chills, and lower abdominal pain. Her symptoms worsened and she presented with fever of 42°C, generalized abdominal pain, and a swollen right knee, with blood pressure 120/80 and pulse 150/min and regular. The date of her last menstrual period, which was described as normal, was 1 week before admission. The patient was oriented as to time, person, and place. Physical examination was unremarkable except for tender abdomen and rigidity, and decreased bowel sounds; the right knee was red, hot, tender, and swollen. Pelvic examination showed some white discharge of the cervical os (Figure 3) A swab was obtained from her cervix for culture (Figure 4). Her right knee was tapped.

Laboratory findings:
- hemoglobin: 12 g dl^{-1}
- white blood cell (WBC) count: 26000 μl^{-1}
- differential: 70 PMN, 5 bands, 25 lymphocytes
- urinalysis: specific gravity, 1.010; protein, 2+; sugar, negative; WBC, 8–10 per HPF; no casts.

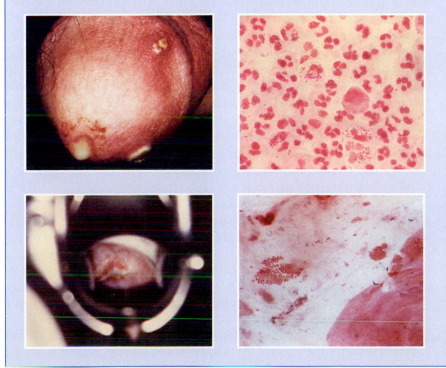

Figure 1 (top left). Gonorrhoeae infection in the male. Creamy purulent discharge from the urinary meatus. In many cases infection is asymptomatic, but may cause painful urination or a purulent discharge, as seen here. In severe cases it may also cause inflammation of the testicles and prostate gland, and infertility. Treatment is with antibiotics.

Figure 2 (top right). Gram stain of penile discharge showing intracellular gram-negative diplococci.

Figure 3 (bottom left). Cervicitis with purulent discharge.

Figure 4 (bottom right). Gram stain of a cervical smear showing extracellular and intracellular gram-negative diplococci.

1. What is the causative agent, how does it enter the body and how does it spread a) within the body and b) from person to person?

Causative agent

The patient has gonorrhea and his girlfriend has gonorrhea-related **pelvic inflammatory disease** (**PID**) and disseminated gonococcal infection (DGI) caused by the bacterium *Neisseria gonorrhoeae* (often termed the gonococcus). *Neisseria* species are **capnophilic**, gram-negative cocci. The cocci are often found in pairs where their adjacent sides are flattened giving them a coffee bean appearance. Their habitat is the mucous membranes of mammals and many species are **commensals** of these surfaces. Two species, however, *N. gonorrhoeae* and *N. meningitidis* (termed the meningococcus), are obligate pathogens of humans. *Neisseria* are oxidase-positive, catalase-positive, and produce acid from a variety of sugars by oxidation.

Neisseria species have a typical gram-negative envelope but the outer leaflet of their outer membrane contains lipooligosaccharide (LOS) instead of **lipopolysaccharide** (**LPS**). Gonococcal and meningococcal LOS lacks the repeating 'O' somatic antigenic side-chains that are present in the *Enterobacteriaceae*, for example. The core polysaccharide of LOS of both the gonococcus and the meningococcus undergoes **antigenic variation** exposing a variety of epitopes on the cell surface. The cell surface of both the gonococcus and the meningococcus (see case study) is decorated with pili (type IV pili), which are hair-like projections composed of polymers of the structural protein **pilin** (PilE). The protein PilC is located at the tip of the **pilus** and is the **adhesin** that mediates initial attachment of the bacterium to the surface of mucosal epithelium. The PilC protein binds CD46, a complement regulatory protein that has cofactor activity that inactivates the complement components C3b and C4b by serum factor I.

The pilin (PilE) protein contains a constant N-terminal domain, a hypervariable C-terminal domain, and several variable regions termed mini-cassettes, which are encoded by genes with varying DNA sequence. The gonococcus has a single complete copy of the pilin gene termed *pilE* but as many as 15 truncated genes with variable DNA sequence. The truncation is at the 5′ end, resulting in lack of the sequence encoding the N-terminal constant domain and promoter elements. These truncated genes are termed *pilS* (silent) and form the *pilS* locus. By recombination of *pilS* sequences into the *pilE* gene the bacterium can express a high number of antigenically distinct pili. In addition to antigenic variation the pili undergo phase variation. In phase variation the bacterium has the ability to turn pilus expression on or off at a high frequency.

Among the outer membrane proteins are a family of opacity-associated proteins (Opa), so named because they give rise to an opaque colony phenotype. Opa proteins are important in the ability of the organism to adhere tightly to epithelia. They also dictate the tissue tropism of the gonococcus and its ability to invade epithelial cells. There are as many as 12 genes encoding Opa proteins and they undergo phase variation such that a neisserial population will contain bacteria expressing none, one or several Opa proteins. There are two hypervariable domains within the extracellular portion of the molecule that give rise to new Opa variants as a result of point mutation and by modular exchange of domains between different Opa proteins. These hypervariable domains are also the sites of interaction with cellular receptors. The Opa proteins of the gonococcus and the meningococcus can be divided into two major groups based on the cellular receptors to which they bind. The Opa$_{HS}$-type proteins bind to heparan sulfate and extracellular matrix molecules such as fibronectin and vitronectin, while the Opa$_{CEA}$-type proteins bind to carcinoembryonic antigen-related cell-adhesion molecule (CEACAM). Opa$_{HS}$ and Opa$_{CEA}$ adhesins mediate internalization of gonococci by various human cells. Either directly through engagement with heparan sulfate proteoglycans on the cell surface or indirectly using fibronectin or vitronectin to bridge Opa$_{HS}$-type proteins and cell surface integrins, the actin cytoskeleton of the epithelial cell is signaled to rearrange to enable engulfment of the bacterial cell. Engagement of Opa$_{CEA}$-type proteins with members of the CEACAM receptor family also triggers actin cytoskeleton rearrangement, permitting internalization by neutrophils, macrophages, and various other types of cells.

Gonococcal porin proteins are named protein IA (PorPIA) and protein IB (PorPIB) and they assemble into trimers that form channels that traverse the gonococcal outer membrane. When gonococci are apposed to the epithelial cell membrane porin PIB appears to be able to insert into the eukaryotic cell membrane and translocate to the cytosol where it participates in the reorganization of the actin cytoskeleton. However, neisserial porins can inhibit neutrophil actin polymerization, degranulation, expression of opsonin receptors, and the respiratory burst.

N. gonorrhoeae and *N. meningitidis* are adept at acquiring iron from their human host from numerous iron sources, which contributes to their ability to replicate on mucosal surfaces, intracellularly, and in the blood.

Both *N. gonorrhoeae* and *N. meningitidis* secrete an IgA1 protease. The enzyme cleaves IgA1 at the hinge region to produce Fab and Fc fragments. While it is logical to assume that IgA1 protease contributes to the **virulence** of the gonococcus by subverting the protective effects of sIgA it should be realized that half of sIgA is of subclass 2 that is resistant to IgA1 protease. Moreover, it has been demonstrated that experimental urethral infections of male volunteers with an IgA1 protease-negative mutant of *N. gonorrhoeae* matching the parent strain in expression of Opa proteins, LOS, and pilin was indistinguishable from that of the parent strain. A role for IgA1 protease may lie in its ability to cleave **lysosome**-associated membrane protein 1 (h-lamp-1). As their name implies h-lamp-1and h-lamp-2 are found in the membranes of mature lysosomes but also in the membranes of **phagosomes**/endosomes. Their functions are not fully understood but they are thought to protect the membrane from the action of degradative enzymes within the lysosome and appear to be required for fusion of lysosomes with phagosomes. It has been shown that gonococcal IgA1 protease can cleave the less glycosylated form of h-lamp-1 found in epithelial cell phagosomes/endosomes, which may enable the bacteria to escape into the cytosol of the cell and prolong their intracellular survival.

Entry and spread within the body

In uncomplicated gonorrhea the bacteria adhere to urethral epithelium of males and to the cervical epithelium and urethral epithelium of females. The apical surface of the epithelium is induced to take up the gonococci via engagement of Opa$_{CEA}$ with CEACAMs. Rearrangement of the actin cytoskeleton facilitates internalization and the bacteria pass through the cells via transcytosis and are released at the basolateral surface into the subepithelial space where infection is established. From this site the gonococci may seed the bloodstream and from there the joints and skin. The up-regulation of CEACAM1 on endothelium brought about by LOS and pro-inflammatory **cytokines** contribute to transgression of the bacteria into blood. In women the cervical infection may ascend to the fallopian tubes (**salpingitis**), which can lead to scarring, **ectopic pregnancy**, sterility, and chronic pelvic pain.

Person to person spread

The gonococcus is a sexually transmitted pathogen and it is acquired and spread horizontally (person to person) by vaginal, anal or oral intercourse. Ejaculation need not occur for the gonococcus to be transmitted or acquired. Gonorrhea can also be spread vertically from mother to baby during delivery. Persons who have had gonorrhea and received treatment may become infected again if they have sexual contact with a person with gonorrhea.

Epidemiology

The US Center for Disease Control and Prevention (CDC) estimates that more than 700 000 persons in the US get new gonorrheal infections each year. By law gonorrhea is a reportable infectious disease in the US – as are chlamydia and syphilis, the other major bacterial sexually transmitted infections (STIs) – but only about half of these infections are reported to the CDC. Gonorrhea is the second most commonly reported bacterial STI in the US after chlamydia (caused by *Chlamydia trachomatis*). Very often individuals are co-infected with both *N. gonorrhoeae* and *Chlamydia trachomatis*.

2. What is the host response to the infection and what is the disease pathogenesis?

Innate immunity in the male and female urinary tract is largely mediated by the anti-microbial proteins β-defensins (HBD), cathelicidin, lactoferrin, and lipocalin. **Tamm-Horsfall protein** (THP) is an abundant urinary glycoprotein that binds specific mannosylated residues and blocks adhesion of certain bacterial urinary tract pathogens such as *Escherichia coli*. The endocervix provides a protective mucous layer rich in HBD-1, HD-5, lysozyme, and lactoferrin for the vagina. Cervical, vaginal, and urethal mucosal IgA antibodies likely contribute to immune protection at these mucosal surfaces. Both the gonococcus and the meningococcus produce an IgA1 protease (see above) that may subvert mucosol IgA antibodies. The vaginal pH varies between 3.5 and 5.0, which restricts the establishment of all but aciduric bacteria. Thus, gonococci must successfully negotiate these defense mechanisms. Gonococci are readily taken up by neutrophils and macrophages in the submucosa. LOS and other components of the outer membrane as well as IgA1 protease are pro-inflammatory and induce release of **tumor necrosis factor-α (TNF-α)**, **interleukin (IL)-1β**, IL-6, and IL-8 from phagocytes, which induces a florid inflammatory reaction.

The complement system serves as the primary mechanism of killing gonococci. This is supported by the fact that individuals with complement deficiencies are at increased risk for disseminated neisserial infections although, interestingly, they have less severe symptomatology than do persons with an intact complement system. Individuals deficient in terminal complement proteins C5, C6, C7, C8, and/or C9 commonly exhibit recurrent, often systemic, neisserial infections. Consistent with the importance of complement, disseminated gonococcal infections are associated predominantly with gonococci that express type PI.A porin molecules that bind factor H, which inhibits the **alternative complement pathway**. Similarly the PI.A porin binds C4-binding protein (C4bp), which results in the inhibition of the classical **complement pathway**. The **membrane attack complex** (MAC) is crucial to the bactericidal action of the complement system and MAC assembly occurs in an aberrant manner on the surface of serum-resistant gonococci.

Gonococcal infection does not appear to result in protective humoral or cellular immunity despite the intense inflammatory reaction engendered by the bacteria. Although local and systemic antibodies can be detected in infected persons they are at low levels and appear not to protect against re-infection. The reason for this is thought to be the extensive and rapid variation of pili, LOS, and the Opas. In addition, both the male and female genital tracts lack inductive mucosal sites. Furthermore, antibodies to the reduction-modifiable protein (Rmp) also termed outer membrane protein (OMP)3, interfere with the ability of antibodies directed against other surface components of the gonococcus to activate complement properly. It has been shown that gonococcal binding of CEACAM1 on **CD4+** T cells can down-regulate the activation and proliferation of these cells, which may contribute to the suboptimal adaptive immune response.

3. What is the typical clinical presentation and what complications can occur?

In the male the most common clinical presentation is infection of the genitourinary tract producing **urethritis**. The most common symptoms are urethral discomfort, dysuria, and discharge of varying severity. If the infection ascends to the epididymis, **epididymitis** presents as unilateral pain and swelling localized posteriorly within the scrotum.

In the female the most common clinical presentation is **endocervicitis**. The most common symptom is a thin, purulent, and unpleasant-smelling vaginal discharge, although many women may be asymptomatic. Women may also have urethritis as well as **cervicitis**, which manifest as dysuria or a slight urethral discharge.

Ascending infection of the endometrium, fallopian tubes, ovaries, and peritoneum manifests as pelvic or lower abdominal pain, which may be in the midline, unilateral, or bilateral. The pain may be accompanied by fever, nausea, and vomiting. Infection of the peritoneum may spread to that covering the liver (peri-hepatitis, Fitz-Hugh-Curtis syndrome) resulting in right upper quadrant pain.

Rectal infection may follow receptive anal intercourse and, in women, by local spread of the gonococcus from the vaginal **introitus**. Often rectal infection is asymptomatic, but pain, **pruritus** (itch), **tenesmus** (the constant feeling of the need to empty the bowel), discharge, and bloody diarrhea may occur.

Oral sex may result in **pharyngitis**, which usually is asymptomatic. Conjunctivitis may follow accidental inoculation of the eye(s) with fingers.

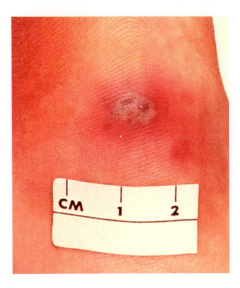

Figure 5. Cutaneous gonococcal lesion due to a disseminated *N. gonorrhoeae* infection.

In neonates vertical transmission may occur from an infected mother to her neonate during vaginal delivery. This results in bilateral conjunctivitis (ophthalmia neonatorum). The symptoms of gonococcal conjunctivitis are pain, redness, and a purulent discharge. Corneal ulceration, perforation, and blindness can occur if treatment is not given promptly. Blindness from neonatal gonococcal infection is a serious problem in developing countries but is uncommon in the United States and Europe where neonatal prophylaxis is routine.

In a few percent of cases gonococci disseminate from the site of mucosal infection via the bloodstream giving rise to systemic disease termed disseminated gonorrheal infection (DGI). Gonococcemia (gonococci in the blood) occurs most frequently in the adolescent and young adult population, with a peak incidence in males aged 20–24 years and females aged 15–19 years.

The clinical manifestations of DGI are biphasic with an early **bacteremic** phase consisting of **tenosynovitis** of the hands, **arthralgias**, and **rash**, followed by a localized phase consisting of localized **septic arthritis** typically of the knee. The rash consists of small **papules** that become pustules on broad **erythematous** bases with necrotic centers (Figure 5) and is usually found below the neck and also may involve the palms and the soles but usually spares the face, scalp, and mouth. The gonococci may also seed the bone (**osteomyelitis**), the central nervous system (**meningitis**) and the heart (**endocarditis**). Other serious complications are **adult respiratory distress syndrome (ARDS)**, and fatal **septic shock**. Pregnant or menstruating women are particularly susceptible to DGI, as are individuals with complement deficiencies, HIV or **systemic lupus erythematosus (SLE)**.

There is a strong association between the acquisition of HIV-1 and other sexually transmitted diseases (STDs) including gonorrhea. An increased HIV load has been found in semen of men with gonorrhea. The mechanisms by which the gonococcus enhances HIV-1 infection are not fully understood but may include the ability of the bacterium to enhance infection and replication of HIV-1 in **dendritic cells**.

4. How is the disease diagnosed and what is the differential diagnosis?

Pathogenic *Neisseria* are fastidious and although they will grow on blood agar as well as chocolate agar (Figure 6) they are usually isolated on selective

Figure 6. A pure culture of *N. gonorrhoeae* growing on (A) blood agar and (B) chocolate agar.

A   B

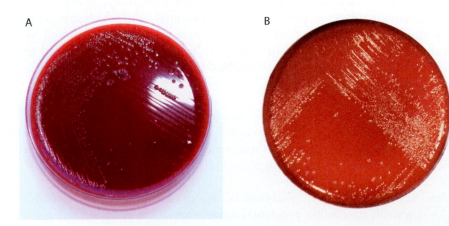

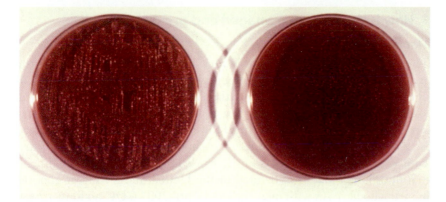

Figure 7. A rectal swab specimen plated onto chocolate agar (left) and the selective chocolate agar-based Thayer-Martin (T-M) agar medium (right). Note the overgrowth of the endogenous rectal microbiota on the chocolate agar plate (left), which suppressed the growth of the gonococcus. On the selective T-M agar shown on the right there is a pure culture of the gonococcus because T-M agar contains antimicrobials that inhibit the growth of organisms other than *N. gonorrhoeae*.

media based on chocolate agar because clinical specimens are frequently obtained from mucosal surfaces and growth of the gonococcus may be inhibited by the attendant commensal **microbiota** in the specimen. Mucosal specimens should be collected using Dacron or Rayon swabs rather than alginate or cotton because both of the latter may be inhibitory for gonococci. Modified Thayer-Martin (MTM) medium is commonly used in the United States and contains various antimicrobial agents to suppress gram-positive and gram-negative bacteria and fungi to allow the selective recovery of *N. gonorrhoeae* and *N. meningitidis* from mucosal surfaces (Figure 7). Plates are incubated at 35–37°C in 3–7% CO_2 in a humid environment. A candle extinction jar is suitable. Identification of the gonococcus and meningococcus is achieved by subjecting oxidase-positive (see *N. meningitidis* case, Figure 4) and catalase-positive (30% H_2O_2) gram-negative diplococci (Figure 8) to a panel of sugars for carbohydrate utilization tests, specific fluorescent antibodies, chromogenic substrates, or DNA probes. Differential utilization of the sugars glucose, maltose, sucrose, and lactose is a simple and common way to speciate the pathogenic *Neisseria*. *N. gonorrhoeae* utilizes glucose only (Figure 9). Because they grow on MTM the selective medium used to isolate *N. gonorrhoeae* and *N. meningitidis*, *Neisseria lactamica* and *Moraxella catarrhalis* must be differentiated from these bacteria. *N. lactamica* can be differentiated because it oxidizes lactose in addition to glucose and maltose, whereas *M. catarrhalis* does not oxidize glucose, maltose, sucrose or lactose. Of the four species *M. catarrhalis* alone produces butyrate esterase and deoxyribonuclease (DNase). In STD clinics a diagnosis of gonococcal urethritis in adult males is usually made by observing intracellular (neutrophils) gram-negative diplococci on smears of urethral discharge (see Figure 2). However, confirmatory tests are required in females and for all extra-genital infections because there can be social and medicolegal issues resulting from the findings.

Differential diagnosis

For uncomplicated gonococcal urethritis and cervicitis the differential diagnosis should include chlamydial genitourinary infections, male and female urinary tract infection, and **vaginitis**.

For ascending infections in the male the differential diagnosis should include testicular torsion and in the female **endometriosis**, **endometritis**, and **ectopic pregnancy**.

For disseminated gonococcal infection meningococcemia should be considered.

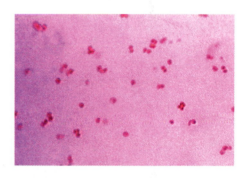

Figure 8. Gram stain of *N. gonorrhoeae* showing small, gram-negative, kidney-shaped diplococci.

Glu Mal Lac Suc

Figure 9. *N. gonorrhoeae* produces acid by oxidation of glucose (Glu) but not from maltose (Mal), sucrose (Suc), or lactose (Lac). The acid turns the pH indicator, phenol red, from red to yellow

5. How is the disease managed and prevented?

Management

The therapy for gonorrhea is antibiotics; however, many gonococcal strains are resistant to penicillins, tetracyclines, spectinomycin, and fluoroquinolones. Since the early 1990s the fluoroquinolones have been widely used for the treatment of gonorrhea because of their efficacy, and convenience as a single-dose, oral therapy. However, fluoroquinolone resistance in *N. gonorrhoeae* has been steadily increasing and is now widespread in the United States and other parts of the world. By 2004, CDC recommended that fluoroquinolones not be used in the United States to treat gonorrhea in men who have sex with men and as of April 2007 CDC no longer recommends the use of fluoroquinolones for the treatment of gonococcal infections and associated conditions such as pelvic inflammatory disease (PID). Currently only the cephalosporins are still recommended for the treatment of gonorrhea. The current treatment regimens are shown in Box 1.

Gonorrhea is a notifiable disease in the US but not in the UK. However, ophthalmia neonatorum, which is caused by *N. gonorrhoeae* or *Chlamydia trachomatis*, is notifiable in the UK.

Prevention

The use of latex condoms to prevent gonorrhea and other STIs is recommended for all sexually active individuals unless in a long-term monogamous relationship in which both partners are proven free of STI.

Box 1. Recommended antibiotic therapy for *Neisseria gonorrhoeae*

Uncomplicated gonococcal infections of the cervix, urethra, and rectum*
Recommended regimens
Ceftriaxone 125 mg in a single intramuscular (i.m.) dose **OR**
Cefixime[†] 400 mg in a single oral dose

PLUS
TREATMENT FOR CHLAMYDIA IF CHLAMYDIAL INFECTION IS NOT RULED OUT

Alternative regimens
Spectinomycin[†] 2 g in a single i.m. dose **OR**
Cephalosporin single-dose regimens[§]

Uncomplicated gonococcal infections of the pharynx*
Recommended regimens
Ceftriaxone 125 mg in a single i.m. dose

PLUS
TREATMENT FOR CHLAMYDIA IF CHLAMYDIAL INFECTION IS NOT RULED OUT

---

*For all adult and adolescent patients, regardless of travel history or sexual behavior. Information regarding management of these infections in patients with documented severe allergic reactions to penicillin or cephalosporins is available at http://www.cdc.gov/std/treatment

[†]Not available in the US.

[§]Other single-dose cephalosporin regimens that are considered alternative treatment regimens against uncomplicated urogenital and anorectal gonococcal infections include ceftizoxime 500 mg i.m.; or cefoxitin 2 g i.m., administered with probenecid 1 g orally; or cefotaxime 500 mg i.m. Some evidence indicates that cefpodoxime 400 mg and cefuroxime axetil 1 g might be oral alternatives.

Information on treatment for other gonococcal infections and associated conditions can be found at http://www.cdc.gov/std/treatment

SUMMARY

1. What is the causative agent, how does it enter the body and how does it spread a) within the body and b) from person to person?

- Gonorrhea is the second most commonly reported bacterial STI in the US after chlamydia (caused by *Chlamydia trachomatis*). Very often individuals are co-infected with both bacteria.

- *Neisseria* species are capnophilic, gram-negative cocci. The cocci are often found in pairs (diplococci) where their adjacent sides are flattened giving them a coffee bean appearance.

- Their habitat is the mucous membranes of mammals and many species are commensals of these surfaces. Two species, however, *N. gonorrhoeae* and *N. meningitidis*, are obligate pathogens of humans.

- *Neisseria* are oxidase-positive, catalase-positive, and produce acid from a variety of sugars by oxidation. *N. gonorrhoeae* utilizes glucose only.

- *Neisseria* species have a typical gram-negative envelope but the outer leaflet of their outer membrane contains lipooligosaccharide (LOS), which lacks repeating 'O' somatic antigenic side-chains.

- The core polysaccharide of LOS of the gonococcus undergoes antigenic variation exposing a variety of epitopes on the cell surface.

- The cell surface of the gonococcus is decorated with pili, which are hair-like projections composed of polymers of the structural protein pillin (PilE).

- Pili mediate initial attachment to mucosal epithelia. Pili are antigenically variable and pilus expression can be turned on or off at a high frequency.

- Opa outer membrane proteins enable the organism to adhere tightly to epithelia and facilitate uptake by various types of host cell.

- *N. gonorrhoeae* and *N. meningitidis* are adept at acquiring iron from their human host.

- Both *N. gonorrhoeae* and *N. meningitidis* secrete an IgA1 protease.

- The gonococcus is a sexually transmitted pathogen and it is acquired and spread horizontally (person to person) by vaginal, anal or oral intercourse.

- Gonorrhea can also be spread vertically from mother to baby during delivery.

- From the subepithelial space where infection is established the gonococci may seed the bloodstream and disseminate to various organs.

2. What is the host response to the infection and what is the disease pathogenesis?

- In the genitourinary tract gonococci encounter various innate antimicrobial factors such as β-defensins, cathelicidin, lactoferrin, lipocalin, lysozyme, lactoferrin, and Tamm-Horsfall protein.

- Low levels of secretory IgA antibodies are induced that are ineffective at preventing re-infection.

- Por proteins in the outer membrane facilitate uptake by epithelial cells and inhibit phagocytosis by neutrophils.

- The complement system serves as the primary mechanism of killing gonococci.

- Gonococci can subvert the alternative and classical complement pathways.

- Gonococcal infection does not result in protective humoral or cellular immunity because of antigenic variation and other mechanisms of immune evasion.

3. What is the typical clinical presentation and what complications can occur?

- In the male the most common clinical presentation is infection of the genitourinary tract producing urethritis.

- In the female the most common clinical presentation is endocervicitis. The most common symptom is a thin, purulent, and unpleasant-smelling vaginal discharge, although many women may be asymptomatic.

- Women may experience ascending infection of the endometrium, fallopian tubes, ovaries, and peritoneum termed pelvic inflammatory disease (PID).

- Rectal infection may follow receptive anal intercourse and, in women, by local spread of the gonococcus from the vaginal introitus.

- Oral sex may result in pharyngitis, which usually is asymptomatic. Conjunctivitis may follow accidental inoculation of the eye(s) with fingers.

- In neonates vertical transmission may occur from an infected mother to her neonate during vaginal delivery. This results in bilateral conjunctivitis (ophthalmia neonatorum), which can lead to blindness.

- Rarely gonococci disseminate via the bloodstream giving rise to systemic disease termed disseminated gonorrheal infection (DGI).

- Bacteremia results in seeding of the joints and skin resulting in tenosynovitis of the hands, arthralgias, and rash.

4. How is the disease diagnosed, and what is the differential diagnosis?

- Pathogenic *Neisseria* are fastidious and are usually isolated on selective media based on chocolate agar.

- Modified Thayer-Martin (MTM) medium is commonly used in the US to allow the selective recovery of *N. gonorrhoeae* from mucosal surfaces. Plates are incubated at 35–37°C in 3–7% CO_2 in a humid environment.

- Identification of the gonococcus is achieved by subjecting oxidase-positive and catalase-positive gram-negative diplococci to carbohydrate utilization tests. *N. gonorrhoeae* oxidizes glucose only.

- Specific fluorescent antibodies, chromogenic substrates, or DNA probes can also be used for identification.

- In STD clinics a diagnosis of gonococcal urethritis in adult males is usually made by observing intracellular (neutrophils) gram-negative diplococci on smears of urethral discharge.

- Confirmatory tests are required in females and for all extra-genital infections because there can be social and medicolegal issues resulting from the findings.

5. How is the disease managed and prevented?

- The use of latex condoms to prevent gonorrhea and other sexually transmitted infections (STIs) is recommended for all sexually active individuals unless in a long-term monogamous relationship in which both partners are proven free of STI.

- Many gonococcal strains are resistant to penicillins, tetracyclines, spectinomycin, and fluoroquinolones.

- Currently only the cephalosporins are recommended for the treatment of gonorrhea.

- Gonorrhea is a notifiable disease in the US but not in the UK. However, ophthalmia neonatorum, which is caused by *Neisseria gonorrhoeae* or *Chlamydia trachomatis*, is notifiable in the UK.

FURTHER READING

Mims C, Dockrell HM, Goering RV, Roitt I, Wakelin D, Zuckerman M. Medical Microbiology, 3rd edition. Mosby, Edinburgh, 2004.

Murphy K, Travers P, Walport M. Janeway's Immunobiology, 7th edition. Garland Science, New York, 2008.

Murray PR, Rosenthal KS, Pfaller MA. Medical Microbiology, 5th edition. Elsevier Mosby, Philadelphia, PA, 2005.

REFERENCES

Boulton IC, Gray-Owen SD. Neisserial binding to CEACAM1 arrests the activation and proliferation of CD4+ T lymphocytes. Nat Immunol, 2002, 3: 229–236.

Chromek M, Slamova, Z, Bergman P, et al. The antimicrobial peptide cathelicidin protects the urinary tract against invasive bacterial infection. Nat Med, 2006, 12: 636–641.

Fichorova RN, Desai PJ, Gibson FC III, Genco CA. Distinct proinflammatory host responses to *Neisseria gonorrhoeae* infection in immortalized human cervical and vaginal epithelial cells. Infect Immun, 2001, 69: 5840.

Ganz T. Defensins in the urinary tract and other tissues. J Infect Dis, 2001, 183(Suppl 1): S41–S42.

REFERENCES

Hauk CR, Meyer F. 'Small' talk: Opa proteins as mediators of *Neisseria*-host-cell communication. Curr Opin Microbiol, 2003, 6: 43–49.

Hedges SR, Mayo MS, Mestecky J, Hook EW 3rd, Russell MW. Limited local and systemic antibody responses to *Neisseria gonorrhoeae* during uncomplicated genital infections. Infect Immun, 1999, 67: 3937–3946.

King AE, Critchley HOD, Kelly RW. Innate immune defences in the human endometrium. Reprod Biol Endocrinol, 2003, 1: 116.

Kirchner M, Meyer TF. The PilC adhesin of the *Neisseria* type IV pilus – binding specificities and new insights into the nature of the host cell receptor. Mol Microbiol, 2005, 56: 945–957.

Massari P, Ram S, Macleod H, Wetzler LM. The role of porins in neisserial pathogenesis and immunity. Trends Microbiol, 2003, 11: 87–93.

Mestecky J, Moldoveanu Z, Russell MW. Immunologic uniqueness of the genital tract: challenge for vaccine development. Am J Reprod Immunol, 2005, 53: 208–214.

Quayle AJ. The innate and early immune response to pathogen challenge in the female genital tract and the pivotal role of epithelial cells. J Reprod Immunol, 2002, 57: 61–79.

Ram S, Cullinane NI, Blom AM, et al. Binding of C4b-binding protein to porin: a molecular mechanism of serum resistance of *Neisseria gonorrhoeae*. J Exp Med, 2001, 193: 281–295.

Ram S, Mackinnon FG, Gulati S, et al. The contrasting mechanisms of serum resistance of *Neisseria gonorrhoeae* and group B *Neisseria meningitidis*. Mol Immunol, 1999, 36: 915–928.

Saemann MD, Horl WH, Weichhart T. Uncovering host defences in the urinary tract: cathelicidin and beyond. Nephrol Dial Transplant, 2007, 22: 347–349.

Zasloff M. Antimicrobial peptides, innate immunity, and the normally sterile urinary tract. J Am Soc Nephrol, 2007, 18: 2810–2816.

Zhang J, Li G, Bafica A, et al. *Neisseria gonorrhoeae* enhances infection of dendritic cells by HIV type 1. J Immunol, 2005, 174: 7995–8002.

WEB SITES

Centers for Disease Control and Prevention, Morbidity and Mortality Weekly, April 13, 2007/56(14);332–336: http://www.cdc.gov/mmwr/preview/mmwrhtml/mm5614a3.htm?s_cid=mm5614a3_e

Centers for Disease Control and Prevention, Division of Sexually Transmitted Diseases Prevention, Atlanta, GA, USA: http://www.cdc.gov/std/Gonorrhea/

eMedicine, Inc., 2007, provider of clinical information and services to physicians and health-care professionals: http://www.emedicine.com/emerg/topic220.htm

eMedicine, Inc., 2007, provider of clinical information and services to physicians and health-care professionals: http://www.emedicine.com/med/topic922.htm

MULTIPLE CHOICE QUESTIONS

The questions should be answered either by selecting True (T) or False (F) for each answer statement, or by selecting the answer statements which best answer the question. Answers can be found in the back of the book.

1. A 24-year old sexually active female presents with dysuria (pain on urination) and cervical discharge. Gram stain of the cervical discharge reveals gram-negative diplococci in pairs. Which one of the following would be the most appropriate therapy to initiate?

   A. Penicillin.

   B. Erythromycin.

   C. Ceftriaxone.

   D. Ceftriaxone plus doxycycline.

   E. Ampicillin.

2. The process of phase and antigenic variation in gonococci involves?

   A. Lysogeny by a bacteriophage.

   B. Plasmid transfer.

   C. Genetic recombination.

   D. Sex pili.

   E. Transduction.

3. A young army recruit presents at a sexually transmitted disease (STD) clinic with acute urethritis. A Gram stain of his urethral exudate reveals neutrophils with intracellular gram-negative diplococci. He is treated with ceftriaxone and sent home. He is requested to return in 1 week so that a urethral culture can be obtained to confirm antibiotic cure. Which ONE of the following culture media should be used for the follow-up culture?

   A. Blood agar.

   B. Chocolate agar.

   C. MacConkey agar.

   D. Modified Thayer-Martin agar.

   E. Mannitol-salt agar.

4. A 24-year-old sexually active female presents with dysuria and a purulent discharge from the cervical canal. A Gram stain of the discharge shows the presence of gram-negative diplococci and numerous neutrophils. The lab reports the isolation of nonhemolytic, gram-negative, oxidase-positive diplococci that utilize glucose. A deletion of the genes responsible for the synthesis of which one of the following would make the organism unable to initially attach to male urethral epithelial cells and endocervical cells?

   A. Capsular polysaccharide.

   B. C-carbohydrate.

   C. O side-chain of lipooligsaccharide.

   D. Pili.

   E. Spore coat protein.

5. A 22-year-old sexually active female from New England presents with fever and right knee swelling. Her joint aspirate reveals gram-negative diplococci. She also has had a recent bout of cervicitis. What is the most likely cause of this patient's septic arthritis?

   A. *Staphylococcus aureus.*

   B. *Streptococcus pyogenes.*

   C. *Neisseria gonorrhoeae.*

   D. *Pseudomonas aeruginosa.*

   E. *Borrelia burgdorferi.*

6. A 16-year-old female comes to the physician because of an increased vaginal discharge. She developed this symptom 2 days ago. She also complains of dysuria. She is sexually active with one partner and uses condoms intermittently. Examination reveals some erythema of the cervix but is otherwise unremarkable. A urinalysis is negative. Sexually transmitted disease testing is performed and the patient is found to have gonorrhea. While treating this patient's gonorrhea infection, treatment must also be given for which one of the following?

   A. Bacterial vaginosis.

   B. Chlamydia.

   C. Herpes.

   D. Syphilis.

   E. Trichomoniasis.

7. All of the following are important in the isolation and laboratory diagnosis of *Neisseria gonorrhoeae* infections, EXCEPT?

   A. Use of selective culture media to suppress the growth of other bacteria and fungi while allowing gonococci to grow.

   B. Positive oxidase test.

   C. The ability of the gonococcus to use only glucose out of the four sugars tested on an acid production panel.

   D. The presence of intracellular gram-negative cocci in penile discharge.

   E. Detection of IgM antibody to pili by ELISA.

8. *Neisseria gonorrhoeae* is capable of all of the following, EXCEPT?

   A. Aerobic metabolism.

   B. Cleavage of secretory immunoglobulin A subclass 1 (sIgA1).

   C. Transcytosis of epithelial cells.

   D. Secretion of hemolytic toxins.

   E. Production of purpuric skin lesions.

9. Which of the following statements about gonorrhea are correct?

   A. Bacteremia is common.

   B. Men are frequently asymptomatic carriers.

   C. Penicillins are no longer recommended for treatment.

   D. Diagnosis by Gram stain in men is unreliable.

   E. Serologic diagnosis is more reliable than culture.

10. All of the following contribute to the virulence of the gonococcus, EXCEPT?

   A. Pili.

   B. Opa proteins.

   C. IgA1 protease.

   D. Capsule.

   E. Porins.

Case 25

Neisseria meningitidis

A 19-year-old college student was in his usual state of health until the evening before admission, when he went to bed with a headache. He told his room-mate that he felt feverish, and on the following morning his room-mate found him in bed, moaning and lethargic. He was taken to the emergency room, where he appeared toxic and drowsy but oriented. His temperature was 40°C, his heart rate was 126/min, and his blood pressure was 100/60 mm Hg. His neck was supple. He had an impressive, nonblanching purpuric rash, most prominent on the trunk, wrists, and legs (Figure 1). His white blood cell count was 26 000 μl^{-1} with 25% band forms. The platelet count was 80 000 μl^{-1}. Blood cultures were obtained and the patient was started on intravenous ceftriaxone. Blood cultures revealed gram-negative diplococci (Figure 2).

It should be noted that in North America this patient would have received a lumbar puncture (LP) as an essential component in the diagnosis of meningococcal disease except where explicitly contraindicated. However, this is not the case in the UK where severe sepsis is a contraindication to lumbar puncture (www.meningitis.org). In the UK many pediatricians and adult physicians feel that LP should not be performed acutely in patients suspected of having meningococcal disease and, if LP is considered to be necessary, it is done once the patient is stable. In the case considered above the cerebrospinal fluid (CSF) glucose, protein, and white blood cell count were normal, and CSF bacterial culture was negative.

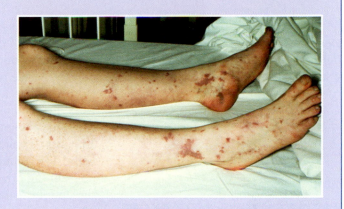

Figure 1. Legs of patient showing a purpuric rash typical of meningococcal septicemia.

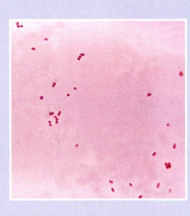

Figure 2. Gram stain of *N. meningitidis*.

1. What is the causative agent, how does it enter the body and how does it spread a) within the body and b) from person to person?

Causative agent

The patient has **bacteremia** (meningococcemia) caused by *Neisseria meningitidis* (often termed the meningococcus). This bacterium is one of the three principal causes of bacterial **meningitis**, the other two being *Streptococcus pneumoniae* (Sp) and *Haemophilus influenzae* serotype b (Hib). However, the introduction of **conjugate vaccines** for immunization of infants has reduced invasive disease caused by Sp and Hib significantly.

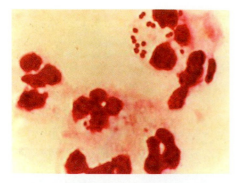

Figure 3. Gram stain of cerebrospinal fluid: _N. meningitidis_, intracellular, gram-negative diplococci.

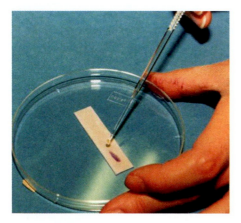

Figure 4. Kovac's oxidase test – positive reaction.

The pathogenic _Neisseria_ species _N. meningitidis_ and _N. gonorrhoeae_ are **capnophilic**, gram-negative cocci. The cocci are found in pairs where their adjacent sides are flattened giving them a coffee bean appearance (Figure 3). _Neisseria_ species are oxidase-positive (Figure 4), catalase-positive, and produce acid from sugars by oxidation. Their habitat is the mucous membranes of mammals and many species are **commensals** of these surfaces. However, _N. meningitidis_ is an obligate human pathogen whose principal habitat is the nasopharynx. However, it has been reported to colonize the mucosae of the endocervix, urethra, and anus.

Neisseria species have a typical gram-negative envelope (Figure 5) but the outer leaflet of their outer membrane contains lipooligosaccharide (LOS) instead of **lipopolysaccharide** (**LPS**). The O antigen is absent. The core polysaccharide of LOS of the meningococcus undergoes **antigenic variation** exposing a variety of epitopes on the cell surface. Phase variation of LOS directs interconversion between invasive and antibody/complement-resistant phenotypes of _N. meningitidis_. LOS variants containing low amounts of sialic acid enter human mucosal cells efficiently whereas those with LOS containing high amounts of sialic acid or expressing capsule do not. On the other hand these variants with low LOS sialylation are highly susceptible to killing by antibody and complement in the bloodstream whereas variants with highly sialated LOS and expressing capsule are serum-resistant. The cell surface is decorated with **pili**, which are hair-like projections composed of polymers of the structural protein **pilin** (PilE). Pili mediate initial attachment to nasopharyngeal epithelium and enable the bacterium to resist **phagocytosis**. The PilC protein present at the tip of the pilus is the **adhesin** that binds a receptor(s) on the host cell surface thought to be CD46.

The pilin (PilE) protein contains a constant N-terminal domain, a hypervariable C-terminal domain, and several variable regions termed mini-cassettes, which are encoded by genes with varying DNA sequence. The meningococcus has a single complete copy of the pilin gene termed _pilE_ but many truncated genes with variable DNA sequence. The truncation is at the 5′ end, resulting in lack of the sequence encoding the N-terminal constant domain and promoter elements. These truncated genes are termed

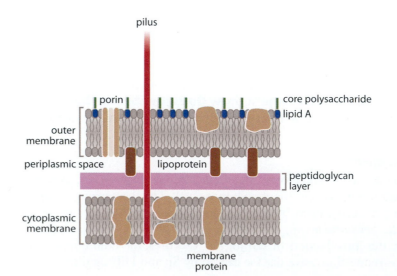

Figure 5. Schematic representation of the _N. meningitidis_ envelope.

pilS (silent) and form the *pilS* locus. By recombination of *pilS* sequences into the *pilE* gene the bacterium can express a high number of antigenically distinct pili. In addition to antigenic variation the meningococcus undergoes phase variation. In phase variation the bacterium has the ability to turn pilus expression on or off at a high frequency.

The major outer-membrane proteins (OMPs) of the meningococcus are subdivided into five classes based on descending molecular weight. Class I, II, and III are porin proteins with class I being PorA, and class II and III, PorB. The class IV protein is the reduction modifiable protein (Rmp) and the class V proteins are the opacity-associated proteins (Opa), so named because they give rise to an opaque colony phenotype. Opa proteins are important in the ability of the organism to adhere tightly to epithelia. They also dictate the tissue tropism of the meningococcus and its ability to invade epithelial cells. There are three or four genes encoding Opa proteins and they undergo phase variation such that a neisserial population will contain bacteria expressing none, one or several Opa proteins. There are two hypervariable domains within the extracellular portion of the molecule that give rise to new Opa variants as a result of point mutation and by modular exchange of domains between different Opa proteins. These hypervariable domains are also the sites of interaction with cellular receptors. The Opa proteins of the meningococcus can be divided into two major groups based on the cellular receptors to which they bind. The Opa_{HS}-type proteins (Opc) bind to heparan sulfate and extracellular matrix molecules such as fibronectin and vitronectin, while the Opa_{CEA}-type proteins (Opa) bind to carcinoembryonic antigen (CEA) and related molecules. Opa_{HS}-type proteins and Opa_{CEA}-type proteins mediate internalization of meningococci by various human cells. Either directly through engagement with heparan sulfate proteoglycans on the epithelial cell surface or indirectly using fibronectin or vitronectin to bridge Opa_{HS}-type proteins and cell surface integrins, the actin cytoskeleton of the epithelial cell is signaled to rearrange to enable engulfment of the bacterial cell. Engagement of Opa_{CEA}-type proteins with CEA-family cell adhesion molecules (CEA-CAM) also triggers actin cytoskeleton rearrangement, permitting internalization by neutrophils, macrophages, and various other types of cells.

The porin proteins that form channels that traverse the outer membrane are thought to play several roles in pathogenesis. For example, when meningococci are apposed to the host cell membrane, PorB appears to be able to translocate into the eukaryotic cell membrane and affect the maturation of **phagosomes**.

An important **virulence** determinant of the meningococcus is the production of a **capsule**. Noncapsulate strains rarely cause invasive disease and are often found colonizing the nasopharynx of asymptomatic carriers. Although there are 13 different types of capsule only five commonly cause invasive disease, A, B, C, Y, and W-135. However, recently there have been outbreaks of invasive disease caused by serogroup X in West Africa. The capsule of serogroup A *N. meningitidis* is composed of *N*-acetyl mannosamine-1-phosphate whereas the capsules of serogroups B, C, Y, and W-135 meningococci are composed completely of poly-sialic acid or sialic acid linked to glucose or galactose. *N. meningitidis* serogroup X synthesizes capsular polymers of $\alpha1{\rightarrow}4$-linked *N*-acetylglucosamine 1-phosphate. The capsule is a major vaccine antigen (see later). Differences in the chemical structure of the capsular polysaccharide determine the meningococcal serogroups whereas the serotype is defined by differences in the class 2/3 OMPs. Subtypes based on class 1 OMP have also been defined. Meningococci can switch the type of capsule they express and this is probably accomplished by transformation and horizontal DNA exchange with other capsule types of *N. meningitidis*.

N. meningitidis is adept at acquiring iron from numerous iron sources in its human host. This property contributes to its ability to replicate on mucosal surfaces, intracellularly, and in the blood.

Entry and spread within the body

The nasopharyngeal epithelium is induced to take up the meningococci by receptor-mediated **endocytosis** and the bacteria pass through the cells into the subepithelial space where infection is established. From this site the meningococci may seed the bloodstream where production of capsule is important for survival and from there to the meninges and many other sites including joints and the skin. Type IV pili mediate adhesion of *N. meningitidis* to endothelial cells and induce Rho- and Cdc42-dependent cortical actin polymerization that results in the formation of membrane protrusions that lead to bacterial uptake. An essential step in the formation of membrane protrusions appears to be the phosphorylation of cortactin because it controls the polymerization of cortical actin. LOS plays an essential role in the induction of membrane protrusions because it induces the recruitment and phosphorylation of cortactin.

Spread from person to person

The meningococcus is spread horizontally (person to person) by respiratory droplets or direct contact. At any given time about 10% of a population harbors the meningococcus in the nasopharynx. The rate of carriage increases with age to a maximum in late adolescence and early adulthood, when as many as a third of individuals may harbor the meningococcus. The carriage rate is related to the degree of crowding such as is encountered in barracks, dormitories, prisons, university dormitories (halls of residence), and many other host and environmental factors.

Epidemiology

For the most part *N. meningitidis* exists in a state of asymptomatic carriage in the human nasopharynx and it has been estimated that as many as 10% of persons are colonized by the organism (see above). Why the meningococcus on rare occasions penetrates the mucosa and invades the blood is not completely understood. Meningococcal meningitis occurs sporadically and unpredictably in small clusters throughout the world. In temperate regions the number of cases increases in winter and spring. In the United States the incidence of meningococcal disease is approximately 1 case per 100 000 population, with a fatality rate of about 10%. Infections are most commonly caused by serogroups B and C. The predominant serogroups in various parts of the world are as follows: Asia and Africa, A, C, W-135, and X; Europe, North and South America, B, C, and Y. In 1995 the number of cases and deaths from meningococcal disease in the UK rose, largely as a result of serogroup C disease. The introduction of a serogroup C conjugate vaccine (see below) reversed this trend. The highest incidence of meningococcal disease occurs in sub-Saharan Africa, in an area that stretches from Senegal in the west to Ethiopia in the east. The population of this area is about 300 million people and it is referred to as the 'meningitis belt.' There are several factors that contribute to the high incidence of disease in this area. Dust winds and cold nights during the dry season are thought to reduce the local immunity of the pharynx and transmission of *N. meningitidis* is facilitated by overcrowded housing, pilgrimages, and traditional markets. Attack rates as high as 1000 cases per 100000 population have been reported.

There is considerable interest in determining why outbreaks of meningococcal disease occur. *N. meningitidis* is a highly polymorphic species and meningococci isolated from asymptomatic carriers are highly diverse. Based on the use of **multilocus enzyme electrophoresis (MLEE)** or **multilocus sequence typing** the carriage population has been found to comprise clonal complexes or lineages. Some of these clonal complexes are disproportionately the source of strains isolated from invasive disease such that about 10 of these lineages, termed 'hyperinvasive lineages,' caused the majority of meningococcal disease during the 20th century.

2. What is the host response to the infection and what is the disease pathogenesis?

The innate secretory immune system in the nasopharynx serves as a particle filter and a sticky microbicidal trap to prevent pathogenic microorganisms from adhering to the nasal mucosa and from reaching the lower respiratory tree. The anatomy of the nasal cavities is designed to create turbulent airflow to maximize the contact of microorganisms with the nasal hairs and the mucous coating overlying the epithelium. Turbulent airflow is created by the turbinate bones located laterally in the nasal cavity. These are covered by ciliated, pseudostratified columnar epithelium supported by a thick, vascular, and erectile glandular **lamina propria**. The inhaled air is warmed and humidified and this may facilitate binding of the polysaccharide capsule of the meningococcus and other capsulate pathogenic bacteria. Microorganisms become trapped in the mucus, which is moved backwards towards the throat by cilial action (mucociliary ladder/blanket) where it can be swallowed and enter the stomach acid bath. Mucus is a viscoelastic gel, which is complemented by an armamentarium of antimicrobial factors such as lysozyme, lactoferrin, and **defensins**.

Acquired humoral immunity in the nasopharynx and at other mucosal surfaces is mediated by secretory immunoglobulin A (SIgA). The principal purpose of SIgA is immune exclusion, that is, prevention of pathogen adherence to, and subsequent invasion of the mucosal epithelium. SIgA can also neutralise exotoxins and microbial enzymes. SIgA is anchored in the mucus where it can participate in pathogen clearance mediated by the mucociliary ladder. It can also agglutinate microorganisms in nasopharyngeal secretions facilitating their clearance. With some viruses polymeric IgA has been shown to play a role in viral clearance from the lamina propria by binding the virions prior to its attachment to the polyIg receptor and transporting them through epithelial cells as the immunoglobulin traffics to the apical surface and ultimately into secretions. In addition, during transport through epithelial cells to the mucosal surface, SIgA is able to intercept and neutralize virions as they make their way in the other direction (from the apical surface to the basal surface of the cell). Whether SIgA can clear meningococci and other pathogenic bacteria from the lamina propria and intercept then within epithelial cells remains to be determined. It is thought that an **IgA1 protease** secreted by *N. meningitidis*, like *N. gonorrhoeae*, plays a role in subverting the host antibody responses. This enzyme cleaves IgA1 at the hinge region to produce Fab and Fc fragments. However, although IgA in plasma is almost all IgA of subclass 1, half of sIgA at mucosal surfaces is sIgA of subclass 2 that is resistant to IgA1 protease.

Once meningococci have transgressed the mucosal barrier by adherence to epithelium and receptor-mediated endocytosis (described earlier), the **mannose-binding lectin** and the **alternative pathways** of the complement system serve as the primary mechanism of meningococcal killing. Individuals with deficiency in the **membrane attack components (C5-9)** of the complement cascade are highly susceptible to invasive disease. Anti-meningococcal bactericidal **IgM** and particularly **IgG** antibodies play the principal role in acquired immune defense. Neonates are highly resistant to meningococcal disease as a result of transplacental transfer of maternal IgG antibodies, but they are extremely susceptible by 6 months of age. Thereafter, antibodies are induced and maintained as a result of intermittent carriage of strains of *N. meningitidis*, although carriage is rare in children under 5 years, and constant exposure to the commensal *N. lactamica*. In addition, *E. coli* strain K1 and *Bacillus pumilis* share structurally and immunologically identical capsules with *N. meningitidis* serogroup B and serogroup A strains, respectively, and may induce cross-reactive antibodies that protect against the meningococcus. The role of cell-mediated immunity in protection against *N. meningitidis* is not well understood.

Pathogenesis

This is a function of the characteristics of the strain of *N. meningitidis*, particularly the propensity to release LOS and the characteristics of host immunity. The feared outcomes of meningococcemia are shock, meningitis, **disseminated intravascular coagulation (DIC)** and myocardial dysfunction. Shock and DIC are interrelated processes that are initiated by LOS. During growth and **lysis** the meningococcus releases LOS in blebs. LOS activates the complement cascade while the lipid A component of LOS binds the CD14/TLR-4/MD-2 receptor complex in the membrane of many cell types, particularly macrophages. Binding induces the release of large amounts of pro-inflammatory **cytokines**, particularly **tumor necrosis factor-α (TNF-α)** but also **interleukin**-1 (**IL**-1), IL-6, IL-8, GM-CSF, IL-10, IL-12, and **interferon-γ (IFN-γ)**, the levels of which are closely correlated with disease severity and risk of death. These cytokines activate vascular endothelium, lymphocytes, and **natural killer (NK)** cells and recruit neutrophils, basophils, and T cells to the site of infection. Furthermore, TNF-α, IL-1, and IL-6 induce fever and the production of acute phase proteins. The release of LOS into the bloodstream brings about the release of TNF-α by macrophages in the liver, spleen, and other sites, causing massive vasodilation resulting in a fall in blood pressure and increased vascular permeability leading to loss of albumin followed by loss of fluid and electrolytes. Initially vasoconstriction of arterial and venous vascular beds attempts to compensate for the decrease in plasma volume but as fluid continues to be lost venous return to the heart is impaired and cardiac output falls. Extravascular fluid accumulates in tissues and organs leading in the lungs to pulmonary **edema** and respiratory failure. Intravascular coagulation occurs as a result of marked **thrombocytopenia** and prolonged coagulation. The end point is organ failure.

3. What is the typical clinical presentation and what complications can occur?

The onset of meningococcemia may be insidious and can result in several outcomes. If the bacteria are rapidly cleared from the blood by the mononuclear phagocyte system the patient manifests a short episode of **febrile**, flu-like symptoms; if not, overt disease develops. Multiplication of meningococci in the blood is associated with chills, fever, and localized or generalized **myalgia**. A hemorrhagic rash develops in about 80% of patients, although half of patients with meningitis do not have a rash. The rash begins as **petechiae** or **maculopapules** that occur initially on the extremities and trunk but may progress to involve any part of the body. As meningococcemia proceeds, pustules, **bullae**, and hemorrhagic lesions with central **necrosis** may develop. From the blood meningococci may seed the brain. Fulminant meningococcal sepsis (FMS), comprising shock and DIC, occurs within a matter of hours, which may or may not be accompanied by meningitis. Gedde-Dahl et al. 1983 (cited in van Deuren et al. 2000, see Extra Learning Resources) divided patients with invasive meningococcal disease into four groups: (1) patients with bacteremia without **shock**; (2) patients with bacteremia and shock but without meningitis (FTS); (3) patients with shock and meningitis; and (4) patients with meningitis alone.

Complications of meningococcal meningitis include **seizures**, increased intracranial pressure, cerebral venous and sagittal sinus **thrombosis**, and **hydrocephalus.** Cerebral herniation is rare. Communicating hydrocephalus can lead to gait difficulty, mental status changes, incontinence, and hearing loss. Fulminant meningococcemia may result in severe DIC leading to bleeding into the lungs, urinary tract, and gastrointestinal tract. Suppurative complications are infrequent but include **septic arthritis**, purulent **pericarditis**, **endophthalmitis**, and **pneumonia**.

4. How is the disease diagnosed and what is the differential diagnosis?

A Gram stain of CSF or material obtained from the characteristic skin lesion is informative and will reveal extracellular and intracellular coffee bean-shaped gram-negative diplococci (Figure 3). Pathogenic *Neisseria* are fastidious and are usually cultured using chocolate agar, although they will grow on blood agar (Figures 6–8). For clinical specimens obtained from mucosal surfaces modified Thayer-Martin (MTM) medium is commonly used in the United States. This medium contains various antimicrobial agents to suppress the growth of commensal gram-positive and gram-negative bacteria and fungi to allow the selective recovery of *N. meningitidis* and *N. gonorrhoeae*. Agar plates are incubated at 35–37°C in 3–7% CO_2 in a humid environment. A candle extinction jar can be used for this purpose.

Figure 6. *N. meningitidis* growing on chocolate agar. The colonies in the lower right quadrant of the plate have been tested for oxidase activity and are positive (the colonies are surrounded by a purple ring).

Figure 7. *N. meningitidis* growing on blood agar.

Figure 8. Overnight growth of *N. meningitidis* on blood agar plate appears as round, moist, glistening, and convex colonies.

Identification of the meningococcus is achieved by subjecting oxidase-positive and catalase-positive (30% H_2O_2) gram-negative diplococci to a panel of sugars for carbohydrate utilization tests, specific fluorescent antibodies, chromogenic substrates, or DNA probes. A flow chart of a simple scheme to identify *N. meningitidis* is shown in Figure 9. Differential utilization of the sugars glucose, maltose, sucrose, and lactose is a simple way to speciate the pathogenic *Neisseria*. *N. meningitidis* utilizes glucose and maltose (Figure 10). Rapid tests are available commercially that detect capsule antigen in CSF and urine by latex agglutination. Serogrouping can be accomplished by agglutination of bacterial cells from a colony with specific antisera. **Polymerase chain reaction (PCR)** on blood and CSF has become routine in the UK.

Differential diagnosis

Bacterial meningitis caused by *Haemophilus influenzae*, *Streptococcus pneumoniae*, *Borrelia burgdorferi*, *Mycobacterium tuberculosis*, *Streptococcus agalactiae*, *Escherichia coli*, *Listeria* species, and *Staphylococcus aureus* and viral meningitis should be considered in the differential diagnosis. In immunocompromised patients such as those with HIV infection, meningitis caused by the fungus *Cryptococcus neoformans*, the protozoan *Toxoplasma gondii*, and the herpes simplex viruses should be considered. In patients presenting with the purpuric rash, **erythema multiforme**, **Henoch-Schönlein purpura**, and hypersensitivity **vasculitis** could be considered. Other conditions to be considered are **Dengue fever**, viral hemorrhagic fever, enterovirus infection, gonococcal infection, infective **endocarditis**, malaria, Rocky Mountain spotted fever, and thrombotic thrombocytopenic purpura.

5. How is the disease managed and prevented?

Management

Because of the rapidity and severity of deterioration in the condition of patients with meningococcal disease, antibiotic therapy should never await the results of laboratory diagnostic procedures. Successful treatment

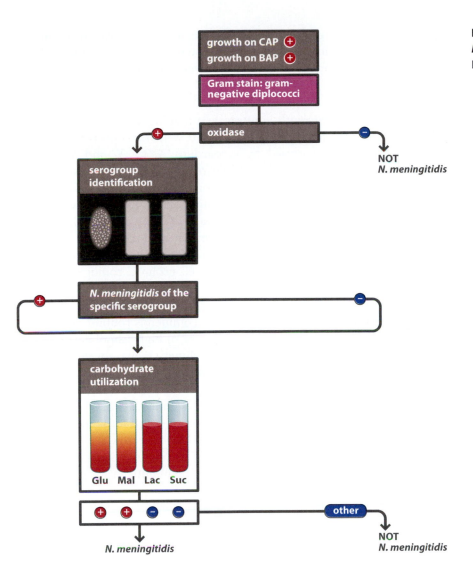

Figure 9. Identification of
N. meningitidis. CAP: chocolate agar plate;
BAP: blood agar plate.

includes early recognition, prompt parenteral antibiotics, aggressive fluid resuscitation in cases of meningococcal sepsis, frequent re-evaluation and, in patients with a poor prognosis, supportive therapy and transfer to an intensive care unit.

Antibiotics for the treatment of meningococcal disease include penicillin G and some cephalosporins (ceftriaxone, cefotaxime, cefuroxime). In patients hypersensitive to β-lactam antibiotics chloramphenicol is used. Decreased susceptibility of meningococci to penicillin has been observed in several countries that results from alteration of penicillin-binding proteins and sometimes plasmid-mediated β-lactamase. Furthermore, reduced susceptibility to rifampicin has resulted in some countries substituting fluoroquinolones for prophylaxis. However, the frequency of fluoroquinolone resistance is quite low.

Individuals with close (kissing) contact during the week before onset of illness are reported to have a 1000-fold increased risk of being infected. Such individuals include household members, day-care contacts, cell-mates, or individuals exposed to infected nasopharyngeal secretions through kissing, mouth to mouth resuscitation or other means of contact with oral secretions.

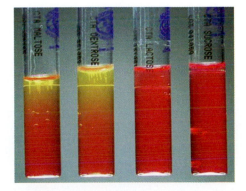

Figure 10. Cystine trypticase agar-sugar reactions differentiating *N. meningitidis* from other *Neisseria* species. Acid production (yellow color) shows oxidative utilization of dextrose (glucose) and maltose with no utilization of lactose and sucrose.

For such individuals the US and several European countries recommend chemoprophylaxis with rifampin, ciprofloxacin or ceftriaxone. Prophylaxis must also be given to the index case to prevent onward spread.

Prevention

As regards vaccines, currently, a tetravalent meningococcal capsular polysaccharide-protein conjugate vaccine (MCV4) that protects against meningococcal disease caused by serogroups A, C, Y, and W-135 is licensed for use in persons aged 11–55 years. In the United States the same vaccine is also available without conjugation of the capsular polysaccharide to a carrier protein (MPSV4). Both vaccines are effective. However, the conjugate vaccine (MCV4) has the advantage of being a T-cell-dependent immunogen that will induce protective immunity in infants, confer long-lasting protection, and provide herd immunity by reducing nasopharyngeal carriage and transmission. In the UK serogroup C conjugate vaccine is given as three doses at 3, 4, and 12 months of age. The Advisory Committee on Immunization Practices (ACIP) of the US Center for Disease Control and Prevention (CDC) recommends routine immunization with MCV4 of persons aged 11–12 years, at high school entry if previously unvaccinated, and of college freshmen living in dormitories. Immunization is also recommended for persons with deficiencies in the complement pathway (C3, C5–9) or who are asplenic. Travelers to Mecca, Saudi Arabia during the annual Hajj and Umrah pilgrimage must have proof of immunization. Immunization is recommended for persons traveling to the meningitis belt in Africa during the dry season, which extends from December through June. A serogroup A conjugate vaccine is under development for the prevention of epidemic meningitis in Africa.

Although serogroup B accounts for about a third of invasive infections and half of the cases in infants in the United States and is a significant cause of disease in developing and developed countries, no licensed vaccine is currently available in the United States. The reason for this is that the serogroup B capsule is composed of repeating units of α2-8 *N*-acetyl neuraminic acid, which is not only a poor immunogen in humans, but is identical to polysialic acid commonly found on human cell surface glycoproteins. Thus, there is the possibility that the vaccine might induce autoreactive antibodies that might result in immunopathology. Research is focusing on OMPs that may serve as candidate vaccine antigens. However, these appear to offer only strain-specific protection.

SUMMARY

1. What is the causative agent, how does it enter the body and how does it spread a) within the body and b) from person to person?

- *Neisseria meningitidis* is one of the three principal causes of bacterial meningitis, the others being *Haemophilus influenzae* and *Streptococcus pneumoniae*.

- *Neisseria* species are capnophilic, gram-negative cocci. The cocci are often found in pairs where their adjacent sides are flattened giving them a coffee bean appearance.

- *N. meningitidis* is oxidase-positive, catalase-positive, and produces acid from glucose and maltose by oxidation.

- *N. meningitidis* is an obligate human pathogen whose principal habitat is the nasopharynx.

- *Neisseria* species have a typical gram-negative envelope but the outer leaflet of their outer membrane contains lipooligosaccharide (LOS) instead of lipopolysaccharide (LPS).

- The core polysaccharide of LOS undergoes antigenic variation exposing a variety of epitopes on the cell surface.

- The cell surface is decorated with pili, which undergo antigenic and phase variation.

- Opacity-associated proteins (Opa) are important in the ability of the organism to adhere tightly to epithelia.

- The most important virulence determinant of the meningococcus is the production of a capsule.

- Meningococci can switch the type of capsule they express.

- *N. meningitidis* is adept at acquiring iron from its human host, which contributes to its ability to replicate on mucosal surfaces, intracellularly, and in the blood.

- *N. meningitidis* secretes an IgA1 protease, which may contribute to virulence.

- Following entry via the nasopharyngeal epithelium, the meningococci may seed the bloodstream and from there the meninges and many other sites including joints and the skin.

- The meningococcus is spread horizontally (person to person) by respiratory droplets or direct contact.

- At any given time about 10% of a population harbors the meningococcus in the nasopharynx.

- The rate of carriage increases with age to a maximum in late adolescence and early adulthood, when as many as a third of individuals may harbor the meningococcus.

- The carriage rate is related to the degree of crowding such as is encountered in barracks, dormitories (halls of residence), prisons, and so forth.

- Meningococcal meningitis occurs sporadically in small clusters throughout the world. In temperate regions the number of cases increases in winter and spring.

- In the United States the incidence of meningococcal disease is approximately 1 case per 100 000 population, with a fatality rate of about 10%.

- Infections are most commonly caused by serogroups B and C.

- The predominant serogroups in various parts of the world are as follows: Asia and Africa, A, C, W135, and X; Europe, North and South America, B, C, and Y.

- The highest incidence of meningococcal disease occurs in sub-Saharan Africa, in an area from Senegal to Ethiopia termed the 'meningitis belt,' where attack rates as high as 1000 cases per 100 000 population have been reported.

2. What is the host response to the infection and what is the disease pathogenesis?

- At the mucosal surface defensins are likely the principal mediator of innate defense and SIgA antibody mediates against humoral immunity.

- In the tissues the innate mannose-binding lectin and the **alternative pathways** of the complement system serve as the primary mechanism of meningococcal killing.

- Anti-meningococcal bactericidal IgM and IgG antibodies play the principal role in acquired immune defense. Antibodies are induced and maintained as a result of intermittent carriage of strains of *N. meningitidis*, and constant exposure to the commensal *N. lactamica*.

- The role of cell-mediated immunity in protection against *N. meningitidis* is not well understood.

- *N. meningitidis* secretes an IgA1 protease, which may subvert the immune response and contribute to virulence.

- Pathogenesis is a function of the characteristics of the strain of *N. meningitidis*, particularly the propensity to release LOS in the form of blebs.

- During growth and lysis the meningococcus releases LOS, which induces the release of large amounts of pro-inflammatory cytokines, particularly tumor necrosis factor-α (TNF-α).

3. What is the typical clinical presentation and what complications can occur?

- The onset of meningococcemia may be insidious and can result in several outcomes ranging from a short episode of febrile, flu-like symptoms to **septicemia** and meningitis.

- A hemorrhagic rash develops in about 80% of bacteremic patients, which begins as petechiae that occur initially on the extremities and trunk but may progress to involve any part of the body.

- As meningococcemia proceeds, pustules, bullae, and hemorrhagic lesions with central necrosis may develop.

- Fulminant meningococcal sepsis (FMS), comprising shock and DIC, occurs within a matter of hours, which may or may not be accompanied by meningitis.

4. How is the disease diagnosed, and what is the differential diagnosis?

- A Gram stain of blood or CSF or material obtained from the characteristic skin lesion is informative and will reveal extracellular and intracellular coffee bean-shaped gram-negative diplococci.

- *Neisseria meningitidis* is fastidious and is usually cultured using chocolate agar although it will grow on blood agar.

- For clinical specimens obtained from mucosal surfaces modified Thayer-Martin (MTM) medium is commonly used in the US.

- Plates are incubated at 35–37°C in 3–7% CO_2 in a humid environment.

- Identification of the meningococcus is achieved by subjecting oxidase-positive and catalase-positive, gram-negative diplococci to a panel of sugars for carbohydrate utilization tests. The meningococcus oxidises glucose and maltose.

- Specific fluorescent antibodies, chromogenic substrates, or DNA probes may be used.

- Rapid tests are available commercially that detect capsule antigen in CSF and urine by latex agglutination and PCR.

- Serogrouping can be accomplished by agglutination of bacterial cells from a colony with specific anti-capsule antisera.

- Bacterial meningitis caused by *H. influenzae* and *Streptococcus pneumoniae* should be considered in the differential diagnosis.

- In patients presenting with the purpuric rash erythema multiforme and hypersensitivity vasculitis are part of the differential diagnosis.

- Other conditions to be considered are Dengue fever, Ebola virus, enterovirus infection, gonococcal infection, infective endocarditis, malaria, Rocky Mountain spotted fever, and thrombotic thrombocytopenic purpura.

5. How is the disease managed and prevented?

- Because of the rapidity and severity of deterioration in the condition of patients with meningococcal disease antibiotic therapy should never await laboratory diagnostic procedures.

- Successful treatment includes early recognition, prompt parenteral antibiotics, frequent re-evaluation and, in patients with a poor prognosis, supportive therapy and transfer to an intensive care unit.

- Antibiotics for the treatment of meningococcal disease include penicillin G and some cephalosporins (ceftriaxone, cefotaxime, cefuroxime). In patients hypersensitive to β-lactam antibiotics chloramphenicol is used.

- Decreased susceptibility of meningococci to penicillin has been observed in several countries.

- Close contacts of the index case should receive chemoprophylaxis with rifampin or ciprofloxacin.

- A tetravalent meningococcal capsular polysaccharide-protein conjugate vaccine (MCV4) that protects against serogroups A, C, Y, and W-135 is given to children aged 11–12 years,

- at high school entry if previously unvaccinated, and to college freshmen living in dormitories.

- Immunization is also recommended for persons with deficiencies in the complement pathway (C3, C5–9) or who are asplenic.

- In the UK serogroup C conjugate vaccine is given as three doses beginning at 3 months of age.

- Immunization is recommended for persons traveling to the meningitis belt in Africa during the dry season, which extends from December through June.

- Currently there is no vaccine available for serogroup B disease.

FURTHER READING

Mims C, Dockrell HM, Goering RV, Roitt I, Wakelin D, Zuckerman M. Medical Microbiology, 3rd edition. Mosby, Edinburgh, 2004.

Murphy K, Travers P, Walport M. Janeway's Immunobiology, 7th edition. Garland Science, New York, 2008.

Murray PR, Rosenthal KS, Pfaller MA. Medical Microbiology, 5th edition. Elsevier Mosby, Philadelphia, PA, 2005.

REFERENCES

Harrison LH. Prospects for vaccine prevention of meningococcal infection. Clin Microbiol Rev, 2006, 19: 142–164.

Hauk CR, Meyer F. 'Small talk': Opa proteins as mediators of *Neisseria*-host-cell communication. Curr Opin Microbiol, 2003, 6: 43–49.

Kirchner M, Meyer TF. The PilC adhesin of the *Neisseria* type IV pilus – binding specificities and new insights into the nature of the host cell receptor. Mol Microbiol, 2005, 56: 945–957.

Pathan N, Faust SN, Levin M. Pathophysiology of meningococcal meningitis and septicaemia. Arch Dis Child, 2003, 88: 601–607.

Stephens DS, Greenwood B, Brandtzaeg P. Epidemic meningitis, meningococcaemia, and *Neisseria meningitidis*. Lancet, 2007, 369: 2196–2210.

Van Deuren M, Brandtzaeg P, van der Meer JWM. Update on meningococcal disease with emphasis on pathogenesis and clinical management. Clin Microbiol Rev, 2000, 13: 144–166.

WEB SITES

CDC Travellers' Health: Yellow Book, Meningococcal Disease: http://wwwn.cdc.gov/travel/yellowBookCh4-Menin.aspx#419

eMedicine, Inc., 2007, provider of clinical information and services to physicians and health-care professionals: Meningococcemia http://www.emedicine.com/derm/topic261.htm; Meningococcemia: http://www.emedicine.com/MED/topic1445 .htm; meningococcal infections: http://www.emedicine.com/med/topic1444.htm

World health Organization, Meningococcal meningitis: http://www.who.int/mediacentre/factsheets/fs141/en/

Guidance for public health management of meningococcal disease in the UK: http://www.hpa.org.uk/infections/topics_az/meningo/meningococcalguidelines.pdf

MULTIPLE CHOICE QUESTIONS

The questions should be answered either by selecting True (T) or False (F) for each answer statement, or by selecting the answer statements which best answer the question. Answers can be found in the back of the book.

1. **A 19-year-old college freshman was brought to the Emergency Room (ER) because of severe headache, fevers, and neck stiffness. She is a resident of a dormitory. In the ER, she was unresponsive, and had a diffuse purpuric/petechial rash; her temperature was 104°F (40°C). Her lumbar puncture was consistent with meningitis. A culture of her cerebrospinal fluid was most likely to yield?**
   A. *Eschericha coli.*
   B. *Neisseria meningitidis.*
   C. Group B *Streptococcus* (GBS).
   D. *Streptococcus pneumoniae.*
   E. *Haemophilus influenzae.*

2. **A frantic mother brought her 10-day-old infant to the Emergency Room (ER) because of fevers and failure to thrive. The baby was diagnosed with meningitis. The cerebrospinal fluid would most likely yield?**
   A. *Streptococcus pyogenes* (group A streptococci).
   B. *Streptococcus pneumoniae.*
   C. *Streptococcus agalactiae* (group B streptococci).
   D. *Staphylococcus aureus.*
   E. *Neisseria meningitidis.*

3. **All of the following are characteristics of *Neisseria meningitidis*, EXCEPT?**
   A. Produces an extracellular polysaccharide capsule.
   B. Gram-negative diplococcus.
   C. Oxidase-positive.
   D. Produces lipooligosaccharide (LOS).
   E. Serogrouped by differences in the outer-membrane proteins.

4. **Which one of the following is the reservoir of *Neisseria meningitidis*?**
   A. Soil.
   B. Salt water.
   C. Human carriers.
   D. Domesticated animals.
   E. Reptiles.

5. **Neisseria meningitidis is spread by which one of the following?**
   A. Respiratory secretions or droplets.
   B. Contaminated fomites.
   C. Consumption of contaminated food or water.
   D. Insect vectors.
   E. Vaginal intercourse.

6. **Neisseria meningitidis is capable of all of the following, EXCEPT?**
   A. Anaerobic metabolism.
   B. Cleavage of secretory immunoglobulin A subclass 1 (sIgA1).
   C. Transcytosis of epithelial cells.
   D. Production of capsule.
   E. Production of purpuric skin lesions.

7. **All of the following are important in protection against meningococcal infection, EXCEPT?**
   A. The alternative complement pathway.
   B. IgM and IgG anti-capsular antibodies.
   C. CD8 cytotoxic T cells.
   D. Polymorphonuclear leukocytes.
   E. The classical complement pathway.

8. **In the treatment and prevention of meningococcal disease all of the following are correct, EXCEPT?**
   A. Antibiotics should not be given until the diagnosis is confirmed by laboratory investigations.
   B. Parenteral penicillin G or ceftriaxone are suitable antibiotics unless the patient is hypersensitive to them.
   C. Close contacts of the index case should receive chemoprophylaxis with rifampin or ciprofloxacin.
   D. Meningococcal capsular polysaccharide-protein conjugate vaccines protect against disease caused by serogroups A, C, Y, and W-135 but not B.
   E. Frequent re-evaluation and, in patients with a poor prognosis, supportive therapy and transfer to an intensive care unit is indicated.

9. **Which of the following statements about the isolation and identification of the meningococcus are CORRECT?**
   A. A Gram stain of blood, CSF or material obtained from the characteristic skin lesion is informative.
   B. Blood or chocolate agar may be used to recover meningococci from the blood and CSF.
   C. *N. meningitidis* oxidizes the sugars glucose and maltose.
   D. Rapid latex agglutination can be used to detect capsular polysaccharide in CSF and urine.
   E. All of the above.

10. **Which of the following contribute to the virulence of the meningococcus?**
    A. Pili.
    B. Opa proteins.
    C. IgA1 protease.
    D. Capsule.
    E. All of the above.

Case 26

Norovirus

A doctor was called to a home for the elderly because two residents had developed vomiting and watery diarrhea within 2 days of each other – an 80-year-old woman followed by an 87-year-old man. On taking a history the doctor found that both patients had flatulence and stomach cramps during the first 24 hours of illness followed by vomiting, diarrhea, aching joints, and neck pains. The doctor examined the patients; the first case had recovered and the symptoms were resolving in the second case. She requested a stool sample from both patients to be sent to the local microbiology laboratory to be tested for bacterial and viral pathogens. On the doctor's return to the home the next day both patients were well and a telephone message from the laboratory reported norovirus detected in both stool samples. The local health authority was notified of a potential outbreak. In response, the community physician advised scrupulous attention to hand-washing to prevent further spread of the infection.

However, the next day another resident became ill with diarrhea and vomiting and a few days later a member of staff developed the same symptoms and had to leave work as she was unwell. Five days afterwards a five further members of staff telephoned to say that they were sick. A nonresident day-patient also reported vomiting 2 days after her last attendance at the home. An outbreak of norovirus-induced vomiting and diarrhea was declared.

Course of the outbreak

Setting

The setting was a home for the elderly that had been recently refurbished. A total of 51 full-time and part-time staff cared for 50 resident and 60 nonresident day-patients who each attended 1 day a week.

Course

The entire outbreak lasted 9 weeks (Figure 1). Most of those affected recovered completely within 72 hours. The first case reported was in a resident who had little contact with the outside community. The first day-patient to be affected became ill 2 days after the initial case. None of the staff became unwell until 9 days after the outbreak began. In all, 34/51 (67%) staff, 27/50 (53%) residents, and 9/60 (15%) day-patients were affected.

Investigation of the outbreak

The time course of the outbreak (Figure 1) suggested serial transfer from one individual to another. After several visits to the home, Environmental Health Officers reported that, although the kitchens and preparation of food were adequate, hand-washing facilities were poor throughout the home, with very few paper towel dispensers. Because of the design of the building soiled bedpans had to be carried though the dining area so that they could be

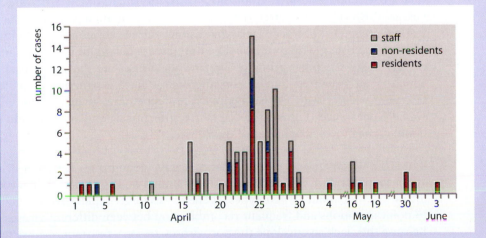

Figure 1. Time course of the norovirus outbreak.

emptied into uncovered sluices located in rooms alongside baths. This meant that the surrounding surfaces in both the dining area and around the sluices would be contaminated with virus particles as a result of air-borne dispersal of fecal material. Thus fecal contamination of the environment in the dining and washing areas accounted at least in part for the continuing person to person spread of norovirus infection.

Control measures
Several measures, which caused considerable practical difficulties for both residents and staff, were instituted in an effort to restrict the spread of infection. The home was closed to new admissions and the meals-on-wheels service to the local community was suspended. However, no attempt was made to isolate affected residents and visiting continued. The transfer of day-patients from the home to a local hospital was discontinued after a secondary outbreak of **gastroenteritis** involving several hospital patients. Three bathrooms in the home were designated sluice rooms, the baths being no longer used, and the remaining three were used for bathing only. All toilets were separate from the bathrooms and some were designated for the exclusive use of those with gastroenteritis. Staff who wished to return to work as soon as their symptoms subsided were advised to remain away

for an additional 48 hours so that they did not reintroduce the infection into the home. Soiled bedpans were placed in polythene bags and then into a closed bin before being transported through the dining room into the sluice room. Hand-washing facilities were improved with the installation of more paper towel dispensers.

Comment
The high attack rate and long duration of the outbreak were related to the vulnerability of the elderly population (average age 84 years) and the difficulties in maintaining high standards of hygiene. Although the home had recently been refurbished and internally redesigned, the planners had not fully considered the facilities necessary for the care of aged and infirm residents, nor had they consulted community physicians or microbiologists at any stage of the planning process. The standards should have been comparable to those of a hospital ward for the care of the elderly. More building work, at considerable cost, was undertaken in order to redesign parts of the building so as to separate sluice rooms from rooms with baths.

Note – This description of an outbreak is adapted from a report of a real one that happened 20 years ago (Gray et al:. see References section) but the principles it illustrates still stand.

1. What is the causative agent, how does it enter the body and how does it spread a) within the body and b) from person to person?

Causative agent

Noroviruses are unenveloped single stranded RNA viruses with an icosahedral capsid, that were formerly known as Norwalk viruses (because of their first identification in Norwalk, USA) or small round structured viruses (SRSVs) – the latter term arose because of their characteristic feathery, ragged appearance lacking a distinct surface structure as seen by electron microscopy (Figure 2A). The noroviruses belong to the family of caliciviruses – the other group of human caliciviruses are the sapoviruses, which when visualized by electron microscopy have a structure distinct from noroviruses with cup-shaped surface depressions giving a 'Star of David' appearance (Figure 2B). Sapovirus infection is endemic in childhood, causing occasional cases of diarrhea.

There are 3 genogroups of human noroviruses (genogroups I, II, and IV; genogroups III and V contain exclusively animal noroviruses) subdivided into altogether about 30 genotypes – genetic variability rapidly evolves due to point mutations and frequent recombination between different viruses. Because they lack an envelope the noroviruses are very resistant to adverse environmental conditions and even to commonly used disinfectants – this

A   B

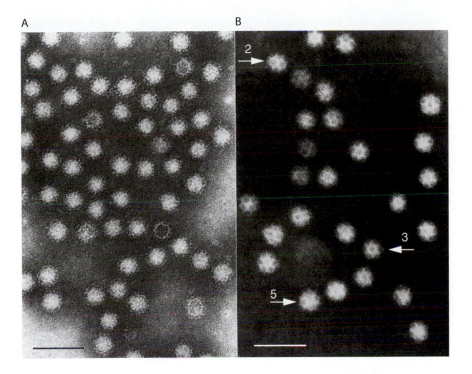

Figure 2. Electron microscopic appearance of human enteric caliciviruses. (A) Noroviruses lack the distinctive surface morphology characteristic of sapoviruses, although a ragged edge is clearly visible surrounding some of the particles. (B) The distinctive surface structures on a sapovirus viewed along the two-, three- and five-fold (indicated) axes of symmetry. Bar: 100 nm for each panel.

property together with genetic variability facilitates their propensity to cause outbreaks (see below).

Entry and spread within the body

After ingestion with food or drink, the virus passes through the stomach and due to its inherent resistance to acid arrives undamaged in the small intestine. Infection is thought to be limited to the small intestine but direct evidence for this is very difficult to obtain in the human host (see Pathogenesis below). The incubation period is 24–48 hours. Virus is found in vomit and shed in feces during the illness and fecal shedding continues for 3 weeks or more after recovery.

Person to person spread

Noroviruses are usually spread via the **fecal–oral** route, either directly from person to person or indirectly. The latter occurs because noroviruses are unenveloped and hence resistant to disinfectants and able to persist in the environment. Indirect transmission occurs after consumption of food (commonly salads and vegetables), or water and ice contaminated by the feces of an infected person. Shellfish are a common source of infection, especially oysters – these are filter feeders and concentrate virus from sewage-polluted water. Swimming in sewage-polluted water also carries the risk of norovirus infection. The infection is highly contagious – aerosolized vomit and transmission by **fomites** facilitates spread during outbreaks.

Epidemiology

Norovirus gastroenteritis is very common, usually, but not always, occurring in the winter and with norovirus infection sweeping through the community with a high attack rate. In the developed world by the fifth decade of life more than 60% of the population have norovirus antibodies, whereas antibodies are acquired much earlier in developing countries.

Noroviruses cause the majority of gastroenteritis cases identified in community-based studies in developed countries and these viruses are the most common cause of outbreaks of nonbacterial gastroenteritis in both children and adults; in the UK since the turn of the 21st century over 2000 norovirus-associated outbreaks have been reported. In developing countries noroviruses are present but less is known about their contribution to the burden of disease.

Figure 3 gives an example of the settings and presumptive modes of transmission for outbreaks of norovirus gastroenteritis in the USA. From this it can be seen that outbreaks are a common problem in hospitals and in residential homes for the elderly. Persistent outbreaks occur on cruise ships and frequently involve multiple routes of transmission including by consumption of food or water, directly from person to person, and from contamination of the environment (fomites). Such outbreaks can only be terminated by closing the ship and deep-cleaning all possibly infected surfaces including carpets and curtains. Interestingly, norovirus was also identified as a cause of outbreaks among American military personnel during the Gulf War.

2. What is the host response to the infection and what is the disease pathogenesis?

Host response to infection

The nature of immunity to norovirus infection is difficult to study because the virus cannot be grown in tissue culture. The limited information available comes from volunteer studies where it has been suggested that the minimal infectious dose is low, being 10–100 **virions**. Immunity to norovirus re-infection persists for up to about 14 weeks in volunteers after previously induced norovirus illness. However, long-term immunity is not maintained. For example, in one volunteer study, 6 of 12 individuals developed gastroenteritis upon exposure to a norovirus and the same 6 individuals became ill again upon rechallenge 27–42 months later. Although pre-existing serum antibodies do not correlate with resistance to re-infection, the importance of the adaptive immune response for the rapid resolution of norovirus infection is apparent because immunocompromised persons may be symptomatic and shed virus for months.

Figure 3. Settings and presumptive modes of transmission for 90 outbreaks of nonbacterial gastroenteritis in the United States from January 1996 to June 1997. Noroviruses were detected in 86 (96%) of the 90 outbreaks by reverse transcription PCR.

Part of the explanation for the lack of resistance to re-infection may lie in the genetic variability of the virus, that is the adaptive immune response

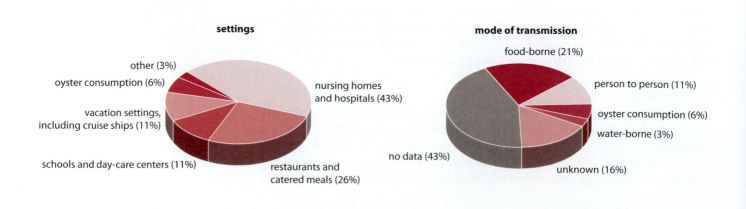

may be specific for a particular genotype of the virus. Indeed a new variant of norovirus genogroup II genotype 4 has recently appeared in Europe and has been responsible for many outbreaks (Lopman et al.: see References). However, intriguingly, the genetic make-up of the individual infected with a norovirus also appears to play a role in susceptibility. For example, about 20% of the population seem to be endowed with long-term resistance to norovirus infection. Evidence has recently emerged that this may result from a genetically determined variation in virus receptors in the intestinal tract involving ABH and Lewis carbohydrate blood group antigens. However, individual noroviruses use different blood group antigens as receptors and thus no one is resistant to all of the noroviruses. It is important to note that genetic variation, either viral or host, cannot solely explain the absence of long-term immunity to noroviruses; the human volunteer studies that first provided evidence for this atypical pattern of immunity involved repeated challenge of the same individual with the identical inoculum of virus. Thus, there was no genetic variation in this case and susceptible individuals still failed to develop lasting protective immunity.

Pathogenesis

As regards the vomiting that is characteristic of norovirus infection, a marked delay in gastric emptying was observed in volunteers who became ill after experimental infection. It has therefore been proposed that abnormal gastric motor function is responsible for nausea and vomiting but the precise mechanism is unknown. Another equally plausible explanation would be inflammation of the pyloric junction between the stomach and the intestine.

The pathogenesis of norovirus-induced diarrhea seems to be noninflammatory or secretory resulting from damage to and blunting of the small intestinal villi (Figure 4) but the exact mechanism is presently unknown. While most enteric viruses replicate in and kill enterocytes, resulting in decreased fluid absorption causing diarrhea with loss of water, sugar, and

Figure 4. Presumed mechanism of norovirus pathogenesis – development of damage to gut mucosa and ensuing diarrhea.

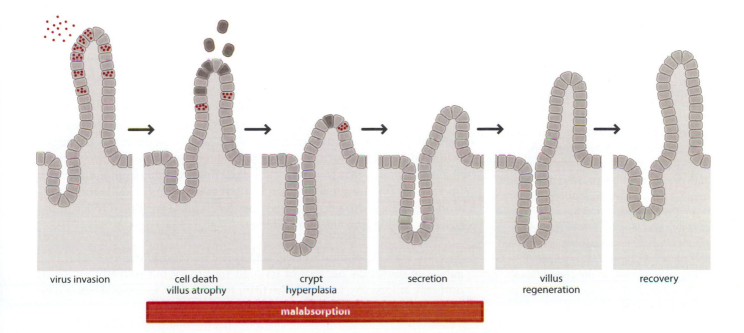

virus invasion | cell death villus atrophy | crypt hyperplasia | secretion | villus regeneration | recovery

malabsorption

salts, there is accumulating evidence that noroviruses do not infect entero-cytes: (1) when intestinal biopsy samples from norovirus-infected volunteers were stained with convalescent serum, no viral antigen could be detected in the enterocytes; (2) recent unpublished work demonstrates instead the detection of viral antigen in cells of the **lamina propria** underlying the enterocytes (M.K. Estes, American Society of Virology conference 2008, keynote address); and (3) a related mouse norovirus replicates in lamina propria cells but not epithelial cells. It is not clear how infection of lamina propria cells could result in diarrhea but there is an intricate interplay between these cells and the overlying enterocytes.

3. What is the typical clinical presentation and what complications can occur?

The typical clinical presentation includes vomiting and because of its seasonal occurrence the illness is known as 'winter vomiting disease.' Symptoms begin abruptly with nausea, abdominal cramps, and vomiting which is characteristically projectile. Diarrhea may be absent, mild or severe but when present it is not bloody, lacks mucus and may be watery. Headache and myalgia are common and low grade fever occurs in 50% of cases. Although infections in the very young and elderly can be quite severe, the disease is usually self-limiting and symptoms subside within 24–48 hours.

Complications are rare but the effects of dehydration (metabolic alkalosis, **hyponatremia**, **hypokalemia**, and renal failure) may occur in the elderly. Immunosuppressed individuals are at risk of prolonged diarrhea, perhaps lasting for months, requiring fluid replacement and in severe cases **enteral** and/or **parenteral** feeding to combat malnutrition and weight loss.

4. How is this disease diagnosed and what is the differential diagnosis?

Sporadic cases are rarely investigated and what follows concerns the differential diagnosis of outbreaks of gastroenteritis. Such diagnosis is based on the characteristic clinical features of the illness, that is vomiting in over 50% of cases and mild diarrhea without blood or mucus with a duration of about 48 hours, together with detection of norovirus in feces. A fecal sample, preferably obtained while the patient is still symptomatic, should be tested by reverse transcription **polymerase chain reaction** (RT-**PCR**), so as to detect viral RNA. Before the development of such highly sensitive molecular testing, electron microscopy was widely used but this technique is limited by its insensitivity as it requires large numbers of virus particles in a stool sample, that is >10^6 m^{-1} for detection. Notably immune-electron microscopy (Figure 5A and B) can be used to prove antibody **seroconversion** in patients with norovirus gastroenteritis and is still used today in the research setting to concentrate virus.

Note – RT-PCR for viral nucleic acid is the method of choice for investigation of outbreaks because of its high sensitivity but nevertheless in an outbreak at least five fecal samples from five different individuals should be tested to ensure detection.

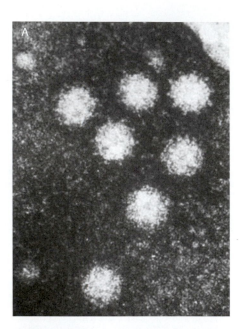

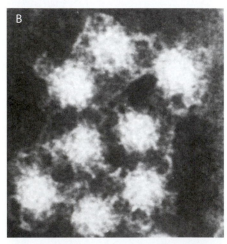

Figure 5. Immune-electron microscopy to show antibody seroconversion.
Noroviruses (size 27nm) from a patient with gastroenteritis visualized by electron microscopy (A) after incubation with serum taken from the patient in the acute phase of illness and (B) after incubation with serum taken from the patient in the convalescent phase of illness. Specific antibody molecules in the convalescent serum of the patient aggregate and coat the virus particles.

Table 1. Viruses causing gastroenteritis and their epidemiological features

Family	Virus	Endemic infection in children	Outbreaks of infection in all ages	Food- or water-borne
Reovirus	Group A rotavirus	Major cause of childhood diarrhea	Occasional in adults – most often among the elderly in hospitals or in residential homes	No
	Group B rotavirus	–	Large outbreaks in China	No
	Group C rotavirus	–	Occur but uncommon	No
Caliciviruses	Noroviruses	–	Major cause of outbreaks	Yes
	Sapoviruses	Less common cause of childhood diarrhea	–	No
Adenovirus	Adenovirus types 40 and 41	Second most common cause of childhood diarrhea	–	No
Astrovirus	Human astrovirus	5–10% of childhood diarrhea cases	Family outbreaks occur	No

Differential diagnosis

The other possible viral causes are given in Table 1 – notably outbreaks of diarrhea are commonly caused by rotavirus in young children and the elderly. Bacterial causes and fecal samples should be tested for *Salmonella*, *Shigella*, and *Campylobacter* spp. In addition, *Clostridium difficile* can cause serious outbreaks with high morbidity and even mortality in hospitals.

5. How is the disease managed and prevented?

Management

The illness is usually short and requires attention only to fluid and electrolyte replacement, except in the case of persistent diarrhea in the immunosuppressed, which may also require enteral or parenteral nutrition.

Prevention

This relies not only on clean drinking water and efficient sewage disposal but also on good standards of personal and food hygiene plus adequate cleaning arrangements in hospitals and residential homes. Raw shellfish should be cooked before consumption and fruit washed if to be eaten raw.

Outbreak control

As illustrated in the example of an outbreak in a residential home given above, this is a difficult issue and outbreaks can grumble on for long periods and often do not end until the majority of susceptible people have been infected. Control measures rely firstly on limiting contact between ill and susceptible persons, for example by isolation of those affected in single rooms in hospital or nursing homes, and using contact (enteric) precautions – gloves, aprons, and scrupulous hand-washing or if more than one case by closing the hospital ward or relevant section of a nursing home to new admissions. Secondly, they rely on measures to prevent water-borne, food-borne, fomite-borne, and person to person spread – transmission is often by more than one route in an outbreak. Thirdly, exclusion of those affected from food-handling is required for 48 hours after recovery. *Note –* although the latest evidence suggests that virus is shed in feces for longer than this after recovery, it should be remembered that the virus load will be decreasing and it is impractical to exclude staff from work for 2 weeks.

Vaccine: Given the lack of understanding of the nature of immunity to norovirus, it is not surprising that there is no vaccine available.

SUMMARY

1. What is the causative agent, how does it enter the body and how does it spread a) within the body and b) from person to person?

- Norovirus – an unenveloped single-stranded RNA virus.
- Noroviruses were formerly known as Norwalk or small round structured viruses.
- Three genogroups, GI, GII, and GIV, of norovirus infect humans.
- Norovirus infection has a seasonal incidence – 'winter vomiting disease.'
- Noroviruses are spread by the fecal–oral route.
- The main source of virus is another infected person, contaminated food, or water.
- Norovirus outbreaks are common in health-care settings and cruise ships.

2. What is the host response to the infection and what is the disease pathogenesis?

- Immunity is short-lived and re-infection is common.
- Very little is known regarding the pathogenesis of norovirus infection.
- The cause of the delay in gastric emptying that is most likely responsible for the high incidence of vomiting episodes is unknown.
- The virus appears to infect lamina propria cells in the intestine instead of enterocytes; it is unclear how this leads to diarrhea.

3. What is the typical clinical presentation and what complications can occur?

- Self-limiting vomiting and diarrhea.
- Main complication is dehydration.
- Causes persistent diarrhea in the immunocompromised.

4. How is this disease diagnosed and what is the differential diagnosis?

- The virus cannot be routinely grown in cell culture so diagnosis relies on RT-PCR to detect viral nucleic acid in feces.
- Differential diagnosis includes infection with *rotavirus*, *Salmonella*, *Shigella*, and *Campylobacter* spp, and in the hospital setting *Clostridium difficile*.

5. How is the disease managed and prevented?

- There is no specific treatment.
- There is no vaccine.
- Control of outbreaks includes reinforcing good hygiene, closure of hospital wards, and isolation of ill persons until 48 hours after symptoms have resolved.

FURTHER READING

Atmar RL, Estes MK. Norwalk virus and related caliciviruses causing gastroenteritis. In: Richman DD, Whitley RJ, Hayden FG, editors. Clinical Virology, 2nd edition. ASM Press, Washington, DC, 2002: Chapter 47.

Desselberger U, Gray JJ. Viruses associated with acute diarrhoeal disease. In: Zuckerman AJ, Banatvala JE, Pattison JR, Griffiths PD, Shaub BD, editors. Principles and Practice of Clinical Virology, 5th edition. Wiley, Chichester, 2004: Chapter 4.

REFERENCES

Gray JJ, Ward KN, Clarke IR. A protracted outbreak of viral gastroenteritis in a residential home for the elderly. Communicable Diseases Report No. 52, 30 December 1988.

Lopman B, Vennema H, Kohli E, et al. Increase in viral gastroenteritis outbreaks in Europe and epidemic spread of new norovirus variant. Lancet, 2004, 363: 682–688.

PHLS Advisory Committee on Gastrointestinal Infections. Preventing person-to-person spread following gastrointestinal infections: guidelines for public health physicians and environmental health officers. Commun Dis Public Health, 2004, 7: 362–384. (Review.)

WEB SITES

Centers for Disease Control and Prevention, Morbidity and Mortality Weekly Report, April 16, 2004/53(RR04);1–33, Atlanta, GA, USA: http://www.cdc.gov/mmwr/preview/mmwrhtml/rr5304a1.htm

Centers for Disease Control and Prevention, National Centre for Immunization and Respitory Diseases, Atlanta GA, USA: http://www.cdc.gov/ncidod/dvrd/revb/gastro/faq.htm

Centre for Infections, Health Protection Agency, HPA Copyright, 2008: http://www.hpa.org.uk/webw/HPAweb&Page&HPAwebAutoListName/Page/1191942172966?p=1191942172966

MULTIPLE CHOICE QUESTIONS

The questions should be answered either by selecting True (T) or False (F) for each answer statement, or by selecting the answer statements which best answer the question. Answers can be found in the back of the book.

1. **Which of the following are true of norovirus?**
   A. This virus is the most usual cause of infantile diarrhea.
   B. Genetic variation arises by reassortment.
   C. It is closely genetically related to rotavirus.
   D. It is closely genetically related to sapovirus.

2. **Which of the following are true concerning the spread of norovirus?**
   A. It is susceptible to adverse environmental conditions.
   B. It can be transmitted via fomites.
   C. It may be acquired by fecal–oral spread.
   D. Infection may be acquired transplacentally.
   E. Only a very small proportion of the population are susceptible.
   F. Re-infection may occur.

3. **Which of the following tests would be helpful in the diagnosis of norovirus gastroenteritis?**
   A. Inoculation of a cell culture with feces from an infected person.
   B. RT-PCR for norovirus protein in feces.
   C. Examination of a fecal sample by electron microscopy.
   D. Fecal microscopy to detect pus cells.
   E. A characteristic appearance of the intestine on endoscopy.

4. **Which of the following are true for the clinical presentation of norovirus infection?**
   A. It presents with dysentery.
   B. After 2 weeks the illness resolves.
   C. A complication may be hypokalemia.
   D. Abdominal pain is severe.

5. **Which of the following are useful in the prevention of transmission of norovirus infection?**
   A. Isolation of infected patients.
   B. Hand-washing.
   C. Chlorine compounds.
   D. Persons with norovirus gastroenteritis may prepare food for others if scrupulous hand-washing is observed.
   E. Ward closure.

Case 27

Parvovirus

A 25-year-old teacher, who was 12 weeks pregnant, went to see her doctor as she had developed an extensive **erythematous** rash on her face, trunk, and limbs (Figure 1). She was also feeling feverish, and had noticed some aching in her wrists.

A blood sample was sent to the laboratory, and the following results were received: rubella **IgG**-positive, **IgM**-negative; parvovirus B19 IgM-positive; parvovirus B19 DNA-positive.

The unequivocal interpretation of these results was that the patient was suffering from an acute infection with parvovirus B19.

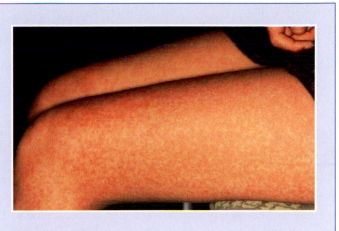

Figure 1. Leg rash.

1. What is the causative agent, how does it enter the body and how does it spread a) within the body and b) from person to person?

Causative agent

Parvovirus B19 belongs to the family of viruses known as the *Parvoviridae*. These are characterized by having a small single-stranded DNA genome. Parvovirus B19 belongs to the *Erythrovirus* genus within the subfamily *Parvovirinae* (and is therefore sometimes referred to as 'erythrovirus'). The suffix 'B19' is somewhat spurious, arising from the coding of the particular serum in which this virus was first identified, that is there are no parvovirus A or B1–18 viruses. There are, however, three recognized genotypes of B19 that differ slightly in their DNA sequence but behave similarly in terms of the clinical consequences of infection.

Entry and spread within the body

Infection is acquired via the respiratory route. About 6 days after infection, virus is found within the bloodstream, that is there is a viremic phase, which lasts about 6 days.

Spread from person to person

About 1 week after infection, at the time of the **viremia**, virus is detectable within the throat, and is shed from the respiratory tract, which represents the common route of spread. However, there is also a possibility of transmission via blood and blood products if blood donation is made at this stage of infection. Virus in the maternal bloodstream will also gain access to, and may cross, the placenta, giving rise to fetal infection *in utero*.

Epidemiology

Infection has been described worldwide. In temperate climates, while infection may occur throughout the year, it is more common in late winter/spring and early summer. Large-scale epidemics have a periodicity of about every 4–5 years. These are often based in primary schools, where over 50% of children may become infected, which also results in infection of susceptible adults such as parents and teachers.

2. What is the host response to the infection and what is the disease pathogenesis?

IgM antibodies to the virus can be detected in the bloodstream about 9 days after infection, but the IgG antibody response is not detectable for 2–3 weeks. The appearance of IgM anti-parvovirus antibodies coincides with a decline in viral titers in blood. The IgM levels peak after a few days, persist for 3 or 4 weeks, and then decline to undetectable levels 2–3 months post infection.

The pathogenesis of disease has been revealed by experiments in which volunteers were inoculated with virus in the nose. An initial fairly nonspecific illness with fever and mild upper respiratory tract symptoms occurred at around 6 days, and is due to the production of inflammatory **cytokines** and the presence of replicating virus in the respiratory tract, respectively. In some individuals there is then a second phase of illness, with a **rash** and joint pains (**arthralgia**) and even swelling (**arthritis**). These latter manifestations arise at the time the IgG antibody response becomes detectable (i.e. at 2–3 weeks post-infection), and are due to the presence of circulating **immune complexes**, which may be deposited in the skin and joints, leading to complement activation and giving rise to acute inflammatory reactions at those sites.

The small DNA genome of the parvoviruses, encoding a very small number of proteins, means that they have to rely on certain host cell processes and enzymatic functions to replicate their genome. Because the host cell has to replicate its DNA during cell division the necessary enzymes are expressed most abundantly in rapidly dividing cells. Therefore parvovirus B19 has a particular predilection for rapidly dividing cells within the bone marrow, especially the erythroid precursors. The basis for this is that the cellular receptor for the virus is the blood group P antigen (also known as globoside), found in large amounts on such cells. Examination of a bone marrow aspirate in acute parvovirus B19 infection reveals abnormal **normoblasts**, (late erythroid precursors) with characteristic intranuclear inclusion bodies, which are the site of production of new virus particles, and hence the cause of the profuse viremia. An important clinical consequence of the bone marrow site of replication of this virus is consequent suppression of normal bone marrow function, which leads to a transient drop in cells from the red cell lineage (particularly their immediate precursors, the **reticulocytes**), white cells, and platelets in the peripheral blood (Figure 2).

A fetus is also an obvious source of rapidly dividing cells. Parvovirus B19 will have access to the placenta during the viremic phase of infection, and there is therefore a risk of fetal damage arising if a pregnant woman (such as the patient in this case) acquires infection during pregnancy.

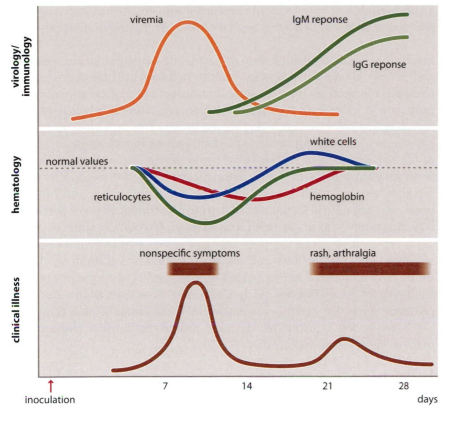

Figure 2. Events following infection with parvovirus B19. Viremia peaks about 8 days following inoculation, while the antibody response becomes detectable at about 14 days (top panel). Maximal bone marrow infection and suppression coincide with the peak viremia (middle panel), as do the nonspecific clinical manifestations of infection, such as fever and malaise. Rash and arthralgia are most likely mediated by the deposition of immune complexes in the skin and joints, which does not occur until the antibody response is generated, that is 2–3 weeks after initial infection (bottom panel).

3. What is the typical clinical presentation and what complications can occur?

Infection with parvovirus B19 can lead to a wide range of clinical manifestations, from trivial to life-threatening, depending on both host (e.g. age, degree of immunocompetence) and viral (e.g. dose) factors. As with most viruses, the most likely outcome of infection is no disease, or asymptomatic **seroconversion,** which occurs in about half of all infections in childhood. Minor illnesses include nonspecific fever and mild respiratory tract symptoms.

The most typical clinical manifestation is the development of a rash (arising from immune complex deposition, see above). In children, this is most prominent on the cheeks (Figure 3), and is hence known as 'slapped cheek syndrome,' although the more medical terminology is **erythema infectiosum,** or fifth disease. On the trunk and limbs, the rash has a more lacy, or reticular appearance. In adults, the malar prominence is not so marked, and the rash is therefore more difficult to distinguish from the many other causes of rash, particularly rubella virus infection.

In some patients, joint symptoms and signs (also arising from immune complex formation) may predominate. Arthralgia/arthritis is unusual in children, but can be very pronounced in adults, especially females. The pattern of joint involvement is reminiscent of that seem in **rheumatoid arthritis,** that is polyarticular and bilateral, mostly in the small joints of the wrist and hands. Indeed, testing of patients attending an early rheumatoid

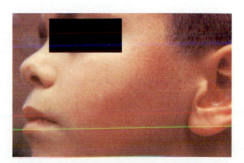

Figure 3. Slapped cheek syndrome.

arthritis clinic will reveal a small percentage who are, in fact, suffering from parvovirus arthritis. Recovery may be prolonged in some individuals.

Infection is associated with a drop in circulating hemoglobin and white cell and platelet counts. In otherwise healthy individuals, these drops are not clinically relevant, and due to their transient nature often go unnoticed. However, in individuals with chronic **hemolytic anemia**, who survive on a very low hemoglobin largely due to the presence of reticulocytes in their peripheral blood, the immediate disappearance of reticulocytes and failure of the bone marrow to replace them throws them into a condition known as an **aplastic** crisis, with insufficient oxygen transportation capacity in the blood. There are many causes of chronic hemolytic anemia, all of which predispose the individual to this life-threatening complication of parvovirus B19 infection, for example **sickle cell disease**, **hereditary spherocytosis.**

In immunocompetent hosts, parvovirus infection is self-limiting, that is, the virus is eliminated after a period of time, and IgG antibodies provide solid protection against any further infection. However, in immunocompromised hosts, for example those with HIV infection, or post-transplantation, parvovirus infection may become chronic, resulting in chronic bone marrow suppression and anemia.

Maternal infection with parvovirus B19 poses a significant risk to the pregnancy, although large-scale studies have shown that in the majority of such pregnancies, a normal infant is delivered at term. Certainly, there is no collection of fetal developmental abnormalities (as there is in rubella-affected pregnancies, the so-called 'congenital rubella syndrome'), which arise as a result of parvovirus infection in pregnancy. However, infection in the second trimester of pregnancy is associated with at least a 10-fold increased risk of spontaneous miscarriage, occurring most frequently about 4–6 weeks after the onset of maternal rash illness. Another well-recognized complication is **hydrops fetalis**. Here, fetal infection results in arrest of red blood cell formation, leading to anemia, heart failure, and consequent fluid accumulation, or **edema**. This complication has only been described in pregnancies affected before 21 weeks of pregnancy.

4. How is this disease diagnosed and what is the differential diagnosis?

Diagnosis of acute infection is made by demonstrating the presence of IgM anti-parvovirus antibodies and/or of parvovirus B19 DNA in a serum sample. The advent of extremely sensitive genome amplification techniques such as the **polymerase chain reaction (PCR)** means that as few as 10 viral particles per ml can be detected. The most common clinical situations where DNA testing is used are firstly in an acute aplastic crisis, where DNA positivity may precede the appearance of specific IgM, and secondly in an immunosuppressed patient with chronic anemia that may be due to chronic parvovirus infection. In the latter case, the patient may have been unable to generate any antibody response at all, and if the infection has become chronic, IgM antibodies would be undetectable. Low levels of parvovirus DNA can persist for many weeks or even months after acute infection, and thus the detection of parvovirus DNA alone is not used for the diagnosis of acute infection in an immunocompetent individual.

Differential diagnosis

There are a large number of causes of a red rash, especially in children, and diagnostic tests are not always performed in this setting. However, in a pregnant woman with a rash, the most important differential diagnosis is acute rubella virus infection. The distinction is made on serological grounds.

Parvovirus arthritis can be confused with rheumatoid arthritis. Again, the distinction can be made by demonstrating the presence of specific antiviral IgM antibodies.

5. How is the disease managed and prevented?

Management

The vast majority of acute parvovirus B19 infections are self-limiting and require no specific therapy. Indeed, there are no effective antiviral agents with activity against parvovirus B19. In pregnancy, it is important to make the diagnosis (and particularly to prove that the infection is not due to rubella virus), as this will allow appropriate counseling of the patient. Parvovirus arthritis may require administration of analgesics and anti-inflammatory agents. Patients with a parvovirus-induced aplastic crisis may require blood transfusion to get them through the acute phase. A similar rationale exists for giving intra-uterine transfusion to babies of infected mothers who present with hydrops fetalis. Some success in the treatment of chronic parvovirus anemia in immunosuppressed subjects has been reported through the use of normal human immunoglobulin, which is a source of potent neutralizing antibodies.

Prevention

There is, as yet, no vaccine available for the prevention of parvovirus B19 infection, although there are reports of experimental vaccines undergoing trials. In theory, susceptible close contacts of parvovirus-infected individuals could be protected by passive immunization with normal human immunoglobulin, and this is certainly worth recommending for contacts who have an underlying chronic hemolytic anemia, or are heavily immunosuppressed. As many of the causes of chronic hemolysis are inherited, this may involve giving prophylaxis to whole families.

SUMMARY

1. What is the causative agent, how does it enter the body and how does it spread a) within the body and b) from person to person?

- Parvovirus B19, also known as erythrovirus, has a single-stranded DNA genome, is nonenveloped, and belongs to the family *Parvoviridae*.

- Infection is acquired by the respiratory route, and virus is subsequently shed from the throat.

- A viremic phase occurs about 6 days post-infection, thus there is the potential for transmission via blood/blood product transfusion, and for transplacental transmission to the fetus from a pregnant mother.

- Infection is described worldwide. Epidemics occur with a periodicity of 4–5 years.

2. What is the host response to the infection and what is the disease pathogenesis?

- IgM antibody responses are detectable about 9 days post-infection, coinciding with a decline in viremia. IgG antibodies may take up to 3 weeks to appear.

- Illness associated with infection is biphasic. The viremic phase may present with fever (due to cytokine release), and mild upper respiratory tract symptoms (due to local inflammatory response).

- After 2–3 weeks, patients may present with skin rash and joint pain and swelling. These manifestations are thought to be mediated by immune complex formation and deposition in skin and joints.

- The cellular receptor for parvovirus B19 is the P blood group antigen, or globoside.

- Most viral replication takes place in the bone marrow, particularly in erythroid precursor cells. Infection is therefore associated with a drop in all cellular elements of the peripheral blood (red cells, white cells, platelets).

3. What is the typical clinical presentation and what complications can occur?

- Most infections are asymptomatic.

- Clinical disease usually presents with a rash. In children, parvovirus causes erythema infectiosum, also known as fifth disease. The rash is most prominent on the cheeks, and hence another name is slapped cheek syndrome. In adults, the rash is more nonspecific, with a lacy, reticular appearance.

- Complications include arthralgia and arthritis. This is more common in women, can be polyarticular, and can take several months to resolve.

- Infection of patients with chronic hemolytic anemia may result in a life-threatening acute aplastic crisis.

- Immunocompromised hosts may fail to eliminate infection, resulting in chronic bone marrow suppression, particularly anemia.

- Infection in pregnancy is associated with an increased risk of spontaneous miscarriage, and fetal anemia leading to fetal hydrops. There is no congenital parvovirus syndrome.

4. How is this disease diagnosed and what is the differential diagnosis?

- Diagnosis is by demonstration of IgM anti-parvovirus antibodies.

- In addition, parvovirus DNA may be detected in blood or tissue (e.g. aborted fetus) by genome amplification.

- In a pregnant woman with a rash, the most important differential diagnosis is acute rubella, which is clinically indistinguishable but has dramatically different consequences for the fetus.

- Parvovirus arthritis may mimic rheumatoid arthritis.

5. How is the disease managed and prevented?

- There are no specific antiviral drugs for the treatment of parvovirus B19 infection.

- Arthritis is managed with appropriate analgesic and anti-inflammatory drugs.

- Acute aplastic crisis may necessitate blood transfusion.

- Intra-uterine blood transfusion may also be life-saving for fetal hydrops.

- Normal human immunoglobulin (containing anti-parvovirus antibodies) may be helpful in the treatment of chronic parvovirus infection in the immunosuppressed.

- There is currently no vaccine available to prevent parvovirus infection.

FURTHER READING

Humphreys H, Irving WL. Problem-orientated Clinical Microbiology and Infection, 2nd edition. Oxford University Press, Oxford, 2004.

Richman DD, Whitley RJ, Hayden FG. Clinical Virology, 2nd edition. ASM Press, Washington, DC, 2002.

Zuckerman AJ, Banatvala JE, Pattison JR, Griffiths PD, Shaub BD. Principles and Practice of Clinical Virology, 5th edition. Wiley, Chicester,2004.

REFERENCES

Broliden K, Tolfvenstam T, Norbeck O. Clinical aspects of parvovirus B19 infection. J Intern Med, 2006, 260: 285–304. (Review.)

Crowcroft NS, Roth CE, Cohen BJ, Miller E. Guidance for control of parvovirus B19 infection in healthcare settings and the community. J Publ Health Med, 1999, 21: 439–446.

de Jong EP, de Haan TR, Kroes AC, Beersma MF, Oepkes D, Walther FJ. Parvovirus B19 infection in pregnancy. J Clin Virol, 2006, 36:1–7. (Review.) Erratum in: J Clin Virol, 2007, 38: 188.

Servey JT, Reamy BV, Hodge J. Clinical presentations of parvovirus B19 infection. Am Fam Physician, 2007, 75: 373–376. (Review.)

Young NS, Brown KE. Parvovirus B19. N Engl J Med, 2004, 350: 586–597.

WEB SITES

All the Virology on the WWW Website, developed and maintained by Dr David Sander, Tulane University: http://www.virology.net/garryfavweb13.html#Parvo

Centers for Disease Control and Prevention, Atlanta, GA, USA: http://www.cdc.gov/ncidod/dvrd/revb/respiratory/parvo_b19.htm

Health Protection Agency UK: http://www.hpa.org.uk/infections/topics_az/parvovirus/menu.htm

Website of Derek Wong, a medical virologist working in Hong Kong: http://virology-online.com/viruses/Parvoviruses.htm

MULTIPLE CHOICE QUESTIONS

The questions should be answered either by selecting True (T) or False (F) for each answer statement, or by selecting the answer statements which best answer the question. Answers can be found in the back of the book.

1. Which of the following statements regarding parvovirus B19 are true?
   A. It contains a double-stranded DNA genome.
   B. The cellular receptor for binding is the blood group B antigen.
   C. It causes annual epidemics in temperate climates.
   D. It preferentially infects erythroid precursor cells in the bone marrow.
   E. It belongs to the same virus family as rubella virus.

2. Which of the following are recognized routes of spread of parvovirus B19 infection?
   A. Blood transfusion.
   B. Contaminated water.
   C. Droplet spread from respiratory tract secretions.
   D. Mother to fetus.
   E. Sexual transmission.

3. Which of the following statements concerning parvovirus B19 infection are true?
   A. Anemia arises because of infection of erythroid precursor cells.
   B. Symptoms of fever and headache arise as a result of immune complex formation.

MULTIPLE CHOICE QUESTIONS (continued)

C. Symptoms of rash and arthralgia occur about 6–9 days after infection.

D. Symptoms are usually less severe than those caused by parvovirus B3 infection.

E. Viral DNA is detectable in the bloodstream before any antibody response is detectable.

4. **Which of the following are recognized manifestations of parvovirus B19 infection?**

   A. The childhood rash known as exanthem subitum (also known as 6th disease).

   B. Polyarticular arthritis.

   C. Meningitis.

   D. Respiratory tract illness.

   E. Pancreatitis.

5. **Which of the following serological markers indicates recent infection with parvovirus B19 virus?**

   A. The presence of parvovirus-specific IgA in a serum sample.

   B. The presence of parvovirus-specific IgD in a serum sample.

   C. The presence of parvovirus-specific IgE in a serum sample.

   D. The presence of parvovirus-specific IgG in a serum sample.

   E. The presence of parvovirus-specific IgM in a serum sample.

6. **In which of the following patients can parvovirus B19 infection be life-threatening?**

   A. Patients with sickle cell anemia.

   B. Hemophiliacs.

   C. Fetuses.

   D. Patients with HIV infection.

   E. Patients with rheumatoid arthritis.

7. **A 25-year-old woman presents with a diffuse morbilliform rash and a small joint polyarthropathy. She is 18 weeks pregnant. Serology reveals the following results:**

   • IgG anti-rubella positive, IgM anti-rubella negative

   • IgG anti-parvovirus positive, IgM anti-parvovirus positive.

   Which of the following adverse events may arise from this in her pregnancy?

   A. Congenital cataracts.

B. Congenital heart disease.

C. Congenital deafness.

D. Hydrops fetalis.

E. Spontaneous miscarriage.

8. **A 25-year-old woman presents with a diffuse morbilliform rash and a small joint polyarthropathy. Her last menstrual period was 8 weeks ago, and a pregnancy test is positive. Serology reveals the following results:**

   • IgG anti-rubella positive, IgM anti-rubella positive

   • IgG anti-parvovirus positive, IgM anti-parvovirus negative.

   Which of the following is the most likely outcome of this pregnancy?

   A. Spontaneous miscarriage.

   B. Hydrops fetalis due to fetal anemia.

   C. A normal healthy baby.

   D. A baby with multiple congenital abnormalities.

   E. A baby with congenital deafness, otherwise normal.

9. **A 25-year-old woman presents with a diffuse morbilliform rash and a small joint polyarthropathy. She is 10 weeks pregnant. Serology reveals the following results:**

   • IgG anti-rubella positive, IgM anti-rubella negative

   • IgG anti-parvovirus positive, IgM anti-parvovirus positive.

   Which of the following is the most likely outcome of this pregnancy?

   A. A normal healthy baby.

   B. A baby with multiple congenital abnormalities.

   C. A baby with congenital deafness, otherwise normal.

   D. A baby with congenital anemia.

   E. A baby with congenital cataracts.

10. **Which of the following drugs/modalities of therapy have been proven useful in the management of patients with parvovirus B19 infection?**

   A. Aciclovir.

   B. Blood transfusion.

   C. Normal human immunoglobulin.

   D. Monoclonal anti-parvovirus antibodies.

   E. Zanamavir.

Case 28

Plasmodium spp.

A 26-year-old model went to see her doctor about 1 week after returning from a job in the Gambia. She complained of an abrupt onset of bouts of shivering and feeling cold, vomiting, rigors, and profuse sweating accompanied by a headache and nausea. On examination she was noted to be pale with a temperature of 39.5°C and had tachycardia. She gave a history of having taken anti-malarial tablets before and during her stay in the Gambia but was admitted to hospital with a provisional diagnosis of malaria.

1. What is the causative agent, how does it enter the body and how does it spread a) within the body and b) from person to person?

Causative agent

The organism causing malaria is *Plasmodium*, a eukaryotic protozoan that infects the erythrocytes of humans. It has the characteristics of eukaryotes, with a nucleus, mitochondria, endoplasmic reticulum, and so forth. Until recently four species of *Plasmodium* were identified as being able to infect humans: *P. falciparum*, *P. ovale*, *P. vivax*, and *P. malariae*. A simian plasmodium, *P. knowlesi*, has been recently proven infective to humans. *P. falciparum* is the most **virulent** species of malaria in humans. All these species have similar life cycles in which the organisms undergo both sexual and asexual reproduction in the vector and host and alternate between intracellular and extracellular forms. The female *Anopheles* mosquito is the vector for malaria. The risk of malaria transmission is therefore restricted to those areas where mosquitoes can breed and where the parasite can develop within the mosquito. The maximum extent of malaria risk is between approximately 60°N and 30°S (except areas higher than around 2500 meters), although this distribution has been reduced dramatically and is currently restricted mainly to the tropics and subtropics – see Epidemiology below.

Entry and spread within the body

The transmission stage of *Plasmodium* is the sporozoite, which is injected into the bloodstream of a human when the female *Anopheles* mosquito takes a blood meal (Figure 1). The detailed life cycle is shown in Figure 2. Following the mosquito bite, at least some of the sporozoites remain in the dermis for some time before entering the bloodstream and some pass into draining lymph nodes. Only a few dozen sporozoites are transmitted during feeding but there is rapid translocation into the liver to begin the first stage of disease.

Figure 1. *Anopheles funestus* mosquito taking a blood meal from its human host. This mosquito species, together with *Anopheles gambiae*, is one of the two most important malaria vectors in Africa. Note the blood passing through the proboscis.

Liver stage (pre-erythrocytic stage)

The blood-borne sporozoites localize in the liver via the sinusoids, where through their surface circumsporoite protein (CSP) they attach to the highly sulfated heparan sulfate proteoglycans (HSPGs) on the surface of the of hepatocytes. Other membrane molecules are important for this binding. The sporozoites actively enter the hepatocytes, (invade – rather than are taken up passively by **endocytosis**) and here they increase in number and develop into schizonts. *P. vivax* and *P. ovale* also produce a

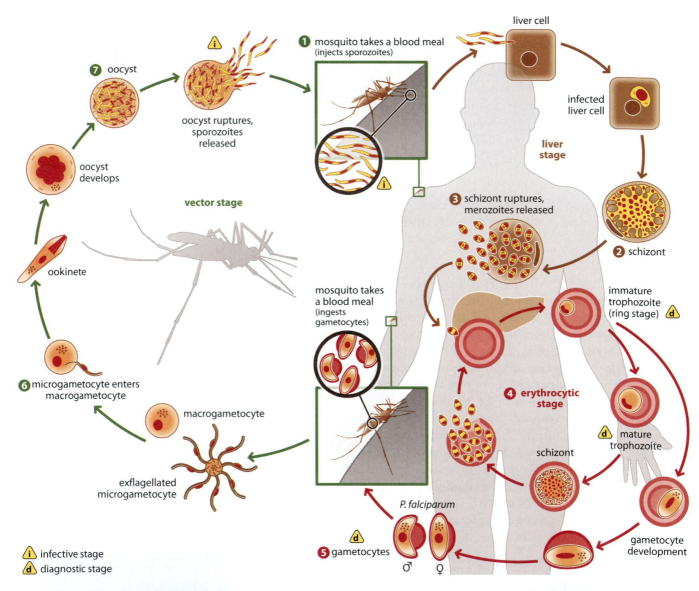

Figure 2. The life cycle of *Plasmodium*. (1) The mosquito injects saliva containing sporozoites as it takes a blood meal and the parasite localizes in the liver (liver stage), where it undergoes a stage of development to produce a schizont in the infected liver cell, which contains merozoites (2). *P. vivax* and *P. ovale* also produce a resting stage within the liver cell called hypnozoites, which can persist in the liver and result in relapses months or even years later. The dead liver cell breaks open and the schizont ruptures (3) releasing merozoites into the bloodstream. These invade erythrocytes (erythrocytic stage) and undergo developmental stages as trophozoites, which mature and produce schizonts, at which stage the erythrocyte bursts rupturing the shizonts to release further merozoites (4). Further cycles of asexual development within uninfected erythrocytes occur, releasing more merozoites to infect further erythrocytes. Differentiation of the immature trophozoite into male and female gametocytes occurs in some erythrocytes (5) and these are ingested when a mosquito takes a blood meal. The male (microgametocyte – exflagellated) fertilizes the female macrogametocyte (6) to form a zygote within the intestine of the mosquito (vector stage) and this becomes an ookinete that invades the intestinal wall where it develops into an oocyte (7). The oocyte matures into sporozoites, which are released and migrate to the salivary gland of the mosquito. Here they will be transmitted to a new human host when the mosquito takes a blood meal and the cycle starts again (1).

resting stage within the liver cell called hypnozoites, which are responsible for the relapses that occur with these forms of malaria (see later). This asexual stage takes up to 2 weeks. Rupture of the liver cells releases the schizonts into the bloodstream as merozoites (with about 10–40 000 being released from the liver).

Erythrocytic stage

These invade and destroy erythrocytes giving rise to symptoms (see complications later). The entry of the merozoites into erythrocytes is achieved through attachment of a number of surface molecules (merozoite surface proteins, MSPs) to structures on the erythrocyte, for example band 3 protein for *P. falciparum*. *P. vivax* has a specific reticular binding protein to enable it to attach and invade **reticulocytes**. In addition, *P. vivax* has surface molecules (**Duffy binding proteins** – DBPs) that bind to Duffy blood group antigens on the erythrocytes. The lack of this antigen in some human populations, mostly West Africans, explains their resistance to *P. vivax*.

Within the erythrocyte, the merozoites undergo further development as a trophozoite (seen as a 'ring' stage – see Figure 2) and then undergo asexual reproduction to produce schizonts, at which stage the erythrocyte bursts releasing merosomes containing 16–32 daughter merozoites into the bloodstream.

Each asexual cycle takes 44–48 hours, and is followed by cell rupture and re-invasion steps that induce periodic waves of fever in the patient (see Figure 3 and Section 3). This erythrocytic cycle may continue for months or years. However, in some erythrocytes the trophozoites differentiate into male and female gametocytes and a mosquito taking a blood meal will take up some of gametocyte-containing erythrocytes, heralding the sexual developmental phase in the vector.

Vector stage

Within the intestine of the insect, male (exflagellated microgametocytes) and female gametes (macrogametocytes) fuse to become a zygote. These become an ookinete, which then invades the intestinal wall where it develops into an oocyte. The oocyte develops into thousands of sporozoites, which then migrate to the mosquito's salivary gland.

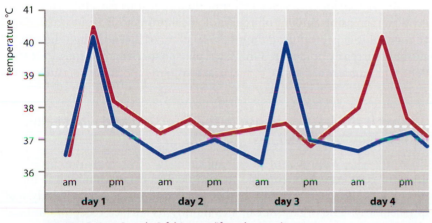

Figure 3. Cyclical fever coincident with the release of merozoites.

Person to person spread

The sporozoites are injected into an individual when an infected female *Anopheles* mosquito feeds and the whole cycle starts again.

Non-mosquito spread

Malaria can also be transmitted through blood transfusion, hypodermic needle sharing or accidents, and from mother to fetus.

Epidemiology

According to the World Health Organization (WHO), malaria is worsening or barely contained in many parts of the world. Global incidence is estimated to be 350–500 million cases of clinical malaria each year, with 300 million carriers of the parasite. Estimates made independently by others using a combination of epidemiologic, geographic, and demographic data have put the overall clinical episodes of malaria at up to 50% higher and 200% higher for areas outside Africa. The higher values are believed to reflect the WHO's reliance on passive national reporting for these countries.

Most malaria infections and deaths occur in sub-Saharan Africa, where it is estimated to account for 80% of all clinical cases and about 90% of all people that carry the parasite. Malaria deaths have been estimated at 800 000 per year in children and a child dies every 30 seconds! Asia, Latin America, the Middle East, and parts of Europe are also affected. With increasing international travel, there continues to be a rise in the number of cases of malaria in travelers returning to nonmalarious areas from countries where malaria is endemic. During the last decade, there has been an average of 1843 cases of malaria in Great Britain each year. The global incidence of *P. falciparum* is shown in Figure 4. Pregnancy has a high risk of malaria. An estimated 10 000 pregnant women and 200 000 of their infants die annually in sub-Saharan Africa as a result of malaria infection during pregnancy. HIV-infected pregnant women are at increased risk.

Human genetic factors that decrease the infection rates of Plasmodium

As already mentioned, the absence of DBPs in most West Africans prevents infection by *P. vivax*, since it uses the Duffy blood group antigen as a means of attachment. Sickle cell trait (heterozygous for HbS with HbA) gives an increasing amount of immunological protection against malaria as young children grow during their first 10 years of life, although the mechanism is currently unknown. Glucose 6 phosphate dehydrogenase (G6PD) deficiency confers resistance to malaria; again, the mechanism is unknown.

2. What is the host response to the infection and what is the disease pathogenesis?

Plasmodium has a number of 'escape mechanisms' that allow it to avoid the immune response. For example, it has numerous morphological forms through its life cycle (Section 1) that are found both extracellularly and intracellularly. The organism can also modify its surface receptors to bind to hepatocytes (sporozoites) and erythrocytes (merozoites). In addition,

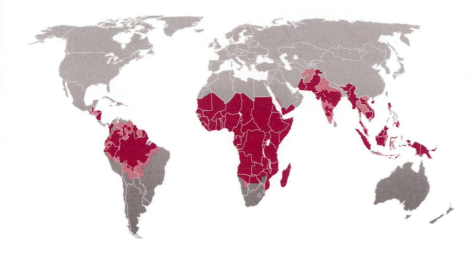

Figure 4. Global incidence of P. falciparum. Dark red areas: ≥0.1 per thousand of the population; pink areas: <0.1 per thousand of the population; gray areas: no risk. The borders of 87 countries where *P. falciparum* is endemic are show. From Guerra et al. 2008, see references.

there is evidence for a direct immunosuppressive effect of sporozoites on **dendritic cell** function and prevention of oxygen burst in **Kupffer cells** of the liver.

The understanding of the immune response to *Plasmodium* is far from complete but there is some information to date. Some immunity does develop to the parasite with continuous infection (especially to blood stage parasites) and children who survive early attacks become resistant to severe disease by about 5 years. Levels of parasites fall progressively until adulthood, when they are low or absent most of the time. This immunity, however, appears to be lost after spending a year away from exposure, presumably due to lack of repeated antigenic stimulation needed for its maintenance. **IgG** antibodies to blood stage parasites are transferred across the placenta and limit parasitemia and severe disease in the neonates. Thus, although immune responses do develop to different stages of the life cycle in infected individuals they are weak and cannot eradicate the parasite.

Principal mechanisms that are thought to be responsible for immunity at different stages of the life cycle of Plasmodium

- Initial entry of the sporozoites into the lymph nodes and bloodstream: antibodies.

- Intra-hepatocyte: there is evidence that both **CD4+** T cells and cytotoxic **CD8+** T cells might be important at this stage, perhaps through direct cytotoxicity but also through production of **tumor necrosis factor-α (TNF-α)**, **interferon-γ (IFN-α)**, and **interleukin (IL)**-1.

- Merozoites in the bloodstream: antibodies.

- Intra-erythrocyte stage: antibodies, **reactive oxygen intermediates**, reactive nitrogen intermediates, eosinophil chemotactic factor, and TNF.

- Male and female gametocytes in the bloodstream: antibodies.

Some innate immunity is also believed to be induced in the mosquito to the gametes and sporozoites of *Plasmodium*.

Understanding the exact mechanisms of immunity against *Plasmodium* will aid in the successful development of vaccines (see below).

Pathogenesis

There are three major pathological mechanisms that account for the pathology and clinical signs and symptoms in patients with malaria (see below). These are as follows.

- The response of the **mononuclear phagocytic system** to the parasite and the **hemozoin** pigment that accumulates in infected erythrocytes causes the release of **cytokines** from the host cells. The cytokines, such as TNF, act as endogenous pyrogens.

- Destruction by infection and defective production of red blood cells probably induced by cytokines, leads to **anemia**. Modification of the erythrocyte membrane makes it less deformable and the cells are removed by the spleen. This leads to **splenomegaly**. Additionally antibody-mediated **lysis** of the erythrocytes occurs as parasite antigens are expressed on the surface of the erythrocytes.

- Obstruction of capillaries caused by parasitized red blood cells. This is also caused by modification of the erythrocyte membrane as the surface molecules bind to adhesion molecules on the endothelium of the capillaries, which is particularly important in the pathogenesis of severe *P. falciparum* malaria (for example, this mechanism leads to development of cerebral malaria).

3. What is the typical clinical presentation and what complications can occur?

In uncomplicated malaria the first symptoms are fever, headache, chills, and vomiting that appear 10–15 days after a person is infected. The pre-erythrocytic stage in the liver is asymptomatic. If not treated promptly with effective medicines, malaria can cause severe illness that is often fatal. Patients may also have splenomegaly due to the sequestering of infected erythrocytes.

With time the fevers take on a periodicity depending on the species of malarial parasite.

Cyclical temperature fluctuations coincide with the rupture of erythrocytes and the release of merozoites into the bloodstream, which during the erythrocytic stage occurs every 48 hours for *P. falciparum*, *P. ovale*, and *P. vivax*, and every 72 hours for *P. malariae*. (see Figure 3).

Complications

In the most severe form of malaria (*P. falciparum*) there can be cerebral complications – **thrombosis** may occur due to occlusion of the cerebral vessels caused by the increased stickiness of the erythrocytes (see above). Multi-organ damage and renal failure can also result from the infection. The latter is uncommon in children with severe malaria who present with prostration, respiratory distress, severe anemia, and/or cerebral malaria.

Blackwater fever can also occur due to intravascular **hemolysis** leading to **hemoglobinuria** and kidney failure. *P. falciparum* also causes pulmonary **edema** and *P. malariae* causes **nephrotic syndrome**. In patients with splenomegaly the spleen may rupture due to its large size.

The pattern of severe malaria is not really understood, although genetic factors, age, and intensity of transmission determine the development of susceptibility to severe anemia and cerebral malaria. Other infectious diseases can interact with malaria and modify susceptibility and/or severity of either disease. HIV infection with malaria increases the risk of both uncomplicated and severe malaria, and *Plasmodium* causes a transient increase in viral load, which might promote transmission of the virus.

Plasmodium and Epstein-Barr virus are concurrent risk factors for **Burkitt's lymphoma** in Africa.

4. How is this disease diagnosed and what is the differential diagnosis?

Clinical diagnosis of malaria can be quite difficult because of the overlap of symptoms with other infectious diseases.

The mainstay of diagnosis is by a blood film. Thick or thin films are prepared from finger-prick blood. These are stained with Giemsa or Fields stain and viewed under the microscope. The degree of parasitemia is noted and the morphological appearance of the trophozoite (or gametocyte) can be used to identify the species of malaria (see diagnostic stages in Figure 2).

- *P. falciparum*: ring forms seen often resemble stereo headphones, may be seen at the edge of the cell. The red cell is of normal size (Figure 5).
- *P. vivax*: irregular large thick ring. The red cells are enlarged.
- *P. ovale*: regular dense ring. The red cell is oval.
- *P. malariae*: dense thick ring or band form. The red cell is normal in size.

Blood films may also be stained with acridine orange to identify the species.

Malaria can also be diagnosed serologically by the detection of antibodies or antigen, the latter being used for rapid identification of acute cases. A number of commercial kits are available for rapid diagnosis of *P. falciparum*. Although many are highly sensitive, some are not as sensitive as a blood film. They detect *P. falciparum*, histine-rich protein (HRP), for example ParaSight F® kits, or lactate dehydrogenase (LDH), for example Kat-Quick® kits).

Diagnosis can also be made by **polymerase chain reaction (PCR)** or DNA probes.

Differential diagnosis

Since clinical diagnosis is difficult the ultimate diagnostic test is the blood film as described above.

5. How is the disease managed and prevented?

Early diagnosis and prompt treatment are the basic elements of malaria control and this is crucial to prevent the development of complications and the majority of deaths from malaria.

The two main approaches to malaria control are:

- drug treatment of the patients with infection and prevention;
- control of the vector.

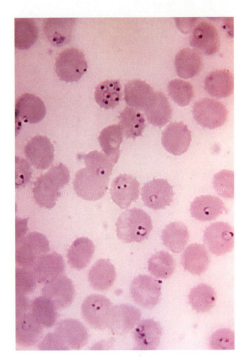

Figure 5. Blood film showing the presence of *P. falciparum* rings in human erythrocytes (×1125). Note that some red blood cells contain multiple parasites, which is more common with *P. falciparum* than other *Plasmodium* spp.

Management

Antimalarial treatment policies vary between countries depending on epidemiology of the disease, transmission, patterns of drug resistance, and political and economic contexts.

Drugs currently used mainly target the erythrocyte stage of the parasite. Chloroquine and quinine have historically been the first-line drugs for treating malaria and are still used for *P. ovale*, *P. vivax*, and *P. malariae*. *P. falciparum* has become increasingly resistant to chloroquine. In areas where resistance developed sulfadoxine-pyrimethamine combinations of drugs have been used but there is now significant resistance to these antifolates. Quinine has some effects on *P. falciparum* and is still used for many cases of severe malaria. Drugs currently used against *P. falciparum* (as well as the ACTs – see below) are: mefloquine, atavaquone/proguanil, and in some cases tetracycline antibiotics.

One drug that targets the pre-erythrocytic phase of the disease is primaquine and this is used after chloroquine to particularly kill the hypnozoites of *P. vivax* and *P. ovale* in the liver.

Resistance of *P. falciparum* (and *P. vivax* in some areas) to chloroquine and sulfadoxine-pyrimethamine has led to the development of new drugs with an emphasis on the artemisinin-based combination therapies (ACTs). Artemisinin was first isolated and developed in China in the 1980s from the plant *Artemisia annua*. Artemisinin derivatives include artesunate, artemether, and dihydro-artemisinin. ACTs combine a derivative of artemisinin, a fast-acting antimalarial endoperoxide, with a longer-lasting partner drug that continues to reduce the parasite numbers after the artemisinin has dropped below therapeutic levels. ACTs act not only on the asexual blood stages to alleviate symptoms but also on the gametes, therefore reducing the spread of the disease.

Following advice from the WHO, most countries in the world now use ACTs as the first line of treatment against malaria. Due to the past history of the ability of the parasite to develop drug resistance, the WHO has asked for the use of oral artemisinin monotherapies to be banned at various levels, including manufacturers, international drug suppliers, national health authorities, and international aid and funding agencies involved in the funding of essential antimalarial medicines.

Sequencing of the parasite genome has aided and will continue to aid in the discovery of new drugs to treat malaria.

Prevention

Indoor residual spraying (IRS) of long-acting insecticide and use of long-lasting insecticidal nets (LLINs) are now the main measures in many parts of the world for controlling the vector. DDT is still used as a long-acting insecticide in those areas where the mosquitoes have not become resistant to it. Unfortunately insecticide resistance is becoming more and more of a problem and the development of new insecticides with active compounds that target different proteins in the insects than DDT is a priority.

Infection is also prevented by controlling the breeding cycle of the vector, which should significantly reduce the number of cases and rate of parasite infection. Larval control is important by limiting breeding sites. Biological control using *Bacillus thuringiensis* toxin is widely used as a larvicide for mosquito larvae. Methoprene kills the larvae, and also the introduction of fish which eat the larvae, into the breeding grounds, is important. These approaches are generally considered to be environmentally friendly methods of mosquito control.

Vaccines

A successful vaccine is not available but several are currently in clinical trials.

Strategies for producing vaccines include using sporozoite antigens, which are expressed in the surface of the infected hepatocytes to enhance cytotoxic T-cell responses against the liver stage. Other strategies target merozoite surface molecules used to attach to and enter erythrocytes. These would not prevent infection but would prevent the symptoms seen during the erythrocytic stage of the disease. It is likely that more than one stage, including the gametes, will have to be targeted if eradication of the parasite is ever to be achieved.

Travelers to malaria endemic areas

With travel to malaria endemic areas now very common a number of drugs are given to prevent infection with the parasite, as well as steps taken to avoid mosquito bites. The drugs that are used in prophylaxis may be the same as those used for clinical cure except in lower doses.

The patient described in this case had been taking anti-malarial tablets as a preventive measure against getting malaria. Possibilities as to why she still got malaria are: a) the patient was taking homeopathic tablets; b) the patient did not take the drugs on a regular basis, or c) the particular malaria species that the patient was infected with was resistant to the prophylactic treatment.

Information as to the appropriate drugs for the area to be visited is available at several websites, for example http://whqlibdoc.who.int/publications/ 2005/9241580364_chap7.pdf

SUMMARY

1. What is the causative agent, how does it enter the body and how does it spread a) within the body and b) from person to person?

- The organism causing malaria is *Plasmodium*, a protozoan with a complex life cycle.

- Four species of *Plasmodium* can infect humans: *P. ovale*; *P. vivax*, and *P. malariae*, with *P. falciparum* being the most virulent species of malaria. A fifth species, *P. knowlesi* of simian origin, has also been shown to infect humans.

- The female *Anopheles* mosquito is the vector for malaria.

- The sporozoite is transmitted from the mosquito into the blood during a blood meal and localizes in the liver. Schizonts are produced through asexual reproduction.

- Liver schizonts rupture and release merozoites into the bloodstream, which invade and destroy erythrocytes giving rise to symptoms.

- Within the erythrocyte the merozoite undergoes another round of schizogony to produce trophozoites.

- Differentiation of the trophozoite into gametocytes occurs in some erythrocytes.

- When another mosquito feeds it takes in the gametocytes, where they fuse to form a zygote, which becomes an oocyte.

- The oocyte matures into sporozoites, which migrate to the salivary gland of the mosquito and are inoculated into another human when the mosquito feeds.

- The global incidence is estimated to be 350–500 million cases of clinical malaria each year, with 300 million carriers of the parasite. This is probably an underestimate.

- Most malaria infections and deaths occur in sub-Saharan Africa where it is estimated to account for 80% of all clinical cases and about 90% of all people that carry the parasite.

- Asia, Latin America, the Middle East, and parts of Europe are also affected.

- With increasing international travel, there continues to be a rise in the number of cases of malaria in travelers returning to nonmalarious areas from countries where malaria is endemic.

- Pregnancy has a high risk of malaria.

- Absence of Duffy blood group antigen in West Africans, sickle cell trait, and glucose 6 phosphate dehydrogenase (G6PD) deficiency are examples of genetic factors leading to reduced infection rates for malaria.

2. What is the host response to the infection and what is the disease pathogenesis?

- A number of 'escape mechanisms' allow it to avoid the immune response. These include different life cycle forms and direct immunosuppression by the parasite.

- Some immunity does develop to the parasite with continuous infection but this is weak and does not eliminate the parasite.

- Immune mechanisms that exist are different against the different stages of the life cycle including: antibodies against the sporozoites, cytotoxic T cells against the hepatocyte stage, antibodies against the merozoites and gametes.

- Pathogenesis occurs due to: (a) cytokine release leading to spikes of fever at regular intervals, (b) destruction of erythrocytes leading to anemia, and (c) obstruction of capillaries due to binding of erythrocytes through parasite encoded surface molecules to endothelium.

3. What is the typical clinical presentation and what complications can occur?

- In uncomplicated malaria the first symptoms are fever, headache, chills, and vomiting that appear 10–15 days after a person is infected.

- Patients can develop splenomegaly.

- Patients can have cerebral complications, severe anemia, multi-organ damage, and renal failure.

- Cyclical temperature fluctuations coincide with the rupture of erythrocytes and the release of merozoites into the bloodstream, which during the erythrocytic stage occurs every 48 hours for *P. falciparum*, *P. ovale*, and *P. vivax*, and every 72 hours for *P. malariae*. This gives rise to fevers of differing periodicity.

- Complications include cerebral malaria, where

thrombosis may occur (with *P. falciparum*, severe anemia and blackwater fever).

- HIV with malaria increases the risk of both uncomplicated and severe malaria, and *Plasmodium* causes a transient increase in viral load, which might promote transmission of the virus.

- *Plasmodium* and Epstein-Barr virus are concurrent risk factors for Burkitt's lymphoma in Africa.

4. How is this disease diagnosed, and what is the differential diagnosis?

- The mainstay of diagnosis is by a blood film.

- The morphologic appearance of the trophozoite or gametocyte can be used to identify the species.

- Malaria can also be diagnosed serologically using the detection of antibodies or antigens.

- Diagnosis can also be made by PCR or DNA probe.

- Clinical diagnosis of malaria can be quite difficult because of the overlap of symptoms with other infectious diseases.

5. How is the disease managed and prevented?

- The two main approaches to malaria control are: drug treatment of the patients with infection and prevention and control of the vector.

- Most drug treatments target the erythrocytic stage of the infection.

- Chloroquine and quinine have been the first-line drugs for treating malaria and are still in use for *P. ovale*, *P. vivax*, and *P. malariae* but *P. falciparum* has become increasingly resistant to chloroquine.

- Drugs currently used against *P. falciparum* in addition to the new artemisinin-based combination therapies (ACTs) are mefloquine, atavaquone/proguanil, and in some cases tetracycline antibiotics.

- Primaquine targets the pre-erythrocytic stage of the disease and is used to kill the hypnozoites of *P. ovale* and *P. vivax* in liver cells.

- ACTs act not only on the asexual blood stages to alleviate symptoms but also on the gametes, therefore reducing the spread of the disease.

- Sequencing of the parasite genome has aided and will continue to aid in the discovery of new drugs to treat malaria.

- Prevention is through controlling the vector using indoor residual spraying (IRS) of long-acting insecticide and long-lasting insecticidal nets (LLINs). DDT is still used as a long-acting insecticide in those areas where the mosquitoes have not become resistant to it. Mosquito larval control is also used.

- A successful vaccine is not currently available but many are being developed targeting the liver stage, erythrocytic stage, and sexual stage (gametes).

- Prophylactic drugs for international travelers depend on drug resistance in the endemic areas to be visited.

FURTHER READING

Goering RV, Dockrell HM, Zuckerman M, et al. Mims Medical Microbiology, 4th edition. Mosby, Elsevier, 2008.

Lydyard P, Lakhani S, Dogan A, et al. Pathology Integrated: An A–Z of Disease and its Pathogenesis. Edward Arnold, 2000: 254–256.

Male D, Brostoff J, Roth DB, Roitt I. Immunology, 7th edition. Mosby, Elsevier, 2007.

Murphy K, Travers P, Walport M. Janeway's Immnunobiology, 7th edition. Garland Science, New York, 2008.

REFERENCES

Chakravarty S, Cockburn IA, Kuk S, Overstreet MG, Sacc JB, Zavala F. CD8+ T lymphocytes protective against malaria liver stages are primed in skin-draining lymph nodes. Nat Med, 2007, 13: 1035–1041.

Good MF, Doolan DL. Malaria's journey through the lymph node. Nat Med, 2007, 13: 1023–1024.

Greenwood BM, Fidock DA, Kyle DE, et al. Malaria: progress, perils and prospects for eradication. J Clin Invest, 2008, 118: 1266–1276.

Guerra CA, Gikandi PW, Tatem AJ, et al. The limits and intensity of *Plasmodium falciparum* transmission: implications for malaria control and elimination worldwide. PLoS Med, 2008, 5: e38.

Hale V, Keasling JD, Renninger N, Diagana TT. Microbially derived artemisinin: a biotechnology solution to the global problem of access to affordable antimalarial drugs. Am J Trop Med Hyg, 2007, 77(6 Suppl): 198–202.

Snow RW, Guerra CA, Noor AM, Myint HY, Hay SI. The global distribution of clinical episodes of *Plasmodium falciparum* malaria. Nature, 2005, 434: 214–217.

WEB SITES

Centers for Disease Control and Prevention, National Center for Zoonotic, Vector-Borne, and Enteric Diseases, Atlanta, GA, USA: http://www.cdc.gov/malaria/history/index.htm

Malaria Atlas Project (MAP). MAP is a joint project between the Malaria Public Health & Epidemiology Group, Centre for Geographic Medicine, Kenya and the Spatial Ecology & Epidemiology Group, University of Oxford, UK with collaborating nodes in America and Asia Pacific region: http://www.map.ox.ac.uk

Malaria Journal Published by Reuters (ISI), MEDLINE and PubMed: http://www.malariajournal.com/

World Health Organization, © Copyright World Health Organization (WHO) 2008. All Rights Reserved: http://www.who.int/mediacentre/factsheets/fs094/en/index.html

World Health Organization, © Copyright World Health Organization (WHO), 2008. All Rights Reserved: http://whqlibdoc.who.int/publications/2005/9241580364_chap7.pdf

MULTIPLE CHOICE QUESTIONS

The questions should be answered either by selecting True (T) or False (F) for each answer statement, or by selecting the answer statements which best answer the question. Answers can be found in the back of the book.

1. **Which of the following transmit *Plasmodium*?**
   A. Ticks.
   B. Fleas.
   C. Mosquitoes.
   D. Bugs.
   E. Mites.

2. **Which of the following cell types are involved in the life cycle of *Plasmodium*?**
   A. Erythrocyte.
   B. Hepatocyte.
   C. Osteoblast.
   D. Lymphocyte.
   E. Neuron.

3. **In what cell/organ does sexual reproduction of *Plasmodium* occur?**
   A. The human erythrocyte.
   B. The mosquito liver.
   C. The human liver.
   D. The mosquito salivary gland.
   E. The mosquito intestine.

4. **Which of the following human genetic markers affect the development of malaria?**
   A. Duffy blood group antigen.
   B. Lewis blood group antigen.
   C. Glucose 6 phosphate dehydrogenase (G6PD) deficiency.
   D. Sickle cell hemoglobin HbS.
   E. Secretor status.

5. **Which of the following are directly important for the symptoms of malaria?**
   A. Hemolysis of erythrocytes.
   B. Sequestration of erythrocytes.
   C. Lysis of liver cells.
   D. Release of TNF.
   E. Destruction of retinal cells.

6. **Which of the following are used in the treatment/prevention of malaria?**
   A. Flucloxacillin.
   B. Artemisinin-based combination therapies (ACTs).
   C. Chloroquine.
   D. Ciprofloxacin.
   E. Proguanil.

7. **Which of the following are clinical indicators of the diagnosis of malaria?**
   A. Cyclical fever.
   B. Splenomegaly.
   C. Iritis.
   D. Anemia.
   E. Skin rash.

8. **Which of the following are complications of malaria?**
   A. Cerebral thrombosis.
   B. Skin necrosis.
   C. Renal failure.
   D. Splenic rupture.
   E. Nephrotic syndrome.

Case 29

Respiratory syncytial virus (RSV)

A 2-month-old child presented to a doctor with a 2-day history of a **febrile** upper respiratory tract infection. His clinical condition deteriorated with worsening cough and **dyspnea**, necessitating admission to hospital. On clinical examination in the hospital emergency department the child had obvious evidence of respiratory distress with severe intercostal retraction and **tachypnea** (>70 breaths per minute) together with **tachycardia**. On **auscultation** of the chest diffuse high pitched wheezing was audible and there were fine inspiratory crackles. Pulse oximetry gave an oxygen saturation of 0.92. Radiologic investigation revealed findings typical of **bronchiolitis**, with hyperinflation of both lung fields (Figure 1). A diagnosis of acute bronchiolitis was made and the child was admitted to hospital for further management. Respiratory syncytial virus (RSV) was detected in nasal secretions (Figure 2); the child was nursed in isolation on oxygen therapy and made an uneventful recovery.

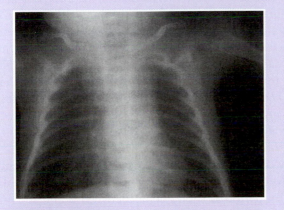

Figure 1. Acute bronchiolitis. Early radiographic appearance – flat diaphragm and hyperlucent lung fields indicate hyperflation.

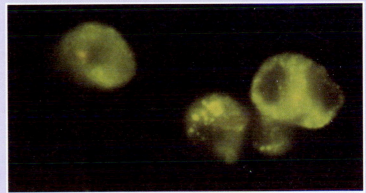

Figure 2. Photomicrograph of respiratory epithelial cells taken from a child infected with respiratory syncytial virus (RSV) and stained with a monoclonal antibody specific for respiratory syncytial virus. The antibody is labeled with fluorescein so that patchy (speckled) green fluorescence is observed under illumination with ultraviolet light. Note cytoplasmic staining and binucleate cell (result of RSV induced cell fusion).

1. What is the causative agent, how does it enter the body and how does it spread a) within the body and b) from person to person?

Causative agent

RSV is responsible for about 80% of cases of bronchiolitis. It is a paramyxovirus that is closely related to the recently discovered Human metapneumovirus. Both are enveloped RNA viruses with a helical **nucleocapsid**. The RSV envelope contains the virus attachment glycoprotein, G, the fusion glycoprotein, F, and a small hydrophobic protein, SH, suspected of contributing to the pathology in the host. There are two major groups of RSV strains, A and B, which circulate concurrently. RSV evolves quickly by random mutations that result in a high degree of variability, particularly in the G protein, which is the major target for the immune response. Thus at any one time there may be multiple sequence variants co-circulating within a population.

Other agents can cause bronchiolitis and these include human metapneumovirus, rhinovirus, adenovirus, influenza and parainfluenza, and enteroviruses.

Entry and spread within the body

The G protein on RSV binds to its receptor on the surface of ciliated epithelial cells in the mucous membranes of the nose and throat. Once the viral and cell membranes have been closely juxtaposed, fusion is induced by means of the fusion peptide of the F protein and the helical viral nucleocapsid is released into the cell's cytoplasm. Viral RNA-dependent RNA polymerase, an integral part of the **virion**, initiates virus replication with the synthesis of viral messenger RNA.

Virus progeny from the first infected cell then spread to neighboring cells (often involving cell to cell fusion and the formation of syncytia, also known as multinucleate giant cells (Figures 3 and 4 – hence the name of the virus) and by shedding from the apical surface of cells with subsequent spread via respiratory secretions to more distant mucosal cells. As the infection progresses cell damage occurs and there is an outpouring of fluid due to inflammation. At this stage the host's phagocytes begin to clear up the cell debris but there is overgrowth on the damaged mucosa by bacterial **commensals** and the mucosal fluid becomes purulent. The virus infection resolves with the production of **interferon** (**IFN**) and antibody.

In infants and the elderly the infection is not limited to the upper respiratory tract and may involve the trachea, bronchi, bronchioles, and alveoli. In all cases the incubation period is short, being between 2 and 8 days, since the infection is restricted to respiratory mucosa and does not become systemic.

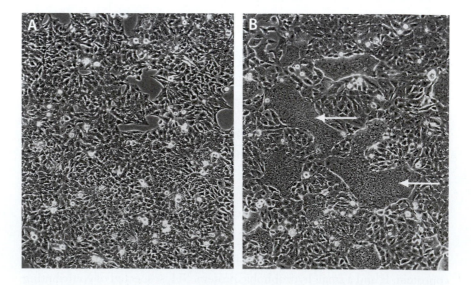

Figure 3. Photomicrograh of RSV-infected respiratory epithelial cells as seen in tissue culture. Cultured respiratory epithelial cells on day 1 (A) and day 3 (B) following infection with the virus. After 1 day, the cells continue to appear healthy. After 3 days, many of the cells have fused, forming large syncytia (arrows), that is a mass of cytoplasm containing several separate nuclei enclosed in a continuous membrane. Magnification: ×400.

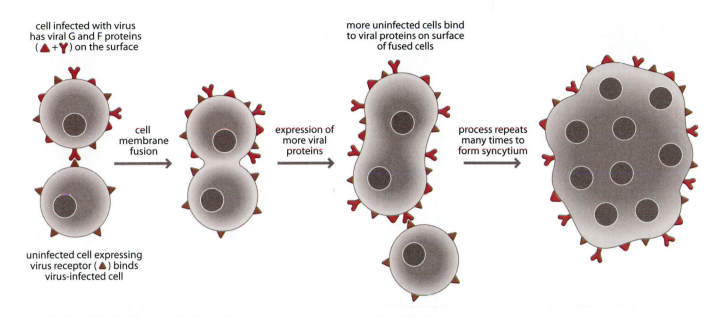

cell infected with virus has viral G and F proteins (▲ + Y) on the surface

cell membrane fusion

uninfected cell expressing virus receptor (▲) binds virus-infected cell

expression of more viral proteins

more uninfected cells bind to viral proteins on surface of fused cells

process repeats many times to form syncytium

Figure 4. Schematic representation of the mechanism of syncytium formation due to RSV infection of respiratory epithelial cells. The virus envelope glycoproteins, G (▲) and F (Y), whose roles on virus particles are in receptor binding and membrane fusion, are expressed on the surface of infected cells. Uninfected epithelial cells bearing the cellular virus receptor (▲) and coming into contact with these infected cells are fused together to form a syncytium.

Person to person spread

RSV infection is highly contagious, being shed in respiratory secretions for several days. Spread is via large droplets (produced for example by sneezing and coughing) but also, and more importantly, by contact with contaminated surfaces. Virus in infected secretions is viable for up to 6 hours on nonporous surfaces, up to 45 minutes on cloth and up to 20 minutes on skin.

Epidemiology

In hospitals, RSV outbreaks are common on pediatric wards; if on a neonatal unit there is a high risk of mortality. Although initially introduced from the community, subsequent spread is usually **nosocomial**. Hospital staff and parents and siblings become infected and while symptomatic are responsible for spreading the virus on the wards via aerosol, **fomites**, and direct contact.

Incidence of bronchiolitis

RSV is the most common cause of lower respiratory tract infection in children <1 year old – worldwide as many as 1 million children may die from RSV infection annually. In the UK and USA two-thirds of infants become infected in the first year of life and of these one-third will develop lower respiratory tract symptoms. In the first 6 months of life, bronchiolitis is an important and life-threatening disease; 2–3% of infants up to 1 year old are admitted to hospital each year with bronchiolitis caused by RSV, and many more will be managed in the community.

Occurrence

Infection occurs in community-wide annual winter outbreaks in temperate climates (Figure 5). Much less is known about the epidemiology in developing countries, although RSV outbreaks are often associated with the rainy season in the tropics.

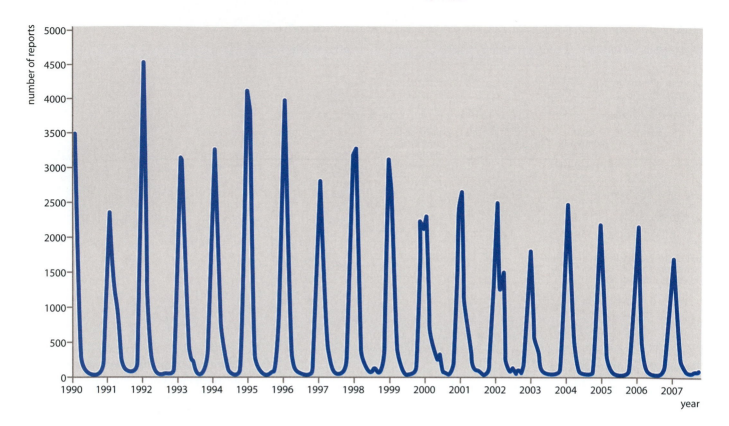

Figure 5. Laboratory reports of RSV infections, England and Wales, 1990–2007 (4 weekly).

2. What is the host response to the infection and what is the disease pathogenesis?

Although first infection is almost always early in life, immunity is short-lived and re-infections occur throughout life.

Antibody responses

After first infection with RSV, **IgM** persists for about 3 months and the **IgG** and **IgA** antibodies that are produced decline to low or absent levels within a year. Re-infection with the virus leads to a rapid increase in all three classes of antibodies. Since RSV infection is restricted to mucosa, antibodies present in upper and lower respiratory tract are important for disease prevention. Antibody, especially IgA, directed against the G protein probably neutralizes the virus by preventing attachment to cells, while antibody to the F protein neutralizes infection by preventing the fusion of the virus with the cell and inhibiting cell to cell spread.

High levels of maternally acquired antibodies protect against lower respiratory tract infection in early life; the level of neutralizing antibody in cord blood correlates directly with the age of first respiratory virus infection and inversely with the severity of illness.

T-cell responses

CD4+ and **CD8+** lymphocytes are critical for the control and clearance of infection. Whereas immunocompetent infants infected with RSV shed virus for 1–21 days, children or adults with impaired cellular immunity shed virus for months and may develop fatal **pneumonia**.

As mentioned earlier, the propensity of RSV to mutate its surface glycoproteins means that new variants are generated on a regular basis. Similar to influenza virus **antigenic drift**, the implication is that the ability to recognize RSV wanes over a period of time, hence re-infection can occur repeatedly throughout an individual's life. This process, coupled with the fact there are two subtypes, results in a complicated epidemiology. This means that in a given geographical region (e.g. the UK) there are multiple forms circulating.

Pathogenesis of bronchiolitis

There is a peribronchiolar lymphocytic infiltration together with **edema** of the bronchiolar walls. This is followed by **necrosis** of the bronchiolar epithelium and necrotic material blocks the lumen of the small airways. Virus infection also causes increased secretion of mucus, which compounds the problem. The infant's bronchioles are very narrow and hence especially prone to obstruction. On inspiration negative intrapleural pressure due to obstructed bronchioles is clinically demonstrated by supraclavicular, intercostal, and subcostal recession of the soft tissues. On expiration positive intrapleural pressure further narrows the bronchiolar lumen and there is resistance to expiration causing hyperinflated lungs. In areas where the bronchiolar lumen becomes completely obstructed the trapped air may become absorbed, causing areas of **atelectasis** (collapse of the lung).

3. What is the typical clinical presentation and what complications can occur?

As with most respiratory virus infections, the clinical consequences vary from asymptomatic or mild afebrile upper respiratory tract illness to severe and fulminating pneumonia.

Bronchiolitis: this is characteristic of the first infection with RSV in infants. There are usually prodromal upper respiratory tract symptoms. As the infection progresses cough is a consistent and prominent finding, together with retraction of the chest wall, dyspnea, and **hypoxia**, with or without **cyanosis**. On auscultation there is a diffuse expiratory wheeze with crackles. Fever is low grade. Hyperinflation is seen on chest X-ray (Figure 1). The duration of illness is between 7 and 12 days. Most hospitalized infants improve clinically after 3–4 days, and most previously normal infants are discharged soon after.

Complications are **otitis media**, respiratory failure, apneic attacks, recurrent wheezing. Interstitial pneumonia is uncommon but carries a bad prognosis.

Pneumonia: RSV is an important cause of lower respiratory tract infection and causes death due to pneumonia in the elderly and the immunocompromised, especially hematopoietic stem cell transplant patients. The signs and symptoms are often indistinguishable from those of influenza.

4. How is this disease diagnosed and what is the differential diagnosis?

Bronchiolitis is defined by two clinical signs, namely wheezing and hyperinflation of the lungs. Nonspecific supportive findings are normal or slightly raised white cell count with lymphocytes predominating. Chest X-ray shows hyperinflation, prominent bronchial wall markings, and multiple areas of collapse and consolidation, and there is **hypoxemia**. However, in hospital, the assistance of a microbiology laboratory is required to confirm the etiological agent; bronchiolitis due to RSV must be distinguished from that due to other respiratory virus infections, especially the closely related human metapneumovirus but also parainfluenza virus type 3, so that appropriate infection control can be instigated.

Samples: a good respiratory sample contains a mixture of respiratory ciliated epithelial cells as well as respiratory secretions. Samples from the upper respiratory tract are sufficient for the diagnosis of RSV infection. These can be nasopharyngeal aspirates or washes, a nasopharyngeal swab or combined nose and throat swab – sampling only the throat mucosa is likely to provide mainly squamous epithelial cells, which do not support respiratory virus replication very well.

Specific diagnosis: this relies on detection of the virus or one of its components. The possibilities for a rapid result, that is within a day, are an immunofluorescence test (Figure 2) or **enzyme immunoassay** for viral antigen, or **polymerase chain reaction (PCR)** for virus nucleic acid. For the most rapid result so-called 'point of care tests' can be used; such tests are based on the detection of RSV antigen and can be used at the bedside rather than in the laboratory. Virus isolation in tissue culture requires viable virus and takes from 3 to 7 days, by which time the patient is usually well on the way to recovery. Likewise diagnosis of RSV by the use of antibody tests is not clinically useful as it takes several weeks for an antibody response to develop.

Differential diagnosis

The differential diagnosis of bronchiolitis includes **croup**, **epiglottitis**, pertussis (whooping cough), and pneumonia (Table 1), and the most likely noninfectious disease is **asthma**.

Table 1. Lower respiratory tract infections in young children, their characteristic clinical features and common causes

Syndrome	Presenting symptoms and signs	Commonest cause
Croup (laryngotracheitis)	Hoarseness, cough, inspiratory stridor with laryngeal obstruction	Parainfluenza viruses (types 1, 2, and 3)
Bronchiolitis	Expiratory wheeze with or without tachypnea, air trapping, and intercostal indrawing	Respiratory syncytial virus Human metapneumovirus
Pneumonia	Crackles on lung auscultation or evidence of pulmonary consolidation on physical examination or chest X-ray	Many different possibilities including respiratory viruses and enteroviruses*

* Whereas croup and bronchiolitis are viral in origin, pneumonia is often caused by bacteria: <1 month old – group B streptococci and *Escherichia coli*; 1–3 months – *Chlamydia trachomatis*; 3 months to 5 years – *Haemophilus influenzae* type b and *Streptococcus pneumoniae*.

5. How is the disease managed and prevented?

Management

Bronchiolitis: Patients most at risk of hospital admission are those with congenital heart disease, any survivor of extreme prematurity, or with any pre-existing lung disease or immunodeficiency. The patient should be admitted if there is a need for oxygen or tube feeding or impending respiratory failure. Management involves the use of humidified oxygen, maintaining oral nutrition, and respiratory support if required. Nebulized ribavirin (an antiviral drug with *in vitro* activity against RSV) has been advocated for children likely to develop severe disease (infants <2 months, those born prematurely or with chronic cardiorespiratory disease). However, despite promising **clinical trials**, in day to day clinical use ribavirin has given disappointing results and it is now rarely used.

Pneumonia: Nebulized ribavirin may be effective as treatment for pneumonia in the immunocompromised but, as above, this is controversial and its use cannot be recommended.

Prevention

Vaccine: There is no vaccine currently available. A formalin-inactivated RSV vaccine was developed in the past but the results were unexpected and disastrous. After immunization subsequent exposure to the virus resulted in worse bronchiolitis and pneumonia than in unimmunized children, with severe morbidity and several deaths. To this day it is not known exactly why this happened. However, animal model studies suggest that the formalin inactivation process revealed novel epitopes on the G protein that primed the immune system to over-react when vaccinees were challenged. The result was a catastrophic inflammatory response that destroyed lung tissue, a process that continued even after viral clearance.

Immunoprophylaxis: A humanized **monoclonal antibody** directed against the F protein of RSV (palivizumab) can be used as passive immunoprophylaxis to protect especially vulnerable infants. However, it is of limited benefit and expensive. For anyone other than an infant immunoprophylaxis with palivizumab is impractical, as the amount of antibody required is prohibitively expensive.

Control of infection: Hospitalized infants or immunocompromised patients infected with RSV pose a risk of infection to neighboring vulnerable patients. Rapid identification of infection by, for example, the 'point of care tests' described above, is of paramount importance to enable timely isolation of infected patients. Scrupulous hand-washing together with cohort nursing are also vital.

SUMMARY

1. What is the causative agent, how does it enter the body and how does it spread a) within the body and b) from person to person?

- The main cause of bronchiolitis is respiratory syncytial virus (RSV) but also human metapneumovirus and parainfluenza viruses.

- RSV attaches to the mucous membranes of the nose and throat, infects the ciliated epithelial cells, and spreads to neighboring cells by the formation of syncytia.

- Shedding of virus from the apical surface of infected cells allows spread via respiratory secretions to more distant mucosal cells in the trachea, bronchi, bronchioles, and alveoli.

- Spread from person to person is via respiratory secretions, either large droplets or contaminated surfaces.

2. What is the host response to the infection and what is the disease pathogenesis?

- Immunity to RSV infection is incomplete and re-infections occur in all age groups. This is because every year or two the virus will have changed enough, antigenically, so that the population is effectively immunologically naïve and hence open to re-infection. This has implications for vaccine development since, as for influenza vaccine, the composition of an RSV vaccine might need to be changed each year.

- RSV causes infection and inflammation of the bronchioles, leading to obstruction of expiratory airflow causing wheeze and hyperinflation and sometimes even collapse of the lung.

3. What is the typical clinical presentation and what complications can occur?

- Bronchiolitis typically presents with tachypnea, expiratory wheeze, and inspiratory intercostal recession.

- Complications are secondary bacterial infection, apnea, hypoxia, and respiratory failure.

4. How is this disease diagnosed and what is the differential diagnosis?

- Bronchiolitis is a clinical diagnosis defined by wheezing and hyperinflation of the lungs.

- Identification of the causal agent (usually RSV) requires a respiratory sample, for example nasopharyngeal aspirate, and a test for the virus, for example enzyme immunoassay for viral antigen.

- The differential diagnosis includes croup, epiglottitis, pertussis, pneumonia, and asthma.

5. How is the disease managed and prevented?

- Admit to hospital if there is a need for oxygen or tube feeding or impending respiratory failure.

- Management involves the use of humidified oxygen, maintenance of oral nutrition, and respiratory support.

- Use of ribavirin is controversial and cannot be recommended.

FURTHER READING

Hall CB. Respiratory syncytial virus and human metapneumovirus. In: Feigin RD, Cherry J, Demmler GJ, Kaplan S. Textbook of Pediatric Infectious Diseases, 5th edition, Vol 2. Elsevier/Saunders, Philadelphia, 2004: Chapter 185A.

Murphy K, Travers P, Walport M. Janeway's Immunobiology, 7th edition. Garland Science, New York, 2008.

Richman DD, Whitley RJ, Hayden FG. Clinical Virology, 2nd edition. ASM Press, Washington, DC, 2002.

Zuckerman AJ, Banatvala JE, Pattison JR, Griffiths PD, Shaub BD. Principles and Practice of Clinical Virology, 5th edition. Wiley, Chicester, 2004.

REFERENCES

Bush A, Thomson AH. Acute bronchiolitis. BMJ, 2007, 335: 1037–1041.

WEB SITES

Centers for Disease Control and Prevention, National Center for Immunization and Respiratory Diseases, Atlanta, GA, USA: http://www.cdc.gov/ncidod/dvrd/revb/respiratory/rsvfeat.htm

Centre for Infections, Health Protection Agency, HPA Copyright, 2008: http://www.hpa.org.uk/infections/topics_az/rsv/default.htm

Scottish Intercollegiate Guidelines Network (SIGN). Bronchiolitis in children (A national clinical guideline), 2006: www.sign.ac.uk

MULTIPLE CHOICE QUESTIONS

The questions should be answered either by selecting True (T) or False (F) for each answer statement, or by selecting the answer statements which best answer the question. Answers can be found in the back of the book.

1. **Which of the following are true of respiratory syncytial virus?**

   A. One of the envelope proteins is a hemagglutinin.

   B. Annual outbreaks of infection occur every summer in the USA.

   C. It is closely related to human metapneumovirus.

   D. The G protein of the virus is hypervariable and accounts for re-infection with the virus throughout life.

   E. The viral genome is segmented.

2. **Which of the following diseases are proven to be caused by RSV?**

   A. Bronchiolitis.

   B. Epiglottitis.

   C. Asthma.

   D. Croup.

   E. Whooping cough.

3. **Which of the following tests would be helpful in the diagnosis of bronchiolitis?**

   A. Polymerase chain reaction for RSV antigen in respiratory secretions.

   B. Immunofluorescence test for respiratory virus antigens in respiratory epithelial cells.

   C. Polymerase chain reaction for human metapneumovirus RNA in respiratory secretions.

   D. Chest X-ray.

   E. Pulse oximetry.

4. **What are the typical signs and symptoms of a patient presenting with bronchiolitis?**

   A. Barrel-shaped chest, prominent neck veins, and downward displacement of the liver.

   B. Inspiratory stridor.

   C. Crackles on auscultation of the lung.

   D. Conjunctivitis.

   E. Expiratory wheezing.

5. **How would you treat a patient with respiratory syncytial virus infection?**

   A. Treat with the monoclonal antibody, palivizumab, administered intramuscularly.

   B. Drain pleural fluid.

   C. Maintain an adequate airway and ensure oxygenation.

   D. Maintain nutrition and hydration.

   E. Treat with nebulized ribavirin.

Case 30

Rickettsia spp.

A university professor returned from a botanical expedition to Kenya feeling generally unwell. He was complaining of a headache. He also noticed a large swollen black lesion on his thigh that was painful (Figure 1). Thinking he had injured himself and that it was now infected, he went to his primary health-care provider who gave him co-amoxiclav. Over the next few days the lesion did not respond and he continued to feel unwell with headache and **myalgia**. He presented to a local travel clinic where the doctor identified the lesion on his thigh as a tick bite. He also noticed that the patient had regional **lymphadenopathy**. Making a provisional diagnosis of rickettsiosis the doctor took blood for **serology** and started the patient on an appropriate antibiotic.

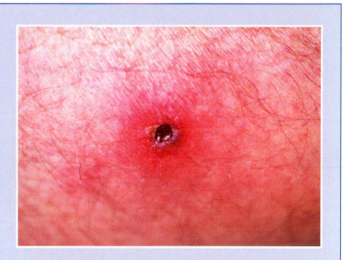

Figure 1. Tick bite showing the black lesion with a central bite.

1. What is the causative agent, how does it enter the body and how does it spread a) within the body and b) from person to person?

Causative agent

The patient has a rickettsial infection. The taxonomy and nomenclature of the *Rickettsiaceae* have undergone a major revision based upon 16S rRNA sequences. Currently the *Rickettsiales* contain the *Rickettsia*, *Orientia*, *Ehrlichia*, *Anaplasma*, *Wolbachia*, and *Neorickettsia*. *Coxiella burnetti* is now included in the γ-proteobacteria along with *Legionella* and *Francisella*. Frequently *Rickettsia* are given species names relating to their original geographic location. There are many species of *Rickettsia* found in nature, only some of which have been linked to illness (Table 1) and present clinically as spotted fevers or typhus (see later).

The *Rickettsia* are small bacteria (0.3 × 0.1 μm) that are obligate intracellular pathogens. They have a typical tri-laminar gram-negative cell wall structure and chemistry, although they stain poorly with the Gram stain. Morphologically the cell wall of *Orientia* (previously included as a *Rickettsia* sp.) has a different cell wall structure with a prominent outer layer compared with the *Rickettsia* spp., which have a prominent inner layer. The *Rickettsia* spp. have a small genome size of about 1.3 Mbp, which is the result of gene decay. To date, three rickettsial genomes have been sequenced: *R. prowazekii*, *R. typhi*, and *R. conori*. Twenty-three genes are

Table 1. Species of *Rickettsia* and *Orientia* causing pathogenic human infection and their arthropod vectors				
Disease	**Tick**	**Flea**	**Louse**	**Mite**
Spotted fevers	R. rickettsii R. conori R. japonica R. sibirica R. australis R. slovaca R. africae R. honei R. helvetica	R. felis		R. akari
Typhus		R. typhi	R. prowazekii	
Scrub typhus				O. tsutsugamushi

common to *R. prowazekii* and *R. typhi*, 15 are common to *R. prowazekii* and *R. conori*, 24 are found only in *R. typhi*, and 775 are common to all three species. Since their cytosolic habitat is rich in nutrients, amino acids, and nucleotides, they lack enzymes for sugar and amino acid metabolism, and for lipid and nucleotide synthesis. Their genome therefore codes for several transport proteins, which enable them to utilize host cell products including ATP, although they may also synthesize ATP. They possess a **type IV secretion system**.

Entry and spread within the body

The bacterium is injected into the host skin by an arthropod vector (see below). It multiplies locally in the dermal tissue, primarily within the endothelial cells of the vasculature, although *R. akari* and *O. tsutsugamushi* infect monocytes. Subsequently the organism is spread by the bloodstream to all body organs where it attaches to and reproduces in endothelial cells of the blood vessels.

Rickettsia attach to and enter the vascular endothelium by inducing **endocytosis** using a zipper-like process whereby there is sequential interaction between ligand and an **adhesin** (Figure 2). Two cell surface proteins

Figure 2. Entry, multiplication, and outcome of *Rickettsia* inside the endothelial cell.

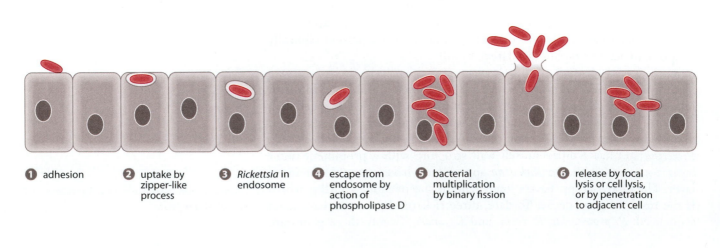

1 adhesion   **2** uptake by zipper-like process   **3** *Rickettsia* in endosome   **4** escape from endosome by action of phospholipase D   **5** bacterial multiplication by binary fission   **6** release by focal lysis or cell lysis, or by penetration to adjacent cell

appear to be involved in the adhesion process, OmpA (outer-membrane protein A) and Sca2, which are both members of an autotransported family of proteins called surface cell antigen (Sca). Other members of this family are Sca1, Sca3, and OmpB. OmpB may also be involved in the adhesion process. A rickettsial phospholipase A2 appears to be involved in the uptake of the rickettsia into the cell. The intracellular signaling events mediating induced endocytosis involve various signaling pathways such as Cdc-42 PI 3 kinase, c-Src, and Arp2/3. Once they have entered the endothelial cell they escape the **endosome** with the aid of bacterial enzymes, for example phospholipase D, and replicate in the cytoplasm. Like many other intracellular organisms, the rickettsia inhibit **apoptosis** of the host cell to allow their replication to proceed. They spread from an infected cell to adjacent cells by polymerizing actin monomers at one pole of the cell, thereby pushing the organism forward. RickA (a group of proteins found in the spotted fever group of rickettsia but not in the typhus group) induces nucleation of actin momomers via the Arp2/3 complex, thereby mediating intracellular movement (Figure 2). Rickettsia are also released into the bloodstream using the polymerizing actin monomers.

Pathogenic rickettsia are spread to humans by different arthropod vectors (Table 1). Some *Rickettsia* spp. have a fairly restricted geographic location, while others can be found on all the continents. The distribution of *Rickettsia* spp. is related to that of their arthropod vectors (Figures 3–6), particularly since ticks, fleas, and lice are ubiquitous. The bacteria are maintained in nature by colonizing/infecting mammalian hosts (dogs, rats, ferrets, deer, etc.), which act as a reservoir for further human infection. In the case of ticks the bacteria are transmitted **trans-stadially** and thus the vector also acts as a reservoir. Ticks transmit the organism during feeding. The tick may feed over several days and may go unnoticed. Transmission of the *Rickettsia* sp. does not occur immediately and may take a few days. It is thus important to scrutinize one's body if one has been hiking in tick-infested locations and carefully remove any ticks without crushing them.

Spread from person to person

Some *Rickettsia* spp. are only found in human hosts and are spread from one human host to the next by one of the arthropod vectors. Table 2 shows the organisms found in a number of geographic locations with their reservoirs.

Epidemiology

The prevalence of infection with *Rickettsia* spp. is probably under-reported. *Rickettsia* spp. are not common in the UK and cases occur in returning travelers. However, *Rickettsia felis* has been detected in 6–12% of the cat fleas in the UK, suggesting that endemic rickettsial disease is a possibility. Between 1990 and 2002 there were 66 reported cases. In the USA, where rickettsia are endemic, between 1993 and 1996 there were 2313 cases. An overall prevalence of 0.5–0.7/100 000 population is recognized. In Spain between 1982 and 1987, 4683 cases of rickettsia were confirmed. Studies in Africa (in 150 individuals) demonstrated a 25% **seropositivity** to the spotted fever group and a 28% seropositivity to the typhus group.

Figure 3. A tick – the vector for some rickettsial diseases.

Figure 4. A flea – the vector for some rickettsial diseases.

Figure 5. A louse – the vector for some rickettsial diseases.

Figure 6. A mite – the vector for some rickettsial diseases.

Table 2. The geographic distribution of *Rickettsia* and *Orientia* and their reservoirs

Organism	Location	Reservoir
R. rickettsii	North, Central, and South America	Dogs, deer, rodents, ticks
R. conori	Africa, India, Middle East, Mediterranean, Russia	Dogs, rodents, ticks
R. japonica	Japan, China	Dogs, rodents, ticks
R. sibirica	Russia, China , France, Africa	Dogs, rodents, ticks
R. australis	Australia	Dogs, rodents, ticks
R. slovaca	Europe	Dogs, rodents, ticks
R. africae	sub-Saharan Africa	Dogs, rodents, ticks
R. honei	Flinders Island Australia	Dogs, rodents, ticks
R. felis	America, Brazil, Europe	Dogs, rodents, ticks
R. akari	Eastern USA, Europe, Korea, South Africa	Mice
R. helvetica	Scandinavia	Dogs, rodents, ticks
R. typhi	Worldwide	Rat
R. prowazekii	Worldwide	Human Flying squirrels in USA
O. tsutsugamushi	Japan, Russia, Australia	Rodent, mites

Seasonal variation of some rickettsial infections may occur, which is related to variation of the tick population. In Australia a seroprevalence study showed that 5.6% of nearly 1000 individuals were positive for rickettsia, with 24% of these positive for scrub typhus. Despite its importance as a global pathogen and indeed as an organism that could be used for bioterrorism, it has been little studied to date. Figure 7 shows the global distribution of different species of the *Rickettsiales*.

Figure 7. Distribution of rickettsial diseases globally, showing location of different species and detection of species from locations where they had not previously been detected (from Walker 2007).

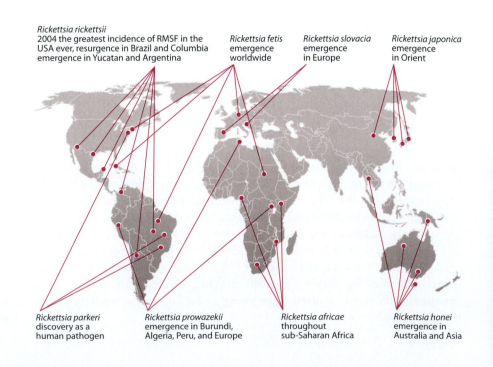

Rickettsia rickettsii
2004 the greatest incidence of RMSF in the USA ever, resurgence in Brazil and Columbia emergence in Yucatan and Argentina

Rickettsia fetis emergence worldwide

Rickettsia slovacia emergence in Europe

Rickettsia japonica emergence in Orient

Rickettsia parkeri discovery as a human pathogen

Rickettsia prowazekii emergence in Burundi, Algeria, Peru, and Europe

Rickettsia africae throughout sub-Saharan Africa

Rickettsia honei emergence in Australia and Asia

2. What is the host response to the infection and what is the disease pathogenesis?

The pathophysiology of rickettsial infection is currently poorly understood and the details may well vary according to the species of *Rickettsia*, particularly the two main groups: the spotted fever group and the typhus group.

Damage to the endothelial cells leads to changes in microvascular permeability. Currently it is believed that the injury is the result of oxidative stress in the endothelial cells mediated by **reactive oxygen species**, which cause lipid peroxidative membrane damage. This is probably the direct effect on the endothelial cells by the infective organism but mostly by cells of the immune system.

Endothelial cell infection induces the production of pro-inflammatory **cytokines** such as **interleukin (IL)-1α** and this causes an up-regulation of E-selectin (*R. rickettsii*) allowing increased adhesion of polymorphs to the vasculature. Following infection with *R. conorii*, **E-selectin, ICAM-1**, and **VCAM -1** are all up-regulated, allowing adhesion of mononuclear cells including macrophages and T cells.

Endothelial damage-mediated changes in microvascular permeability result in **hypovolemia**, hypotension, and pulmonary **edema**. Membrane leakiness induced by nitric oxide, which is a host response by the endothelial cells and macrophages, can also lead to interstitial **pneumonia**, **myocarditis**, perivascular lesions in the brain and other organs, and in surface peripheral blood vessels leads to a **rash** . In skin cells from *R. conorii*-infected patients the host responds by up-regulation of **tumor necrosis factor-α (TNF-α)**, **interferon-γ (IFN-γ)**, IL-10, RANTES, and inducible nitric oxide synthase (iNOS), and the levels of the cytokines correlate with severity of disease. The **vasculitis** leads to increased local consumption of platelets with a **thrombocytopenia** in a proportion of patients and a pro-coagulation state, although **disseminated intravascular coagulation (DIC)** and **thromboses** are rare.

Immune response

The immune response, as well as playing a role in pathogenesis, is eventually able to control the infection. Production of an enzyme **indoleamine 2,3-dioxygenase** by the endothelial cells following the effect of pro-inflammatory cytokines, limits the replication of rickettsia by metabolizing tryptophan, an essential amino acid for rickettsia.

Early in infection it is thought that **natural killer (NK)** cells inhibit growth of rickettsia in endothelial cells through release of IFN-γ. **CD4** T cells and **CD8+** T cells are believed to produce IFN-γ and also RANTES that enhance intracellular killing of rickettsia via nitric oxide production and hydrogen peroxide. CD8+ T cells are able to kill rickettsia-infected endothelial cells via a **perforin**-dependent cytotoxic mechanism, which contributes significantly to recovery from infection. There is also an antibody response generated to some of the rickettsial adhesion molecules, such as OmpA and OmpB, but these antibodies appear in significant amounts only after the infection has resolved. They are therefore thought to be protective against re-infection.

3. What is the typical clinical presentation and what complications can occur?

Rickettsial disease presents in one of two ways: (a) as a spotted fever or (b) as **typhus**, depending upon the infecting organism (Tables 1 and 3). The typical presentation of rickettsial disease is abrupt onset of fever, headache, myalgia, and a rash.

Spotted fevers

The incubation period is about 7 days but may be up to 14 days. Patients present with fever, headache, myalgia, and a **maculopapular rash**.

The rash typically begins around the wrists and ankles but can also cover the trunk and may also appear on the palms and soles. In 10% of patients with Rocky Mountain spotted fever (RMSF) caused by *R. rickettsii* the rash may be absent. Following infections with *R. slovaca* and *R. helvetica* typically there is no rash and a rash is uncommon with *R. africae*.

About 25% of patients may show signs of central nervous system (CNS) involvement with encephalitis or **seizures**. Gastrointestinal symptoms of nausea, vomiting, abdominal pain, and diarrhea may be present.

Despite the low platelets and prolonged coagulation time, bleeding and DIC are uncommon. In infections with *R. conori* a pro-coagulation state exists and thromboses may occur in 10% of patients.

A tick bite may be visible as a black eschar (tache noire), although this is uncommon in RMSF, or as multiple eschars in infections with *R. africae*. Following infections with *R. slovaca*, the tick bite is commonly in the scalp and may lead to **alopecia**.

Complications include coma, renal failure, **adult respiratory distress syndrome (ARDS)**, **myocarditis**, and death in about 10% of patients.

Table 3. Common eponyms for some rickettsial diseases

	Species	*Disease*
Spotted fever	R. rickettsii	Rocky Mountain spotted fever (RMSF)
	R. conori	Mediterranean spotted fever, Kenya tick typhus, Indian tick typhus, Boutonneuse fever
	R. sibirica	N. Asian tick typhus, Siberian tick typhus
	R. japonica	Japanese tick typhus
	R. australis	Queensland tick typhus
	R. honei	Flinders Island spotted fever
	R. africae	African tick bite fever
	R. akari	Rickettsial pox
Typhus	R. prowazekii	Epidemic typhus, Brill-Zinsser disease (recrudescence)
	R. typhi	Murine (or endemic) typhus
	O. tsutsugamushi	Scrub typhus

Fulminant infection is more common in patients with glucose 6 phosphate dehydrogenase (G6PD) deficiency.

Laboratory investigations show a **thrombocytopenia** in about 50% of patients, **anemia**, abnormal liver function tests, increased urea, **hyponatremia** and **hypoalbuminemia**. Prolonged coagulation times may be observed and increased levels of creatine kinase. A chest X-ray may reveal pulmonary edema.

Rickettsial pox

This presents with a fever, headache, and a rash. The rash is vesicular and starts as a **papule**, becoming vesicular, and scabs leaving a black eschar. The rash may also appear in the oral cavity as well as the palms and soles. Regional lymphadenopathy is present. The patients may also complain of profuse sweating, rigors, sore throat, and **photophobia**. Laboratory results indicate a **leukopenia** although liver function tests, urea, electrolytes, and hemoglobin are usually normal.

Typhus

Epidemic typhus

Epidemic typhus is associated with poor hygiene, homelessness, overcrowding, war, poverty, and natural disasters. The incubation period is about 10 days. The patient presents with fever, headache, myalgia, and a macular rash that is found on the trunk although not usually on the palms and soles. Respiratory and CNS symptoms may also be present. The case fatality rate is 15%. The disease acquired from flying squirrels is less severe.

Laboratory results indicate a leukopenia and thrombocytopenia. Elevated liver function tests, creatine kinase, and blood urea are found. A chest X-ray may show interstitial shadowing.

Once infected and after apparently successful treatment, a patient may suffer a recrudescence of typhus (**Brill-Zinsser disease**) with the same signs and symptoms. This is frequently induced by stress or immune suppression.

Murine (endemic) and scrub typhus

For both endemic and scrub typhus the incubation period is 7–14 days and the patient presents with the typical symptoms of abrupt onset of fever, headache, and myalgia. A maculopapular rash on the trunk develops during the course of the illness in the majority of cases. Hepatosplenomegaly may be present in both endemic and scrub typhus. In scrub typhus the rash is pale and transient and may be missed. Gastrointestinal (nausea, vomiting), respiratory (cough), and CNS symptoms (altered consciousness, seizures) may also occur. Complications similar to RMSF may occur and the case fatality rate is about 5%.

4. How is this disease diagnosed and what is the differential diagnosis?

Diagnosis of rickettsial disease relies on **serology** and **polymerase chain reaction** (**PCR**) because of the difficulty in culturing these organisms in a routine setting. Reference and specialist laboratories may offer a culture service where the organism is grown on cell culture using a shell-vial assay or in animals. An accurate diagnosis of the infecting species is frequently not possible using serology because of the many cross-reactions between the rickettsia and other bacteria, particularly *Proteus* sp., which is the basis of the **Weil-Felix test**. Immunohistochemistry can be used on biopsy material from any rash and can help distinguish rickettsial disease from other causes of a rash. **Immunofluorescence** or **ELISA** for serum antibodies are not useful in making a diagnosis of acute disease. The Weil-Felix agglutination test using Proteus OX-2 and OX-19 has a low sensitivity and specificity. More accurate diagnosis can be made using PCR combined with restriction fragment length polymorphism (RFLP) or sequencing to confirm the identity of any amplicon detected. RT-PCR and real-time PCR assays are also available. PCR is most useful using skin biopsies of the rash and is less helpful for blood samples. The primers are based on a number of different genes including those for rOmpA, rOmpB, a 17 kDa genus-specific protein (*htrA*), and citrate synthase (*gltA*).

Differential diagnosis

Rickettsial disease presents with a fever and a rash and thus in the differential diagnosis any other infection with a similar presentation, including other rickettsial illnesses, should be considered. Other infectious causes that should be considered are measles, rubella, varicella, arboviral infections, meningococcal and disseminated gonococcal disease, typhoid, secondary syphilis, anthrax, and leptospirosis. Of the noninfectious causes **idiopathic thrombocytopenic purpura** and **immune complex** disease should be considered.

5. How is the disease managed and prevented?

Management

The mainstay of treatment is doxycycline. Other tetracyclines may also be used. An alternative although less effective treatment is chloramphenicol, which can be used during pregnancy or in cases of allergy to the tetracyclines. Azithromycin may also be an effective antibiotic in some cases. Quinolones and rifampicin have *in vitro* activity but poor clinical response. β-lactams, aminoglycosides, and sulfonamides are ineffective

Prevention

Preventive measures include the use of insect repellents and protective clothing. Currently there is no vaccine available.

SUMMARY

1. What is the causative agent, how does it enter the body and how does it spread a) within the body and b) from person to person?

- *Rickettsia* and *Orientia* spp. are obligate intracellular gram-negative bacteria.

- They have a small genome size and restricted metabolic activity but the genome codes for several autotransporter proteins.

- The bacteria are spread by the bite of an arthropod vector: tick, flea, louse, mite.

- The bacteria are spread by the bloodstream and infect vascular endothelial cells systemically.

- Adhesins of rickettsia for attachment to endothelial cells belong to an autotransporter family of proteins.

- Induced endocytosis of rickettsia uses a zipper-like action.

- Intracellular signaling events depend upon Arp2/3 complex.

- Rickettsia escape from the endosome and replicate in the cytoplasm.

- Infection with rickettsia up-regulates adhesion molecules on the endothelial cell, inducing adhesion of leukocytes.

- The bacteria are maintained in nature by infecting mammals such as dogs and rodents, which act as a reservoir of infection.

- Some *Rickettsia* spp. are confined to humans.

- The rickettsia and orientia are found on all continents although some have a more restricted geographical distribution.

2. What is the host response to the infection and what is the disease pathogenesis?

- Damage to the endothelial cells leads to changes in microvascular permeability through oxidative stress in the endothelial cells mediated by the infection but also by cells of the immune system.

- Membrane leakiness induced by nitric oxide, which is a host response by the endothelial cells and macrophages, can also lead to interstitial pneumonia, myocarditis, or perivascular lesions in the brain.

- The vasculitis leads to increased local consumption of platelets with a thrombocytopenia in a proportion of patients.

- Cellular damage is caused by free radicals.

- Cell damage results in increased vascular permeability and edema.

- Host indoleamine 2,3-dioxygenase limits replication of rickettsia.

- CD8+ T cells are important in recovery from rickettsial infection.

3. What is the typical clinical presentation and what complications can occur?

- Rickettsial disease presents with abrupt onset of fever, headache, myalgia, and rash.

- The incubation period is usually 7–14 days.

- The rash is maculopapular but in rickettsial pox it is vesicular.

- Gastrointestinal, respiratory, and CNS symptoms may occur.

- Evidence of an insect bite in the form of a black eschar may be found.

- Complications include renal failure, ARDS, and death in 5–15% of cases.

- Although damage occurs directly to endothelium, DIC and major vascular incidents are rare.

4. How is this disease diagnosed and what is the differential diagnosis?

- Culture of *Rickettsia* spp. is not part of routine diagnosis but is performed in specialist or reference laboratories.

- Detection of serum antibodies is more useful as a retrospective diagnosis.

- Accurate species diagnosis by serology is difficult because of cross-reactions within the genus.

- Immunofluorescence and enzyme immunoassay are frequently used serological assays.

- The Weil-Felix agglutination reaction has low sensitivity and specificity.

- PCR combined with RFLP or sequencing can give an accurate species identification, particularly on biopsy material from rashes.

- Differential diagnosis includes many illnesses that present with a fever and a rash.

5. How is the disease managed and prevented?

- The mainstay of treatment is doxycycline.

- Alternative treatments include chloramophenicol and azithromycin.

- Preventative measures include insect repellents and protective clothing.

FURTHER READING

Cimolai N. Laboratory Diagnosis of Bacterial Infections. Marcel Dekker, New York, 2001: 823–860.

Heymann DL. Control of Communicable Disease Manual. American Public Health Association, Washington, DC, 2004: 459–464, 583–590.

Mandell GL, Bennet JE, Dolin R. Principles & Practice of Infectious Diseases, 6th edition, Vol 3. Elsevier, Philadelphia, 2005: 2284–2310.

Mims C, Dockrell HM, Goering RV, Roitt I, Wakwlin D, Zuckerman M. Medical Microbiology, 3rd edition. Mosby, Edinburgh, 2004: 386–389.

Murphy K, Travers P, Walport M. Janeway's Immunobiology, 7th edition. Garland Science, New York, 2008.

REFERENCES

Carl M, Tibbs CW, Dobson ME, Paparello S, Dasch GA. Diagnosis of acute typhus using the polymerase chain reaction. J Infect Dis, 1990, 161: 791–793.

Dignat-George F, Teysseire N, Mutin M, et al. *Rickettsia conorii* infection enhances vascular cell adhesion molecules-1 and intercellular adhesion molecule 1-dependent mononuclear cell adherence to endothelial cells. J Infect Dis, 1997, 175: 1142–1152.

Eremeeva ME, Dasch GA, Silverman DJ. Evaluation of a PCR assay for quantitation of *Rickettsia rickettsii* and closely related spotted fever group Rickettsia. J Clin Microbiol, 2003, 41: 5466–5472.

Hechemy KE, Oteo JA, Raoult DA, Silverman DJ, Blanco JR. A century of rickettsiology: emerging, re-emerging rickettsioses, clinical, epidemiologic, and molecular diagnostic aspects and emerging veterinary rickettsioses. Ann N Y Acad Sci, 2006, 1078: 1–14.

Kenny MJ, Birtles RJ, Day MJ, Shaw SE. *Rickettsia felis* in the United Kingdom. Emerg Infect Dis, 2003, 9: 1023–1024.

Li H, Walker DH. rOmpA is a critical protein for the adhesion of *Rickettsia rickettsii* to host cells. Microb Pathog, 1998, 24: 289–298.

Martinez JJ, Cossart P. Early signalling events involved in the entry of *Rickettsia conorii* into mammalian cells. J Cell Sci, 2004, 117: 5097–5106.

Olano JP. Rickettsial infection. Ann N Y Acad Sci, 2005, 1063: 187–196.

Raoult D, Fournier PE, Fenollar F, et al. *Rickettsia africae*, a tick-borne pathogen in travelers to sub-Saharan Africa. N Engl J Med, 2001, 344: 1504–1510.

Treadwell TA, Holman RC, Clarke MJ, Krebs JW, Paddock CD, Childs JE. Rocky mountain spotted fever in the United States, 1993–1996. Am J Trop Med Hyg, 2006, 63: 21–26.

Walker DH. Rickettsiae and rickettsial infection: the current state of knowledge. Clin Infect Dis, 2007, 45: S39–S44.

WEB SITES

All the Virology on the WWW Website, developed and maintained by Dr David Sander, Tulane University: http://textbookofbacteriology.net/Rickettsia.html

Centers for Disease Control and Prevention, Division of Viral and Rickettsial Diseases, Atlanta, GA, USA: http://www.cdc.gov/ncidod/dvrd/

Health Protection Agency, Centre for Infections, UK. © Health Protection Agency: http://www.hpa.org.uk/infections/topics_az/zoonoses/

Microbiology and Immunology Online, School of Medicine, University of South Carolina: http://pathmicro.med.sc.edu/mayer/ricketsia.htm

MULTIPLE CHOICE QUESTIONS

The questions should be answered either by selecting True (T) or False (F) for each answer statement, or by selecting the answer statements which best answer the question. Answers can be found in the back of the book.

1. **Which of the following are true of the genus *Rickettsia*?**

   A. They can be cultured on artificial media.

   B. They have a large genome.

   C. *Coxiella* is a not member of the *Rickettsiales*.

   D. They have a gram-positive cell wall.

   E. *Salmonella typhi* is one of the *Rickettsiales*.

2. **Which of the following are true concerning the spread of *Rickettsia* spp.?**

   A. *Rickettsia* are spread by arachnidae.

   B. Animal reservoirs maintain the bacteria in nature.

   C. Some species of *Rickettsia* only infect humans.

   D. Many *Rickettsia* have a global distribution.

   E. *Rickettsia* typically infect epithelial cells.

3. **Which of the following are known to be important in the pathogenesis of rickettsial disease?**

   A. Extracellular location of the organism.

   B. Induced endocytosis.

   C. Leukocyte adhesion to vascular endothelium.

   D. Increased vascular permeability.

   E. Infection of epithelial cells.

4. **Which of the following are true of *Rickettsia rickettsii*?**

   A. It causes Brill-Zinsser disease.

   B. It is transmitted by a flea bite.

   C. It presents with fever, headache, myalgia, and rash.

   D. The disease is more severe in patients with G6PD deficiency.

   E. It produces a vesicular rash.

5. **Which of the following are true statements concerning the diagnosis of rickettsial disease?**

   A. The Gram stain of biopsy material is useful.

   B. Serology provides an accurate diagnosis of acute disease.

   C. Immunofluorescence is frequently used in laboratory diagnosis.

   D. PCR can give an accurate species identification.

   E. PCR primers that are useful include those based on OmpA and citrate synthase.

6. **Which of the following are used in the treatment of rickettsial disease?**

   A. Ampicillin.

   B. Chloramphenicol.

   C. Doxycycline.

   D. Oxytetracycline.

   E. Ciprofloxacin.

Case 31

Salmonella typhi

A 53-year-old lady returned from a visit to Lahore complaining of feeling generally unwell. She had a temperature and a cough. She attended a clinic where she was seen by a doctor who confirmed that she had a temperature of 38°C and he noticed a rash on the upper chest. She was admitted to hosptial for investigation, which included a thick and thin film for malaria, a full blood count, urea and electrolytes, a chest X-ray, and blood cultures. The malaria investigation was negative, the chest X-ray showed patchy basal consolidation, the full blood count revealed a relative **lymphocytosis** and gram-negative bacilli were seen in the blood culture (Figure 1). A provisional diagnosis of enteric fever was made and she was started on appropriate antibiotics. The diagnosis was confirmed by isolation of *Salmonella typhi* from the blood cultures.

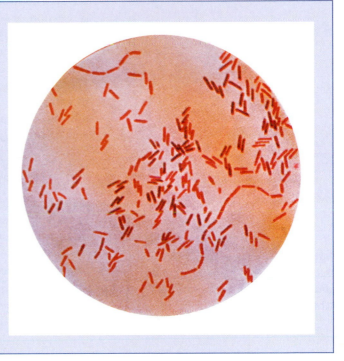

Figure 1. Gram-negative bacilli in a blood culture.

1. What is the causative agent, how does it enter the body and how does it spread a) within the body and b) from person to person?

Causative agent

The nomenclature of the genus *Salmonella* has undergone a number of revisions leading to two systems of validly published names, the latest version of which (post-2002) has the disadvantage of not highlighting important human pathogens such as *Salmonella typhi* or *Salmonella enteritidis* by not giving them succinct names. In the current version of the nomenclature the cause of enteric fever and the organism that is one of the important causes of gastroenteritis would be *S. enterica* subsp. *enterica* serovar Typhi and S. *enterica* subsp. *enterica* serovar Enteritidis, respectively. For pragmatic reasons, this nomenclature will not be used in this text but will be shortened to *Salmonella* Typhi. There are only three species of which *Salmonella enterica* is further subdivided into subspecies and **serovars**.

Salmonella organisms are motile nonsporing gram-negative facultative anaerobic rods measuring 2–3 × 0.4–0.6 μm. The genome of *Salmonella* Typhi and *Salmonella* Typhimurium have both been sequenced and contain about 4.8 million base pairs with 4000–5000 coding sequences, of which 98% are homologous between the two strains. Several **pathogenicity** islands (**PAIs**) are present in *Salmonella enterica* (SPI-1, which codes for a **type III secretion system** and another, SPI-2) and *Salmonella* Typhi has an additional SPI-7 coding for the Vi antigen. Frequent genome rearrangements associated with the PAI lead to outgrowth of strains that are better adapted to different environmental circumstances. Horizontal gene transfer is also an important factor in the evolution of the genus. A large plasmid is carried in *Salmonella* Typhi (pHCM1) that encodes drug resistance. **Virulence** plasmids are also carried in nontyphoidal strains. There are over 2500 serovars in the genus and they are grouped according to the possession of somatic O antigens, flagellar H antigens, and surface virulence (Vi) antigens in the Kauffman-White scheme. K (capsular) antigens are also present (Figure 2). Most *Salmonella* spp. express two types of flagella made up of different proteins (H antigen) and they switch from one phase to the next at characteristic frequencies. Thus the H antigens occur in two phases: phase I and II. The O antigens are designated by arabic numerals, the phase I antigens by letters a–z, and the phase II antigens by arabic numerals. Thus a specific serovar may be designated [9,12,Vi/d/–] = *Salmonella* Typhi (*Salmonella* Typhi does not have a phase II antigen) or [1,4,5,12/i/1,2] = *Salmonella* Typhimurium. For epidemiological purposes a number of typing methods have been used. These include **phage typing**, plasmid profiling, **ribotyping**, **pulse field gel electrophoresis** (**PFGE**), variable number of tandem repeats (**VNTR**), and enterobacterial repetitive intergenic consensus-**polymerase chain reaction** (ERIC-**PCR**).

Entry and spread within the body

Salmonella are ingested in food or water. The infecting dose is uncertain but has been reported to vary between 10^2 and 10^6. Reduced gastric acidity is

Figure 2. The relationship between the O (cell wall) and H (flagella) antigens and the Kaufmann-White serotyping scheme.

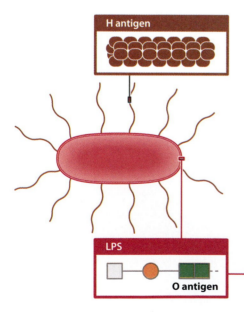

group	serovar	O	H1	H2
A	*paratyphi A*	1, 2, 12	a	(1, 5)
B	*typhimurium*	1, 4, (5), 12	i	1, 2
	brandenburgh	4, 12	i, v	enz$_{15}$
	saint paul	1, 4, (5), 12	e, h	1, 2
C	*choleraesuis*	6, 7	c	1, 5
	montevideo	6, 7	g, m, s	(1, 2, 7)
D	*typhi*	9, 12 , Vi	d	-
	enteritidis	9, 12	g, m	(1, 7)
	panama	9, 12	l, v	1, 5

lipid A	core area	O antigen variable region

repeat unit

key	
H	haunch (breath) bacteria which produce a thin film on the growth medium similar to the effect of breath on a glass, only motile bacteria do this because they have flagellae, H antigens can occur as two types, H1 and H2
O	ohne (without) that is without breath i.e. do not produce a thin film on the growth medium because they do not have flagellae and are nonmotile
Vi	virulence

a predisposing risk factor and colonization by *Helicobacter pylori* may be an important co-morbidity through its reduction in acid output in cases of pan-gastritis. Adhesion to epithelial cells occurs via **fimbriae**. Both *Salmonella* Typhi and nontyphoidal strains enter via the M cells and/or **dendritic cells** and enterocytes; *Salmonella* Typhi adheres to the **cystic fibrosis transmembrane conductance receptor** (**CFTM**) on gastrointestinal epithelial cells. Mutations in this protein may lead to lack of adhesion and thus resistance to infection. Once adherent, the bacteria invade by a process called bacteria-mediated **endocytosis** where the type III secretion system is involved. Adhesion of the bacterium induces cytoskeletal changes in the epithelial cell with membrane ruffling that encloses the organism into a vacuole. Once endocytosed a proportion of **vesicles** will fuse with the basolateral membrane and enter the **lamina propria**. During adhesion *Salmonella* Typhi induces up-regulation of the host CFTM receptor and an increased level of translocation to the lamina propria. Here the organism enters macrophages and dendritic cells where it survives and replicates. *Salmonella* Typhi surviving and replicating within the monocytic lineage are released and pass to the mesenteric lymph nodes and via the thoracic duct to the general circulation where they localize in the cells of the **mononuclear phagocyte system** (in liver, spleen, and bone marrow). The bacteria continue to replicate and are shed into the bloodstream with the onset of clinical illness and are circulated in the blood to all body organs and induce organ specific signs and symptoms. Movement of *Salmonella* Typhi from the gastrointestinal tract may also occur when dendritic cells and macrophages carrying the intracellular organisms migrate to mesenteric lymph nodes.

Nontyphoid *Salmonella*

These salmonella penetrate intestinal epithelial cells but rarely disseminate round the body by the bloodstream and are more prone to cause local disease. Co-infection with HIV is a risk factor for severe nontyphoidal infection (but not for enteric fever, which emphasizes important differences in the pathophysiology of the diseases) and is a significant cause of mortality.

Spread from person to person

Salmonella Typhi

Salmonella Typhi (and *Salmonella* Paratyphi) is a strictly human pathogen and it is **endemic** in several countries (see below). Endemicity of typhoid fever is associated with a poor social hygiene infrastructure (inadequate sewage disposal, inadequate potable water supplies, and inadequate food hygiene practices) and social upheaval.

Salmonella Typhi is acquired either directly or indirectly from another human or carrier by ingestion. Direct acquisition from a person is uncommon but can occur associated with certain sexual practices. More usually infection is acquired from fecally contaminated water or food and less commonly in laboratories handling clinical specimens. Outbreaks have been linked to food-handlers who are carriers of the organism – the most notorious being 'Typhoid Mary'. Mary Mallon was a cook for a New York banker, who infected most of his family with typhoid because she was a healthy carrier excreting *Salmonella* Typhi. Investigations by a civil engineer employed by the banker to identify the source revealed that outbreaks of typhoid had occurred in seven families for whom Mary had been the cook.

The organism is endemic in India, Africa, South America, and South-East Asia. In Africa the incidence is about 900/100 000 population. In Indonesia enteric fever is one of the commonest causes of death in children, with 20 000 deaths each year. In nonendemic countries such as the UK and the USA the incidence is about 0.2/100 000 population. In the UK the number of cases in 1980 was 336 (caused by *Salmonella* Typhi, *Salmonella* Paratyphi A and B) and in 1996 the number of cases was 525.

A significant proportion of cases of typhoid in industrialized countries are acquired abroad: in the UK in 2006 this comprised over 50% of the cases and in the USA, 72% of cases between the years 1985 and 1994.

Nontyphoid *salmonella*

Nontyphoid *Salmonella* are found in a variety of animal species. Infection is acquired by ingestion from contaminated food, particularly eggs, poultry, and dairy produce. The sudden increase in *Salmonella* Enteritidis PT4 in the UK was linked to contaminated eggs. Infection may also be acquired from direct contact with infected animals, particularly exotic pets, for example snakes.

Epidemiology

The number of cases of gastroenteritis caused by the 2500 different serovars of salmonella (e.g. *Salmonella* Enteritidis, *Salmonella* Typhimurium, etc.) has increased over the years. In the USA about 1.4 million cases occur annually with an incidence of 17/100 000 population, with *Salmonella* Typhimurium and *Salmonella* Enteritidis being the most common serovars causing infection. In the UK in 1981 the number of reported cases was 8939, which increased to 12 375 in 2006. The most frequent serovar causing disease was *Salmonella* Typhimurium, until 1988 when there was a sudden increase in *Salmonella* Enteritidis, which became the predominant serovar. The predominant strain of *Salmonella* Enteritidis causing disease is phage type 4 (PT4) and in 1988 *Salmonella* Enteritidis accounted for 12 302 cases, outnumbering all the other salmonella types combined. It remains the predominant phage type, although the number of isolates is now less than the combined number of all the other phage types.

2. What is the host response to the infection and what is the disease pathogenesis?

Low gastric acid, immunosuppression, and certain HLA polymorphisms predispose to infection.

Innate immunity

Infection with *Salmonella* Typhi induces a monocytic response with little diarrhea, due in part to the immunosuppressive effect of the **Vi antigen**. On the other hand, nontyphoid salmonella induces an acute inflammatory reaction with the secretion of **interleukin (IL)-8**, the recruitment of neutrophils, and pronounced diarrhea. The diarrhea is induced in part by translocated proteins and in part by disruption of the epithelial barrier.

In the lamina propria salmonella will enter epithelial cells and macrophages by bacteria-mediated endocytosis or **phagocytosis**, respectively. Once inside the macrophage, the bacterium is protected from the humoral immune system and it survives intracellular killing by the macrophage by up-regulating numerous genes required for intracellular survival and replication, including the Vi antigen, which is principally anti-phagocytic. Changes occur in the cell wall of salmonella making it resistant to bactericidal peptides and **reactive oxygen species** and cause less inflammation. Additionally, products coded by the pathogenicity island SPI-2 are secreted into the macrophage cytoplasm, which inhibit phagolysosome fusion by modulation of the actin cytoskeleton surrounding the vacuole and by incorporating proteins into the membrane of the vacuole.

In response to bacterial invasion, **pathogen-associated molecular patterns (PAMPs)** are recognized by **Toll-like receptors** (TLR2, lipoprotein; TLR4, **lipopolysaccharide (LPS)**; TLR5, flagellin) and initiate an inflammatory response in the gastrointestinal tract. Pro-inflammatory **cytokines** are induced by the salmonella, particularly IL-1, **interferon (IFN)**, IL-2, IL-6, and the **chemokine** IL-8, which recruits neutrophils. The epithelial monolayer is an important target and mainly secretes the IL-8 in response to the bacterium. The production of IL-8 is a result of the activation of NFκB within the epithelial cells and is dependent on the type III secretion system in the bacterium, which injects bacterial proteins into the cells thereby inducing the cytokine response. Killing by macrophages is also important as defects in macrophage function, for example by lack of IFN-γ, leads to severe disease.

Adaptive immunity

Eradication of the infection and the production of immunity require **CD4+** helper T cells and the production of specific antibodies by B cells. The role of cytotoxic CD8 cells in effective eradication of the organism in less clear.

Pathogenesis

The pathogenesis of disease occurs in several stages. Initial entry into the body and intracellular survival are followed by dissemination round the body to all organ systems by the bloodstream. Organs particularly affected are the cells of the mononuclear phagocytic system in the liver and spleen, bone marrow, gallbladder, and importantly, the Peyer's patches in the intestine. **Kupffer cells** in the liver are a major defense mechanism of the host but surviving salmonella that invade hepatocytes induce **apoptosis** of the cells. At this stage the patient will show systemic signs of inflammation: fever, **jaundice**, hepatosplenomegaly, **myalgia**, headache, and in some cases mental confusion. Later on in the infection, **rose spots** may appear and if left untreated by the third week there is an intense monocytic infiltration of Peyer's patches, which may rupture leading to signs of perforation.

3. What is the typical clinical presentation and what complications can occur?

Salmonella *Typhi*

Enteric (or typhoid) fever is a serious infection with an appreciable mortality if left untreated. It is caused by *Salmonella* Typhi or *Salmonella* Paratyphi. The incubation period is from 5 to 21 days depending on the size of the

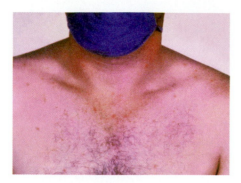

Figure 3. Rose spots on the upper trunk in a case of enteric fever.

ingested inoculum and host co-morbidity, for example immunosuppression. In adults the disease may present as a nonspecific **febrile** illness (**pyrexia of unknown origin – PUO**) or it may present with fever and gastrointestinal symptoms. In children there may be a much more nonspecific presentation. Typical symptoms include fever, headache, abdominal pain and tenderness, constipation or diarrhea, and delirium. On examination there may be a relative **bradycardia** (where the pulse rate is less than expected from the temperature of the patient), a pinkish **maculopapular rash** found on the trunk (rose spots – Figure 3), and hepatosplenomegaly. Laboratory investigations (see later) demonstrate **anemia**, **leukopenia**, classically, but not always, with a relative lymphocytosis, abnormal liver function tests, and elevated creatine phosphokinase (CPK). Risk factors for infection include immunosuppression and sickle cell anemia. Complications include **cholecystitis**, intestinal perforation and hemorrhage, **osteomyelitis**, and **endocarditis**. *Salmonella* Paratyphi (paratyphoid fever) causes a milder form of the illness compared with *Salmonella* Typhi. Following infection about 1–5% of patients may become long-term gastrointestinal carriers of the organism. These individuals are at greater risk of developing hepatobiliary and intestinal carcinoma.

Nontyphoid salmonella

Nontyphoid *Salmonella* are one cause of a self-limiting gastroenteritis that is indistinguishable from other bacterial causes. The incubation period is from 12 to 48 hours following ingestion of contaminated food. It presents with fever, abdominal pain, and diarrhea, which may occasionally be blood-stained. Complications include **paralytic ileus** leading to **toxic megacolon**.

4. How is this disease diagnosed and what is the differential diagnosis?

Diagnosis of enteric fever is made by isolation of *Salmonella* Typhi or Paratyphi from a blood culture, feces, or bone marrow of a symptomatic patient. The specimen providing the highest diagnostic yield is the bone marrow. Sampling of duodenal secretions using the string test may add to the diagnostic yield. The organism may also be isolated from other specimens such as rose spots, urine, sputum, and cerebrospinal fluid (CSF). Nontyphoid *Salmonella* are usually isolated from only the feces but occasionally may also be isolated from the blood.

One of a number of selective media may be used for fecal specimens, including xylose-lysine-deoxycholate (XLD), deoxycholate citrate agar (DCA), bismuth sulfite agar, Salmonella Shigella (SS) agar, and Hektoen agar. The general principle of these media is that they include sugars and a pH indicator to differentiate *Salmonella* spp. (and *Shigella* spp.) from nonenteric pathogens; a source of sulfur and iron as the *Salmonella* spp. generate hydrogen sulfide, which precipitates the iron to give black colonies; and a selective agent such as bile to which *Salmonella* spp. are resistant (Figure 4).

The organism is identified biochemically and serologically by testing for slide agglutination with O and H antibodies and by reference to the Kaufmann-White scheme. In the UK, enteric fever is a notifiable disease, as is food poisoning (of any microbiological cause). Although most industrialized countries have a reporting system for infectious diseases, many countries do not.

Figure 4. Growth of *Salmonella* on DCA medium. Note the black center in some colonies indicating precipitation of FeS.

Serologically, the **Widal test** has been used specifically for *Salmonella* Typhi but is not recommended for diagnosis (in the UK) as it does not have sufficient sensitivity; as many as 50% of patients with culture-confirmed enteric fever are negative in the Widal test. Newer serological assays such as a dot **ELISA** test format (Typhidot) detecting **IgG** and **IgM** in studies has shown a sensitivity and specificity of over 90%. A rapid latex agglutination inhibition test (Tubex) detecting only IgM to the somatic O9 antigen is also available. PCR and DNA probes are available but not currently in routine use, particularly in resource-poor countries.

Differential diagnosis

The differential diagnosis includes a wide range of causes of PUO such as brucellosis, malaria, typhus, and **Dengue fever** as well as causes of fever with intra-abdominal pathology such as **abscess**, amebic dysentery, and abdominal tuberculosis.

5. How is the disease managed and prevented?

Management

Salmonella Typhi and Paratyphi

Because of the increasing resistance to antibiotics including chloramphenicol, co-trimoxazole, ampicillin, and ciprofloxacin in India and South-East Asia, the treatment depends on the geographic location of the infecting strain. Similar antibiotic resistance also occurs in nontyphoid *Salmonella* spp. and, particularly in the UK, the emergence of resistance to ciprofloxacin was temporally associated with the use of a veterinary fluoroquinolone – enrofloxacin. The current recommended treatment in the UK for patients from India and South-East Asia is azithromycin in moderate illness and ceftriaxone for severe disease. For patients from other locations the antibiotic of choice is ciprofloxacin.

Enteric fever can be prevented by vaccination. Three main types of vaccine exist including an oral live attenuated strain (Vivotif-Ty21a), a parenteral vaccine based on the Vi capsular antigen (Typhim Vi), and a **conjugate vaccine** based on the Vi antigen and recombinant exotoxin A of *Pseudomonas aeruginosa* (Vi-rEPA). An acetone-inactivated parenteral vaccine also exists. The protection afforded by these vaccines ranges from 60 to 90%.

Nontyphoid Salmonella spp.

Generally gastroenteritis does not require antibiotic treatment as the illness is self-limiting. However, in severe disease or if there is evidence of systemic spread then depending on the sensitivities an appropriate antibiotic such as a cephalosporin or ciprofloxacin can be used. An unwanted effect of antibiotics is to increase the carriage rate of the organism.

Prevention

Prevention of gastroenteritis relies on control of critical points along the food chain from farm to dining room, which includes infection-free animals, sanitary animal transport, hygienic processes in abattoirs, and correct storage and cooking conditions in the kitchen.

SUMMARY

1. What is the causative agent, how does it enter the body and how does it spread a) within the body and b) from person to person?

- *Salmonella* spp. are motile gram-negative rods.

- There are over 2500 serovars, they are grouped in the Kauffman-White scheme based upon the O and H antigens.

- *Salmonella* Typhi is a strictly human pathogen and is acquired from food or water contaminated with organisms from another case or a carrier.

- Nontyphoid *Salmonella* spp. are found in animals and infection is acquired from contaminated food or pets.

- *Salmonella* Typhimurium and *Salmonella* Enteritidis are the two most frequent causes of nontyphoid *Salmonella* spp. gastroenteritis.

- *Salmonella* Typhi penetrates the gastrointestinal epithelium and is disseminated by the bloodstream.

- Nontyphoid *Salmonella* spp. generally remain within the gastrointestinal tract, although some serovars may disseminate in the blood.

2. What is the host response to the infection and what is the disease pathogenesis?

- Low gastric acid, immunosuppression, and certain HLA polymorphisms predispose to infection.

- Survival in macrophages occurs by modulating phagolysosome fusion and the bacterial cell wall to resist inimical bactericides.

- *Salmonella* Typhi induces a monocytic response with few gastrointestinal symptoms.

- Nontyphoid *Salmonella* spp. induce a granulocyte response with pronounced gastrointestinal symptoms.

- *Salmonella* Typhi localizes to cells of the monocyte/macrophage system where it replicates.

- *Salmonella* Typhi is distributed to all body organs by the blood, giving rise to organ-specific disease.

3. What is the typical clinical presentation and what complications can occur?

- Enteric fever presents as a PUO and abdominal symptoms.

- The incubation period is 5–21 days.

- A rash (rose spots) may be present.

- There may be a relative bradycardia and a leukopenia.

- Hepatosplenomegaly may be present.

- Complications include cholecystitis, intestinal perforation, endocarditis, and osteomyelitis.

- Nontyphoidal gastroenteritis is self-limiting.

- The incubation period is 12–48 hours.

- It presents with fever, abdominal pain, and diarrhea.

4. How is this disease diagnosed and what is the differential diagnosis?

- Enteric fever is diagnosed by isolation and identification of the organism.

- The highest yield of organism is from the bone marrow.

- Identification of *Salmonella* spp. depends on biochemical reactions and serological reactions using the Kaufmann-White scheme.

- In many countries including the UK food poisoning and enteric fever are notifiable diseases.

5. How is the disease managed and prevented?

- There are increasing levels of antibiotic resistance in *Salmonella*.

- Treatment is with azithromycin or ceftriaxone in patients from India and SE Asia.

- Treatment is with ciprofloxacin in patients from other areas.

- Protection from enteric fever is provided by vaccination.

- Antibiotics are not usually required in cases of gastroenteritis caused by nontyphoid *Salmonella*.

- Prevention of gastroenteritis is by adequate hygiene standards at all points in the food chain.

FURTHER READING

Cimolai N. Laboratory Diagnosis of Bacterial Infections. Marcel Dekker, New York, 2001: 423–497.

Lydyard P, Lakhani S, Dogan A, et al. Pathology Integrated: An A–Z of Disease and its Pathogenesis. Edward Arnold, London, 2000: 257.

Mandell GL, Bennet JE, Dolin R. Principles & Practice of Infectious Diseases, 6th edition, Vol 2. Elsevier, Philadelphia, 2005: 2636–2654.

Murphy K, Travers P, Walport M. Janeway's Immunobiology, 7th edition. Garland Science, New York, 2008.

Mims C, Dockrell HM, Goering RV, Roitt I, Wakwlin D, Zuckerman M. Medical Microbiology, 3rd edition. Mosby, Edinburgh, 2004: 300–302.

REFERENCES

Baker S, Dougan G. The genome of *Salmonella enterica* serovar Typhi. Clin Infect Dis, 2007, 45(Suppl 1): S29–S33.

Bhan MK, Bahl R, Bhatnager S. Typhoid and paratyphoid fever. Lancet, 2005, 366: 749–762.

Cooke FJ, Wain J. The emergence of antibiotic resistance in typhoid fever. Travel Med Infect Dis, 2004, 2: 67–74.

Everest P, Wain J, Roberts M, Rook G, Dougan G. The molecular mechanisms of severe typhoid fever. Trends Microbiol, 2001, 9: 316–320.

Fang FC, Vazques-Torres A. *Salmonella* selectively stops traffic. Trends Microbiol, 2002, 10: 391–392.

Gentschev I, Spreng S, Sieber H, et al. Vivotif – a magic shield for protection against typhoid fever and delivery of heterologous antigens. Chemotherapy, 2007, 53: 177–180.

Guzman CA, Borsutzky S, Griot-Wenk M, et al. Vaccines against typhoid fever. Vaccine, 2006, 24: 3804–3811.

Haraga A, Ohlson MB, Miller SI. *Salmonella* interplay with host cells. Nat Rev Microbiol, 2008, 6: 53–66.

Huang DB, DuPont HL. Problem pathogens: extraintestinal complications of *Salmonella enterica* serovar Typhi infection. Lancet Infect Dis, 2005, 5: 341–348.

Kam KM, Luey KY, Chiu AW, et al. Molecular characterization of *Salmonella enterica* serotype Typhi isolates by pulsed-field gel electrophoresis in Hong Kong 2000-2004. Foodborne Pathog Dis, 2007, 4: 41–49.

Liu GR, Liu WQ, Johnston RN, et al. Genome plasticity and ori-ter rebalancing in *Salmonella typhi*. Mol Biol Evol, 2006, 23: 365–371.

Parry CM, Hien TT, Dougan G, White NJ, Farrar JJ. Typhoid fever. N Engl J Med, 2002, 347: 1770–1782.

Raffetellu M, Chessa D, Wilson RP, Tukel C, Akcelik M, Baumler AJ. Capsule mediated immune evasion: a new hypothesis explaining aspects of typhoid fever pathogenesis. Infect Immun, 2006, 74: 19–27.

Roumagnac P, Weill FX, Dolecek C, et al. Evolutionary history of *Salmonella typhi*. Science, 2006, 314: 1301–1304.

Srikantiah P, Vafokulov S, Luby SP, et al. Epidemiology and risk factors for endemic typhoid fever in Uzbekistan. Trop Med Int Health, 12: 838–847.

Vazques-Torres A, Fang FC. Cellular routes of invasion by enteropathogens. Curr Opin Microbiol, 2000, 3: 54–59.

Zhang XL, Jeza VT, Pan O. *Salmonella typhi*: from a human pathogen to a vaccine vector. Cell Mol Immunol, 2008, 5: 91–97.

WEB SITES

LPSN, List of Prokaryotic names with Standing in Nomenclature, Salmonella nomenclature page: http://www.bacterio.cict.fr/salmonellanom.html

Centre for Infections, Health Protection Agency, HPA Copyright, 2008: http://www.hpa.org.uk/infections/topics_az/typhoid/menu.htm

eMedicine, Inc., 2007, provider of clinical information and services to physicians and health-care professionals: http://www.emedicine.com/MED/topic2331.htm

MULTIPLE CHOICE QUESTIONS

The questions should be answered either by selecting True (T) or False (F) for each answer statement, or by selecting the answer statements which best answer the question. Answers can be found in the back of the book.

1. **Which of the following are true of the genus *Salmonella*?**

    A. They are motile.

    B. They can be grouped serologically into a small number of serovars.

    C. The Kauffman-White scheme of classification is based upon the O, H, and Vi antigens.

    D. Phase variation occurs with the O antigen.

    E. Phage typing may be used for epidemiological purposes.

2. **Which of the following are true concerning *Salmonella* Typhi?**

    A. *Salmonella* Typhi is widespread in animals.

    B. Infection is usually acquired by ingestion of contaminated food or water.

    C. The global burden of disease is an estimated 16 million cases per year.

    D. A significant proportion of cases are linked to a food-handler who is a carrier of the organism.

    E. In industrialized countries many cases are acquired abroad from an endemic country.

3. **Which of the following are true statements concerning nontyphoid *Salmonella*?**

    A. They are widespread in animals.

    B. They are one of the commonest causes of gastroenteritis.

    C. *Salmonella* Typhimurium is one of the commonest serovars causing illness.

    D. In Europe and the USA infection with *Salmonella* Enteritidis PT4 is associated with keeping exotic animals.

    E. The number of reported cases is decreasing.

4. **Which of the following is true of *Salmonella* Typhi?**

    A. It adheres to the cystic fibrosis transmembrane receptor.

    B. It produces a significant secretory response.

    C. Large numbers of granulocytes are induced.

    D. It inhibits phagolysosome fusion.

    E. It induces bacteria-mediated endocytosis.

5. **Which of the following are true of the clinical presentation of disease caused by *Salmonella* Typhi?**

    A. It often presents with a PUO.

    B. Disease is usually restricted to the gastrointestinal tract.

    C. A rash may be a presenting feature.

    D. Gastrointestinal perforation is a complication.

    E. The carrier state is unknown.

6. **Which of the following is useful in the diagnosis of *Salmonella*?**

    A. Gram stain.

    B. Culture of bone marrow.

    C. Serology.

    D. PCR.

    E. String test.

7. **Which of the following are used in the treatment or prevention of *Salmonella* Typhi infection?**

    A. Vaccination.

    B. Azithromycin.

    C. Erythromycin.

    D. Enrofloxacin.

    E. Ceftriaxone.

Case 32

Schistosoma spp.

A 25-year-old male was admitted to hospital vomiting blood. On examination of the abdomen both the liver and the spleen were enlarged. A full blood count revealed a reduced hemoglobin level of 9 g dl⁻¹ and the white cell differential showed a raised eosinophil count of 1.1×10^9 L⁻¹ (normal range $<0.45 \times 10^9$ L⁻¹). Endoscopic examination revealed dilated **esophageal varices** (Figure 1). **Cirrhosis** of the liver was suspected, but the patient denied any excessive drinking of alcohol and tests showed that he was negative for hepatitis B and C. A liver biopsy was performed and confirmed an abnormal liver, with extensive granulomatous change and **fibrosis**. On further questioning it transpired that he had spent much of his childhood living in a small village in Kenya. His parents had been working there on a farm. He would regularly paddle and swim in a nearby lake. He recalled no particular illness while living in Africa. Urine examination was negative but stool microscopy demonstrated the presence of *Schistosoma mansoni* eggs. A rectal biopsy obtained at **sigmoidoscopy** showed **granulomas**, containing *Schistosoma* eggs. Schistosomal **serology** was also positive. He was treated with praziquantel and his stool was monitored for the disappearance of *Schistosoma* eggs. He was referred to a hepatologist for further management of his esophageal varices. Serial ultrasounds were performed to monitor the liver fibrosis.

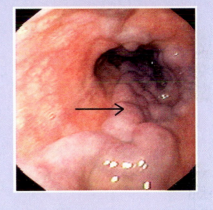

Figure 1. Endoscopic view of large esophageal varices (arrowed).

1. What is the causative agent, how does it enter the body and how does it spread a) within the body and b) from person to person?

Causative agent

Human schistosomiasis, also known as bilharzia, is mainly caused by three species of *Schistosoma*. *S. mansoni* occurs in Africa, parts of the Arabian peninsula, the Caribbean, and South America. *S. haematobium* occurs in Africa and parts of the Arabian peninsula. *S. japonicum* is now found in China, the Philippines, and Indonesia. *S. intercalatum* occurs in pockets in west and central African and *S. mekongi* is restricted to small pockets in Cambodia and Laos.

Schistosomes are blood flukes, which are also known as trematodes. Adult worms are less than 2 cm long (Figure 2). Males have a longitudinal groove in which the female worm resides. They live in the veins around the bladder in the case of *S. haematobium* or in the mesenteric veins of the intestine in the case of *S. mansoni* and *S. japonicum*.

Entry and spread within the body

The life cycle of schistosomiasis is shown in Figure 3. Humans are infected when they enter water containing *Schistosoma* cercariae (Figure 4). These

Figure 2. Adult schistosome worms. The female (on the left) is slender and fits into a groove on the ventral surface of the male.

Figure 3. Life cycle of *Schistosoma* spp.
Eggs shed in urine or feces (1) from human hosts release miracidia on contact with fresh water (2) and then enter intermediate snail hosts (3). After multiplication in the snail hosts (4) cercariae arise (5), which penetrate human skin on water contact (6). They develop into schistosomula (7), which mature into adult worms and live within blood vessels (10) around the bladder or intestine (A, B, C). Adult male and female pairs shed eggs, about half of which exit from the body and the other half are deposited in tissues, causing pathology.

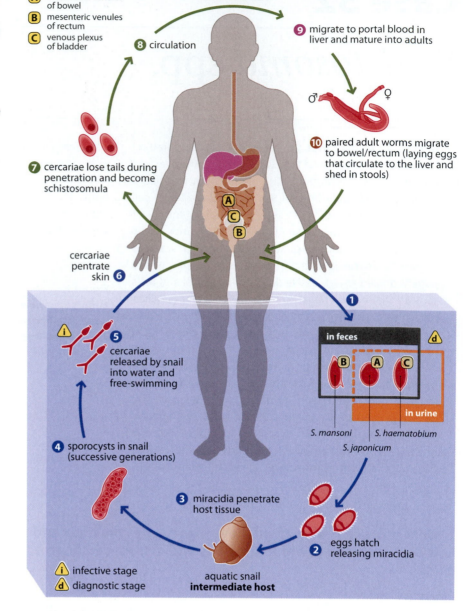

larvae penetrate the skin. The cercariae transform into a larval stage called the schistosomulum. These enter blood vessels and pass round the venous circulation, through the lungs and then enter the arterial circulation. They reach the liver. The schistosomula mature over about a month to adult worms in the portal vein. Males and females pair and then migrate distally to the veins around the bladder and the intestine. They usually live for 3–5 years. However, there are reports of individuals who have left endemic areas for several years and show signs of active infection many years later. From such reports it is thought that occasionally adult pairs may survive for up to 30 years. The case history illustrates this prolonged survival.

Person to person spread

Schistosomiasis does not spread directly from person to person but the life cycle involves a snail intermediate host. *S. mansoni* and *S. haematobium*

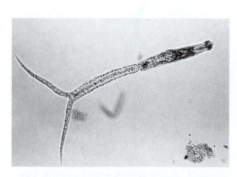

Figure 4. A Schistosome cercaria, which emerges from snails and penetrates human skin.

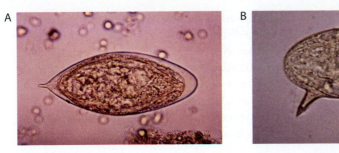

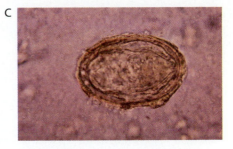

females lay between 20 and 300 eggs per day. The figure can be considerably higher for *S. japonicum*, with an estimated 500–3500 eggs per day. About half of these eggs succeed in passing through the wall of the bladder or intestine to emerge in urine or feces. Otherwise eggs get stuck in the bladder or intestinal wall or embolize in the venous circulation to the liver. In advanced, heavy infections collateral circulation develops and eggs disseminate into the lungs, brain, and other organs.

Eggs have a spinous process. The shape of the egg and the position of the spine is characteristic for each species of *Schistosoma* (Figure 5). *S. haematobium* has a terminal spine. *S. mansoni* and *S. japonicum* have lateral spines, with *S. japonicum* having smaller-sized eggs.

The eggs contain a ciliated larval form called the miracidium, which is released when the eggs come into contact with fresh water in lakes, rivers, and streams. Miracidia go on to infect specific aquatic snails. Within the snails they multiply into sporocysts from which arise the cercariae that close the life cycle. The species of snails differs between the species of *Schistosoma*. *S. mansoni* is transmitted by *Biomphalaria* snails, *S. haematobium* by *Bulinus*. These are aquatic snails, while *S. japonicum* is transmitted by an amphibious snail called *Oncomelania*.

Figure 5. Eggs of (A) *S. haematobium* **with a terminal spine, (B)** *S. mansoni* **with a lateral spine, and (C)** *S. japonicum,* **which are smaller and rounder** (courtesy of Public Health Image Library, Centers for Disease Control & Prevention, USA).

Epidemiology

Schistosomiasis is endemic in many tropical and subtropical countries and affects more than 200 million people worldwide. It is estimated that about 20 million people have severe illness and that there are at least 14 000 deaths per annum. Some even estimate mortality reaching 200 000 per annum.

2. What is the host response to the infection and what is the disease pathogenesis?

A host immune inflammatory response may be directed against the antigens of the schistosomula larvae, the adult worms or the eggs. There is evidence that the host response provides some level of protective immunity to re-infection. Rates of re-infection have been studied in endemic areas when adults or children are treated to clear infection. The rate of re-infection can only partly be explained by behavior and intensity of water contact. Correcting for the extent of water contact, adults who have had a greater lifetime exposure to schistosomiasis are less likely to be re-infected than children.

In vitro studies have shown that schistosomula are more susceptible to immune attack than adult worms. Immunity correlates with levels of **IgE** directed against schistosomula or adult worm antigens. Key antigens include schistosomal glutathione-*S*-transferase and glyceraldehyde-3-phosphate. The levels of antibodies rise with age. *In vitro*, IgE mediates killing of schistosomula by **opsonization** to eosinophils. The eosinophils degranulate to release toxic molecules such as eosinophil cationic protein. However, immunity is inversely correlated with levels of **IgG₂** and IgG₄ antibody directed against egg polysaccharide antigens. It is thought that these IgG isotypes also bind to schistosomula and block the binding of effector IgE antibodies.

Larval, adult, and egg antigens are sometimes released in such quantities that together with significant levels of circulating IgG, an **immune complex** disease develops (**type III hypersensitivity**). This occurs early in infection and is referred to as **Katayama syndrome**. Some observations also implicate **tumor necrosis factor-α (TNF-α)** in severe symptomatology.

Once adults mature there seems to be a ceiling to the number of adults present in the vasculature. This may be because of the physical space available. Adults become coated with host proteins, including self HLA molecules, and this 'host mimicry' may confuse the immune response, leading the host to believe the worm is 'self.' Very little, if any, inflammation is seen around adult worms.

A considerable immune and inflammatory response occurs against the eggs that lodge in tissues. Blood **eosinophilia** is often observed in association with considerable egg deposition. These eggs release soluble antigens that present a strong stimulus to **CD4+** T-helper lymphocytes, which orchestrate an inflammatory response. Granulomas form around the eggs with an influx of lymphocytes and macrophages. Fibroblasts deposit collagen and extracellular matrix proteins. The extent of resulting fibrosis depends on the balance of activities between matrix metalloproteinases (MMPs) and their tissue inhibitors (TIMPs). In turn these enzymes are affected by the balance of a variety of cytokines. **Transforming growth factor-β (TGF-β)**, , **interleukin** (IL)-1, IL-4, IL-13, and TNF-α promote fibrosis, while **interferon (IFN)**-γ is associated with reduced fibrosis. The actions of these **cytokines** in turn are affected by other regulatory cytokines such as IL-10 and IL-12. About 5–10% of patients develop extensive fibrosis (Figure 6). Numerous granulomas with surrounding fibrosis constitute the major part of tissue pathology. In time there can be remodelling of granulomas with some resorption of fibrotic tissue.

3. What is the typical clinical presentation and what complications can occur?

Clinical features may occur at different stages of infection. When cercariae first penetrate the skin there can be local irritation, which may appear immediately or within a few days. Raised, red spots may last a few days. Similarly, a swimmers' itch can be caused by trematodes of animal origin, particularly in temperate climatic zones.

A type III hypersensitivity reaction to high antigen release is referred to as Katayama syndrome, named after a region in Japan. This may occur between 2 and 12 weeks after infection. There is an abrupt onset of fever,

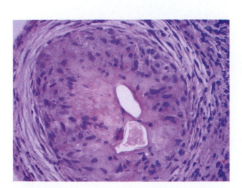

Figure 6. Granuloma formed around a schistosome egg.

malaise, aches, and tiredness. The lung becomes involved, with patchy shadowing on a chest X-ray. This is accompanied by a dry cough. There is also abdominal pain with diarrhea and both liver and spleen can be enlarged. On blood tests there is a high eosinophil count.

No symptomatology is attributed to the adult worms. Even once eggs are being deposited in tissues, symptoms may be absent or mild. Nonspecific features include malaise and tiredness. When considerable numbers of eggs settle in the intestine, bladder, liver, lung or other tissues problems arise as described below. These take some time to develop.

With *S. haematobium* granulomatous inflammation of the bladder wall can also result in ulceration (Figure 7). Bleeding into urine may be microscopic or macroscopic. Chronic lesions can progress to fibrosis and bladder calcification. This can block the ureters, with back pressure dilating the ureters and compromising kidney function. In the long term there is a risk that the damaged mucosal epithelium undergoes malignant change, resulting mainly in squamous cell carcinomas rather than transitional cell carcinomas. This is more common in smokers.

With *S. mansoni* and *S. japonicum* granulomatous inflammation of the intestine is principally found in the rectum and distal bowel. The mucosa may ulcerate. Abdominal pain is accompanied by diarrhea that may contain blood. From the peri-intestinal blood vessels eggs laid by adults embolize through the portal vein to the liver. In heavily infected or genetically predisposed individuals the granulomatous reaction in the liver results in considerable fibrosis; this periportal fibrosis – described as 'clay pipe stem' fibrosis – develops along the portal veins (Figure 8). The liver can enlarge and can feel hard and uncomfortable in the right upper abdomen. As blood flow through the liver is impaired by fibrosis, **portal hypertension** develops. This enlarges the spleen, which causes discomfort in the left upper abdomen. Gastro-esophageal varices develop and can bleed, as illustrated in the case history. Eventually damage to the liver may be so extensive that liver failure results in death.

In 5–10% of infected individuals portal hypertension becomes well established and collateral blood vessels open up around the liver. Eggs from the peri-intestinal blood vessels pass toward the liver but then bypass through these dilated blood vessels into the lung. This results in granulomatous inflammation in lung tissue. Eggs may reach any part of the body. *S. japonicum* worms lay more eggs than the other species and in the Far East granulomas may develop in the central nervous system and be a cause of epilepsy.

4. How is the disease diagnosed and what is the differential diagnosis?

The diagnosis of schistosomiasis can be made by microscopic analysis through finding eggs in urine or stool. Urine collected at the end of micturition provided at mid-day is the best specimen in which to look for *S. haematobium* eggs. This terminal specimen is likely to have a higher concentration of eggs. Egg shedding is also maximal about mid-day. Multiple stool specimens may be required to find eggs, as egg shedding is

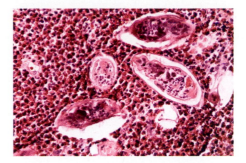

Figure 7. Intense eosinophilic inflammation around *S. haematobium* eggs in the bladder wall (courtesy of Public Health Image Library, Centers for Disease Control & Prevention, USA).

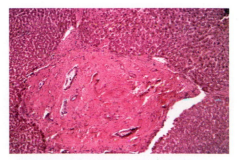

Figure 8. Normal liver architecture is seen at the edge of the slide. In the center there is considerable fibrosis around schistosome eggs (courtesy of Public Health Image Library, Centers for Disease Control & Prevention, USA).

variable. Stool specimens can be suspended and concentrated. In endemic areas, the so-called **Kato-Katz technique** (microscopic examination of 42 mg thick stool smears prepared on microscope slides) is the most widely used diagnosis. If eggs are not seen in urine or stool, endoscopic examination of bowel or bladder may show inflammatory changes and biopsies may yield granulomas containing eggs. Ultrasonography can be used to detect complications of infection such as changes in the bladder, ureters, kidneys, and liver.

In endemic areas the presence of **IgM**, IgG and IgE antibodies in serum does not distinguish current from past infection. In returned travelers who have not previously been to a schistosomiasis-endemic area positive serology may be diagnostic of exposure. Serology does not become positive till about 12 weeks after infection, and occasionally longer. Eosinophilia is suggestive of ongoing infection in the absence of another explanation.

Differential diagnosis

The differential diagnoses to be considered depend on the presentation (Table 1). In people who have traveled from endemic areas one may have to consider differential diagnoses for hematuria, diarrhea with or without blood, eosinophilia, and granulomas on biopsy. Diarrhea with blood is also known as **dysentery**. The case history illustrates late stage presentation when one also has to think of cirrhosis of the liver, due for instance to alcohol or chronic viral hepatitis B or C. In endemic areas it is possible to get both schistosomiasis and chronic viral hepatitis.

5. How is the disease managed and prevented?

Management

Standard treatment is monotherapy with oral praziquantel in a single dose or two doses 4–6 hours apart. In mass chemotherapy programmes single doses are more practical. Praziquantel acts against the adult worms, causing paralysis and damage to their surface. It does not act on schistosomula and cannot be used very soon after infection, before adults have matured. Artemisinin derivatives have activity against schistosomula. Praziquantel cures up to 90% of patients. Success is gauged by the disappearance of egg excretion 4–6 weeks after treatment. Repeat courses of praziquantel may be given. Increasing resistance of the worms to praziquantel is a concern in some regions, but may be overcome by more prolonged courses with higher doses.

Prevention

Vaccines against schistosomiasis are under investigation. While we wait in hope for an effective vaccine, prevention depends on health education, control of intermediate host snail populations, reducing contaminated water contact, and mass chemotherapy to treat infected individuals and reduce egg shedding. Environmental management to reduce intermediate host snail populations is difficult, as is separating humans from water contact. The building of dams for irrigation and hydroelectric purposes has had mainly negative effects on the incidence of infection. The provision of safe water supply and sanitation infrastructure are very important in reducing episodes of contaminated water contact. Some countries have undertaken population-wide mass chemotherapy with repeated cycles of treatment. This has reaped success in countries such as Brazil, China, and Egypt. The World Health Organization (WHO) wishes to extend this approach to many more countries. A Schistosomiasis Control Initiative is now under way in selected countries in Africa.

Table 1. Differential diagnoses to be considered with different presentations of schistosomiasis

Hematuria (blood in urine)
Urinary tract infection
Bladder or ureteric stones
Tumor of bladder or prostate in adults
Glomerulonephritis
Polycystic kidneys
Coagulopathy

Dysentery (bloody diarrhea)
Infections due to
  Campylobacter
  Shigella
  Enterohemorrhagic *Escherichia coli*
  Entamoeba histolytica
Colonic tumor
Inflammatory bowel disease, particularly Crohn's disease, which also causes granulomas

Eosinophilia
Infections due to
  Strongyloides
  Filaria (e.g. *Wuchereria, Onchocerca*)
  Migrating larval stages of intestinal nematodes (e.g. *Ascaris*, hookworms)
Atopic/allergic reactions
  Asthma
  Eczema
  Drug reactions
Vasculitis
Malignancy

Granulomas
Infections due to
  Mycobacteria
  Brucella
  Chlamydia granulomatosis
  Francisella tularensis
  Listeria monocytogenes
  Nocardia spp.
  Histoplasma
  Coccidiodes
  Fasciola hepatica
  Paragonimus
Sarcoidosis
Crohn's disease
Vasculitis
Lymphoma
Primary biliary cirrhosis
Chronic granulomatous disease
Berylliosis

SUMMARY

1. What is the causative agent, how does it enter the body and how does it spread a) within the body and b) from person to person?

- Schistosomiasis in humans is caused by three main species of blood flukes (trematodes) – *S. mansoni*, *S. haematobium*, and *S. japonicum* (two further species are of regional importance only).

- Adult male and female worms lodge together in blood vessels around the bladder (*S. haematobium*) or the intestine (*S. mansoni* and *S. japonicum*). Females lay eggs, about half of which pass into urine (*S. haematobium*) or feces (*S. mansoni* and *S. japonicum*) and others disseminate into various organs.

- Upon fresh water contact in lakes, rivers, and streams eggs release a larva called a miracidium that infects specific aquatic snails (*Biomphalaria* in the case of *S. mansoni*, *Bulinus* for *S. haematobium*) or the amphibious snail *Oncomelania* (in the case of *S. japonicum*).

- Cercariae arise from snails and transcutaneously infect humans when they are in contact with water. The first larval stage in humans is the schistosomulum.

2. What is the host response to the infection and what is the disease pathogenesis?

- Schistosomula are susceptible to immune attack by IgE and eosinophils. The binding of IgE may be blocked by IgG_2 and IgG_4 antibody against egg antigens.

- There is virtually no inflammation around adult worms living in blood vessels.

- Granulomas form around eggs that deposit in tissues. These granulomas are subsequently surrounded by fibrosis. Disease is largely due to granulomatous inflammation and extensive post-granulomatous fibrosis.

- Very occasionally excessive antigen release results in an immune complex disease.

3. What is the typical clinical presentation and what complications can occur?

- An itch and rash may occur when cercariae first penetrate the skin, particularly after primary infection.

- The occasional immune complex disease is called Katayama syndrome and comprises high fever, aches, cough, and later diarrhea.

- Disease of the bladder can cause bleeding into urine (hematuria) and later fibrosis can block the flow of urine with ureteral dilatation and renal failure. There is a risk of developing bladder cancer.

- Disease of the intestine can cause diarrhea with (dysentery) or without blood.

- Severe granulomatous change and fibrosis in the liver lead to portal hypertension with the risk of life-threatening bleeding from esophageal varices. The liver and spleen become enlarged.

- When collateral circulation develops granulomas can appear elsewhere such as in the lung or the central nervous system.

4. How is the disease diagnosed and what is the differential diagnosis?

- Diagnosis is based on visualizing characteristic eggs in urine, feces or biopsies.

- Serological tests are also available and are most useful in returned travelers.

- On routine blood testing eosinophilia may be found.

- The differential diagnosis includes other causes of hematuria, dysentery, eosinophilia, and granulomas.

- Ultrasonography is useful for identifying structural changes in the renal tract or liver.

5. How is the disease managed and prevented?

- Praziquantel is the mainstay of treatment.

- Prevention requires health education, environmental management to control intermediate host snail populations, and decreasing opportunities for water contact (through improved access to clean water and adequate sanitation).

- The provision of a safe water supply and sanitation infrastructure are important for schistosomiasis and many other diseases.

- Some countries undertake population-wide mass chemotherapy.

FURTHER READING

Davis A. Schistosomiasis. In: Cook GC, Zumla AI. Manson's Tropical Diseases, 21st edition. Elsevier, Edinburgh 2003, 1431–1469.

Murphy K, Travers P, Walport M. Janeway's Immunobiology, 7th edition. Garland Science, New York, 2008.

REFERENCES

Booth M, Mwatha JK, Joseph S, et al. Periportal fibrosis in human *Schistosoma mansoni* infection is associated with low IL-10, low IFN-γ, high TNF-α, or low RANTES, depending on age and gender. J Immunol, 2004, 172: 1295–1303.

Gryseels B, Polman K, Clerinx J, Kestens L. Human schistosomiasis. Lancet, 2006, 368: 1106–1118.

Ross AG, Bartley PB, Sleigh AC, et al. Schistosomiasis. N Engl J Med, 2002, 346: 1212–1220.

Ross AG, Vickers D, Olds GR, Shah SM, McManus. Katayama syndrome. Lancet Infect Dis, 2007, 7: 218–224.

Steinmann P, Keiser J, Bos R, Tanner M, Utzinger J. Schistosomiasis and water resources development: systematic review, meta-analysis and estimates of people at risk. Lancet Infect Dis, 2006, 6: 411–425.

Whetham J, Day JN, Armstrong M, Chiodini PL, Whitty CJ. Investigation of tropical eosinophilia; assessing a strategy based on geographical area. J Infect, 2003, 46: 180–185.

Whitty CJ, Mabey DC, Armstrong M, Wright SG, Chiodini PL. Presentation and outcome of 1107 cases of schistosomiasis from Africa diagnosed in a non-endemic country. Trans R Soc Trop Med Hyg, 2000, 94: 531–534.

WEB SITES

Center for Disease Control, Atlanta, GA, USA: www.cdc.gov/

Health Protection Agency: www.hpa.org.uk

Schistosomiasis Control Initiative: www.schisto.org

World Health Organization: www.who.int

MULTIPLE CHOICE QUESTIONS

The questions should be answered either by selecting True (T) or False (F) for each answer statement, or by selecting the answer statements which best answer the question. Answers can be found in the back of the book.

1. **Which of the following statements are true for the causative agent of schistosomiasis?**
   A. It is a trematode or blood fluke.
   B. It is restricted to North and South America.
   C. It lives in adult form in the human intestinal lumen.
   D. It reaches lengths of 20 cm in adult form.
   E. It is associated with snail populations.

2. **Which of the following statements are true for the life cycle of *Schistosoma*?**
   A. There has to be human contact with fresh water.
   B. Humans are infected by a larval form called a miracidium.
   C. Within humans the first larval form is called the schistosomulum.
   D. Adult male and female worms of *S. haematobium* normally live in blood vessels around the intestine.
   E. Adult worms usually live for 3–5 months.

3. **Which of the following statements are true about schistosome eggs?**
   A. *S. mansoni* females lay about 3500 eggs per day.
   B. Eggs of *S. japonicum* normally pass through the bladder wall into urine.
   C. Granulomas form around eggs deposited in tissue.
   D. *S. haematobium* eggs have a terminal spine.
   E. *S. japonicum* eggs are larger than the other species.

4. **Which of the following statements are true for the host immune response in schistosomiasis?**

   A. It is not thought to provide effective immunity to re-infection.

   B. It is capable of killing schistosomula through IgE-mediated mechanisms.

   C. It can lead to a hypersensitivity reaction with immune complex formation.

   D. It includes IgG$_2$- and IgG$_4$-mediated killing of adult worms.

   E. It is responsible for the pathology of disease.

5. **Which of the following are true for tissue granulomas in schistosomiasis?**

   A. They develop around schistosomula and adult worms.

   B. They are purely due to a Th1 response in schistosomiasis.

   C. Cytokines including IFN-γ and TNF-α are present.

   D. They eventually dismantle and disappear without fibrosis.

   E. They are associated with eosinophilia.

6. **Which of the following are clinical features of schistosomiasis?**

   A. A skin itch after swimming in water.

   B. Katayama syndrome with dry cough and changes on chest X-ray.

   C. Problems with adult worms blocking blood vessels.

   D. Nonspecific malaise and tiredness.

   E. Blood in the urine.

7. **Which of the following are true regarding the complications of schistosomiasis?**

   A. They are due to the inflammatory and fibrotic reaction around eggs.

   B. They include scarring of the bladder wall and blockage of ureters.

   C. They develop very soon after first infection.

   D. They include enlarged esophageal varices that arise from adult worms living in these blood vessels.

   E. They include epilepsy.

8. **Which of the following are true regarding the diagnosis of schistosomiasis?**

   A. It depends on finding characteristic eggs in a blood film.

   B. It depends on finding eggs in urine usually provided first thing in the morning in the case of *S. haematobium*.

   C. It requires serological testing of returned travelers 10–14 days after water exposure.

   D. It is supported by the presence of eosinophilia.

   E. It may need to be differentiated from Crohn's disease in the case of *S. mansoni*.

9. **Which of the following statements are true for the treatment of schistosomiasis?**

   A. A 6-week course of praziquantel is required.

   B. It depends on effective killing of eggs by praziquantel.

   C. It involves the use of praziquantel immediately after water exposure.

   D. A combination of drugs is usually required.

   E. It cures up to 90% of patients after the first treatment course.

10. **Which of the following statements are true for the control and prevention of schistosomiasis?**

    A. It has resulted in elimination from South East Asia.

    B. It is being achieved in some countries by population-wide mass chemotherapy.

    C. It has been facilitated by the construction of dams and irrigation canals.

    D. It is not yet possible through vaccination.

    E. It is easily achieved by killing intermediate host snail populations.

Case 33

Staphylococcus aureus

A 44-year-old male presented to the emergency room with complaints of chest pain and was found to have suffered a myocardial infarction. His past medical history was significant for hypertension, noninsulin-dependent diabetes, **hypercholesterolemia**, and a history of heavy smoking (two packs per day for 15 years). A cardiac catheterization on hospital day 3 showed three vessel coronary artery disease, and he underwent triple coronary artery bypass graft surgery on hospital day 5. On hospital day 7, he developed **septic shock** with acute renal and respiratory failure requiring intubation. At that time, he had a fever of 39.3°C, his arterial blood gas revealed a pO$_2$ of 89 mmHg on 100% O$_2$, and he had a white blood cell count of 27 000/mm^3. Two blood cultures were obtained. A chest radiograph showed a left lower lobe infiltrate with pleural **effusion**. A chest tube was placed to drain the effusion. On hospital day 11, pus was noted to be seeping from his sternal wound. His wound was debrided and a rib biopsy was performed. Blood, drainage from his chest tube, tracheal aspirates, pus from his sternal wound, and a rib biopsy were cultured. The bacterium isolated from the clinical specimens was a gram-positive, catalase-positive, coagulase-positive coccus that was resistant to methicillin.

1. What is the causative agent, how does it enter the body and how does it spread a) within the body and b) from person to person?

Causative agent

The patient has an infection caused by methicillin (oxacillin)-resistant *Staphylococcus aureus* (MRSA). Bacteria in the genus *Staphylococcus* are gram-positive cocci that grow in grape-like clusters and produce the enzyme catalase (see Section 4). The genus *Staphylococcus* comprises some 35 species, many of which are members of the endogenous **microbiota** of the skin and mucous membranes of the gastrointestinal and genitourinary tracts of humans. Interestingly, a number of these species have defined habitats on the human body. Three species account for most human disease, namely *S. aureus*, *S. epidermidis*, and *S. saprophyticus*. Of these, *S. aureus* is the most pathogenic. *S. aureus* has a typical gram-positive cell wall that features a thick peptidoglycan layer extensively cross-linked with pentaglycine bridges. The extensive cross-linking makes the cell very resistant to drying so that staphylococci can survive on **fomites** (inanimate objects) for long periods of time. The wall also contains covalently bound teichoic acid and lipoteichoic acid, which is anchored in the cytoplasmic membrane. There are a number of molecules that are exposed on the cell surface. These are either anchored in the cytoplasmic membrane and traverse the cell wall to the outside or they are anchored in the wall and extend from it (Figure 1). Among the important cell surface-exposed molecules of *S. aureus* are various microbial surface components that recognize adhesive matrix molecules (**MSCRAMMs**). These components recognize and bind various extracellular matrix proteins such as fibronectin, collagen, and fibrinogen (clumping factor), and the Fc region of mammalian **IgG** (protein A). In addition to wall-associated clumping factor that binds solid-phase fibrinogen, *S. aureus*

**Figure 1. Schematic of the envelope of
S. aureus.**

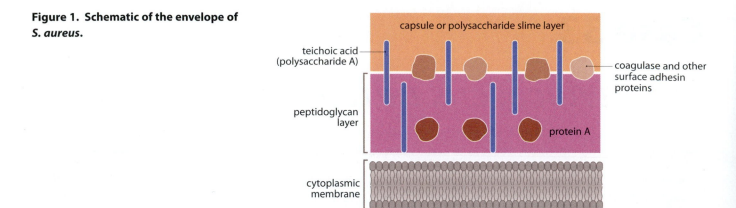

secretes coagulase, which binds soluble fibrinogen. Coagulase binds pro-thrombin in a 1:1 molar ratio to form a complex termed staphylothrombin, which converts fibrinogen to fibrin. Because *S. aureus* is the only staphylo-coccal species that possesses coagulase and protein A their detection serves as a method to identify the bacterium (see Section 4). MSCRAMMs facili-tate invasion of keratinocytes and endothelial cells by *S. aureus*. Once taken up by cells, the bacteria survive in vacuoles and free in the cytoplasm where some persist for several days. Bacterial invasion results in cytotoxicity and apoptotic cell death.

Almost all *S. aureus* clinical isolates produce capsular polysaccharides, which have been divided into 11 serotypes. Serotype 1 and 2 staphylococci are heavily capsulated and produce mucoid colonies. The remaining serotypes produce 'microcapsules' and form nonmucoid colonies. Interestingly, most clinical isolates are microcapsule-producing serotypes 5 and 8.

Almost all strains of *S. aureus* secrete a group of enzymes and cytotoxins that includes four **hemolysins** (alpha, beta, gamma, and delta), nucleases, proteases, lipases, hyaluronidase, and collagenase. Alpha-toxin is a pore-forming toxin that acts on many types of human cells. It is dermonecrotic and neurotoxic. Beta-toxin is a sphingomyelinase C whose role in patho-genesis remains unclear. Delta-toxin is able to damage the membrane of many human cells, possibly by inserting into the cytoplasmic membrane and disordering lipid chain dynamics, although there are suggestions that it may form pores. Two types of bicomponent toxin are made by *S. aureus*, gamma-toxin and Panton-Valentine leukocidin (PVL). While gamma-toxin is produced by almost all strains of *S. aureus*, PVL is produced by only a few percent of strains. The toxins affect neutrophils and macrophages. PVL has pore-forming activity and has been associated with necrotic lesions involving the skin and with severe necrotizing **pneumo-nia** in community-acquired *S. aureus* infections. Also, *S. aureus* produces a staphylokinase, which is a potent activator of plasminogen, the precursor of the fibrinolytic protease plasmin.

S. aureus produces several families of **exotoxins** that have **superantigen** activity. These include heat-stable **enterotoxins** (A–E, G–I) that cause food poisoning, exfoliative (epidermolytic) toxins A (chromosome encoded) and B (plasmid encoded) (ETA and ETB) that are the cause of staphylococcal scalded skin syndrome (SSSS) occuring predominantly in children, and **toxic shock syndrome** toxin 1 (TSST-1) that causes staphylococcal toxic shock.

Entry and spread within the body

S. aureus requires a breach in the skin or mucosa such as produced by a catheter, a surgical incision, a burn, a traumatic wound, ulceration or viral skin lesions to facilitate its entry into the tissues. Furthermore, staphylococcal infections are frequently associated with reduced host resistance brought about by **cystic fibrosis**, immunosuppression, diabetes mellitus, viral infections, and drug addiction. *S. aureus* has several enzymes that aid in invasion of the barrier epithelia and serve as spreading factors. Among these are lipases, hyaluronidase, and fibrinolysin.

Person to person spread

Because they colonize the surface of the skin staphylococci are readily shed and easily transferred horizontally by direct contact or by contact with fomites such as bed linen, medical instruments, and so forth.

Epidemiology

Infections with *S. aureus* are worldwide and do not show a seasonal prevalence. Transient colonization of the umbilical stump and skin, particularly in the perineal region, is common in neonates. About 15% of children and adults are persistent carriers of *S. aureus* in the anterior nares. There is a high rate of carriage in hospital personnel, IV drug users, diabetics, and dialysis patients, and MRSA and MSSA (methicillin-sensitive *S. aureus*) colonization is common in the population. In fact, studies suggest that as many as one-third of the US population currently is colonized in the anterior nares by this organism. MRSA is identified as a **nosocomial** pathogen throughout the world. Established risk factors for MRSA infection include recent hospitalization or surgery, residence in a long-term-care facility, dialysis, and indwelling percutaneous medical devices and catheters. Recently, however, cases of MRSA have been documented in healthy community-dwelling persons without established risk factors for MRSA acquisition. Because they are apparently acquired in the community, these infections are referred to as community-acquired (CA)-MRSA. CA-MRSA infections have been reported in North America, Europe, Australia, and New Zealand. A collection of 117 CA-MRSA strains from these countries shared a type IV SCC*mec* cassette and the PVL locus, whereas the distribution of other toxin genes was quite specific to the strains from each continent. Within each continent, the genetic background of CA-MRSA strains did not correspond to that of the hospital-acquired MRSA.

2. What is the host response to the infection and what is the disease pathogenesis?

S. aureus targets those with a breach of the skin and/or mucosal barriers (burns, traumatic wounds, surgical incisions, ulceration, viral skin lesions) together with those with reduced host resistance (cystic fibrosis, immunosuppression, diabetes mellitus, drug addiction). Intact barrier epithelia together with innate immune factors such as α-**defensins** and **cathelicidin** contribute to preventing invasion of *S. aureus*. If *S. aureus* breaches the epithelial barrier then the bacteria are combated by soluble innate antimicrobial proteins in the intercellular spaces and by neutrophils. α-Defensins and cathelicidin are also secreted by epithelial cells and neutrophils as well as being constituents of phagocyte **lysosomes**. However, staphylokinase and the metalloproteinase, aureolysin, can inactivate these defense factors. *S. aureus* also is resistant to lysozyme, a muramidase found in external secretions and in lysosomal vacuoles. Neutrophil **phagocytosis** is critical in defense against *S. aureus*. *S. aureus* is able to inhibit chemotaxis of neutrophils and monocytes in response to C5a and formylated bacterial peptides by secreting a chemotactic inhibitory protein (CHIP). In addition, nuclease produced by the bacterium are able to degrade the neutrophil extracellular traps (NETs) that are comprised of chromatin and microbicidal granule proteins. Initially, phagocytosis is aided by the innate pathways of the complement cascade (lectin and alternative pathways) and later by opsonic **IgM** and IgG antibodies. Consistent with the importance of opsonophagocytosis in host defense against *S. aureus*, the bacterium has several methods by which to avoid **opsonization** and phagocytosis and death within the phagolysosome. The bacterial **capsule** and protein A impair attachment and internalization. Protein A binds IgG antibodies via the Fc region, which prevents their specific binding that would otherwise lead to opsonization and complement activation resulting in further opsonization by C3b. In addition, plasminogen bound on the surface of *S. aureus* can be cleaved by staphylokinase into plasmin, which can then cleave IgG and C3b. Furthermore, *S. aureus* is able to inhibit the C3 convertase. *S. aureus* resists the bactericidal environment of the phagolysosome by the production of **catalase** and carotenoids. Carotenoids give *S. aureus* colonies their golden color. Both molecules neutralize singlet oxygen and superoxide produced as a result of the respiratory burst. Furthermore, several of the exotoxins produced by *S. aureus* kill neutrophils and monocytes/macrophages as well as other types of cells. The **pus** formed in these infections is a mixture of dead organisms and dead phagocytes and the release of their lysosomal contents contributes to the observed tissue damage. Neutralizing IgG antibodies are important in defense against TSST-1.

Pathogenesis

S. aureus is a versatile pathogen that causes a wide range of diseases. Disease may be mediated by invasion, bacterial multiplication leading to formation of **abscesses**, and destruction of a variety of tissues, as well as by a range of exotoxins, some of which are superantigens. Furthermore, *S. aureus* and other staphylococcal species are broadly antibiotic-resistant (see below). Factors contributing to the **virulence** of *Staphylococcus aureus* are presented in Table 1.

Skin infections range from those that are superficial such as impetigo (Figure 2) (see also *Streptococcus pyogenes*), infections of hair follicles (folliculitis) (Figure 3), and boils (furuncles) (Figure 4) to deeper infections like carbuncles (Figure 5), which occur when furuncles unite and extend deeper into subcutaneous tissues forming sinus tracts. From subcutaneous sites the bacteria can seed the bloodstream and spread to other tissues.

S. aureus is a frequent cause of **bacteremia** and about half of these instances are nosocomial, associated with surgical wound infection and foreign bodies such as indwelling catheters, or sutures. From the blood *S. aureus* may seed many organs including the heart, lungs, bones, and joints. *S. aureus* **endocarditis** has a high mortality and may throw off septic emboli that

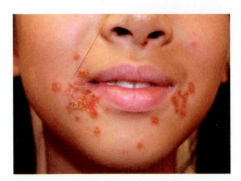

Figure 2. Impetigo around the mouth of an 11-year-old girl. This is a highly contagious bacterial skin infection caused by *S. aureus*.

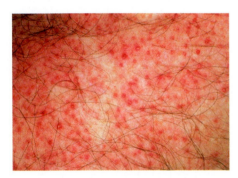

Figure 3. Folliculitis caused by *S. aureus*.

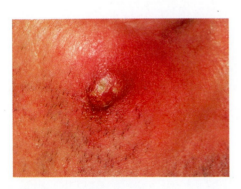

Figure 4. Furuncle (boil) on the cheek caused by *S. aureus*.

Table 1. Factors contributing to virulence of *Staphylococcus aureus*	
Factor	*Action*
Cell wall-associated	
Capsule/slime	Adherence; antiphagocytic
Protein A	Prevents opsonization; anticomplementary
Clumping factor	Binds to fibrinogen and platelet–fibrin clots
Secreted products – toxins	
Exfoliative toxins (ETA and ETB)	Serine proteases that disrupt intercellular adhesion in the stratum granulosum; possibly superantigens
α, β, δ, γ, cytotoxins	Pore-forming or detergent-like cytotoxins that are toxic for many types of cell
Panton-Valentine leukocidin	Leukotoxic
Toxic shock syndrome toxin-1	Superantigen
Enterotoxins	Family of nine heat-stable, tasteless proteins that cause food poisoning and are superantigens; proposed biothreat agents
Secreted products – enzymes	
Catalase	Detoxifies H_2O_2 in the phagolysosome of neutrophils
Coagulase	Induces fibrin clots
Hyaluronidase	Aids in tissue spread by hydrolyzing the hyaluronic acid intercellular matrix
Lipases	Hydrolyze skin lipids aiding invasion
Fibrinolysin	Aids spreading by dissolving fibrin clots
Nucleases	Hydrolyzes DNA liquifying pus; degrades neutrophil extracellular traps (NETs)

may infect other organs. Pneumonia may result if bacteria reach the lungs via the bloodstream or as a result of aspiration. Acute and chronic **osteomyelitis** and **septic arthritis** may result from hematogenous spread or from a skin infection.

Enterotoxins (A–E, G–I) produced by *S. aureus* are the most common cause of food poisoning, with enterotoxin A causing most cases. Food poisoning results from the production of toxin in the foodstuff and its subsequent ingestion. The most frequently contaminated foods are salted meats, potato salad, custards, and ice cream. Generally, the food is contaminated by *S. aureus* on the hands of food preparers. Holding the food at room temperature allows the staphylococci to proliferate and produce toxin. The enterotoxins are heat-stable and tasteless so that, although heating of the food will kill the staphylococci, the enterotoxins are unaffected. The intoxication has a rapid onset, usually within 4 hours, and the symptoms are nausea, vomiting, diarrhea, and abdominal pain. Symptoms usually resolve within 24 hours. The mechanism(s) of action of the enterotoxins are not understood but may include mast cell activation and release of inflammatory mediators and neuropeptide **substance P** in the gastrointestinal tract and elsewhere.

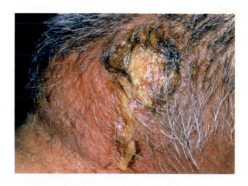

Figure 5. Carbuncle on the neck caused by *S. aureus*.

S. aureus secretes several toxins, some of which function as superantigens. These include TSST-1, enterotoxins, and the exfoliative (epidermolytic) toxins. Superantigens bind to both MHC class II molecules on the surface

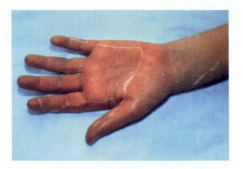

Figure 6. Staphylococcal toxic shock syndrome showing desquamation of the left palm.

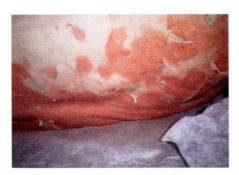

Figure 7. Staphylococcal scalded skin syndrome showing peeling of the skin on the back. Reprinted by permission from Macmillan Publishers Ltd: *Journal of Investigative Dermatology*, Volume 124: Pages 700 – 703, Figure S1, 2005.

of macrophages and specific Vβ regions of the T-cell receptor on the surface of CD4 T cells. Binding results in the activation of both cell types, leading to excessive production of pro-inflammatory **cytokines** and T-cell proliferation, causing fever, a diffuse macular **rash**, desquamation that includes the palms (Figure 6) and the soles, hypotension, and **shock**. Although toxic shock syndrome (TSS) was first reported in 1928, this syndrome came to prominence in 1980 with the introduction of superabsorbent tampons. It was found that *S. aureus* in the vagina could multiply rapidly in these tampons and that tampon fibers acted as an ion exchanger, reducing the level of magnesium ions in the tampon. Under conditions of magnesium limitation the production of TSST-1 by *S. aureus* is markedly increased. The toxin was then absorbed though the vaginal wall and entered the bloodstream. TSS may also follow localized growth of *S. aureus* in a wound. Interestingly, the vast majority of adults have antibodies that neutralize TSST-1, so only those lacking antibody are at risk.

Staphylococcal scalded skin syndrome (SSSS) is mediated by the exfoliative toxins A and B (ETA, ETB) produced by certain strains of *S. aureus*. The disease is seen mainly in neonates and young children following an infection of the mouth, nasal cavities, throat, or umbilicus. It begins as redness and inflammation around the mouth that extends to the whole body. The superficial layers of skin slip free from the lower layers with a slight rubbing pressure (Nikolsky's sign). Large areas of the skin blister and peel away, leaving wet, red, and painful areas. The blisters or **bullae** do not contain staphylococci or leukocytes. The exotoxins, ETA and ETB, are thought to bind to desmoglein 1 in **desmosomes** in the stratum granulosum, causing it to break down thus releasing the superficial skin layers (Figure 7). Healing typically occurs within 1–2 weeks. SSSS can occur in a localized form termed bullous impetigo. The presentation is one of localized bullae (large blisters) without Nikolsky's sign. In contrast to SSSS the bullae contain *S. aureus* and, therefore, the disease is highly contagious.

3. What is the typical clinical presentation and what complications can occur?

Because *S. aureus* is such a versatile pathogen and causes a wide range of disease via invasion and destruction of tissue and by production of toxins, it is difficult to give a 'typical' clinical presentation. Recently, a retrospective chart review was conducted of patients coded for *S. aureus* sepsis (SAS) requiring admission to a pediatric intensive care unit (PICU) in Auckland, New Zealand over a 10-year period (see References: Miles et al., 2005). Fifty-eight patients were identified with SAS over the 10-year study period. The age distribution of the children was bimodal with peaks at 0–2 and 12–14 years. Almost all of the children had community-acquired SAS; only three had nosocomial SAS. Musculoskeletal symptoms dominated (79%). Pneumonia and/or **empyema** were diagnosed in 78% of children on admission. Two-thirds of the children either presented with multiple site involvement or secondary sites developed during hospitalization that included pneumonia, septic arthritis, osteomyelitis, and soft tissue involvement (**cellulitis, fasciitis**, abscess). A few children had epidural abscesses, vascular **thrombosis**, or endocarditis. Minor limb trauma within the preceding month was reported in half of the children. Osteomyelitis was also diagnosed in about half of the children, and septic arthritis in slightly less.

A pre-existing medical condition was present in eight children and included acute and chronic **myeloid leukemia** and diabetes.

The mortality rate in children with SAS was 8.6% compared with an overall mortality of 6.0% for other patients admitted to the pediatric neonatal intensive care unit. Five children who presented with refractory septic shock died. Complications of SAS occurred in 20 surviving children (38%), including multi-system organ failure, respiratory disease, **paraplegia** from epidural abscesses, and extensive venous thrombosis of the inferior vena cava or femoral veins.

4. How is the disease diagnosed and what is the differential diagnosis?

Diagnosis

Microscopy: Depending on the disease and the type of clinical specimen *S. aureus* may be observed by Gram's stain. From clinical specimens the staphylococci appear as single cells or small clusters of gram-positive cocci (Figure 8). However, as staphylococci are **commensals** of the skin and mucous membranes Gram staining may not be useful unless the specimen is from a normally sterile body site, but even here as in the case of blood, few bacteria may be present and culture of the specimen is indicated.

Culture: *S. aureus* and other staphylococci grow well on blood agar after 24–48 hours of incubation at 37°C. Staphylococci are **facultative anaerobes**. Colonies of *S. aureus* are large, smooth, glossy domes with an entire edge. If held at room temperature the colonies will change color from white to a golden yellow (*ergo* – *S. aureus*, the golden staphylococcus). Almost all strains of *S. aureus* are β-hemolytic due to their secretion of cytotoxins (Figure 9). Mannitol-salt agar containing 7.5% sodium chloride and mannitol as the carbon source is a useful selective and differential medium to recover staphylococci from clinical specimens, for example skin, that contain a mixed **microbiota**. Most bacteria other than staphylococci are inhibited by this concentration of NaCl and *S. aureus* can ferment mannitol whereas most other staphylococcal species cannot. Incorporation of the pH indicator, phenol red, reveals colonies fermenting mannitol because acid produced by the colony changes the color of the agar from pink to yellow (Figure 10).

Determinative tests: *S. aureus* are gram-positive cocci growing in grape-like clusters that are positive for the presence of the enzyme catalase. Catalase reduces hydrogen peroxide to water and molecular oxygen and the presence

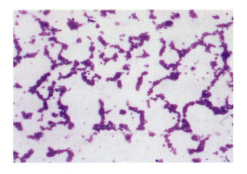

Figure 8. Gram stain of *S. aureus*.

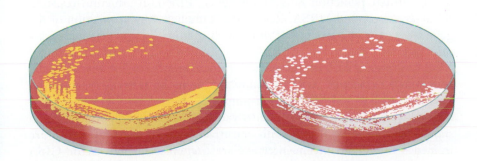

Figure 9. *S. aureus* growing on blood agar. Left, *S. aureus* showing typical golden yellow colonies; right, white colonies of a coagulase-negative *Staphylococcus*.

Figure 10. *S. aureus* growing on mannitol-salt agar.

A

B

Figure 11. Catalase test: A, positive; B, negative. Reprint permission kindly given by Jay Hardy, President of Hardy Diagnostics, www.Hardy.Diagnostics.com.

of the enzyme can be readily detected by placing a drop of H_2O_2 on top of staphylococcal cells smeared on a microscope slide. Contact between H_2O_2 and the cells results in the immediate evolution of O_2 that can be seen as bubbles (Figure 11). Also, *S. aureus* is coagulase-positive and ferments mannitol (see above). The production of coagulase can be observed by inoculating 0.5 ml of rabbit plasma containing **EDTA** with a few colonies of *S. aureus* and incubating for up to 4 hours at 35–37°C. Coagulase will cause the plasma to clot (Figure 12). This test can also be performed by emulsifying a colony of *S. aureus* in a drop of rabbit plasma on a microscope slide where the bacterial cells will clump if they produce coagulase. In fact, species of staphylococci other than *S. aureus* are frequently termed 'CNS' (coagulase-negative staphylococci). It should be noted that there are two other genera of catalase-positive, gram-positive cocci that are opportunistic pathogens that may be confused with staphylococci, particularly CNS. Members of the genus *Micrococcus* are transients of the skin and mucous membranes while *Stomatococcus mucilaginosus* (weakly catalase-positive) is a commensal of the **oropharynx**. Both have been reported to cause opportunistic infections. *Micrococcus* species may produce bright yellow colonies similar to *S. aureus*, but they are not β-hemolytic, nor do they produce coagulase. *Stomatococcus mucilaginosus* colonies are usually clear to white, mucoid (because of the presence of capsule), and adherent to the surface of the agar. They do not grow in the presence of 5% NaCl.

Commercial kits for the rapid detection of clumping factor (cell-bound coagulase) and protein A based on latex agglutination are widely available. In these agglutination tests latex particles coated with fibrinogen and IgG are rapidly agglutinated by *S. aureus* cells.

Differential diagnosis

The differential diagnoses of *S. aureus* infections are shown in Table 2.

5. How is the disease managed and prevented?

Management

Currently less than 10% of staphylococcal isolates are susceptible to penicillin. Resistance to this antibiotic is mediated by a plasmid-encoded β-lactamase (penicillinase), which hydrolyzes the β-lactam ring of the molecule. Furthermore, about half of *S. aureus* and CNS isolates are resistant to semi-synthetic penicillins (methicillin, nafcillin, oxacillin) that are resistant to β-lactam ring hydrolysis. Methicillin acts by binding to and competitively inhibiting the transpeptidase used to cross-link the peptide D-alanyl-alanine used in peptidoglycan synthesis. Methicillin is a structural analog of D-alanyl-alanine, and the transpeptidases that are bound by it are termed penicillin-binding proteins (PBPs). Methicillin-resistant *S. aureus* are termed MRSA. Resistance is mediated by acquisition of the gene *mec*A that encodes a PBP for which methicillin and related antibiotics lack affinity. Initially, MRSA were limited to hospitals but recently MRSA outbreaks have been reported in the community. Interestingly, the community-associated MRSA strains appear to be clonally related, but unrelated to hospital MRSA. Alarmingly, *S. aureus* and CNS have been isolated with low-level resistance to vancomycin, an antibiotic to which staphylococci had been uniformly sensitive. Such strains of *S. aureus* are termed VISA (vancomycin-intermediate *Staphylococcus aureus*) or GISA (glycopeptide-intermediate *Staphylococcus aureus*). There appear to be two different mechanisms of resistance to glycopeptides. One seems to be related to increased thickness of the cell wall, which prevents penetration of the antibiotic to its target in the cytoplasmic membrane. The other results from the acquisition of the *van*A gene from vancomycin-resistant enterococci (VRE). Glycopeptide antibiotics such as vancomycin act by binding to the D-alanyl-D-alanine terminus of peptidoglycan precursors preventing cell wall synthesis. However, the *van*A gene specifies the synthesis of peptidoglycan precursors terminating in the depsipeptide

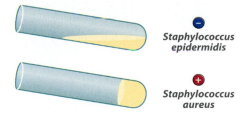

Figure 12. Positive and negative coagulase tests. Of the medically important species of *Staphylococcus* only *S. aureus* produces coagulase. Thus, other staphylococcal species are termed coagulase-negative staphylococci (CNS).

Table 2. Differential diagnosis of *Staphylococcus aureus* infections

Staphylococcus aureus infection	Differential diagnosis
Impetigo	None
Bullous impetigo	Pemphigus Pemphigoid Burn Stevens-Johnson syndrome Dermatitis herpetiformis
Scalded skin syndrome (Ritter disease)	Nonaccidental injury Scalding Abrasion trauma Sunburn Erythema multiforme Toxic epidermal necrolysis
Bone and joint infections	Bone infarction (in patients with sickle cell disease) Toxic synovitis Leukemia
Septic arthritis	Trauma Deep cellulitis Henoch-Schönlein purpura Slipped capital femoral epiphysis Legg-Calvé-Perthes disease Leukemia Toxic synovitis Metabolic diseases affecting joints (ochronosis*)
Endocarditis – bacteremia	None
Toxic shock syndrome	Staphylococcal scalded skin syndrome Meningococcemia Rubeola Adenoviral infections Dengue fever Severe allergic drug reactions

Taken from: Tolan RW Jr, Baorto EP, 2006. *Staphylococcus aureus* infection. Emedicine (from WebMD) http://www.emedicine.com/PED/topic2704.htm

*Ochronosis: bluish-black discoloration of certain tissues, such as the ear cartilage and the ocular tissue, seen with alkaptonuria, a metabolic disorder.

D-alanyl-D-lactate, to which vancomycin does not bind. Serious infections with methicillin-sensitive *S. aureus* (MSSA) infections are treated with parenteral penicillinase-resistant penicillins or first- or second-generation cephalosporins with clindamycin. Vancomycin is reserved for staphylococcal strains that are resistant to penicillinase-resistant penicillins and clindamycin. Mupirocin may be used to treat superficial or localized skin infections caused by *S. aureus* (i.e. impetigo). This antibiotic works by inhibiting isoleucyl-tRNA synthetase, which is required for protein synthesis. However, it should be noted that most community-acquired MRSA strains are already resistant or are acquiring resistance to mupirocin.

In addition to antibiotic therapy, temporary intravascular devices such as catheters should be promptly removed if infection is suspected. Long-term intravascular devices should be removed if infection with *S. aureus* is confirmed. Abscesses must be drained and, usually, if the infection involves a prosthetic joint it should be removed.

Prevention
Prevention of staphylococcal infection relies on the practice of good hygiene. The infectious dose of *S. aureus* required to initiate a wound infection in a healthy person is high unless the wound contains a foreign body of some sort. The Centers for Disease Control and Prevention (CDC) recommend keeping one's hands clean by washing thoroughly with soap and water or using an alcohol-based hand sanitizer; keeping cuts and scrapes clean and covered with a bandage until healed; avoiding contact with other people's wounds or bandages; and avoiding sharing personal items such as towels or razors. In the hospital setting attempts have been made to eliminate nasal carriage by hospital personnel by the use of antibiotic regimens.

SUMMARY

1. What is the causative agent, how does it enter the body and how does it spread a) within the body and b) from person to person?

- Staphylococci are gram-positive cocci that grow in grape-like clusters and produce the enzyme catalase.

- The genus *Staphylococcus* comprises some 35 species, many of which are members of the endogenous microbiota of the skin and mucous membranes of the gastrointestinal and genitourinary tracts of humans.

- *S. aureus* and other staphylococcal species are readily isolated on blood agar incubated aerobically at 37°C for 24–48 hours.

- *S. aureus* is unique among staphylococci because it produces a cell-associated and secreted coagulase, which converts fibrinogen to fibrin and it expresses protein A on its surface, which binds the Fc region of human IgG. These properties are utilized in rapid ID tests for *S. aureus*.

- Staphylococci are very tolerant of salt and will grow in the presence of 7.5% NaCl.

- These properties serve as the bases for a selective and differential medium termed mannitol-salt agar that is useful in isolating staphylococci from skin and mucosal surfaces. Mannitol-salt agar differentiates *S. aureus* from coagulase-negative staphylococci (CNS) species that do not ferment mannitol.

- *S. aureus* has a typical gram-positive cell wall that features a thick peptidoglycan layer extensively cross-linked with pentaglycine bridges. The extensive cross-linking makes the cell very resistant to drying so that the bacterium can survive on inanimate objects for long periods of time. The cell envelope contains membrane-anchored lipoteichoic acids and wall-anchored teichoic acids.

- The anterior nares are the reservoir of *S. aureus*. From this site it is readily transferred to the hands.

- Because they also colonize the surface of the skin staphylococci are readily shed and easily transferred horizontally (person to person) by direct contact (bacteria on the hands) or by contact with fomites (inanimate objects) such as bed linen, medical instruments, and so forth.

- There is a high rate of carriage of methicillin-resistant *S. aureus* (MRSA) in hospital personnel, IV drug users, diabetics, and dialysis patients.

- As many as one-third of the US population are colonized in the anterior nares by this organism.

- Transient colonization of the umbilical stump and skin, particularly in the perineal region, is common in neonates.

- *S. aureus* (and other species of staphylococci) produce an extracellular polysaccharide capsule or slime that enables them to form a biofilm on catheters and other prostheses.

- *S. aureus* requires a breach in the skin or mucosa such as produced by a catheter, a surgical incision, a burn, a traumatic wound, ulceration or viral skin lesions to facilitate its entry into the tissues.

- Staphylococcal infections are frequently associated with reduced host resistance brought about by cystic fibrosis, immunosuppression, diabetes mellitus, viral infections, and drug addiction.

- *S. aureus* has several enzymes that aid in invasion of the barrier epithelia and serve as spreading factors. Among these are lipases, hyaluronidase, and fibrinolysin.

2. What is the host response to the infection and what is the disease pathogenesis?

- Intact barrier epithelia together with innate immune factors at these surfaces are effective in preventing invasion of *S. aureus*.

- If *S. aureus* breaches the epithelial barrier the bacteria are phagocytosed by neutrophils.

- Phagocytosis is aided by opsonization via the innate pathways of the complement cascade (lectin and alternative pathways) and later by IgM and IgG antibodies.

- *S. aureus* has various mechanisms to resist phagocytosis and intracellular killing.

- Neutralizing IgG antibodies are important in defense against TSST-1.

3. What is the typical clinical presentation and what complications can occur?

- *S. aureus* is a versatile pathogen that causes a wide range of disease via invasion and destruction of tissue and by production of toxins.

- Impetigo begins as a small area of **erythema** that progresses into fluid-filled bullae that rupture and heal with the formation of a honey-colored crust. Impetigo is highly contagious.

- Bullous impetigo is a localized form of staphylococcal scalded skin syndrome (SSSS) that presents as bullae without surrounding erythema. The bullae contain staphylococci and rupture, leaving a denuded area. Bullous impetigo is highly contagious.

- Scalded skin syndrome begins with peri-oral erythema that extends over the whole body in 1–2 days. Friction to the skin causes it to be displaced (positive Nikolsky's sign) leading to the formation of fragile bullae that burst, leaving a tender base. This is followed by desquamation of the skin. The patient often is **febrile** and occasionally has a **mucopurulent** eye discharge.

- Folliculitis is a tender pustule that involves the hair follicle.

- Furuncles are small abscesses that involve both the skin and the subcutaneous tissues in areas with hair follicles, such as the neck, axillae, and buttocks.

- Carbuncles are collections of inter-connected furuncles with several sinus tracts opening to the skin surface. Often they are the source of systemic infection via bacteremic spread of the staphylococci indicated by fever and chills.

- Osteomyelitis often presents with a sudden onset of fever and bone tenderness or limp. There may be little pain or pain can be throbbing and quite severe.

- Septic arthritis presents with decreased range of motion, warmth, erythema, and tenderness of the joint together with constitutional symptoms and fever. However, in infants these signs may be absent.

- Endocarditis presents with high fever, chills, sweats, and malaise. The condition can deteriorate rapidly with reduced cardiac output and septic emboli manifested by **Osler nodes**, subungual hemorrhages, Janeway lesions, and **Roth spots**.

- Toxic shock syndrome (TSS) is a rapidly progressing, life-threatening intoxication. It begins abruptly with fever of 38.9°C or higher, a diffuse macular, **erythematous** rash, and hypotension. Conjunctival and/or vaginal **hyperemia** may be present. Multiple organ systems are affected such as gastrointestinal, musculature, renal, hepatic, hematologic, and nervous. Also, the entire skin including the palms of the hands and the soles of the feet desquamates.

- Staphylococcal pneumonia is a rapidly progressive disease observed primarily at the extremes of age in persons with predisposing systemic conditions such as **chronic obstructive pulmonary disease** (**COPD**), **bronchiectasis**, cystic fibrosis or influenza infection. There are no unique clinical or radiographic findings but patients may also have prominent gastrointestinal tract symptoms.

4. How is the disease diagnosed and what is the differential diagnosis?

- Examination of appropriate clinical specimens by microscopy will reveal gram-positive cocci in grape-like clusters and individual cells.

- Clinical specimens are cultured on blood agar at 37°C for 24–4 hours in air. Colonies are round, glossy, domed with an entire edge, and golden yellow in color. Most *S. aureus* strains are β-hemolytic. Mannitol-salt agar can be used to isolate *S. aureus* from skin or mucosal surfaces. *S. aureus* colonies will be yellow in color and surrounded by a yellow zone indicating fermentation of mannitol.

- Determinative tests include: production of catalase and coagulase; rapid latex agglutination test to detect clumping factor and protein A; mannitol fermentation.

- Bullous impetigo: the differential diagnosis would include pemphigus, pemphigoid, and burns.

- Scalded skin syndrome: the differential diagnosis would include scalding, sunburn, and trauma.

- Toxic shock syndrome: SSSS, meningococcemia, rubeola, adenovirus infections, **Dengue fever**, severe allergic reactions.

5. How is the disease managed and prevented?

- Prevention of staphylococcal infection relies on the practice of good hygiene. The infectious dose of *S. aureus* required to initiate a wound infection in a healthy person is high unless the wound contains a foreign body of some sort.

- The Centers for Disease Control and Prevention (CDC) recommendations are: keeping one's hands clean by washing thoroughly with soap and water or using an alcohol-based hand sanitizer; keeping cuts and scrapes clean and covered with a bandage until healed; avoiding contact with other people's wounds or bandages; and avoiding sharing personal items such as towels or razors.

- In the hospital setting attempts have been made to eliminate nasal carriage of *S. aureus* by hospital personnel by the use of antibiotic regimens.

- Intravascular devices such as catheters should be promptly removed if infection is suspected. Long-term intravascular devices should be removed if infection with *S. aureus* is confirmed. Abscesses must be drained and, usually, if the infection involves a prosthetic joint it should be removed.

- Antibiotic therapy is problematic for staphylococcal infections because of broad antibiotic resistance.

- Currently less than 10% of staphylococcal isolates are susceptible to penicillin. Resistance to this antibiotic is mediated by a plasmid-encoded β-lactamase (penicillinase), which hydrolyzes the β-lactam ring of the molecule.

- About half of *S. aureus* and CNS isolates are resistant to semi-synthetic penicillins (methicillin, nafcillin, oxacillin) that are resistant to β-lactam ring hydrolysis. Methicillin-resistant *S. aureus* are termed MRSA.

- Serious infections with methicillin-sensitive *S. aureus* (MSSA) infections are treated with parenteral penicillinase-resistant penicillins or first- or second-generation cephalosporins with clindamycin.

- Vancomycin is reserved for staphylococcal strains that are resistant to penicillinase-resistant penicillins and clindamycin. Mupirocin may be used to treat superficial or localized skin infections caused by *S. aureus*.

- Low-level resistance to vancomycin, an antibiotic to which staphylococci had been uniformly sensitive, has been recognized.

FURTHER READING

Fischetti VA, Novick RP, Ferretti JJ, Portnoy DA, Rood JI. Gram-Positive Pathogens. ASM Press, American Society for Microbiology, Washington, DC, 2000.

Mims C, Dockrell HM, Goering RV, Roitt I, Wakelin D, Zuckerman M. Medical Microbiology, 3rd edition, Mosby, Edinburgh, 2004.

Murphy K, Travers P, Walport M. Janeway's Immunobiology, 7th edition. Garland Science, New York, 2008.

Murray PR, Rosenthal KS, Pfaller MA. Medical Microbiology, 5th edition, Elsevier Mosby, Philadelphia, PA, 2005.

REFERENCES

Chambers HF. Methicillin resistance in staphylococci: molecular and biochemical basis and clinical implications. Clin Microbiol Rev, 1997, 10: 781–791.

Dinges MM, Orwin PM, Schlievert PM. Exotoxins of Staphylococcus aureus. Clin Microbiol Rev, 2000, 13: 16–34.

Foster TJ. Immune evasion by staphylococci. Nat Rev Microbiol, 2005, 3: 948–958.

Fournier B, Philpott DJ. Recognition of Staphylococcus aureus by the innate immune system. Clin Microbiol Rev, 2005, 18: 521–540.

Iwatsuki K, Yamasaki O, Morizane S, Oono T. Staphylococcal cutaneous infections: invasion, evasion and aggression. J Dermatol Sci, 2006, 42: 203–214.

Kluytmans J, van Belkum A, Verbrugh H. Nasal carriage of Staphylococcus aureus: epidemiology, underlying mechanisms, and associated risks. Clin Microbiol Rev, 1997, 10: 505–520.

Ladhani S, Joannou CL, Lochrie DP, Evans RW, Poston SM. Clinical, microbial, and biochemical aspects of the exfoliative toxins causing staphylococcal scalded-skin syndrome. Clin Microbiol Rev, 1999, 12: 224–242.

Massey RC, Horsburgh MJ, Lina G, Hook M, Recker M. The evolution and maintenance of virulence in Staphylococcus aureus: a role for host-to-host transmission? Nat Rev Microbiol, 2006, 4: 953–958.

Miles F, Voss L, Segedin E, Anderson BJ. Review of Staphylococcus aureus infections requiring admission to a paediatric intensive care unit. Arch Dis Child, 2005, 90: 1274–1278.

O'Riordan K, Lee JC. Staphylococcus aureus capsular polysaccharides. Clin Microbiol Rev, 2004, 17: 218–234.

Rooijakkers SHM, van Kessel KPM, van Strijp JAG. Staphylococcal innate immune evasion. Trends Microbiol, 2005, 13: 596–601.

Srinivasan A, Dick JD, Perl TM. Vancomycin resistance in staphylococci. Clin Microbiol Rev, 2002, 15: 430–438.

Vandenesch F, Naimi T, Enright MC, et al. Community-acquired methicillin-resistant Staphylococcus aureus carrying Panton-Valentine leukocidin genes: worldwide emergence. Emerg Infect Dis, 2003, 9: 978–984.

Weber JT. Community-associated methicillin-resistant Staphylococcus aureus. Clin Infect Dis, 2005, 41(Suppl 4): S269–S272.

WEB SITES

Centers for Disease Control and Prevention (CDC), National Center for Preparedness, Detection, and Control of Infectious Diseases, 2008, United States Public Health Agency: http://www.cdc.gov/ncidod/dhqp/ar_mrsa_ca_public.html

eMedicine, Inc., 2007, provider of clinical information and services to physicians and health-care professionals: http://www.emedicine.com/PED/topic2704.htm

eMedicine, Inc., 2007, provider of clinical information and services to physicians and health-care professionals: http://www.emedicine.com/PED/topic1172.htm

eMedicine, Inc., 2007, provider of clinical information and services to physicians and health-care professionals: http://www.emedicine.com/PED/topic1677.htm

eMedicine, Inc., 2007, provider of clinical information and services to physicians and health-care professionals: http://www.emedicine.com/PED/topic2269.htm

MULTIPLE CHOICE QUESTIONS

The questions should be answered either by selecting True (T) or False (F) for each answer statement, or by selecting the answer statements which best answer the question. Answers can be found in the back of the book.

1. A young woman recently underwent rhinoplasty to correct a deviated septum and her nose was packed postoperatively. She experienced a 2–3-day period of low-grade fever, muscle aches, chills, and malaise. This was followed by the development of a sunburn-like rash that covered most of her body. The packing was removed and a nasal swab grew gram-positive cocci that were catalase-positive and produced colonies on mannitol-salt agar that were surrounded by a yellow halo. The bacteria clumped when mixed with rabbit plasma. What is your diagnosis?

    A. Scalded skin syndrome.

    B. Toxic shock syndrome.

    C. Carbunculosis.

    D. Erysipelas.

    E. Impetigo.

2. Which one of the following mechanisms of action of toxic shock syndrome toxin-1 (TSST-1) produced by *Staphylococcus aureus* produces toxic shock syndrome?

    A. ADP-ribosylates a G protein of the adenylate cyclase complexes.

    B. ADP-ribosylates elongation factor 2 (EF2) needed for protein synthesis.

    C. Binds to MHC class II molecules on antigen-presenting cells and to the CD4 T-cell receptor.

    D. Prevents release of glycine and other inhibitory neurotransmitters.

    E. Membrane action results in host cell lysis.

Refer to the case history below to answer questions 3–5.

A man presents to the emergency room with high fever and malaise. He claims to have felt this way for about 3 days, worse every day. He appears underweight and very ill. Physical examination reveals needle marks in both antecubital fossae. Listening for heart sounds, you hear a distinctive systolic heart murmur. You order blood cultures and make a presumptive diagnosis of acute bacterial endocarditis. Cultures confirm your initial diagnosis and the patient begins to respond to the drug regimen you ordered.

3. Which one of the following antibiotics is most likely to be effective in this case?

    A. Nafcillin.

    B. Methicillin.

    C. Cephalosporin.

    D. Penicillin.

    E. Vancomycin.

4. Which of the following molecules contribute to the virulence of this organism?

    A. M protein.

    B. Polyglutamic acid capsule.

    C. IgA1 protease.

    D. Protein A.

    E. Exotoxin A.

5. Which one of the following infections would MOST commonly be associated with the same organism that caused the man's infection?

    A. Community-acquired bacterial pneumonia.

    B. Bacterial meningitis.

    C. Urinary tract infection.

    D. Toxic shock syndrome.

    E. Erysipelas.

Refer to the case history below to answer questions 6–8.

A 1-year-old boy is brought to his pediatrician with a severely inflamed cut on his knee. He has had four ear infections in the past year, and several additional wound infections that generally respond to topical therapy. Culture of the current wound reveals gram-positive cocci that are catalase-positive and coagulase-positive.

6. Protein A, a determinant of pathogenesis for these bacteria is?

    A. Required for attachment to epithelial tissue.

    B. A superantigen.

    C. Binds to the Fc portion of immunoglobulin G.

    D. A cytotoxin.

    E. An exo-enzyme that aids in spreading of bacteria through tissue.

7. Given the history, which of the following immune abnormalities would be MOST LIKELY?

    A. B-cell deficiency.

    B. T-cell deficiency.

    C. Complement deficiency.

    D. Eosinophil deficiency.

    E. Mast cell deficiency.

MULTIPLE CHOICE QUESTIONS (continued)

8. Scalded skin syndrome may develop in this child. Which of the following statements presents the MOST LIKELY reason why this may not occur?

   A. The bacterial strain causing this infection does not produce exfoliative toxin.

   B. The bacteria causing this infection are not the correct species to cause this disease.

   C. Underlying disease in this child will prevent the expression of the syndrome.

   D. Development of the syndrome requires infection of a pre-existing burn, not just a cut.

   E. Maternally derived IgG antibodies will neutralize the toxin.

9. Six hours after consumption of potato salad at a picnic, several family members presented to the ER with the abrupt onset of nausea, vomiting, and diarrhea. Which one of the following is the MOST LIKELY causative pathogen?

   A. *Bacillus cereus.*

   B. *Enterococcus faecalis.*

   C. *Staphylococcus aureus.*

   D. *Escherichia coli.*

   E. *Clostridium perfringens.*

10. An outbreak of sepsis caused by *Staphylococcus aureus* has occurred in the newborn nursery. You are asked to investigate. What is the most likely source of the organism?

    A. Scalp.

    B. Nose.

    C. Oropharynx.

    D. Vagina.

    E. Axilla.

Case 34

Streptococcus mitis

A 60-year-old male presented at the Emergency Room with a 2-week history of low-grade fevers and fatigue. The patient had a past medical history significant for hypertension, **aortic stenosis**, and hyperlipidemia. He stated that he has had daily temperature elevations up to 37.8–38.5°C, which were relieved with acetaminophen. He also noted generalized fatigue, **myalgia**, and **arthralgia**. He denied any headache, cough, dyspnea, chest pain, nausea, vomiting, or diarrhea. He has recently noticed a painful nodule on the tip of his left index finger. He reported that he had a tooth extracted 4 weeks ago. No antibiotic prophylaxis was provided by his dentist.

- On physical examination, the patient was pale and appeared fatigued. His temperature was 38.6°C; pulse 90; BP 160/86.
- Head, eyes, ears, nose, and throat examination revealed **Roth spots** (retinal hemorrhages with a white center) on funduscopic examination (Figure 1A). There was a conjunctival hemorrhage in the left eye (Figure 1B).
- Lungs: clear to **auscultation**.
- Heart: tachycardic with a grade III/VI harsh systolic murmur heard best at the right second intercostal space, and radiating to the neck.
- Abdomen: benign.
- Extremities: an **Osler node** was present on the right index finger and thumb. There were splinter hemorrhages in the nail beds of both hands (Figure 1C). Splinter hemorrhages are characteristic of infectious **endocarditis**.
- A transesophageal **echocardiogram** revealed a vegetation on the aortic valve.
- Blood cultures at 24 hours revealed gram-positive cocci growing in chains (Figure 2).

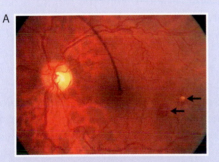
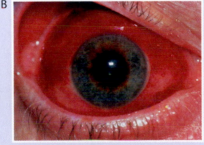
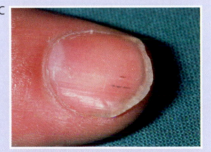

Figure 1 (above). (A) Fundus photograph of the right temporal retina demonstrates multiple white-centered hemorrhages (Roth spots). (B) Close-up of eye: subconjunctival hemorrhage. (C) Splinter hemorrhage on fingernail in endocarditis. Close-up of a fingernail showing splinter hemorrhage: dark, thin lines under the nail caused by bleeding. This is a sign of subacute bacterial endocarditis (SBE), in which the membrane (endocardium) lining a valve in the heart becomes infected with bacteria and inflamed. SBE can smolder undetected for months, causing mild symptoms (such as weakness and vague aches and pains) but doing serious damage to the heart. If untreated it can lead to heart failure and death. Treatment includes high doses of intravenous antibiotics and, in some cases, surgery to replace the damaged heart valve.

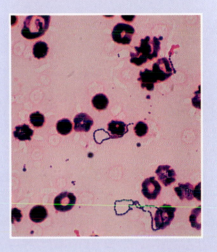

Figure 2 (right). Photomicrograph of *Streptococcus mitis* grown in a blood culture. Viridans streptococci are responsible for approximately half of all cases of bacterial endocarditis. These bacteria are commensals in the oropharynx.

1. What is the causative agent, how does it enter the body and how does it spread a) within the body and b) from person to person?

Causative agent

The patient has subacute bacterial endocarditis (SBE) caused by *Streptococcus mitis*. *S. mitis* is one of about 20 species of α-hemolytic streptococci, also termed viridans streptococci, that are normal inhabitants of the **oropharynx**. In fact, collectively, viridans streptococci are the principal cause of SBE. They comprise a large fraction of the **commensal** microbiota of the oropharynx and many of them, particularly those that produce extracellular polysaccharide, can cause subacute bacterial endocarditis. Clinicians tend to use the trivial term 'Streptococcus viridans' to describe them and the clinical laboratory generally does not identify them to the species level. However, the term 'Streptococcus viridans' is without taxonomic standing and should not be used. As α-hemolytic streptococci are emerging as causes of head and neck infection, and because they are demonstrating increased antibiotic resistance, it is likely that these bacteria will warrant speciation in the future.

S. mitis is a pioneer species in the oropharynx and is numerically dominant on mucosal surfaces, but also colonizes teeth. It should be noted that *S. mitis* and another species, *S. oralis*, are genetically very closely related to *S. pneumoniae* at the level of 16S ribosomal RNA. In fact the assignment of α-hemolytic streptococci to *S. mitis* and *S. oralis* is difficult and both are members of the 'mitis group' that includes *S. pneumoniae*, *S. sanguinis*, *S. parasanguinis*, *S. gordonii*, *S. cristatus*, *S. infantis*, and *S. peroris*. Although *S. mitis* does not form a capsule like *S. pneumoniae*, it shares antigens with the pneumococcus and about two-thirds of *S. mitis* isolated from the mouth produce IgA1 protease. Streptococci are gram-positive and catalase-negative. The **catalase** test is used to distinguish streptococci from staphylococci, which are the other principal, medically important genus of gram-positive cocci (see Figure 11 in the *Staphylococcus aureus* case for the catalase test). *S. mitis*, in common with other species of *Streptococcus*, grows readily on blood agar incubated at 37°C in the presence of 5% CO_2 for 24–48 hours. On blood agar *S. mitis* forms gray-green, punctuate, rough, dry colonies that are surrounded by a zone of incomplete (greening) hemolysis (**α-hemolysis**) (Figure 3). The term 'viridans' streptococci, therefore, is used to describe α-hemolytic streptococci because they turn erythrocytes in blood agar green. Commensal α-hemolytic streptococci are distinguished from *S. pneumoniae* which is often carried transiently in the pharynx, by the distinct cellular morphology (bullet-shaped diplococci) of the latter, but also because the pneumococcus is susceptible to optochin while all other α-hemolytic streptococci are resistant (Figure 4). As mentioned above, it is difficult to distinguish between certain species of optochin-resistant, α-hemolytic streptococci, and to compound this problem there are differences in the ways in which they are speciated by different groups of workers in the field.

S. mitis has a typical gram-positive-type cell wall that features a thick peptidoglycan layer with covalently bound teichoic acid and cell membrane-anchored lipoteichoic acid, except that they lack ribitol or glycerol phosphate unlike classical teichoic acids. The cell wall of *S. mitis* contains C-polysaccharide like the pneumococcus. The C-polysaccharide of *S. mitis* contains phosphocholine. Many species of viridans streptococci

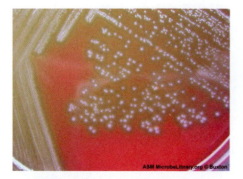

Figure 3. Viridans group streptococci growing on blood agar. Note that the colonies are surrounded by α-hemolysis (incomplete or greening hemolysis).

produce extracellular glucan slime although this is not a feature of *S. mitis*. *S. mitis* is a commensal bacterium that functions as an opportunistic pathogen. Other than the production of IgA1 protease virtually nothing is known about potential **virulence** determinants.

Streptococcus bovis (Lancefield group D), an inhabitant of the alimentary tract, is another important cause of SBE. Any patient with *S. bovis* bacteremia, with or without endocarditis, should be examined for GI tract malignancy, inflammatory bowel disease or liver disease because of association between the presence of this bacterium and these conditions. *S. bovis* has also been isolated more frequently from the stools of patients with malignancies of the colon.

Entry and spread within the body

The organisms colonize the oropharynx at birth. The commensal *S. mitis* is an inhabitant of the shedding and nonshedding surfaces of the oropharynx. If there is a breach in the integrity of the mucosal barrier as a result of tooth extraction, scaling, periodontal surgery or other minor oral surgery *S. mitis* can gain access to the bloodstream. It is worth remembering that even normal chewing results in a transient **bacteremia** containing *S. mitis*. Following tooth extractions bacteremia may last as long as 30 minutes but is usually shorter. From the bloodstream the organism may seed the heart or the central nervous system.

Person to person spread

S. mitis causes endogenous infections, that is, the person who is colonized by the bacterium as a member of their endogenous microbiota becomes infected when *S. mitis* translocates from the mucosal surface. There is no horizontal spread of the bacterium from one person to another.

2. What is the host response to the infection and what is the disease pathogenesis?

During colonization of the oropharynx at birth protection is initially mediated by the mucous membranes acting as barriers and the large number of innate immune factors present in saliva (see Further Reading: Cole & Lydyard, 2006). Secretory **IgA** antibodies secreted at low concentrations from birth are effective barriers to the penetration of *S. mitis*. Host phagocytes are a second line of defense against streptococcal invasion. Streptococci are opsonized by activation of the **alternative** and lectin innate **complement pathways** and the **classical pathway** in the presence of **IgM** and **IgG** antibodies in the plasma and tissue fluid. Viridans streptococci entering the bloodstream are attracted to damaged endocardium or endothelial tissue. The streptococci commonly adhere to congenitally malformed or **prosthetic** heart valves, with the mitral valve usually the most commonly affected. However, the bacteria may colonize a septal defect or attach to other areas of the heart. The bacteria form a vegetation (an accumulation of bacteria, platelets, and fibrin). Alternatively, streptococci in the blood can adhere to thrombi, leading to increased platelet and fibrin deposition, such that these vegetations grow by accretions of layers of fibrin and platelets with bacterial colonies sandwiched between them. These vegetations are friable (fragile, crumbly) and portions may detach forming emboli that may seed many organs including spleen, kidneys, bowel, or brain, causing infection and **infarction**.

Figure 4. Optochin susceptibility test for identification of viridans streptococci.
α-Hemolytic, gram-positive, catalase-negative cocci are streaked in bands across a blood agar plate and an optochin disc is placed on each streak. After incubation for 24–48 hours in 5% CO_2 in air at 37°C growth around the discs is examined. The strain in the top streak has grown up to the disc and is resistant to optochin and, therefore, is not *S. pneumoniae* but is another species of viridans streptococci. The strains in the center and lower streaks show a zone of inhibition, indicating that they are susceptible to optochin and are pneumococci.

Pathogenesis

The mechanism by which the organisms cause pathogenesis is unclear. However, experiments in rabbits suggest that for endocarditis to occur there needs to be an event that induces an intravascular pro-inflammatory response (release of **interleukin (IL)**-1) a few hours before the bacteremia. It appears that IL-1 acts on endocardium to release epithelial tissue factor, a potent profibrotic factor, which is implicated in fibroblast proliferation, angiogenesis, and extracellular matrix (ECM) synthesis. It also causes the release of thrombin. *S. mitis* might contribute to this by inducing IL-1 release itself on the endocardium.

3. What is the typical clinical presentation and what complications can occur?

Persons at greatest risk for subacute bacterial endocarditis are those who have had previous infective endocarditis or rheumatic heart disease, those with prosthetic heart valves, congenital heart disease, other malformations of the heart, and those who use intravenous drugs. Persons with **periodontal disease** and poor oral hygiene are also at higher risk. The signs and symptoms of bacterial endocarditis resemble a nonspecific flu-like illness. Almost all patients are **febrile** and may have chills, sweats, anorexia, malaise, cough, headache, myalgia and/or arthralgia, and confusion. Fever is usually low-grade, rarely exceeding 39°C, remittent, and usually not associated with rigors. In about one-third of patients there may be neurologic abnormalities that include **stroke**, intracerebral hemorrhage, and subarachnoid hemorrhage. Peripheral symptoms include **petechiae** on the conjunctiva, buccal or palatal mucosa, and the extremities. There may be splinter and subungual hemorrhages in the nail beds of the fingers and toes and Osler nodes in the pulp of the digits. In China *S. mitis* has been reported to cause a scarlet fever-like **pharyngitis** and about half the cases developed a streptococcal toxic shock-like syndrome. This finding indicates that some *S. mitis* strains must express superantigenic activity.

4. How is the disease diagnosed and what is the differential diagnosis?

Diagnosis of infective endocarditis (IE) uses the Duke criteria (see References: Durack et al., 1994).

Detection and isolation of viridans streptococci
- *Microscopy*: Examination of blood stained by Gram's method for gram-positive cocci growing in chains can provide a rapid preliminary diagnosis.
- *Culture*: Specimens should be plated onto blood agar, which is incubated for 24–48 hours at 37°C in the presence of 5% CO_2.
- *Determinative tests*: *S. mitis* and other commensal α-hemolytic streptococci are resistant to **lysis** when exposed to bile or optochin (ethylhydrocupreine dihydrochloride) in contrast to *S. pneumoniae*, which is sensitive (Figure 4).

Differential diagnosis
- Tuberculosis, salmonellosis, gastrointestinal and genitourinary infections, and other disorders causing fever of undetermined origin.

- Juvenile **rheumatoid arthritis, polymyalgia rheumatica**.
- Acute rheumatic fever.
- Nonbacterial thrombotic endocarditis.
- Polyarteritis nodosa.
- **Systemic lupus erythematosus** with antiphospholipid antibody syndrome.
- Cardiac myxoma (a benign neoplasm arising within the heart).
- Neoplasms.

5. How is the disease managed and prevented?

Since dental treatment is considered the principal factor predisposing to bacterial endocarditis, which has high morbidity and mortality, antibiotic prophylaxis was recommended before beginning dental procedures likely to produce bacteremia in all patients with congenital heart disease, prosthetic heart valves and those with a history of rheumatic fever. However, in 2007 the American Heart Association in participation with the American Dental Association published new guidelines redefining the need for prophylactic antibiotics before dental treatment. The guidelines were also endorsed by the Infectious Diseases Society of America and by the Pediatric Infectious Diseases Society (http://www.ada.org/prof/resources/topics/infective_endocarditis.asp). The guidelines are based on accumulating evidence that taking antibiotics before dental treatment does not prevent IE in patients who are at risk of developing a heart infection. The reason for this is that all of us generate a bacterial shower into the bloodstream many times a day whenever we chew, brush or floss our teeth. Studies suggest that IE is more likely to result from these everyday activities than from dental treatment. Added to this is the fact that the risks of taking prophylactic antibiotics, such as adverse reactions and generation of antibiotic resistance, outweigh the benefits for the majority of patients. Therefore, patients with the following conditions who have taken prophylactic antibiotics routinely in the past but no longer require them include:

- mitral valve prolapse
- rheumatic heart disease
- bicuspid valve disease
- calcified aortic stenosis
- congenital heart conditions such as ventricular septal defect, atrial septal defect, and hypertrophic cardiomyopathy.

Prophylactic antibiotics before dental treatment are advised for patients with:

1. artificial heart valves
2. a history of infective endocarditis
3. certain specific, serious congenital (present from birth) heart conditions, including:
   - unrepaired or incompletely repaired cyanotic congenital heart disease, including those with palliative shunts and conduits
   - a completely repaired congenital heart defect with prosthetic material or device, whether placed by surgery or by catheter intervention, during the first 6 months after the procedure

- any repaired congenital heart defect with residual defect at the site or adjacent to the site of a prosthetic patch or a prosthetic device

4. a cardiac transplant that develops a problem in a heart valve.

When antibiotic prophylaxis is recommended before beginning dental procedures the standard regimen is 2 g of amoxicillin orally 1 hour before the procedure or 2 g of ampicillin given intravenously or intramuscularly within 30 minutes of the procedure. In patients allergic to penicillin, clarithromycin, cephalexin, cefadroxil or clindamycin may be given 1 hour before the procedure or cefazolin or clindamycin may be given intravenously 30 minutes before the procedure.

The viridans streptococci are increasingly important causes of sepsis and **pneumonia** in neutropenic persons and sepsis and **meningitis** in neonates. Their portal of entry is the oral mucosa and oral mucositis is a predisposing factor as are profound **neutropenia** and administration of trimethoprim-sulfamethoxazole or quinolines. Treatment is the administration of appropriate antibiotics and blood cultures are usually negative after 24 hours of therapy.

It should be noted that antibiotic resistance among viridans streptococci is increasing, with *S. mitis* being more resistant than other viridans species. There is evidence of horizontal gene transfer from viridans streptococci to *S. pneumoniae*. Commensal bacteria can be reservoirs of antibiotic resistance genes. Within the viridans streptococci there is species-related variability of susceptibility – especially to penicillin, macrolides, and tetracycline, with *S. mitis* and *S. oralis*, in particular, displaying high rates of resistance to penicillin and macrolides. The difference in susceptibilities between species of viridans streptococci indicates the importance of their accurate identification rather than considering them as a group, that is 'S. viridans'.

SUMMARY

1. What is the causative agent, how does it enter the body and how does it spread a) within the body and b) from person to person?

- *S. mitis* is one of about 20 species of commensal α-hemolytic streptococci. These bacteria are the most common cause of subacute bacterial endocarditis.

- *S. mitis* is a gram-positive, catalase-negative coccus that is resistant to optochin and bile.

- *S. mitis* grows readily on blood agar incubated at 37°C in the presence of 5% CO_2 for 24–48 hours, forming small, punctate colonies surrounded by a zone of incomplete hemolysis (α-hemolysis).

- *S. mitis* has a typical gram-positive-type cell wall that features a thick peptidoglycan layer with covalently bound teichoic acid.

- Like *S. pneumoniae*, to which it is closely related, *S. mitis* has a C-polysaccharide and its teichoic acid contains choline.

- Although *S. mitis* does not form extracellular polysaccharide, extracellular slime is produced by many other species of commensal viridans streptococci.

- About two-thirds of *S. mitis* isolates produce IgA1 protease to subvert mucosal immunity, but little else is known about their determinants of pathogenesis.

- *S. mitis* is an endogenous pathogen. It is an opportunist that causes disease when it is translocated to the bloodstream as a result of trauma (surgical or otherwise) in the mouth. There is no horizontal spread of the bacterium from person to person.

- *S. mitis* and other commensal, α-hemolytic streptococci adhere to damaged endocardium and valves.

2. What is the host response to the infection and what is the disease pathogenesis?

- The barrier epithelium, in concert with innate immune factors at these surfaces, is an effective barrier to the penetration of *S. mitis*.

- Secretory immunoglobulin A (sIgA) antibodies in saliva are important in acquired specific immunity in the mouth.

- Host phagocytes are a second line of defense against *S. mitis* that breaches the mucosal barrier.

- *S. mitis* are opsonized by activation of the alternative and lectin innate complement pathways and the classical pathway in the presence of IgM and IgG antibodies in the plasma and tissue fluid.

- The pathogenic potential of *S. mitis* likely results from its ability to adhere to damaged endocardium and valves and to initiate vegetations.

3. What is the typical clinical presentation and what complications can occur?

- Subacute bacterial endocarditis with signs and symptoms resembling a nonspecific flu-like illness.

- Persons at greatest risk are those who have had previous infective endocarditis or rheumatic heart disease, those with prosthetic heart valves, congenital heart disease, other malformations of the heart, and those who use intravenous drugs.

- Almost all patients are febrile and may have chills, sweats, anorexia, malaise, cough, headache, myalgia and/or arthralgia, and confusion.

- Fever is usually low-grade, rarely exceeding 39°C.

- There may be neurologic abnormalities that include stroke, intracerebral hemorrhage, and subarachnoid hemorrhage.

- Peripheral symptoms include petechiae on the conjunctiva, buccal or palatal mucosa, and the extremities.

- There may be splinter and subungual hemorrhages in the nail beds of the fingers and toes and Osler nodes in the pulp of the digits.

4. How is the disease diagnosed and what is the differential diagnosis?

- Microscopy: examination of positive blood bottles by Gram's method can provide a rapid preliminary diagnosis by revealing gram-positive cocci growing in chains.

- Culture: positive blood bottles are plated onto blood agar. Colonies are small, gray-green and surrounded by a zone of α-hemolysis.

- Determinative tests: resistance to optochin or bile.
- The Duke criteria (see References).

5. How is the disease managed and prevented?

- Maintain good oral hygene to eliminate dental plaque.

- Ampicillin is the antibiotic of choice.
- Clarithromycin, cephalexin, cefadroxil or clindamycin may be used in patients allergic to penicillin.

FURTHER READING

Cole MF, Lydyard PM. Oral microbiology and the immune response. In: Lamont RJ, Burne RA, Lantz MS, LeBlanc DJ, editors. Oral Microbiology and Immunology. ASM Press, Washington, DC, 2006: Chapter 10.

Mims C, Dockrell HM, Goering RV, Roitt I, Wakelin D, Zuckerman M. Medical Microbiology, 3rd edition. Mosby, Edinburgh, 2004.

Murphy K, Travers P, Walport M. Janeway's Immunobiology, 7th edition. Garland Science, New York, 2008.

Murray PR, Rosenthal KS, Pfaller MA. Medical Microbiology, 5th edition, Elsevier Mosby, Philadelphia, PA, 2005.

REFERENCES

Cole MF, Bryan S, Evans MK, et al. Humoral immunity to commensal oral bacteria in human infants: salivary secretory immunoglobulin A antibodies reactive with *Streptococcus mitis* biovar 1, *Streptococcus oralis*, *Streptococcus mutans*, and *Enterococcus faecalis* during the first two years of life. Infect Immun, 1999, 67: 1878–1886.

Cole MF, Evans M, Fitzsimmons S, et al. Pioneer oral streptococci produce immunoglobulin A1 protease. Infect Immun, 1994, 62: 2165–2168.

Durack DT, Lukes AS, Bright DK. New criteria for diagnosis of infective endocarditis: utilization of specific echocardiographic findings. Duke Endocarditis Service. Am J Med, 1994, 96: 200–209.

Herzberg MC. Platelet-streptococcal interactions in endocarditis. Crit Rev Oral Biol Med, 1996, 7: 222–236.

Kilan M, Poulsen K, Blomqvist T, Havarstein LS, Bek-Thomsen M, Tettelin H, Sorensen UB. Evolution of the *Strepococcus pneumoniae* and its close commensal relatives. PLoS One (electronic resource). 3(7):e2683, 2008.

Kirchherr JL, Bowden GH, Richmond DA, Sheridan MJ, Wirth KA, Cole MF. Clonal diversity and turnover of *Streptococcus mitis* bv. 1 on shedding and nonshedding oral surfaces of human infants during the first year of life. Clin Diagn Lab Immunol, 2005, 12: 1184–1190.

Lu HZ, Weng XH, Zhu B, et al. Major outbreak of toxic shock-like syndrome caused by *Streptococcus mitis*. J Clin Microbiol, 2003, 41: 3051–3055.

Pearce C, Bowden GH, Evans M, et al. Identification of pioneer viridans streptococci in the oral cavity of human neonates. J Med Microbiol, 1995, 42: 67–72.

Tak T, Dhawan S, Reynolds C, Shukla SK. Current diagnosis and treatment of infective endocarditis. Expert Review of Antiinfective Therapy, 2003, 1: 639–654.

Teng L-J, Hsueh P-R, Chen Y-C, Ho S-W, Luh K-T. Antimicrobial susceptibility of viridans group streptococci in Taiwan with an emphasis on the high rates of resistance to penicillin and macrolides in *Streptococcus oralis*. J Antimicrob Chemother, 1998, 41: 621–627.

WEB SITES

American Family Physician, published by the American Academy of Family Physicians, 'Preventing Bacterial Endocarditis,' February 01, 1998: http://www.aafp.org/afp/980201ap/taubert.html

American Heart Association website: http://www.americanheart.org/presenter.jhtml?identifier=4436

eMedicine, Inc., 2007, provider of clinical information and services to physicians and health-care professionals: http://www.emedicine.com/MED/topic671.htm

MULTIPLE CHOICE QUESTIONS

The questions should be answered either by selecting True (T) or False (F) for each answer statement, or by selecting the answer statements which best answer the question. Answers can be found in the back of the book.

1. **Which of the following distinguish the commensal viridans streptococci from the pneumococcus?**

   A. Susceptibility to optochin.

   B. Hemolysis.

   C. Cellular morphology revealed by Gram staining.

   D. A and B.

   E. A and C.

2. **Which one of the following bacteria is the most common cause of subacute bacterial endocarditis (SBE)?**

   A. *Streptococcus pneumoniae.*

   B. *Streptococcus pyogenes.*

   C. *Streptococcus agalactiae.*

   D. *Streptococcus mitis.*

   E. *Enterococcus faecium.*

3. **All of the following are characteristics of *Streptococcus mitis*, EXCEPT?**

   A. Gram-positive.

   B. α-Hemolytic.

   C. Produce an IgA1 protease.

   D. Optochin resistant.

   E. Uniformly susceptible to penicillin.

4. ***Streptococcus mitis* is the most common cause of which one of the following infections that are caused by streptococci**

   A. Neonatal meningitis.

   B. Necrotizing fasciitis.

   C. Streptococcal toxic shock syndrome (STSS).

   D. Subacute bacterial endocarditis.

   E. Lobar pneumonia.

5. **All of the following statements about *S. mitis* or the disease it causes are correct, EXCEPT?**

   A. It causes endogenous infections.

   B. It is an opportunistic pathogen.

   C. It is a member of the oropharyngeal commensal microbiota.

   D. Disease results from horizontal transmission via respiratory droplets.

   E. Most strains of *S. mitis* produce IgA1 protease.

6. ***S. mitis* has been reported to cause all of the following diseases, EXCEPT?**

   A. Impetigo.

   B. Subacute bacterial endocarditis (SBE).

   C. Intracerebral hemorrhage.

   D. Subarachnoid hemorrhage.

   E. Toxic shock syndrome.

7. **Which of the following conditions predispose towards bacterial endocarditis caused by *S. mitis*?**

   A. Intravenous drug use.

   B. Rheumatic fever.

   C. Prosthetic heart valve.

   D. Congenital heart disease.

   E. All of the above.

Case 35

Streptococcus pneumoniae

A 45-year-old male with a known history of alcohol abuse presented at the Emergency Room with a 1-day history of fevers, shaking chills, and productive cough. He stated that he was coughing up blood-tinged sputum that had a rusty appearance. The patient also remarked that he had some left-sided chest pain that was worse when he breathed in or coughed (**pleuritic pain**) and some **dyspnea** (shortness of breath).

He denied any significant past medical history, and was not taking any medications. He reported a 20-pack per-year history of smoking, and drinking at least one six-pack of beer per day.

On physical exam he was a well-developed male in moderate distress. His temperature was 40°C, pulse 110 beats per minute, blood pressure 140/80 mmHg, respiratory rate 24, and O_2 saturation 90% on room air. His lung examination revealed decreased breath sounds and dullness to percussion in the left lower lung base. There were some **rales** and **egophony** present. A sputum sample and blood were collected and a chest X-ray was ordered. The results of the investigations were as follows.

- Laboratory findings: WBC 18 000, 80% PMNs, 10% bands, 10% lymphocytes.
- Chest X-ray: dense consolidation and air bronchograms in the left lower lobe (Figure 1).
- Sputum Gram stain: gram-positive, bullet-shaped diplococci and inflammatory cells (Figure 2).
- Blood cultures at 24 hours revealed gram-positive diplococci. (Figure 3).

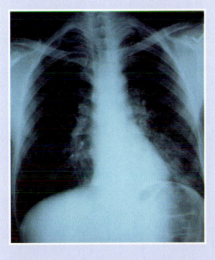

Figure 1. AP chest X-ray shows pneumonia of the left lower lobe with early consolidation. This disease is caused by the bacterium *Streptococcus pneumoniae*. The alveoli (the air sacs of the lungs, too small to be seen) become blocked with pus, forcing air out and causing the lung to solidify. Symptoms include coughing and chest pain.

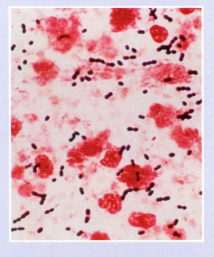

Figure 2. Gram stain of *S. pneumoniae* in sputum: note the bullet-shaped diplococci.

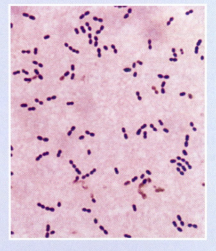

Figure 3. Gram stain of *S. pneumoniae* from blood culture broth showing gram-positive bullet-shaped diplococci.

Figure 4. A mucoid strain of S. pneumoniae growing on blood agar. The mucoid appearance results from production of capsule.

1. What is the causative agent, how does it enter the body and how does it spread a) within the body and b) from person to person?

Causative agent

The patient is infected with *Streptococcus pneumoniae*, frequently termed the pneumococcus. *S. pneumoniae* is one of the main causes of respiratory tract infections worldwide, killing over 1 million individuals, mostly children, every year. Pneumococcal infection is the sixth leading cause of disease in the US and is responsible for about 6 million cases of **otitis media**, about 200 000 hospitalizations from **pneumonia** and invasive disease, and at least 40 000 deaths annually, which are more deaths than acute infection with any other bacteria. *S. pneumoniae* is a gram-positive, **catalase**-negative diplococcus. The bacterial cells are bullet- or lancet-shaped when observed by Gram's stain (Figure 3). The catalase test is used to distinguish streptococci from staphylococci, which are the other medically important genus of gram-positive cocci (see the *Staphylococcus aureus* case for an image of the catalase test). *S. pneumoniae* grows readily on blood agar incubated at 37°C in the presence of 5% CO_2 for 24–48 hours and forms small to medium size colonies with a varying mucoid appearance. The colonies are gray-green in color and are surrounded by a zone of incomplete (greening) **hemolysis** (α-hemolysis) (Figure 4). Older colonies of pneumococci collapse in the center due to autolysis and are termed 'draughtsman-like' colonies.

The most important component of the *S. pneumoniae* outer surface is the polysaccharide **capsule**, since this is the principal **virulence** determinant of the bacterium. The capsule impedes **phagocytosis** primarily by inhibiting deposition of the opsonic complement component, C3b, on the bacterial surface thus impairing the immune response to *S. pneumoniae*. The structure of the capsule is complex and there are some 91 distinct antigenic types. *S. pneumoniae* has a typical gram-positive-type cell wall that features six layers of peptidoglycan with covalently bound teichoic acid (C-polysaccharide) and cell membrane-anchored lipoteichoic acid (Forssman antigen) (Figure 5). Both forms of teichoic acid have identical carbohydrate structures and contain choline. Although the vast majority of

Figure 5. Schematic representation of the S. pneumoniae envelope. Teichoic acids (TAs) and lipoteichoic acids (LTAs) (both in blue) are carbohydrate phosphate polymers rich in choline (red spheres). TAs are linked to the peptidoglycan via a phosphodiester linkage, whereas LTAs are linked to the cell membrane via a C-terminal fatty acyl group. Choline-binding proteins are linked to cell wall TAs or LTAs via choline-binding domains (CBDs). Pneumococcal surface antigen (PsaA) is located beneath the peptidoglycan layer and is attached to the cell membrane; penicillin-binding proteins (PBPs) are located in the 'periplasmic space' (the zone beneath the peptidoglycan and above the cytoplasmic membrane) and interact with the peptidoglycan. Hyaluronate lyase (Hyl) is tethered to the peptidoglycan. Taken and adapted from Figure 2 in: Di Guilmi AM & Dessen A. 2002. New approaches towards the identification of antibiotic and vaccine targets in *Streptococcus pneumoniae*. EMBO Reports 3(8):728–734.

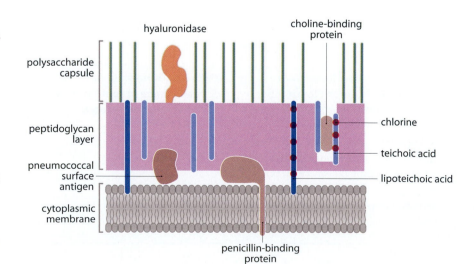

the teichoic acid is covered by the capsule, anti-phosphocholine antibody can protect against experimental pneumococcal infection and the acute phase reactant, **C-reactive protein**, can interact with choline residues in the cell wall. On the outer surface of the cell wall is a family of 12 choline-binding proteins (CBPs) that are noncovalently bound to the phosphorylcholine moiety of the wall. These include: the major autolysin of the pneumococcus, LytA; two other cell wall hydrolases, LytB and LytC, suggested to be involved in daughter cell separation as well as in bacterial uptake of DNA; PcpA, which is thought to be involved in protein–protein and protein–lipid interactions; PspA that decreases complement deposition on the bacterial surface during sepsis; and CbpA (SpsA), the most abundant of the CBPs, that functions as a cell-surface **adhesin** in the nasopharynx. CbpA has been shown to bind the secretory component of **IgA** and the complement component, C3. The pneumococcus contains a cholesterol-dependent pore-forming cytotoxin termed pneumolysin, which is stored intracellularly in most strains and is released upon cell **lysis** mediated by LytA. This toxin is important in the causation of **meningitis** because it damages ependymal cilia that line the ventricles of the brain and induces brain cells to undergo **apoptosis**. Also, H_2O_2 produced by the pneumococcus is a potent **hemolysin**. Finally, the bacterium secretes an **IgA1 protease**, which is able to subvert the activity of IgA1 by cleaving the molecule at the hinge region, and a neuraminidase that cleaves terminal sialic acid from glycoconjugates thereby uncovering **epitopes** for pneumococcal adherence.

Entry and spread within the body

S. pneumoniae enters the body through the oral–nasal route via respiratory droplets and adheres to the oropharyngeal epithelium via the adhesin molecules listed in Table 1. Initial binding is thought to involve carbohydrate epitopes in the respiratory mucosa through an as yet unidentified adhesin. Pneumococci then use CbpA to bind to the poly-Ig

Table 1. Factors contributing to virulence of *Streptococcus pneumoniae*

Factor	Action
Capsule	Anti-phagocytic
PspA	Anti-phagocytic
Pneumolysin	Cytotoxin
Hydrogen peroxide	Hemolysin
Neuraminidase	Removes terminal sialic acid exposing glycoconjugates for adhesion
Teichoic acid	Adhesin (binds fibronectin), immunomodulatory
CbpA	Adhesin and invasion (binds platelet-activating factor [PAF] receptor and poly-Ig [pIg] receptor)
PsaA	Mn^{2+} binding protein, required for adherence of bacteria to host cells
PcpA	Adhesin (protein–protein and protein–lipid interactions)
IgA1 protease	Cleaves sIgA1 in respiratory secretion

(pIg) receptor and phosphocholine to bind to the platelet-activating factor (PAF) receptor, enabling the pneumococcus to invade the epithelium. *S. pneumoniae* transiently colonizes the pharynx and carriage of the bacterium in healthy people has been reported to be as high as 75% depending on detection method and population. Colonization is more common in infants and children than in adults. Repeat episodes of carriage are associated with different capsule serotypes and the duration of carriage decreases with each successive colonization episode, in part because of the induction of serotype-specific antibodies. Disease results from the spread of bacteria from the nasopharynx to the sinuses, middle ear, **meninges**, and lungs (as it did in this patient). Bacteremic spread may also occur.

Person to person spread

Disease occurs as a result of horizontal spread of the organism by respiratory droplets, particularly in crowded settings such as hospitals, day-care centers, and jails.

2. What is the host response to the infection and what is the disease pathogenesis?

Immune response

Table 1 lists the various factors that contribute to the virulence of the organism and some of these include avoidance mechanisms that help to prevent the organism from being eliminated by the host defenses. The respiratory mucosa with it muco-ciliary blanket together with innate immune factors and secretory IgA antibodies are effective barriers to the invasion of *S. pneumoniae* through the epithelium. As for most extracellular, capsulate bacteria, phagocytosis aided by complement and opsonic **IgM** and **IgG** antibodies are the principal modalities of host defense after invasion. Thus, pneumococci are opsonized by activation of the **alternative** and lectin innate **complement pathways** and the **classical** pathway in the presence of anti-capsular antibodies in the plasma and tissue fluid. The pneumococcus targets those with impaired innate and acquired immune systems, for example, persons who lack a spleen, are immunodeficient, or lack early or late complement components. Such individuals are generally the very young and very old who have low levels of anti-capsular antibody and those with impaired bacterial clearance. However, neonates are protected by transplacental transfer of maternal IgG anti-capsular antibodies and infants are protected against pneumococcal disease through antibacterial components of human milk. Impaired bacterial clearance is observed in persons with defective epiglottal and/or cough reflexes, antecedent viral respiratory infections, chronic pulmonary disease, chronic cardiovascular disease, alcoholism, renal dysfunction, and diabetes, as well as individuals with certain hereditary or acquired immunodeficiencies.

Pathogenesis

The host immune response plays a major role in pathogenesis of the pneumococcus. Resolution of infection occurs when the bacteria are rapidly cleared without neutrophil recruitment. This is accomplished by both humoral and cellular components of the immune system. Lung tissue injury results from rapid bacterial growth with attendant increase in pro-inflammatory **cytokines** and neutrophil recruitment. Thus, the hallmark of pneumococcal disease is the intense inflammatory infiltrate at the site of infection. This results from the action of pneumolysin and hydrogen peroxide and the pro-inflammatory nature of the bacterial cell wall components such as teichoic acid and peptidoglycan fragments. These are released from the wall by the autolysins, LytA, B, and C. Pneumococci can be induced to undergo autolysis by exposure to β-lactam antibiotics (see Management below), hypertonic media, detergents or extended incubation in the stationary phase of growth. The wall fragments activate the complement cascade releasing the anaphylatoxins, C3a and C5a. In addition, **interleukin** (**IL**)-1 and tumor necrosis factor-α (TNF-α) are released from activated leukocytes leading to the recruitment of additional inflammatory cells, amplifying tissue damage and causing fever.

3. What is the typical clinical presentation and what complications can occur?

Pneumococcal pneumonia typically presents with an abrupt onset of fever to 41°C and shaking chills. These symptoms may follow a viral infection of the respiratory tract. The majority of patients have a productive cough with blood-tinged sputum and **pleurisy**. The mortality from pneumococcal pneumonia is about 5% in the United States. Among the over 90 capsule types, capsule type 3 organisms are the most virulent.

S. pneumoniae is also a frequent cause of **sinusitis** and otitis media. From the middle ear or sinuses the bacteria can seed the central nervous system (CNS). CNS infection can also follow **bacteremia**. Meningitis is most commonly observed in children and adults and less frequently in the elderly. Of the bacteria causing meningitis, *S. pneumoniae* infection has the highest mortality and results in the most severe neurological impairment.

4. How is the disease diagnosed and what is the differential diagnosis?

Microscopy

Examination of sputum, blood, cerebrospinal fluid, and other specimens stained by Gram's method for gram-positive, bullet-shaped diplococci can provide a rapid preliminary diagnosis. A Gram stain of sputum is shown in Figure 2.

Culture

Specimens should be plated onto blood agar and incubated for 24–48 hours at 37°C in the presence of 5% CO_2. Colonies of various strains vary in the amount of capsule they produce. Highly mucoid, α-hemolytic colonies are shown in Figure 4; however, pneumococcal colonies frequently produce little to no capsule (nonmucoid). Examples of colonies of a nonmucoid strain of *S. pneumoniae* are shown in Figure 6.

Determinative tests

S. pneumoniae is lysed by exposure to bile or optochin (ethylhydrocupreine dihydrochloride) while commensal α-hemolytic streptococci in the nasopharynx are resistant to both. An optochin disc placed on a blood agar

A

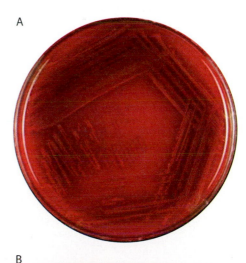

B

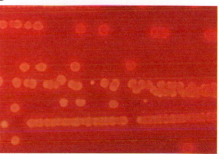

Figure 6. (A) A nonmucoid strain of *S. pneumoniae* growing on a blood agar plate. (B) Enlargement of (A).
S. pneumoniae appear as small grayish colonies with a greenish zone of α-hemolysis surrounding them on the blood agar plate.

Figure 7. Optochin susceptibility test for identification of *S. pneumoniae*.
α-Hemolytic, gram-positive, catalase-negative cocci are streaked in bands across a blood agar plate and an optochin disc is placed on each streak. After incubation for 24–48 hours in 5% CO_2 in air at 37°C growth around the discs is examined. The strain in the top streak has grown up to the disc and is resistant to optochin and, therefore, is not a pneumococcus. The strains in the center and lower streaks show a zone of inhibition indicating that they are susceptible to optochin and are pneumococci.

plate inoculated with *S. pneumoniae* will show a wide zone of growth inhibition (Figure 7).

Serology

The capsule type of the *S. pneumoniae* isolate can be determined by the **Quellung test** in which bacteria are mixed into a drop of capsule-specific antiserum and observed under the microscope. Capsule swelling and increased refractility result from the interaction of the antibody with the capsule for which it is specific.

Differential diagnosis

The clinical presentation and findings suggest pneumonia of bacterial etiology because of the acute onset and severity of symptoms. Pneumonia caused by viruses has a more gradual onset and is generally less severe. Symptoms of viral pneumonia may include fever, dry cough, headache, and muscle pain. Within 12–16 hours, other symptoms may appear, such as shortness of breath, sore throat, increased cough, and mucus with cough. Bacteria to include in the differential diagnosis would be *Haemophilus influenzae*, *Moraxella catarrhalis*, *Klebsiella pneumoniae*, *Staphylococcus aureus*, *Legionella pneumophila*, *Mycoplasma pneumoniae* and *Chlamydophila pneumoniae*.

5. How is the disease managed and prevented?

Management

S. pneumoniae, once highly susceptible to penicillin, has now acquired resistance to this and many other antibiotics. Penicillin resistance results from the generation of penicillin-binding proteins (PBPs) with decreased affinity for penicillin. PBPs are involved in the assembly of the cell wall. It is likely that these altered PBP genes arose by interspecies recombination in which segments of the PBPs' structural genes were replaced by regions derived from PBP genes of oral streptococci. These altered PBP genes of penicillin-resistant pneumococci can be spread horizontally to sensitive pneumococci by transformation. About one-third of the strains isolated in the United States are resistant to penicillin and higher rates of resistance have been observed in other countries. Although there are over 90 *S. pneumoniae* serotypes, over 90% of penicillin-resistant strains are found within seven serotypes (6A, 6B, 9V, 14, 19A, 19F, and 23F). Unfortunately, these are the same serotypes that cause the vast majority of infections in children. Cefotaxime, ceftriaxone, and clindamycin are effective antibiotics for treating pneumonia caused by penicillin-resistant pneumococcal isolates that are susceptible to these antibiotics. Clindamycin or vancomycin is recommended when a pneumococcal isolate is resistant to cefotaxime or ceftriaxone.

Prevention

Two vaccines are available to prevent pneumococcal disease. Both consist of the purified capsular polysaccharide. A 23-valent vaccine consisting of capsular polysaccharide from the 23 serotypes that are most commonly isolated from infected patients is recommended for use in children above the age of 2 years and in adults. This vaccine is only 60% effective. Polysaccharides are T-cell-independent antigens and infants below the age of 2 years respond poorly to them. Therefore, for children younger than 2 years of age, a 7-valent **conjugate vaccine** is recommended. In this vaccine the capsular polysaccharide is conjugated to a protein carrier to render it T-cell-dependent. The protein carrier used is either tetanus toxoid or diphtheria toxoid, themselves vaccine antigens. The 7-valent conjugate vaccine reduces invasive infection in children, yet has little effect on the incidence of otitis media and colonization. The long-term efficacy and benefit of this vaccine are, therefore, unknown.

SUMMARY

1. What is the causative agent, how does it enter the body and how does it spread a) within the body and b) from person to person?

- *S. pneumoniae* is the most common cause of community-acquired bacterial pneumonia.

- *S. pneumoniae* is a gram-positive, catalase-negative, bullet-shaped diplococcus.

- *S. pneumoniae* grows readily on blood agar incubated at 37°C in the presence of 5% CO_2 for 24–48 hours, forming small, mucoid colonies surrounded by a zone of incomplete hemolysis (α-hemolysis).

- *S. pneumoniae* has a typical gram-positive-type cell wall that features a thick peptidoglycan layer with covalently bound teichoic acid.

- *S. pneumoniae* undergoes autolysis releasing cell wall fragments that are pro-inflammatory.

- The capsule is the most important virulence factor of *S. pneumoniae* and there are over 90 antigenically distinct types.

- *S. pneumoniae* has many determinants of pathogenesis including a capsule that is anti-phagocytic, adhesins, cytotoxins (pneumolysin and H_2O_2), pro-inflammatory cell wall fragments (teichoic acid, lipoteichoic acid, and peptidoglycan fragments), and IgA1 protease to subvert mucosal immunity.

- *S. pneumoniae* is spread person to person via respiratory droplets.

- There is a high rate of carriage in healthy persons, particularly children.

- *S. pneumoniae* adheres to and invades respiratory epithelial cells from where it can seed the bloodstream and central nervous system.

2. What is the host response to the infection and what is the disease pathogenesis?

- The barrier epithelium, in concert with innate immune factors at these surfaces, is an effective barrier to the penetration of *S. pneumoniae*.

- Secretory immunoglobulin A (sIgA) antibodies in respiratory mucus are important in acquired specific immunity at the mucosal surface.

- Host phagocytes are a second line of defense against *S. pneumoniae* invasion.

- *S. pneumoniae* are opsonized by activation of the alternative and lectin innate complement pathways and the classical pathway in the presence of anti-capsular antibodies in the plasma and tissue fluid.

- The virulence of *S. pneumoniae* results from its ability to subvert sIgA antibodies, adhere to and invade host cells, and to avoid opsonization and phagocytosis by means of is capsule.

- The cytotoxins pneumolysin and H_2O_2 lyse host cells.

- *S. pneumoniae* readily undergoes autolysis releasing pneumolysin but also fragments of the cell wall (teichoic acid, lipoteichoic acid, and peptidoglycan) that are strongly pro-inflammatory.

3. What is the typical clinical presentation and what complications can occur?

- Pneumonia with a sudden onset comprising shaking chills, fever to 41°C, productive cough with blood-tinged or purulent sputum, and pleurisy.

- *S. pneumoniae* is also a common cause of sinusitis and otitis media.

- Bacteremia and meningitis are serious complications.

4. How is the disease diagnosed and what is the differential diagnosis?

- Microscopy: examination of sputum, blood, cerebrospinal fluid or aspirates stained by Gram's method can provide a rapid preliminary diagnosis by revealing gram-positive, bullet-shaped diplococci.

- Culture: clinical specimens are plated onto blood agar to grow the characteristic small to medium, mucoid colonies surrounded by a zone of α-hemolysis.

- Determinative tests: susceptible to optochin or bile.

- Antigen detection: capsule typing can be performed using type-specific antisera.

5. How is the disease managed and prevented?

- If the strain is susceptible, penicillin is the antibiotic of choice in patients that are not hypersensitive.

- Over the last decade *S. pneumoniae* has become increasingly resistant to penicillin and other antibiotics.

- Cefotaxime, ceftriaxone, and clindamycin are effective antibiotics for treating pneumonia caused by penicillin-resistant pneumococci that are susceptible to these antibiotics.

- Clindamycin or vancomycin is recommended for cefotaxime- or ceftriaxone-resistant pneumococci.

- There are currently two vaccines approved for use in the US. The unconjugated vaccine for adults is only 60% effective in the elderly and is ineffective in children. The newly developed 7-valent conjugate vaccine reduces invasive infection in children, yet has little effect on the incidence of otitis media and colonization.

- Pneumococcal disease is complicated by an increasing incidence (currently 30% in the US) of clinical isolates resistant to commonly used antibiotics.

FURTHER READING

Mims C, Dockrell HM, Goering RV, Roitt I, Wakelin D, Zuckerman M. Medical Microbiology, 3rd edition. Mosby, Edinburgh, 2004.

Murphy K, Travers P, Walport M. Janeway's Immunobiology, 7th edition. Garland Science, New York, 2008.

Murray PR, Rosenthal KS, Pfaller MA. Medical Microbiology, 5th edition. Elsevier Mosby, Philadelphia, PA, 2005.

Tuomanen E, editor. The Pneumococcus. ASM Press, Washington, DC, 2004.

REFERENCES

Bergmann S, Hammerschmidt S. Versatility of pneumococcal surface proteins. Microbiology, 2006, 152(Pt 2): 295–303.

Di Guilmi AM, Dessen A. new approaches towards the identification of antibiotic and vaccine targets in *Streptococcus pneumoniae*. EMBO Reports. 3(8) 728–734, 2002 Aug.

Durbin WJ. Pneumococcal infections. Pediatr Rev, 2004, 25: 418–424.

File TM Jr. Clinical implications and treatment of multiresistant *Streptococcus pneumoniae* pneumonia. Clin Microbiol Infect, 2006, 12(Suppl 3): 31–41.

Hammerschmidt S. Adherence molecules of pathogenic pneumococci. Curr Opin Microbiol, 2006, 9: 12–20.

Hausdorff WP, Feikin DR, Klugman KP. Epidemiological differences among pneumococcal serotypes. Lancet Infect Dis, 2005, 5: 83–93.

Kadioglu A, Andrew PW. The innate immune response to pneumococcal lung infection: the untold story. Trends Immunol, 2004, 25: 143–149.

Kilan M, Poulsen K, Blomqvist T, Havarstein LS, Bek-Thomsen M, Tettelin H, Sorensen UB. Evolution of the *Strepococcus pneumoniae* and its close commensal relatives. PLoS One (electronic resource). 3(7):e2683, 2008.

Lee C-J, Lee LH, Gu X-X. Mucosal immunity induced by pneumococcal glycoconjugate. Crit Rev Microbiol, 2005, 31: 137–144.

Lopez R, Garcia E. Recent trends on the molecular biology of pneumococcal capsules, lytic enzymes, and bacteriophage. FEMS Microbiol Rev, 2004, 28: 553–580.

Lynch JP 3rd, Zhanel GG. Escalation of antimicrobial resistance among *Streptococcus pneumoniae*: implications for therapy. Semin Respir Crit Care Med, 2005, 26: 575–616.

Mitchell TJ. The pathogenesis of streptococcal infections. Nat Rev Microbiol, 2003, 1: 219–230.

Ortqvist A, Hedlund J, Kahn M. *Streptococcus pneumoniae*: epidemiology, risk factors, and clinical features. Semin Respir Crit Care Med, 2005, 26: 563–574.

Paterson GK, Mitchell T. Innate immunity and the pneumococcus. Microbiology, 2006, 152(Pt 2): 285–293.

WEB SITES

Centers for Disease Control & Prevention, National Center for Immunization and Respiratory Diseases, Atlanta, GA, USA: http://www.cdc.gov/NCIDOD/DBMD/DISEASEINFO/streppneum_t.htm

World Health Organization, Division of Immunizations, Vaccines and Biologicals, © 2003 WHO/OMS: http://www.who.int/vaccines/en/pneumococcus.shtml

MULTIPLE CHOICE QUESTIONS

The questions should be answered either by selecting True (T) or False (F) for each answer statement, or by selecting the answer statements which best answer the question. Answers can be found in the back of the book.

1. **Which one of the following statements concerning optochin discs is correct?**
   A. Used to determine whether a β-hemolytic streptococcus is *S. pyogenes*.
   B. Inhibits viridans streptococci in respiratory specimens.
   C. Selects for *S. pneumoniae* in respiratory specimens.
   D. Distinguishes *S. pneumoniae* from other species of viridans streptococci.
   E. Tests the antibiotic sensitivity of gram-positive cocci.

2. **Which one of the following statements concerning the pneumococcal conjugate vaccine is correct?**
   A. Composed of capsular polysaccharide from 23 serotypes of *S. pneumoniae*.
   B. Nonimmunogenic in children under the age of 2 years.
   C. Composed of seven capsular polysaccharide serotypes linked to a protein carrier.
   D. A T-independent antigen.
   E. A conjugate of pneumococcal surface protein A (PspA) and tetanus toxoid.

3. **All of the following statements concerning the capsule of pneumococci are correct, EXCEPT?**
   A. It is an important virulence determinant.
   B. It is immunogenic in children over 2 years of age.
   C. It is anti-phagocytic.
   D. It is antigenically diverse.
   E. All capsule types of pneumococci are equally invasive.

4. **All of the following statements concerning the envelope of *S. pneumoniae* are correct, EXCEPT?**
   A. Autolyses when treated with penicillin.
   B. Pneumolysin is a principal cell wall protein.
   C. Fragments initiate inflammation.
   D. Peptidoglycan is covalently linked to choline-containing teichoic acid.
   E. Contains several choline-binding adhesins.

5. **Which one of the following bacteria is the most common cause of community-acquired pneumonia?**
   A. *Streptococcus pneumoniae*.
   B. *Legionella* species.
   C. *Mycoplasma* species.
   D. *Chlamydophila pneumoniae*.
   E. *Pseudomonas aeruginosa*.

6. **All of the following are characteristics of *Streptococcus pneumoniae*, EXCEPT?**
   A. Gram-positive.
   B. α-Hemolytic.
   C. Produce an IgA1 protease.
   D. Bullet-shaped diplococci.
   E. Uniformly susceptible to penicillin.

7. ***Streptococcus pneumoniae* is the most common cause of bacterial meningitis in all of the following patients, EXCEPT?**
   A. A 14-day-old neonate.
   B. A 33-year-old male who developed penetrating head trauma in a motor vehicle accident.
   C. A 45-year-old alcoholic.
   D. A 64-year-old with diabetes mellitus.
   E. A previously healthy 70-year-old with a recent bout of sinusitis.

8. **All of the following are features of the host–pneumococcus relationship, EXCEPT?**
   A. Transmitted person to person by respiratory droplets.
   B. Vertical transmission from mother to neonate during delivery.
   C. High rate of carriage in healthy children.
   D. Repeat episodes of carriage are associated with different capsule serotypes.
   E. Duration of carriage decreases with each successive colonization episode.

9. **The pneumococcus causes all of the following diseases, EXCEPT?**
   A. Pneumonia.
   B. Otitis media.
   C. Sinusitis.
   D. Meningitis.
   E. Necrotizing fasciitis.

10. **Which of the following conditions predispose towards pneumococcal disease?**
    A. Antecedent viral respiratory infection.
    B. Chronic pulmonary disease.
    C. Cigarette smoking.
    D. Chronic renal disease.
    E. All of the above.

Case 36

Streptococcus pyogenes

A 7-year-old boy was well until yesterday when he developed **dysphagia**, painful anterior lymph nodes, and a fever to 40°C. The patient vomited once and complained of a headache. Physical examination showed an acutely ill patient with a temperature of 39°C and a pulse of 104 beats per minute. Examination of his head, eyes, ear, nose, and throat revealed bilateral tonsillar **hypertrophy** with grayish-white exudates and punctate hemorrhages (Figure 1). Bilateral tender submandibular lymph nodes were palpated. The remainder of the physical examination was within normal limits. A throat swab was obtained. Laboratory findings were: hemoglobin, normal; hematocrit, normal; WBC count, 19 000 mm^3; differential, 80% PMN, 4% bands, 15% lymphocytes.

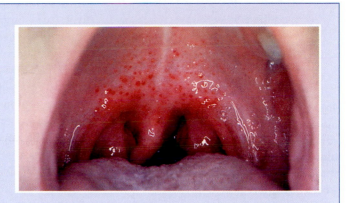

Figure 1. Pharynx of the patient showing punctate hemorrhages and inflamed tonsils caused by *Streptococcus pyogenes*.

1. What is the causative agent, how does it enter the body and how does it spread a) within the body and b) from person to person?

Causative agent

The patient has acute **pharyngitis**. The vast majority of cases of acute pharyngitis are caused by viruses, but the principal causative bacterial agent is *Streptococcus pyogenes*, also called the group A streptococcus or GAS. *S. pyogenes* is a gram-positive, catalase-negative coccus that grows in chains. A Gram stain of streptococci is shown in Figure 2. As for other streptococci, the **catalase** test is used to distinguish them from staphylococci, which are the other medically important genus of gram-positive cocci (see Figure 11 in the *Staphylococcus aureus* case for the catalase test). Streptococci grow readily on blood agar incubated at 37°C in the presence of 5% CO_2 for 24–48 hours. GAS form small, gray opalescent colonies and unlike *S. pneumoniae* and *S. mitis*, which show α- (incomplete) **hemolysis**, GAS colonies are surrounded by a large zone of complete hemolysis (β-hemolysis) (Figure 3). GAS has a typical gram-positive-type cell wall that features a thick peptidoglycan layer with covalently bound teichoic acid (Figure 4). There are a number of molecules that are exposed on the cell surface. These are either anchored in the cytoplasmic membrane and traverse the cell wall to the outside or they are anchored in the wall and extend from it. Among the membrane-anchored molecules are M protein and lipoteichoic acid. M protein is the most important **virulence** factor of GAS. M protein is an α-helical coiled-coil fibrillar protein. The amino acid sequence of the extracellular portion of the molecule

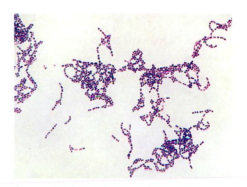

Figure 2. Gram stain of *Streptococcus pyogenes*. The streptococci form long chains and appear as purple beads. This form of growth is most obvious when the bacteria are obtained from liquid samples.

Figure 3. *Streptococcus pyogenes* inoculated onto trypticase soy agar containing 5% defibrinated sheep's blood, that is blood agar plate (BAP) that had been 'streaked,' and 'stabbed' with a bacteriological loop. A clear, colorless region surrounding each colony reflects the fact that the red blood cells in the blood agar medium have been destroyed, or hemolyzed, and indicates that these bacteria are β-hemolytic. Hemolysis is caused by two exotoxins termed streptolysin O and streptolysin S. Streptolysin O is oxygen-labile and so the inoculum is stabbed below the surface of the agar where the oxygen tension is reduced and both hemolysins are active. Stabs are shown at the 12:00 and 2:00 positions.

is highly variable, giving rise to over 100 serotypes of M protein (see immune response to *S. pyogenes*). The various serotypes are referred to as 'M' types. There are two classes of M protein, class I and II. M types with class I M proteins share surface-exposed antigenic epitopes and do not produce opacity factor (OF), whereas M types with class II M proteins lack these shared epitopes and produce OF. The importance of the division of M types into two classes is in determining their propensity to cause secondary complications. While both classes cause **suppurative infections** and **glomerulonephritis**, only strains with class I M proteins cause rheumatic fever (see complications later). Another surface appendage important in typing GAS strains is the T antigen of which there are about 20 types. T antigens form the backbone of pilus-like structures that extend from the cell surface and may be involved in adhesion and invasion. It is interesting that the genes encoding the pili are found on a pathogenicity island (large genomic regions 10 - 200 kb that are found in pathogenic bacteria but not in non-pathogenic bacteria of the same or closely related species which are acquired by horizontal gene transfer) that also encodes the genes for other extracellular matrix binding adhesins such as fibronectin-binding protein. Lipoteichoic acid is an important molecule that is pro-inflammatory and contributes to adherence of GAS by binding fibronectin on host cells. This together with a variety of cell surface molecules including the **adhesins**, F protein (fibronectin-binding protein), glyceraldehyde 3-phosphate dehydrogenase (G3-PD), and enolase (binds plasminogen) are involved in attachment of the bacteria to the epithelium and therefore contribute to pathogenesis (Table 1). Also located on the cell surface is the group-specific carbohydrate on which is based the Lancefield typing system for β-hemolytic streptococci. Some GAS strains produce a capsule composed of hyaluronic acid. The structure of the capsule is identical to that of the mammalian intercellular matrix, thus disguising the organism.

Entry and spread within the body

GAS enters the body via the **oropharynx** and is very adept at adhering to epithelial cells via the adhesin molecules described above. GAS transiently colonizes the pharynx and skin and carriage of the bacterium is reported to be as high as 20%. GAS only really causes pathogenesis when the organism invades and penetrates the epithelial surface or localizes in deeper

Figure 4. Schematic of the GAS envelope.

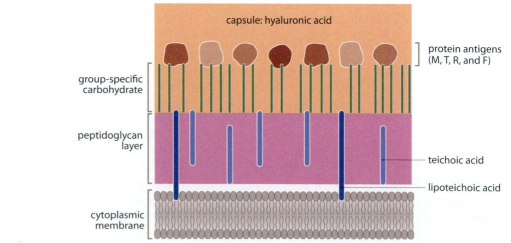

Table 1. Factors contributing to virulence of *Streptococcus pyogenes*

Factor	Action
Capsule	Antiphagocytic
M protein	Antiphagocytic, adhesin, degrades complement C3b,
T protein	Adhesin
Lipoteichoic acid	Adhesin
F protein	Adhesin (ligand fibronectin)
G3-PD	Adhesin
Enolase	Adhesin
Serum opacity factor	Adhesion (ligand fibronectin)
M-like proteins	Bind IgM and IgG, antiphagocytic
Pyrogenic exotoxins	Fever, superantigen, enhances endotoxin activity, shock, scarlet fever
Streptolysin S	Cytotoxin
Streptolysin O	Cytotoxin
Streptokinase	Fibrinolysis
DNase	Hydrolyzes DNA; degrades neutrophil extracellular traps (NETS)
C5a peptidase	Inactivates chemotactic factor C5a
Hyaluronidase	Spreading factor

G3-PD, glyceraldehyde 3-phosphate dehydrogenase.

subcutaneous tissues (see complications later). The adhesins, particularly M protein and F protein, enable GAS to invade epithelial surfaces.

GAS can seed the lymphatics and bloodstream. What causes the organism to become invasive following a local infection is still not fully understood. However, it has recently been suggested that clotting factors and the level of the pyrogenic **exotoxin** SpeB might be important.

Person to person spread

GAS is spread person to person via respiratory droplets. Crowded environments such as classrooms and day-care centers facilitate spread. In the case of skin infections, GAS is spread via the hands or by **fomites**. Spread via the hands can result in auto-inoculation, that is spread of the organism to additional parts of the body as well as spread to other persons. This is well illustrated in the case of **pyoderma (impetigo)**, a highly contagious superficial skin infection seen in young children in day-care or kindergarten settings.

2. What is the host response to the infection and what is the disease pathogenesis?

The virulence of GAS results from its ability to adhere to and invade host cells and to avoid **opsonization** and **phagocytosis**. The skin and mucous membranes in concert with innate immune factors at these surfaces are effective barriers to the penetration of GAS. However, GAS are able to invade epithelial cells. It appears the fibronection-binding proteins and M protein are important co-operative invasins, but it is clear that other surface adhesion molecules listed in Table 1 are implicated. Fibronectin may serve as a bridging molecule between the bacterial surface and the α5b1 integrin on the host cell membrane. Adhesion of GAS to the epithelial cell surface induces cytoskeletal reorganization and cell membrane ruffling and streptococci appear to be taken up by a zippering mechanism and enter the endosomal pathway. However, they are able to escape the early **endosome**, perhaps as a result of the action of the pore-forming cytolysin, streptolysin O. Within the cytosol GAS are combated by being enveloped by autophagosomes and killed when autophagosomes fuse with **lysosomes**. Thus, **autophagy** may be a mechanism by which nonphagocytic epithelial cells protect themselves against GAS invasion.

Host phagocytes are a second line of defense against streptococcal invasion. Streptococci are opsonized by activation of the **alternate** and lectin innate **complement pathways** and the **classical** pathway in the presence of anti-M protein antibodies in the plasma and tissue fluid. The hyaluronic **capsule** is poorly immunogenic, antiphagocytic, and serves to mask cell surface antigens from host immunity. Only M protein extends beyond the capsule. M protein binds factor H, a regulatory protein of the alternative pathway of complement, which degrades the complement component C3b, which is a potent opsonin. Also, by binding fibrinogen, M protein blocks deposition of C3b. In addition, GAS produces a serine protease that inactivates the complement component C5a, which is a powerful chemoattractant for neutrophils and macrophages.

GAS secretes several exotoxins and enzymes that play a role in pathogenesis and are listed in Table 1. The pyrogenic exotoxins (SpeA, B, C, F) are phage-specified **superantigens** that bring about the release of **cytokines (interleukin (IL)-1**, IL-2, IL-6, **tumor necrosis factor-α (TNF-α)**, **TNF-β**, **interferon-γ (IFN-γ)**) from macrophages and CD4 T cells. These cytokines mediate **shock** and organ failure characteristic of streptococcal **toxic shock syndrome** and give rise to the **rash** associated with scarlet fever. Streptolysins S and O are **hemolysin**s that are responsible for the β-hemolysis of GAS when it is grown on blood agar. However, they also lyse leukocytes, platelets, and likely other host cells. GAS produces three enzymes that facilitate the spread of the organism; hyaluronidase hydrolyzes the intercellular matrix, DNase depolymerizes extracellular DNA reducing the viscosity of **pus** and degrade the neutrophil extracellular traps (NETs), and streptokinase lyses blood clots aiding bloodstream spread. Finally, the immunoglobulin-binding M-like proteins function in blocking phagocytic activity and also degrade complement C3b.

As mentioned above it is clear that GAS has several strategies to invade nonphagocytic cells, such as the utilization of the fibronectin-binding protein to bind fibronectin, which is used as a bridging molecule to bind to the α5b1 integrin cellular receptor. However, the invasive phenotype of GAS as seen in the highly virulent subclone of serotype M1T1, a major cause of **necrotizing fasciitis**, may be related to the accumulation of plasminogen, fibrinogen, and streptokinase on the cell surface. Down-regulation of the pyrogenic exotoxin SpeB appears to favor cell surface accumulation of these factors. The proteolytic activity of SpeB cleaves host proteins and other GAS determinants of pathogenesis among which are plasminogen and streptokinase. So, although SpeB is an important determinant of pathogenesis early in the infection cycle, its down-regulation favors invasion of GAS.

3. What is the typical clinical presentation and what complications can occur?

GAS is a versatile pathogen capable of causing a wide spectrum of diseases. It is convenient to divide them into local infections with GAS and its products, invasive infections, and post-group A streptococcal disease.

Local infections

Pharyngitis occurs 24–48 hours post-exposure with sudden onset of sore throat, malaise, fever, and headache. The pharynx and tonsils may be **erythematous** with creamy, yellow exudates. There may be cervical **lymphadenopathy**. Complications of streptococcal pharyngitis are scarlet fever and acute rheumatic fever (see Figure 1). If the infecting GAS strain is lysogenized by a temperate **bacteriophage** that specifies production of pyrogenic exotoxin an erythematous rash beginning on the trunk and spreading to the extremities develops 1–2 days following the onset of pharyngitis (Figure 5A). The rash spares the palms and soles and the skin

A
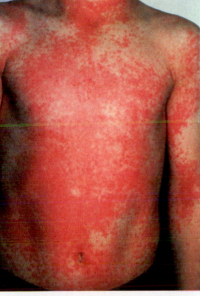

B
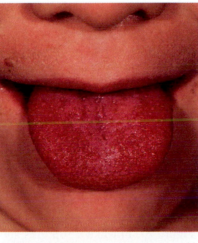

C
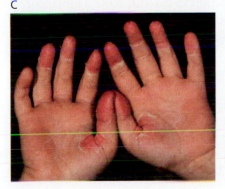

D
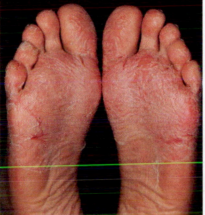

Figure 5. (A) Scarlet fever rash on a boy's torso and arms. (B) Strawberry tongue characteristic of scarlet fever. (C) Scarlet fever showing desquamation of palms and fingers. (D) Scarlet fever showing desquamation of the soles of the feet.

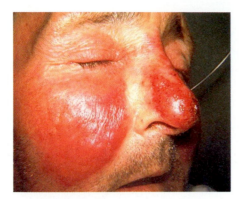

Figure 6. Facial erysipelas manifested as severe malar and nasal erythema and swelling. Erysipelas is a dermatologic condition, which involves the inoculation of the skin and subcutaneous tissue with GAS causing edema and bright red erythema of the affected areas. Erysipelas can be differentiated from cellulitis by its characteristically raised advancing edges and sharply demarcated borders, reflecting its more superficial nature. Cellulitis has no lymphatic component and exhibits nondiscrete margins.

around the mouth (circumoral pallor). Initially the tongue is covered with a white coating, which is lost to reveal a red, raw surface termed 'strawberry tongue' (Figure 5B). After about a week the rash fades and is replaced by desquamation (Figure 5C and D). It has been found that about three-quarters of pharyngeal isolates of GAS from children are positive for the pyrogenic exotoxins, *speA* and *speC*. Most invasive pediatric GAS strains are identical to acute pharyngitis strains; thus, childhood pharyngitis serves as a major reservoir for strains with invasive potential.

Acute rheumatic fever, which can be a nonsupperative sequela of GAS pharyngitis is considered below.

Pyoderma (*impetigo*) is a highly contagious, superficial infection of exposed skin, typically the face, arms and legs, seen most frequently in young children. It is often a mixed infection of GAS and *Staphylococcus aureus*. GAS enters the subcutaneous tissue via a breach in the integrity of the skin. The course of infection begins as **vesicles**, which progress to pustules that rupture and are replaced by honey-colored crusts (see *Staphylococcus aureus* case, Figure 2). A complication of impetigo is acute glomerulonephritis. This will be discussed below.

Invasive infections

Erysipelas is an acute infection of the skin accompanied by lymphadenopathy, fever, chills, and leukocytosis. The painful, inflamed skin is raised and clearly demarcated from the surrounding healthy skin (Figure 6). Although it can occur on any part of the body the legs are a frequent site of infection because of venous insufficiency and stasis ulcerations.

Cellulitis is an infection similar in nature to erysipelas except that it involves not only the skin but the connective tissues (Figure 7).

Necrotizing fasciitis is a deep infection of the connective tissue that spreads along fascial planes and destroys muscle and fat. The bacterium enters through a trivial break in the skin. The course of the infection is rapid, often beginning with severe pain without evidence of injury or wound. Over a matter of hours there is swelling and the appearance of a spreading red or dusky blue skin discoloration, often with fluid-filled **bullae** (Figure 8). Flu-like symptoms such as diarrhea, nausea, fever, confusion, dizziness, and weakness are also apparent.

Streptococcal toxic shock syndrome (STSS) often follows necrotizing fasciitis with the disease progressing to shock and organ failure. Patients with STSS are **bacteremic** in contradistinction to patients with staphylococcal toxic shock syndrome. M1 and M3 serotypes are strongly associated with necrotizing fasciitis and STSS. These M types are heavily capsulate and produce SpeA and SpeC.

Post-streptococcal sequelae

Acute rheumatic fever (ARF) is an inflammatory disease of the connective tissue, particularly the heart, joints, blood vessels, subcutaneous tissue, and central nervous system that occurs about 3 weeks after GAS pharyngitis in susceptible subjects. The attack rate is about 3% in untreated pharyngitis. All layers of the heart are involved and there is

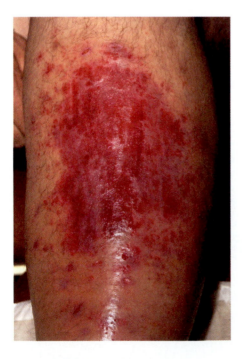

Figure 7. Cellulitis of the lower leg caused by GAS.

chronic damage to the valves, particularly the mitral valve. Also, there is a migratory **arthritis**, subcutaneous nodules, a serpiginous, flat, painless rash (erythema marginatum), and chorea (Sydenham's chorea). ARF is an autoimmune disease in which antibodies induced against streptococcal antigens cross-react with human tissues. The basis for this cross-reactivity lies in the coiled-coil nature of M protein and its homology with tropomyosin, myosin, keratin, laminin, desmin, vimentin, and other coiled-coil proteins. Also, there may be cross-reactivity with sarcolemmal membranes. ARF is associated with M types 1, 3, 5, 6, and 18.

Acute glomerulonephritis (AGN) is a second autoimmune disease that can follow either pharyngeal or skin infection with certain GAS M types. The target organ is the kidney, resulting in inflammation of the glomeruli, **edema**, hypertension, **hematuria**, and proteinuria. The M types causing AGN are termed nephritogenic and the pharyngeal nephritogenic M types are different from the skin nephritogenic M types. In the case of AGN the autoantigens are glomerular heparin-sulfate proteoglycan and basement membrane type IV collagen and laminin. In addition, **immune complexes** of streptococcal antigens deposit in the glomeruli activating the complement cascade.

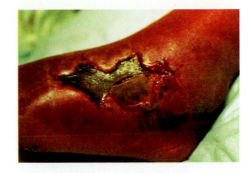

Figure 8. Necrotizing fasciitis. The course of the infection is rapid, often beginning with severe pain without evidence of injury or wound. Over a matter of hours there is swelling and the appearance of a spreading red or dusky blue skin discoloration, often with fluid-filled bullae.

4. How is the disease diagnosed and what is the differential diagnosis?

Diagnosis

Microscopy: Examination of infected tissues or body fluids stained by Gram's method can provide a rapid preliminary diagnosis.

Culture: Swabs from the posterior pharynx, undisturbed crusted pustules, tissue, pus, blood, and so forth, are plated onto blood agar, which is incubated for 24–48 hours at 37°C in the presence of 5% CO_2. The first sector of the inoculated plate should be stabbed to promote β-hemolysis (see Figure 3).

Determinative tests: GAS are uniformly susceptible to the antibiotic bacitracin. A bacitracin disc placed on a blood agar plate inoculated with GAS will show a wide zone of growth inhibition (Figure 9). Lancefield group C and G streptococci may also exhibit a large zone of β-hemolysis; however, GAS can be distinguished from these groups because it produces the enzyme *L*-pyrrolidonyl arylamide (PYR) that can be detected by a rapid (2 minute) disc test (Figure 10).

Antigen detection: Rapid tests are available to directly detect GAS on throat swabs from an individual suspected of having streptococcal pharyngitis. These tests are based on immunological detection of the Lancefield group A wall carbohydrate antigen. The antigen is extracted from the swab using acid or an enzyme and is detected by antibody immobilized on latex beads (latex agglutination) or on a membrane (Figure 11). These tests are highly specific and sensitive if performed correctly. However, a negative rapid test must be confirmed by culture.

Serology: Individuals infected with GAS produce antibodies to many components of the bacterial envelope and proteins secreted by the organism. Immunity is mediated by antibodies to M protein and is M type-specific.

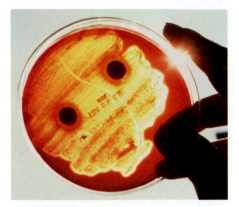

Figure 9. Bacitracin sensitivity of S. pyogenes. The image shows a blood agar plate inoculated with *S. pyogenes* on which two bacitracin disks have been placed. Growth of the GAS is shown by the clear area of the agar in which hemolysins produced by GAS have lysed the red blood cells. GAS are uniformly sensitive to bacitracin. This is shown by the zone of blood agar surrounding each disk in which the red blood cells are intact.

Figure 10. PYR test. Hydrolysis of 4-pyrrolidonyl-a-naphthylamide (PYR) is a presumptive test for identification of group A streptococci and *Enterococcus* species. In the disk test shown, colonial growth is applied to the moistened disk containing the PYR substrate. After 2 minutes, dimethylaminocinnamaldehyde reagent is applied to the disk to detect the free α-naphthylamine that is released on hydrolysis of PYR (red color). Reprint permission kindly given by Jay Hardy, President of Hardy Diagnostics, www.Hardy.Diagnostics.com.

Antibody against streptolysin O (the ASO titer) is a useful marker for confirming acute rheumatic fever or acute glomerulonephritis resulting from GAS pharyngitis. Because streptolysin O is inhibited by cholesterol in the skin, patients with AGN following GAS pyoderma do not exhibit an elevated ASO titer. In this case elevated antibodies against DNase B are a useful indicator.

Differential diagnosis

Pharyngitis
- Viral pharyngitis: rhinovirus, adenovirus, herpes simplex virus, influenza virus, parainfluenza virus, coronavirus, enterovirus, respiratory syncytial virus (RSV), cytomegalovirus, HIV.
- Infectious mononucleosis – Epstein-Barr virus (EBV).
- Peritonsillar abscess.
- Pharyngitis caused by groups C and G streptococci.

Pyoderma (impetigo)
- Atopic dermatitis.
- Bullous pemphigoid.
- Candidiasis, cutaneous.
- Contact dermatitis, allergic and irritant.
- Herpes simplex.
- Insect bites.
- Pemphigus foliaceus.
- Pemphigus vulgaris.
- Scabies.
- Staphylococcal scalded skin syndrome.
- Thermal burns.
- Tinea pedis.

Erysipelas
- None.

Cellulitis
- Angioedema.
- Chemical burns.
- Dermatitis: atopic, contact, and exfoliative.
- Erythema multiforme.
- Gas gangrene.
- Impetigo.

Necrotizing fasciitis
- Cellulitis.
- Gas gangrene.
- Toxic shock syndrome.

Streptococcal toxic shock syndrome
- Staphylococcal infections (staphylococcal toxic shock syndrome).
- Cellulitis.
- Clostridial gas gangrene.
- Erythema multiforme (Stevens-Johnson syndrome).
- Meningococcemia.

Acute rheumatic fever
- Aortic regurgitation.
- Atrial fibrillation.
- Endocarditis.
- Huntington chorea.
- Juvenile rheumatoid arthritis.
- Leukemia.
- Mitral regurgitation.
- Mitral stenosis.
- Myocarditis.
- Pediatrics: Kawasaki disease, scarlet fever.
- Scarlet fever.

Acute glomerulonephritis
- Post-infectious bacterial etiologies: *S. pneumoniae*, *Staphylococcus* species, *Mycobacterium* species, *Salmonella typhosa*, *Brucella suis*, *Treponema pallidum*, *Corynebacterium bovis*, *Actinobacillus* species.
- Post-infectious viral etiologies: cytomegalovirus, coxsackievirus, Epstein-Barr virus, hepatitis B, rubella, rickettsial scrub typhus, mumps.
- Post-infectious fungal and parasitic etiologies: *Coccidioides immitis* and the following parasites: *Plasmodium malariae*, *Plasmodium falciparum*, *Schistosoma mansoni*, *Toxoplasma gondii*, filariasis, trichinosis, and trypanosomes.

A

B

Figure 11. A latex agglutination kit used in a hospital pathology laboratory for grouping (identifying) hemolytic streptococci.

- Systemic causes of glomerular nephritis:
  - Wegener granulomatosis.
  - Hypersensitivity vasculitis.
  - Cryoglobulinemia.
  - Systemic lupus erythematosus.
  - Polyarteritis nodosa.
  - Henoch-Schönlein purpura.
  - Goodpasture syndrome.
  - Renal diseases.
  - Membranoproliferative glomerulonephritis.

5. How is the disease managed and prevented?

Management

GAS remains sensitive to penicillin and this antibiotic is the treatment of choice in patients who are not hypersensitive. In this case erythromycin or a cephalosporin can be substituted. However, it should be noted that during the last few years, erythromycin-resistant *S. pyogenes* has been reported in different parts of the world. Streptococcal pharyngitis is self-limiting, however, antibiotic therapy is indicated because it prevents the development of rheumatic fever but, interestingly, it does not appear to prevent the development of acute glomerulonephritis. Serious soft tissue infections require drainage and surgical debridement as the first line of therapy.

Prevention

In April 2007, the American Heart Association updated its guidelines for prevention of endocarditis and concluded that there is no convincing evidence linking dental, gastrointestinal or genitourinary tract procedures with the development of endocarditis. The prophylactic use of antibiotics prior to a dental procedure is now recommended only for those patients with the highest risk of adverse outcome resulting from endocarditis, such as patients with a prosthetic cardiac valve, previous endocarditis, or those with specific forms of congenital heart disease. The guidelines no longer recommend prophylaxis prior to a dental procedure for patients with rheumatic heart disease unless they also have one of the underlying cardiac conditions listed above. Antibiotic prophylaxis solely to prevent bacterial endocarditis is no longer recommended for patients who undergo a gastrointestinal or genitourinary tract procedure. (See *Streptococcus mitis* case, Case 34, for prophylatic recommendations for patients with ARF).

SUMMARY

1. What is the causative agent, how does it enter the body and how does it spread a) within the body and b) from person to person?

- *S. pyogenes*, also known as the group A streptococcus (GAS), is the most common cause of bacterial pharyngitis.

- The vast majority of cases (about 70%) of pharyngitis are caused by viruses.

- GAS is a gram-positive, catalase-negative coccus that grows in chains.

- GAS grows readily on blood agar incubated at 37°C in the presence of 5% CO_2 for 24–48 hours, forming small, gray opalescent colonies surrounded by a large zone of complete hemolysis (β-hemolysis).

- GAS has a typical gram-positive-type cell wall that features a thick peptidoglycan layer with covalently bound teichoic acid.

- M protein is the most important virulence factor of GAS and there are about 100 types.

- GAS has many determinants of pathogenesis that include lipoteichoic acid (pro-inflammatory), adhesins, superantigens (pyrogenic exotoxins), cytotoxins (streptolysin O and S), and a capsule.

- GAS is spread person to person via respiratory droplets, the hands, and fomites.

- GAS adheres to and invades epithelial cells from where it can invade subcutaneous tissues and seed the lymphatics and bloodstream.

2. What is the host response to the infection and its pathogenesis?

- The barrier epithelia in concert with innate immune factors at these surfaces are effective barriers to the penetration of GAS.

- Host phagocytes are a second line of defense against GAS invasion.

- GAS is opsonized by activation of the alternate and lectin innate complement pathways and the classical pathway in the presence of anti-M protein antibodies in the plasma and tissue fluid.

- The virulence of GAS results from its ability to adhere to and invade host cells and to avoid opsonization and phagocytosis by means of capsule, M protein, and C5a peptidase.

- The hemolysins, streptolysin S and O, are cytotoxins that can lyse erythrocytes, leukocytes, and platelets and likely other host cells.

- The pyrogenic exotoxins (erythrogenic toxins) are superantigens that result in the release of pro-inflammatory cytokines that mediate shock and organ failure characteristic of streptococcal toxic shock syndrome and give rise to the rash associated with scarlet fever.

3. What is the typical clinical presentation and what complications can occur?

- *Pharyngitis* occurs 24–48 hours post-exposure with sudden onset of sore throat, malaise, fever, headache, and erythematous pharynx and tonsils with creamy, yellow exudates and cervical lymphadenopathy.

- *Scarlet fever* is caused by GAS strains lysogenized by a temperate bacteriophage that specifies production of pyrogenic exotoxin resulting in an erythematous rash, circumoral pallor, and strawberry tongue.

- *Pyoderma (impetigo)* is a highly contagious, superficial infection of exposed skin, typically the face, arms, and legs, characterized by vesicles that progress to pustules that rupture and are replaced by honey-colored crusts.

- *Erysipelas* is an acute infection of the skin, most frequently the legs, accompanied by lymphadenopathy, fever, chills, and leukocytosis in which the infected skin is painful, inflamed, raised, and clearly demarcated.

- *Cellulitis* is an infection similar in nature to erysipelas except that it involves not only the skin but the connective tissues.

- *Necrotizing fasciitis* is a deep infection of the connective tissue that spreads along fascial planes and destroys muscle and fat.

- *Streptococcal toxic shock syndrome* (STSS) often follows necrotizing fasciitis; the disease is characterized by shock and organ failure.

- *Acute rheumatic fever* (ARF) is an autoimmune inflammatory disease of the connective tissue, particularly the heart, joints, blood vessels, subcutaneous tissue, and central nervous system that occurs about 3 weeks after GAS pharyngitis in susceptible subjects.

- *Acute glomerulonephritis* (*AGN*) is an autoimmune disease that can follow either pharyngeal or skin infection causing inflammation of the glomeruli, edema, hypertension, hematuria, and proteinuria.

4. How is the disease diagnosed and what is the differential diagnosis?

- *Microscopy*: Examination of infected tissues or body fluids stained by Gram's method can provide a rapid preliminary diagnosis.

- *Culture*: Swabs from the posterior pharynx, undisturbed crusted pustules, tissue, pus, blood, and so forth, are plated onto blood agar to grow the characteristic colonies surrounded by a large zone of β-hemolysis.

- *Determinative tests*: Susceptible to bacitracin and PYR-positive.

- *Antigen detection*: Rapid tests based on immunological detection of the Lancefield group A wall carbohydrate antigen are available to directly detect GAS on throat swabs; a negative rapid test must be confirmed by culture.

- *Serology*: Anti-streptolysin O titer (the ASO titer) is a useful marker for confirming acute rheumatic fever or acute glomerulonephritis resulting from GAS pharyngitis; anti-DNase B for AGN following skin infection.

5. How is the disease managed and prevented?

- Penicillin is the antibiotic of choice in patients who are not hypersensitive.

- Erythromycin or a cephalosporin can be used in patients allergic to penicillin.

- Antibiotic treatment of GAS pharyngitis prevents the development of rheumatic fever.

- Serious soft tissue infections require drainage and surgical debridement as the first line of therapy.

- Prophylactic antibiotic therapy is required for several years in individuals who have had rheumatic fever.

- Any person with pre-existing damage to their heart valves should receive prophylactic antibiotics before any procedure likely to result in transient bacteremia, such as dental treatment. (See *Sreptococcus mitis* case, Case 34.)

FURTHER READING

Mims C, Dockrell HM, Goering RV, Roitt I, Wakelin D, Zuckerman M. Medical Microbiology, 3rd edition. Mosby, Edinburgh, 2004.

Murphy K, Travers P, Walport M. Janeway's Immunobiology, 7th edition. Garland Science, New York, 2008.

Murray PR, Rosenthal KS, Pfaller MA. Medical Microbiology, 5th edition. Elsevier Mosby, Philadelphia, PA, 2005.

REFERENCES

Bisno AL, Brito MO, Collins CM. Molecular basis of group A streptococcal virulence. Lancet Infect Dis, 2003, 3: 191–200.

Carapetis JR, Steer AC, Mulholland EK, Weber M. The global burden of group A streptococcal diseases. Lancet Infect Dis, 2005, 5: 685–694.

Cole JN, McArthur JD, McKay FC, et al. Trigger for group streptococcal M1T1 invasive disease. FASEB J, 2006, 20: 1745–1747.

Courtney HS, Hasty DL, Dale JB. Molecular mechanisms of adhesion, colonization, and invasion of group A streptococci. Ann Med, 2002, 34: 77–87.

Cunningham MW. Pathogenesis of group A streptococcal infections. Clin Microbiol Rev, 2000, 13: 470–511.

Cunningham MW. Autoimmunity and molecular mimicry in the pathogenesis of post-streptococcal heart disease. Front Biosci, 2003, 8: s533–543.

Desjardins M, Delgaty KL, Ramotar K, Seetaram C, Toye B. Prevalence and mechanisms of erythromycin resistance in group A and group B *Streptococcus*: implications for reporting susceptibility results. Clin Microbiol, 2004, 42: 5620–5623.

Kreikemeyer B, McIver KS, Podbielski A. Virulence factor regulation and regulatory networks in *Streptococcus pyogenes* and their impact on pathogen-host interactions. Trends Microbiol, 2003, 11: 224–232.

Jarva H, Jokiranta TS, Wurzner R, Meri S. Complement resistance mechanisms of streptococci. Mol Immunol, 2003, 40: 95–107.

Molinari G, Chhatwal GS. Streptococcal invasion. Curr Opin Microbiol, 1999, 2: 56–61.

Stevens DL. Streptococcal toxic shock syndrome associated with necrotizing fasciitis. Annu Rev Med, 2000, 51: 271–288.

Stollerman GH. Rheumatic fever in the 21st century. Clin Infect Dis, 2001, 33: 806–814.

WEB SITES

Centers for Disease Control and Prevention, Department of Health and Human Services, United States Government: http://www.cdc.gov/ncidod/dbmd/diseaseinfo/groupastreptoccal_g.htm

Department of Health, New York State Government: http://www.nyhealth.gov/nysdoh/communicable_diseases/en/gas.htm

National Institute of Allergy and Infectious Diseases, National Institutes of Health, Department of Health and Human Services, United States Government: http://www.niaid.nih.gov/factsheets/strep.htm

MULTIPLE CHOICE QUESTIONS

The questions should be answered either by selecting True (T) or False (F) for each answer statement, or by selecting the answer statements which best answer the question. Answers can be found in the back of the book.

1. **All of the following are components of the GAS cell wall, EXCEPT?**

   A. Thick peptidoglycan layer.

   B. Teichoic acid.

   C. Group-specific polysaccharide.

   D. Lipopolysaccharide.

   E. Lipoteichoic acid.

2. **All of the following are characteristics of GAS, EXCEPT?**

   A. β-Hemolytic.

   B. Resistant to bacitracin.

   C. Capnophilic, i.e. grows best in the presence of CO_2.

   D. Catalase-negative.

   E. PYR-positive.

3. **Which of the following laboratory tests are important in the diagnosis of GAS disease?**

   A. Microscopy.

   B. Culture.

   C. Serology.

   D. Antigen detection.

   E. All of the above (A–D) are important.

4. **All of the following statements concerning the M protein of *Streptococcus pyogenes* are correct, EXCEPT?**

   A. M protein is anti-phagocytic.

   B. Patients can have multiple episodes of *S. pyogenes* infection since there are multiple antigenic types of M protein.

   C. M protein is the most important virulence determinant.

   D. All M types of *S. pyogenes* are equally virulent.

   E. Epitopes on M protein cross-react with those on several human tissues.

5. **Which one of the following statements concerning the capsule of S. pyogenes is true?**

   A. It is associated with strains causing severe systemic infection.

   B. It is composed of hyaluronic acid.

   C. It masks the organism from the humoral immune system.

   D. A and B.

   E. A, B, and C.

6. **All of the following statements about streptococcal pyrogenic exotoxins are correct, EXCEPT?**

   A. They are superantigens.

   B. They are encoded by bacteriophage.

   C. They are responsible for the rash of scarlet fever.

   D. They bind IgG via the Fc region.

   E. They are associated with severe infections, including shock.

7. **All of the following are mechanisms by which GAS avoids host defenses, EXCEPT?**

   A. Inhibition of complement activation.

   B. Resist phagocytosis.

   C. Bind antibody via the Fc region.

   D. Cause unfocused activation of CD4 helper T cells.

   E. Produce proteases that cleave antibody molecules.

8. **All of the following may result from infection with GAS, EXCEPT?**

   A. Acute pharyngitis.

   B. Pyoderma.

   C. Necrotizing fasciitis.

   D. Meningitis.

   E. Cellulitis.

9. **Prior infection with GAS can be demonstrated in patients with acute rheumatic fever by which one of the following?**

   A. Blood culture.

   B. Skin culture.

   C. Culture of a heart valve in a patient with carditis.

   D. A high titer of antibody against the hyaluronic acid capsule.

   E. A high titer of antibody against streptolysin O.

10. **All of the following statements about antibiotic treatment of GAS infections are correct, EXCEPT?**

    A. Penicillin is the antibiotic of choice unless the patient is allergic.

    B. Penicillin need not be given to patients with GAS pharyngitis because the disease is self-limiting.

    C. Penicillin is required for patients with GAS pharyngitis because it prevents the development of acute rheumatic fever.

    D. Penicillin alone is inadequate treatment for necrotizing fasciitis.

    E. Patients with a history of rheumatic fever require long-term antibiotic prophylaxis.

Case 37

Toxoplasma gondii

A 32-year-old Nigerian female was admitted to hospital having had two **fits**. She reported a headache over the previous few weeks. Over about 6 months she had been obtaining over the counter treatment from the local pharmacy for oral thrush. She had a slight fever on admission, no focal neurological signs or **papilledema** on **funduscopy**. On a full blood count she had a slight **pancytopenia**. A **CT scan** of her head showed one small ring enhancing lesion in the cerebrum (Figure 1). After deliberation it was considered safe and prudent to perform a lumbar puncture. There were 10 lymphocytes (normal <5), a protein of 940 mg L^{-1} (normal range 150–450), but no organisms. She was started on phenytoin to control fits and ceftriaxone and metronidazole to treat bacterial brain **abscesses.** Later in her admission the possibility of HIV was considered. After counseling she tested positive for HIV. This changed the differential diagnosis. There had been no improvement in her fever or headache and she was presumptively diagnosed with toxoplasmic **encephalitis**. Treatment was changed to pyrimethamine and sulfadiazine plus folinic acid to help her bone marrow. **Serology** for *Toxoplasma* was **IgG**-positive and **IgM**-negative. Her cerebrospinal fluid (CSF) was positive for *Toxoplasma* by **polymerase chain reaction (PCR)**. Over the next 3 months serial CT scans showed resolution of her lesion on treatment.

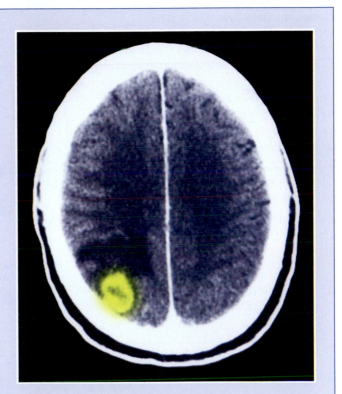

Figure 1. A single toxoplasmic encephalitic lesion is seen on the left-hand side of this CT scan of the head. The injection of contrast highlights the lesion, in this case coloured yellow. There is surrounding edema, which gives a black appearance.

1. What is the causative agent, how does it enter the body and how does it spread a) within the body and b) from person to person?

Causative agent

Toxoplasma gondii is a protozoan intracellular parasite. The definitive host is the cat and other felines. Many warm-blooded animals can serve as an intermediate host including, for example, poultry, rodents, cattle, sheep, and pigs. Humans can become accidentally infected. *T. gondii* is found throughout the world but human infection is more common in some countries than others.

Entry and spread within the body

T. gondii oocysts (Figure 2) are shed in the feces of cats and may survive up to 18 months in the environment. During this time they may be ingested

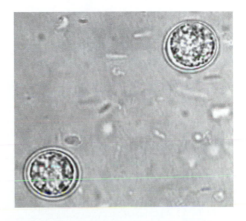

Figure 2. *Toxoplasma gondii* **oocysts.**

Figure 3. The life cycle of *Toxoplasma gondii*. Cats shed unsporulated oocysts onto the soil (1). These sporulate (2) and are ingested by the intermediate host when they graze on the ground (3). Humans may ingest oocysts, which contaminate foods or water (4) or eat cysts in infected meat (5). Within the intestine of the intermediate host the oocysts release sporozoites that invade the intestinal epithelium. They mature into tachyzoites that form tissue cysts at various sites in the body (6). These may then be ingested by cats to continue the life cycle (7). In humans tissue cysts develop in the same way as in the intermediate hosts.

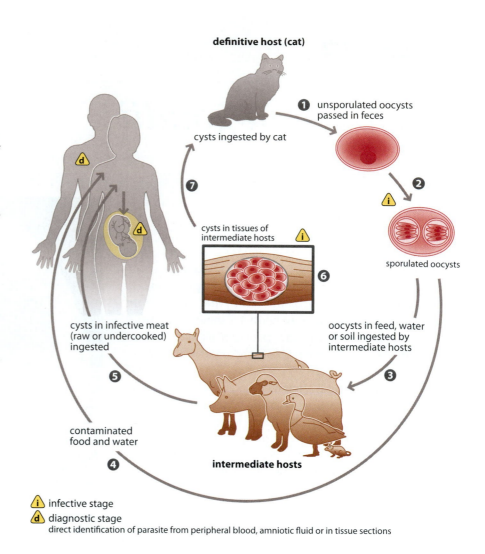

by other animals such as rodents (Figure 3). Within the intestine of these intermediate hosts, the oocysts release sporozoites that invade the intestinal epithelium and multiply to form tachyzoites (Figure 4). Tachyzoites may spread to local cells or pass via the lymphatics to regional lymph nodes and then to elsewhere in the body. They can infect any cell in the body. However, there is a predilection for lymph nodes, skeletal and cardiac muscle, the brain, and the eye. At each tissue site tachyzoites enter cells and multiply to form tissue **cysts**, which contain bradyzoites (Figure 5). Bradyzoites in tissue cysts multiply more slowly and may pass unnoticed in tissues for prolonged periods. Humans are infected when they ingest bradyzoites in tissue cysts within uncooked or poorly cooked meat from the intermediate hosts or eat food including vegetables contaminated with cat feces and oocysts (Figure 3).

Tissue cysts are destroyed by proper cooking or freeze-thawing. The practice of eating undercooked meats is more common in some countries, such as France, where the prevalence of toxoplasmosis is also higher. Toxoplasmosis observed in vegetarians is presumably through eating vegetables contaminated by oocysts in the soil, which are not completely washed off. Oocysts can also enter the water supply but if the water is properly filtered oocysts will be excluded. In the developed world outbreaks of

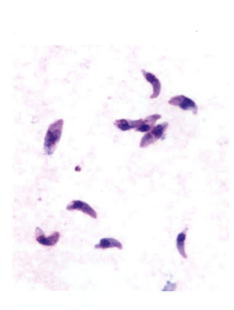

Figure 4. *Toxoplasma gondii* tachyzoites.

toxoplasmosis have occurred when filtration procedures have been disrupted. In the developing world the quality of filtration and water supplies cannot be guaranteed.

As infection with *T. gondii* can be silent its true prevalence is not known. Some indication of prevalence comes from studies on seroprevalence performed in the United States and Western Europe. Seropositivity rises with age. In the United States about 15% of women of child-bearing age are **seropositive**, whilst in Western Europe this figure is about 50%.

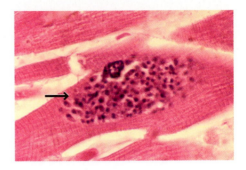

Figure 5. Tissue cyst of *Toxoplasma gondii* in a cardiac myocyte.

Person to person spread

Humans can pass infection vertically from pregnant mother to fetus but are not a source of horizontal transmission. Nonimmune cats may get infected by ingesting oocysts lying around in the soil. Alternatively they could eat animals, for example rodents, harboring tissue cysts. Bradyzoites then emerge and infect the feline intestinal epithelium. *T. gondii* undergoes sexual reproduction in the intestinal epithelium of the cat. Oocysts shed in cat feces are then a source for human ingestion directly or via an intermediate host.

2. What is the host response to the infection and what is the disease pathogenesis?

Targets for the host immune response are tachyzoites or bradyzoites, which are extracellular, or intracellular organisms within phagocytes or tissue cells. An antibody response is mounted and antibodies persist lifelong. On binding to extracellular tachyzoites the antibodies fix complement and cause target **lysis**.

Within macrophages *T. gondii* produces a **parasitophorous vacuole**. Somehow lysosomal fusion does not occur with this vacuole and *T. gondii* is spared lysosomal attack. There are other microbicidal and microbistatic mechanisms elaborated by macrophages. *T. gondii* is susceptible to some of these, such as **reactive oxygen species**, reactive nitrogen intermediates, and tryptophan depletion. To enable these mechanisms macrophages require activation by a type 1 **cytokine** pattern. Interferon-γ (IFN-γ) is central to this pattern. In an experimental mouse model of toxoplasmosis IFN-γ is produced by **natural killer (NK)** cells, and **CD4+** and **CD8+** T lymphocytes. Other cytokines that can promote the production of IFN-γ are **interleukin (IL)**-1, IL-2, IL-12, and IL-15. Downstream **tumor necrosis factor-α (TNF-α)** released from macrophages enhance their own activation.

Macrophages have MHC class II molecules on their surface, which are recognised by CD4+ T-helper cells. Tissue cells only have MHC class I molecules, and can only be recognized by CD8+ T cytotoxic cells. If *T. gondii* antigen is processed to appear on the tissue cell surface in the context of MHC class I molecules the cells are recognized by *T. gondii* reactive CD8+ T cells and killed.

After primary infection there is a lag before the immune response is induced. During this lag period *T. gondii* spreads throughout the body and the immune response has to chase and curtail infection. The eye and the central nervous system represent relatively immunologically privileged sites. It is harder for immune and inflammatory responses to operate. Lymphocytes do traffic into the CSF and enter the brain substance. However, in the brain, astrocytes and **microglial cells** serve a phagocytic function and are seen to proliferate during acute infection. In the majority of subjects the immune response brings infection under control without any symptoms. In some cases the local inflammatory response may cause problems in lymph nodes (**lymphadenopathy**), heart muscle (**myocarditis**), skeletal muscle (myositis), and the retina (**chorioretinitis**). Symptomatic disease is more likely if there is an exuberant immune response. IL-10 acts as a counter-regulatory cytokine to damp down the type I cytokine response.

Bradyzoites in chronic infection are intermittently reactivating. Local tissue spread and infection is prevented through immune surveillance. However, in the immunocompromised host, immune surveillance is diminished and bradyzoite reactivation goes unchecked leading to local tissue inflammation.

A

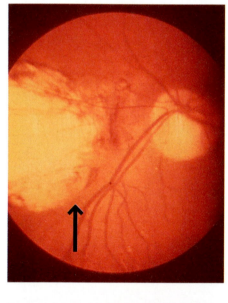

B

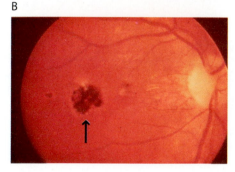

Figure 6. Two different stages and appearances of _Toxoplasma_ chorioretinitis.

3. What is the typical clinical presentation and what complications can occur?

If pregnant mothers acquire _T. gondii_ infection during their pregnancy there is a risk of transplacental transfer of tachyzoites. The chance of this does not depend on whether the mother is symptomatic or not. If the infection had been acquired before pregnancy there is the possibility that reactivation of tissue cysts will release tachyzoites during pregnancy. However, an immune response will be present and circulating tachyzoites will be quickly destroyed by antibody and complement, thus preventing any fetal infection. The situation may be different in the immunocompromised individual. Severe fetal infection may result in abortion. The greatest risk of congenital abnormality is with infections between 10 and 24 weeks gestation. Infection of the brain may cause lesions around the channels for CSF. Blockage of these channels causes **hydrocephalus** with swelling of the ventricles. Lesions within the brain substance may subsequently calcify. Disturbance of brain development can stunt its growth resulting in **microcephaly**. Fetal infection at a later stage of infection may not cause any immediate problems. Tissue cysts are established that may reactivate after birth.

Following congenital infection the commonest manifestation of later reactivation is in the eye with inflammation of the retina called chorioretinitis (Figure 6). Reactivation probably occurs elsewhere in skeletal muscle but this goes unnoticed. Chorioretinitis may represent a very small volume of inflammation, but becomes very noticeable and troublesome if it affects the visual fields. Understandably central lesions are more noticeable in causing a patch of visual loss, a **scotoma**. After initial inflammation and **necrosis** there is scarring and pigmentary change in the retina. More than one episode of chorioretinitis can occur as further tissue cysts reactivate.

Primary infection in children and adults remains asymptomatic in about 80% of subjects. There may be a constitutional illness with fever, malaise, fatigue, and then an organ-specific problem. The most common manifestation is lymphadenopathy. This could be anywhere in the body, and be single or multiple. If in the neck it may be similar to that seen in infectious mononucleosis. Illness can subside without treatment, but the size of lymph nodes can fluctuate for a few months but will invariably settle down.

Immunocompromise that may cause reactivation of toxoplasmosis may arise from hematological malignancy, transplant immunosuppression, and HIV infection. Prophylaxis with co-trimoxazole, diminishes, but does not preclude reactivation. In HIV the commonest site for lesions is in the brain. Toxoplasmic encephalitis represents a progressive, focal disease process with central necrosis, surrounded by vascular inflammation. Lesions may be single or multiple and present with headaches, focal fits, and focal neurological signs depending on the location of lesions. In immunocompromised individuals many other manifestations are possible including myocarditis, myositis, and **pneumonitis**.

4. How is the disease diagnosed and what is the differential diagnosis?

The main diagnostic test used is serology. Histology of an excised lymph node or CT and MRI scans of the head may be highly suggestive of toxoplasmosis but supportive evidence is usually obtained. PCR of amniotic fluid, CSF or samples from the eye has high specificity but variable sensitivity.

Currently **ELISA** techniques are used to assay for anti-*Toxoplasma* IgG and IgM. Formerly live or formalin-fixed tachyzoites were used in neutralization, **immunofluorescence**, and agglutination assays. The **Sabin-Feldman dye test** is a neutralization assay of live tachyzoites. It is the standard against which other assays are compared. Serological tests are useful in excluding *T. gondii* infection if negative on two occasions 3 weeks apart. However, a positive test does not necessarily imply that the current illness is due to toxoplasmosis. IgG antibodies appear 2 weeks after infection, peak at 2 months, and then wane after 2 years, but remain lifelong. The persisting titer of antibody varies between individuals and bears no relationship to disease severity. IgM antibodies in many other infections just appear in the acute phase and are a good indicator of acute infection. In toxoplasmosis IgM can persist for a few years and is of limited use in defining acute infection.

Toxoplasmosis has to be considered in the differential diagnosis for congenital abnormalities, lymphadenopathy, chorioretinitis, and ring enhancing CNS lesions. It is a less common explanation for myocarditis or myositis. '**TORCH**' is a useful mnemonic for key congenital infections that may produce overlapping clinical features. It stands for **T**oxoplasmosis, **R**ubella, **C**ytomegalovirus, and **H**erpes virus. When lymphadenopathy is in the neck toxoplasmosis may present with similar symptoms to that of infectious mononucleosis. Infectious mononucleosis is most commonly due to Epstein-Barr virus (see Case 9), but can also be due to cytomegalovirus and toxoplasmosis. Differential diagnoses for lymphadenopathy, chorioretinitis, and toxoplasmic encephalitis are shown in Table 1.

Table 1. Differential diagnoses for various manifestations of toxoplasmosis

Lymphadenopathy	Chorioretinitis	Encephalitis
Pyogenic bacteria	*Infections:*	Focal CNS lesions:
Epstein-Barr virus	Tuberculosis	*Infection:*
Cytomegalovirus (CMV)	Toxocariasis	Brain abscess
Mycobacteria	Histoplasmosis	Tuberculoma
HIV	Syphilis	Cryptococcoma
Syphilis	Cytomegalovirus	Focal CMV encephalitis
Bartonella (cat scratch disease)	Varicella-zoster virus	Nocardia
Kikuchi's syndrome		
Lymphoma	*Connective tissue disease:*	*Neoplasia:*
Sarcoidosis	e.g. rheumatoid arthritis	Primary or secondary brain tumors
Various other infections that cause lymphadenopathy may have accompanying diagnostic features, e.g. a measles rash	*Sarcoidosis*	Lymphoma

5. How is the disease managed and prevented?

Management

Pyrimethamine and sulfadiazine are effective in killing tachyzoites, but do not clear tissue cysts. Pyrimethamine is a folate antagonist and causes severe bone marrow suppression, unless the bone marrow is spared with folinic acid supplements. Pyrimethamine and sulfadiazine are of proven efficacy in treating toxoplasmic encephalitis in HIV patients. In immunocompetent subjects disease will self-limit anyway and administering potentially toxic therapy is of no added benefit. One exception may be for large chorioretinitis lesions. Small lesions will probably settle before assessment, diagnosis, and commencement of treatment.

Management of toxoplasmosis in pregnancy is challenging and practice varies from country to country. Mothers who have not previously been infected may be totally asymptomatic if they acquire infection during the pregnancy. If infection is diagnosed on the basis of a **seroconversion**, from being seronegative early in pregnancy to being seropositive, anti-*Toxoplasma* treatment may commence after the fetus has already been infected. Conducting trials on the efficacy of treatment to prevent fetal infection is understandably difficult. Contrasting results have come from various small trials. The drug used in pregnant mothers is spiramycin. This does not cross the placenta and is only used to kill tachyzoites before they can transfer to the fetus. Some of the various trials have shown a benefit with spiramycin. Once the fetus has been infected, perhaps confirmed by amniotic PCR, pyrimethamine and sulfadiazine may be used after the first trimester. In the first trimester treatment-associated folate antagonism runs a significant risk of fetal malformation.

Prevention

Prevention of infection requires care with cooking and hand hygiene. Avoidance of undercooked meats is essential in pregnancy. Adequate cooking of meat to high temperatures eliminates the risk. Hands can get contaminated in the garden or handling a cat litter. Careful hand-washing or wearing gloves will reduce risk. Prophylaxis with co-trimoxazole can reduce the risk of disease in those who are already infected and are immunocompromised. The prime indication for co-trimoxazole is actually prevention of *Pneumocystis* pneumonia. If patients develop allergic reactions to co-trimoxazole they may be switched to other prophylactics that do not prevent *T. gondii* infection.

SUMMARY

1. What is the causative agent, how does it enter the body and how does it spread a) within the body and b) from person to person?

- *Toxoplasma gondii* is a protozoan parasite.
- The definitive host is the cat and other felines. From multiplication in the intestinal epithelium oocysts are shed in the feces.
- Other animals ingest the oocysts. Within their intestine invasion and multiplication yields tachyzoites, which pass around the body.
- In various tissues tachyzoites invade cells and produce tissue cysts.
- Humans may be infected by ingesting oocysts from the environment or by eating tissue cysts in undercooked meat.

2. What is the host response to the infection and what is the disease pathogenesis?

- Antibody and complement can kill circulating tachyzoites.
- Macrophages can phagocytose tachyzoites and when activated by IFN-γ kill them through reactive oxygen or nitrogen species and tryptophan depletion.
- Tissue cysts are attacked by CD8+ lymphocytes with the help of CD4+ lymphocytes.
- The inflammatory response created by attacking tissue cysts may be low grade and asymptomatic. However, if extensive, with numerous tachyzoites invading cells locally, there can be organ-specific problems such as chorioretinitis, myositis, myocarditis, and encephalitis.

3. What is the typical clinical presentation and what complications can occur?

- Congenital infection can result in fetal loss or severe congenital abnormalities involving the brain.
- *In utero* infection may also be silent with later problems as infection reactivates, most commonly in the eye as chorioretinitis.
- Primary infection in children or adults is asymptomatic in the majority but if symptomatic can cause lymphadenopathy.
- In the immunocompromised infection can reactivate and cause encephalitis. This is most often seen in HIV patients.
- Reactivation may also affect skeletal muscle, the heart, and the lung.

4. How is the disease diagnosed and what is the differential diagnosis?

- Antibody to *Toxoplasma* can be measured by a variety of serological tests.
- IgM antibodies can persist for prolonged periods and are not useful for distinguishing acute from chronic infection.

- Antibodies persist lifelong and their presence does not necessarily mean that the current illness is due to toxoplasmosis. The absence of antibody is sometimes more useful in excluding infection.

- PCR for toxoplasmosis has high specificity, but variable sensitivity.

- Histology or the clinical picture can be highly suggestive of toxoplasmosis.

5. How is the disease managed and prevented?

- Treatment of toxoplasmosis is usually only required in the immunocompromised and for the infected fetus. Pyrimethamine and sulfadiazine, with folinic acid, are used to kill tachyzoites, although they have no effect on the tissue cysts.

- Immunocompetent individuals will usually have self-limiting disease.

- The management of toxoplasmosis in pregnancy is controversial and practice varies from country to country.

- Avoidance of undercooked meat, thorough cooking, and hand hygiene are important in prevention.

FURTHER READING

Montoya JG, Kovacs JA, Remington JS. *Toxoplasma gondii*. In: Mandell GL, Bennett JE, Dolin R, editors. Principles and Practice of Infectious Diseases, 6th edition. Elsevier, Philadelphia, 2005: 3170–3198.

Murphy K, Travers P, Walport M. Janeway's Immunobiology, 7th edition. Garland Science, New York, 2008.

REFERENCES

Cook AJC, Gilbert RE, Buffolano W, et al. Sources of toxoplasma infection in pregnant women: European multicentre case-control study. BMJ, 2000, 321: 142–147.

Dunn D, Wallon M, Peyron F, et al. Mother-to-child transmission of toxoplasmosis: risk estimates for clinical counselling. Lancet, 1999, 353: 1829–1833.

WEB SITES

Center for Disease Control, Atlanta, GA, USA: www.cdc.gov/

Health Protection Agency: www.hpa.org.uk

MULTIPLE CHOICE QUESTIONS

The questions should be answered either by selecting True (T) or False (F) for each answer statement, or by selecting the answer statements which best answer the question. Answers can be found in the back of the book.

1. **Which of the following are true of the causative agent of toxoplasmosis?**

   A. *Toxoplasma gondii* is a protozoan intracellular pathogen.

   B. The definitive host is the dog and other canids.

   C. In highly endemic areas humans are a key reservoir for infection.

   D. The rapidly dividing form is called a tachyzoite.

   E. Bradyzoites are the only extracellular stage.

2. **Which of the following are true of the transmission of T. *gondii*?**

   A. Infection may be acquired by the ingestion of oocysts in the soil.

   B. Vegetarians do not acquire toxoplasmosis.

   C. Transmission through water supplies is possible.

   D. Tissue cysts in meat are destroyed by thorough cooking, but not by freezing.

   E. Measures to reduce transmission are particularly important during pregnancy.

3. **Which of the following are true of the host response to T. *gondii*?**

   A. Extracellular killing of tachyzoites is mediated by antibody and complement fixation.

   B. Infected macrophages activate CD4+ T lymphocytes through antigen presented in the context of MHC class I molecules.

   C. Macrophages can kill phagocytosed tachyzoites by tryptophan depletion of the parasitophorous vacuole.

   D. IFN-γ is produced by NK cells during infection.

   E. TNF-α helps to activate infected macrophages.

4. **Which of the following are true of the pathogenesis of toxoplasmosis?**

   A. Infection spreads through the body mostly after an immune response has been mounted.

   B. Highly infected tissues include cardiac muscle because it is an immunologically privileged site.

   C. Pathology is entirely due to tissue damage from dividing tachyzoites.

   D. IL-10 promotes the tissue inflammatory response.

   E. Bradyzoites intermittently reactivate and cause disease when immune surveillance is compromised.

5. **Which of the following are true of the clinical features of toxoplasmosis in immunocompetent individuals?**

   A. Infection in pregnancy is always symptomatic.

   B. Fetal infection results in congenital abnormalities mainly when it occurs in the first trimester (0–13 weeks).

   C. The most common late manifestation of congenital toxoplasmosis is chorioretinitis.

   D. Primary infection in adults is symptomatic in 80%.

   E. Lymphadenopathy is the most common manifestation of primary infection in adults.

6. **Which of the following are true of the clinical features of toxoplasmosis in immunocompromised individuals?**

   A. In HIV the most common site for reactivation is in the brain.

   B. Toxoplasmic encephalitis is always manifest as a single focal lesion.

   C. *Toxoplasma* pneumonitis can occur in immunocompromised individuals.

   D. Cardiac failure is possible in toxoplasmosis.

   E. Congenital infection is more likely in the immunocompromised mother.

7. **Which of the following are true of the diagnosis of toxoplasmosis?**

   A. Live tachyzoites are used in the Sabin-Feldman neutralization assay for antibodies.

   B. Two negative serological tests are useful in ruling out infection.

   C. A positive IgM assay indicates recent infection.

   D. PCR for *T. gondii* has high specificity, but variable sensitivity.

   E. Histology of lymph nodes or infected tissue is always diagnostic.

8. **Which of the following are true of the differential diagnosis of toxoplasmosis?**

   A. Congenital *Toxoplasma* infection may have to be differentiated from cytomegalovirus (CMV) infection.

   B. *Toxoplasma* cervical lymphadenopathy may have to be differentiated from glandular fever due to Epstein-Barr virus (EBV).

   C. *Toxoplasma* chorioretinitis may have to be differentiated from sarcoidosis.

   D. The radiological appearance of toxoplasmic encephalitis is characteristic, with few other possible alternatives.

MULTIPLE CHOICE QUESTIONS (continued)

E. *T. gondii* is the most common of the pathogens responsible for the mononucleosis syndrome.

9. **Which of the following are true of the treatment of toxoplasmosis?**

A. Pyrimethamine and sulfadiazine are used to treat toxoplasmosis in immunocompetent hosts.

B. Pyrimethamine and sulfadiazine are used to treat all episodes of chorioretinitis.

C. Pyrimethamine must be co-administered with folinic acid to prevent bone marrow suppression.

D. Spiramycin is used to treat fetal infection.

E. Treatment of fetal infection in the first trimester runs the risk of significant fetal malformation.

10. **Which of the following are true of the control and prevention of toxoplasmosis?**

A. Pets can be treated before purchase to prevent infection in the household.

B. Adequate filtration of water supplies is necessary to exclude oocysts.

C. Undercooked meats should not be eaten in pregnancy.

D. Gardens are safe for homes that do not have cats.

E. Prophylaxis with co-trimoxazole in immunocompromised individuals reduces the chances of reactivation of latent infection.

Case 38

Trypanosoma spp.

A 32-year-old female from Brazil presented to her local hospital with a sudden onset of left leg, arm, and facial weakness. She was able to speak and reported being in good health in the past except that she got short of breath running after her 4-year-old son. She grew up as a child in Brazil but came to the UK in her early twenties. A CT scan of her brain showed an acute **stroke**. The ECG tracing of her heart was abnormal with broadened QRS complexes. A chest X-ray showed enlargement of the heart (Figure 1). An **ECHO scan** of the heart showed a dilated left ventricle with an apical **aneurysm**, in which there was a small thrombus. An **ELISA** test for *Trypanosoma cruzi* antibodies was performed because of her Brazilian origin and proved positive. She was diagnosed with Chagas' disease with cardiomyopathy and an embolic stroke. She was treated with intensive rehabilitation, anticoagulated, and commenced on ACE inhibitors. For her infection she was treated with benznidazole.

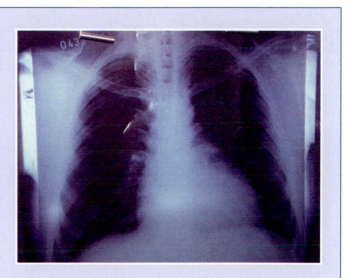

Figure 1. **A chest X-ray showing enlargement of the heart due to Chagas' disease.**

1. What is the causative agent, how does it enter the body and how does it spread a) within the body and b) from person to person?

Causative agent

Trypanosomes are flagellated protozoan parasites. In Africa the species of *Trypanosoma brucei* has two subspecies that infect humans (Figure 2). In Central and West Africa this is *T. brucei gambiense* and in East and Southern Africa it is *T. brucei rhodesiense*. The species in South America is *Trypanosoma cruzi* (Figure 3). It is estimated that in the course of evolution *T. brucei* and *T. cruzi* diverged from each other about 100 million years ago. The life cycle and clinical features arising from these two species differ.

Entry and spread within the body

When *T. brucei* is inoculated into a new human host by tsetse flies local multiplication occurs under the skin. There is spread to lymph nodes and then entry into the bloodstream. Multiplication occurs in blood. Within a few weeks *T. brucei rhodesiense* passes from the bloodstream, probably through the choroid plexus, into the central nervous system (CNS). This CNS invasion takes several months with *T. brucei gambiense*.

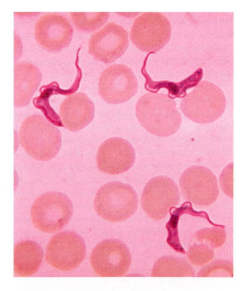

Figure 2. *Trypanosoma brucei* **spp. in a blood film. The trypanosomes are slender with an undulating membrane leading to the flagellum.**

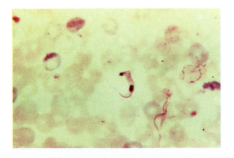

Figure 3. *Trypanosoma cruzi* in a blood film. It often appears as a 'C' shape.

Figure 4. *Triatoma infestans*, which transmits *Trypanosoma cruzi*.

Figure 5. Tsetse fly, *Glossina morsitans*, the vector for African trypanosomiasis.

T. cruzi is transmitted by triatomine bugs (Figure 4). Local inflammation occurs at the site of inoculation. Again there is spread to local lymph nodes and then entry into the bloodstream. Various tissues may be invaded but key targets are the heart and the gastrointestinal tract. Multiplication occurs in the tissues. The intracellular life form is the amastigote, which lacks a flagellum. Replication of the amastigote produces a pseudocyst within a cell. The extracellular life form is the trypomastigote. *T. cruzi* emerges from infected cells and passes onto other cells.

Person to person spread

African trypanosomiasis is spread by tsetse flies belonging to the genus *Glossina* (Figure 5). The geographical distribution of African trypanosomiasis is determined by the ecological requirements of tsetse flies. This is patchy in countries between the sub-Saharan region and the Kalahari and Namib deserts. Infection of humans can be person to person but also a **zoonosis** with tsetse flies transmitting trypanosomes from a reservoir of ungulates (see below). *Glossina morsitans morsitans* usually transmits *T. brucei rhodesiense* and *G. palpalis* usually transmits *T. brucei gambiense*. Trypanosomes ingested from an animal host pass through the mid-gut of the tsetse fly, undergo developmental changes and reach the salivary glands. Here they are referred to as metacyclic trypomastigotes. When tsetse flies bite humans their saliva passes into the bites bearing these trypomastigotes. This is called **salivarian transmission**. The life cycle of African trypanosomiasis is shown in Figure 6.

In South America various animals can serve as a reservoir, examples include the armadillo and the opossum. Triatomine bugs ingest trypanosomes when biting infected animals. The trypanosomes remain within the intestine of the bug. If they next feed on humans they bite the skin and defecate at the same time. Humans reflexly scratching in the vicinity of the bite rub feces bearing metacyclic trypomastigotes into the open wound. This is called **stercorarian transmission**, the Latin root *sterco* referring to feces. The life cycle of South American trypanosomiasis is shown in Figure 7.

Regrettably another form of transmission of trypanosomiasis is through blood transfusion. In resource-poor settings blood screening may not be

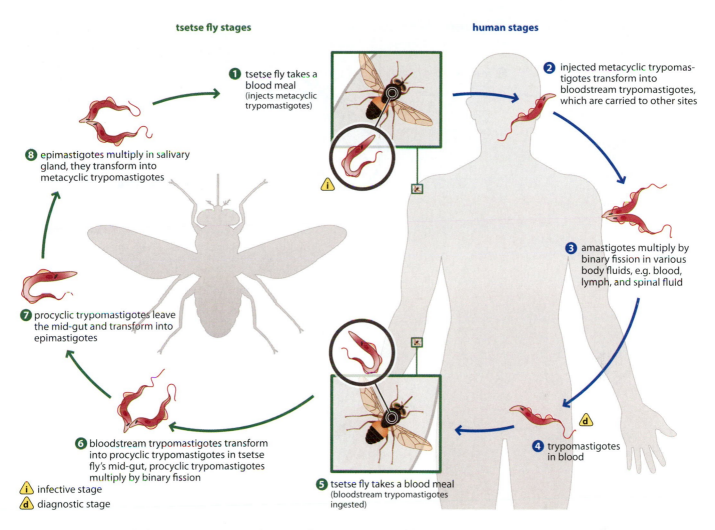

Figure 6. Life cycle of African trypanosomiasis. Tsetse flies inoculate humans with metacyclic trypomastigotes when they take a blood meal (1). Trypomastigotes multiply in tissue fluid, then in lymph nodes, and then enter the bloodstream (2). In the bloodstream they continue multiplying with antigenic variation to avoid the host response. Eventually they enter the central nervous system (3). Circulating parasite may be engulfed by tsetse flies when they take a blood meal (4). There is then development within the mid-gut of the tsetse fly before migration to the salivary glands (6–8).

feasible. However, transfusion-related transmission has been described in the USA from blood donated by South American immigrants. Understandably antibody screening of blood is not routine in nonendemic countries.

Epidemiology
Despite control efforts WHO estimated in 2004 that African trypanosomiasis has a prevalence of about 0.5 million and accounts for 48000 deaths per annum, while complicated Chagas' disease afflicts about 5 million in South America and is responsible for about 14000 deaths per annum. These are very likely to be underestimates as there are no robust mechanisms for reporting cases. African trypanosomiasis is present in 36 countries and South American trypanosomiasis is present in 18 countries.

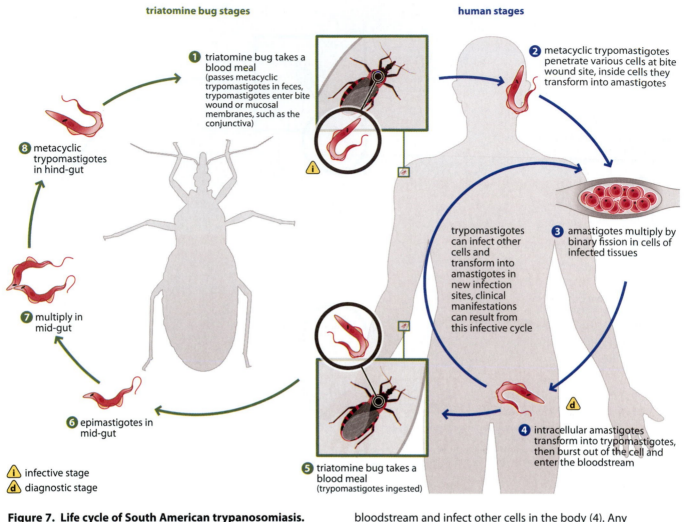

triatomine bug stages

human stages

① triatomine bug takes a blood meal (passes metacyclic trypomastigotes in feces, trypomastigotes enter bite wound or mucosal membranes, such as the conjunctiva)

② metacyclic trypomastigotes penetrate various cells at bite wound site, inside cells they transform into amastigotes

⑧ metacyclic trypomastigotes in hind-gut

③ amastigotes multiply by binary fission in cells of infected tissues

trypomastigotes can infect other cells and transform into amastigotes in new infection sites, clinical manifestations can result from this infective cycle

⑦ multiply in mid-gut

⑥ epimastigotes in mid-gut

④ intracellular amastigotes transform into trypomastigotes, then burst out of the cell and enter the bloodstream

⑤ triatomine bug takes a blood meal (trypomastigotes ingested)

ⓘ infective stage
ⓓ diagnostic stage

Figure 7. Life cycle of South American trypanosomiasis. Humans are infected by stercorarian transmission from triatomine bugs (1). At the site of infection metacyclic trypomastigotes enter cells (2), transform into amastigotes and multiply (3). Trypomastigotes arise from infected cells, enter the bloodstream and infect other cells in the body (4). Any circulating parasite may be engulfed by triatomine bugs when they take a blood meal (5). Further development occurs in the mid-gut of the bugs (6-8).

2. What is the host response to the infection and what is the disease pathogenesis?

There are innate and adaptive host responses to trypanosomal infection. Humans may be exposed to nonpathogenic species of trypanosomes. However, in normal human serum, apolipoprotein L-1 binds to the trypanosome surface and is endocytosed. Within the trypanosomal cytoplasm, apoliprotein L-1 reaches and forms pores in **lysosomes**. Release of lysosomal contents causes trypanosomal killing. Species of trypanosomes that are usually nonpathogenic may only cause infection in humans with apolipoprotein L-1 deficiency. The pathogenic species *T. brucei rhodesiense* possesses a serum resistance-associated protein (SRA), which strongly binds to apolipoprotein L-1, inhibiting its toxic action. The other pathogenic species, *T. brucei gambiense*, does not possess SRA and how it might resist the action of apolipoprotein L-1 is not clear.

T. brucei remains extracellular in the bloodstream and is exposed to the host immune response. Specific antibodies appear against the surface glycoprotein and lyse the trypanosomes through activation of complement. However, in *T. brucei rhodesiense* there are an estimated one thousand different variants of the surface glycoprotein. *T. brucei* switches the gene from one variant to another (**antigenic variation**). Each new antibody response is met with a gene switch and the escaping, new variants of trypanosomes multiply in successive waves (Figure 8).

The variant specific glycoproteins (VSG) stimulate B lymphocytes to produce **IgM** in a T-cell independent manner. Eventually host serum has an excess of polyclonal IgM antibodies. The polyclonal activation of B cells compromises their ability to respond to other pathogens (another escape mechanism, like the antigenic variation to fool the host immune system). Other trypanosomal antigens pass through antigen-presenting cells and stimulate **CD4+** lymphocytes to mount a T-cell-dependent antibody response. Trypanosomes directly release factors that stimulate **CD8+** T lymphocytes and macrophages. Through this cellular activation various **cytokines** and mediators appear. **Tumor necrosis factor-α (TNF-α)** contributes to the weight loss of chronic infection. It seems to inhibit trypanosomal growth, but conversely **interferon-γ (IFN-γ)** seems to help trypanosomal proliferation.

Dysregulated antibody production leads to the appearance of autoantibodies. **Immune complexes** damage vascular endothelium and on binding to the surface of red blood cells cause **hemolysis**. Chronic infection suppresses bone marrow function, probably through cytokine effects. This, added to hemolysis, causes **anemia**. Equally, platelet numbers may fall and disturbance of clotting may lead to hemorrhage or conversely **thrombosis**.

Inflammation occurs in tissues containing *T. brucei*. This ranges from the skin at the site of local inoculation, to lymph nodes, to organs seeded from

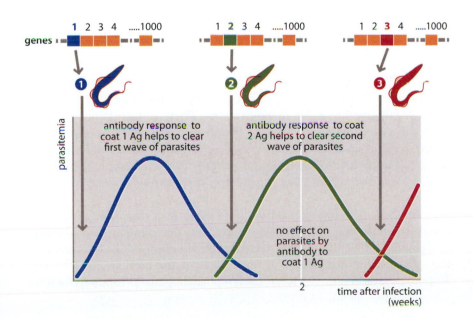

Figure 8. Antigenic variation of *T. brucei*. An antigen-specific response is made against the surface coat antigen, which gradually removes the parasites from the bloodstream (coat 1). During this time some of the parasites escape by switching their surface coat gene and the antibody response has to start from scratch. Antibodies to the second coat antigen (coat 2) are made, which remove these new parasites from the circulation. Some of the parasite will then escape again by switching their gene usage to coat 3 and so on. In *T. brucei rhodesiense* it is estimated that there are a thousand different variants of the glycoprotein that makes up the surface coat. (For simplicity, individual genes are depicted in a linear sequence whereas they are actually scattered about in the genome of the organism.)

the bloodstream such as the heart, the kidneys, and eventually the CNS. Within the CNS there is progressive inflammation of the **leptomeninges.** The inflammatory response, immune complexes, local cytokines, and **prostaglandins** all contribute to CNS pathology, which eventually involves disruption of the **choroid plexus** and the blood–brain barrier.

While *T. brucei* is extracellular *T. cruzi* is principally intracellular. The pseudocysts within tissues do not excite an inflammatory response. This only occurs once they rupture from cells and release antigens. *T. cruzi* is coated with a mucin-type glycoprotein that is anchored to the surface by a glycosylphosphatidylinositol (GPI). There are several hundred surface mucin genes, but antigenic variation in *T. cruzi* has not been established. GPI is well recognized as a potent macrophage activator and is thereby pro-inflammatory. Cell-mediated immune responses occur in infected tissue. In mice, CD8+ lymphocytes are essential for survival during the early stages of experimental infection. There is a polyclonal activation of B cells and it is conceivable that autoantibodies arise through dysregulation. Various *T. cruzi* antigens are also cross-reactive with host antigens found in vascular endothelium, muscle interstitium, and cardiac and nervous tissue. Autoimmune pathology is possible as passive transfer of serum or lymphocytes can cause pathology in recipients. However, it is not clear whether chronic pathology is caused by long-term infection-stimulated tissue inflammation or an autoimmune process or both.

3. What is the typical clinical presentation and what complications can occur?

In African trypanosomiasis local inflammation at the site of tsetse fly inoculation is sometimes apparent and referred to as a trypanosomal **chancre.** This may occur in half of *T. brucei rhodesiense* infections but is rare with *T. brucei gambiense*. It appears after 1–2 weeks, may last for 2–3 weeks, and may reach a diameter of 3–5 cm. There may be regional **lymphadenopathy**.

Symptoms occur once trypanosomes multiply and circulate in the bloodstream. These commence about 1–3 weeks after infection. There are fevers, headache, malaise, and loss of appetite. The fevers follow a cyclical pattern as the VSGs change. More generalized lymphadenopathy, **hepatomegaly**, and **splenomegaly** appear. There may be skin **rashes**.

Invasion of the CNS represents the second stage of infection. There is progressive mental deterioration culminating in coma and death, with a variety of intervening CNS manifestations. There is disturbance of motor function, co-ordination, behavior, and sleep. African trypanosomiasis is called sleeping sickness. Patients may sleep in the daytime and be awake at night. *T. brucei rhodesiense* infection progresses more rapidly, with death within 9 months, while *T. brucei gambiense* may take a few years.

In South American trypanosomiasis inflammation at the site of inoculation is called a **chagoma.** The disease was originally described by Carlos Chagas in Brazil in 1907 and the disease is also called Chagas' disease. Sometimes the inflammation occurs around the eyelid and the local swelling is called **Romana's sign** (Figure 9). Regional lymphadenopathy is followed by fever,

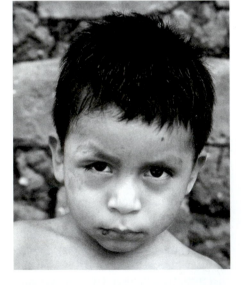

Figure 9. Romana's sign in a boy. There is swelling by the right eyebrow. The boy will have been infected in that region by a triatomine bug.

headache, malaise, hepatomegaly, splenomegaly, and rash. In the acute phase, inflammation in seeded organs results in **myocarditis**, diarrhea, and vomiting or **meningoencephalitis**.

The clinical features of the acute phase last 1–2 months and features may not return despite continued infection. Up to a third of patients suffer chronic complications after one or two decades. These principally affect the heart and the gastrointestinal tract.

Cardiac muscle weakens and thins. Impaired cardiac function results in heart failure. Aneurysms can develop, typically at the apex of the left ventricle. Clots formed within the aneurysm can embolize. Cardiac conduction is altered with ECG abnormalities. **Dysrhythmias** or complete heart block may cause sudden death.

In Brazil, more than elsewhere, gastrointestinal involvement results in dilatation and loss of peristalsis. This results in **megaesophagus** and **megacolon**, with accompanying **dysphagia** and constipation (Figure 10). The inability to eat causes **cachexia** and death.

4. How is the disease diagnosed and what is the differential diagnosis?

Unfortunately many individuals suffer from trypanosomiasis distant from diagnostic facilities. Clinical features are nonspecific. In Chagas' disease Romana's sign and local lymphadenopathy are suggestive.

In African trypanosomiasis the simplest test is the **C**ard **A**gglutination **T**est for **T**rypanosomiasis (**CATT**). A drop of heparinized whole blood is placed on a card containing antigen and the presence of agglutination is observed over 5 minutes. This test has high sensitivity but false positive results can occur. Direct visualization of parasite may be observed in lymph node aspirates, blood films, and cerebrospinal fluid (CSF). If present in reasonable numbers the motile trypanosomes are easily observed. However, the level of parasitemia may be low in later stages of infection making parasitological diagnosis difficult. Examination of the CSF is essential for staging infection.

In Chagas' disease trypanosomes may be seen in a blood film during the acute phase. The main diagnostic test is **serology**, which may employ an ELISA. Once there are organ complications, ECGs or ECHOs of the heart reveal structural abnormalities and imaging of the esophagus and colon reveal dilatation.

Differential diagnosis

Mega syndromes do not have alternative explanations of note. The differential diagnosis for the cardiac complications includes other cardiomyopathies. Otherwise the systemic **febrile** phase of early Chagas' disease and African trypanosomiasis has a long differential diagnosis. In Africa the CNS phase may have to be distinguished from **encephalitis**, cerebral tuberculosis, HIV-related neurological disease, and cerebral tumors.

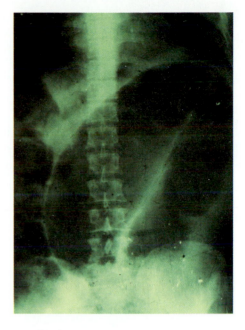

Figure 10. A plain abdominal X ray.
There is marked dilatation of the colon, megacolon, which is easily apparent on the X ray.

5. How is the disease managed and prevented?

Management

The treatments available for African trypanosomiasis are shown in Table 1. The year in which these agents were first developed is shown in brackets. The drugs are toxic and need to be bettered. Melarsoprol is an arsenical drug and causes a reactive encephalopathy in 20% on treatment and mortality in 2–12%.

Drugs for Chagas' disease include nifurtimox and benznidazole. Again they are both toxic and it is uncertain whether they should be used in the

Table 1. Drugs used to treat trypanosomiasis	
Pathogen	**Drugs and comments**
T. brucei gambiense	**Early stage**
	Pentamidine (1940)
	Intravenous or intramuscular
	Can cause drop in blood pressure, and changes in renal and hepatic function and blood glucose levels
	Late stage
	Melarsoprol (1949)
	Intravenous
	Can cause encephalopathy, with convulsions and coma, headache, thrombophlebitis, rash
	Eflornithine (1990)
	Intravenous
	Can cause bone marrow suppression and gastrointestinal upset
	Alternative to melarsoprol if available
T. brucei rhodesiense	**Early stage**
	Suramin (1920)
	Intravenous
	Can cause reversible nephrotoxicity, bone marrow suppression, and rash
	Late stage
	Melarsoprol (see above)
T. cruzi	Benznidazole
	Oral
	Can cause gastrointestinal upset, headache, joint pains, itch, and rash
	Nifurtimox
	Oral
	Can cause gastrointestinal upset, headache, joint and muscle pains

asymptomatic phase of infection before complications arise. Potentially individuals will die with, rather than from, *T. cruzi* infection. Toxicity seems to be less in children than in adults. There are now studies of ben‑znidazole in children showing that a 60-day course is reasonably well tol‑erated with disappearance of *T. cruzi* antibodies in almost 60%.

Prevention

The control of trypanosomiasis is hampered by its rural occurrence. When possible it is important to actively diagnose cases and offer treatment for or before complications. Otherwise vector control is the key strategy employed. Insecticide sprays are used to kill tsetse flies. Insecticide-impregnated traps are positioned in key locations (Figure 11). Tsetse pop‑ulations have been reduced in some localities by the release of sterilized male flies. While tsetse flies are principally outdoors the triatomine bugs of South America live within cracks in walls or in thatched roofs. Improved housing is important in reducing bug populations. Insecticide spraying has also been very successful in eliminating transmission in many areas. In South America universal blood screening is also important.

Figure 11. Tsetse fly traps in rural Africa.

SUMMARY

1. What is the causative agent, how does it enter the body and how does it spread a) within the body and b) from person to person?

- Trypanosomes are flagellated protozoan parasites.
- African trypanosomiasis is caused by *Trypanosoma brucei rhodesiense* or *T. brucei gambiense*.
- South American trypanosomiasis is caused by *Trypanosoma cruzi*.
- Tsetse flies transmit infection in Africa from other humans or an animal, ungulate reservoir.
- Triatomine bugs transmit infection in South America from an animal reservoir.
- After an insect bite trypanosomes spread from local tissue to lymph nodes, then enter the bloodstream, and finally invade tissues.
- In African trypanosomiasis invasion of the central nervous system occurs after weeks to several months.
- In South American trypanosomiasis an aflagellate, intracellular life form, the amastigote, forms a pseudocyst within cells.

2. What is the host response to the infection and what is the disease pathogenesis?

- Nonpathogenic trypanosomes are killed by normal human serum through the action of apolipoprotein L-1. Pathogenic trypanosomes neutralize this through a serum resistance-associated protein (SRA).
- In the bloodstream *T. brucei* species are killed by the action of antibody and complement.
- *T. brucei* species undergo antigenic variation and evade the antibody response.
- Antigenic stimulation by *T. brucei* causes polyclonal antibody production by B cells.
- Trypanosome-derived lymphocyte triggering factor (TLTF) and trypanosomal macrophage activating factor (TMAF) stimulate CD8+ T cells and macrophages, respectively.
- In African trypanosomiasis various cytokines are produced, and among these TNF-α can cause weight loss.

- Invasion of the central nervous system and leptomeningeal inflammation cause a 'sleeping sickness.'

- *T. cruzi* sheds glycosylphosphatidylinositol (GPI), which is a potent macrophage activator.

- Inflammation occurs in *T. cruzi*-infected tissue and this may lead to long-term pathology, particularly of cardiac muscle and gastrointestinal smooth muscle.

- Autoantibodies also appear in *T. cruzi* infection but their relative contribution to pathology is unclear.

3. What is the typical clinical presentation and what complications can occur?

- Swelling may occur at the site of the insect bite.

- As trypanosomes enter the bloodstream there are fevers, headache, malaise, and loss of appetite.

- Once the central nervous system is invaded in African trypanosomiasis there is progression to coma and death with a variety of intervening neurological problems.

- After clinical features from acute *T. cruzi* infection up to a third of patients suffer complications in the heart or gastrointestinal tract.

- Cardiac muscle can become weak and aneurysmal.

- Weakening of intestinal smooth muscle leads to dilatation and megaesophagus or megacolon.

4. How is the disease diagnosed, and what is the differential diagnosis?

- The simplest test for African trypanosomiasis is a card agglutination test for antigen.

- Trypanosomes may be seen on a blood film or in a tissue sample.

- South American trypanosomiasis may be diagnosed by a serological test.

5. How is the disease managed and prevented?

- Drugs for trypanosomes are toxic.

- Melarsoprol is used for late stage African trypanosomiasis but can cause a fatal encephalopathy.

- For chronic South American trypanosomiasis treatment with benznidazole is probably beneficial in children but of uncertain value in adults.

- Tsetse fly numbers can be reduced by outdoor insecticide-impregnated traps.

- Improving housing conditions reduces the population of triatomine bugs, which would otherwise live in cracks in walls or thatched roofs.

FURTHER READING

Burri C, Brun R. Human African trypanosomiasis. In: Cook GC, Zumla AI. Manson's Tropical Diseases, 21st edition. Elsevier, Edinburgh, 2003, 1303–1323.

Miles MA. American trypanosomiasis (Chagas disease). In: Cook GC, Zumla AI. Manson's Tropical Diseases, 21st edition. Elsevier, Edinburgh, 2003, 1325–1337.

Murphy K, Travers P, Walport M. Janeway's Immunobiology, 7th edition. Garland Science, New York, 2008.

REFERENCES

Barrett MP, Burchmore RJS, Stich A, et al. The trypanosomiases. Lancet, 2003, 362: 1469–1480.

De Andrade ALSS, Zicker F, de Oliviera RM, et al. Randomised trial of efficacy of benznidazole in treatment of early *Trypanosoma cruzi* infection. Lancet, 1996, 348: 1407–1413.

Pentreath VW. Trypanosomiasis and the nervous system. Trans R Soc Trop Med Hyg, 1995, 89: 9–15.

Vanhollebeke B, Truc P, Poelvoorde P, et al. Human *Trypanosoma evansi* infection linked to a lack of apolipoprotein L-1. N Engl J Med, 2006, 355: 2752–2756.

WEB SITES

Center for Disease Control, Atlanta, GA, USA: www.cdc.gov/
Health Protection Agency: www.hpa.org.uk

World Health Organization: www.who.int

MULTIPLE CHOICE QUESTIONS

The questions should be answered either by selecting True (T) or False (F) for each answer statement, or by selecting the answer statements which best answer the question. Answers can be found in the back of the book.

1. **Which of the following are true about *Trypanosoma*?**
   A. They are flagellated protozoan parasites.
   B. *T. cruzi* is responsible for African trypanosomiasis.
   C. They have animal reservoirs.
   D. *T. brucei* species cause infections in Asia.
   E. *T. cruzi* remains extracellular.

2. **Which of the following are true about the life cycle of trypanosomiasis?**
   A. *T. brucei* species are transmitted by sandflies.
   B. *T. cruzi* is transmitted by triatomine bugs.
   C. *T. cruzi* is inoculated into a new human host through saliva from the biting vector.
   D. Trypanosomiasis can be transmitted by blood transfusion.
   E. Infections never pass from human to human.

3. **Which of the following are true about the spread of trypanosomes?**
   A. They are inoculated directly into the bloodstream.
   B. Invasion of the central nervous system occurs in the first few days.
   C. A key target for *T. cruzi* is the reticuloendothelial system including the liver, spleen, and bone marrow.
   D. *T. cruzi* invades tissues and multiplies between cells rather than within cells.
   E. *T. brucei gambiense* invades the heart tissue.

4. **Which of the following are true of the host response to *Trypanosoma brucei*?**

   A. Antibody and complement lyse *T. brucei* species.

   B. CD4+ T lymphocytes help B lymphocytes to produce anti-trypanosomal antibodies.

   C. Tumour necrosis factor-α inhibits trypanosomal growth.

   D. *T. brucei* species must inhibit apolipoprotein L-1 to avoid lysosomal killing.

   E. *T. brucei rhodesiense* evades host antibody responses by switching between about 100 variant surface glycoproteins.

5. **Which of the following are true of the host response to *Trypanosoma cruzi*?**

   A. Autoantibodies appear in the course of infection.

   B. Antigenic variation is a clearly established immune evasion mechanism.

   C. Tissue pseudocysts excite a tissue inflammatory reaction.

   D. CD8+ T lymphocytes can kill cells bearing *T. cruzi* antigen.

   E. Macrophages are suppressed by *T. cruzi*.

6. **Which of the following statements are true about the clinical features of African trypanosomiasis?**

   A. A trypanosomal chancre may be seen in up to half of *T. brucei rhodesiense* infections.

   B. Lymphadenopathy is not observed in *T. brucei gambiense* infection.

   C. Fevers commence within a month of infection.

   D. Death follows invasion by trypanosomes of the central nervous system.

   E. Progression to death is more rapid with *T. brucei gambiense* than with *T. brucei rhodesiense*.

7. **Which of the following are true of the clinical features of South American trypanosomiasis?**

   A. Inflammation at the site of inoculation is called a chagoma.

   B. In the acute phase of infection the liver and spleen may be enlarged.

   C. Long-term complications occur in all those infected.

   D. Infection of the heart results in thickening of heart muscle.

   E. Weakening of the gastrointestinal smooth muscle occurs with infection in all geographical areas.

8. **Which of the following are true of the diagnosis of trypanosomiasis?**

   A. South American trypanosomiasis, Chagas' disease, can always be diagnosed by characteristic clinical features.

   B. The card agglutination test for African trypanosomiasis tests for trypanosomal antigen in the bloodstream.

   C. The card agglutination test for trypanosomiasis has a low sensitivity, but a high specificity.

   D. For African trypanosomiasis diagnosis may be based on microscopy of blood or cerebrospinal fluid.

   E. A serological test for antibody is available for Chagas' disease.

9. **Which of the following are true for the treatment of trypanosomiasis?**

   A. The use of melarsoprol for African trypanosomiasis is associated with encephalopathy in about 20%.

   B. Benznidazole is better tolerated by children than adults with *T. cruzi* infection.

   C. All individuals with *T. cruzi* infection should be treated to prevent complications.

   D. Pentamidine is used in the early stage of *T. brucei gambiense* infection.

   E. Late stage African trypanosomiasis is treated with benznidazole.

10. **Which of the following are true of the control of trypanosomiasis?**

   A. A vaccine is available.

   B. Treatment of infected individuals is a key part in eliminating the reservoir of infection.

   C. Insecticide-treated traps are used to control tsetse fly populations.

   D. Corrugated roofs are preferable to thatched roofs to limit triatomine bug populations.

   E. Screening of blood before transfusion is ineffective in preventing transmission.

Case 39

Varicella-zoster virus

A young girl was taken to see her doctor by her mother. She had developed an itchy blistering **rash** that first appeared as red spots on her forehead 2 days previously and had now spread over her whole body. She had been previously well, and was not on any medications.

On examination, the rash was indeed extensive. The girl had a temperature of 38.1°C. The doctor made a clinical diagnosis of chickenpox.

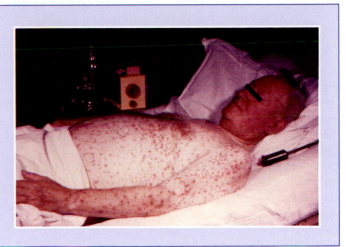

Figure 1. An adult male with chickenpox. Note the global distribution of the rash all over the body.

1. What is the causative agent, how does it enter the body and how does it spread a) within the body and b) from person to person?

Causative agent

The causative agent of chickenpox is varicella-zoster virus (VZV). This belongs to the *Herpesviridae* family of viruses, the characteristics of which are that the genome is double-stranded DNA, surrounded by a protein coat or **capsid** with icosahedral symmetry, and the **nucleocapsid** in turn is surrounded by a loosely fitting, irregular **lipid envelope** (Figure 2). The double-barrelled name of the virus – VZV – is because initial infection with this virus gives rise to the disease varicella (the proper name for chickenpox), while reactivation of infection (explained below) gives rise to the disease zoster (also known as **shingles**). There are currently eight herpesviruses that can cause human disease (see Table 1 in the Epstein-Barr virus case, Case 9).

Entry and spread within the body

Varicella-zoster virus infection is acquired through entry of droplets containing viral particles into the upper respiratory tract. After an initial phase of replication at the site of entry, the virus spreads to the regional lymph nodes and organs of the **mononuclear phagocyte system** where further replication takes place. About 10–12 days after initial infection, virus is released from the lymph nodes into the bloodstream (**viremia**), and then is deposited in the skin, where the final phase of replication occurs, giving rise to the characteristic vesicular rash of chickenpox, the clinical manifestation of primary infection with this virus.

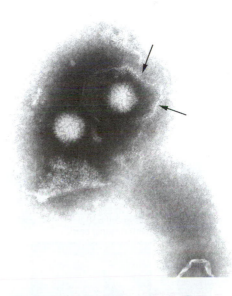

Figure 2. Electron micrograph of a herpesvirus, showing the geometrically regular nucleocapsid, which contains the viral DNA, and the loosely fitting irregular lipid envelope (arrows).

Like all herpesviruses, VZV establishes latency (see also HSV-1 and HSV-2 cases). Replicating virus within the skin gains entry into the nerve terminals supplying the skin, and travels in a retrograde direction up the nerve axons to the nerve cell bodies, the site of **latency** for this virus. In latently infected cells, the viral genome is present, but there is no replication or production of virus particles, and few, if any, viral proteins are synthesized (Figure 3). As there has been viremic spread of the virus around the body, all nerve cell bodies in the spinal cord and those of the cranial nerves may be latently infected.

Once an individual is infected with VZV, they remain infected for life. The virus remains in a latent state and does not cause any damage to the infected nerve cell bodies. However, under certain circumstances, latent virus may become reactivated, resulting in viral replication and production of new virus particles. The precise molecular mechanisms whereby virus establishes or breaks out of latency, are not fully understood, but a number of clinical factors that predispose to reactivation of latent virus are well recognized (Table 1). Newly formed virus particles travel back down the axon to reach the skin, resulting in a vesicular rash similar to that of chickenpox, but characterized by having a distribution limited to the area of skin supplied by the nerve in which reactivation has taken place – a so-called **dermatomal** distribution. This reactivated infection (also referred to as secondary or recurrent infection) is known as herpes zoster, zoster, or shingles.

Spread from person to person

An individual with chickenpox sheds large amounts of virus in droplet form from the nose and throat, and VZV is transmitted via an air-borne route. The clinical attack rate among exposed susceptible individuals is in the range 70–90%, indicating that an individual with chickenpox is highly infectious.

Figure 3. VZV latency. (A) During primary infection, bloodstream spread of virus results in the characteristic rash of chickenpox appearing all over the body. Some virus enters the nerve terminals of the nerves supplying the skin, and ascends up the nerve axon to reach the nerve cell body, the site of latency for this virus. (B) Latent virus sits within the nerve cell bodies. In this state, there is no viral replication, and no damage to the cell. (C) This latent virus may become reactivated out of latency, and start to replicate. Mature virus particles then descend down the nerve axon to reach that area of skin supplied by that particular nerve, thereby resulting in the characteristic dermatomal distribution of the zoster (or shingles) rash.

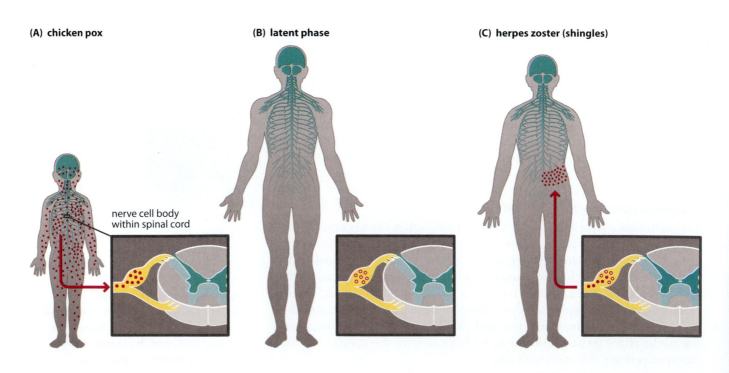

(A) chicken pox

nerve cell body within spinal cord

(B) latent phase

(C) herpes zoster (shingles)

Table 1. Factors predisposing to reactivation of latent VZV
Increasing age
Trauma (including surgery, dentistry)
Stress (including illness)
Sunburn
Immunocompromise
HIV infection
Post transplantation
Underlying malignancy*
Radiotherapy, chemotherapy
High dose steroids

* Particularly of the lymphoid and myeloid systems, e.g. leukemia, lymphoma

It is possible to catch chickenpox from someone who has an attack of shingles, as there is plenty of viable virus in the vesicular rash of shingles. However, such individuals are much less infectious than patients with chickenpox, as they are not shedding large amounts of virus from the nasopharynx.

Incidence of infection

Infection with VZV is extremely common worldwide, with some geographic variations in epidemiology. In temperate countries, infection is usually acquired in childhood, with a peak age of infection less than 5 years old. In the UK, by adulthood, over 90% of the population have antibodies to VZV, and are therefore latently infected (and at risk of developing shingles). In tropical regions, infection usually occurs at a much older age, with a mean age of infection above 20 years old. Reasons for this are not clear, but may reside around different patterns of social mixing between infected cases and susceptible contacts.

The average annual incidence of herpes zoster, the reactivated or secondary form of infection, is 0.2%. Disease is very much age-related, with rates of disease in the over 65s around 10 times higher than in the under 50s – reflecting declining cellular immune function with increasing age.

2. What is the host response to the infection and what is the disease pathogenesis?

Immune response

In an immunocompetent host, VZV infection stimulates both humoral and cellular arms of the immune response. Antibodies to VZV, initially of the **IgM** class, but then of the **IgG** class, can be detected about 14 days after initial infection. These antibodies have neutralizing activity, and therefore prevent the access of virus into uninfected cells. However, the development of cellular immune responses is more important in controlling initial (and reactivated) infection. This involves both **natural killer (NK)** cells and **CD8+** cytotoxic T cells.

The adaptive immune responses are important in maintaining latency of VZV – as evidenced by the fact that immunosuppressed individuals, especially those with defects in cellular, rather than humoral, immune responses, are more likely to develop shingles.

Pathogenesis

The pathogenesis of chickenpox has been described above. Herpes zoster represents reactivation of virus in a nerve cell body within a single sensory ganglion (a collection of nerve cell bodies), with subsequent transport of the virus down the axon to infect the skin supplied by that nerve. Zoster can occur anywhere on the body. As indicated in Table 1, herpes zoster is much more common in patients with underlying immunodeficiency, which may arise through a number of different causes. Malignant disorders of the reticuloendothelial system such as **lymphoma**, **leukemia**, and **multiple myeloma**, are associated with abnormalities in T-cell function and hence development of zoster, which may be unusually prolonged or aggressive. An attack of zoster in a younger-than-expected patient may be the presenting feature of underlying HIV infection. On a more subtle level, it is not unusual for patients to present with zoster rashes at times of stress, for example bereavement or divorce, suggesting the existence of as yet ill-defined links between central nervous system function and the immune system.

3. What is the typical clinical presentation and what complications can occur?

Many primary infections with VZV do not result in clinical disease. When symptomatic, primary infection results in varicella, also known as chickenpox. After an incubation period of 13–17 days, disease usually presents with its most striking feature, the vesicular rash (Figure 1), in a centripetal distribution (i.e. more evident centrally than peripherally), with lesions usually on the trunk and face (including within the mouth). Prodromal symptoms, for example a flu-like illness, are relatively unusual, especially in children. The lesions evolve from **macules** (flat red patches in the skin) to **papules** (raised lesions) to **vesicles** (fluid-filled blisters) on a red (**erythematous**) base in a few hours (Figure 4). The vesicles become pustular, and then dry out and scab over the next 2–4 days, followed by regeneration of the underlying epithelium. Cropping occurs over several days, so that lesions of differing ages may be present. The patient is also usually **febrile** once the rash appears, although constitutional symptoms are usually mild, particularly in children. The rash is itchy in some, but not all patients.

Complications of chickenpox are unusual, especially in an immunocompetent child. Secondary bacterial infection of the vesicles may occur, which very occasionally may lead to **septicemia**, and also may result in scarring once the lesions heal. The disease is much more debilitating in adults, especially smokers, who are also at significant risk of varicella **pneumonia**, the commonest life-threatening complication. This presents with shortness of breath and cough about 2–3 days after onset of the rash. **Encephalitis** is uncommon, but may present about 7–10 days after onset. The most likely pathogenesis of this is a post-infectious encephalitis, that is, not due to the presence of virus itself within the brain substance, but arising through aberrant immune responses to infection that damage the

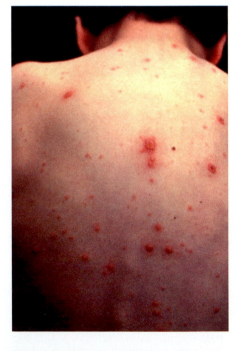

Figure 4. Close-up of the varicella-zoster rash. Note the fluid-filled vesicles on an erythematous base.

brain tissue. A cerebellar syndrome is most often seen in children, but **hemiplegia** is also described. Recovery is usual, but long-term sequelae and even fatalities may occur.

Chickenpox in an immunosuppressed individual, whether child or adult, is altogether a more serious and life-threatening disease. Varicella pneumonia is much more common in this patient group, as is hemorrhagic varicella, a severe complication often associated with **thrombocytopenia**. Bleeding occurs into the skin lesions, from mucous membranes, and into unaffected skin, and the prognosis is poor.

Chickenpox in pregnancy is of particular concern. Adult pregnant females are more likely to die from varicella pneumonia than adult nonpregnant females. In addition, virus may cross the placenta and damage the developing fetus, giving rise to congenital varicella or varicella embryopathy, the characteristic features of which are extensive areas of skin scarring, and failure of limb bud development. This occurs in about 1% of pregnancies where maternal varicella occurs within the first 20 weeks of pregnancy. If the mother acquires infection late in pregnancy, then there is a risk that the baby will be born and acquire infection before the mother has had time to generate protective IgG anti-VZV, which can cross the placenta. Neonatal chickenpox may be very severe, with mortality rates of up to 30% reported.

Shingles

The clinical features of shingles are protean, and depend to a large extent on the location of the nerve cell body in which reactivation arises. The shingles rash is acutely painful, and occurs in a dermatomal distribution (see above, and Figure 5). The commonest and most distressing complication is **post-herpetic neuralgia** (PHN), defined as pain persisting for at least 3 months (although the precise time period may vary between different studies) in the area of a zoster rash after the rash itself has disappeared. The pathogenesis is poorly understood, but it can persist for months or even years, and can have a significantly detrimental effect on the patient's quality of life. The risk of PHN increases with increasing age of the patient, and is also related to the intensity of the shingles rash (i.e. the number of lesions). Other complications depend on the anatomic location of the rash. Ophthalmic zoster (occurring in the ophthalmic branch of the trigeminal nerve) is particularly serious, as virus may gain access to the cornea, giving rise to an ulcerating zoster **keratitis**. While mostly thought of as a disease of sensory nerves, motor nerve damage can also arise in zoster, giving rise to lower motor neuron paralysis, for example zoster of the facial nerve (seventh cranial nerve) may give rise to a **Bell's palsy** (facial nerve paralysis).

Both primary and secondary VZV infections can be life-threatening in patients with underlying immunodeficiencies, particularly involving the T-cell arm of the immune response, for example patients with HIV infection, or transplant recipients on potent immunosuppressive drugs. The presence of virus within the bloodstream (which is a routine stage of the pathogenesis of primary infection, and may arise in secondary infection) means that virus may seed visceral organs, and the absence of an appropriate immune response to contain this infection may result in, for instance, fatal varicella **pneumonitis**, **hepatitis**, or encephalitis.

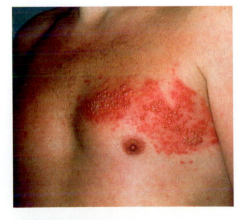

Figure 5. A zoster rash. This arises due to reactivation of VZV from its site of latency within the nerve cell body. The rash is dermatomal in distribution, i.e. affects only that area of skin supplied by the affected nerve. In this patient, the rash continued onto his back, stopping in the midline.

4. How is this disease diagnosed and what is the differential diagnosis?

Diagnosis of primary infection is usually made on clinical grounds, as the vesicular rash of chickenpox is so distinctive. There are very few differential diagnoses, especially since the elimination of smallpox. The smallpox rash, in contradistinction to chickenpox, is usually centrifugal, and the lesions are rounder and deeper than the more irregular and superficial chickenpox ones. More importantly, the lesions of smallpox evolve at a similar rate and, therefore, are not present in all stages as occurs with chickenpox.

Zoster can be a difficult diagnosis to make, especially in the early stages where the patient may present with acute pain before the rash has appeared. However, once the rash appears, again, a clinical diagnosis is usually reliable, the main differential diagnosis being herpes simplex virus (HSV) infection.

Laboratory confirmation of the diagnosis can be made in a number of ways. A vesicle swab should be sent to the laboratory, where the virus may be isolated in tissue culture, or more rapidly, cells shaken off the swab can be stained with immunofluorescent anti-VZV antibodies (antigen detection). Vesicle fluid contains sufficient virus for the particles to be visible by electron microscopy, although this does not allow differentiation between VZV and HSV, both of which appear as typical herpesvirus particles. The presence of viral genome in vesicle fluid can be demonstrated using a genome amplification technique such as the **polymerase chain reaction (PCR)**. Finally, a serum sample from a patient with chickenpox will be IgM anti-VZV positive – however, this serological approach is not as reliable in the diagnosis of zoster, as this is a reactivated infection and IgM responses may not be as clear-cut.

5. How is the disease managed and prevented?

Management

VZV is treated with aciclovir and its derivatives. VZV is able to phosphorylate aciclovir using the virally encoded enzyme thymidine kinase, and therefore replication of VZV is sensitive to aciclovir, and derivatives of aciclovir such as valaciclovir (a valine ester) or famciclovir (see HSV-1 case for more detail on the action of aciclovir). However, aciclovir therapy is not usually indicated in otherwise uncomplicated childhood chickenpox. In the USA, aciclovir is licensed for that indication, but the justification for its use is an economic, not a medical one – antiviral therapy results in a speedier recovery of the child, which allows the adult carers to go back to work sooner. It is reasonable to treat chickenpox in an adult with aciclovir, as the disease is inherently more severe in this age group. Although aciclovir is not licensed for use in pregnancy, it can be life-saving if varicella pneumonia develops.

It is not necessary to treat every case of shingles. Most guidelines recommend treatment if the patient has any underlying immunodeficiency, regardless of the cause (risk of internal organ dissemination and death), if the patient is over 50 years old (increased risk of PHN), or if the rash is in the territory of the ophthalmic branch of the trigeminal nerve (risk of zoster keratitis) or the facial nerve (risk of motor nerve damage). If therapy is indicated, it must be started within 72 hours of onset of the rash to be most effective.

Prevention

There is a licensed vaccine against VZV – this consists of a live attenuated virus known as the OKA strain. It was initially developed for use in children with severe underlying disease such as leukemia, who would otherwise be at risk of life-threatening chickenpox. Thus, it has an excellent safety record, even in patients with impaired immune systems. As this is a live virus, occasionally vaccinees may develop vaccine-associated chickenpox, although this is usually mild. Vaccine virus may also go latent in the spinal cord and reactivate in the form of shingles, but this is rare and not usually severe. In the USA, this is recommended as a routine vaccination of childhood. However, in the UK, this vaccine is currently only licensed for use in health-care workers shown to be susceptible to infection (i.e. anti-VZV-negative). There is a theoretical risk that vaccination in childhood may paradoxically increase the burden of disease associated with VZV infection. If vaccine-induced immunity is not long-lasting, then vaccination in childhood may result in older adolescents and adults acquiring primary infection, with a consequent increase in severity of disease. Also, it has been suggested that exposure to children with chickenpox serves to boost VZV immunity in adults, which in turn allows continued suppression of VZV reactivation. Elimination of disease in small children may therefore lead to an increase in herpes zoster in adults. Decisions on the use of VZV vaccine in the UK are therefore awaiting the results of the widespread use of vaccine in the USA. Recently published evidence suggests that vaccination of elderly individuals may decrease the risk of herpes zoster.

SUMMARY

1. What is the causative agent, how does it enter the body and how does it spread a) within the body and b) from person to person?

- VZV, a herpesvirus.

- Infection acquired in the throat, initial viral replication in the local lymph nodes, followed by viremic (i.e. blood-borne) spread taking virus to the skin, giving the characteristic widespread vesicular rash of varicella or chickenpox.

- Virus replicating in the skin gains access to sensory nerve terminals and travels up the nerve axon to the nucleus of the nerve cell body in the sensory ganglion.

- Nerve cell body is the site of virus latency.

- Reactivation from latency results in virus particles traveling down to the nerve terminals, giving rise to the rash of herpes zoster, or shingles.

- Patients with chickenpox shed large amounts of virus from the throat, as well as from the skin rash.

- Patients with herpes zoster are also infectious as the skin lesions contain viable virus.

2. What is the host response to the infection and what is the disease pathogenesis?

- An antibody response develops that is first detectable about 7 days after onset of the rash, initially IgM only, followed by IgG.

- T-cell responses are also key to preventing the development of life-threatening internal organ dissemination of virus.

3. What is the typical clinical presentation and what complications can occur?

- Primary infection, when symptomatic, presents as chickenpox (varicella), a widespread vesicular rash, with fever and generalized lymphadenopathy.

- The rash evolves from macule to papule to vesicle to crust in 4 days.

- Crops of lesions appear, firstly on face and trunk, and then spread to scalp and limbs.

- Complications in childhood are unusual.

- In adults, varicella is a more severe disease, and varicella pneumonia is a life-threatening complication.

- Maternal varicella in the first 20 weeks of pregnancy may result in fetal damage in 1% of cases. Maternal varicella in late pregnancy may give rise to neonatal varicella. Pregnant women with chickenpox are also at increased risk of varicella pneumonia.

- Herpes zoster (shingles) is a painful vesicular rash occurring in a dermatomal distribution.

- Complications of zoster include post-herpetic neuralgia, zoster keratitis, motor nerve damage, and dissemination in immunosuppressed hosts.

4. How is this disease diagnosed and what is the differential diagnosis?

- Diagnosis of both chickenpox and shingles can be made on clinical grounds, on the basis of the characteristics of the rash.

- Laboratory confirmation can be made by virus isolation, electron microscopy, antigen detection, or genome amplification techniques.

- Antibody tests are only of use in primary infection, where an IgM anti-VZV response can be demonstrated.

5. How is the disease managed and prevented?

- Treatment is possible with aciclovir (an antiviral drug that is a deoxyguanosine analog) or derivatives such as valaciclovir or famciclovir.

- Treatment is not usually indicated in otherwise uncomplicated chickenpox in children.

- Treatment is recommended for chickenpox in adults.

- In zoster, treatment is recommended if the patient is immunodeficient, aged over 50 years, or zoster is in ophthalmic or facial nerves. It must be initiated within 72 hours of onset of the rash.

- Prevention is by means of a live attenuated vaccine.

- In the UK, varicella vaccine is only licensed for VZV-susceptible health-care workers.

- In the USA, varicella vaccine is licensed for routine use in childhood.

FURTHER READING

Humphreys H, Irving WL. Problem-orientated clinical micro-biology and infection, 2nd edition. Oxford University Press, Oxford, 2004.

Murphy K, Travers P, Walport M. Janeway's Immunobiology, 7th edition. Garland Science, New York, 2008.

Richman DD, Whitley RJ, Hayden FG. Clinical Virology, 2nd edition. ASM Press, Washington, DC, 2002.

Zuckerman AJ, Banatvala JE, Pattison JR, Griffiths PD, Shaub BD. Principles and Practice of Clinical Virology, 5th edition. Wiley, Chicester, 2004.

REFERENCES

Gnann JW, Whitley RJ. Herpes Zoster. N Engl J Med, 2002, 347: 340–346.

Heininger U, Seward JF. Varicella. Lancet, 2006, 368: 1365–1376.

Oxman MN, Levin MJ, Johnson GR, et al. A vaccine to prevent herpes zoster and postherpetic neuralgia in older adults. N Engl J Med, 2005, 352: 2271–2284.

Vazquez M, LaRussa PS, Gershon AA, Steinberg SP, Freudigman K, Shapiro ED. The effectiveness of the varicella vaccine in clinical practice. N Engl J Med, 2001, 344: 955–960.

WEB SITES

All the Virology on the WWW Website, developed and maintained by Dr David Sander, Tulane University: http://www.virology.net/garryfavweb12.html#Herpe

Centers for Disease Control and Prevention, National Center for Immunization and Respiratory Diseases, Atlanta, GA, USA: http://www.cdc.gov/ncidod/diseases/list_varicl.htm

Centre for Infections, Health Protection Agency, HPA Copyright ©, 2008: http://www.hpa.org.uk/infections/topics_az/chickenpox/menu.htm

Website of Derek Wong, a medical virologist working in Hong Kong: http://virology-online.com/viruses/VZV.htm

MULTIPLE CHOICE QUESTIONS

The questions should be answered either by selecting True (T) or False (F) for each answer statement, or by selecting the answer statements which best answer the question. Answers can be found in the back of the book.

1. **Which of the following are true of VZV?**
   A. It is an enveloped virus.
   B. It is a member of the *Poxviridae* (pox virus family).
   C. It carries a single-stranded DNA genome.
   D. It undergoes latency in nerve cell bodies.
   E. It has a capsid of helical symmetry.

2. **Which of the following are true of VZV infection?**
   A. Infection is usually acquired via the gastrointestinal tract.
   B. During the incubation period, VZV is spread around the body in the bloodstream.
   C. The rash of chickenpox is due to the deposition of immune complexes into the skin.
   D. It is possible for a susceptible host to acquire infection by exposure to a patient with shingles (herpes zoster).
   E. The incubation period from infection to appearance of rash is of the order of 6–8 days.

3. **Which of the following statements are true?**
   A. The average age at primary infection is higher in tropical countries than in temperate ones.
   B. In the UK, at least one-third of individuals greater than 20 years old have no evidence of prior infection with VZV infection.

MULTIPLE CHOICE QUESTIONS (continued)

C. Hemorrhagic chickenpox usually arises as a result of a vigorous immune response to infection.

D. The attack rate of infection in susceptible individuals following exposure to a case of chickenpox is less than 50%.

E. During primary infection, virus goes latent in nerve cell bodies.

4. **Which of the following statements concerning chickenpox are true?**

A. Infection in an adult usually results in more severe disease than in a child.

B. In adults, varicella encephalitis is the commonest life-threatening complication.

C. The absence of itch means that the rash is not likely to be due to VZV infection.

D. New lesions may continue to appear for a few days after first appearance of the rash.

E. Lesions are usually more abundant at the extremities (i.e. scalp, hands, and feet) than over the rest of the body.

5. **With regard to chickenpox infection in pregnancy, which of the following are true?**

A. Varicella embryopathy may arise from maternal chickenpox at any stage during the pregnancy.

B. Over 10% of babies born to mothers who acquire chickenpox in pregnancy may suffer developmental defects.

C. Failure of limb bud development is one of the features of varicella embryopathy.

D. Maternal shingles carries a significant risk that virus may cross the placenta and damage the developing fetus.

E. A baby whose mother develops chickenpox 2 days after delivery is not at risk of severe neonatal chickenpox.

6. **Which of the following factors increase the likelihood of reactivation of latent VZV infection?**

A. Increasing age.

B. HIV infection.

C. Antibody deficiency.

D. Hodgkin's lymphoma.

E. Eczema.

7. **Which of the following are recognized complications of herpes zoster (shingles) in an immunocompetent patient?**

A. Ulceration of the cornea.

B. Motor nerve paralysis.

C. Persistence of pain after the rash has resolved.

D. Hepatitis.

E. Pneumonia.

8. **Which of the following statements regarding diagnosis of varicella-zoster virus infection are true?**

A. Laboratory confirmation should be sought for all cases of clinically diagnosed chickenpox.

B. VZV can be isolated in tissue culture.

C. Antigen detection by immunofluorescence on cells scraped from the base of a vesicular lesion allows distinction between reactivated herpes simplex virus infection and reactivated VZV infection.

D. Electron microscopy of vesicle fluid allows distinction between reactivated herpes simplex virus infection and reactivated VZV infection.

E. Genome detection by polymerase chain reaction assay allows distinction between reactivated herpes simplex virus infection and reactivated VZV infection.

9. **Which of the following patients should be offered antiviral therapy with aciclovir (or derivatives)?**

A. A 45-year-old man with two vesicular lesions on the left side of his forehead.

B. A 40-year-old woman with an uncomplicated shingles rash on her trunk which first appeared 36 hours ago.

C. A 28-year-old renal transplant recipient with three vesicular lesions just below his right nipple.

D. A 58-year-old woman with a painful vesicular rash on her left side that first appeared 5 days ago.

E. A 27-year-old male known to have HIV infection with a shingles rash in the distribution of the 10th thoracic nerve (i.e. at the level of his umbilicus).

10. **Which of the following statements regarding varicella vaccine are true?**

A. The vaccine consists of a live attenuated virus.

B. It should not be given to children with leukemia.

C. Susceptible health-care workers who work in areas where they may be exposed to VZV should be vaccinated.

D. Vaccinated patients may develop vaccine-associated shingles.

E. Vaccine may be used as a routine vaccination in childhood.

Case 40

Wuchereria bancrofti

A 24-year-old man from Southern India presented with swelling of his left foot. The swelling diminished overnight. He was otherwise well with no fevers or systemic upset. Initially there was concern that he had a deep vein **thrombosis**, but his D-dimers were not raised. There were a few enlarged lymph nodes in his groin, and on closer questioning he reported previous episodes with the nodes becoming painful and swollen. In his home town in India he recalled seeing individuals with grossly swollen legs. His full blood count revealed an **eosinophilia** of $1.4 \times 10^9 \ L^{-1}$ (normal range 0.04–0.4). His filarial **serology** was positive. He was treated with a prolonged course of doxycycline. The eosinophil count fell to $0.4 \times 10^9 \ L^{-1}$ and his swelling subsided (Figure 1).

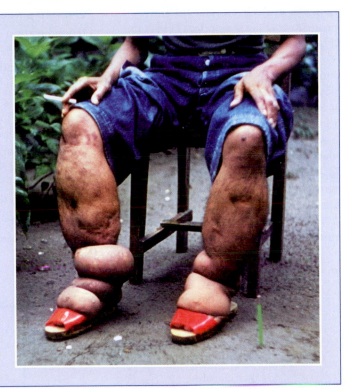

Figure 1. This man has more advanced swelling of his legs than in the case history. His bilateral lymphedema is more advanced and will not resolve with treatment. (Courtesy of Public Health Image Library, Centers for Disease Control & Prevention, USA.)

1. What is the causative agent, how does it enter the body and how does it spread a) within the body and b) from person to person?

Causative agent

Wuchereria bancrofti is the causative agent of Bancroftian filariasis. It is a round worm, or nematode, which in the adult form lives in lymphatic vessels. Females are about 4–10 cm long and males about 2–4 cm. However, they are very slender, being 0.25 and 0.1 mm in width, respectively. *W. bancrofti* is found in Asia, Africa, South America, and the Caribbean. Other filarial nematodes that cause a similar illness are *Brugia malayi* and *B. timori*, which are confined to South East Asia.

Entry and spread within the body

Infection occurs with the bite of various species of mosquito vector (see below). Larvae enter the tissues and are thought to migrate along the lymphatics. They commonly reach the lymphatics draining to lymph nodes in the groin and sometimes the armpit. Adult worms mature, mate, and after about 8 months females release microfilariae from the ova in their uterine bodies. These circulate in the bloodstream and are available to be ingested when a mosquito takes a blood meal (Figure 2). The life

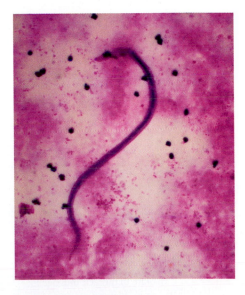

Figure 2. *Wuchereria bancrofti* **microfilaria in a blood film** (courtesy of Public Health Image Library, Centers for Disease Control & Prevention, USA).

cycle of *W. bancrofti* is shown in Figure 3. In areas with nocturnally periodic transmission microfilariae appear in the bloodstream in maximum numbers around midnight, which coincides with peak local vector biting activity. In other areas, where the local vectors bite during the day, microfilariae levels remain relatively constant in the bloodstream throughout the day. Microfilariae have a lifespan of about 1 year. Adult worms are capable of living up to 8 years, but may not survive that long.

Person to person spread

Depending on geographical location *W. bancrofti* is transmitted by *Anopheles*, *Aedes*, *Culex* or *Mansonia* mosquitoes. When mosquitoes ingest microfilariae-laden blood the microfilariae pass into the midgut, shed their sheath, and penetrate the wall of the midgut to enter the body cavity. They migrate to muscles in the thorax and undergo two molts. The third stage larvae are the infective form and pass onto the mouth parts of the mosquito. Upon blood feeding, the larvae are deposited onto the skin surface and enter the host via the puncture made by the mosquito. The reservoir for infection is the local microfilaremic human population.

Epidemiology

In 2000 the World Health Organization (WHO) estimated that there were 120 million cases of lymphatic filariasis in 80 countries, with 40 million

Figure 3. Life cycle of *Wuchereria bancrofti*. L3 larvae escape from the mosquito onto the skin surface at the time of taking a blood meal from humans (1). The larvae then enter the human host through the puncture made by the mosquito. The larvae mature into adults, which settle in lymphatics (2). Male and female adult worms mate and then females release ensheathed microfilariae larvae, which circulate in the blood (3). These may be ingested by mosquitoes feeding on humans (4). In the mosquito mid-gut the sheaths are shed (5). Larvae then mature into the L3 form, which passes to the mosquito mouthparts (6-8).

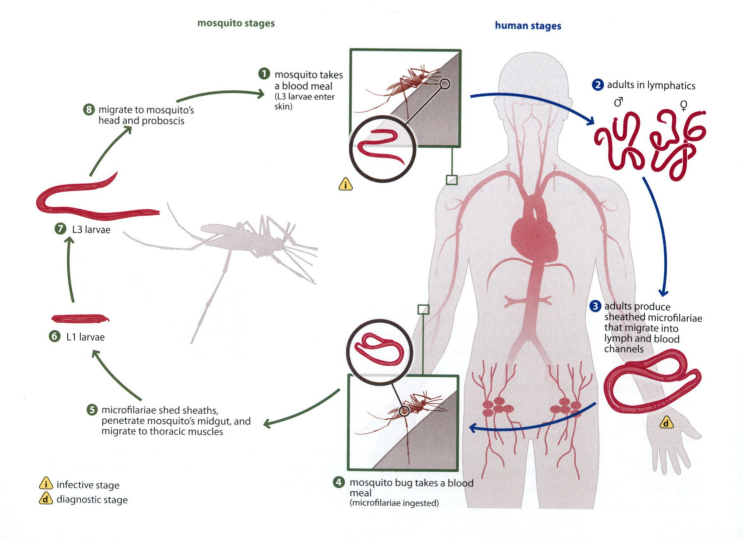

mosquito stages   **human stages**

1. mosquito takes a blood meal (L3 larvae enter skin)
2. adults in lymphatics ♂ ♀
8. migrate to mosquito's head and proboscis
7. L3 larvae
6. L1 larvae
5. microfilariae shed sheaths, penetrate mosquito's midgut, and migrate to thoracic muscles
4. mosquito bug takes a blood meal (microfilariae ingested)
3. adults produce sheathed microfilariae that migrate into lymph and blood channels

ⓘ infective stage
ⓓ diagnostic stage

individuals being severely incapacitated. About one-third of all cases are in India, one-third in Africa, and the remainder in other parts of South East Asia, Pacific Islands, South America, and the Caribbean.

2. What is the host response to the infection and what is the disease pathogenesis?

The immune response may be directed against infecting L3 larvae, adult worms or microfilariae and may involve antibody, lymphocytes, and other mononuclear cells. *In vitro* microfilariae can be killed by **IgE**-mediated degranulation of eosinophils and mast cells. This killing mechanism may be modified.

Repeated mosquito bites expose individuals to the third stage larvae (L3). These go on to mature into adult worms, which mate and in turn release microfilariae. The immune response to each of these stages seems to be segregated. Some individuals will progressively stimulate a protective immune response to the L3 larvae. This may become so effective by adult-hood that new infections can no longer occur. Adult worms present within lymphatics will eventually die and so in an endemic community there will be some who are immune and clear of infections.

In endemic areas there will be mothers who have filariasis at the time of pregnancy. Microfilariae have been seen in cord blood and newborns have **IgM** and IgE antibodies to microfilarial antigen. The latter **antibody iso-types** do not cross the placenta and their production implies *in utero* microfilarial antigen exposure. It is postulated that this early antigen exposure modifies the subsequent immune response when children are exposed to microfilariae. In some studies in children anti-microfilarial IgE-mediated degranulation of eosinophils and mast cells does not occur. It is thought that high titers of **IgG4** block anti-microfilarial IgE in an antigen-specific way. This enables the survival of microfilariae. Peripheral blood lymphocytes taken from teenagers born to microfilaremic mothers do not proliferate *in vitro* and produce **interferon-γ (IFN-γ)** to the same extent as controls. There is a dominance of **Th2** responses over **Th1** responses, with potentially **interleukin** 10 (IL-10) and **transforming growth factor-β (TGF-β)** down-regulating Th1 responses. At the molecular level there is down-regulation of the expression of **Toll-like receptors** on T cells (TLR1, TLR2, TLR4, and TLR9), which means that they are less responsive to stimulation through these receptors. Thus by various possible means the immune response to infection is down-regulated through childhood and adolescence. This results in persistent infection and microfilaremia.

Immune responses do not appear to be effective in killing adult worms. However, cumulative exposure to adult worms, alive or dying, stimulates immune responses to adult worm antigens. These may reach a stage where intense inflammatory reactions occur around adult worms in lymphatics. This could contribute to lymphatic pathology. This will not be the only contributory mechanism. Adult worms themselves can cause dilatation of the lymphatics, which is called **lymphangiectasia**, without an accompanying inflammatory response. Secondary bacterial infections have been shown to contribute to the maintenance of lymphatic pathology.

A small minority of individuals experience a battle between microfilariae and the immune response which takes place in the lung. The clinical condition of tropical pulmonary eosinophilia is the result.

There is variation between individuals and populations in the pathogenesis of disease. There is a genetic component to immune responses, with familial aggregation of certain forms of the clinical disease. People who migrate into endemic areas do not have the possible benefit of a modified immune response from childhood. They experience stronger immune responses and more lymphatic pathology. The same applies to those in endemic areas not born to microfilaremic mothers.

3. What is the typical clinical presentation and what complications can occur?

There is a spectrum of clinical manifestations. A majority of individuals, about 70%, remain asymptomatic while others experience complications related to the adult worms or the microfilariae.

The most common clinical manifestation of lymphatic filariasis is **hydrocoele**, whereby adult *W. bancrofti* localize in the scrotal lymphatics and lymphatics of the spermatic cord. Adult worms can also cause acute **adenolymphangitis**, accompanied by fever. Episodes of acute adenolymphangitis probably relate to the death of adult worms and settle over a week, but can recur. The lymphatic channels become damaged and dilate (lymphangiectasia). Poor lymphatic drainage predisposes to secondary bacterial infection. When bacteria inflame afferent lymphatics red streaks appear along their course (**lymphangitis**) and further lymphatic damage occurs. Eventually **lymphedema** can develop in the limbs or genitals, which can lead to **elephantiasis** (Figures 1 and 4). Lymphedema stretches the skin and causes it to thicken, with underlying **fibrosis**. Pushing a finger into the swollen leg does not create an indented pit. Cracks in the skin can lead to further secondary bacterial infection, aggravating the damage already done to the lymphatics. Sometimes blocked lymphatics within the abdomen can result in lymph discharging into the urinary tract. Urine takes on a milky appearance due to the fat content of lymph. This is called **chyluria**.

The presence of microfilariae in the bloodstream is in general asymptomatic but may rarely be associated with intermittent, nocturnal fevers. A strong immune response may be manifest in the lung. As microfilariae pass through the lung they cause an inflammatory reaction, with fever, cough, and wheeze and widespread inflammatory infiltrates on a chest X-ray. This is most pronounced at night. The eosinophil count is raised and there is a high titer of antifilarial antibodies. This respiratory manifestation is called tropical pulmonary eosinophilia (TPE) and is usually observed in Asia rather than Africa or South America. TPE is actually quite rare and affects about 1% of infected subjects. In TPE microfilariae are actually scanty in the peripheral blood and are not found on blood films. Recurrent bouts of TPE damage the lungs with fibrosis. Bouts can be curtailed with a microfilaricidal drug, diethylcarbamazine (see below).

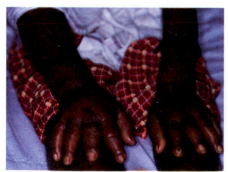

Figure 4. Lymphedematous arms.

4. How is the disease diagnosed and what is the differential diagnosis?

Elephantiasis once established is unmistakable, but earlier stages of infection have differential diagnoses and can warrant specific investigation.

Adult worms are small and are rarely found in tissue. Careful ultrasound of the scrotum may sometimes reveal wriggling adults in lymphatic vessels. Microfilariae may be found in chyluria, but parasitological diagnosis is usually based on finding microfilariae in a peripheral blood film. This requires nocturnal blood sampling in areas where lymphatic filariasis is nocturnally periodic. One can provoke microfilariae to appear in the bloodstream during the day by giving the drug diethylcarbamazine (DEC), and taking blood 30–60 minutes later. The DEC provocation test is as sensitive as a nocturnal blood film. As filariasis is a chronic infection microfilariae may be present in the blood over prolonged periods. In endemic areas microfilariae may be present in the blood when subjects succumb to another infection, such as malaria. Thus a positive microfilaria blood film does not necessarily mean that the current illness is due to filariasis.

Serological tests for antifilarial antibodies have suffered from poor sensitivity and specificity. In endemic areas it is difficult to distinguish between current and past infection. Tests have suffered from cross-reactions with other parasites. There are now tests for filarial antigen. If positive they indicate current infection. Antigen is present in the blood in the day and the night and there is no need for nocturnal blood sampling. Current tests are highly sensitive and specific for Bancroftian filariasis. There are two commercial antigen tests. One is an **enzyme-linked immunosorbent assay** (**ELISA**) and the other an immunochromatographic card test.

Across the clinical spectrum of filariasis there are differentials for different manifestations, including intermittent fevers, **lymphadenopathy**, **edema**, cough, wheeze, and eosinophilia. **Febrile** illness may be due to a long list of infections. Those that do not settle spontaneously include malaria, tuberculosis, and HIV. The differential diagnosis of lymphadenopathy is discussed with the *Toxoplasma* case. A swollen limb may be due to local **cellulitis**, a deep vein thrombosis, heart failure or low protein states from **nephrotic syndrome** or **cirrhosis**. Edema once it has reached the stage of elephantiasis in most endemic areas is unmistakable. In certain parts of Africa walking barefoot introduces silicates and minerals into the skin. When these pass along the lymphatics and to lymph nodes there is a fibrotic reaction that can cause lymphatic obstruction and gross swelling. This is called **podoconiosis** and can be as severe as elephantiasis. Intermittent cough, wheeze, and eosinophilia have to be differentiated from asthma, some connective tissue diseases, **allergic alveolitis**, and lung migratory phases of intestinal nematode infections. Eosinophilia is discussed with the *Schistosoma* case.

5. How is the disease managed and prevented?

Management

There are four key drugs that have been used to treat filariasis, and each has differing effects on microfilariae and on adult worms. Drug efficacy can be monitored by blood films, ultrasound examination for adult worms, and antigen tests.

DEC has been used for several decades. It kills about 70% of microfilariae and about 50% of adult worms. It has been the backbone of attempts to eradicate filariasis. It has to be used in repeated annual cycles to reduce transmission and adult worm numbers. Ultimately eradication of infection requires death of all adult worms, either naturally or through treatment. DEC can cause headaches and gastrointestinal upset. Dying adult worms excite episodes of adenolymphangitis. DEC is used to abort episodes of TPE due to microfilariae in the lung.

Albendazole can be used with DEC or ivermectin. Ivermectin is very good at killing microfilariae in the short term but has no effect on adult worms. Microfilariae therefore reappear in the blood rapidly.

The most recent and exciting development in the treatment of filariasis comes from the use of doxycycline. Living within adult *W. bancrofti* are bacteria endosymbionts called *Wolbachia*. *Wolbachia* can be killed with the antibiotic doxycycline and if this antibiotic is used in patients with filariasis adult *W. bancrofti* die. In a trial of an 8-week course of doxycycline there was an 80% reduction in adult worms and a loss of microfilariae in the blood 14 months after treatment. Doxycycline is avoided in pregnancy, breast-feeding mothers, and children under 12 years because of its effects on bones and teeth. These features preclude the use of doxycycline in mass treatment programs but a search for alternative antibiotics active against *Wolbachia* is being pursued. The use of doxycycline in the case history is not typical of treatment worldwide.

Beyond specific antifilarial drugs subjects with lymphedema need a lot of help and support in the care of their swollen limb(s) (Figure 5). Secondary bacterial infections need to be avoided. This requires attention to skin care, elevation of the leg to reduce swelling if possible, and prompt use of antibiotics if infection occurs. Sometimes surgery has been employed, but with limited success for swollen legs. The most effective treatment for patients suffering from hydrocoele is surgery.

Prevention

There is now a collaboration of public and private parties to eliminate lymphatic filariasis. This is called the Global Alliance to Eliminate Lymphatic Filariasis (GAELF). The mainstay of elimination efforts is periodic mass chemotherapy. Entire, defined populations are given chemotherapy irrespective of whether they are microfilaremic or amicrofilaremic. This is simpler than testing every individual by blood films. Repeated administration at annual intervals is intended to reduce levels of microfilaremia so that it is less likely for mosquitoes to transmit infection. A high level of population coverage is required for this to work. Administration must also be repeated for a number of years. How many

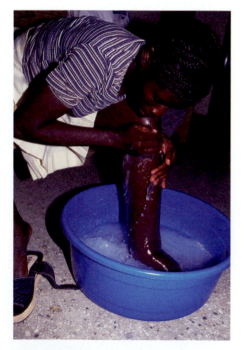

Figure 5. Woman carefully washing her lymphedematous leg to reduce the chances of secondary bacterial infection.

years depends on the efficacy of the drugs used in killing adult worms, but should be at least 4–6 years. Traditionally DEC has been the backbone of mass chemotherapy, to which now albendazole may be added. In areas where **onchocerciasis** is present DEC causes severe reactions in individuals co-infected with *Onchocerca volvulus*. Therefore in these areas ivermectin is used in preference to DEC.

Mass chemotherapy alone is unlikely to eliminate lymphatic filariasis transmission. Mosquito control also has a part to play in the elimination of lymphatic filariasis. This has a broader context in the prevention of malaria and other mosquito-borne diseases. Mosquitoes only live for several weeks. Without the use of insecticides, disease-bearing mosquitoes will die naturally. As long as there is another generation of mosquitoes some will become vectors for continued transmission of disease. Prevention of mosquito breeding has to be highly effective to avoid this new generation of vectors. They will not become vectors if there is no human reservoir of disease from which to feed. However, rapid, complete eradication of the microfilaremic human reservoir is not possible and efforts in mosquito control and mass chemotherapy work better combined than in isolation.

SUMMARY

1. What is the causative agent, how does it enter the body and how does it spread a) within the body and b) from person to person?

- Bancroftian, lymphatic filariasis is caused by *Wuchereria bancrofti*, a tissue-dwelling round worm or nematode.

- Male and female adult worms live in lymphatics.

- Female worms release larvae called microfilariae which enter the bloodstream.

- Various species of mosquitoes ingest microfilariae in the blood and pass infection onto others.

- *Wuchereria* is only found in humans, and there is no animal reservoir.

2. What is the host response to the infection and what is the disease pathogenesis?

- In endemic areas pregnant mothers may have chronic filarial infection and the fetus is exposed to filarial antigen. This early exposure may modify the subsequent immune response in children.

- IgG4 antibodies may block the effects of IgE in an antigen-specific manner.

- Th1 and Th2 responses can be down-regulated in children and adolescents.

- Repeated exposure to third stage larvae stimulates immunity, which may prevent further infection in older subjects.

- Some adult individuals also develop immune responses to worms that cause lymphatic pathology.

- In Asia there are occasional individuals who mount a strong anti-microfilaremic response, which causes intermittent pulmonary pathology with a marked peripheral eosinophilia.

3. What is the typical clinical presentation and what complications can occur?

- Infection is often asymptomatic.

- Adult worm death can cause intermittent inflammatory episodes of lymph nodes and lymphatics.

- Lymphatic damage leads to lymphedema and potentially gross swelling of dependent parts.

- Microfilariae in the blood can cause intermittent fevers.

- Tropical pulmonary eosinophilia is the clinical manifestation of a strong immune response to microfilariae with fever, cough, wheeze, and infiltrates on a chest X-ray.

4. How is the disease diagnosed and what is the differential diagnosis?

- The presence of microfilariae in a blood film indicates current infection.

- Microfilariae may be present in the blood over long periods and therefore care must be taken in attributing the current illness to filariasis.

- Filarial antigen tests, if positive, also indicate current infection.

- In endemic areas gross lymphedema and hydrocoele is usually diagnostic of lymphatic filariasis.

5. How is the disease managed and prevented?

- Diethylcarbamazine (DEC) and ivermectin kill microfilariae. These are often used in conjunction with albendazole to maximize killing of adult worms.

- Doxycycline can kill adult worms through its effect on the endosymbiont bacteria, *Wolbachia*. When these bacteria are killed the adult worms die.

- Patients complicated with gross lymphedema require careful attention to their swollen limbs and prevention of secondary infection.

- Control programs have involved repeated, annual community-wide use of DEC or ivermectin (in onchocerciasis areas) to reduce microfilarial numbers for transmission by mosquitoes.

FURTHER READING

Nutman TB. Lymphatic Filariasis. Tropical Medicine: Science and Practice. Imperial College Press, London, 2000.

REFERENCES

Babu S, Blauvelt CP, Kumaraswami V, Nutman TB. Diminished T cell TLR expression and function modulates the immune response in human filarial infection. J Immunol, 2006, 176: 3885–3889.

Babu S, Blauvelt CP, Kumaraswami V, Nutman TB. Regulatory networks induced by live parasites impair both Th1 and Th2 pathways in patent lymphatic filariasis: implications for parasite prevalence. J Immunol, 2006, 176: 3248–3256.

Taylor MJ, Makunde WH, McGarry HF, et al. Macrofilaricidal activity after doxycycline treatment of *Wuchereria bancrofti*: a double-blind, randomised placebo controlled trial. Lancet, 2005, 365: 2116–2121.

WEB SITES

A-WOL Consortium: www.a-wol.com/

Centers for Disease Control & Prevention, Atlanta, GA, USA: www.cdc.gov/

Global Programme to Eliminate Lymphatic Filariasis: www.filariasis.org/

Health Protection Agency: www.hpa.org.uk

NGO Amaury Coutinho: www.amaurycoutinho.org.br

World Health Organization: www.who.int

MULTIPLE CHOICE QUESTIONS

The questions should be answered either by selecting True (T) or False (F) for each answer statement, or by selecting the answer statements which best answer the question. Answers can be found in the back of the book.

1. **Which of the following are true statements about *W. bancrofti*?**

    A. It is a tapeworm (cestode).

    B. Adult worms live in lymphatics.

    C. Larvae (microfilariae) released from adult worms are ingested from lymphatics by mosquitoes.

    D. Circulating microfilarial numbers are greatest around mid-day.

    E. Adult worms have a lifespan that can reach 8 years.

2. **Which of the following are true statements about the transmission of *W. bancrofti*?**

    A. There is an animal reservoir.

    B. Only female *Anopheles* mosquitoes are capable of transmission.

    C. Microfilariae have to molt in the mosquito before transmission is possible.

    D. Transmission from humans can only occur 8 months or longer after infection.

    E. Microfilariae are inoculated into blood vessels when mosquitoes take a blood meal.

3. **Which of the following are true statements about the immune response in filariasis?**

    A. Repeated exposure to infectious mosquito bites dampens the immune response.

    B. There is an absence of a cell-mediated immune response.

    C. High IgG4 levels may block potentially beneficial IgE-mediated degranulation of eosinophils and mast cells.

    D. A weak immune response results in a failure to control infection and considerable symptoms and pathology.

    E. Down-regulation of Toll-like receptors decreases stimulation of T lymphocytes.

4. **Which of the following are true statements about protective immunity to filariasis?**

   A. Immunity to infection is directed against the L3 larvae.

   B. Strong immune responses against adult worms can cause more pathology than protection to hosts.

   C. Strong immune responses against microfilariae reduce the probability of mosquitoes ingesting microfilariae.

   D. There is a genetic component to protective immunity.

   E. Children born to microfilaremic mothers are protected against infection with L3 larvae.

5. **Which of the following are true about the clinical features of filariasis?**

   A. The death of adult worms results in resolution of lymphedema.

   B. Adult worms cause lymphedema through inflammatory damage to lymphatics.

   C. Lymphedema is an inevitable consequence of all infections.

   D. Chyluria occurs when lymph appears in the urine.

   E. Secondary bacterial infections can complicate lymphedema.

6. **Which of the following are true of tropical pulmonary eosinophilia (TPE)?**

   A. There is a low titer of antifilarial antibodies.

   B. Fever, cough, and wheeze are accompanied by infiltrates on chest X-ray.

   C. This manifestation is mainly seen in Asia.

   D. During exacerbations high numbers of microfilariae are seen on a blood film.

   E. Long-term lung damage occurs with repeated bouts of TPE.

7. **Which of the following are true of the diagnosis of filariasis?**

   A. The presence of microfilariae in a blood film is conclusive evidence that filariasis is responsible for a bout of febrile illness.

   B. Adult worms may be visualized in lymphatic vessels, especially in the scrotum.

   C. Microfilariae are preferably found in blood films.

   D. Diethylcarbamazine can provoke the appearance of microfilariae in the blood 30–60 minutes after administration.

   E. Serological tests for antibody are highly sensitive and specific.

8. **Which of the following are true of the differential diagnosis of filariasis?**

   A. A lymphedematous leg may also arise from walking barefoot in certain parts of Africa.

   B. In endemic areas gross lymphedema, elephantiasis, is invariably due to filariasis.

   C. Intermittent cough and wheeze in TPE have to be differentiated from asthma.

   D. Lymphedema due to filariasis can be distinguished from other causes because it is always unilateral.

   E. Lymphadenopathy in the neck is usually due to other diagnoses.

9. **In the treatment of filariasis which of the following are true?**

   A. Diethylcarbamazine does not kill microfilariae.

   B. A single dose of diethylcarbamazine kills all adult worms.

   C. Albendazole acts synergistically with diethylcarbamazine to kill adult worms.

   D. Ivermectin only kills microfilariae and reduces numbers in the short term.

   E. Prompt treatment of secondary bacterial infections has little effect on long-term prognosis.

10. **Which of the following are true of the control of filariasis?**

    A. Mass chemotherapy is used for individuals who are microfilariae blood film positive.

    B. In onchocerciasis-endemic areas diethylcarbamazine is preferred to ivermectin for mass chemotherapy.

    C. Mass chemotherapy is performed in repeated annual cycles.

    D. Mosquito control measures must be combined with mass chemotherapy for maximum effect.

    E. Doxycycline may now be used safely for future mass chemotherapy programs.

Glossary

α/β T cells
the main T cell type that recognises peptide antigens in association with major histo-compatibility antigens (HLA) class I or HLA class II; uses two chains for the T cell receptor – α and β chains

α-defensins
see entry for defensins

γ/δ T cells
T cells that use γ and δ chains for their receptors; these cells do not recognise processed peptides presented on classical MHC molecules but have been reported to recognise whole protein structures, stress proteins, small molecules produced by some bacteria and even lipids

abscess
a localized collection of pus in any part of the body that is surrounded by swelling (inflammation)

acidophile
a bacterium that grows well under acid conditions

acrodermatitis chronica atrophicans
unusual, progressive, fibrosing skin process due to the effect of continuing active infection with *Borrelia afzelii*

acute rheumatic fever (ARF)
an inflammatory disease of the connective tissue, particularly the heart, joints, blood vessels, subcutaneous tissue and central nervous system that occurs about three weeks following GAS pharyngitis in susceptible subjects

acute-phase proteins
a group of proteins (produced largely by the liver) that appear in the blood in increased amounts shortly after the onset of infection or tissue damage; they include C-reactive protein, serum amyloid, fibrinogen, mannose-binding protein, proteolytic enzyme inhibitors, and transferrin

Addison's disease
a disorder that occurs when the adrenal glands produce less of their hormones; caused by damage to the adrenal cortex through infection, autoimmunity and tumours

adenitis
inflammation of a gland

adenocarcinoma
a malignant tumor originating in glandular tissue: carcinoma – derived from epithelial cells

adenolymphangitis
inflammation of the lymph nodes and the lymphatics

adenopathy
large or 'swollen' lymph nodes: see lymphadenopathy entry

adhesins
these are bacterial proteins that promote adherence to host-cell membranes. Some are simply anchored in the bacterial membrane, whereas others are placed at the tips of pili. Adhesion is a prerequisite for entry into the body and effective adhesion may be a virulence factor

adult respiratory distress syndrome (ARDS)
a diffuse pulmonary parenchymal injury associated with noncardiogenic pulmonary edema and resulting in severe respiratory distress and hypoxemic respiratory failure. The pathologic hallmark is diffuse alveolar damage (DAD)

allergic alveolitis
inflammation of the small airways or alveoli caused by an immunological reaction to an inhaled bioaerosol (a mist or dust of biological particles), or some organic chemicals

alopecia
baldness

alternative complement pathway
complement activation pathway not initiated by antibody but directly through a variety of antigens including bacterial lipopolysaccharide and some virus components

alveolar macrophages
macrophages in the lung alveoli; part of the mononuclear phagocyte system

amyloidosis
a disorder that results from the abnormal deposition of a particular protein (amyloid), in various tissues of the body; amyloid proteins include immunoglobulin light chains in patients with myeloma

anaphylaxis
an immediate (within 30 min of allergen challenge) severe immune reaction mediated primarily by IgE antibodies, resulting in vasodilation and constriction of smooth muscle that may lead to death

anemia
strictly defined as a decrease in red blood cell (RBC) mass; three main classes include excessive blood loss (acutely such as a hemorrhage or chronically through low-volume loss), excessive blood cell destruction (hemolysis) or deficient red blood cell production in the bone marrow (ineffective hemopoiesis)

aneurysm
an aneurysm occurs when part of an artery swells. It is caused by a damaged blood vessel or a weakness in the blood vessel wall. The pressure of blood in the artery causes it to 'balloon' out at the weak point. The swelling can be small and spherical (berry-sized), normally occurring near blood vessel branches, or it can be larger and balloon-like

anhidrosis
the inability to sweat in response to heat

antibody isotypes
can be heavy chain or light chain isotypes; often used synonymously for classes of antibody (IgM, IgG, IgA, IgD and IgE; each isotype is encoded by separate immunoglobulin constant region genes μ, γ, α, δ and ε for heavy chains and κ and λ for light chains

antigenic shift
a sudden shift in the antigenicity of a virus resulting from the recombination of the genomes of two viral strains; seen only with influenza A viruses it usually results from the replacement

of the hemagglutinin with a novel subtype that has not been present in human influenzaviruses for a long time; the consequences of the introduction of a new hemagglutinin into human viruses is usually a pandemic, or a worldwide epidemic

antigenic variation
changes in cell surface antigenic determinants of a microorganism in order to avoid the host immune response

antimicrobial peptides
small molecular weight proteins with a broad spectrum antimicrobial activity; produced by a variety of cell types, as part of the innate immune response

aortic coarctation
narrowing of the aorta between the upper-body artery branches and the branches to the lower body (at the insertion of the ductus arteriosus)

aortic stenosis
narrowing of the aortic valve in the heart

aplastic anemia
a syndrome of bone marrow failure characterized by peripheral pancytopenia (reduced red blood cells but also white blood cells and platelets) and marrow hypoplasia

apoptosis
programmed death of a cell by 'suicide': involves enzymatic digestion of the DNA into packages: governed by complex signaling pathways and marked by a well-defined sequence of morphological changes, resulting in small bodies that are phagocytosed by other cells

arthralgia
joint pain

arthritis
a term used to describe conditions affecting the joints; includes degeneration of the cartilage (osteoarthritis), inflammatory autoimmune arthritis (rheumatoid arthritis) and directly induced by microbes (septic arthritis)

Arthus reaction
a Type III hypersensitivity reaction; are immune complex-mediate antibody complexes mainly in the vascular walls, serosa (pleura, pericardium, synovium), and glomeruli

aspergilloma
a clump of fungus which populates a lung cavity (also known as a mycetoma or fungus ball)

asthma
also called hyper-reactive airway disease; a chronic inflammatory disorder characterized by generalized but reversible bronchial airways obstruction; can be precipitated by non-specific factors, irritants including physical exertion, cold, smoke, scents and pollution; specific factors that can precipitate asthma include allergens in the form of pollen, dust, animal fur, mold and some kinds of food. Some viruses and bacteria, chemical fumes or other substances at the workplace and certain medicines, e.g. aspirin and other NSAIDS

atelectasis
the collapse of part or all of a lung; caused by a blockage of the bronchus or bronchioles or by pressure on the lung

atopic asthma
asthma associated with specific IgE antibodies

atopic dermatitis
a common chronic skin disease (also called eczema) usually occurring in individuals with a hereditary predisposition to all atopic disorders; frequently seen in association with a high serum IgE level

atrophicus
a chronic skin disease that is characterized by the eruption of flat white hardened papules with central hair follicles often having black keratotic plugs

atrophy
wasting of tissues, organs, or the entire body, as from death and reabsorption of cells, diminished cellular proliferation, decreased cellular volume, pressure, ischemia, malnutrition, lessened function, or hormonal changes

auscultation
listening for sounds made by internal organs, such as the heart and lungs, to aid in the diagnosis of certain disorders

autophagy
the process by which cells digest parts of their own cytoplasm; allows for both recycling of macromolecular constituents under conditions of cellular stress and remodeling the intracellular structure for cell differentiation

azoospermia
absence of live spermatozoa in the semen

bacteremia
the presence of bacteria in the bloodstream

bacteriophage
a virus that infects bacteria, also called a phage

bacteriuria
bacteria in the urine: a number of automated laboratory tests are available to detect/quantitate bacteriuria, these include turbidometric, bioluminescence, electrical impedance, flow cytometry and radiometric tests

Behcet's syndrome
a rare autoimmune disorder that causes ulcers and skin lesions

Bell's palsy
an idiopathic unilateral facial nerve paralysis, usually self-limiting

bilirubin
bilirubin is a product that results from the breakdown of hemoglobin: hemoglobin is broken down to heme and globin. Heme is converted to bilirubin, which is then carried by albumin in the blood to the liver

Blackwater fever
a complication of malaria characterized by intravascular hemolysis, hemoglobinuria and kidney failure

Bornholm's disease
infection of the chest wall musculature giving rise to severe chest pain

bradycardia
a resting heart rate of under 60 beats per minute in adults, though it is seldom symptomatic until the rate drops below 50 beat/min; resting heart rate under 100 beats per minute for children

Brill-Zinsser disease
reappearance of epidemic typhus years after the initial attack. The agent that causes epidemic typhus (*Rickettsia prowazekii*) remains viable for many years and then when host defenses are down, it is reactivated causing recurrent typhus

bronchiectasis
a chronic lung disease characterized by permanent, irreversible

progressive, abnormal dilitation and destruction of the bronchi that results in poor clearance and pooling of mucus. This disorder also makes a person more susceptible to infections caused by bacteria

bronchiolitis
inflammation of the bronchioles, the airways that extend beyond the bronchi and terminate in the alveoli. Bronchiolitis is due to viral infections such as parainfluenza, influenza, adenovirus and, especially, respiratory syncytial virus (RSV)

bronchoalveolar lavage
a diagnostic procedure of washing a sample of cells and secretions from the alveolar and bronchial airspaces

bronchoscopy
the examination of the bronchi (the main airways of the lungs) using a flexible tube (bronchoscope). Bronchoscopy helps to evaluate and diagnose lung problems, assess blockages, obtain samples of tissue and/or fluid, and/or to help remove a foreign body

bruit
a sound, especially an abnormal one: a bruit may be heard over an artery reflecting turbulence of flow. Listening for a bruit in the neck is a simple, safe, and inexpensive way to screen for stenosis (narrowing) of the carotid artery

bullus/bullae
a blister of more than 5 mm diameter with thin skin and full of fluid

Burkitt's lymphoma
an uncommon type of Non-Hodgkin Lymphoma (NHL); commonly affects children; it is a highly aggressive type of B-cell lymphoma; associated with Epstein-Barr virus

cachexia
weight loss, wasting of muscle, loss of appetite, and general debility that can occur during a chronic disease

candidiasis
an infection caused by the fungus *Candida* (also called thrush)

capnophilic
term used to describe bacteria that grow best in an atmosphere of increased carbon dioxide (5–10% CO_2 in air)

capsid
the outer protein coat of a virus surrounding and protecting the nucleic acid; made up of subunits called capsomeres; the outer proteins of the capsid also mediate attachment of the virus to specific receptors on the host cell surface

cardia
the opening of the esophagus into the stomach; also the part of the stomach adjoining this opening

cardiomegaly
enlargement of the heart

carditis
inflammation of the heart; can be endocarditis (inflammation of the endocardium), myocarditis (inflammation of the heart muscle) or pericarditis (inflammation of the pericardium)

caseation
necrosis with conversion of damaged tissue into a soft cheesy substance

catalase
an enzyme that catalyzes the decomposition of hydrogen peroxide to water and oxygen

CATT
Card Agglutination Test for Trypanosomiasis

cavitation
the formation of cavities in a body tissue or an organ, especially those cavities that form in the lung as a result of tuberculosis

CD11/CD18
leukocyte adhesion molecules (β 2 integrins)

CD4+ T cells
helper T cells; T cells that recognize specific antigenic peptides when presented to them on MHC class II molecules on antigen presenting cells; helper T cells

CD59
also called lymphocyte function antigen - 3 (LFA-3); it binds to CD2

CD8+ T cells
cytotoxic T cells; T cells that recognize specific antigenic peptides when presented to them on MHC class I molecules on antigen presenting cells/target cells

cellulitis
an infection of the deep subcutaneous tissue of the skin

cerebellar ataxia
a disorder of coordination due to a lesion affecting the cerebellum; a symptom or syndrome rather than a diagnosis, with many possible causes

cerebritis
inflammation of the brain

cervicitis
inflammation of the cervix; a common infection of the lower genital tract

chagoma
a swelling resembling a tumor that appears at the site of infection in Chagas' disease

chancre
painless ulceration formed during the primary stage of syphilis; also associated with sleeping sickness

chancroid
a sexually transmitted infectious disease characterized by painful ulcers, bubo formation, and painful inguinal lymphadenopathy; caused by *Haemophilus ducreyi*, a gram-negative coccoid-bacillary rod

Charcot-Leyden crystals
crystals in the shape of elongated double pyramids, formed from eosinophils, found in the sputum in bronchial asthma and in other exudates or transudates containing eosinophils

chemokines
a class of cytokines; the main families of the four described thus far include the CC and the CXC group; they have both chemotactic and activating properties

cholecystitis
inflammation of the gallbladder that occurs most commonly because of an obstruction of the cystic duct

chorioretinitis
an exudative inflammatory process involving the retinal vessels usually caused by congenital viral, bacterial, or protozoan infections in neonates; congenital toxoplasmosis is the most common cause of infectious chorioretinitis in immunocompetent children; congenital cytomegalovirus (CMV) infection is the second

most common etiology; fungal infections are more frequently identified, and emergent pathogens, including West Nile virus and lymphocytic choriomeningitis virus, have been described. In rare instances, chorioretinitis is part of a systemic noninfectious process

choroid plexus
a vascular proliferation of the cerebral ventricles that serves to regulate intraventricular pressure by secretion or absorption of cerebrospinal fluid

chronic fatigue syndrome
also called myalgic encephalomyelitis (or encephalopathy; ME); a disorder of unknown etiology that probably has an infectious basis

chronic granulomatous disease
an inherited disorder of phagocytic cells, resulting from an inability of phagocytes to undergo a respiratory burst to produce bactericidal superoxide anions (O_2^-); leads to recurrent life-threatening bacterial and fungal infections

chronic interstitial lung disease
a heterogeneous group of disorders that distort the architecture of the lung parenchyma with inflammation and fibrosis; includes idiopathic pulmonary fibrosis, sarcoidosis, hypersensitivity pneumonitis, eosinophilic pneumonias, inorganic pneumoconioses, drug-induced lung diseases, connective tissue diseases affecting the lungs, pulmonary vasculitides, the sequelae of acute respiratory distress syndrome, and bronchiolitis, among others

chronic interstitial nephritis
a large and heterogeneous group of disorders characterized primarily by interstitial fibrosis with mononuclear leukocyte infiltration and tubular atrophy; there are many causes including urinary obstruction, chronic bacterial infections, drugs (lithium, Chinese herbs) and radiation

chronic obstructive pulmonary disease (COPD)
a disease state characterized by the presence of airflow obstruction due to chronic bronchitis or emphysema; airflow obstruction is usually progressive, may be accompanied by airway hyperreactivity, and may be partially reversible

chyluria
the presence of lymph and fat in the urine, giving it a milky appearance. Due to obstruction of lymph flow, which causes rupture of lymph vessels

cirrhosis
scarring of the liver involving the formation of fibrous (scar) tissue associated with the destruction of the normal architecture of the organ

clade
a group of biological taxa (as species) that includes all descendants of one common ancestor

colectomy
the surgical removal of the colon or part of the colon (partial colectomy)

colitis
inflammation of the large intestine (the colon). There are many forms of colitis, including ulcerative, Crohn's, infectious, pseudomembranous, and spastic

collagen vascular disease (CVD)
a heterogeneous group of autoimmune disorders of unknown etiology; many CVDs involve the lungs directly or involve them as a complication of treatment of the CVD. CVDs include systemic lupus erythematosus, rheumatoid arthritis, progressive systemic sclerosis or scleroderma, dermatomyositis and polymyositis, ankylosing spondylitis, Sjögren syndrome and mixed connective-tissue disease

commensal
the normal bacterial microbiota of the body that, unless displaced from the skin or mucosal surfaces, generally do not cause disease and may actually help protect against colonization with pathogenic bacteria. If commensals transgress the epithelial barriers they may become opportunistic pathogens

complement pathway
pathway leading to activation of complement; three pathways defined, the 'classical' involving antibody, the 'alternative' pathway that activates complement without the use of antibody (i.e. via innate immune system) and the 'lectin' pathway which activates complement via mannose binding protein

conjugate vaccine
vaccines that are made by coupling the protective 'weak' antigen (e.g. pneumococcal carbohydrate antigens) with a strong antigen (carrier protein – that can be recognised by the T helper cells) e.g. diptheria toxoid

conjunctivitis
an inflammation of the conjunctivae

corticosteroids
a class of drugs that have potent anti-inflammatory effects; derivatives of the steroid hormones that are produced in the adrenal cortex

CR2
complement receptor 2; also called CD21; found on follicular dendritic cells (where it attaches immune complexes) and B cells where it amplifies activation via the surface antibody receptor

C-reactive protein(CRP)
an acute phase protein produced in the liver; quantity found in the serum gives a measurement of systemic inflammation

Crohn's disease
a chronic inflammation of the digestive tract that can affect any part of the digestive tract from the mouth to the anus but usually involves the terminal part of the small intestine, the beginning of the large intestine (cecum), and the area around the anus

croup
also termed laryngotracheitis or laryngotracheobronchitis, is a viral respiratory tract infection; manifests as hoarseness, a seal-like barking cough and a variable degree of respiratory distress

CT scan
computed tomography scan

cyanosis
bluish discoloration of the skin and mucous membranes due to the presence of deoxygenated hemoglobin in blood vessels near the skin surface

cyst
an abnormal closed sac-like structure containing liquid, gaseous or semisolid substance that can occur in any part of the body e.g. sebaceous, ovarian, kidney or liver

cystic fibrosis
a disease of exocrine gland function involving multiple organ systems that mainly results in chronic respiratory infections, pancreatic enzyme insufficiency, and associated complications in untreated patients. Pulmonary involvement occurs in 90% of patients surviving the neonatal period. End-stage lung disease is the principal cause of death

cystic fibrosis transmembrane conductance receptor (CFTM)
a protein that functions as a chloride channel and is regulated by cyclic adenosine monophosphate (cAMP). Mutations in the CFTR gene give rise to cystic fibrosis that results in abnormalities of chloride transport across epithelial cells on mucosal surfaces. This leads to decreased secretion of chloride and increased reabsorption of sodium and water across epithelial cells. The resulting reduced height of epithelial lining fluid and decreased hydration of mucus results in mucus that is stickier to bacteria, which results in infection and inflammation

cystitis
inflammation of the bladder

cytokine storm
unregulated release of proinflammatory cytokines, such as TNF-α and IL-1, into the blood by monocytes and macrophages in respone to lipopolysaccharide (LPS) or lipooligosaccharide (LOS) from gram-negative bacteria

cytokines
small molecular weight proteins produced by different cell types; families classified in different ways but include the interleukins, colony stimulating factors, tumour necrosis factors, chemokines and interferons

cytolethal distending toxin
a bacterial toxin produced by a variety of pathogenic bacteria; it enters eukaryotic cells and breaks double-stranded DNA resulting in the arrest of the cell cycle and finally leads to apoptosis of the target cell

cytopathic
pertaining to or characterized by pathological changes in cells e.g. cytopathic viruses

decay accelerating factor
also called CD55; a 70kD protein present on the surface of nucleated body cells; prevents complement mediated lysis and acts as a receptor for some bacteria e.g. *Helicobacter pylori* on stomach epithelial cells

defensins
small cationic antimicrobial proteins that insert in lipid bilayers and form pores. Alpha defensins are produced by neutrophils, macrophages and Paneth cells, and β-defensins are produced by many kinds of leucocytes and epithelial cells

dendritic cells (DCs)
several families of cells with multiple processes (resembling dendrite processes of a neuron) to enable them to interact with several other cells at the same time

dengue fever
Dengue (DF) and dengue hemorrhagic fever (DHF) are primarily diseases of tropical and sub-tropical areas and are caused by four closely related but antigenically distinct serotypes of the genus *Flavivirus* (DEN1-4); transmitted to humans by the mosquito *Aedes aegypti*; produces a spectrum of clinical illness ranging from a non-specific viral syndrome to severe and fatal hemorrhagic disease

dermatome/dermatomal
localized area of skin that has its sensation via a single nerve from a single nerve root of the spinal cord

desmosomes
specialized cell junction characteristic of epithelial tissues into which intermediate filaments (tonofilaments of cytokeratin) are inserted; particularly present in tissues such as the skin that have to withstand mechanical stress

diathesis/diasthesia
a hereditary predisposition of the body to a disease, a group of diseases, or another disorder

disseminated intravascular coagulation (DIC)
not an illness in its own right but rather a complication or an effect of progression of other illnesses; a complex systemic thrombohemorrhagic disorder involving the generation of intravascular fibrin and the consumption of procoagulants and platelets resulting in intravascular coagulation and hemorrhage; most commonly observed in severe sepsis and septic shock (see entry for septic shock)

diuresis
increased formation of urine by the kidney

diverticulitis
inflammation around a diverticulum; diverticula are more commonly seen in the small or large intestine and consist of herniation of the mucosa and submucosa through the muscularis, usually at the site of a nutrient artery

Donovan body
intracytoplasmic, nonflagellated form of certain intracellular flagellates (e.g. *Leishmania*); also called amastigotes

Duffy binding proteins
parasite proteins that bind to Duffy antigens on erythrocyte surfaces and facilitate the invasion process in *Plasmodium* infections

dysasthesia
neuropathy

dysentery
diarrhea with blood

dyspepsia
indigestion

dysphagia
difficulty in swallowing

dyspnea
shortness of breath

dysrhythmias
irregularities in heart beats or brain waves

dysuria
burning micturition

ECHO scan
scan using ultrasound

echocardiogram
an ultrasound heart scan

ectopic pregnancy
also known as tubal pregnancy; when the fertilized egg implants in any tissue other than the uterine wall

edema
abnormal accumulation of fluid outside the blood vessels; may be a manifestation of cardiac failure

EDTA
ethylenediaminetetraacetic acid; a chelating agent that sequesters Ca⁺⁺ and Mg⁺⁺ ions

effusion
escape of fluid from the blood vessels or lymphatics into the tissues or a cavity

egophony
increased resonance of voice sounds

electrocardiogram (EEG)
the electrocardiogram (ECG or EKG) measures and records the electrical activity of the heart allowing the diagnosis of a wide range of heart conditions

electroencephalogram
test that measures and records the electrical activity of the brain

elephantiasis
the gross (visible) enlargement of the arms, legs, genitals or breasts to elephantoid size

ELISpot test
enzyme-linked immunosorbent spot test; an enzyme linked antibody test to enumerate antigen-specific cells usually by measuring a cytokine produced

embolism
an obstruction of a blood vessel, usually by a blood clot but can also be by an air bubble, amniotic fluid, a clump of bacteria, a chemical substance e.g. talc, and drugs

emphysema/emphysematous
a progressive lung disease in which there is abnormal permanent enlargement of air spaces distal to the terminal bronchioles, accompanied by the destruction of the walls and without obvious fibrosis

empyema
an inflammatory fluid (pus) and debris in the pleural space resulting from an untreated pleural-space infection that progresses from free-flowing pleural fluid to a complex collection in the pleural space

encephalitis
an inflammation of the brain parenchyma; presents as diffuse and/or focal neuropsychological dysfunction; caused by infection and autoimmune mechanisms

endemic
restricted or peculiar to a locality or region

endocarditis
inflammation or infection of the endocardium, which is the inner lining of the heart muscle and, most commonly, the heart valves usually caused by bacterial infection

endocervicitis
inflammation of the mucous lining of the uterine cervix

endocytosis
the process by which a cell engulfs a large particle (bacteria, viruses, etc.) and contains it in a vescicle in the cytoplasm

endometriosis
the presence of normal endometrial mucosa (glands and stroma) that implants at sites other than the uterine cavity; this tissue responds as normal endometrium resulting in microscopic internal bleeding, with the subsequent inflammatory response, neovascularization, and fibrosis formation

endometritis
infection of the endometrium or decidua, with extension into the myometrium and parametrial tissues; the most common cause of fever during the postpartum period

endophthalmitis
inflammation of the intraocular cavities (i.e., the aqueous or vitreous humor); usually caused by infection

endoscopy
a procedure where the inside of the body is investigated using an endoscope

endotoxin
a toxin that is part of the outer cell wall (outer membrane) of gram-negative bacteria; also called lipopolysaccharide

enteral
the route by which food or drugs are given via the mouth

enterocolitis
inflammation of the small intestine and colon

enterotoxin
a bacterial toxin that exerts its effects on the intestine e.g. staphylococcal enterotoxin B (SEB)

enzyme-linked immunosorbent assay (ELISA)/enzyme immunoassay
an enzyme-based system used to quantitate antigens or antibodies; enzymes coupled to antibodies include alkaline phosphatase, and peroxidase

eosinophilia
an abnormal increase in the number of eosinophils in the blood

epididymitis
inflammation of the epididymis can be acute (<6 wk) or chronic and is most commonly caused by infection

epididymo-orchitis
acute inflammatory reaction of the epididymis and testis secondary to infection

epiglottitis
inflammation of the epiglottis

episome
a portion of genetic material that can exist independent of the main body of genetic material (i.e. chromosome) at some times, while at other times is able to integrate into the chromosome

epitopes
an alternative name for antigenic determinants that are the smallest parts of antigens detected by antibodies or T cell receptors

erysipelas
an acute bacterial infection of the skin accompanied by lymphadenopathy, fever, chills and leukocytosis

erythema infectiosum
in children is a normal response to infection by human parvovirus-B19; acute infection in a host who is immunocompetent results in formation of immune complexes that are probably responsible for the deposition of the immune complexes in the skin and joints of individuals with this condition; also refered to as 'slapped cheek' disease or Fifth disease

erythema migrans
a bulls-eye rash classically 5–6.8 cm following infection with *Borrelia burgdorferi* delivered via a tick bite

erythema multiforme
a rash that results from an allergic response, most often secondary to a drug

erythema nodosum
a disorder characterized by the formation of tender, red nodules on the front of the legs; erythema nodosum mainly affects women and has been associated with certain infections and thought to be due to a hypersensitivity reaction

erythema/erythematous
skin redness produced by congestion of the capillaries, which may result from a variety of causes; see erythema nodosum etc.

erythrocyte sedimentation rate (ESR)
measures the time taken for erythrocytes to settle in blood plasma; test used to estimate the level of inflammation in patients with some autoimmune diseases

E-selectin
also called ELAM-1 and CD62E; a member of the selectin family of adhesion molecules; recognizes sialylated carbohydrate groups related to Lewis x or Lewis a family; used by T cells to enter inflamed skin

esophageal varices
abnormally enlarged veins in the lower esophagus and upper part of the stomach; they develop when normal blood flow in the hepatic portal vein is blocked; separate blood backs up into smaller, more fragile blood vessels in the esophagus, and often stomach or rectum, causing the vessels to swell; esophageal varices can rupture, causing a life-threatening condition

exocytosis
the process by which cellular products within vesicles are secreted from a cell when the vesicle membrane fuses with the plasma membrane

exotoxin
bacterial protein toxin that is secreted by pathogenic bacteria and contributes to infectious disease: examples include diptheria toxin, tetanus toxin

facultative anaerobe
an organism that can grow well in both the absence and the presence of oxygen equivalent to that of normal air (21%)

feco-oral route
transmission of infection from feces via the oral route

farmer's lung
also called extrinsic allergic alveolitis; an allergic disease resulting from deposition of immune complexes (type III hypersensitivity) in the lung following repeated inhalation of antigenic materials from actinomycete fungi (found in moldy hay)

fasciitis
inflammation of the fascia (lining tissue under the skin)

febrile
pertaining to or marked by fever; feverish

fibromyalgia
a common disorder and a syndrome composed of different signs and symptoms; now recognized as one of many central pain-related common syndromes; pain associated with fibrous tissue

fibrosis
formation of excessive fibrous tissue as the result of repair or a reactive process (e.g. during inflammation which becomes chronic); a synonym for scar formation

fimbriae
thin, hair-like appendages, 1 to 20 microns in length and present on the cells of gram-negative bacteria; unlike flagella, they are non-motile and possess antigenic and hemagglutinating properties; some fimbriae mediate the attachment of bacteria to cells via adhesins

fits
the clinical expression of an abnormal cerebral discharge – a dysrhythmia; may be convulsive (stiffening or jerking) or non-convulsive (as in absence or complex partial seizures)

flagella typing
typing of flagellar H antigens by serology or PCR

Flavivirus
a genus of the family *Flaviviridae*; genus includes the West Nile virus, dengue virus, Tick-borne Encephalitis Virus, Yellow Fever Virus, and several other viruses which may cause encephalitis

fomites
inanimate objects that carry microbes leading to spread of infections e.g. cups, spoons, surgical instruments etc.

fundoscopy
examination of the optic disc, retina and blood vessels using an ophthalmoscope

G6PD deficiency
glucose-6-phosphatase dehydrogenase (G6PD) deficiency is an enzyme deficiency of erythrocytes; inherited as an X-linked disorder, G6PD deficiency affects 400 million people worldwide (about 10% of the world population); leads to lysis of erythrocytes – hemolytic anemia; it confers protection against malaria, which probably accounts for its high gene frequency; affects all races

gastritis
inflammation or irritation of the stomach mucosa; can be acute or chronic; can be caused by alcohol, aspirin and *Helicobacter pylori* (chronic)

gastroenteritis
a nonspecific term for various pathological states of the gastrointestinal tract; primary manifestation is diarrhea, but may be accompanied by nausea, vomiting, and abdominal pain

G-CSF
granulocyte colony stimulating factor; a cytokine produced by macrophages, fibroblasts and other cells; have an effect on stem cells; used to enhance the development of neutrophils when given with stem cell transplants; also injections used to drive CD34+ stem cells into the blood stream for collection for stem cell donation

Glasgow coma scale score
the measurement used to quantify the level of consciousness following traumatic brain injury

glaucoma
a nonspecific term used for several ocular diseases that ultimately result in increased intraocular pressure (IOP) and decreased visual acuity resulting from the effect of the pressure on the optic nerve

GM-CSF
granulocyte-monocyte colony stimulatng factor; a cytokine produced by T cells, macrophages and endothelial cells in response to infection/inflammation; results in specific increase in granulocytes and monocytes in the bone marrow from their stem cells to aid in fighting the infection

Gortex tube
Gortex is a biomaterial; tubes made of Gortex are used to carry blood in heart surgery

gout
a common disorder of uric acid metabolism (found in 1% of the population) that can lead to deposition of monosodium urate (MSU) crystals in soft tissue, recurrent episodes of debilitating joint inflammation, and, if untreated, joint destruction and renal damage

granulocytopenia
a reduced number of granulocytes in the circulation – neutrophils, eosinophils and basophils

granuloma
an accumulation of macrophages, epithelioid cells, with or without lymphocytes; can result from macrophages being activated by chronic bacterial infections such as tuberculosis, leprosy and syphilis; similar but not identical structures are induced in some parasitic infections such as schistosomiasis and by non-infectious agents such as asbestos

granulysin
a protein present in the acidic granules of human NK cells and cytotoxic T cells with antimicrobial activity against a broad spectrum of microbial pathogens; when released acts directly on the microbial cell wall/membrane (e.g. of mycobacteria) leading to increased permeability and lysis

granzymes
granule-associated enzymes in cytotoxic T cells and NK cells that are involved in inducing apoptosis of virus infected cells

guarding (abdominal)
tensing of the abdominal wall muscles to guard inflamed organs within the abdomen from the pain when abdominal wall is pressed; characteristic finding in the physical examination for appendicitis or diverticulitis

Guillain-Barré syndrome
a heterogeneous group of immune-mediated processes generally characterized by motor, sensory, and autonomic dysfunction. Classically an acute inflammatory demyelinating polyneuropathy characterized by progressive symmetric ascending muscle weakness, paralysis, and hyporeflexia with or without sensory or autonomic symptoms; in severe cases, muscle weakness may lead to respiratory failure

gynecomastia
a benign enlargement of the male breast resulting from a proliferation of the glandular component of the breast

hemagglutinin
a substance that causes agglutination of erythrocytes; an integral envelope protein of influenza virus; responsible for binding of the virus to the host cell and subsequent fusion of viral and host membranes in the endosome after the virus has been taken up by endocytosis

hematuria
the presence of erythrocytes in the urine

hemoglobinuria
the presence of free hemoglobin in the urine

hemolysin
exotoxins produced by some bacteria that damage or destroy cells; alpha hemolysin is a sphingomyelinase produced by staphylococci. It produces partial hemolysis of sheep and cattle erythrocytes and appears to have little pathogenic effect; β hemolysin (produced by streptococcus) results in complete lysis of the erythrocytes around the bacteria colonies

hemolysis
used in the empirical identification of microorganisms based on the ability of bacterial colonies grown on agar plates to break down red blood cells in blood agar plates; on growing the organism it can be classified as to whether or not it has caused hemolysis in the red blood cells (RBCs) incorporated in the medium; of particular importance in the classification of streptococcal species that produce β hemolysin

hemolytic anemia
an autoimmune disease where autoantibodies bind to the patient's erythrocytes and cause lysis (mediated by complement) and removal via opsonization

hemolytic ureamic syndrome (HUS)
a disease mainly of infancy and early childhood; characterized by the triad of microangiopathic hemolytic anemia, thrombocytopenia, and acute renal failure; the most common cause of acute renal failure in children; diarrhea and upper respiratory infection are the most common precipitating factors of HUS

hemophagocytosis
pathologic finding of activated macrophages, engulfing erythrocytes, leukocytes, platelets, and their precursor cells

hemozoin
produced during the intra-erythrocytic phase of *Plasmodium* infection by digestion of hemoglobin; hemozoin may play a role in malaria-associated immunosuppression, affecting both the antigen processing and immunomodulatory functions of macrophages

hemiparesis
muscle weakness on one side of the body

hemiplegia
paralysis of one side of the body

hemoptysis
coughing up blood

Henoch-Schölein purpura
a small-vessel vasculitis characterized by immunoglobulin A (IgA), C3, and immune complex deposition in arterioles, capillaries, and venules; affects mostly children and involves the skin and connective tissues, gastrointestinal tract, joints, and scrotum as well as the kidneys

hepatitis
inflammation of the liver; may be acute or chronic; causative agents of acute hepatitis include hepatitis viruses A–E, CMV, some bacterial infections, some medicines e.g. paracetamol and toxins such as alcohol; chronic hepatitis can also be caused by hepatitis B, C and D viruses, some drugs, toxins and can be autoimmune

hepatocellular carcinoma
a primary malignant proliferation of hepatocytes

hepatoma
primary liver cancer beginning in the liver; most are hepatocellular carcinomas (HCC)

hepatomegaly
enlargement of the liver

hereditary spherocytosis
a familial hemolytic disorder with marked heterogeneity of clinical features, ranging from an asymptomatic condition to fulminant hemolytic anemia; the morphological hallmark of HS is the microspherocyte, which is caused by loss of membrane surface area, and an abnormal osmotic fragility in vitro

herpangina
painful oral ulceration

heterophile antibodies
antibodies that react with an antigen entirely different from and phylogenetically not related to the antigen responsible for their production e.g. antibodies that agglutinate sheep red blood cells in patients with infectious mononucleosis; caused by EBV that polyclonally activates B cells independent of their specificity; usually IgM

highly active antiretroviral treatment (HAART)
a combination of three or more active drugs to effectively treat patients with HIV

histiocytes
tissue macrophages

hydatid cyst
a cyst formed in the liver, or, less frequently, elsewhere, by the larval stage of *Echinococcus*

hydrocephalus
a disturbance in the formation, flow, or absorption of cerebrospinal fluid (CSF) that leads to an increase in fluid in the central nervous system

hydrocoele
enlargement of the scrotum as the result of accumulation of fluid in the coat around the testis

hydronephrosis
defined as a dilation of the renal pelvis and calyces; physiological response to interuption of the flow of urine that can be caused by infection

hydrops fetalis (fetal hydrops)
a serious condition of the fetus defined as abnormal accumulation of fluid in two or more fetal compartments, including ascites, pleural effusion, pericardial effusion, and skin edema

hypercholesterolemia
a condition where there is a high level of cholesterol in the blood

hyperemia
an increase in blood flow to any part of the body ususally due to vasodilation

hypergastrinemia
excessive secretion of gastrin

hypersplenism
increased activity of the spleen; can be caused by tumors, anemia, malaria, tuberculosis, and various connective tissue and inflammatory diseases

hypertension
abnormally high blood pressure

hypertrophy
an increase in cell size

hypoesthesia
partial loss of sensitivity to sensory stimuli; diminished sensation

hypoalbuminemia
decreased levels of albumin in the blood; can be caused by various conditions, including nephrotic syndrome, hepatic cirrhosis, heart failure, and malnutrition; however, most cases of hypoalbuminemia are caused by acute and chronic inflammatory responses

hypochlorhydria
a condition characterized by an abnormally low level of hydrochloric acid in the stomach

hypochondrium
the upper lateral region of the abdomen on either side of the epigastrium and below the lower ribs

hypogammaglobulinemia
an immunodeficiency disease marked by abnormally low levels of all classes of immunoglobulins, with increased susceptibility to infectious diseases; may be primary (congenital), or secondary (acquired)

hyponatremia
an abnormally low concentration of sodium in the blood

hypovolemia
an abnormal decrease in blood volume

hypoxemia
decreased amount of oxygen in the blood

hypoxia
deficiency in the amount of oxygen reaching tissues

ICAM-1
intercellular adhesion molecule 1; an adhesion molecule important in many immune processes; the receptor for rhinoviruses that cause the common cold

icterus
another term for jaundice; yellowing of the skin and the whites of the eyes caused by an accumulation of bilirubin in the blood; can be a symptom of gallstones, liver infection or anemia

idiopathic thrombocytopenic purpura (ITP)
also known as primary immune thrombocytopenic purpura and autoimmune thrombocytopenic purpura; defined as isolated thrombocytopenia with normal bone marrow and the absence of other causes of thrombocytopenia

IFN-γ
a type of interferon produced by NK cells and T cells; involved mostly in signaling between cells of the immune system

IgA
immunoglobulin A; secreted across epithelial surfaces into the the tracts of the body, e.g. gastrointestinal; protects the body by prevention of attachment of bacteria and viruses

IgA nephropathy
also known as IgA nephritis and Berger's disease; kidney disease characterized by IgA deposition in the glomerular mesangium; now the most common cause of glomerulonephritis in the world

IgA1 protease
a proteolytic enzyme that cleaves specific peptide bonds in the hinge region sequence of immunoglobulin A1 (IgA1); secreted by several species of pathogenic bacteria at mucosal sites of infection to destroy the structure and function of human IgA1 thereby eliminating an important aspect of host defense

IgE
immunoglobulin E; an antibody involved in enhancing acute inflammation and the cause of type 1 hypersensitivity (allergy) reactions

IgG
immunoglobulin G; the most common class of antibody in the blood; only antibody class to pass across the placenta to protect the neonate

IgM
immunoglobulin M; the largest antibody molecule and because of its size is found mainly in the blood stream

Immune complexes
antibodies and antigens bound in various ratios; may also contain complement components; responsible for type III hypersensitivity reactions

immune-paresis
defined as reduction in the levels of all immunoglobulin classes or reduction in all classes of immunoglobulins other than that of the paraprotein in multiple myeloma

immunofluorescence assay (IFA)
an assay that uses specific antibodies labeled with fluorescent dyes to detect antigens on or in cells. The cells are examined under a fluorescence microscope which excites the dyes to give off light. Different dyes emit light of different wavelengths and thus different colors

in situ hybridization
a technique for localizing and detecting specific mRNA sequences in morphologically preserved tissues sections or cell preparations by hybridizing the complementary strand of a nucleotide probe to the sequence of interest

indoleamine-2 3dioxygenase (IDO)
an enzyme that breaks down tryptophan; may play a role in immunosuppression since T cell function is dependent on this amino acid

infarction
an area of tissue death due to the local lack of oxygen

inflammatory bowel disease
collective term for ulcerative colitis (UC) and Crohn's disease (CD); inflammatory diseases of the large and small intestines

interleukin (IL)
member of the cytokine family; communication molecules between leukocytes; to date up to 33 different ILs identified (IL1 to IL33)

interstitial nephritis
also called tubulointerstitial disease; inflammatory disease of the renal interstitial cells between the nephrons and sometimes tubules (as opposed to glomeruli – glomerulonephritis)

intrathecal
introduced into the space under the brain arachnoid membrane or the spinal cord

intravenous immune globulin (IVIG)
pooled serum immunoglobulins given originally to treat primary antibody deficiencies; now used to treat a variety of diseases including some secondary immunodeficiencies and autoimmune diseases

introitus
an entrance that normally goes into a canal or hollow organs (e.g. vagina)

ischemia
deprivation of oxygen to a tissue leading to damage; usually caused by vasoconstriction, thrombosis or embolism

ischemic necrosis
necrosis caused by hypoxia (see entry for hypoxia)

isoenzyme typing
a technique involving electrophoresis of enzymes to type strains of bacteria within a particular species, e.g., multilocus enzyme electrophoresis (MLEE)

Jarisch-Herxheimer reaction
a transient inflammatory reaction following treatment of syphilis with antibiotics; includes fever, chills, and worsening of the skin rash or chancre

jaundice
a yellow discoloration of the skin, most easily seen by looking at the sclera of the eyes; due to excess of circulating bilirubin

kala-azar
synonymous with leishmaniasis

kallikrein
a serine protease; produced by renal tubules; causes dilation of blood vessels; part of the kinin system that plays a role in inflammation, blood pressure, coagulation and pain

Kaposi sarcoma
a connective tissue tumor caused by Human Herpes Virus 8 (HHV8), also known as Kaposi's Sarcoma-associated Herpes Virus (KSHV). It can affect people with a weakened immune system, including people with HIV (Human Immunodeficiency Virus) and AIDS

Katayama syndrome
a hypersensitivity reaction to schistosomula antigens with immune complex formation

Kato-Katz technique
a (semi) quantitative assay using thick stool smears for microscopic examination

keratitis
infection of the cornea

keratoconjunctivitis
inflammation of the cornea and conjunctiva

Kupffer cells
part of the mononuclear phagocyte system, macrophages of the liver

kyphosis
an abnormal rounding forward of the shoulders and upper back; also refers to the type of curve formed by the sacrum and thoracic spine

lamina propria
a vascular layer of connective tissue under the basement membrane lining a layer of epithelium e.g. in the intestine where it contains many kinds of immune cells

latent/latency
the ability of a pathogenic virus to lie dormant within a cell. A latent viral infection is a type of persistent viral infection

Legionnaires disease
caused by *Legionella* species that are obligate, aerobic, gram-negative bacilli: pneumonia is the most common presenting problem

leptomeninges
the two innermost layers of the tissue that cover the brain and spinal cord, called the arachnoid mater and pia mater

leukemia
a cancer of the blood or bone marrow characterized by an abnormal proliferation of blood cells, usually white blood cells

leukopenia
an abnormally low white blood cell count

lichen sclerosis
a chronic inflammatory dermatosis that results in white plaques with epidermal atrophy; more commonly seen on the vulva or penis but can occur extragenitally; has recently been associated with autoantibodies to the glycoprotein extracellular matrix protein 1 (ECM1)

lipid envelope
the host-derived outer lipid layer of enveloped viruses

lipopolysaccharide
the 'endotoxin' in the wall of gram-negative bacteria made of a sugar backbone with lipid A attached; potent activator of the immune system

lymphadenitis
inflammation of a lymph node

lymphadenopathy
swollen or enlarged lymph nodes; may be generalized or local

lymphangiectasia
dilatation of lymphatic vessels

lymphangitis
inflammation of the lymphatic vessels

lymphocytosis
an increase in the number of lymphocytes in the blood caused mainly by infection or inflammation

lymphoedema
the result of accumulation of lymph in interstitial spaces; usually occurs where the lymphatic system becomes damaged following surgery and/or radiotherapy (as in cancer treatment) or as a result of infection, severe injury, burns or trauma

lymphoma
a cancer in which lymphocytes are the cells of origin; over 40 kinds of lymphoma with the two major subgroups being Hodgkins lymphoma (also called Hodgkins disease) and Non Hodgkins lymphoma of which there are about 30 subtypes; treatment is different for the different patient subgroups

lysis
disintegration of a cell resulting from destruction of its membrane; e.g. bacterial cells by antibody and complement; certain viruses cause lysis; also some chemical substance e.g. enzymes

lysosomes
cytoplasmic, membrane-bound organelles that contain hydrolytic enzymes

macules
flat red patches on the skin

maculopapular rash
a rash that contains both macules – a discolored spot or area on the skin that is not elevated above the surface – and papules, small, solid, usually inflammatory raised areas in the skin without pus

maculopathy
disease of the macula (of the retina), such as age-related macular degeneration

madarosis
loss of eyelashes or eyebrows

magnetic resonance imaging (MRI) scan
a computer based imaging system using a powerful magnetic field and radio frequency pulses to produce detailed pictures of organs, soft tissues, bone and virtually all other internal body structures

mannose binding lectin
also called mannose binding protein; a calcium dependent acute phase protein; member of the collectin family; activates the complement system when it binds to pathogen surfaces

mastocytosis
a disorder characterized by mast cell proliferation and accumulation within various organs, most commonly the skin

megacolon
dilatation of the colon that is not caused by mechanical obstruction; can be acute or chronic

megaesophagus
chronic dilitation of the esophagus; usually associated with asynchronous function of the esophagus and the caudal esophageal sphincter; usually a congenital condition, causing accumulation of food and saliva and regurgitation and aspiration pneumonia from an early age; may also be secondary to systemic disease, particularly general neuromuscular disorders such as myesthenia gravis

membrane attack complex
components of complement (C5 to C9) that lead to pore formation and membrane damage

meninges
the three membranes (pia mater, arachnoid mater, and dura mater) that surround the brain and spinal cord

meningitis
inflammation of the meninges; can be acute or chronic; infectious and non-infectious causes eg. bacteria and viruses and NSAIDS and antibiotics

meningoencephalitis
inflammation affecting both the brain parenchyma and the meninges; often caused by virus infection

metaplasia
cellular adaptation to physiological or mechanical stress whereby one type of differentiated tissue is converted into another type of differentiated tissue; benign and reversible

micro-aerobic
related to microorganisms requiring only small amounts of oxygen

microaerophilic
related to microorganisms requiring oxygen for growth but at lower concentration than is present in the atmosphere

microbiota
the microorganisms that typically inhabit a particular region of the body

microcephaly
a neurodevelopmental disorder in which the circumference of the head is smaller than normal

microglial cells
phagocytic non-neuronal cells of the brain; part of the mononuclear phagocytic system and derived from myeloid precursors

miliary
characterized by lesions resembling millet seeds; miliary tuberculosis marked by appearance of many small lesions

Miller Fisher syndrome
a rare variant of Guillain-Barré syndome that typically presents with the classic triad of ataxia, areflexia, and ophthalmoplegia

MIP-1β
macrophage inflammatory protein 1β; a C-C chemokine (class of cytokines); potent chemoattractant for monocytes

Mollaret's meningitis
a rare form of meningitis; benign recurrent aseptic meningitis

monoclonal antibodies
antibodies produced by a B cell clone and therefore are all identical with regard to specificity; used as laboratory reagents for cell phenotyping and in immunotherapy e.g. herceptin for treating breast cancer

mononuclear phagocytic system
also called the reticuloendothelial system; term for all the phagocytic cells of myeloid origin that are found in the different organs and and blood stream e.g. kupffer cells of liver, alveolar macrophages of the lung, microglial cells of the brain etc.

MSCRAMMs
microbial surface components that recognize adhesive matrix molecules

muco-ciliary escalator
the co-ordinated movement of the cilia, hair-like structures on the surface of respiratory epithelial cells

mucopurulent
pertaining to a discharge that combines pus and mucus

multi-locus enzyme electrophoresis
a molecular sub-typing technique (see enzyme typing above)

multi-locus sequence typing
procedure for characterizing isolates of bacterial species using the sequences of internal fragments of (usually) seven house-keeping genes

multiple myeloma
monoclonal plasma cell malignancy; multiple because there can be several growth foci of abnormal plasma cells (myeloma cells) in the bone marrow; can lead to bone erosion, anemia, kidney damage and immunosuppression

myalgia
muscle pain

mycosis
a disease caused by fungi

myelitis
inflammation of the spinal cord

myeloid
refers to the nonlymphocytic groups of white blood cells, including the granulocytes, monocytes and platelets

myelopathy
a disorder in which the tissue of the spinal cord is diseased or damaged; can also mean pathological change in the bone marrow

myiasis
infestation of the body by fly larvae (usually through a wound or other opening) or a disease resulting from the infestation

myocarditis
collection of diseases with toxic, infectious and autoimmune etiologies characterized by inflammation of the myocardium

myopericarditis
inflammation of both the myocardium and pericardium

NADPH oxidase-dependent killing
killing of microorgansims through the oxygen-dependent antimicrobial mechanism of macrophages and neutrophils; relies on activation of NADPH oxidase that catalyzes the transfer of electrons from NADPH to molecular oxygen, resulting in the formation of superoxide anion and a number of downstream toxic oxidants

natural killer (NK) cell
a cell of the innate immune system that kills virus infected cells; morphologically, large granular lymphocytes

necrosis
death in response to cell or tissue damage, which ends in the release of the intracellular content and the onset of inflammation; there are different kinds of necrosis e.g. haemorrhagic etc.

necrotizing fasciitis
a deep bacterial infection of the connective tissue that spreads along fascial planes and destroys muscle and fat

nephrotic syndrome
a clinical entity characterized by massive loss of urinary protein (primarily albuminuria) leading to hypoproteinemia (hypoalbuminemia) and its result, edema

neuropathy
a functional disturbance or pathological change in the peripheral nervous system

neurotensin
a 13-residue peptide transmitter, sharing significant similarity in its 6 C-terminal amino acids with several other neuropeptides; distributed throughout the central nervous system; highest levels in the hypothalamus, amygdala and nucleus accumbens. It induces a variety of effects, including: analgesia, hypothermia and increased locomotor activity. It is also involved in regulation of dopamine pathways. In the periphery, neurotensin is found in endocrine cells of the small intestine, where it leads to secretion and smooth muscle contraction

neutropenia
an abnormally low neutrophil count in the blood

neutrophilia
an increased number of neutrophils in the blood

nitric oxide synthase (NOS)
produces nitric oxide (NO) and citrulline from arginine, molecular oxygen, and NADPH. NO plays an important role as a mechanism of cytotoxicity for macrophages and as a signaling molecule involved in neurotransmission

nonsteroidal anti-inflammatory drugs (NSAIDS)
used to treat inflammation in a number of diseases; includes aspirin, ibuprofen and indomethacin

normoblasts
myeloid cell precursors of erythrocytes found in the bone marrow

nosocomial (infection)
relates to an infection acquired or taking place in a hospital

nucleocapsid
the nucleic acid core and surrounding capsid of a virus

oliguric
defined as having a urine output of less than 1 mL/kg/h in infants, less than 0.5 mL/kg/h in children, and less than 400 mL/d in adults; one of the clinical hallmarks of renal failure

onchocerciasis
a parasitic disease caused by the filarial worm *Onchocerca volvulus*; transmitted through the bites of infected female blackflies of *Simulium* species, which carry immature larval forms of the parasite from human to human; larvae die causing a variety of conditions, including blindness, skin rashes, lesions, intense itching and skin depigmentation

opportunistic infections
infections that take advantage of weakness in the immune defenses e.g. those that occur in AIDS patients

opsonization
the coating of microbes with antibodies or complement (called opsonins) to enhance phagocytosis

ophthalmia neonatorum
conjunctivitis that is contracted by newborns during delivery; usually caused by the gonococcus

orchitis
testicular inflammation: it may occur idiopathically or it may be associated with infections such as mumps, gonorrhoeae, filarial disease, syphilis or tuberculosis

oro-oral
oral–oral route of transmission of infection e.g. EBV

oropharynx
the back of the mouth where it meets the throat

Osler node
a small, raised, and tender cutaneous lesion characteristic of subacute bacterial endocarditis, usually appearing in the finger pads

osteomyelitis
acute or chronic infection of the bone usually caused by pyogenic bacteria or mycobacteria; can result in destruction of the bone and deformity if not treated

otitis media
a build-up of fluid or mucus in the middle ear: mucus can become infected with bacteria resulting in inflammation

otomycosis
a fungal infection of the external auditory canal characterized by inflammation, pruritus and scaling

oxygen free radicals
reactive oxygen species (ROS) that are ions or very small molecules that include oxygen ions, free radicals, and peroxides, both inorganic and organic. They are highly reactive due to the presence of unpaired valence shell electrons

pancreatitis
acute or chronic inflammation of the pancreas

pancytopenia
an abnormal reduction in all three types of cells in the blood, including erythrocytes (anemia), white blood cells (neutrophils; neutropenia) and platelets (thrombocytopenia); often the result of changes in the bone marrow

papilledema
an optic disc swelling, secondary to elevated intracranial pressure with underlying infectious, infiltrative, or inflammatory etiologies

papular
referring to one or more small solid rounded lesions rising from the skin that are each usually 0.5 cm or less in diameter (less than a half inch across)

papule
a skin lesion that is small, solid, and raised, usually not more than 0.5 cm in diameter

paralytic ileus
functional non-mechanical 'obstruction' of intestinal flow

paraplegia
the loss of movements and/or sensation following damage to the nervous system

parasitophorous vacuole
a vacuole formed by layers of endoplasmic reticulum around an intracellular parasite which may serve to isolate the parasite and enclose it for lysozymal attack

parasthesiae
abnormal skin sensations (as tingling or tickling or itching or burning), with no apparent physical cause; usually associated with peripheral nerve damage

parenteral
administration of a drug etc through any route other than the digestive tract; usually via injection

paronychia
an infection of the skin which is just next to a nail (the nail-fold): the nail itself may become infected or damaged if a nail-fold infection is left untreated

pathogen associated molecular patterns (PAMPs)
evolutionarily conserved stuctures (patterns) of microorganisms but not humans that are the danger signals detected by Pattern Recognition Receptors (PRRs) such as Toll-like receptors on the surface of epithelial cells, phagocytes and antigen-presenting cells

pathogenicity islands (PAIs)
discrete genetic units flanked by direct repeats, insertion sequences or tRNA genes, which are sites for recombination into the DNA. They are incorporated in the genome of pathogenic microorganisms but are usually absent from those non-pathogenic organisms of the same or closely related species. PAIs carry genes encoding one or more virulence factors: adhesins, toxins, invasins, etc. The G+C content of PAIs often differs from that of the rest of the genome

pattern recognition receptors (PRRs)
receptors such as Toll-like receptors that bind PAMPs and tranduce a signal resulting in the release of proinflammatory cytokines and chemokines

pelvic inflammatory disease
infection of the womb and fallopian tubes

peptidoglycan
the rigid, shape-determining complex polymer that forms the cell wall of bacteria. It is made up of chains of heteropolysaccharides linked by tetrapeptides

perforin
a protein that can polymerize to form the membrane pores that are an important part of the killing mechanism of NK and cytotoxic T cells

pericarditis
inflammation of the outer lining of the heart

periodontal disease
a group of chronic inflammatory diseases that affect the tissues that support and anchor the teeth. If untreated, periodontal disease would result in the destruction of the gums, alveolar bone and the outer layer of the tooth root

peripheral nerve neuropathy
disorders of the peripheral nerves; in its most common form it causes pain and numbness in hands and feet, and tingling burning pain; caused by traumatic injuries, infections, metabolic problems and exposure to toxins; one of the most common causes of the disorder is diabetes

petechiae
pinpoint-sized hemorrhages of small capillaries in the skin or mucous membranes

phage typing system
Infection of bacteria by bacteriophages is highly specific; this system uses a panel of bacteriophages (also called phages) to identify unknown bacterial strains

phagocytosis
the process of ingesting large particles such as bacteria and viruses by specialized immune cells of the mononuclear phagocyte system

phagosome
the vacuole containing the large particle(s) ingested by the process of phagocytosis

pharyngitis
painful inflammation of the pharynx (sore throat)

photophobia
sensitivity or intolerance to light

picornavirus
viruses in the family *Picornaviridae*; non-enveloped, genome of a single stranded, positive-sense RNA viruses with an icosahedral capsid; important human pathogens

pilin
a class of proteins found in the pilus structures of bacteria

pillus/pilli
smaller than flagella; used in the exchange of genetic material during bacterial conjugation; see also fimbriae

pinocytosis
the ingestion of dissolved materials by endocytosis

placentitis
inflammation of the placenta

pleocytosis
an abnormal increase in the number of cells (as lymphocytes) in a bodily fluid e.g. the cerebrospinal fluid; indicative of an inflammatory, infectious or malignant condition

pleural thickening
thickening and hardening of the pleura – two layered protective membrane surrounding the lung

pleurisy/pleuritic pain
inflammation of the pleura leading to pain on breathing (pleuritic pain)

pneumocytes
alveoli are lined by two types of pneumocytes – Type 1 and Type 2; type 1 responsible for gaseous exchange and type 2 for production and secretion of surfactant, a molecule that reduces the surface tension of pulmonary fluids and contributes to the elastic properties of the lungs

pneumonia
inflammation of the lung parenchyma; usually caused by several species of bacteria, viruses, fungi or parasites but can have a non-infectious etiology

pneumonitis
a general term referring to inflammation of the lung tissue

pneumotropic
the affinity of a microbe for lung tissue; infectious agents that cause pneumonia generally have specific adhesion molecules that allow them to bind to lung epithelia

podoconiosis
nonfilarial, noninfective 'elephantiasis', usually crystalline blockage of the limb lymphatics, almost always affecting the lower limbs and especially the feet

polycistronic
mRNA is said to be polycystronic when it contains the coding sequences for multiple gene products, each of which is independently translated from the mRNA; usual for prokaryotes but not eukaryotes

polymerase chain reaction (PCR)
a technique used to copy and amplify specific DNA sequences over a millionfold in a simple enzyme reaction; DNA corresponding to genes, or gene fragments can be amplified from chromosomal DNA; the amplified DNA can be used for the analysis and manipulation of genes

polymorphisms
silent mutations that have no effect on the encoded protein that tend to accumulate in the DNA

polymyalgia rheumatica
a disorder of the muscles and joints characterized by severe aching and stiffness of the neck, shoulder girdle and pelvic girdle; classified as a rheumatic disease but has unknown etiology

post-herpetic neuralgia
pain arising or persisting in areas affected by herpes zoster at least three months after the healing of the skin lesions

Pott's disease
partial destruction of the vertebrae, usually caused by a tuberculous infection and often producing curvature of the spine

preauricular lymphadenopathy
swollen/enlarged lymph nodes draining the eyelids and conjunctivae, temporal region and pinna

proctitis
inflammation of the rectal mucosa; many causes

proctocolitis
inflammation of the rectum and the colon

pro-inflammatory mediators
molecules that initiate inflammatory reactions

prostaglandins
unsaturated carboxylic acids, synthesized from the fatty acid, arachidonic acid via the cyclooxygenase pathway. They are extremely potent mediators of a diverse group of physiological processes

prostatitis
inflammation of the prostate gland; may be bacterial or non-bacterial and in the case of prostatodynia the complaints are consistent with prostatitis but no evidence of prostatic inflammation

prosthesis
an artificial substitute for a body part

protean
very variable, easily changeable or continually changing

protein losing enteropathy
abnormal loss of protein from the gastrointestinal tract, or the inability of the digestive tract to take in proteins

provirus
a form of a virus that is integrated into the genetic material of a host cell and by replicating with the integrated form can be transmitted from one cell generation to the next without causing lysis

pruritis
itch

pseudomembranous colitis
sometimes called antibiotic-associated colitis or *C. difficile* colitis; inflammation of the colon that occurs in some people who have received antibiotics; almost always associated with an overgrowth of the bacterium *Clostridium difficile* (*C. difficile*), although in rare cases, other organisms can be involved

psoriatic arthritis
an inflammatory joint disease associated with psoriasis

pulse field gel electrophoresis
a technique used to separate especially long strands of DNA by length in order to tell differences among samples. It operates by alternating electric fields to run DNA through a flat gel matrix of agarose

purpuric
having small hemorrhages in the skin, mucous membranes, or serosal surfaces

pus
contains fluid, living and dead leucocytes, dead tissue, and bacteria or other foreign substances

pyelonephritis
infection of the upper part of the urinary tract; infection of the renal pelvis and often the renal parenchyma

pyoderma (impetigo)
a highly contagious, superficial infection of exposed skin by *S. pyogenes* or *Staph. aureus*, typically the face, arms and legs; seen most frequently in young children

pyogenic infection
infection characterized by severe local inflammation, usually with pus formation, generally caused by one of the pyogenic bacteria

pyrexia
fever, when the body's temperatue rises above normal

pyrexia of unknown origin-PUO
defined as a temperature greater than 38.3°C on several occasions, more than 3 weeks' duration of illness, and failure to reach a diagnosis despite one week of inpatient investigation; caused by infections (30–40%), neoplasms (20–30%), collagen vascular diseases (10–20% and miscellaneous diseases (5–10%).

pyuria
the presence of pus in the urine; abnormal numbers of white blood cells in the urine indicative of infection

Quellung test
bacteria are mixed into a drop of capsule-specific antiserum and observed under the microscope. Capsule swelling and increased refractility result from the interaction of the antibody with the capsule for which it is specific

radiculomyelopathy
irritation of the sacral nerve roots

radionuclide brain scanning
an imaging technique that uses a small dose of a radioactive chemical (isotope) called a tracer that can detect cancer, trauma, infection or other disorders – in this case in the brain

rales
wet, crackly lung noises heard on inspiration

rashes
macular, petechial, scarlatiniform, urticarial, or erythema multiforme

RAST
radioallergosorbent test; test for specific IgE antibodies to a variety of allergens including cow's milk, nuts, eggs etc.

reactive oxygen species (ROS)/reactive oxygen intermediates (ROI)
ROS form as a natural byproduct of the normal metabolism of oxygen and have important roles as bacteriocidal agents and in cell signaling: release by neutrophils and macrophages are thought to play a role in tissue damage

recto-vaginal fistulae
an abnormal connection between the rectum and vagina; may result from injury during childbirth, complications following surgery, cancer or inflammatory bowel disease such as Crohn's disease; may resolve spontaneously but most often requires surgery

regulatory T cells
T cells that regulate cellular and antibody responses and are thought to prevent autoimmunity; mostly of the helper phenotype i.e. CD4+ but also expressing high levels of CD25 and FoxP3

Reiter's syndrome
now known as reactive arthritis; refers to acute nonpurulent arthritis complicating an infection elsewhere in the body; can be triggered following enteric or urinogenital infections and is often associated with HLA-B27

restriction fragment length polymorphism (RFLP)
a technique by which organisms may be differentiated by analysis of patterns derived from cleavage of their DNA

reticulocytes
newly differentiated erythrocytes without a nucleus: they retain a fine network of endoplasmic reticulum that stains with reticulocyte stains

retrovirus
retrovirus is an enveloped virus possessing an RNA genome. It contains an enzyme reverse transcriptase which is used to perform the reverse transcription of its genome from RNA into DNA. This can then be integrated into the host's genome with an integrase enzyme. The virus then replicates as part of the cell's DNA.

Reye's syndrome
sudden (acute) brain damage (encephalopathy) and liver function problems of unknown cause, often associated with the use of aspirin to treat chickenpox or influenza in children

rheumatoid arthritis
a chronic inflammatory autoimmune disease affecting the synovial joints that can manifest itself systemically. Systems affected can include the lung, etc.

rheumatoid factor (RF)
antibodies with a specificity for the Fc region of IgG; they can be of the IgM, IgG or IgA class; found in around 75% of patients with rheumatoid arthitis but also in some patients having a number of other autoimune diseases such as systemic lupus erythematosus; IgM RF commonly found in low levels in normal individuals

rhinitis
a group of conditions that are caused by inflammation of the lining of the nose; can be allergic (mediated by IgE) or non-allergic through bacterial or viral infections

rhinorrhoea
a discharge from the nasal mucous membrane, especially if excessive

ribotyping
a technique used to determine genetic and evolutionary relationships between organisms; oligonucleotide probes targeted to highly conserved domains of coding sequences of ribosomal RNA are amplified and the products are visualized by gel electrophoresis banding patterns and compared with known species and strains to determine organism relatedness

Romana's sign
conjunctivitis and swelling around the eye resulting from infection via the eye with *T. cruzi*

rosaceae
a chronic (long-term) disease that affects the skin and sometimes the eyes; characterized by redness, pimples, and, in advanced stages, thickened skin; usually affects the face with other parts of the upper body only rarely involved

rose spots
red macular lesions, 2–4 millimeters in diameter occurring in patients suffering from *Salmonella typhi* and *Salmonella paratyphi*; rose spots can also occur following invasive non-typhoid salmonellosis

Roth spot
a round white retina spot surrounded by hemorrhage in bacterial endocarditis, and in other retinal hemorrhagic conditions

Sabin-Feldman dye test
a serological test for the diagnosis of toxoplasmosis; measures antibodies to toxoplasma

salivarian transmission
transmission of the African trypanosome via the saliva of the fly vector

salpingitis
an infection and inflammation in the fallopian tubes

saprophytic
feeding or growing on decaying organic matter

sarcoidosis
a disease characterized by non-caseating granulomas; associated with a number of immune system aberrations; commonly involves the lungs and the lymph nodes (especially the intrathoracic nodes), the skin, the eyes, the liver;,the heart, and the nervous, musculoskeletal, renal, and endocrine systems; of unknown etiology

scotoma
an area of diminished vision within the visual field

seizure
a common, nonspecific manifestation of neurologic injury and disease

septic arthritis
also known as infectious arthritis; direct infection of the joint, mainly bacterial

septic shock
sepsis is a systemic inflammatory response to an infectious agent that releases endotoxin or superantigen into the blood; likely mediated by TNFα, IL-1 and IFNγ; life threatening condition

septicemia
the active multiplication of bacteria in the bloodstream that results in an overwhelming infection: can be caused by emboli

seroconversion
the development of antibodies in response to infection or immunization

serology
the study of serum, often for the presence of specific antibodies to microbes

seropositive
the presence of specific antibodies to a microorganism following infection or immunization

serovar
a sub-division of microorganisms on the basis of antigenic structure

severe combined immunodeficiency
a primary immunodeficiency that results from any of a heterogenous group of genetic conditions affecting the immune system, leading to severe T- and B-cell dysfunction

shingles
a viral disease caused by Herpes zoster and characterized by a painful skin rash with blisters in a limited area on one side of the body, often in a stripe – see Case 39

sickle cell disease
a group of genetic disorders that affects hemoglobin

sigmoidoscopy
procedure used to examine the rectum and the sigmoid colon (using a sigmoidoscope)

sinusitis
an inflammation of the paranasal sinuses, which may or may not be as a result of infection, from bacterial, fungal, viral, allergic or autoimmune issues

spirochetes
motile, unicellular, spiral-shaped organisms which are morphologically distinct from other bacteria; they range from obligate anaerobes to aerobes and from free-living forms to obligate parasites

splenomegaly
enlargement of the spleen

stercorarian transmission
transmission of the american trypanosome *T. cruzi* (causing Chagas disease) via the feces of the insect vector

Stevens-Johnson syndrome
an immune-complex–mediated hypersensitivity that is a severe expression of erythema multiforme; typically involves the skin and the mucous membranes; caused by many drugs, viral infections and malignancies; cell death results causing separation of the epidermis from the dermis

Still's disease
a form of juvenile idiopathic arthritis that affects large joints and bone growth may be retarded; characterized by inflammation with high fever spikes, fatigue and salmon-colored rash

stroke
the rapidly developing loss of brain functions due to a disturbance in the blood vessels supplying blood to the brain

substance P
a small neuropeptide involved in regulation of the pain threshold

superantigen
antigens, mostly from bacterial toxins, that interact with a subset of T lymphocytes expressing products of the VβT cell recep-

tor genes; as a consequence, superantigens activate large numbers of T cells independent of their specificity

suppurative infections
pus producing infections

sylvatic
referring to diseases only occurring in wild animals

syncope
a brief loss of consciousness caused by a temporary deficiency of oxygen in the brain

systemic lupus erythematosus
a chronic inflammatory disease that affects multiple organ systems; an autoimmune disease characterized by antinuclear antibodies especially those to double stranded DNA

tachycardia
a rapid heart rate, especially one above 100 beats per minute in an adult

tachypnea
abnormally rapid breathing or respiration

Tamm-Horsfall protein
also called uromodulin; a mucoprotein derived from the secretion of renal tubular cells and a normal constituent of urine; may play a role in preventing attachment of type 1 fimbriated *Escherichia coli* to the urinary tract thus contributing to innate immunity in that tract

tegument
a natural outer covering; in relation to many enveloped viruses is a cluster of non-essential and essential proteins that line the space between the envelope and nucleocapsid

tenesmus
the constant feeling of a need to empty the bowels

tenosynovitis
inflammation of the sheath that surrounds a tendon

terminal ileitis
inflammation of the lower section of the small intestine

Th1
a subset of helper T cells (expressing CD4) that are involved primarily in producing pro-inflammatory cytokines like IFN-γ, IL-2, and TNF-α and help the development of cell mediated immune responses

Th2
a subset of helper T cells (expressing CD4) that are involved primarily in producing cytokines like IL-4, IL-5, IL-9, IL-13 that are primarily involved in the development of antibody mediated immune responses

thrombocytopenia
a disorder where the platelet count is below the normal lower limit (usually 150×10^9/L)

thrombosis
the formation of a blood clot (thrombus) inside a blood vessel that obstructs the flow of blood through the circulatory system

thyroiditis
inflammation of the thyroid gland; causes include autoimmune eg. Hashimoto's thyroiditis and thyrotoxicosis, infections (bacterial or viral) and drugs

tinnitus
the name given to the condition of noises 'in the ears' and/or 'in the head' with no external source. Tinnitus noises are described variously as ringing, whistling, buzzing and humming

TNF-α
tumour necrosis factor alpha; a pro-inflammatory cytokine produced mainly by macrophages; has a multitude of functions – kills some tumors (hence name) but enhances growth of others, has effects on blood vessel dilation and hence acute inflammation etc.

Toll-like receptors
a group of pattern recognition receptors that are found on many cells of the body but particularly on macrophages and polymorphs. Important in early recognition of invading microbes.

TORCH
Toxoplasmosis, **R**ubella, **C**ytomegalovirus and **H**erpes virus

toxic megacolon
an acute toxic colitis with dilatation of the colon; may complicate any number of colitides, including inflammatory, ischemic, infectious, radiation, and pseudomembranous

toxic shock syndrome
usually caused by an enterotoxin from staphylococcal infections of the vagina entering the circulation; this acts as a superantigen (see entry above) and stimulates a subset of VβT cells releasing large amounts of cytokines resulting in a life threatening condition

tracheostomy
a surgical procedure to create an opening (stoma) in the trachea; the opening itself can be called a tracheostomy

trachoma
an infection of the eye caused by *Chlamydia trachomatis*; the most common cause of blindness in the developing world after cataracts

transforming growth factor
one member of a group of cytokines that were identified by their ability to promote the growth of fibroblasts but are also generally immunosuppressive

trans-stadially
the passage of a microbial parasite, such as a virus or rickettsia, from one developmental stage (stadium) of the host to its subsequent stage or stages, particularly as seen in mites

transthoracic echocardiography (TTE)
also known as a standard echocardiography where an echocardiogram is made; see entry for echocardiogram

trichiasis
an uncomfortable condition in which the eye lashes are misdirected toward the eye ball and scratch its surface, the cornea; can be caused by infection, inflammation, autoimmune conditions, and trauma such as burns

tularemia
an acute, febrile, granulomatous, infectious zoonosis caused by the aerobic gram-negative pleomorphic bacillus *Francisella tularensis*; infection of humans is through introduction of the bacillus by inhalation, intradermal injection, or oral ingestion

Type I hypersensitivity
immediate type hypersensitivity, allergy; mediated by IgE antibodies that bind to mast cells and the allergen causes degranulation giving rise to an excessive acute inflammatory reaction

Type II hypersensitivity
mediated by non IgE antibodies usually IgM or IgG and/or complement; leads to tissue damage including enhanced phagocytosis and lysis

Type III hypersensitivity
immune-complex mediated hypersensitivity; tissue damage caused by enzymes released from neutrophils and complement activation e.g. streptococcal glomerulonephritis

Type III secretion apparatus
one of 4 types of secretion systems of gram negative bacteria that require ATP and mediate transport and injection of pathogenicity molecules into target eukaryotic cells e.g. shigella

Type IV hypersensitivity
T cell mediated hypersensitivity; tissue damage caused by granuloma formation in mycobacterial infections

Type IV secretion apparatus
as well as delivering toxic molecules into cells, type IV secretion systems are also used to transport DNA or protein-DNA complexes. One such process is bacterial conjugation whereby two mating bacteria exchange genetic material; used in pathogenisity by *Helicobacter*

typhus
an infectious disease that is spread by lice or fleas resulting in acute febrile illness; caused by rickettsial organisms

uremic
pertaining to uremia which is the accumulation in the blood of nitrogenous waste products (urea) that are usually excreted in the urine

urethritis
inflammation of the urethra

urolithiasis
the process of forming stones in the kidney, bladder and or urethra (urinary tract)

urticaria
commonly referred to as hives; raised areas of erythema and edema involving the dermis and epidermis; may be acute (often due to type I hypersensitivity) or chronic; acute caused by drugs, foods, insect venoms and parasites; also physical causes include pressure and cold or light

uveitis
inflammation of one or all parts of the uveal tract. Components of the uveal tract include the iris, the ciliary body, and the choroid

uvula
a pendant, fleshy mass: palatine uvula is the small, fleshy mass hanging from the soft palate above the root of the tongue

vaginitis
inflammation of the vagina characterized by discharge, odour, irritation, and/or itching

vasculitis
inflammation of the blood vessel wall; may be caused by the immune system

VCAM -1
vascular cell adhesion molecule 1; present on endothelial cells; ligand for VLA-4, an integrin present on lymphocytes and some phagocytes; involved in adhesion of leukocytes to inflamed endothelium and therefore their movement into sites of inflammation

vesicles
fluid filled blisters on the skin

vesicoureteric reflux
the abnormal flow of urine from the bladder back into the ureters

Vi antigen
virulence antigen; determinant of pathogenesis

viremia
the presence of viruses in the blood

virion
a complete virus particle with its DNA or RNA core and protein coat as it exists outside a cell; also called viral particle

virulent/virulence
the ability of a microorganism to cause disease

virus envelope
glycoprotein coat surrounding the nucleocapsid; composed of two lipid layers interspersed with protein molecules; contains lipids of host origin as well as proteins of viral origin; many viruses also develop glycoprotein spikes on their envelope to help them attach to specific cell surfaces

Weil's disease
synonymous with Leptospirosis; a bacterial zoonotic disease caused by spirochaetes of the genus *Leptospira*

Weil-Felix test
an agglutination test for rickettsial infection based on production of nonspecific agglutinins in the blood of infected patients, and using various strains of *Proteus vulgaris* as antigen

Western blot
also Western immunoblot, a technique for identification of antigens in a mixture by electrophoresis, blotting onto nitrocellulose and labeling with enzyme or radio-labeled antibodies

white cell casts
casts in urine made up of white blood cells; reason called casts because they acquire the shape of the renal tubule

Widal test
a serological test for the diagnosis of typhoid and other salmonella infections

Zollinger-Ellison syndrome
the triad of non beta islet cell tumors of the pancreas (gastrinomas), hypergastrinemia, and severe ulcer disease

zoonosis
an animal disease that can be transmitted to humans

Medicine Net. com
http://www.medterms.com/script/main/art.asp?articlekey=2533

The free dictionary
http://www.thefreedictionary.com/

The Ohio State University Medical Center
http://medicalcenter.osu.edu/

Emedicine
http://www.emedicine.com

Trip database
http://www.tripdatabase.com

Online medical dictionary
http://cancerweb.ncl.ac.uk

Merriam-Webster dictionary
http://medical.merriam-webster.com/medical
http://www.geneed.com/website/catalog/glossary
http://www.nature.com/nrg/journal/v3/n6/glossary

Answers to Multiple Choice Questions

CASE 1: *Aspergillus fumigatus*

1. **Which of the following are characteristics of *A. fumigatus*?**
   A. *A. fumigatus* can only grow at low temperatures.
      FALSE: it is thermotolerant and grows at temperatures ranging from 15°C to 55°C.
   B. Its natural habitat is soil.
      TRUE: it grows on organic debris, decaying vegetation, and bird excreta.
   C. *A. fumigatus* produces large spores.
      FALSE: it produces small conidia that are easily airborne.
   D. Aspergillus is a filamentous fungus with branching hyphae.
      TRUE: *Aspergillus* hyphae are 3–5 µm in diameter and form dichotomous branches.
   E. Wounds are the main port of entry of *A. fumigatus*.
      FALSE: conidia are inhaled and the respiratory tract is the main port of entry.

2. **Which of the following are risk factors for the development of invasive aspergillosis (IA)?**
   A. Aggressive chemotherapy of leukemia and lymphoma patients.
      — Key for all
      TRUE: immunodeficiency caused by chemotherapy can lead to IA.
   B. Intensive and prolonged treatment with steroids.
      TRUE: corticosteroids predispose to the establishment of IA.
   C. Bone marrow transplantation.
      TRUE: BMT patients are at increased risk for the development of IA.
   D. Alzheimer's disease.
      FALSE: does not predispose to IA.
   E. Graft-versus-host reaction.
      TRUE: GVH patients are at increased risk for the development of IA.

3. **Which of the following are the most frequent clinical presentations of ABPA?**
   A. Severe joint pain.
      FALSE: not characteristic for ABPA.
   B. Wheezing, cough, fever, malaise, weight loss.
      TRUE: general symptoms characteristic for both ABPA and asthma patients.
   C. Skin rash.
      FALSE: not characteristic for ABPA.
   D. High levels of serum IgE.
      TRUE: characteristic for ABPA and asthma patients.
   E. Recurrent pneumonia.
      TRUE: additional clinical condition associated with ABPA.

4. **What are the normal immune responses to *A. fumigatus* infection?**
   A. Innate immunity plays a leading role in the disposal of this fungus.
      TRUE: innate immune mechanisms are essential in early clearance of the fungus.
   B. Neutrophils are more efficient in killing of conidial forms while resident macrophages dispose of the hyphae stage of *A. fumigatus*.
      FALSE: macrophages attack early conidial forms of *A. fumigatus*, while neutrophils are important in the killing of later invasive hyphal forms.
   C. Killing of the fungal cells by CD8+ cytotoxic T cells is essential.
      FALSE: the role of CD8+ CTLs does not appear to be important.
   D. Specific antibodies play an important protective role against *A. fumigatus*.
      FALSE: the protective role of specific antibodies in aspergillosis is believed to be poor.
   E. Type I pro-inflammatory cytokines help to kill this pathogen.
      TRUE: Th1 pro-inflammatory signals recruit neutrophils into sites of infection and provide resistance to *A. fumigatus*.

5. **Which of the following are true of aspergilloma?**
   A. It can be reliably diagnosed only by CT scan or radiography.
      TRUE: aspergilloma can only be diagnosed using these methods.
   B. Aspergilloma pathogenesis is based on acute immune inflammation.
      FALSE: aspergilloma pathogenesis is not associated with acute inflammation.
   C. Aspergilloma mostly develops in pre-existing lung cavities.
      TRUE: the cavities are a result of pre-existing diseases such as tuberculosis.
   D. There could be multiple aspergillomas.
      TRUE: some cavities may contain multiple aspergillomas.
   E. It requires intensive antifungal therapy.
      FALSE: antifungal therapy is not efficient for the treatment of aspergilloma.

6. **Which of the following are true of ABPA?**
   A. ABPA patients have specific symptoms and the condition is easy to diagnose.
      FALSE: symptoms are nonspecific and ABPA is difficult to diagnose.

B. Secondary ABPA is particularly frequent in patients with diabetes and SLE.
FALSE: it is frequent in patients with asthma and cystic fibrosis.

C. ABPA pathogenesis is based on type II hypersensitivity reactions.
FALSE: ABPA pathogenesis is based on type I and type III hypersensitivity.

D. It is associated with blood eosinophilia.
TRUE: blood eosinophilia is characteristic for ABPA.

E. It presents as bronchial asthma.
TRUE: nonspecific symptoms are similar to asthma.

7. **Which of the following tests are used for the diagnosis and monitoring of IA?**
A. Microscopy of sputum.
TRUE: *Aspergillus* hyphae sometimes associated with conidia can be found in sputum.

B. Sputum or bronchoalveolar lavage cell cultures.
TRUE: specimen is cultured on a special agar for up to 6 months at 30°C.

C. Neutrophil cell counts.

TRUE: extremely important for prognosis and monitoring of IA.

D. CT scan.
TRUE: is used for the diagnosis of IA and other aspergillosis.

E. ELISA.
TRUE: currently developed for the confirmation of IA.

8. **Which of the following drugs are used in the treatment of IA?**
A. Amphotericin B.
TRUE: AmB is still the first choice drug for the treatment of IA.

B. Erythromycin.
FALSE: not used for the treatment of IA.

C. Itraconazole.
TRUE: is used for the treatment of IA, particularly in patients with nephropathy.

D. Voriconazole.
TRUE: is effective for the treatment of IA.

E. Tetracycline.
FALSE: not used for the treatment of IA.

CASE 2: *Borrelia burgdorferi* and related species

1. A 23-year-old woman presents on a Monday morning after having spent the weekend camping. She asks to have a tick removed that became affixed to her lower leg during a long hike the day before. Which one of the following is the most appropriate treatment after removing the tick?
A. Cefuroxime axetil.
FALSE: appropriate treatment for a confirmed case of Lyme disease.

B. Doxycycline.
FALSE: appropriate treatment for a confirmed case of Lyme disease.

C. Amoxicillin.
FALSE: appropriate treatment for a confirmed case of Lyme disease.

D. Any of the above antibiotics would be effective to treat a confirmed case of Lyme disease.
FALSE.

E. None of the above.
TRUE: in the absence of signs and symptoms consistent with Lyme disease antibiotics are not indicated.

2. Which one of the following is the least likely clinical manifestation of Lyme disease?
A. Skin rash.
TRUE: erythema migrans is seen in about 80% of persons with Lyme disease.

B. Arthritis.
TRUE: arthritis is a frequent manifestation of Lyme disease.

C. Nausea, vomiting, diarrhea.
FALSE: nausea, vomiting, diarrhea are not features of Lyme disease.

D. Myocarditis.
TRUE: myocarditis and A/V block are common in Lyme disease.

E. Malaise and fatigue.
TRUE: malaise and fatigue are the most common findings after the skin lesion in early disease and are

reported by as many as 80% of persons in the US but by less than 35% of persons in Europe.

3. A 5-year-old girl travels with her family to Maryland in June for vacation. A week later, the father finds a 2 mm black spot behind her ear that he thought was a scab. Four days later, a red enlarging circular lesion appears around the same ear and fades within a week. They return to Minnesota and visit their pediatrician who notes that the child's smile is not quite symmetrical. Which one of the following is the most likely diagnosis?
A. Tinea capitis.
FALSE: tinea capitis is a superficial fungal infection of the skin of the scalp, eyebrows, and eyelashes, with a propensity for attacking hair shafts and follicles.

B. Hypersensitivity to a mosquito bite.
FALSE: the features of a hypersensitivity reaction are discomfort, erythema, tenderness, warmth, and edema of tissues surrounding the bite site.

C. Lyme disease.
TRUE: this is the most common neurologic feature in the US and Europe in children with Lyme disease. More than half of children with neurologic symptoms have a facial palsy, which may be bilateral.

D. Erythema infectiosum.
FALSE: erythema infectiosum (fifth disease) is a common childhood exanthem caused by human parvovirus B19, which features a three-phased cutaneous rash. The exanthem begins with a classic slapped-cheek appearance, which may appear like a sunburn and typically fades over 2–4 days. At 1–4 days later there is a an erythematous macular-to-morbilliform eruption occurring primarily on the extremities. After several days this eruption fades into a lacy pattern. It should be noted that B19 infection may give rise to an acute polyarthropathy that can last for days to months.

E. Enteroviral exanthema.
FALSE: enteroviruses are a common cause of exanthems and enanthems in children. For example, in herpangina

and hand, foot, and mouth disease the rash is an enanthem found in the oral cavity. In hand, foot, and mouth disease skin lesions appear together with or shortly after the oral lesions and vary in number from a few to more than 100. They begin as erythematous macules or papules, which quickly become small, gray, oval or linear vesicles surrounded by a red halo. The hands are more commonly involved than the feet. Lesions usually occur on the lateral aspects of the fingers and toes, especially around the nails, but they may be seen in the digital flexures and on the palms and soles.

4. **During June in a suburb of Trenton, NJ, a patient presents to her doctor concerned about her risk for Lyme disease. She lives near a wooded area, and 3 days previously, she pulled off a small tick from behind her knee. Currently, she is asymptomatic and her physical exam is normal. Which one of the following is the correct course of action?**
   A. Obtain a Lyme serologic test, begin empiric doxycycline therapy and repeat the serologic testing in 6 weeks.
   FALSE: serology and/or antibiotic therapy are not indicated in the absence of signs and symptoms of Lyme disease.
   B. Begin empiric doxycycline therapy and obtain a Lyme serologic test in 6 weeks.
   FALSE: antibiotic therapy is not indicated in the absence of signs and symptoms of Lyme disease.
   C. Do nothing unless clinical signs of early Lyme disease develop over the next weeks.
   TRUE: diagnosis of Lyme disease should be based upon patient history and clinical findings.
   D. Obtain a Lyme serologic test. If positive, begin therapy with doxycycline. If negative, give no therapy.
   FALSE: serologic tests are not indicated in the absence of signs and symptoms of Lyme disease.
   E. Explain to the patient that Lyme disease has never been reported in New Jersey, and she has nothing to worry about.
   FALSE: New Jersey has a high incidence of Lyme disease.

5. **A 17-year-old white male goes to his doctor worried about Lyme disease. He recently returned from a camp in the Upper Peninsula of Michigan where he went hiking in the woods. He recalls no tick bite, but about 1 week after returning home he developed a low-grade fever, myalgia, and fatigue. He had had these symptoms for 1 week when he presented to his doctor. A physical examination is completely normal. What is the single most important diagnostic clue in establishing the diagnosis of Lyme disease?**
   A. New-onset bundle-branch block.
   FALSE: this is a late manifestation when an alternate explanation is not found.

B. Nuchal rigidity compatible with meningitis.
   FALSE: this is a late manifestation when an alternate explanation is not found.
   C. High, spiking fevers.
   FALSE: high spiking fevers are not a feature of Lyme disease.
   D. Erythema migrans.
   TRUE: the rash of erythema migrans is a reliable physical clue. The rash is present in four out of five persons with symptomatic Lyme disease.
   E. Acute arthritis of a large joint.
   FALSE: although arthritis can occur on its own it is not diagnostic of Lyme disease.

6. **All of the following are features of *B. burgdorferi sensu lato*, EXCEPT?**
   A. Microaerophilic.
   TRUE: the bacterium is microaerophilic.
   B. Difficult to culture because of their exacting nutrient requirements.
   TRUE: can be cultivated on specialized media but isolation and culture are not a practical means of diagnosis.
   C. Usually detected in the blood of infected patients by microscopy.
   FALSE: *B. burgdorferi* infection is detected by serology.
   D. They have endoflagella.
   TRUE: borreliae have their flagella located in the periplasmic space.
   E. They have a linear chromosome and contain circular and linear plasmids.
   TRUE: borreliae are unusual in the fact that they have a linear chromosome and not a covalently closed loop.

7. **All of the following are features of the pathogenesis of *B. burgdorferi* in Lyme disease, EXCEPT?**
   A. *B. burgdorferi* is resistant to complement-mediated lysis.
   TRUE: the bacteria, bind factor H, which inactivates C3b.
   B. IgM and IgG antibodies play the principal role in the clearance of *B. burgdorferi*.
   TRUE: humoral immunity is the principal mechanism of host defense.
   C. Arthritis is linked to HLA-DR4 and HLA-DR2.
   TRUE: arthritis has been linked to these HLA haplotypes.
   D. CD8 T cells are important in combating intracellular borreliae.
   FALSE: cytotoxic T cells are not involved in host defense against *B. burgdorferi*.
   E. *B. burgdorferi* displays antigenic variation.
   TRUE: outer surface proteins (Osps) are modulated on and off the surface of the bacterium.

CASE 3: *Campylobacter jejuni*

1. **How can *C. jejuni* be typed?**
   A. Based on the lipopolysaccharide.
   TRUE: there are over 60 serovars in the Penner serotyping system, which detects heat-stable antigens as opposed to the less frequently used Lior typing system using heat labile antigens.
   B. By the Widal reaction.

FALSE: this is a serological test (no longer of value) for *Salmonella typhi* infection.
   C. By phage typing.
   TRUE: and give a further level of discrimination when combined with serotyping.
   D. By the isoprenoid content.
   FALSE: even though the isoprenoid content of

Campylobacter spp. is unusual and is used taxonomically to separate it from *Helicobacter*.
  E. By MLEE.
  TRUE: such as RFLP and MLEE.

2. **Which of the following are true of *C. jejuni*?**
  A. The organism grows in 48 hours on agar.
  TRUE: it grows more slowly than most organisms as it is microaerophilic.
  B. It is a strict anaerobe.
  FALSE: it is a microaerophile.
  C. It is acquired by the oral route.
  TRUE: it is usually food-related although feco–oral spread can occur from animals.
  D. Infection occurs in childhood.
  TRUE: but individuals at any age can be infected, peaks occurring in children and young adults.
  E. It is a rare infection.
  FALSE: *C. jejuni* is the commonest cause of gastroenteritis in the UK, with an incidence of about 350/100 000 and globally it is one of the commonest causes of travelers' diarrhea.

3. **Which of the following are correct statements concerning the host response to *C. jejuni*?**
  A. Few granulocytes are recruited to the area.
  FALSE: there is a strong granulocyte response and an acute inflammatory response is initiated.
  B. There is a strong IgA response.
  TRUE: but it takes several days to develop.
  C. *C. jejuni* is resistant to complement.
  FALSE: it is complement-sensitive but other *Campylobacter* spp., e.g. *C. fetus* var *fetus*, are resistant.
  D. Epithelial ulceration occurs.
  TRUE.
  E. Crypt abscesses occur.
  TRUE.

4. **Which of the following are virulence factors of *C. jejuni* involved in pathogenesis of disease?**
  A. Type IV secretion apparatus.
  FALSE: a flagella secretion system is involved.
  B. Cytolethal distending toxin.
  TRUE: it induces cell cycle arrest and apoptosis.
  C. Arabinogalactan.
  FALSE: this is a component of the mycobacterial cell wall.
  D. Lipopolysaccharide.
  TRUE: and anti-LPS antibodies cross-react with myelin in the nerves.
  E. VacA.
  FALSE: this is found in *Helicobacter*.

5. **Which of the following may be due to infection with *C. jejuni*?**
  A. Megacolon.
  TRUE: but a rare complication in severe disease.
  B. Miller Fisher syndrome.
  TRUE: an autoimmune complication caused by cross-reacting antibodies.
  C. Enterocolitis.
  TRUE: it may present as a mild diarrhea or severe dysentery.
  D. Keratitis.
  FALSE: this may be caused by *Pseudomonas* or viruses.
  E. Septicemia.
  TRUE: it occurs rarely with *C. jejuni* but is more common with *C. fetus* var *fetus*.

6. **Which of the following are typical signs and symptoms of *C. jejuni* infection?**
  A. Colicky lower abdominal pain.
  TRUE.
  B. Fever.
  TRUE.
  C. Bloody diarrhea.
  TRUE.
  D. Skin rash.
  FALSE.
  E. Splenomegaly
  FALSE: this can occur with *S. typhi*.

7. **Which of the following are important in the routine diagnosis of *C. jejuni* infection?**
  A. Culture.
  TRUE: on selective medium containing charcoal incubated for 48 hours in a microaerobic environment.
  B. Urea breath test.
  FALSE: this is a test for *H. pylori*.
  C. Serology.
  FALSE: not in routine diagnosis but may be used in cases of Guillain-Barré syndrome to confirm a recent *Campylobacter* infection.
  D. PCR.
  FALSE: it can be used and may be used increasingly but currently is not part of routine diagnosis.
  E. String test.
  FALSE: this is used for *Giardia* or even *Helicobacter*.

8. **Which of the following antibiotics are used to treat gastroenteritis caused by *C. jejuni*?**
  A. Erythromycin.
  TRUE: but resistance is an increasing problem.
  B. Metronidazole.
  FALSE: although it is used for *Helicobacter*.
  C. Spiromycin.
  FALSE: this is used for toxoplasmosis.
  D. Gentamicin.
  FALSE: but may be used in cases of septicemia caused by *C. jejuni*.
  E. Ciprofloxacin.
  TRUE: but resistance is a problem caused by use of similar fluoroquinolones in veterinary practice.

9. **Which one of the following complications of infection affects the nervous system?**
  A. Guillain-Barré syndrome.
  TRUE: caused by several infections including *C. jejuni* – an ascending paralysis starting in the legs.
  B. Hemolytic uremic syndrome.
  FALSE: a complication of *E. coli* O157 or *Shigella* infection leading to hemolysis and renal failure.
  C. Reiter syndrome.
  FALSE: a complication of *Shigella* infection, among others, leading to arthritis.
  D. Miller Fisher syndrome.
  TRUE: a complication of *C. jejuni* infection affecting the cranial nerves.
  E. Anti-Tourette syndrome.
  FALSE: first reported by the lead singer of Spinal Tap in reference to gorillas, although Tourette syndrome may be related to a previous infection with *Streptococcus pyogenes* and leads to uncontrollable swearing.

CASE 4: *Chlamydia trachomatis*

1. **Which of the following are true?**
   A. Chlamydiaceae are obligatory intracellular pathogens.
      TRUE: they are obligatory intracellular human and animal pathogens.
   B. *Chlamydia trachomatis* is the only known human pathogen in the Chlamydiaceae Family.
      FALSE: *Chlamydophila pneumoniae* is the other human pathogen from this Family.
   C. *Chlamydia* species are characterized by tissue tropism.
      TRUE: different species of *Chlamydia* infect different tissues.
   D. Their principle membrane protein is 40 kDa major outer-membrane protein (MOMP).
      TRUE: this is the main immunodominant outer-membrane protein of *Chlamydia*.
   E. *Chlamydia* species express high density of surface proteoglycans (PGs).
      FALSE: although the gene for the PG was found in the genome of *Chlamydia*, surface expression of PG has not been documented.

2. **What is *Chlamydia trachomatis* characterized by?**
   A. A biphasic developmental cycle.
      TRUE: *Chlamydia* developmental cycle comprises two phases: elementary body (EB) and reticulate body (RB).
   B. Three human biological variants (biovars): trachoma, genital, and lymphogranuloma venereum (LGV).
      FALSE: there are only two human biovars: trachoma and LGV.
   C. One serological variant (serovar).
      FALSE: there are several serovars of *C. trachomatis*.
   D. High spread worldwide.
      TRUE: some 85 million people are affected by active trachoma worldwide, and 50 million new cases of STD caused by *C. trachomatis* are detected every year.
   E. Production of cytotoxins.
      TRUE: *C. trachomatis* like other bacterial pathogens produce bacterial toxins.

3. **Is the following true for the life cycle of *C. trachomatis*?**
   A. The infectious EB form of *C. trachomatis* infects host epithelial cells and phagocytes.
      TRUE: the EB form infects nonciliated columnar epithelial cells and phagocytes.
   B. The entry of the EB form occurs through phagocytosis, endocytosis, and pinocytosis forming endosomes called inclusions.
      TRUE: all these entry mechanisms are employed by *C. trachomatis*.
   C. EB-containing endosomes fuse with host cell lysosomes.
      FALSE: the fusion is prevented by *C. trachomatis* to avoid digestion.
   D. EB forms replicate in the inclusion.
      FALSE: the EB form transforms into the RB form, which replicates in the inclusion and then transforms back into the EB form.
   E. Progeny EB forms are released from the host cell by lysis, reverse endocytosis or apoptosis.
      TRUE: all these mechanisms are used by *C. trachomatis* for freeing newly produced EB forms.

4. **How is *C. trachomatis* spread?**
   A. By blood transfusion (ocular infection).
      FALSE: *C. trachomatis* is not transmitted by blood.
   B. Coughing and sneezing (ocular infection).
      TRUE: the ocular infection can be transmitted by droplets.
   C. By fingers, shared cloths or towels (ocular infection).
      TRUE: transmitted from eye to eye by fingers, shared cloths or towels.
   D. Through vaginal intercourse (genital infection).
      TRUE: *C. trachomatis* is the leading bacterial cause of sexually transmitted infections.
   E. From mother to child during birth.
      TRUE: *C. trachomatis* can be passed through an infected birth canal.

5. **Which of the following is true about the host response to *C. trachomatis* infection?**
   A. Immune response is insufficient and evokes chronic inflammation.
      TRUE: protective immunity to *C. trachomatis* infection is limited and results in inflammatory response and subsequent tissue damage.
   B. The most powerful T-cell-mediated immune response is production of IL-4.
      FALSE: the most powerful anti-chlamydial T-cell-mediated immune function is the production of IFN-γ.
   C. Specific IgG and IgM antibodies are directed towards bacterial lipopolysaccharide.
      FALSE: circulatory IgM and IgG antibodies are mostly directed to the immunodominant antigen MOMP.
   D. *C. trachomatis* blocks expression of MHC molecules on host cells.
      TRUE: this is one of the mechanisms used by the organism to evade the immune response.
   E. Interferon-γ enhances phagocytosis of *C. trachomatis*.
      TRUE: one of the ways IFN-γ limits chlamydial infection.

6. **Is the following true of urogenital infection caused by *C. trachomatis* in women?**
   A. The symptoms develop immediately after infection.
      FALSE: incubation period usually comprises 1–3 weeks.
   B. There is a cervical discharge in women.
      TRUE: these are important early symptoms.
   C. Dysuria and pyuria occur.
      TRUE: these symptoms also develop early.
   D. Cervicitis and urethritis can occur.
      TRUE: these symptoms develop later.
   E. Even treated infection always leads to infertility.
      FALSE: over 95% of women with effectively treated chlamydial infection will not develop infertility.

7. **Which of the following laboratory tests are used for the diagnosis of chlamydial infection?**
   A. Nucleic acid amplification tests (NAATs).
      TRUE: NAATs are now increasingly used for the diagnosis and differential diagnosis of chlamydial infections by the detection of chlamydial ribosomal RNA (rRNA) or DNA.
   B. Blood cell count.
      FALSE: *Chlamydia* cannot be diagnosed in blood.
   C. Microscopy of biopsies.
      FALSE: biopsies are not used for the diagnosis, but swab material is used instead.
   D. Susceptible cell cultures.
      TRUE: infected cultures of susceptible McCoy or HeLa cells are used.
   E. Polymerase chain reaction (PCR).

TRUE: PCR analysis for the presence of rRNA is used for diagnosis and differential diagnosis.

8. **Which of the following drugs are used for the treatment of chlamydial infections?**
   A. Erythromycin and azithromycin.
      TRUE: are used for the treatment of ocular and urogenital infections.
   B. Tetracyclines.
      TRUE: are used for the treatment of ocular infections.
   C. Amoxicillin.
      TRUE: can be used for the treatment of chlamydial infection in pregnant women.
   D. Rifampicin.
      FALSE: not used for the treatment of chlamydial infection.
   E. Doxycycline.
      TRUE: is used for the treatment of urogenital infections.

CASE 5: *Clostridium difficile*

1. **Which of the following are true of *Clostridium difficile*?**
   A. It is an anaerobic bacterium.
      TRUE: it is a gram-positive rod.
   B. It produces spores.
      TRUE: they can remain in the environment for long periods and act as a source of infection.
   C. It produces a zinc metalloprotease as the only toxin.
      FALSE: it produces three toxins: TcdA, TcdB, and a binary toxin. TcdA/B are glucosyltransferases and the binary toxin is an ADP ribosyltransferase.
   D. Toxin production is controlled by a sigma factor.
      TRUE: TcdD (positive regulator, sigma factor) and TcdC (negative regulator).
   E. A phage typing system is most useful in differentiating strains.
      FALSE: typing systems include ribotyping, PFGE, and MLEE.

2. **Which of the following are true concerning the epidemiology of *Clostridium difficile*?**
   A. Asymptomatic carriage in hospital is uncommon.
      FALSE: it can be up to 30%.
   B. Commonly associated with outbreaks in hospital.
      TRUE: although community-acquired infection is becoming increasingly recognized.
   C. Rarely causes extra-intestinal disease.
      TRUE: usually abscess, bacteremia.
   D. A recognized hypervirulent strain is ribotype 027.
      TRUE: although severe disease can also be caused by other ribotypes.
   E. Neonates may be colonized but do not acquire disease.
      TRUE: colonization can occur in up to 70% and it is thought that the neonate does not have the receptor for the toxin.

3. **Which of the following are known to be important in the pathogenesis of *Clostridium difficile*-associated disease?**
   A. Attaching and effacing lesions.
      FALSE: this occurs in *E. coli*.
   B. Production of glucosyltransferases by the organism.
      TRUE: they inactivate Rho, Rac Cdc42.
   C. Vacuolating cytotoxin.
      FALSE: this is produced by *H. pylori*.
   D. Taking antibiotics.
      TRUE: this is a risk factor, as is age > 60 years and hospital admission.
   E. Disruption of the actin cytoskeleton.
      TRUE: this causes loss of the tight junction holding cells together and cell death.

4. **Which of the following are important in the differential diagnosis of *Clostridium difficile*-associated disease?**
   A. Leptospirosis.
      FALSE: this presents with meningitis, jaundice or renal failure.
   B. Toxic megacolon.
      TRUE: this is a complication of CDAD and requires colectomy.
   C. *Giardia* gastroenteritis.
      TRUE: any infectious cause of diarrhea is in the initial differential diagnosis, although in an elderly patient in hospital on antibiotics CDAD is the most likely diagnosis.
   D. Inflammatory bowel disease.
      TRUE.
   E. Duodenal ulcer.
      FALSE: this presents with epigastric pain or evidence of perforation but not diarrhea and is caused by *Helicobacter pylori* or taking NSAIDs/steroids.

5. **Which of the following are true statements concerning the diagnosis of *Clostridium difficile*?**
   A. The Gram stain of fecal material is useful.
      FALSE: Gram stain of feces has no use.
   B. Detection of toxins in the blood is the main way of diagnosis.
      FALSE: detection of the toxin in the feces is the routine way of making a diagnosis.
   C. Culture of the organism is useful for epidemiological investigation.
      TRUE: as the organism can then be typed.
   D. Colonoscopy is important.
      TRUE: to detect pseudomembranes in PMC.
   E. Ribotyping is important.
      FALSE: it is not used in diagnosis of the illness but for epidemiological studies once the diagnosis has been made.

6. **Which of the following are used in the treatment of *Clostridium difficile*-associated disease?**
   A. Ampicillin.
      FALSE: it may cause CDAD.
   B. Cephalosporins.
      FALSE: it is one of the commonest antibiotic groups to cause CDAD.
   C. Metronidazole.
      TRUE.
   D. Vancomycin.
      TRUE.
   E. Clindamycin.
      FALSE: it may cause CDAD.

CASE 6: *Coxiella burnetti*

1. **Which of the following are true of *C. burnetii*?**
   A. *C. burnetii* is an obligate intracellular pathogen.
      TRUE: *C. burnetii* is an obligate intracellular bacterium.
   B. It only infects humans.
      FALSE: it is a zoonosis, is found in almost all animal species and humans are infected through animals.
   C. *C. burnetii* species has limited genetic variability and only consists of one genovar.
      FALSE: there are several varieties of *C. burnetii* genovars that are closely associated with their virulence and geographic distribution.
   D. It is characterized by two so-called phase variations.
      TRUE: *C. burnetii* consists of two phase variations based on the structure of LPS: virulent phase I variation and avirulent phase II variation.
   E. *C. burnetii* has a simple life cycle.
      FALSE: it has a complex life cycle and exists in two distinct forms: small cell variant (SCV) and large cell variant (LCV).

2. **Is *C. burnetii* characterized by?**
   A. Wide geographic distribution.
      TRUE: it is characterized by wide geographic distribution, except of New Zealand.
   B. Aerosol route of transmission.
      TRUE: this is the main route of SCV transmission from animals to humans.
   C. Frequent sexual transmission in humans.
      FALSE: not common.
   D. Ability to live and multiply in flies.
      FALSE: not proven. *C. burnetii* multiplies in the gut cells of ticks and is shed with their feces onto the skin of the animal host.
   E. Dissemination through blood to other organs.
      TRUE: *C. burnetii* is disseminated mostly via the blood monocytes to other organs, including the liver where they reside in the Kupffer cells.

3. **Is the following true for the life cycle of *C. burnetii*?**
   A. The SCV form of *C. burnetii* infects alveolar macrophages and monocytes.
      TRUE: the SCV form infects alveolar macrophages by attaching to integrin-associated proteins (IAPs), TLR4, and complement receptor 3 molecules.
   B. It survives in phagosomes and phagolysosomes with an alkaline pH.
      FALSE: *C. burnetii* is an acidophilic bacterium that can multiply in an acidic environment (pH = 4.7–5.2).
   C. SCVs divide by binary fission inside phagolysosomes and transform into LCVs.
      TRUE: after phagolysosomal fusion, the SCVs are activated, multiply, and may transform into LCVs, which is the metabolically active intracellular form of *C. burnetii*.
   D. LCVs do not divide and transform back into SCVs.
      FALSE: LCVs first divide also by binary fission prior to the transformation back into SCVs.
   E. SCVs are released from the host cell exclusively by macrophage apoptosis.
      FALSE: SCVs are released through cell lysis and exocytosis.

4. **Which of the following is true about the host response to *C. burnetii* and the ways the bacterium evades immune responses?**
   A. In most cases immune response is insufficient and evokes chronic inflammation.
      FALSE: immune responses are sufficient to eradicate *C. burnetii* infection in 98% of cases.
   B. Specific IgG, IgM, and IgA represent a powerful immune response to the infection.
      FALSE: production of antibodies is mostly of a diagnostic value.
   C. *C. burnetii* can release enzymes degrading reactive oxygen intermediates (ROIs).
      TRUE: *C. burnetii* produces enzymes superoxide dismutase and catalase, which degrade ROIs.
   D. *C. burnetii* inhibits maturation of phagosomes.
      TRUE: *C. burnetii* inhibits maturation of phagosomes into phagolysosomes.
   E. Interferon-γ produced by T cells enhances phagocytosis of *C. burnetii*.
      TRUE: IFN-γ is the main Th1 cytokine enhancing effective eradication of *C. burnetii*.

5. **Which of the following are the symptoms of acute Q fever?**
   A. Incubation period is about 14–26 days.
      TRUE: the incubation period of acute Q fever ranges from 14 to 26 days.
   B. It presents as a flu-like febrile illness.
      TRUE: acute Q fever presents as a flu-like febrile illness with headache, myalgia, and high-grade fever with chills and sweating.
   C. Q fever presents with typical pneumonia.
      TRUE: typical pneumonia develops as a result of *C. burnetii* penetration in the lungs.
   D. Q fever hepatitis is a common symptom.
      TRUE: dissemination of *C. burnetii* can cause hepatitis with hepatomegaly and with granulomas in more severe cases.
   E. Meningoencephalitis is a common symptom.
      FALSE: meningoencephalitis is a rare complication.

6. **Which of the following are the symptoms or predisposing factors of chronic Q fever?**
   A. The disease course is about 3 months.
      FALSE: chronic Q fever is defined by a disease course of more than 6 months.
   B. Chronic endocarditis is the main symptom.
      TRUE: chronic endocarditis is present in 60–70% of all chronic Q fever cases.
   C. Underlying cardiac valve defects are predisposing factors.
      TRUE: congenital, rheumatic, syphilitic or degenerative cardiac valve defects are found in over 90% of Q fever endocarditis patients.
   D. Immunosuppression and pregnancy are predisposing factors.
      TRUE: immunosuppression, immunodeficiency, and pregnancy are predisposing conditions.
   E. Doughnut-type granulomas develop in the liver and in the bone marrow.
      FALSE: they are characteristic of acute, but not chronic Q fever hepatitis.

7. **Which of the following laboratory tests are used for the diagnosis of infection with *C. burnetii*?**
   A. Blood cell count changes are highly characteristic for Q fever.
      FALSE: not of diagnostic value for Q fever.
   B. Duke's criteria are used for the clinical diagnosis of Q fever endocarditis.
      TRUE: Duke's criteria are used worldwide to calculate a diagnostic score for infective endocarditis.
   C. High levels of IgG and IgM antibodies to phase I *C. burnetii* indicate acute Q fever.
      FALSE: IgM followed by IgG to phase II *C. burnetii* is indicative of acute Q fever.
   D. Excess or equal titers of IgA and IgM anti-phase I antibodies as compared to anti-phase II antibodies are highly indicative of chronic Q fever.
      TRUE: these are used for the serologic diagnostic criteria for chronic Q fever.
   E. For microscopic observations tissue samples are stained with the Gram technique.
      FALSE: since *C. burnetii* does not have the dipicolinic acid and a spore coat with cysteine it cannot be stained by the Gram technique, and the Gimenez method is used instead.

8. **Which of the following drugs are used for the treatment of Q fever?**
   A. Doxycycline.
      TRUE: the drug of first choice for the treatment of Q fever.
   B. Tetracycline.
      TRUE: also used for the treatment of Q fever.
   C. Erythromycin and azithromycin.
      TRUE: used for the treatment of Q fever in patients with gastric intolerance to tetracyclines.
   D. Chloroquine.
      TRUE: is used for the treatment of chronic Q fever.
   E. Penicillin.
      FALSE: not used for the treatment of Q fever.

CASE 7: Coxsackie B virus

1. **Which of the following viruses are classified as enteroviruses?**
   A. Coxsackie B5 virus.
      TRUE.
   B. Echovirus 33.
      TRUE.
   C. Norwalk virus.
      FALSE: although this virus does indeed cause gastroenteritis, it belongs to the *Caliciviridae*.
   D. Poliovirus type 2.
      TRUE.
   E. Rotavirus.
      FALSE: although this virus does indeed cause gastroenteritis, it belongs to the *Reoviridae*.

2. **Which of the following clinical syndromes may arise in an immunocompetent 5-year-old child as a result of enterovirus infection?**
   A. Conjunctivitis.
      TRUE: particularly from EV70 virus infection.
   B. Herpangina.
      TRUE: this describes the occurrence of painful mouth blisters and ulcers.
   C. Foot and mouth disease.
      FALSE: foot and mouth disease is an infection of cattle and sheep caused by foot and mouth disease virus, which belongs to the picornaviruses, whereas hand, foot, and mouth disease is a disease of children caused by enterovirus infection (usually a coxsackie A virus).
   D. Gastroenteritis.
      FALSE: despite their name, there is no evidence that enteroviruses cause disease in the enteric tract.
   E. Lower motor neuron paralysis.
      TRUE: the disease poliomyelitis, caused by one of the three serotypes of poliovirus, gives rise to lower motor neuron paralysis.

3. **Which of the following features in a cerebrospinal fluid sample are suggestive of a viral cause of meningitis?**
   A. Cloudy fluid.
      FALSE: this is much more suggestive of a bacterial meningitis, with a florid polymorphonuclear white cell infiltrate giving rise to the cloudiness.
   B. A mononuclear cell infiltrate.
      TRUE: this is the typical response to a viral infection.
   C. CSF sugar less than half the value of the blood sugar.
      FALSE: CSF sugar is unaffected in viral meningitis, whereas metabolism of the glucose by bacteria in the CSF gives rise to a low value in bacterial meningitis.
   D. The presence of intracellular gram-negative diplococci.
      FALSE: these are obviously bacteria.
   E. CSF protein level at the upper limit of normal.
      TRUE: the usual finding in viral meningitis.

4. **Which of the following viruses are well-recognized causes of meningitis?**
   A. Coxsackie A9.
      TRUE.
   B. Echovirus 33.
      TRUE: this is another enterovirus.
   C. Herpes simplex virus.
      TRUE: usually in the setting of a young woman with genital herpes.
   D. Measles virus.
      FALSE: the feared complication of measles virus infection is encephalitis.
   D. Mumps virus.
      TRUE: before widespread vaccination, this was indeed the commonest cause of viral meningitis.

5. **Which of the following statements regarding enteroviral meningitis are true?**
   A. Enteroviruses account for up to 30% of cases of viral meningitis.
      FALSE: with the advent of vaccination against mumps meningitis, enteroviruses now account for well over 50% of cases of viral meningitis.
   B. Poor hygiene is a significant factor in the spread of illness.
      TRUE: results in fecal contamination of food and water.
   C. Most cases occur in children under the age of 14 years.
      TRUE.

D. Infection most often occurs in the winter months in temperate climates.
FALSE: enteroviral infections are more common in summer months.
E. Long-term sequelae including nerve deafness are common.
FALSE: the prognosis of enteroviral meningitis is very good – long-term sequelae are almost unheard of, unless the host is immunosuppressed.

6. **Which of the following laboratory assays can be used in the diagnosis of enteroviral meningitis?**
A. Demonstration of a rise in enterovirus-specific antibodies in paired acute and convalescent blood samples.
TRUE.
B. Electron microscopy of a cerebrospinal fluid sample.
FALSE: there are not sufficient amounts of virus in this condition to allow visualization by electron microcopy (which requires a particle density of at last 10^5 ml^{-1} of fluid).
C. Gram staining of a cerebrospinal fluid sample.
FALSE: this is used to identify bacterial causes of meningitis.
D. Reverse transcriptase polymerase chain reaction assay on a sample of cerebrospinal fluid.
TRUE: the RNA genome is reverse transcribed, and then amplified by PCR. This is by far the most sensitive diagnostic assay for this condition.
E. Virus isolation in cell culture from a fecal sample taken 10 days after the onset of illness.
TRUE: enteroviruses may be excreted in feces for 2–3 weeks following the onset of meningitis.

7. **Which of the following antiviral drugs have activity against enteroviruses?**
A. Aciclovir.
FALSE: this is a DNA synthesis inhibitor, enteroviruses

are RNA viruses, and also do not have the requisite thymidine kinase enzyme necessary to phosphorylate this drug and thereby make it active.
B. Amantadine.
FALSE: this inhibits uncoating of influenza A viruses only.
C. Oseltamivir.
FALSE: this is an influenza virus neuraminidase inhibitor.
D. Ritonavir.
FALSE: this is an HIV protease inhibitor.
E. Zidovudine.
FALSE: this is an HIV reverse transcriptase inhibitor.

8. **Poliovirus type 2 is isolated from a stool sample from a 20-year-old male nurse. The nurse recalls being vaccinated against poliovirus infection as a child with Sabin (live attenuated vaccine), and receiving a booster dose 4 years ago before traveling to India. Investigation of this person's immune system is most likely to reveal which of the following?**
A. No abnormality.
UNLIKELY: otherwise immunocompetent individuals may excrete live vaccine-derived poliovirus for perhaps 2–3 months after vaccination, but not for 4 years.
B. Antibody deficiency.
THE CORRECT ANSWER: it is highly likely that this individual is hypogammaglobulinemic.
C. T-cell (cell-mediated) deficiency.
NO: T-cell deficiency is associated with an increased risk of many infectious diseases, but not of chronic carriage of enteroviruses.
D. Abnormalities of phagocytosis.
NO: these are normally associated with recurrent bacterial infections.
E. Complement deficiency.
NO: some complement deficiencies do predispose to recurrent meningitis due to infection with *Neisseria meningitidis*, but not to chronic enterovirus carriage.

CASE 8: *Echinococcus* spp.

1. **Which of the following are true about *Echinococcus*?**
A. *Echinococcus* is a cestode tapeworm.
TRUE.
B. *Echinococcus granulosus* is restricted in its geographical distribution to the Northern Hemisphere.
FALSE: it is *Echinococcus multilocularis* that is restricted to the Northern Hemisphere.
C. Adult worms live in the intestine of humans, sheep or cattle.
FALSE: adult worms live in the intestine of dogs.
D. The life cycle for *Echinococcus granulosus* involves dogs.
TRUE.
E. *Echinococcus multilocularis* does not cause human infections.
FALSE.

2. **Which of the following are true about the life cycle of *Echinococcus* infection?**
A. Humans are infected by swallowing eggs.
TRUE.
B. The larval form that arises from the eggs is called an oncosphere.
TRUE.

C. Cysts usually develop from the oncosphere in the intestinal wall.
FALSE: this is not the usual site for cyst development. Instead they are more commonly found in the liver and/or lung.
D. Humans spread infection when they pass eggs in their feces.
FALSE: with rare exceptions humans are an end host. As adults do not live in the human intestine no eggs are produced to shed in human feces.
E. For *E. granulosus* dogs continue the life cycle when they eat sheep or cattle offal.
TRUE.

3. **Which of the following are true about the epidemiology of *Echinococcus* infection?**
A. Infection is principally an urban disease.
FALSE: infection is usually rural and associated with farms or nomadic herding communities.
B. The high incidence of infection observed in the Turkana district of Kenya is due to local fox populations.
FALSE: it is due to the close relationship of the Turkana people with their domesticated dogs.

C. Infection due to *E. multilocularis* can occur in arctic regions.
TRUE: this can be spread by the arctic fox.
D. Red fox populations are diminishing with a corresponding decline in *E. multilocularis*.
FALSE: conversely red fox populations are increasing, with an increase in *E. multilocularis* infection in some areas.
E. *E. granulosus* infection is disappearing worldwide.
FALSE: there is no sign that this infection has come under control.

4. **Which of the following components are included in the host response to *Echinococcus*?**
A. A Th2-type response has been clearly established against the invading oncospheres.
FALSE: very little is actually known about the host response to the oncospheres when they first penetrate the intestinal mucosa.
B. Complement activation occurs as the oncosphere matures into cysts.
TRUE.
C. Both CD4+ and CD8+ T lymphocytes are found around the cysts.
TRUE.
D. An inflammatory response occurs that continues unabated around the cysts.
FALSE: the inflammatory response abates once the acellular layer is established. In fact immunomodulatory processes come into operation.
E. Anaphylactic reactions occur in response to cysts that rupture and leak.
TRUE.

5. **Which of the following statements are true about hydatid cysts?**
A. Cysts are always symptomatic.
FALSE: small cysts, in noncritical positions may cause no symptoms.
B. The spleen is the commonest location for cysts.
FALSE: 90% of cysts occur in the liver and/or the lung, with the preponderance being in the liver.
C. Cysts are usually limited in size and do not enlarge to make organs palpable.
FALSE: cysts may grow so large that the liver becomes palpable.
D. More than one cyst can develop.
TRUE.
E. Cysts can cause pain.
TRUE: depending on size and position.

6. **What are the recognized complications of infection?**
A. Rupture of cysts leading to secondary daughter cysts.
TRUE.
B. Secondary infection with bacteria.
TRUE.
C. Obstructive jaundice when the biliary tree is compressed by a cyst.
TRUE.
D. Increased blood pressure during an anaphylactic reaction to cyst rupture.
FALSE: in fact in anaphylaxis blood pressure falls, and this may be associated with an urticarial rash and swelling of mucous membranes of the lips, tongue, vocal chords, and bronchial mucosa.
E. Death from progressive infection.
TRUE: this is particularly the case for *E. multilocularis*.

7. **What is the diagnosis of *Echinococcus* infection based on?**
A. A very specific serological test.
FALSE: the test is actually not very specific and is used as supportive evidence once a cyst is visualized.
B. A highly sensitive serological test.
FALSE: the test is not highly sensitive, with reports ranging from 50% for lung cysts and 80% for liver cysts.
C. Recognition of characteristic cyst features on ultrasound or CT scanning.
TRUE.
D. The presence of eosinophilia in the blood.
FALSE: eosinophilia is not the rule for *Echinococcus* infection, but may be seen when cysts leak.
B. A diagnostic aspiration of cysts to look for protoscoleces.
FALSE: aspiration is used more in the context of treatment in the procedure called PAIR, rather than in diagnosis. There is a risk of anaphylaxis if the cyst is ruptured by puncture.

8. **Which of the following should be considered in the differential diagnosis of *Echinococcus* infection?**
A. Benign cysts of the liver.
TRUE: although these will not have a thick, fibrous capsule.
B. Developmental cysts in the lung.
TRUE: these can have various radiological appearances.
C. Liver tumors.
TRUE: but once detailed radiology is obtained tumors will be shown to be solid structures.
D. Cirrhosis of the liver.
FALSE: this can cause liver enlargement through macronodular cirrhosis, but once detailed radiology is obtained there is no cystic change.
E. Amebic liver abscess.
TRUE: although like a benign cyst this will not have a thick, fibrous capsule and radiologists will report the lack of defined structures within.

9. **Which of the following are relevant for the treatment of *Echinococcus* infection?**
A. Surgical resection is always indicated.
FALSE: sometimes surgery is technically not feasible and for very small, asymptomatic cysts may be unnecessary.
B. Antiparasitic drug treatment before surgery is beneficial.
TRUE: this reduces cyst viability and makes surgery safer.
C. For drug treatment albendazole is always combined with praziquantel.
FALSE: at the moment there is insufficient evidence to say this must be done for all cases.
D. Drug treatment is used for 10 days.
FALSE: treatment is continued for months, often 3 months.
E. Cysts can also be killed by the injection of 95% ethanol.
TRUE: As part of the procedure called PAIR used in specialist units.

10. **Which of the following are true for the control of *Echinococcus* infection?**
A. It is helped by regular drug treatment of dogs to clear infection.
TRUE: dogs have been treated with arecoline or praziquantel.
B. It requires the avoidance of feeding farm dogs with offal.
TRUE: this helps to break the life cycle.
C. It requires the routine use of vaccination in animals.

TRUE: at the moment vaccination has been trialed but is not routine.
D. It is helped by vaccination of humans.
FALSE: to date vaccination has not become practice for humans.

E. It is helped by good hand hygiene measures in farm workers.
TRUE.

CASE 9: Epstein-Barr virus

1. **Which of the following are true of Epstein-Barr virus?**
   A. It is an RNA virus.
   FALSE: it is a DNA virus – a gamma herpesvirus.
   B. It has a helical capsid.
   FALSE: the capsid is icosahedral.
   C. It mainly infects T cells.
   FALSE: it infects mainly B cells but probably also squamous pharyngeal epithelial cells.
   D. It is a zoonosis.
   FALSE: it is a human virus that is passed on by direct intimate contact between individuals.
   E. It always causes infectious mononucleosis.
   FALSE: only about 30% of adolescents and young adults undergoing primary EBV infection develop the symptoms of infectious mononucleosis.

2. **Which of the following diseases are associated with EBV?**
   A. X-linked lymphoproliferative disease.
   TRUE: this rare familial condition affecting young boys is characterized by its extreme sensitivity to EBV.
   B. Graves' disease.
   FALSE: this is an autoimmune disease without any known infection origin.
   C. Burkitt's lymphoma.
   TRUE: this malignant tumor associated with EBV is endemic to central parts of Africa and New Guinea.
   D. Cervical cancer.
   FALSE: there is a strong association of this with human papillomavirus which is a DNA virus.
   E. Nasopharyngeal cancer.
   TRUE: this is relatively rare except in populations in southern China.

3. **Which of the following laboratory tests would be helpful in the diagnosis of infectious mononucleosis?**
   A. Paul-Bunnell test.
   TRUE: this test measures heterophil antibodies in the serum of suspected infectious mononucleosis patients, using agglutination of nonhuman erythrocytes such as horse cells. However, these are absent from about 10% of patients with adolescents and young adults.
   B. Monospot test.
   TRUE: the classic Monospot test uses horse erythrocytes but newer tests measure heterophil antibodies using latex agglutination.
   C. Electron microscopy of pharyngeal cell scraping.
   FALSE: not used.

   D. Culture of the virus on fibroblasts.
   FALSE: this virus will not grow on fibroblasts.
   E. Presence of serum IgM VCA antibodies.
   TRUE: the most useful serological assays detect IgM, which appears early in the illness and declines rapidly over the following 3 months. Usually measured by ELISA but can be detected using immunofluorescence methods.

4. **What are the typical symptoms of a patient presenting with infectious mononucleosis?**
   A. Pharyngitis, pyrexia, and cervical lymphadenopathy with tonsillar enlargement.
   TRUE: around 98% of patients presenting with clinical infectious mononucleosis have this triad of symptoms.
   B. Pharyngeal inflammation and transient palatal petechiae.
   TRUE: this presentation is typical in adolescents.
   C. Encephalitis.
   FALSE: this is rare in infectious mononucleosis.
   D. Osteomyelitis.
   FALSE: this is caused by bacterial infections such as *Staphylococcus aureus*.
   E. Hepatitis and jaundice in young patients.
   FALSE: but it is seen in about 9% of older patients.

5. **How would you treat a patient with infectious mononucleosis?**
   A. Just for symptoms.
   TRUE: in most cases there is nothing to give the patient apart from helping with pain, dehydration, etc. (however, see option D in this question).
   B. With aciclovir.
   FALSE: although this might have some effect on the lytic phase of the infection, it cannot work when the virus is latent in B cells.
   C. Aspirin.
   TRUE: patients can be treated with nonsteroidal drugs for fever and myalgias. However, aspirin should not be given to children because it may cause Reye's syndrome.
   D. Steroids.
   TRUE: corticosteroids are recommended in patients with significant pharyngeal edema that might cause respiratory compromise.
   E. Ampicillin.
   FALSE: this is effective against streptococcal infections but causes severe rashes in patients with EBV infection.

CASE 10: *Escherichia coli*

1. **Which of the following are true of *E. coli*?**
    A. It is a significant cause of urinary tract infections.
        TRUE: it is responsible for about 80% of cases.
    B. It is part of the normal flora of the intestine.
        TRUE: it is found in the feces.
    C. It lacks endotoxin in the cell wall.
        FALSE: it has a typical gram-negative cell wall with lipopolysaccharide (endotoxin).
    D. It is a significant cause of gastroenteritis.
        TRUE: and there are different strains – EPEC, ETEC, EIEC, EHEC, EAggEC.
    E. It is gram-positive
        FALSE: it is gram-negative.

2. **Which of the following are correct statements concerning urinary tract infection?**
    A. It is an exogenous infection.
        FALSE: it is an endogenous infection.
    B. In the neonate, it is more common in boys than in girls.
        TRUE: but in all other age groups it is more common in females.
    C. *Staphylococcus saprophyticus* is a common cause in women.
        TRUE: *Staphylococcus saprophyticus* is a cause of cystitis in women.
    D. Asymptomatic bacteriuria may have severe consequences in some individuals.
        TRUE: in pregnant women and children because of renal scarring.
    E. It is not very common in hospital.
        FALSE: it is the most common hospital-acquired infection owing to patients being catheterized.

3. **Which of the following are virulence factors of uropathogenic *E. coli* (UPEC)?**
    A. Cytolethal distending toxin.
        FALSE: this is produced by *Campylobacter*, the *E. coli* toxin is cytotoxin necrotizing factor.
    B. Type I fimbriae.
        TRUE: and are found in most *E. coli* strains.
    C. P-fimbriae.
        TRUE: and bind to the P-blood group antigen expressed on uroepithelium.
    D. S-layer.
        FALSE: this is found in *Campylobacter* and is part of the cell wall.
    E. Serum resistance.
        TRUE: and is due to the capsular polysaccharide, which prevents the binding of complement.

4. **The host responds in which of the following ways in a case of *E. coli* pyelonephritis?**
    A. Poor antibody response.
        FALSE: there is a good antibody response but its role in protection is unclear.
    B. Recruitment of granulocytes.
        TRUE: due to the secretion of IL-8 by uroepithelium.
    C. Development of a temperature.
        TRUE: due to the production of Th1 cytokines.
    D. Inefficient immune defenses in the renal medulla.
        TRUE: due to the high osmolarity and ammonium.
    E. Secretion of IL-6 and IL-8.
        TRUE: both are secreted and the levels of IL-6 correlate with severity of disease.

5. **Which of the following are typical of pyelonephritis?**
    A. Frequency of micturition.
        TRUE: also with cystitis.
    B. Dysuria.
        TRUE: also with cystitis.
    C. Rigors.
        TRUE: uncommon in cystitis.
    D. Loin pain.
        TRUE: usually a dull ache as opposed to the sharp, colicky pain of urolithiasis that radiates to the groin.
    E. Rash.
        FALSE.

6. **Which of the following are important in the routine diagnosis of urinary tract infection?**
    A. Microscopy.
        FALSE: although in some circumstances microscopy can be useful, e.g. in identifying sterile pyuria.
    B. Culture.
        TRUE: the mainstay of diagnosis.
    C. Significant bacteriuria.
        TRUE: greater than 10^5 bacteria per ml of a single species, although UTI can occur with fewer bacteria and in 5% of cases with more than one species.
    D. Automated methods of assessing bacteria may be used.
        TRUE: several methods exist and are commercially available.
    E. Antibiotic sensitivity testing.
        TRUE: as the organisms are variable in their resistance pattern, although initial treatment is often empirically based on the local resistance pattern.

7. **Which of the following are likely to be correct to treat an extended-spectrum β-lactamase (ESBL)-producing *Enterobacter* causing pyelonephritis?**
    A. Three days of cefuroxime.
        FALSE: as it will be resistant to β-lactam and it is too short a course.
    B. A carbapenem.
        TRUE: although carbapenem-resistant ESBL organisms have been reported.
    C. Two weeks of nitrofurantoin.
        FALSE: although 2 weeks is correct, nitrofurantoin does not penetrate tissue and is only useful for cystitis.
    D. Gentamicin.
        TRUE: for 2 weeks.
    E. A fluoroquinolone.
        TRUE: for 2 weeks.

CASE 11: *Giardia lamblia*

1. **Which of the following statements are true of *Giardia*?**
   A. The feco–oral route is a common route of infection.
   TRUE: poor sanitation leading to contamination of food/water and direct contact with fomites, etc. are the common routes of infection.
   B. Infection is through ingestion of eggs.
   FALSE: there is no egg stage in the *Giardia* life cycle. The infectious stage is the cyst, although immediate ingestion of trophozoites after they leave the body can result in infection.
   C. The upper part of the small intestine is the site of attachment of trophozoites.
   TRUE: after leaving the stomach the cyst wall breaks down (excystation) in the duodenum and releases trophozoites, which attach to the upper part of the small intestine, particularly the mid-jejunum.
   D. Trophozoites invade the epithelium and spread throughout the body.
   FALSE: trophozoites very rarely penetrate the epithelium and spread.
   E. Encystation takes place in the duodenum.
   FALSE: excystation takes place here. Encystation takes place lower down in the colon.

2. **What is the usual clinical presentation of giardiasis?**
   A. Watery diarrhea.
   TRUE: this is often seen within a few days of infection.
   B. Pulmonary infection.
   FALSE: this protozoan does not infect pulmonary epithelium and does not spread from the intestine.
   C. Malabsorption.
   TRUE: this is a common presentation in patients with chronic *Giardia*. Loss of weight is often seen.
   D. Trachoma.
   FALSE: trachoma is caused by *Chlamydia trachomatis*. *Giardia* trophozoites only attach to epithelial cells in the upper part of the small intestine.
   E. Blood in the stools of infected individuals.
   FALSE: although attachment of the trophoblasts results in some change in the intestinal epithelium (with some cell death) there is no penetration of the epithelial cell layer and little inflammation. Bloody stools are seen with other gastrointestinal infections such as *Campylobacter* and *Salmonella* that produce exotoxins.

3. **Which of the following statements are true of *Giardia*?**
   A. Most infections produce serious symptoms.
   FALSE: in most cases the disease is asymptomatic but the stools can still contain large numbers of cysts.
   B. It is the most common cause of traveler's diarrhea.
   FALSE: other microbes are responsible for the majority of cases of traveler's diarrhea. These include enterotoxigenic *E. coli* (commonest cause, 40% of cases), *Shigella*, *Campylobacter jejuni*. No pathogens are found in 20% of cases.
   C. It is only found in humans.
   FALSE: *Giardia* is found in a variety of animal species, including mammals, amphibians, and birds. The species of *Giardia* that infects humans is called *G. duodenalis*, *lamblia* or *intestinalis*. There is little evidence that there is significant transmission from animals to humans (i.e. that *Giardia* is a common zoonosis).
   D. It has an incubation period of 8 weeks.

   FALSE: generally symptoms appear within 1–3 weeks of infection.
   E. It shows antigenic variation.
   TRUE: *Giardia* has around 150 genes coding for VSP antigens. Since each trophozoite has only one of these genes expressed it can change expression from one to another VSP, thus helping to avoid the immune response.

4. **Which of the following are used for the treatment of giardiasis?**
   A. Amoxicillin.
   FALSE: this is a β-lactam antibiotic that inhibits the synthesis of the bacterial cell wall of a wide range of gram-positive and some gram-negative bacteria.
   B. Prednisone.
   FALSE: this is a synthetic corticosteroid used to treat inflammatory diseases including some autoimmune diseases, e.g. rheumatoid arthritis.
   C. Metronidazole (nitroimidazoles).
   TRUE: this is the main drug used to treat giardiasis.
   D. Mepacrine.
   TRUE: this is an acridine compound that can be used to treat giardiasis.
   E. Furazolidone.
   TRUE: this is a nitrofuran derivative used for treating giardiasis.

5. **How is *Giardia* infection diagnosed?**
   A. Identification of cysts in the stools
   TRUE: this is done on at least three stool samples. Fresh slide mounts stained with iodine identify the cysts.
   B. Identification of trophozoites in urine.
   FALSE: trophozoites do not invade. They can also be detected in the stool samples.
   C. With fluorescent antibodies to the cysts.
   TRUE: commercially available fluorescent antibodies can be used to stain cysts in stools.
   D. The 'rope' test.
   FALSE: this should be the 'string test,' which is used to diagnose giardiasis.
   E. The breath test.
   FALSE: this test is used for detecting urea in *Helicobacter* infections.

6. **Which of the following statements are true of *Giardia*?**
   A. Infection is more serious in immunocompromised individuals.
   TRUE: more serious infections occur in patients with HIV/AIDS, cancer, transplant patients, and the elderly.
   B. Trophozoites possess four pairs of flagella.
   TRUE: a trophozoite is pear-shaped, has four pairs of flagella, two nuclei, a ventral 'sucking' disk, and median bodies.
   C. Cysts can be boiled and still remain viable.
   FALSE: although cysts are hardy and can remain viable for up to 3 months in the environment, they are killed in boiling water. This is used to purify drinking water.
   D. An effective human vaccine is available.
   FALSE: to date no human vaccine has been produced for *Giardia* although one is commercially available for dog giardiasis.
   E. A prophylactic drug is available for travelers to warm climates.

FALSE: unlike prophylactic drugs used for malaria, no chemoprophylactic drugs are available to prevent giardiasis. Avoidance of contaminated water or food and good hygiene are important in preventing infection.

7. **Which of the following complications are associated with giardiasis?**
   A. Malabsorption.
      TRUE: this is seen in cases of chronic giardiasis and can lead to loss of weight.
   B. Blind loop syndrome.
      FALSE: this occurs when part of the intestine becomes blocked, so that digested food slows or stops moving

through the intestines. This leads to bacterial overgrowth with subsequent absorption problems. Caused by surgery to the intestine, also seen with Crohn's disease and systemic sclerosis.
   C. Failure to thrive.
      TRUE: children with giardiasis often show failure to thrive.
   D. Lactose intolerance.
      TRUE: some patients develop lactose intolerance.
   E. Intussusception.
      FALSE: this is the telescoping or prolapse of one portion of the bowel into an immediately adjacent segment. The cause is unknown.

CASE 12: *Helicobacter pylori*

1. **Which of the following statements are true of the cell wall of *Helicobacter*?**
   A. It increases somatostatin levels.
      FALSE: the LPS blocks the secretion of somatostatin.
   B. It has high endotoxin activity.
      FALSE: it has exceptionally low endotoxin activity due to the phosphorylation levels.
   C. The lipopolysaccharide has low levels of phosphorylation.
      TRUE.
   D. It does not contain lipopolysaccharide.
      FALSE: it is a typical gram-negative cell wall.
   E. It is gram-positive.
      FALSE: it is gram-negative.

2. **Which of the following are true of *H. pylori*?**
   A. The organism grows in 24 hours on agar.
      FALSE: it takes 5 days for primary isolation.
   B. It is a strict anaerobe.
      FALSE: it is microaerobic.
   C. It is acquired by the oral route.
      TRUE: either oro–oral or feco–oral.
   D. Infection occurs in childhood.
      TRUE.
   E. Over 50% of the global population are affected.
      TRUE: making it the commonest infection worldwide except for microbes causing periodontal disease.

3. **Which of the following statements are correct concerning the host response to *Helicobacter*?**
   A. Few granulocytes are recruited to the area.
      FALSE: there is a vigorous response.
   B. There is a strong IgG response.
      TRUE: but the antibody is ineffective due to escape mechanisms of *H. pylori*.
   C. *Helicobacter* is sensitive to complement *in vitro*.
      TRUE: but complement activity is inhibited *in vivo*.
   D. There is a poor IgA response.
      FALSE: there is a good IgA response both locally and systemically.
   E. *Helicobacter* resists phagocytosis.
      TRUE: binding of *Helicobacter* fails to stimulate phagocytosis.

4. **Which of the the following virulence factors of *Helicobacter* are involved in pathogenesis of disease?**
   A. A type IV secretion apparatus.
      TRUE: it both delivers the CagA protein into the host

cell and stimulates IL-8 secretion.
   B. Cytolethal distending toxin.
      FALSE: this is a toxin produced by *Campylobacter*. The toxin produced by *Helicobacter* is vacuolating cytotoxin.
   C. CagA protein.
      TRUE: this is transmitted into the host cell where is becomes phosphorylated and interferes with cell signaling.
   D. Lipopolysaccharide.
      TRUE: it inhibits the production of somatostatin.
   E. Teichoic acid.
      FALSE: it does not occur in the cell wall of *H. pylori*.

5. **Which of the following may be due to infection with *Helicobacter pylori*?**
   A. Gastric ulcer.
      TRUE: and duodenal ulcer.
   B. Idiopathic thrombocytopenic purpura.
      TRUE: and may be due to an autoimmunity stimulated by the *Helicobacter*.
   C. Gastric adenocarcinoma at the cardia.
      FALSE: *Helicobacter* is related to distal gastric carcinoma.
   D. Food poisoning.
      FALSE: this can be caused by *Salmonella* or *Campylobacter*.
   E. Septicemia.
      FALSE: although the organism has been detected in the blood on one occasion.

6. **Which of the following are typical signs or symptoms of duodenal ulcer?**
   A. Lower abdominal pain.
      FALSE: epigastric pain.
   B. Fever.
      FALSE: *H. pylori* does not induce a systemic inflammatory response.
   C. Pain relieved by food.
      TRUE: typically.
   D. Pain occurs in the early hours of the morning.
      TRUE: and classically is a localized pain.
   E. Diarrhea.
      FALSE: this is associated with pathogens such as *Salmonella*, *Campylobacter*, *Shigella*.

7. **Which of the following are important in the ideal routine diagnosis of *Helicobacter pylori*?**
   A. Culture.
      FALSE: not often performed except in referral centres or after failed eradication regimens.

B. Urea breath test.
TRUE: with a sensitivity and specificity of 98%.
C. Serology.
FALSE: as it does not give evidence of current infection.
D. Pepsinogen I/II ratio.
FALSE: but useful for assessing the degree of atrophy.
E. Fecal antigen test.
TRUE: gives evidence of current infection.

8. **Which of the following antibiotics are used to eradicate *Helicobacter pylori*?**
A. Erythromycin.

FALSE: does not penetrate to the epithelial surface of the stomach where the organism is located.
B. Metronidazole.
TRUE: but resistance is becoming a problem.
C. Clarithromycin.
TRUE: the mainstay of treatment.
D. Gentamicin.
FALSE: *Helicobacter* is resistant to this antibiotic.
E. Amoxicillin.
TRUE: and the prevalence of resistance is very low.

CASE 13: Hepatitis B virus

1. **Which of the following viruses contains DNA?**
A. Hepatitis A virus.
FALSE: contains positive single-stranded RNA.
B. Hepatitis B virus.
TRUE: contains partially double-stranded DNA.
C. Hepatitis C virus.
FALSE: contains positive single-stranded RNA.
D. Hepatitis D virus.
FALSE: contains RNA.
E. Hepatitis E virus.
FALSE: contains positive single-stranded RNA.

2. **Which of the following are recognized routes of transmission of hepatitis B virus?**
A. Fecal–oral.
FALSE: HBV is a blood-borne virus. HAV and HEV are transmitted by the fecal–oral route.
B. Via contaminated needles.
TRUE: injecting drug users who share needles/syringes are therefore at risk of infection, as are health-care workers who suffer needlestick injuries.
C. Via blood or blood products.
TRUE: most countries now screen blood donors for HBsAg to prevent this.
D. From mother to baby.
TRUE: the most important route of transmission, arises from exposure to an infected birth canal and infected maternal blood at the time of birth.
E. Sexual.
TRUE: chronic HBV carriers will have virus in their genital tracts.

3. **Which of the following viruses are usually transmitted by contaminated food and water?**
A. Hepatitis A virus.
TRUE: this can give rise to intra-household spread as an infected individual may contaminate food and water.
B. Hepatitis B virus.
FALSE: HBV is blood-borne.
C. Hepatitis C virus.
FALSE: HCV is blood-borne.
D. Hepatitis D virus.
FALSE: HDV is blood-borne and can only infect individuals who are also infected with HBV.
E. Hepatitis E virus.
TRUE: huge outbreaks arise through fecal contamination of water supplies.

4. **Infection with which of the following viruses carries an increased risk of development of hepatocellular carcinoma?**
A. Epstein-Barr virus.
FALSE: EBV is associated with malignant disease, e.g. certain lymphomas, but not with HCC.
B. Hepatitis A virus.
FALSE: there is no evidence linking HAV infection with any malignant disease.
C. Hepatitis C virus.
TRUE: chronic HCV infection, together with chronic HBV infection, accounts for the vast majority of HCC in the world.
D. Hepatitis E virus.
FALSE: there is no evidence linking HEV infection with any malignant disease.
E. Yellow fever virus.
FALSE: there is no evidence linking YFV infection with any malignant disease.

5. **Which of the following statements relating to hepatitis viruses is/are true?**
A. The risk of transmission of hepatitis B virus from mother to baby is greater than that of hepatitis C virus.
TRUE: vertical transmission rates are about 90% and 5%, respectively.
B. The chances of an adult becoming a chronic carrier are greater following infection with hepatitis B virus than with hepatitis C virus.
FALSE: chronic carriage rates following acute infection in an adult are 5–10% and 75–80%, respectively.
C. The risk of acute liver failure (fulminant hepatitis) is greater for infections with hepatitis B virus than it is for hepatitis A virus.
TRUE: about 1% of HBV infections are fulminant. The figure is much smaller (although not zero) for HAV.
D. The mortality associated with acute hepatitis E virus infection is greater than that with acute hepatitis A virus infection.
TRUE: although these two viruses share many epidemiological and clinical features, HEV-associated mortality is higher than HAV, especially in pregnant women.
E. Hepatitis D virus can only replicate in cells already infected with hepatitis C virus.
FALSE: HDV requires the presence of HBV (not HCV) as a helper virus to allow it to complete its replication cycle.

6. **In a chronic carrier of hepatitis B virus, which of the following markers may be present in the patient's serum?**
   A. Antibody to hepatitis B core antigen.
   TRUE: these would be of the IgG class.
   B. Antibody to hepatitis B e antigen.
   TRUE: some carriers eliminate HBeAg, and become anti-HBe antigen-positive. Nevertheless, they remain HBsAg-positive, and therefore, by definition, are still HBV carriers.
   C. Antibody to hepatitis B surface antigen.
   FALSE: it is usually not possible to detect this antibody, as the patients have a vast excess of its cognate antigen, HBsAg. Thus any antibody that is present is bound up in immune complexes and cannot be detected.
   D. Hepatitis B surface antigen.
   TRUE: indeed, this is the defining characteristic of a hepatitis B carrier.
   E. HBV DNA.
   TRUE: in a chronic carrier, virus is replicating in the liver, and new virus particles are formed which are released into the bloodstream. These will contain HBV DNA.

7. **Which of these statements, concerning hepatitis B e antigen (HBeAg) and antibody to HBeAg (anti-HBe), is/are true?**
   A. The presence of HBeAg in serum means the patient is at increased risk of serious chronic liver disease.
   TRUE: the presence of HBeAg in peripheral blood is a surrogate marker for very extensive virus replication within the liver. There is therefore an increased risk of liver cell death (and replacement by fibrous tissue) due to CTL responses.
   B. The presence of anti-HBe in serum means that the patient is of low infectivity.
   TRUE: in some HBV carriers, virus replication slows down, HBeAg is lost, and shortly afterwards, anti-HBe becomes detectable. This is therefore a surrogate marker of lower levels of virus replication and therefore less infectivity.
   C. The e antigen is a breakdown product of the surface antigen.
   FALSE: HBeAg is derived from the same gene as the core antigen.
   D. There are mutants of HBV that do not synthezise e antigen.
   TRUE: in these patients, the absence of HBeAg from serum does NOT correlate with lower infectivity.
   E. The presence of HBeAg in a serum sample indicates that infection with hepatitis B virus has taken place within the last 3 months.
   FALSE: HBeAg can persist for many years in an HBV carrier, therefore its presence does not provide any information about when infection took place.

8. **In relation to the treatment of patients with chronic HBV infection, which of the following statements are correct?**
   A. IFN-α therapy works primarily by rendering uninfected cells resistant to virus replication.
   FALSE: IFN-α in chronic HBV infection is believed to act as an immunomodulatory agent, boosting the host cytotoxic T-cell response, rather than as an antiviral agent.
   B. IFN therapy of chronic HBV infection is less likely to succeed in patients who are co-infected with HIV.
   TRUE: as the mechanism of action of IFN is mediated

through the host immune system, then it will not be as efficacious in individuals whose immune system is not intact.
   C. The main mode of action of lamivudine is to prevent assembly of mature virus particles.
   FALSE: while virus will not be able to assemble in lamivudine-treated cells, the primary mode of action of lamivudine is to inhibit RNA-dependent DNA synthesis.
   D. Polymerase mutations leading to lamivudine resistance arise in the majority of patients treated for more than 3 years.
   TRUE: these resistance rates have been reported in patients receiving lamivudine monotherapy. It is possible that resistance rates would not be so high if patients were treated with combination therapy involving multiple agents.
   E. Treatment with HBV polymerase inhibitors usually leads to elimination of HBV from an infected liver.
   FALSE: essentially, the RT inhibitors suppress virus replication without altering the survival or otherwise of infected hepatocytes. When treatment is stopped, there is often a reappearance of viral replication.

9. **Which ONE of the following statements about hepatitis B vaccine is true?**
   A. Current vaccines are derived from human plasma.
   FALSE: early HBV vaccines were indeed purified from human sera, but nowadays the vaccine is prepared by recombinant DNA technology.
   B. It is a live attenuated vaccine.
   FALSE: it is a subunit (and therefore nonlive) vaccine consisting entirely of HBsAg.
   C. Multiple doses have to be given to ensure an adequate immune response.
   TRUE: this is a feature of many nonlive vaccines – multiple doses are required to generate an immune response of sufficient potency to offer protection.
   D. Response can be monitored by measuring antibody to hepatitis B core antigen in the recipient's serum.
   FALSE: the only immune response generated by the vaccine is anti-HBs, not anti-HBc.
   E. It will also provide protection against infection by hepatitis E virus.
   FALSE: it will provide protection against HDV infection, as HDV carries the same HBsAg surface protein as does HBV, which will therefore also be neutralized by anti-HBs.

10. **Which of the following groups is at increased risk of HBV infection, and should therefore be targeted in a selective HBV vaccination policy?**
   A. Health-care workers.
   TRUE: all HCWs should be vaccinated, as they may be exposed to HBV from infected patients.
   B. Sex workers.
   TRUE: HBV is sexually transmitted, therefore these individuals are at high risk of acquiring infection.
   C. Frequent travelers.
   FALSE: it is not the traveling *per se* that puts an individual at risk – but the behavior of the traveler once he/she has reached his destination.
   D. Sexual partners of known hepatitis B carriers.
   TRUE: HBV is sexually transmitted, therefore these individuals are at high risk of acquiring infection.
   E. Babies of HBV carrier mothers.
   TRUE: mother to baby transmission of infection can be prevented by vaccinating the baby at birth.

CASE 14: Herpes simplex virus 1 (HSV-1)

1. **Which of the following are features of herpesviruses?**
   A. They carry their genome in the form of single-stranded DNA.
      FALSE: herpesvirus genomes are composed of double-stranded DNA.
   B. The genome size is of the order of 150–200 kb.
      TRUE.
   C. They are nonenveloped viruses.
      FALSE: herpesviruses are enveloped.
   D. They have a helical capsid.
      FALSE: herpesviruses have an icosahedral capsid.
   E. There is a reverse transcriptase step in their replication cycle that renders them sensitive to reverse transcriptase inhibitors.
      FALSE: herpesviruses have a DNA polymerase enzyme, but no reverse transcriptase.

2. **Which of the following may be outcomes of herpes simplex virus infection of a cell?**
   A. Acute cytolytic infection.
      TRUE: cell death arises in 24–48 hours.
   B. Chronic infection.
      FALSE: this has not been described.
   C. Integration of the viral genome into a host cell chromosome.
      FALSE.
   D. Latent infection.
      TRUE: the site of latency of HSV is the nerve cell body.
   E. Transformation.
      FALSE: this has not been described. However, other members of the herpesvirus family are transforming viruses, e.g. Epstein-Barr virus transforms peripheral blood B lymphocytes.

3. **Which of the following statements regarding the pathogenesis of herpes simplex encephalitis are true?**
   A. About one-third of cases of HSE arise as a result of primary HSV infection.
      TRUE: in the remaining two-thirds of cases, antibodies to HSV are already present, and thus the pathogenesis is believed to be from reactivated or secondary infection.
   B. Brain cell loss occurs through both virus-induced cell death and the action of host cytotoxic T cells.
      TRUE: virus is cytolytic resulting in cell death. There is also evidence that the host immune response is responsible for some part of the cell death – in patients who are immunodeficient, the disease course is less acute, but progressively deteriorating.
   C. Hemorrhage is a typical pathological feature.
      TRUE: hemorrhagic necrosis is the principal pathological finding.
   D. The predominant cellular infiltrate in affected brain is composed of polymorphonuclear leukocytes.
      FALSE: the predominant cellular infiltrate is with mononuclear cells.
   E. Visually normal brain tissue adjacent to affected areas of brain is also microscopically normal.
      FALSE: microscopic evidence of disease extends beyond what is visible macroscopically.

4. **Which of the following are common clinical manifestations evident in a patient with herpes simplex encephalitis (i.e. occur in > 50% of patients)?**
   A. Bilateral muscle weakness.
      FALSE: hemiparesis, i.e. paralysis of one side of the body, is not unusual, but objective bilateral weakness is.
   B. Diarrhea.
      FALSE: gastrointestinal upset is not a feature of this disease, although a significant number of patients may give a history of vomiting.
   C. Drowsiness.
      TRUE: altered levels of consciousness are a cardinal feature.
   D. Headache.
      TRUE: this is a cardinal feature.
   E. Personality change.
      TRUE: this is a cardinal feature.

5. **Which of the following statements are true of the CSF findings in patients with herpes simplex encephalitis?**
   A. A normal CSF excludes the diagnosis of HSE.
      FALSE: the CSF is normal on first evaluation in about 5–10% of cases of HSE.
   B. Red cells are commonly found in the CSF.
      TRUE: this reflects the hemorrhagic nature of the pathology within the brain.
   C. The typical white cell response is of mononuclear cells.
      TRUE.
   D. The CSF protein is usually below the lower limit of normal.
      FALSE: the protein is usually raised, and continues to rise as disease progresses.
   E. The CSF sugar is usually less than half that of the blood sugar.
      FALSE: low CSF sugar in comparison to blood sugar is usually indicative of a bacterial infection, not a viral one.

6. **Which of the following statements regarding the use of neuroimaging techniques in the diagnosis of herpes simplex encephalitis are correct?**
   A. A focal lesion in the temporofrontal region is the most common finding.
      TRUE: although HSE can arise anywhere within the brain, this is the most common site.
   B. Bilateral lesions may be seen.
      TRUE: particularly if the disease has been present for a few days without therapy being initiated.
   C. CT scanning is the most sensitive neuroimaging technique for the diagnosis of HSE.
      FALSE: lesions are seen first on MRI scans.
   D. EEG is not usually abnormal in a patient with HSE.
      FALSE: spike and slow-wave activity with periodic epileptiform discharges arising from the temporal lobe are frequently seen.
   E. The detection of diffuse cerebral inflammation on an MRI scan confirms the diagnosis of HSE.
      FALSE: HSE is a focal disease, and if diffuse changes across the brain substance are seen, this renders HSE an unlikely diagnosis.

7. **Aciclovir works through which of the following mechanisms?**
   A. DNA polymerase inhibition.
      TRUE: aciclovir triphosphate is a potent inhibitor of viral DNA polymerase.
   B. Neuraminidase inhibition.

FALSE: oseltamivir and zanamavir are neuraminidase inhibitors, used in the treatment of influenza virus infection.

C. Protease inhibition.

FALSE: protease inhibitors such as ritonavir, indinavir, are used in the treatment of HIV infection.

D. Reverse transcriptase inhibition.

FALSE: AZT (zidovudine) and 3TC (lamivudine) are examples of reverse transcriptase inhibitors, which are used in the treatment of HIV infection.

E. Viral entry inhibition.

FALSE: enfuvirtide is a drug that inhibits the entry of HIV into target cells.

8. **Aciclovir is a selective antiviral agent for which of the following reasons (more than one may apply)?**

A. It becomes concentrated specifically in virally infected cells.

TRUE: because within HSV-infected cells, free aciclovir is phosphorylated into aciclovir monophosphate, which therefore reduces the intracellular concentration of aciclovir, and hence more free drug will diffuse into the infected cell.

B. It binds to a target enzyme which is not present in host cells.

FALSE: it binds to DNA polymerase enzymes, and host cells do possess such an enzyme.

C. It binds to a viral enzyme with much greater affinity than it does to the corresponding cellular enzyme.

TRUE: this is one of the reasons why aciclovir is such a nontoxic drug.

D. It requires activation, the first step of which cannot be performed by the host cell and therefore requires the presence of virus within the cell.

TRUE: another reason why aciclovir is a safe drug – it

does not become activated in noninfected cells.

E. It is only able to diffuse into cells that are infected with virus.

FALSE: the drug will diffuse into all cells, but will only progress down the pathway to triphosphorylation in virally infected cells, as the first step in this pathway requires a virally encoded enzyme not present in host cells (thymidine kinase).

9. **Which of the following drugs will inhibit the replication of herpes simplex virus?**

A. Azidothymidine.

FALSE: this is a reverse transcriptase inhibitor that is used in the treatment of HIV infection.

B. Foscarnet.

TRUE: this is a pyrophosphate analog that has activity against a wide range of DNA polymerase enzymes, including that of HSV. However, it is much more toxic than aciclovir and its derivatives, so foscarnet is usually only indicated in the treatment of HSV-associated disease if the virus is known to be aciclovir-resistant.

C. Ganciclovir.

TRUE: however, as with foscarnet, this is a much more toxic drug than aciclovir and its derivatives, and is therefore not routinely used as a first-line agent in the treatment of HSV disease.

D. Penciclovir.

TRUE: the antiviral spectrum of activity and mode of action of penciclovir are very similar to those of aciclovir. It is marketed as famciclovir, a complex ester that is hydrolyzed after absorption into penciclovir.

E. Valaciclovir.

TRUE: this is a valine ester of aciclovir that is hydrolyzed to aciclovir after absorption. It is much better absorbed orally than aciclovir.

CASE 15: Herpes simplex virus 2 (HSV-2)

1. **Which of the following viruses belong to the herpesvirus family?**

A. Coxsackie B virus.

FALSE: this is an enterovirus, family *Picornaviridae*.

B. Cytomegalovirus.

TRUE.

C. Epstein-Barr virus.

TRUE.

D. Varicella-zoster virus.

TRUE.

E. Variola major virus.

FALSE: this is the cause of smallpox, and is a poxvirus.

2. **All herpesviruses undergo the phenomenon of latency. Which of the following statements regarding latency are correct?**

A. Viral nucleic acid is found in the cell nucleus.

TRUE.

B. Viral capsid proteins are found on the plasma membrane of the cell.

FALSE: in latency there is very little, if any, production of viral proteins, and no production of viral structural components.

C. There is continuous release of new viral particles from the cell.

FALSE: there is no formation of viral particles.

D. Viral particles are located in the cell cytoplasm but are not released from the cell.

FALSE: there is no formation of viral particles.

E. The presence of latent virus inhibits the specialized functioning of the cell.

FALSE: there is no evidence for this.

3. **Which of the following are recognized routes of spread of herpes simplex virus?**

A. Air-borne.

FALSE.

B. Direct contact.

TRUE.

C. Mother to baby.

TRUE: resulting in neonatal herpes.

D. Sexual.

TRUE.

E. Water-borne.

FALSE.

4. **Viruses that undergo latency may cause both primary and secondary (or reactivated) infections. Which of the following statements regarding the clinical manifestations of herpes simplex virus infection are true?**

A. Inguinal lymphadenopathy is a common feature of recurrent genital herpes.

FALSE: this may be a feature of primary infection, but is very unusual in recurrent disease.

B. Infection in neonates who acquire infection from mothers undergoing a primary genital infection late in pregnancy is usually confined to the skin and mucous membranes.
FALSE: in only a small minority of cases of neonatal herpes is virus confined to the skin and mucous membranes. In the vast majority of such infections, there is internally disseminated infection, whether or not there is evidence of infection on the skin and mucous membranes.

C. Person to person spread of genital herpes most likely arises from a source patient with asymptomatic genital shedding of virus.
TRUE.

D. Primary infection may give rise to systemic, as well as local, symptomatology.
TRUE: expressed as regional lymphadenopathy and a febrile response.

E. The majority of primary infections in the genital area result in severe symptomatic disease.
FALSE: most people with antibodies to HSV-2 give no history of genital (or oral) disease. Thus, their primary infections must have been asymptomatic.

5. **Recognized complications of genital HSV infection include which of the following?**
   A. Eczema.
   FALSE: although in patients with eczema, primary or recurrent HSV infection can give rise to eczema herpeticum, a life-threatening complication.
   B. Increased risk of acquisition of HIV infection.
   TRUE.
   C. Cholecystitis.
   FALSE.
   D. Meningitis.
   TRUE.
   E. Urinary retention.
   TRUE: either via involvement of autonomic nerves, or if there are lesions near to the urethral meatus.

6. **Which of the following components of the immune system is the most important in controlling the outcome of reactivated genital HSV infection?**
   A. B cells.
   FALSE.
   B. Complement.
   FALSE.
   C. Macrophages.
   FALSE.
   D. NK cells.
   FALSE.
   E. T cells.
   TRUE: patients with T-cell deficiencies, e.g. through HIV infection, or transplant recipients, are very much more likely to get frequent, severe recurrences than are immunocompetent individuals.

7. **Regarding the diagnosis of genital HSV infection, which of the following statements are correct?**
   A. Genome detection assays demonstrating the presence of HSV DNA in a vesicle swab are the most sensitive assays available.
   TRUE: genome detection assays are extremely sensitive, so it is possible for patients to be culture-negative, PCR-positive.
   B. The clinical features of recurrent genital herpes are so distinctive that laboratory confirmation is not usually required.
   FALSE: there are many causes of genital ulceration, and proof that a given ulcer is due to HSV infection requires laboratory confirmation.
   C. The diagnosis can be made by demonstrating the presence of antibodies specific for HSV-2 in a serum sample.
   FALSE: the presence of such antibodies indicates that the patient has been infected with HSV-2, but does not prove that infection was genital (as opposed to oral) or that a given ulcer is due to HSV reactivation.
   D. The diagnosis can be made by immunofluorescent antigen detection using cells scraped from the base of a genital ulcer.
   TRUE: although this is not as sensitive as PCR-based detection.
   E. The virus can be isolated in tissue culture from vesicle fluid.
   TRUE.

8. **Which of the following drugs are routinely used to treat genital HSV infection?**
   A. Amantadine.
   FALSE: this drug is used in the management of influenza virus infection.
   B. Azidothymidine.
   FALSE: this is a reverse transcriptase inhibitor used in the treatment of HIV infection.
   C. Famciclovir.
   TRUE: this is the oral prodrug of penciclovir.
   D. Ganciclovir.
   FALSE: although HSV is sensitive to ganciclovir, this drug has to be given by i.v. injection, and therefore is reserved for possible use in patients with aciclovir-resistant HSV.
   E. Valaciclovir.
   TRUE: this is the oral prodrug of aciclovir.

9. **Resistance to aciclovir can arise in HSV infection through which of the following mechanisms?**
   A. Acquisition of an aciclovir efflux pump.
   FALSE.
   B. Deletion of the viral thymidine kinase gene.
   TRUE: TK-negative mutants of HSV do exist in nature and are selected for when a patient is treated with aciclovir. However, loss of the TK gene is also associated with loss of virulence, so such strains are not a clinical problem.
   C. Mutations in the UL97 gene.
   FALSE: these are associated, in cytomegalovirus, with resistance to ganciclovir.
   D. Mutations in the viral DNA polymerase gene.
   TRUE: such that the binding of aciclovir triphosphate to the viral DNA polymerase is much reduced. These DNA pol mutants remain fully pathogenic.

CASE 16: *Histoplasma capsulatum*

1. **Which of the following are characteristic of *H. capsulatum*?**
   A. Exclusive growth at 37°C temperature.
      FALSE: *H. capsulatum* grows as a saprophytic mold at 25°C and as a parasitic yeast at 37°C.
   B. Ability to infect both birds and bats.
      FALSE: bats are infected, but not birds, although both enrich the soil with excretions and support the growth of the fungus.
   C. Only one variety: *H. capsulatum* var. *capsulatum*.
      FALSE: *H. capsulatum* has three varieties: *H. capsulatum* var. *capsulatum*, which causes the common histoplasmosis, *H. capsulatum* var. *duboisii*, which causes African histoplasmosis, and *H. capsulatum* var. *farciminosum*, which causes lymphangitis in horses.
   D. Ability to produce microconidia as well as macroconidia.
      TRUE: fungal mycelium produces micro- and macroconidia.
   E. Being endemic to the Ohio, Missouri, and Mississippi River valleys.
      TRUE: *H. capsulatum* can be found in temperate climates, predominantly in river valleys in North and Central America.

2. **What are the main characteristics of the interaction between *H. capsulatum* and the host immune system?**
   A. *H. capsulatum* undergoes transformation from mold to yeast form inside alveolar macrophages.
      TRUE: *H. capsulatum* undergoes transformation into the pathogenic yeast and replicates inside macrophages.
   B. In many case *H. capsulatum* prevents formation of phagolysosome.
      TRUE: *H. capsulatum* prevents fusion of phagosomes and lysosomes and increases its pH.
   C. T-cell immunity plays a leading role in eradication of *H. capsulatum*.
      TRUE: T lymphocytes produce cytokines that activate phagocytes, with IFN-γ and TNF-α.
   D. Specific antibodies play an important protective role against *H. capsulatum*.
      FALSE: the protective role of antibodies appears insignificant.
   E. IL-4 and IL-10 are crucial for anti-histoplasma immune responses.
      FALSE: type 1 cytokines are important to fight the infection (IFN-γ and TNF-α).

3. **Which of the following is true for infection with *H. capsulatum*?**
   A. Microconidia and mycelial fragments of *H. capsulatum* are inhaled.
      TRUE: *H. capsulatum* infects by inhalation of air-borne microconidia and mycelial fragments.
   B. It is mostly asymptomatic or self-limiting in immunocompetent hosts.
      TRUE: in immunocompetent hosts infection is asymptomatic, self-limiting or benign.
   C. Progressive disseminated histoplasmosis is a mild disease.
      FALSE: this is a potentially fatal disease developing in immunocompromised individuals.
   D. Chronic pulmonary histoplasmosis develops in patients with underlying pulmonary diseases.
      TRUE: this form develops in patients with emphysema and can co-exist with tuberculosis, actinomycoses, and other lung diseases.
   E. Histoplasmosis never disseminates to the central nervous system.
      FALSE: progressive disseminated histoplasmosis disseminates to the CNS.

4. **Which are the most frequent symptoms of acute progressive disseminated histoplasmosis?**
   A. Oral and bowel ulcers.
      TRUE: the result of dissemination to the gastrointestinal system.
   B. Shrinking of the adrenal glands.
      FALSE: the adrenal glands are enlarged.
   C. Anemia, leukopenia, and thrombocytopenia.
      TRUE: associated pathology of the bone marrow.
   D. Lung cavities.
      FALSE: not characteristic with this form of histoplasmosis.
   E. Meningitis and cerebritis.
      TRUE: the result of dissemination to the CNS.

5. **Which of the following is true about *H. capsulatum* var. *duboisii*?**
   A. Its cells are smaller than those of *H. capsulatum* var. *capsulatum*.
      FALSE: *H. capsulatum* var. *duboisii* usually grows as a large yeast (7–15 μm in diameter).
   B. It causes African histoplasmosis.
      TRUE: infection with var. *duboisii* causes African histoplasmosis (histoplasmosis duboisii).
   C. Strong T-cell responses have been detected.
      FALSE: no appreciable cellular responses to var. *duboisii* have been detected.
   D. Host response involves giant cells and macrophages.
      TRUE: large numbers of giant cells (up to 80 μm), macrophages, and neutrophils are detected.
   E. It often involves the lungs.
      FALSE: it rarely involves the lungs but infects cutaneous, liver, lung, lymphatic, subcutaneous, and bony tissues.

6. **Which of the forms of histoplasmosis can be detected by chest radiography?**
   A. Acute pulmonary histoplasmosis.
      FALSE: usually does not show any irregularities.
   B. Chronic pulmonary histoplasmosis.
      TRUE: the cavitations are found in almost 90% of patients.
   C. Acute progressive disseminating histoplasmosis.
      TRUE: the chest radiography becomes informative when the disease progresses.
   D. Chronic progressive disseminating histoplasmosis.
      FALSE: does not present with chest radiography changes.
   E. Presumed ocular histoplasmosis syndrome.
      FALSE: cannot be diagnosed by chest radiography.

7. **Which of the following laboratory tests are the most informative for the diagnosis of histoplasmosis?**
   A. Cell culture.
      TRUE: sputum and blood are usually used for cultures.
   B. Histology.
      TRUE: direct microscopy evidence of the histological

specimen is a serious proof of infection.
C. Lymphocyte cell counts.
FALSE: not particularly informative.
D. CT scan.
TRUE: helps to detect subacute progressive disseminated and cerebral histoplasmosis.
E. Serology.
TRUE: the complement fixation method is used for diagnosis.

8. **Which of the following drugs are used for the treatment of progressive disseminating histoplasmosis?**
A. Clarithromycin.
FALSE: not used for the treatment of histoplasmosis.

B. Itraconazole.
TRUE: a fungistatic drug, can be used to treat chronic pulmonary histoplasmosis or chronic and subacute progressive disseminated histoplasmosis.
C. Penicillin.
FALSE: not used for the treatment of histoplasmosis.
D. Amphotericin B.
TRUE: the antifungal drug of first choice for the treatment of acute pulmonary histoplasmosis, chronic pulmonary histoplasmosis, all forms of progressive disseminated histoplasmosis.
E. Tetracycline.
FALSE: not used for the treatment of histoplasmosis.

CASE 17: Human immunodeficiency virus (HIV)

1. **Which of the following statements regarding the human immunodeficiency virus are correct?**
A. It belongs to the flavivirus family.
FALSE: HIV is a retrovirus.
B. It carries a double-stranded RNA genome.
FALSE: it carries two copies of ssRNA.
C. It is an enveloped virus.
TRUE: this makes it relatively easy to inactivate, by stripping off the envelope.
D. The genome encodes both a reverse transcriptase and a protease.
TRUE.
E. Infection of humans has arisen through transfer of simian immunodeficiency viruses into the human population.
TRUE: at least four different trans-species transmissions must have occurred in the last 50 years.

2. **Which of the following are recognized routes of infection of human immunodeficiency virus?**
A. Fecal–oral.
FALSE.
B. Droplet spread.
FALSE.
C. Percutaneous.
TRUE: intact skin is resistant to infection, but penetration of the skin, e.g. via a needle, is an efficient way of transmitting infection.
D. Sexual.
TRUE.
E. Transplacental.
TRUE: although it is believed that most babies who acquire infection from their mothers do so at the time of birth, rather than antenatally.

3. **Which of the following statements regarding the entry of HIV into target cells are correct?**
A. The initial binding event occurs between viral gp120 and cellular CD8 molecules.
FALSE: CD4 molecules.
B. CCR5 is a co-receptor for HIV binding and entry.
TRUE: found on CD4+ T cells (especially memory cells).
C. Infection of CD4+ T cells can arise through interaction with dendritic cells.
TRUE: often through HIV bound to DC-Sign on the dendritic cells.

D. Viral entry proceeds through fusion of the viral envelope with the cell membrane.
TRUE.
E. Viral entry can be inhibited by anti-gp120 antibodies.
TRUE.

4. **Which of the following statements relating to the HIV replication cycle are correct?**
A. Reverse transcription results in production of a double-stranded DNA copy of the virus.
TRUE.
B. The RNA genome of HIV can exist in a latent state inside infected cells.
FALSE: the virus is reverse transcribed into DNA, which then integrates into host chromosomes.
C. The viral genes are polycistronic, resulting in the translation of polyproteins.
TRUE.
D. The virus requires the availability of cellular enzymes to integrate its genome into host cell chromosomes.
FALSE: this is performed by a viral integrase.
E. Maturation of viral particles requires the presence of a viral protease enzyme.
TRUE.

5. **Which of the following are true about HIV disease?**
A. The virus causes immunodeficiency by mainly killing cytotoxic T cells.
FALSE: CD4+ helper T cells are killed by the virus.
B. Killing of helper T cells can be through apoptosis.
TRUE: CD4+ helper T cells die by 'cell suicide' through apoptosis either following direct virus infection or via interaction with CD8+ cytotoxic T cells that recognize that the CD4+ T cells are infected with the virus because of the presentation of HIV peptides in their MHC class I surface molecules.
C. HIV causes mortality of the patient directly.
FALSE: patient mortality is mainly through opportunistic infections.
D. Lymphadenopathy results from increased proliferation of B cells and cytotoxic T cells.
TRUE: it is likely that this is mediated through cytokines and bacterial products derived from the gut.
E. Macrophages and dendritic cells act as a reservoir of HIV.
TRUE: these cells can bind intact virus via DC-Sign and may also allow some intracellular virus replication.

6. **Which of the following are true for laboratory diagnosis of HIV?**
   A. A negative anti-HIV result excludes the possibility that the patient is infected with HIV.
   FALSE: could be in the 'window' period when HIV is present in blood but not antibody.
   B. It is necessary to confirm a positive HIV test result in a number of different assays.
   TRUE: the diagnosis is of such importance that it is essential to be sure that the positive result is genuine.
   C. Cell culture of HIV is routinely performed to assess drug sensitivities.
   FALSE: resistance testing is now based on gene rescue by PCR and analysis by direct sequencing.
   D. Detection of anti-HIV antibodies in a 3-month-old baby indicates that mother to baby transmission of infection has occurred.
   FALSE: this represents transplacentally acquired maternal antibody.
   E. In late stage disease, anti-HIV antibodies may disappear, leading to a false-negative test result.
   FALSE: anti-HIV antibodies are present throughout infection.

7. **Which of the following are AIDS-defining illnesses?**
   A. Herpes zoster.
   FALSE.
   B. Recurrent cold sores.
   FALSE.
   C. Esophageal candidiasis.
   TRUE.
   D. Cytomegalovirus disease.
   TRUE: often seen as retinitis.
   E. Kaposi's sarcoma.
   TRUE: as seen in the patient described here.

8. **Which of the following are recognized manifestations of AIDS?**
   A. Cerebral toxoplasmosis.
   TRUE.
   B. Oral hairy leukoplakia.
   TRUE: often the first manifestation of AIDS: can occur in other immunosuppressed patients.
   C. Nasopharyngeal carcinoma.
   FALSE: this is no more common in HIV-infected than in noninfected persons.
   D. Progressive multifocal leukoencephalopathy.
   TRUE: reactivation of JC virus in an immunocompromised host.
   E. *Pneumocystis jiroveci* infection.
   TRUE: as seen in the patient described here.

9. **Which of the following are true of seroconversion illness?**
   A. It occurs in only 10% of infected individuals.
   FALSE: it probably occurs in most.
   B. It has many features of a glandular fever-like illness.
   TRUE.
   C. Anti-HIV antibodies are always detectable at this stage of infection.
   FALSE: antibodies and antigens may be in equivalence at this time, so it is not always possible to detect the antibodies.
   D. Patients are not infectious at this stage of infection.
   FALSE: they are most infectious since they still have live virus in their blood and body fluids.
   E. It is often followed by a long period when the patient is asymptomatic.
   TRUE: this can be up to 10 or more years.

10. **Which of the following statements regarding therapy are correct?**
   A. In the early stages of infection, monotherapy with a nucleoside analog reverse transcriptase inhibitor is recommended.
   FALSE: monotherapy should never be recommended because it encourages the emergence of drug resistance. Exceptions are in the management of mother to baby transmission, where a single dose of nevirapine has a role.
   B. Drug resistance occurs through the accumulation of point mutations in the genes encoding the target enzyme.
   TRUE.
   C. HAART should be given until such time as the viral load becomes undetectable, and then stopped, until the viral load exceeds 10 000 copies per ml.
   FALSE: interruptions of therapy have been tried, but clinical outcome is worse.
   D. Monitoring of therapy is by CD4 count measurement.
   FALSE: monitoring of therapy is by viral load measurement but CD4 counting is likely to be used in addition.
   E. Successful therapy results in restoration of immune function.
   TRUE.

CASE 18: Influenza virus

1. **Which of the following statements regarding influenza viruses are true?**
   A. Their genome consists of eight segments of double-stranded RNA.
   FALSE: the genome does indeed consist of eight segments, but the RNA is a negative single-strand RNA.
   B. They are classified into types A, B, and C on the basis of the nature of their internal proteins, particularly the nucleoprotein.
   TRUE.
   C. Type A influenza viruses are further subdivided into subtypes on the basis of the nature of their matrix proteins.
   FALSE: subtyping of influenza A viruses is on the basis of the two surface glycoproteins, HA and NA.
   D. Type B influenza viruses are further subdivided into subtypes on the basis of the nature of their surface proteins.
   FALSE: there are no subtypes of influenza B virus.
   E. There is no animal reservoir of type B influenza viruses
   TRUE: this is quite possibly why there are no subtypes of this virus.

2. **The emergence of new pandemic strains of influenza virus may arise from which of the following processes?**
   A. Trans-species transfer of an avian influenza virus directly to humans.

TRUE: a possibility, but any virus that makes a trans-species jump is likely to have to acquire several adaptive mutations to enable it to replicate efficiently in human cells.

B. Spontaneous mutations in the genes encoding the surface glycoproteins.
FALSE: these are the basis for the phenomenon of antigenic drift, not shift, and therefore lead to interpandemic epidemics.

C. Reassortment of avian and human influenza viruses within a single host.
TRUE: this is believed to have been the mechanism whereby the 1956 and 1968 influenza pandemics emerged. The 'single host,' or mixing vessel, is thought to have been the pig.

D. Use of neuraminidase inhibitors resulting in mutations in the neuraminidase gene.
FALSE: while it is true that mutations in neuraminidase arise following use of this drug, these do not confer the properties necessary for the virus to cause a pandemic.

E. Natural selection of viral variants in an immunized host population.
FALSE: this is the next step, following random mutations in the genes encoding the surface glycoproteins, that lead to antigenic drift, not shift.

3. **Which of the following statements regarding the epidemiology of influenza viruses is/are correct?**
A. Influenza B viruses undergo antigenic shift but not antigenic drift.
FALSE: influenza B viruses undergo drift, not shift.

B. Antigenic shift in influenza viruses gives rise to global pandemics of influenza.
TRUE.

C. Antigenic drift in influenza viruses gives rise to interpandemic epidemics of influenza.
TRUE.

D. Antigenic shift describes the emergence of new influenza A virus subtypes.
TRUE.

E. Antigenic drift results in amino acid changes clustered within key epitopes of the viral nucleoprotein.
FALSE: the important epitopes that drift are located in the surface glycoproteins, not the nucleoprotein.

4. **Which of the following statements regarding disease associated with influenza virus infection is/are true?**
A. Influenza virus infection of respiratory epithelial cells results in transformation of those cells.
FALSE: the interaction between influenza virus and its target cells is cytolytic, not transforming.

B. Systemic manifestations of influenza virus infection (e.g. fever, myalgia, headache) arise from the presence of virus circulating in the bloodstream.
FALSE: this symptomatology arises from the presence of circulating cytokines, particularly interferon-α, of which influenza viruses are potent inducers.

C. Pneumonia arising as a complication of influenza virus infection is usually due to secondary bacterial invasion.
TRUE: at present, with the current circulating strains of virus. This may not always pertain, e.g. the high mortality associated with the 1918–19 pandemic is thought in part to have been due to primary viral pneumonia.

D. Influenza-related mortality is higher in patients with pre-existing cardiac disease.
TRUE: hence this group is targeted for vaccination.

E. Influenza-related encephalitis arises through cross-reactivity of the immune response to infection with the patient's brain tissue.
TRUE: at least it is not possible to detect virus or viral antigens within the brain substance, so this is the best working hypothesis of the pathogenesis.

5. **Which of the following statements regarding the avian H5N1 influenza virus are true?**
A. Infection of humans results in death in 10–20% of cases.
FALSE: current mortality rates are over 50%.

B. The pathogenesis of disease arises from explosive release of cytokines within the respiratory tract.
TRUE: this high mortality arises from induction of a cytokine storm within the lung.

C. Infection can be prevented by vaccination with vaccines containing antigens derived from A/H1N1 and A/H3N2 viruses.
FALSE: protective vaccines will need to be derived from H5N1 virus.

D. This virus is always sensitive to oseltamivir.
FALSE: there are reports of oseltamivir resistance in these viruses, usually arising in the context of suboptimal therapy.

E. This virus is a possible candidate for the next influenza pandemic.
TRUE: although there is no way to be sure that this will be the one! Certainly, if an H5N1 virus was able to adapt to high levels of replication in a human host, leading to efficient person to person spread, then a pandemic will emerge, as effectively none of the population on the planet has any protective immunity to this virus.

6. **Which of the following statements regarding diagnostic tests is/are true?**
A. Virus isolation in cell culture is a rapid diagnostic technique.
FALSE: most viruses – including influenza viruses – take several days in cell culture before their presence is revealed by the development of a cytopathic effect.

B. Antigen detection techniques are dependent on the presence of viable virus in the sample sent to the laboratory.
FALSE: these techniques rely upon the presence of viral antigens within an infected cell. It does not matter whether the cell/virus remains viable on its journey to the laboratory.

C. Genome detection techniques are the most sensitive assays for diagnosis of virus infections.
TRUE: genome amplification techniques theoretically can detect a single copy of target nucleic acid.

D. Genome amplification assays cannot be used for RNA viruses.
FALSE: there are a number of ways of amplifying RNA – perhaps the simplest is to copy the RNA into DNA as a first step using reverse transcriptase, and then amplify the DNA copy. This is referred to as reverse transcriptase PCR – not to be confused with real-time PCR, which describes something different.

E. Demonstration of high antibody titers to the H5 hemagglutinin in serum samples taken from acutely ill patients will be the mainstay of diagnosis of human infection with avian H5N1 influenza virus.
FALSE: given the need for rapid diagnosis, genome amplification assays such as reverse transcriptase PCR will be the preferred technology.

7. **Which of the following drugs has proven efficacy against influenza A viruses?**
   A. Aciclovir.
      FALSE: this drug is a DNA synthesis inhibitor that needs to be activated by viral thymidine kinase. Influenza viruses are RNA viruses, and do not possess a thymidine kinase enzyme.
   B. Foscarnet.
      FALSE: this drug inhibits DNA synthesis.
   C. Indinavir.
      FALSE: this is an HIV protease inhibitor.
   D. Zanamavir.
      TRUE: this is a neuraminidase inhibitor.
   E. Zidovudine.
      FALSE: this is a reverse transcriptase inhibitor.

8. **With regard to anti-influenza drugs, which of the following statements are true?**
   A. Amantadine is effective as a prophylactic agent against influenza B virus.
      FALSE: amantadine has no activity against influenza B viruses. It does work as prophylaxis against many (but not all) influenza A viruses.
   B. The mode of action of amantadine involves blockage of an ion channel and prevention of viral uncoating.
      TRUE.
   C. Neuraminidase inhibitors have no activity against influenza B virus.
      FALSE: these drugs work against all known influenza neuraminidases.
   D. Resistance to oseltamivir has not been described in influenza A viruses.
      FALSE: it has been described, particularly in patients infected with avian H5N1 virus.
   E. Zanamavir should not be used in patients with a history of egg allergy.
      FALSE: influenza vaccine may be contraindicated in these individuals, in which case, prophylaxis with a neuraminidase inhibitor is a sensible alternative strategy.

9. **Which of the following statements regarding influenza are true?**
   A. Influenza re-infections occur despite the presence of high levels of serum antibodies.
      TRUE: this arises because antigenic drift produces viruses not fully neutralizable by pre-existing circulating antibodies.
   B. Influenza pandemics have only occurred since the beginning of the 20th century.
      FALSE: while the virus itself was only first isolated in 1933, historical records suggest that influenza pandemics have been a natural phenomenon for many centuries.
   C. Human influenza viruses only infect humans.
      FALSE: pigs are also susceptible to infection with human influenza viruses.
   D. Epidemics of influenza occur in the winter months in the northern hemisphere.
      TRUE.
   E. Infections with influenza B virus are less severe than those with influenza C virus.
      FALSE: influenza C virus is not an important human pathogen.

10. **With regard to influenza vaccines, which of the following statements are true?**
    A. Vaccine-induced immunity is clinically useful for at least 10 years.
       FALSE: not least because antigenic drift allows the virus to escape pre-existing immunity in a vaccinated host. Hence at-risk individuals need to be vaccinated each year.
    B. Universal vaccination against influenza is currently recommended.
       FALSE: most countries, including the UK and USA, have selective vaccination policies, i.e. vaccine is targeted at specific risk groups only, although in the USA, such risk groups constitute a majority of the population.
    C. Inactivated influenza vaccines are contraindicated in immunosuppressed individuals.
       FALSE: there is no problem in using inactivated vaccines in an immunocompromised host – although, of course, the immune response to the vaccine may well not be optimal.
    D. Influenza vaccines contain antigens derived from A/H1N1, A/H3N2, and B viruses.
       TRUE: all three viruses co-circulate, therefore the aim is provide protection against all three.
    E. Live attenuated influenza vaccines are administered by subcutaneous injection.
       FALSE: they are administered by intranasal inoculation.

CASE 19: *Leishmania* spp.

1. **Which of the following are true about the causative agent of leishmaniasis?**
   A. The extracellular form of *Leishmania* is called the amastigote.
      FALSE: this is the intracellular form. The flagellated promastigote is extracellular.
   B. *Leishmania* can be taxonomically subdivided by isoenzyme typing.
      TRUE.
   C. The form of clinical disease can depend on the species of *Leishmania*.
      TRUE.
   D. There are both animal and human reservoirs of infection.
      TRUE.
   E. Spread within the body occurs inside macrophages.
      TRUE.

2. **Which of the following are true of the transmission of leishmaniasis?**
   A. The domestic dog can serve as a reservoir of infection.
      TRUE.
   B. The insect vector is the mosquito.
      FALSE: the vector is the sandfly.
   C. The infective form when the vector bites a new host is the amastigote.
      FALSE: it is the promastigote.
   D. Infection can be transmitted between intravenous drug users with shared needles and equipment.
      TRUE.
   E. Individuals with post kala-azar dermal leishmaniasis (PKDL) can serve as a human reservoir of infection.
      TRUE.

3. **Which of the following are true of the host response to leishmaniasis?**
   A. A Th1 response is associated with susceptibility.
   FALSE: Th1 is protective.
   B. Transforming growth factor-β activates macrophages to increase leishmanicidal functions.
   FALSE: interferon-γ activates macrophages.
   C. In different strains of mice infected with *L. major* there is either a Th1 or Th2 response.
   TRUE.
   D. Neutrophils fail to kill phagocytosed *Leishmania*.
   TRUE.
   E. A granulomatous inflammatory response develops in tissues.
   TRUE.

4. **Which of the following are true of the pathogenesis of leishmaniasis?**
   A. *Leishmania* may enter macrophages in apoptotic neutrophil fragments.
   TRUE.
   B. *Leishmania* are susceptible to lysosomal attack.
   FALSE: they resist lysosomal attack.
   C. *Leishmania* homolog of activated C kinase receptor (LACK) steers the host to a Th1 response.
   FALSE: it steers to a Th2 response.
   D. Dendritic cell antigen presentation is down-regulated by *Leishmania*.
   TRUE.
   E. *Leishmania* suppresses the production of reactive oxygen and nitrogen intermediates.
   TRUE.

5. **Which of the following are true of cutaneous and mucocutaneous leishmaniasis?**
   A. Cutaneous lesions are usually found in moist areas such as the armpit and groins.
   FALSE: they are usually found in exposed areas of the body where the sandfly bites.
   B. Lesions are manifest 2–6 days after a bite.
   FALSE: usually 2–6 weeks.
   C. Lesions are always painful.
   FALSE: lesions may be painless, but are more likely to become painful when secondarily infected.
   D. *L. braziliensis* and *L. panamensis* are the key species responsible for mucocutaneous disease.
   TRUE.
   E. In mucocutaneous disease the mucosal lesions appear several years after an initial cutaneous lesion.
   TRUE.

6. **Which of the following are true of visceral leishmaniasis?**
   A. Massive enlargement of the liver and spleen are possible.
   TRUE.
   B. As a reaction to infection there is a rise in the number of platelets.
   FALSE: usually platelet numbers fall due to a combination of bone marrow suppression and hypersplenism.
   C. In the blood there is a hypergammaglobulinemia.
   TRUE.
   D. Post kala-azar dermal leishmaniasis occurs as an immediate reaction to treatment for visceral leishmaniasis.
   FALSE: this is not a drug reaction, but a later relapse of

incompletely treated infection.
   E. In HIV-infected patients usually cutaneous species of *Leishmania* can cause visceral disease.
   TRUE.

7. **Which of the following are true of the diagnosis of leishmaniasis?**
   A. Mucocutaneous disease can be diagnosed clinically in endemic areas.
   TRUE.
   B. On staining tissue samples the characteristic appearance of the organism is called Donovan bodies.
   TRUE.
   C. The sensitivity of splenic aspirates is about 50%.
   FALSE: it is > 90%.
   D. PCR has a high sensitivity and specificity.
   TRUE.
   E. Serological tests can distinguish recent acute infection from past infection.
   FALSE.

8. **Which of the following are true of the differential diagnosis of leishmaniasis?**
   A. Cutaneous tuberculosis may cause similar cutaneous lesions.
   TRUE.
   B. Mucocutaneous leishmaniasis is the only infection that destroys the nasal septum.
   FALSE: other infections include syphilis, leprosy, and paracoccidiodomycosis.
   C. Visceral leishmaniasis may have to be differentiated from malaria.
   TRUE.
   D. Clinical features alone are sufficient to differentiate visceral leishmaniasis from other diagnoses.
   TRUE.
   E. Noninfectious diseases can cause fever like visceral leishmaniasis.
   TRUE.

9. **Which of the following are true of the treatment of leishmaniasis?**
   A. All cutaneous lesions need to be treated to prevent disease progression.
   FALSE: cutaneous lesions can self-cure.
   B. All treatments are parenteral (i.e. need to be given by injection).
   FALSE: miltefosine is oral.
   C. Treatment courses can last up to 28 days.
   TRUE.
   D. Liposomal amphotericin B can be used to treat visceral leishmaniasis.
   TRUE.
   E. Miltefosine is less effective than other treatments for visceral leishmaniasis.
   FALSE: it is equi-effective with amphotericin B.

10. **Which of the following are true of the control and prevention of leishmaniasis?**
    A. Mosquito nets are sufficient to prevent sandfly bites.
    FALSE: the mesh is not always small enough to exclude sandflies.
    B. The distribution of mosquito nets to control malaria has reduced the incidence of leishmaniasis.
    FALSE: there is no evidence for this and in fact the incidence of leishmaniasis appears to be rising.

C. Sandflies are intrinsically resistant to pyrethroid insecticides.
FALSE.
D. There is no effective vaccine to date.
TRUE.

E. Treatment of individuals with post kala-azar dermal leishmaniasis is sufficient to eliminate the reservoir of infection.
FALSE: there is still an animal reservoir.

CASE 20: *Leptospira* spp.

1. **Which of the following statements are true of *Leptospira interrogans*? It:**
   A. Is gram-positive.
   FALSE: it has a gram-negative cell wall structure.
   B. Has axial flagella.
   TRUE: it has two flagella that are located beneath the outer membrane in the periplasmic space.
   C. Can be detected by dark ground microscopy.
   TRUE: as the organism is very thin it is best seen using dark ground microscopy.
   D. Has only one serovar.
   FALSE: it has over 250 serovars.
   E. Is nonmotile.
   FALSE: it is motile.

2. **Which of the following are true concerning *Leptospira*?**
   A. They are typically spread from person to person.
   FALSE: infection is a zoonosis.
   B. They are found in a wide range of animals.
   TRUE: leptospires are found in a wide range of animals.
   C. They are spread by contact with animal urine.
   TRUE: water or soil contaminated with infected animal urine is the source of most human infection, although direct contact with animals can also lead to infection.
   D. All *leptospira* are pathogenic.
   FALSE: only some of the species are pathogenic, others are saprophytes.
   E. Infection is associated with certain occupations.
   TRUE: e.g. veterinarians, sewage workers, and farmers.

3. **Which of the following are true for disease caused by *Leptospira*?**
   A. Vasculitis is a common occurrence.
   TRUE: and is the hallmark of the pathology.
   B. Pulmonary hemorrhages are frequent.
   TRUE: and characteristically they are intra-alveolar.
   C. Acute tubular necrosis may occur.
   TRUE: caused by the inflammatory infiltrate.
   D. It is a granulomatous disease.
   FALSE: granulomas are not formed.
   E. Thrombocytopenia is common.
   TRUE: and is either primary due to thrombopoiesis or secondary to DIC.

4. **Which of the following are true of the clinical presentation of leptospirosis?**
   A. All infection presents in the same fashion.
   FALSE: the clinical presentation may vary depending on the organ principally affected, e.g. meningitis or pneumonia or liver and renal failure.
   B. Any serovar can cause any clinical presentation.
   TRUE: but some serovars are typically linked with certain signs or symptoms, e.g. *L. autumnalis* infection often presents with a rash.
   C. Leptospirosis is a biphasic illness.
   TRUE: an initial nonspecific febrile illness followed by

organ-specific disease.
   D. Leptospirosis may be confused with Legionnaires disease, viral hepatitis or Hantavirus infection.
   TRUE: depending on the clinical presentation of leptospirosis.
   E. Jaundice and renal failure with conjunctival inflammation is a common presentation with *L. interrogans* serovar Icterohaemorrhagia.
   F. TRUE: it occurs in about 80% of cases on infection with this serovar.

5. **Which of the following are complications of leptospirosis?**
   A. Myocarditis.
   TRUE: and may lead to cardiac arrhythmia and sudden death.
   B. Stevens Johnson syndrome.
   FALSE: this may occur following some antibiotics.
   C. Weil's disease.
   FALSE: not a complication, this is the eponymous name given to severe leptospirosis.
   D. Guillain-Barré syndrome.
   TRUE: associated with a myelitis.
   E. Renal failure.
   TRUE: as part of Weil's disease.

6. **Which of the following can be used in the diagnosis of leptospirosis?**
   A. Gram stain.
   FALSE: the organism is too thin to be easily seen but can be detected by dark ground microscopy.
   B. Culture.
   TRUE: but culture uses specialized media, takes several weeks, and is not routine.
   C. Microscopic agglutination test.
   TRUE: although it is group-specific not serovar-specific and has several cross-reactions.
   D. IgM ELISA.
   TRUE: and with the MAT is the usual routine diagnostic approach.
   E. PCR.
   TRUE: but is not serovar-specific.

7. **Which of the following are used in the treatment or prevention of leptospirosis?**
   A. Vaccination.
   TRUE: animals are vaccinated and a human vaccine exists but is only routinely used in France.
   B. Benzylpenicillin.
   TRUE: the standard treatment of leptospirosis.
   C. Gentamicin.
   FALSE: this is not used in treatment of leptospirosis.
   D. Personal protective equipment.
   TRUE: e.g. for sewage workers.
   E. Mosquito nets.
   FALSE: leptospirosis is not transmitted by insects.

CASE 21: *Listeria monocytogenes*

1. **Which of the following are true of *Listeria* species?**
   A. They are gram-positive.
   TRUE: *Listeria* are nonsporing gram-positive bacilli.
   B. *L. monocytogenes* is nonmotile.
   FALSE: *Listeria* are motile and *L. monocytogenes* has flagella and a characteristic tumbling motility at 25°C.
   C. *L. monocytogenes* can grow at 4°C.
   TRUE: and this means that refrigeration does not prevent bacterial multiplication. Also this attribute can be used to enrich for *Listeria*.
   D. All *Listeria* species are pathogenic for humans.
   FALSE: *L. monocytogenes* causes most human disease.
   E. *L. monocytogenes* is alpha-hemolytic.
   FALSE: *L. monocytogenes* produces a narrow zone of beta-hemolysis surrounding the colony.

2. **Which of the following are true concerning *Listeria monocytogenes*?**
   A. It is ubiquitous in the environment.
   TRUE: it is found on vegetation, in water, and colonizes a wide variety of animals.
   B. Infection is a zoonosis.
   TRUE: infection can be acquired from animal sources.
   C. Soft cheese is a low risk food that is unlikely to be contaminated with large numbers of *L. monocytogenes*.
   FALSE: some soft cheeses have high counts of *L. monocytogenes*.
   D. Infection may be acquired transplacentally.
   TRUE: giving rise to fetal disease.
   E. Infection is mostly associated with certain at-risk groups.
   TRUE: pregnant women, neonates, immunocompromised subjects.

3. **Which of the following are true for *Listeria*?**
   A. The bacteria are rapidly killed by antibodies.
   FALSE: as *Listeria* is an intracellular pathogen.
   B. *Listeria* has a tropism for the CNS.
   TRUE: and causes meningoencephalitis.
   C. *Listeria* induces uptake into nonphagocytic cells.
   TRUE: and is taken up into a phagosome from which it escapes with the aid of listeriolysin O.
   D. *Listeria* hydrolyzes actin in the cytoplasm.
   FALSE: it polymerizes actin, which it uses to move from one cell to an adjacent cell.
   E. *Listeria* infection is controlled by cytotoxic CD8 cells.
   TRUE: and is also killed by activated macrophages.

4. **Which of the following are true of the clinical presentation of diseases caused by *L. monocytogenes*?**
   A. Granulomatosis infantiseptica may occur.
   TRUE: following transplacental transmission.
   B. Infection in pregnant women may lead to spontaneous abortion.
   TRUE: as the placenta becomes infected and the fetus develops widespread abscesses.
   C. It is typically associated with necrotizing fasciitis.
   FALSE: this is typically caused by *Streptococcus pyogenes*.
   D. Meningitis is common in pregnant women.
   FALSE: meningitis is uncommon in pregnant women.
   E. Outbreaks are uncommon.
   TRUE: although a number or outbreaks of listeriosis have been reported in association with pate, coleslaw, chocolate milk, compared with many other food-borne pathogens such as *Salmonella* and *E. coli* O157, outbreaks caused by *Listeria* are uncommon.

5. **Which of the following are useful in the diagnosis of *Listeria*?**
   A. Gram stain.
   TRUE: it can be helpful but in some specimens the organism may be regarded as a diphtheroid or appear as a coccus.
   B. Culture.
   TRUE: producing a narrow zone of beta-hemolysis of blood agar.
   C. Serology.
   FALSE: it is not helpful.
   D. PCR.
   FALSE: it is not part of the routine clinical diagnosis but is available.
   E. Gas-liquid chromatography.
   FALSE: this is used to detect anaerobic infection in pus.

6. **Which of the following are used in the treatment or prevention of *Listeria*?**
   A. Vaccination.
   FALSE: a vaccine does not yet exist.
   B. Co-trimoxazole.
   TRUE: it can be used in patients with penicillin hypersensitivity.
   C. Gentamicin.
   TRUE: it is used with ampicillin as part of the routine first-line treatment.
   D. Education leaflets.
   TRUE: as part of the education of pregnant women not to eat high risk foods.
   E. Good kitchen practices.
   TRUE: as they can prevent contamination of cooked food from raw food and correct cooking temperatures will kill the organism.

CASE 22: *Mycobacterium leprae*

1. **Which of the following are true of *M. leprae*?**
   A. Its cell wall contains species-specific phenolic glycolipid I (PGL-I).
   TRUE: this is a species-specific bacterial adhesin.
   B. It binds to the target cells via genus-specific lipoarabinomannan.
   FALSE: binds to the target cells by PGL-I.
   C. The genome of *M. leprae* is decayed compared with *M. tuberculosis*.
   TRUE: the genome contains fewer structural genes and more pseudogenes and inactivated genes.
   D. Nine-banded armadillos are natural hosts of *M. leprae*.

TRUE: *M. leprae* grows in these animals naturally and they are used for the propagation of the pathogen for experimental and diagnostic purposes.
E. Is highly contagious.
FALSE: is not very pathogenic and most infections do not result in leprosy.

2. **Which of the following are characteristic of the physiology of *M. leprae*?**
   A. Slow growth.
      TRUE: requires several weeks to grow.
   B. A requirement of 37°C for optimal growth.
      FALSE: grows in the cooler parts of the body: 30–34°C.
   C. Being disseminated by the digestive route.
      FALSE: uses an air-borne route.
   D. Its growth inside macrophages and Schwann cells.
      TRUE: infects these cells via PGL-I.
   E. Its preference for growth in the peritoneal cavity.
      FALSE: is mostly found in peripheral nerves.

3. **In which of the following regions of the world is leprosy endemic?**
   A. The Pacific region.
      TRUE: 3–5/10 000 cases.
   B. Southern Europe.
      FALSE: few cases, mostly in immigrants from the endemic regions.
   C. Southern America.
      TRUE: 3–15/10 000 cases.
   D. East–Southern Asia.
      TRUE: 3–15/10 000 cases.
   E. Africa.
      TRUE: 3–15/10 000 cases.

4. **Which of the following characterize the tuberculoid (TT) form of leprosy?**
   A. The prevalence of the production of Th1 over Th2 cytokines.
      TRUE: mostly Th1 cytokines IFN-γ, IL-2, IL12, and IL-15 are produced.
   B. The predominantly T-cell-mediated immune responses.
      TRUE: Th1 responses favor T-cell-mediated reactions, mostly of the delayed type.
   C. The development of erythema nodosum leprosum (ENL).
      FALSE: this is a feature of the LL form of leprosy.
   D. A possibility of a spontaneous resolution.
      TRUE: some of the TT cases can resolve spontaneously.
   E. Belonging to the paucibacillary form of leprosy.
      TRUE: extremely few bacilli if any can be found in the tissues.

5. **Which of the following characterize the lepromatous form (LL) of leprosy?**
   A. Prevalence of the production of Th1 over Th2 cytokines.
      FALSE: mostly Th2 cytokines IL-4, IL-5, and IL-10 are produced.
   B. Predominantly antibody-mediated immune responses.
      TRUE: a specific IgM to PGL-I represents the leading immunological feature.
   C. The possibility to revert to the less severe form.
      FALSE: LL is the only form of leprosy that never reverts.
   D. Belonging to the multibacillary form of leprosy.
      TRUE: bacilli can be easily identified in biopsies.
   E. More than five skin lesions.
      TRUE: this is a characteristic diagnostic feature for the LL form.

6. **Which of the following clinical tests are important for the differential diagnosis of leprosy?**
   A. Hypopigmentation of the skin.
      TRUE: one of the gold standard clinical symptoms.
   B. Reddish patches of the skin.
      TRUE: another gold standard clinical symptom.
   C. Loss of sensation.
      TRUE: one of the earliest and important symptoms of leprosy.
   D. Thickening of peripheral nerves.
      TRUE: important later symptom.
   E. Abscesses.
      FALSE: not specific for the diagnosis of leprosy.

7. **Which of the following laboratory tests are used for the establishment or confirmation of the diagnosis of leprosy?**
   A. Bacillary cell culture.
      FALSE: not possible to grow in cultures due to the extremely low rate of proliferation.
   B. Microscopy of biopsies.
      TRUE: used for the diagnosis, particularly of the MB forms.
   C. Serological identification of the specific antibodies.
      TRUE: can be used for the confirmation of diagnosis of the LL and BL forms.
   D. Polymerase chain reaction.
      TRUE: is now implemented in the reference laboratories.
   E. Lepromin test.
      FALSE: this test helps to establish the form of the leprosy (TT), but should not be used for its diagnosis.

8. **Which of the following are included in the multidrug therapy regime (MDT) of leprosy?**
   A. Clofazimine.
      TRUE: one of the main components.
   B. Benzylpenicillin.
      FALSE: not used for the treatment of leprosy.
   C. Isoniazid.
      FALSE: not used for the treatment of leprosy.
   D. Rifampicin.
      TRUE: one of the main components.
   E. Dapsone.
      TRUE: one of the main components.

CASE 23: *Mycobacterium tuburculosis*

1. **Which of the following are true of the cell wall of mycobacteria?**
   A. It contains mycolic acid.
      TRUE: the waxy outer layer.
   B. It contains peptidoglycan.
      TRUE: as occurs in most bacteria.
   C. It stains green with the Ziehl-Neelsen stain.
      FALSE: it stains red.
   D. It is highly permeable to antibiotics.
      FALSE: it has a waxy impermeable cell wall.
   E. It contains lipopolysaccharide.
      FALSE: only found in gram-negative bacteria.

2. **Which of the following are true of *M. tuberculosis*?**
   A. The organism grows in 3 days.
      FALSE: the organism may take 2–4 weeks to grow.
   B. The organism can only be grown in armadillos.
      FALSE: *M. tuberculosis* is grown on Lowenstein–Jensen medium. *M. leprae* cannot be grown on artificial media but grows when injected into armadillos.
   C. The organism is disseminated by the aerial route.
      TRUE: it is disseminated by droplet nuclei.
   D. Each infected person infects on average 20 other individuals.
      TRUE: and the rate of infection in close contacts is proportional to the number of times a patient coughs per time period.
   E. About one-third of the world's population is infected.
      TRUE: and the situation is becoming worse due to antibiotic resistance and co-morbidity with HIV.

3. **Which of the following are associated with the host response to *M. tuberculosis*?**
   A. Macrophages.
      TRUE: these are the typical cells associated with infection by mycobacteria and of chronic inflammation. They are professional phagocytic cells and antigen-presenting cells.
   B. Granulomas.
      TRUE: this is the typical host response to chronic inflammation and can be caused by many stimuli other than mycobacteria, e.g. *Brucella* infection, beryllium or silica dust or schistosomiasis. The granuloma is a collection of macrophages, which may fuse to produce giant cells and the whole entity is surrounded by epithelioid cells and lymphocytes. The center of the granuloma undergoes necrosis. The antigen that induces granuloma formation by *M. tuberculosis* is primarily trehalose mycolate (also called 'cord factor,' as on staining the organism it has the appearance of cords).
   C. Destructive antibody response.
      FALSE: the damage is caused by a T-cell response and neutrophils.
   D. Lipopolysaccharide (LPS) binding protein.
      FALSE: this is associated with gram-negative bacteria, which bind LPS and deliver it to CD14 and TLR 4 on macrophages, triggering the production of TNF.
   E. Cell-mediated immunity.
      TRUE: as the organism is an intracellular pathogen the key defense is cell-mediated.

4. **Which of the following are ways in which *M. tuberculosis* avoids the host defenses?**
   A. Remains a superficial infection.
      FALSE: *M. tuberculosis* is an intracellular pathogen.
   B. Adopts an intracellular location.
      TRUE: the organism is able to survive in macrophages.
   C. Prevents phagolysosome fusion.
      TRUE: the organism inhibits the fusion of the phagosome with the lysosome, which contains destructive cationic proteins that may damage the cell wall of the organism.
   D. Resists acidification of the phagolysosome.
      FALSE: see above.
   E. Is resistant to TNF-α.
      FALSE: TNF-α is produced by macrophages in response to mycobacterial uptake and is related to the weight loss (**cachexia**) experienced by the individual.

5. **Which of the following may be due to infection with *M. tuberculosis*?**
   A. Osteomyelitis.
      TRUE: may cause vertebral collapse if in the spine (Pott's disease) and consequent neurologic complications.
   B. Meningitis.
      TRUE: the onset is chronic with a predominant number of lymphocytes in the cerebrospinal fluid compared with polymorphs that are found in meningitis caused by, e.g. *Neisseria meningitidis*.
   C. Abscess.
      TRUE: and it is called a 'cold' abscess as it is apparently sterile because bacteria that commonly cause abscesses, e.g. *Staphylococcus aureus*, cannot be grown from it. It is not sterile, however, as *M. tuberculosis* may be grown, it just takes longer.
   D. Food poisoning.
      FALSE: although *M. tuberculosis* may affect the gastrointestinal tract it does not present with the symptoms of food poisoning – acute onset vomiting/diarrhoea/pain.
   E. Pyelonephritis.
      TRUE: and give a 'sterile' pyuria. *M. tuberculosis* may be grown but an early morning urine sample is needed (rather than a midstream specimen of urine) and takes longer to grow.

6. **Which of the following are typical signs of pulmonary tuberculosis?**
   A. Chronic cough.
      TRUE.
   B. Seizures.
      FALSE: this may suggest tuberculoma in the brain, i.e. tuberculosis of the central nervous system.
   C. Skin rash.
      FALSE: although *M. tuberculosis* may infect the skin it presents as lupus vulgaris or scrofuloderma – both granulomatous skin lesions.
   D. Weight loss.
      TRUE: probably due partly to the TNF-α produced during the disease.
   E. Low grade fever.
      TRUE.

7. **Which of the following are important in the diagnosis of tuberculosis?**
   A. Culture.
      TRUE: and the medium used is Lowenstein–Jensen,

which contains egg yolk and a dye (malachite green) that inhibits the growth of other more rapidly growing bacteria. The colonies may take 2–4 weeks to become visible.

B. Microscopy.
TRUE: the stain is the ZN stain with mycobacteria staining red and all else staining green.

C. Serology.
FALSE: this is not currently used as a diagnostic test.

D. Polymerase chain reaction.
TRUE: and this is useful for detecting drug-resistant isolates and speeds up the time to diagnosis as it is more rapid than culture.

E. Histopathology.
TRUE: as granuloma may be seen and the tissue section when stained for mycobacteria (ZN stain) may reveal the presence of the organism.

8. **Which of the following could be confused with extrapulmonary tuberculosis?**

A. Staphylococcal abscess of the spine.
TRUE: and may be differentiated by isolation of the organism or evidence of mycobacterial disease at other locations, e.g. chest X-ray.

B. Addison's disease.
TRUE: and one of the causes of adrenal failure may be tuberculosis by destroying the adrenal glands. Addison's disease may also have other causes.

C. Ringworm.
FALSE: this is caused by a dermatophyte fungus and has a characteristic appearance of a spreading erythematous ring lesion.

D. Glioma.
TRUE: as this is a tumor of the CNS and may present with signs and symptoms similar to the presence of a tuberculoma, e.g. seizures.

E. Cryptococcal meningitis.
TRUE: as this is a chronic slow onset meningitis caused by *Cryptococcus neoformans* (a yeast) and has similar signs and symptoms to tuberculous meningitis. It can be distinguished by seeing the organism in the

cerebrospinal fluid using a stain or detecting it by a latex agglutination test for cryptococcal antigen (CRAG). The host cell response is also lymphocytes, similar to tuberculosis.

9. **Which of the following antibiotics are used to treat tuberculosis?**

A. Benzylpenicillin.
FALSE: this has no activity against mycobacteria.

B. Isoniazid.
TRUE: part of the triple regimen.

C. Rifampicin.
TRUE: part of the triple regimen.

D. Pyrazinamide.
TRUE: part of the triple regimen.

E. Vancomycin.
FALSE: this is an antibiotic that acts only against gram-positive bacteria.

10. **Which of the following are true of tuberculosis?**

A. It is a notifiable disease in the UK.
TRUE.

B. Contacts should be started on empirical antituberculous therapy.
FALSE: close contacts are screened with a chest X-ray and followed up; the GP is informed. If the person is susceptible, e.g. immunocompromised, then they may be given isoniazid.

C. DOTS means the patient visits a treatment center for their drugs.
FALSE: DOTS means that the person is observed by a health-care worker actually taking their drugs.

D. Disease prevalence is correlated with social deprivation.
TRUE: tuberculosis is frequently a disease of social deprivation and the homeless. It can of course infect anyone of any social class.

E. The is no effective vaccine.
FALSE: there is an effective vaccine – BCG, although its efficacy has a wide geographic variation from 0 to 80% worldwide.

CASE 24: *Neisseria gonorrhoeae*

1. **A 24-year-old sexually active female presents with dysuria (pain on urination) and cervical discharge. Gram stain of the cervical discharge reveals gram-negative diplococci in pairs. Which one of the following would be the most appropriate therapy to initiate?**

A. Penicillin.
FALSE: gonococci are resistant to penicillin.

B. Erythromycin.
FALSE: gonococci are resistant to erythromycin.

C. Ceftriaxone.
FALSE: while cephalosporins are the antibiotic of choice for the treatment of gonorrhea it is usual to treat for *Chlamydia* infection, too.

D. Ceftriaxone plus doxycycline.
TRUE: ceftriaxone plus doxycycline is the recommended treatment for gonorrhea/chlamydia co-infections.

E. Ampicillin.
FALSE: gonococci are resistant to ampicillin.

2. **The process of phase and antigenic variation in gonococci involves?**

A. Lysogeny by a bacteriophage.
FALSE: integration of phage DNA into the bacterial chromosome is not involved in phase or antigenic variation.

B. Plasmid transfer.
FALSE: extrachromosomal DNA is not involved in phase or antigenic variation.

C. Genetic recombination.
TRUE: recombination of gene sequences gives rise to phase or antigenic variation.

D. Sex pili.
FALSE: bacterial conjugation is not required for phase or antigenic variation.

E. Transduction.
FALSE: transfer of viral and/or bacterial DNA via bacteriophage is not involved in phase or antigenic variation.

3. **A young army recruit presents at a sexually transmitted disease (STD) clinic with acute urethritis. A Gram stain of his urethral exudate reveals neutrophils with intracellular gram-negative diplococci. He is treated with ceftriaxone and sent home. He is requested to return in 1 week so that a urethral culture can be obtained to confirm antibiotic cure. Which ONE of the following culture media should be used for the follow-up culture?**
   A. Blood agar.
      FALSE: gonococci grow poorly on blood agar and would be overgrown by the commensal microbiota in the urethral culture.
   B. Chocolate agar.
      FALSE: gonococci grow well on chocolate agar but would be overgrown by the commensal microbiota in the urethral culture.
   C. MacConkey agar.
      FALSE: MacConkey agar is designed for the isolation of enterobacteriacae.
   D. Modified Thayer-Martin (MTM) agar.
      TRUE: MTM is chocolate agar containing antimicrobials to inhibit commensal bacteria and fungi and is the medium of choice for the isolation of gonococci.
   E. Mannitol-salt agar (MSA).
      FALSE: MSA is a selective and differential medium for isolation of staphylococci.

4. **A 24-year-old sexually active female presents with dysuria and a purulent discharge from the cervical canal. A Gram stain of the discharge shows the presence of gram-negative diplococci and numerous neutrophils. The lab reports the isolation of nonhemolytic, gram-negative, oxidase-positive diplococci that utilize glucose. A deletion of the genes responsible for the synthesis of which one of the following would make the organism unable to initially attach to male urethral epithelial cells and endocervical cells?**
   A. Capsular polysaccharide.
      FALSE: the gonococcus does not produce extracellular polysaccharide.
   B. C-carbohydrate.
      FALSE: the C-carbohydrate or C-polysaccharide is a wall antigen of *Streptococcus pneumoniae*.
   C. O side-chain of lipooligsaccharide.
      FALSE: LOS lacks the O side-chain.
   D. Pili.
      TRUE: pili are involved in the initial adhesion of gonococci to epithelium.
   E. Spore coat protein.
      FALSE: gram-negative bacteria do not form spores.

5. **A 22-year-old sexually active female from New England presents with fever and right knee swelling. Her joint aspirate reveals gram-negative diplococci. She also has had a recent bout of cervicitis. What is the most likely cause of this patient's septic arthritis?**
   A. *Staphylococcus aureus*.
      FALSE: staphylococci are gram-positive cocci growing in clusters.
   B. *Streptococcus pyogenes*.
      FALSE: streptococci are gram-positive cocci growing in chains.
   C. *Neisseria gonorrhoeae*.
      TRUE: the gonococcus is a gram-negative diplococcus.

   D. *Pseudomonas aeruginosa*.
      FALSE: *P. aeruginosa* is a motile gram-negative rod.
   E. *Borrelia burgdorferi*.
      FALSE: *B. burgdorferi* is a spirochete that stains weakly gram-negative.

6. **A 16-year-old female comes to the physician because of an increased vaginal discharge. She developed this symptom 2 days ago. She also complains of dysuria. She is sexually active with one partner and uses condoms intermittently. Examination reveals some erythema of the cervix but is otherwise unremarkable. A urinalysis is negative. Sexually transmitted disease testing is performed and the patient is found to have gonorrhea. While treating this patient's gonorrhea infection, treatment must also be given for which one of the following?**
   A. Bacterial vaginosis.
      FALSE: no treatment for bacterial vaginosis should be given unless there is evidence of this condition.
   B. *Chlamydia*.
      TRUE: persons with gonorrhea are assumed to be co-infected with chlamydia and vice versa.
   C. Herpes.
      FALSE: no treatment for herpes should be given unless there is evidence of this infection.
   D. Syphilis.
      FALSE: no treatment for syphilis should be given unless there is evidence of this infection.
   E. Trichomoniasis.
      FALSE: no treatment for trichomoniasis should be given unless there is evidence of this infection.

7. **All of the following are important in the isolation and laboratory diagnosis of *Neisseria gonorrhoeae* infections, EXCEPT?**
   A. Use of selective media to suppress the growth of other bacteria and fungi while allowing gonococci to grow.
      TRUE: Modified Thayer Martin agar, a chocolate agar containing antibiotics and an antifungal agent is the medium of choice for isolating gonococci.
   B. Positive oxidase test.
      TRUE: *Neisseria* species are oxidase-positive.
   C. The ability of the gonococcus to use only glucose out of the five sugars tested on an acid production panel.
      TRUE: *N. gonorrhoeae* produces acid from glucose.
   D. The presence of intracellular gram-negative diplococci in penile discharge.
      TRUE: in the male the presence of intracellular gram-negative diplococci is pathognomonic for gonorrhea.
   E. Detection of IgM antibody to pili by ELISA.
      FALSE: serology is of no value in the diagnosis of gonorrhea.

8. ***Neisseria gonorrhoeae* is capable of all of the following, EXCEPT?**
   A. Aerobic metabolism.
      TRUE: *Neisseria* untilize sugars oxidatively.
   B. Cleavage of secretory immunoglobulin A subclass 1 (sIgA1).
      TRUE: *N. gonorrhoeae* produces an IgA1 protease.
   C. Transcytosis of epithelial cells.
      TRUE: *N. gonorrhoeae* induces uptake by epithelial cells.
   D. Secretion of hemolytic toxins.
      FALSE: *N. gonorrhoeae* does not produce hemolysin.
   E. Production of purpuric skin lesions.
      TRUE: a feature of disseminated *N. gonorrhoeae* infection is a purpuric skin rash.

9. **Which of the following statements about gonorrhea are correct?**
   A. Bacteremia is common.
      FALSE: disseminated infection is rare.
   B. Men are frequently asymptomatic carriers.
      FALSE: infection in the male is almost always symptomatic.
   C. Penicillins are no longer recommended for treatment.
      TRUE: *N. gonorrhoeae* is resistant to penicillin.
   D. Diagnosis by Gram stain in men is unreliable.
      FALSE: Gram stain of penile discharge is an effective way of detecting gonorrhea.
   E. Serologic diagnosis is more reliable than culture.
      FALSE: serology plays no part in diagnosis.

10. **All of the following contribute to the virulence of the gonococcus, EXCEPT?**
    A. Pili.
       TRUE: pili are the means of initial attachment to epithelium.
    B. Opa proteins.
       TRUE: Opa proteins are involved in tight adhesion and internalization of the bacterium.
    C. IgA1 protease.
       TRUE: plays a role in inducing inflammation if not immune evasion.
    D. Capsule.
       FALSE: gonococci are not capsulate.
    E. Porins.
       TRUE: porins induce apoptosis in host cells.

CASE 25: *Neisseria meningitidis*

1. **A 19-year-old college freshman was brought to the Emergency Room (ER) because of severe headache, fevers, and neck stiffness. She is a resident of a dormitory. In the ER, she was unresponsive, and had a diffuse purpuric/petechial rash; her temperature was 104°F (40°C). Her lumbar puncture was consistent with meningitis. A culture of her cerebrospinal fluid was most likely to yield?**
   A. *Escherichia coli*.
      FALSE: *Escherichia coli* is a cause of meningitis in the immediate post-partum period.
   B. *Neisseria meningitidis*.
      TRUE: *N. meningitidis* causes meningitis in this setting and age of patient. The purpuric rash is an important clinical sign but occurs in less than 50% of patients.
   C. Group B *Streptococcus* (GBS).
      FALSE: GBS is not a cause of meningitis in this age group.
   D. *Streptococcus pneumoniae*.
      FALSE: the incidence of meningitis caused by *S. pneumoniae* is highest in children and the elderly and it does not present with a purpuric rash.
   E. *Haemophilus influenzae*.
      FALSE: meningitis caused by *H. influenzae* is most common in pediatric patients and it does not present with a purpuric rash. Invasive disease has virtually been eradicated in industrialized countries by immunization.

2. **A frantic mother brought her 10-day-old infant to the Emergency Room (ER) because of fevers and failure to thrive. The baby was diagnosed with meningitis. The cerebrospinal fluid would most likely yield?**
   A. *Streptococcus pyogenes* (group A streptococci).
      FALSE: the group A streptococcus is an unusual cause of meningitis.
   B. *Streptococcus pneumoniae*.
      FALSE: except in rare cases neonates are protected against meningitis caused by *S. pneumoniae* by transplacental transfer of maternal anti-capsular IgG antibodies.
   C. *Streptococcus agalactiae* (group B streptococci).
      TRUE: the group B streptococcus is the most common cause of meningitis in this age group. Meningitis caused by the GBS at 10 days post-partum is termed 'late onset disease.'
   D. *Staphylococcus aureus*.
      FALSE: *S. aureus* is an unusual cause of meningitis.

   E. *Neisseria meningitidis*.
      FALSE: except in rare cases neonates are highly resistant to *N. meningitidis* infection because they are protected by transplacental transfer of anti-capsule IgG antibodies.

3. **All of the following are characteristics of *Neisseria meningitidis*, EXCEPT?**
   A. Produces an extracellular polysaccharide capsule.
      TRUE: the capsule is a major virulence determinant.
   B. Gram-negative diplococcus.
      TRUE: *N. meningitidis* is a gram-negative coffee bean-shaped diplococcus.
   C. Oxidase-positive.
      TRUE: *N. meningitidis* produces cytochrome oxidase.
   D. Produces lipooligosaccharide (LOS).
      TRUE: *N. meningitidis* produces LOS, which lacks or has very short O-antigenic sugar chains.
   E. Serogrouped by differences in the outer-membrane proteins.
      FALSE: *N. meningitidis* is serogrouped on the basis of immunological differences in the capsular polysaccharides.

4. **Which one of the following is the reservoir of *Neisseria meningitidis*?**
   A. Soil.
      FALSE: *N. meningitidis* sole habitat is the human oropharynx.
   B. Salt water.
      FALSE: *N. meningitidis* sole habitat is the human oropharynx.
   C. Human carriers.
      TRUE: *N. meningitidis* sole habitat is the human oropharynx.
   D. Domesticated animals.
      FALSE: *N. meningitidis* sole habitat is the human oropharynx.
   E. Reptiles.
      FALSE: *N. meningitidis* sole habitat is the human oropharynx.

5. ***Neisseria meningitidis* is spread by which one of the following?**
   A. Respiratory secretions or droplets.
      TRUE: *N. meningitidis* is spread by respiratory secretions.

B. Contaminated fomites.
FALSE: *N. meningitidis* inhabits the human mucous membranes, principally the oropharynx.
C. Consumption of contaminated food or water.
FALSE: *N. meningitidis* inhabits the human mucous membranes, principally the oropharynx.
D. Insect vectors.
FALSE: *N. meningitidis* inhabits the human mucous membranes, principally the oropharynx.
E. Vaginal intercourse.
FALSE: *N. meningitidis* inhabits the human mucous membranes, principally the oropharynx. However, because the meningococcus has been shown to colonize the genitourinary tract the bacterium conceivably could be transmitted by receptive oral sex.

6. ***Neisseria meningitidis* is capable of all of the following, EXCEPT?**
A. Anaerobic metabolism.
FALSE: *N. meningitidis* is an obligate aerobe.
B. Cleavage of secretory immunoglobulin A subclass 1 (sIgA1).
TRUE: *N. meningitidis* produces an IgA1 protease.
C. Transcytosis of epithelial cells.
TRUE: *N. meningitidis* induces uptake by epithelial cells and is able to transcytose.
D. Production of capsule.
TRUE: the capsule is the principal determinant of pathogenesis.
E. Production of purpuric skin lesions.
TRUE: a characteristic of *N. meningitidis* infection is the production of a purpuric rash.

7. **All of the following are important in protection against meningococcal infection, EXCEPT?**
A. The alternative complement pathway.
TRUE: all pathways of complement are important in defense against the meningococcus.
B. IgM and IgG anti-capsular antibodies.
TRUE: bactericidal IgM and IgG are important in defense against the meningococcus.
C. CD8 cytotoxic T cells.
FALSE: cytotoxic T cells are not important in defense against the meningococcus.
D. Polymorphonuclear leukocytes.
TRUE: the PMN is an important cell in phagocytosis of meningococci.
E. The classical complement pathway.
TRUE: all pathways of complement are important in defense against the meningococcus.

8. **In the treatment and prevention of meningococcal disease all of the following are correct, EXCEPT?**
A. Antibiotics should not be given until the diagnosis is confirmed by laboratory investigations.
FALSE: parenteral antibiotics should be given immediately.

B. Parenteral penicillin G or ceftriaxone are suitable antibiotics unless the patient is hypersensitive to them.
TRUE: these are the recommended antibiotics for therapy.
C. Close contacts of the index case should receive chemoprophylaxis with rifampin or ciprofloxacin.
TRUE: this is the recommended course of action for close contacts.
D. Meningococcal capsular polysaccharide-protein conjugate vaccines protect against disease caused by serogroups A, C, Y, and W-135 but not B.
TRUE: currently there is no vaccine available for serogroup B disease.
E. Frequent re-evaluation and, in patients with a poor prognosis, supportive therapy and transfer to an intensive care unit is indicated.
TRUE: this is the correct management for a patient with invasive meningococcal disease.

9. **Which of the following statements about the isolation and identification of meningococcus are CORRECT?**
A. A Gram stain of blood, CSF or material obtained from the characteristic skin lesion.
TRUE: a Gram stain will show intracellular gram-negative diplococci.
B. Blood or chocolate agar may be used to recover meningococci from the blood and CSF.
TRUE: *N. meningitidis* will grow on either medium.
C. *N. meningitidis* oxidizes the sugars glucose and maltose.
TRUE: oxidation of these sugars defines *N. meningitidis*.
D. Rapid latex agglutination can be used to detect capsular polysaccharide in CSF and urine.
TRUE: this is a sensitive test for the detection of capsule.
E. All of the above.
TRUE: all of the above are suitable methods for detecting and isolating *N. meningitidis*.

10. **Which of the following contribute to the virulence of the meningococcus?**
A. Pili.
TRUE: pili are important in the initial adhesion to nasopharyngeal epithelium.
B. Opa proteins.
TRUE: Opa proteins are important in tight adhesion to nasopharyngeal epithelium.
C. IgA1 protease.
TRUE: IgA1 protease contributes to the virulence of *N. meningitidis*.
D. Capsule.
TRUE: the capsule is the principal virulence determinant.
E. All of the above.
TRUE: all of the above contribute to the virulence of *N. meningitidis*.

CASE 26: Norovirus

1. **Which of the following are true of norovirus?**
   A. This virus is the most usual cause of infantile diarrhea.
      FALSE: the usual cause of infantile diarrhea is rotavirus.
   B. Genetic variation arises by reassortment.
      FALSE: noroviruses are single-stranded RNA viruses in which genetic variation arises by point mutation and recombination but not reassortment, which requires a segmented genome.
   C. It is closely genetically related to rotavirus.
      FALSE: rotavirus is a reovirus, namely a double-stranded segmented RNA virus.
   D. It is closely genetically related to sapovirus.
      TRUE: both are caliciviruses. sapovirus is an infrequent cause of infantile diarrhea.

2. **Which of the following are true concerning the spread of norovirus?**
   A. It is susceptible to adverse environmental conditions.
      FALSE: it persists in the environment mainly due to the lack of a viral envelope.
   B. It can be transmitted via fomites.
      TRUE: norovirus persists in the environment and may be transmitted via fomites, i.e. inanimate objects such as door handles, carpets or writing utensils that may be contaminated with infectious agents.
   C. It may be acquired by feco–oral spread.
      TRUE: most infections are acquired via this route, whereas infection is only rarely acquired from aerosolized vomit.
   D. Infection may be acquired transplacentally.
      FALSE: the infection is thought to be limited to the gut.
   E. Only a very small proportion of the population are susceptible.
      FALSE: immunity is short-lived and at any one time a large proportion of the population are susceptible.
   F. Re-infection may occur.
      TRUE: immunity is short-lasting and strain-specific.

3. **Which of the following tests would be helpful in the diagnosis of norovirus gastroenteritis?**
   A. Inoculation of a cell culture with feces from an infected person.
      FALSE: the virus cannot be grown in cell culture.
   B. RT-PCR for norovirus protein in feces.
      FALSE: RT-PCR does not detect viral protein. However RT-PCR to detect viral RNA is the mainstay of diagnosis.
   C. Examination of a fecal sample by electron microscopy.
      TRUE: the virus is seen by electron microscopy as a small round structured virus. However, this method is very insensitive as there have to be at least a million particles per ml of feces for the virus to be visualized.
   D. Fecal microscopy to detect pus cells.
      FALSE: norovirus infection causes a secretory diarrhea and pus cells are absent.
   E. A characteristic appearance of the intestine on endoscopy.
      TRUE: the intestinal villi are blunted. Nevertheless this method is not recommended for routine diagnosis as the disease is short-lived and also rotavirus infection gives the same appearance.

4. **Which of the following are true for the clinical presentation of norovirus infection?**
   A. It presents with dysentery.
      FALSE: the typical presentation is with vomiting and watery not bloody diarrhea.
   B. After 2 weeks the illness resolves.
      FALSE: the illness lasts about 2 days in the immunocompetent, although it may persist for weeks to months in organ transplant recipients.
   C. A complication may be hypokalemia.
      TRUE: if part of dehydration.
   D. Abdominal pain is severe.
      FALSE: abdominal cramps are characteristically mild.

5. **Which of the following are useful in the prevention of transmission of norovirus infection?**
   A. Isolation of infected patients.
      TRUE: patients with norovirus gastroenteritis should be nursed in an appropriate single room using precautions to prevent transmission of norovirus by contact, e.g. use of gloves and aprons. This is one of the main ways of limiting the spread of infection in the hospital setting.
   B. Hand-washing.
      TRUE: the virus may be spread on hands.
   C. Chlorine compounds.
      TRUE: these agents are effective when used as disinfectants to clean contaminated surfaces.
   D. Persons with norovirus gastroenteritis may prepare food for others if scrupulous hand-washing is observed.
      FALSE: those infected should not prepare food until at least 48 hours after their symptoms. The infectious dose is very low and hand-washing is insufficient to prevent transmission on hands. There is also a major risk of contamination of food and the surrounding environment with vomit.
   E. Ward closure.
      TRUE: norovirus outbreaks are notoriously difficult to control in the hospital setting. It is usually necessary to close an affected ward to new admissions and wait for the outbreak to burn out.

CASE 27: Parvovirus

1. **Which of the following statements regarding parvovirus B19 are true?**
   A. It contains a double-stranded DNA genome.
      FALSE: the *Parvoviridae* have a single-stranded DNA genome.
   B. The cellular receptor for binding is the blood group B antigen.
      FALSE: the receptor is globoside, or the P blood group antigen.
   C. It causes annual epidemics in temperate climates.
      FALSE: while infections occur every year, there is a periodicity to major epidemics of 4–5 years. This is

presumably to allow time for accumulation of sufficient susceptible individuals in a population to support an epidemic.

D. It preferentially infects erythroid precursor cells in the bone marrow.
TRUE: these express abundant P antigen.

E. It belongs to the same virus family as rubella virus.
FALSE: rubella virus has a positive single-stranded RNA genome and belongs to the virus family *Togaviridae*.

2. **Which of the following are recognized routes of spread of parvovirus B19 infection?**
A. Blood transfusion.
TRUE: blood donors may have an asymptomatic viremia, which results in the blood being infectious.
B. Contaminated water.
FALSE: there is no evidence for fecal–oral transmission of this virus.
C. Droplet spread from respiratory tract secretions.
TRUE: the most common route of spread, as virus is shed from the throat.
D. Mother to fetus.
TRUE: parvovirus has access to, and may cross, the placenta from the maternal bloodstream. The effects of this vary according to the stage of pregnancy, but can give rise to erythrogenic arrest, leading to fetal hydrops.
E. Sexual transmission.
FALSE: there is no evidence of sexual transmission of this virus.

3. **Which of the following statements concerning parvovirus B19 infection are true?**
A. Anemia arises because of infection of erythroid precursor cells.
TRUE: infection of the bone marrow is followed by the disappearance of reticulocytes, and a transient drop in circulating hemoglobin levels.
B. Symptoms of fever and headache arise as a result of immune complex formation.
FALSE: these symptoms most likely arise through cytokine release.
C. Symptoms of rash and arthralgia occur about 6–9 days after infection.
FALSE: these symptoms usually appear at least 2 weeks after infection, coincident with the generation of an IgG antibody response.
D. Symptoms are usually less severe than those caused by parvovirus B3 infection.
FALSE: there is no such virus as parvovirus B3.
E. Viral DNA is detectable in the bloodstream before any antibody response is detectable.
TRUE: viremia will precede an antibody response, and can be detected with the very sensitive genome amplification assays now available.

4. **Which of the following are recognized manifestations of parvovirus B19 infection?**
A. The childhood rash known as exanthem subitum (also known as 6th disease).
FALSE: that syndrome is caused by infection with human herpesvirus 6 (HHV6). Parvovirus causes erythema infectiosum, also known as 5th disease.
B. Polyarticular arthritis.
TRUE: this can be quite severe, prolonged, and can mimic rheumatoid arthritis. It is more common in women.
C. Meningitis.
FALSE: this is not a typical manifestation of parvovirus

B19 infection.
D. Respiratory tract illness.
TRUE: usually mild.
E. Pancreatitis.
FALSE.

5. **Which of the following serological markers indicates recent infection with parvovirus B19 virus?**
A. The presence of parvovirus-specific IgA in a serum sample.
B. The presence of parvovirus-specific IgD in a serum sample.
C. The presence of parvovirus-specific IgE in a serum sample.
D. The presence of parvovirus-specific IgG in a serum sample.
E. The presence of parvovirus-specific IgM in a serum sample.
Answer: E. IgM antibodies are the first to appear and, more importantly, the IgM response is transient due to class switching to IgG production. Thus, the presence of specific IgM antibodies in a serum sample is indicative of a recent infection with the virus of interest.

6. **In which of the following patients can parvovirus B19 infection be life-threatening?**
A. Patients with sickle cell anemia.
TRUE: such patients will already have low hemoglobin levels, much of which is due to the presence of circulating reticulocytes. Acute parvovirus infection can therefore result in an acute aplastic crisis.
B. Hemophiliacs.
FALSE: hemophiliacs are more likely to have been infected with parvovirus B19, through exposure in transfused factor VIII, but there is nothing unusual about the infection in these patients.
C. Fetuses.
TRUE: *in utero* parvovirus infection may lead to spontaneous miscarriage or fetal heart failure.
D. Patients with HIV infection.
TRUE: failure to eliminate infection can result in a chronic anemia.
E. Patients with rheumatoid arthritis.
FALSE: parvovirus infection may lead to a rheumatoid-like polyarthritis, but there is nothing special about infection in RA patients.

7. **A 25-year-old woman presents with a diffuse morbilliform rash and a small joint polyarthropathy. She is 18 weeks pregnant. Serology reveals the following results:**
- IgG anti-rubella positive, IgM anti-rubella negative
- IgG anti-parvovirus positive, IgM anti-parvovirus positive.
- **Which of the following adverse events may arise from this in her pregnancy?**
A. Congenital cataracts.
B. Congenital heart disease.
C. Congenital deafness.
D. Hydrops fetalis.
E. Spontaneous miscarriage.
Answer: D, E. The serological results indicate a recent parvovirus infection. Both D and E are recognized complications of maternal parvovirus infection at this stage in pregnancy. Options A, B, and C are well-recognized features of the congenital rubella syndrome and therefore not likely to affect this patient.

8. **A 25-year-old woman presents with a diffuse morbilliform rash and a small joint polyarthropathy. Her last menstrual period was 8 weeks ago, and a**

pregnancy test is positive. Serology reveals the
following results:
- IgG anti-rubella positive, IgM anti-rubella positive
- IgG anti-parvovirus positive, IgM anti-parvovirus
  negative.

Which of the following is the most likely outcome of
this pregnancy?
A. Spontaneous miscarriage.
B. Hydrops fetalis due to fetal anemia.
C. A normal healthy baby.
D. A baby with multiple congenital abnormalities.
E. A baby with congenital deafness, otherwise normal.

*Answer: D. The serological results indicate a maternal acute
rubella infection. At this stage in pregnancy, such infection
almost invariably results in multiple severe developmental
defects in the fetus, the so-called congenital rubella syndrome.*

9. **A 25-year-old woman presents with a diffuse
morbilliform rash and a small joint polyarthropathy.
She is 10 weeks pregnant. Serology reveals the
following results:**
- IgG anti-rubella positive, IgM anti-rubella negative
- IgG anti-parvovirus positive, IgM anti-parvovirus
  positive.

Which of the following is the most likely outcome of
this pregnancy?
A. A normal healthy baby.
B. A baby with multiple congenital abnormalities.
C. A baby with congenital deafness, otherwise normal.
D. A baby with congenital anemia.
E. A baby with congenital cataracts.

*Answer: A. The serological results indicate a maternal acute
parvovirus infection. The most likely outcome among those
listed is a normal healthy baby. There is also a much increased
risk of spontaneous miscarriage. However, in contrast to
rubella infection in pregnancy, there is no such thing as a
congenital parvovirus syndrome.*

10. **Which of the following drugs/modalities of therapy
have been proven useful in the management of patients
with parvovirus B19 infection?**
A. Aciclovir.
FALSE: this drug inhibits viral DNA synthesis, but
requires prior activation by viruses which can only be
performed in cells infected with a virus that possesses a
thymidine kinase enzyme. Parvovirus B19 does not.
B. Blood transfusion.
TRUE: this can be life-saving for patients with chronic
hemolytic anemia and a parvovirus-induced aplastic
crisis, and for fetuses with parvovirus-induced heart
failure (when it has to be given intra-uterine).
C. Normal human immunoglobulin.
TRUE: this may be helpful in the management of
chronic parvovirus infection in an immunosuppressed
host.
D. Monoclonal anti-parvovirus antibodies.
FALSE: there have been no clinical studies of such
antibodies.
E. Zanamavir.
FALSE: this is a neuraminidase inhibitor used in the
treatment of influenza virus infection. Parvovirus B19
does not possess a neuraminidase enzyme.

CASE 28: *Plasmodium* spp.

1. **Which of the following transmit *Plasmodium*?**
A. Ticks.
FALSE: ticks transmit, e.g. Lyme disease.
B. Fleas.
FALSE: fleas transmit, e.g. plague.
C. Mosquitoes.
TRUE: the vector for malaria.
D. Bugs.
FALSE: bugs transmit, e.g. trypanosomes, a blood
protozoa.
E. Mites.
FALSE: mites transmit, e.g. Rickettsiae.

2. **Which of the following cell types are involved in the
life cycle of *Plasmodium*?**
A. Erythrocyte.
TRUE: the merozoites replicate in the erythrocyte.
B. Hepatocyte.
TRUE: the sporozoites replicate in the hepatocyte.
C. Osteoblast.
FALSE: not affected in malaria.
D. Lymphocyte.
FALSE: not affected in malaria.
E. Neuron.
FALSE: not involved in the life cycle.

3. **In what cell/organ does sexual reproduction of
Plasmodium occur?**
A. The human erythrocyte.
FALSE: schizogony occurs in the erythrocyte.

B. The mosquito liver.
FALSE: does the mosquito have a liver?
C. The human liver.
FALSE: replication of the sporozoites occurs in the
hepatocytes.
D. The mosquito salivary gland.
FALSE: sporozoites are found in the salivary gland.
E. The mosquito intestine.
TRUE: the gametocytes fuse to form the zygote in the
enterocytes of the mosquito gut.

4. **Which of the following human genetic markers affect
the development of malaria?**
A. Duffy blood group antigen.
TRUE: receptor for the adhesin of *Plasmodium vivax*.
B. Lewis blood group antigen.
FALSE: not involved in malaria pathogenesis or
replication.
C. Glucose 6 phosphate dehydrogenase (G6PD) deficiency.
TRUE: restricts the development of the malarial parasite.
D. Sickle cell hemoglobin HbS.
TRUE: restricts the development of the malarial parasite.
E. Secretor status.
FALSE: not involved in malaria pathogenesis or
replication.

5. **Which of the following are directly important for the
symptoms of malaria?**
A. Hemolysis of erythrocytes.
TRUE: causes anemia and results in 'blackwater fever.'

B. Sequestration of erythrocytes.
TRUE: by blocking vessels leads to the symptoms/signs of cerebral malaria.

C. Lysis of liver cells.
FALSE: does not lead to symptoms but involved in the life cycle.

D. Release of TNF.
TRUE: responsible for fever.

E. Destruction of retinal cells.
FALSE: retinal cells are not affected.

6. **Which of the following are used in the treatment/prevention of malaria?**

A. Flucloxacillin.
FALSE: an antibiotic affecting bacteria.

B. Artemisinin-based combination therapies (ACTs).
TRUE: being used more and more and being promoted by WHO for first-line therapy worldwide.

C. Chloroquine.
TRUE: also used in prophylaxis. However, this is no longer a highly recommended treatment in most areas.

D. Ciprofloxacin.
FALSE: an antibiotic affecting bacteria.

E. Proguanil.
TRUE: used in the prophylaxis of malaria.

7. **Which of the following are clinical indicators of the diagnosis of malaria?**

A. Cyclical fever.
TRUE: initally the temperature may be irregular but settles down to a characteristic periodicity.

B. Splenomegaly.
TRUE: due to destruction of erythrocytes and should be palpated with care as it may rupture.

C. Iritis.
FALSE: iritis is not a feature of malaria.

D. Anemia.
TRUE: due to hemolysis.

E. Skin rash.
FALSE: a rash is not a feature of malaria.

8. **Which of the following are complications of malaria?**

A. Cerebral thrombosis.
TRUE: due to blocking of capillaries caused by infected erythrocytes.

B. Skin necrosis.
FALSE: does not occur in malaria.

C. Renal failure.
TRUE: due to blackwater fever.

D. Splenic rupture.
TRUE: and may lead to massive intra-abdominal hemorrhage.

E. Nephrotic syndrome.
TRUE: assosiated with *P. malariae* infections.

CASE 29: Respiratory syncytial virus (RSV)

1. **Which of the following are true of respiratory syncytial virus?**

A. One of the envelope proteins is a hemagglutinin.
FALSE: unlike influenza and parainfluenza viruses where the virus attachment protein is a hemagglutinin, the RSV G protein is not a hemagglutinin.

B. Annual outbreaks of infection occur every summer in the USA.
FALSE: in temporal regions such as the USA there are annual winter outbreaks.

C. It is closely related to human metapneumovirus.
TRUE: both RSV and metapneumoviruses are pneumoviruses, a subfamily of paramyxoviruses.

D. The G protein of the virus is hypervariable and accounts for re-infection with the virus throughout life.
TRUE: the RSV G protein is hypervariable and emergence of new variants means individuals are immunologically naïve, hence re-infection.

E. The viral genome is segmented.
FALSE: none of the paramyxoviruses have a segmented genome. A segmented genome is a characteristic of the orthomyxoviruses, i.e. influenza virus.

2. **Which of the following diseases are proven to be caused by RSV?**

A. Bronchiolitis.
TRUE: RSV is the most common cause of bronchiolitis.

B. Epiglottitis.
FALSE: epiglottitis is caused by *Haemophilus influenzae* type b. The bacterium causes swelling and obstruction of the epiglottis and consequent respiratory obstruction.

C. Asthma.
FALSE: asthma may follow RSV infection but it is unclear whether children who get bronchiolitis have a pre-existing predisposition to asthma or whether the relationship is causal.

D. Croup.
TRUE: a small proportion of cases are caused by RSV; more likely etiological agents are parainfluenza (and influenza) viruses.

E. Whooping cough.
FALSE: whooping cough (also known as pertussis) is caused by a bacterium, *Bordetella pertussis*.

3. **Which of the following tests would be helpful for the diagnosis of bronchiolitis?**

A. Polymerase chain reaction for RSV antigen in respiratory secretions.
FALSE: the polymerase chain reaction detects viral nucleic acid rather than antigen. Since RSV is an RNA virus an extra step is required, namely the use of reverse transcriptase to convert RNA to cDNA to allow amplification of the latter in the polymerase chain reaction.

B. Immunofluorescence test for respiratory virus antigens in respiratory epithelial cells.
TRUE: the detection of RSV antigen in respiratory epithelial cells collected by nasopharyngeal aspiration is a frequently used test.

C. Polymerase chain reaction for human metapneumovirus RNA in respiratory secretions.
TRUE: human metapneumovirus is a known cause of bronchiolitis.

D. Chest X-ray.
TRUE: hyperinflation of the lungs is characteristic of bronchiolitis.

E. Pulse oximetry.
TRUE: children with bronchiolitis may have hypoxemia and respiratory failure.

4. **What are the typical signs and symptoms of a patient presenting with bronchiolitis?**
A. Barrel-shaped chest, prominent neck veins, and downward displacement of the liver.
TRUE: these are all consequences of the hyperinflation of the lungs that is induced by bronchiolitis.
B. Inspiratory stridor.
FALSE: this is characteristic of croup and a result of laryngeal obstruction.
C. Crackles on auscultation of the lung.
FALSE: this is characteristic of the pneumonia that may rarely accompany bronchiolitis.
D. Conjunctivitis.
FALSE: adenovirus respiratory infections are sometimes accompanied by conjunctivitis. Conjunctivitis is also seen in measles due to infection of the ocular mucosa by the virus.
E. Expiratory wheezing.
TRUE: this finding is characteristic of bronchiolitis and

is due to obstruction of the bronchioles as a result of viral infection, usually RSV.

5. **How would you treat a patient with respiratory syncytial virus infection?**
A. Treat with the monoclonal antibody, palivizumab, administered intramuscularly.
FALSE: this antibody is only indicated to protect vulnerable infants, such as those with bronchopulmonary dysplasia, from severe RSV infection.
B. Drain pleural fluid.
FALSE: pleural effusions are not a feature of bronchiolitis.
C. Maintain an adequate airway and ensure oxygenation.
TRUE: supportive care is of paramount importance in bronchiolitis.
D. Maintain nutrition and hydration.
TRUE: supportive care is of paramount importance in bronchiolitis.
E. Treat with nebulized ribavirin.
TRUE: although the initial clinical trials appeared promising, results of treatment in clinical practice proved disappointing. Now very rarely used.

CASE 30: *Rickettsia* spp.

1. **Which of the following are true of the genus *Rickettsia*?**
A. They can be cultured on artificial media.
FALSE: they are obligate intracellular bacteria.
B. They have a large genome.
FALSE: they have a small genome size and restricted metabolic activity.
C. *Coxiella* is not a member of the *Rickettsiales*.
TRUE: *Coxiella* used to be included in the *Rickettsiales* but is now included in the γ-proteobacteria.
D. They have a gram-positive cell wall.
FALSE: they are gram-negative and do not stain well with the Gram stain.
E. *Salmonella typhi* is one of the *Rickettsiales*.
FALSE: *Salmonella typhi* is a member of the *Enterobacteriaceae*, not to be confused with *Rickettsia typhi*, which is a member of the *Rickettsiales*.

2. **Which of the following are true concerning the spread of *Rickettsia* spp.?**
A. *Rickettsia* are spread by arachnidae.
FALSE: they are spread by arthropods such as ticks, fleas, lice, and mites.
B. Animal reservoirs maintain the bacteria in nature.
TRUE: usually dogs, deer, rodents.
C. Some species of *Rickettsia* only infect humans.
TRUE: *R. prowazekii* and also *O. tsutsugamushi*.
D. Many *Rickettsia* have a global distribution.
TRUE: such as *R. prowazekii*, while other such as *R. honei* have only thus far been detected in Australia.
E. *Rickettsia* typically infect epithelial cells.
FALSE: they typically infect vascular endothelial cells.

3. **Which of the following are known to be important in the pathogenesis of rickettsial disease?**
A. Extracellular location of the organism.
FALSE: the organism is an obligate intracellular bacterium.
B. Induced endocytosis.

TRUE: the organism stimulates endothelial cells (and macrophages in some species) to take it up using a zipper mechanism.
C. Leukocyte adhesion to vascular endothelium.
TRUE: infection with *Rickettsia* up-regulates cellular adhesion molecules ICAM, VCAM, and E-selectin on the vascular endothelium.
D. Increased vascular permeability.
TRUE: lipid peroxidation and nitric oxide damage the endothelial cells and increase vascular permeability leading to edema.
E. Infection of epithelial cells.
FALSE: *Rickettsia* infect endothelial cells.

4. **Which of the following are true of *Rickettsia rickettsii*?**
A. It causes Brill-Zinsser disease.
FALSE: this is the recrudescent form of epidemic typhus caused by *R. prowazekii*. *R. rickettsii* causes Rocky Mountain spotted fever.
B. It is transmitted by a flea bite.
FALSE: it is transmitted by a tick bite. Fleas transmit *R. felis* and *R. typhi*.
C. It presents with fever, headache, myalgia, and rash.
TRUE: this is the classical presentation of rickettsial disease.
D. The disease is more severe in patients with G6PD deficiency.
TRUE: because the erythrocytes are more susceptible to damage.
E. It produces a vesicular rash.
FALSE: this occurs in rickettsial pox. A maculopapular rash beginning around the wrists and ankles is characteristic of RMSF.

5. **Which of the following are true statements concerning diagnosis of rickettsial disease?**
A. The Gram stain of biopsy material is useful.

FALSE: the bacterium is barely visible by light microscopy.
B. Serology provides an accurate diagnosis of acute disease.
FALSE: serology is more important in making a retrospective diagnosis.
C. Immunofluorescence is frequently used in laboratory diagnosis.
TRUE: an alternative test is the ELISA format.
D. PCR can give an accurate species identification.
TRUE: particularly if combined with sequencing or RFLP of the amplicon.
E. PCR primers that are useful include those based on OmpA and citrate synthase.
TRUE.

6. **Which of the following are used in the treatment of rickettsial disease?**
A. Ampicillin.
FALSE: as this is a β-lactam.
B. Chloramphenicol.
TRUE: it can be used in cases of tetracycline hypersensitivity or pregnancy.
C. Doxycycline.
TRUE: the most effective treatment.
D. Oxytetracycline.
TRUE: as an alternative to doxycycline.
E. Ciprofloxacin.
FALSE: although it has *in vitro* activity.

CASE 31: *Salmonella typhi*

1. **Which of the following are true of the genus *Salmonella*?**
A. They are motile.
TRUE: they are motile.
B. They can be grouped serologically into a small number of serovars.
FALSE: they can be grouped into about 2500 serovars but there are only 3 species.
C. The Kauffman-White scheme of classification is based upon the O, H, and Vi antigens.
TRUE: the scheme divides up *Salmonella* into subgroups based on the antigenicity of the various surface antigens.
D. Phase variation occurs with the O antigen.
FALSE: phase variation occurs with the H antigen.
E. Phage typing may be used for epidemiological purposes.
TRUE: but molecular techniques such as PFGE and ERIC-PCR are also used.

2. **Which of the following are true concerning *Salmonella* Typhi?**
A. *Salmonella* Typhi is widespread in animals.
FALSE: *Salmonella* Typhi is a strictly human pathogen.
B. Infection is usually acquired by ingestion of contaminated food or water.
TRUE: usually because of poor personal hygiene practices or a poor public health infrastructure.
C. The global burden of disease is an estimated 16 million cases per year.
TRUE: and most cases occur in Africa, India, and South-East Asia.
D. A significant proportion of cases are linked to a food-handler who is a carrier of the organism.
TRUE: and in some countries carriers of *Salmonella* Typhi are prohibited from being food-handlers.
E. In industrialized countries many cases are acquired abroad from an endemic country.
TRUE: about 40–50% of cases.

3. **Which of the following are true statements concerning nontyphoid *Salmonella*?**
A. They are widespread in animals.
TRUE: and are found commonly in poultry and cattle as well as pets.
B. They are one of the commonest causes of gastroenteritis.
TRUE: second only to *Campylobacter jejuni*.
C. *Salmonella* Typhimurium is one of the commonest serovars causing illness.

TRUE: although all of the 2500 or so serovars have caused illness at some point in time.
D. In Europe and the USA infection with *Salmonella* Enteritidis PT4 is associated with keeping exotic animals.
FALSE: it is linked to eating eggs.
E. The number of reported cases is decreasing.
FALSE: the number of cases is increasing year by year.

4. **Which of the following is true of *Salmonella* Typhi?**
A. It adheres to the cystic fibrosis transmembrane receptor.
TRUE: and it may penetrate through the M cell.
B. It produces a significant secretory response.
FALSE: this is caused by nontyphoid salmonella.
C. Large numbers of granulocytes are induced.
FALSE: this is caused by nontyphoid salmonella; it induces a monocytic response.
D. It inhibits phagolysosome fusion.
TRUE: by modulating the actin cytoskeleton and the surface of the vacuole.
E. It induces bacteria-mediated endocytosis.
TRUE: but is also taken up by macrophages by phagocytosis.

5. **Which of the following are true of the clinical presentation of disease caused by *Salmonella* Typhi?**
A. It often presents with a PUO.
TRUE: the presentation may be nonspecific but may also present with abdominal symptoms.
B. Disease is usually restricted to the gastrointestinal tract.
FALSE: it is a systemic disease with the organism spreading to many organs. Diseases caused by nontyphoid *Salmonella* spp. are often restricted to the gastrointestinal tract.
C. A rash may be a presenting feature.
TRUE: a reddish-pink maculopapular rash on the trunk called rose spots.
D. Gastrointestinal perforation is a complication.
TRUE: as are cholecystitis, endocarditis, and osteomyelitis.
E. The carrier state is unknown.
FALSE: about 10% of individuals become carriers.

6. **Which of the following is useful in the diagnosis of *Salmonella*?**
A. Gram stain.
FALSE: as there are many gram-negative rods.

B. Culture of bone marrow.
   TRUE: this specimen provides the highest yield of isolation.
C. Serology.
   FALSE: the Widal test has a poor sensitivity and is not used in many countries.
D. PCR.
   TRUE: but currently not part of routine diagnosis.
E. String test.
   TRUE: in combination with other tests such as blood culture it can increase the diagnostic yield.

7. **Which of the following are used in the treatment or prevention of *Salmonella* Typhi infection?**
   A. Vaccination.
      TRUE: and can be given as an oral vaccine or a parenteral vaccine and provides about 60–90% protection.
   B. Azithromycin.
      TRUE: and is used in moderately severe cases in patients from India and SE Asia.
   C. Erythromycin.
      FALSE: this is not used to treat any *Salmonella* species.
   D. Enrofloxacin.
      FALSE: this is a veterinary antibiotic. Ciprofloxacin is the human equivalent, although the use of enrofloxacin in veterinary practice may be linked to the emergence of resistance to ciprofloxacin.
   E. Ceftriaxone.
      TRUE: it is used in severe cases of enteric fever, although sensitivities should be routinely checked as resistance to third-generation cephalosporins has been reported.

CASE 32: *Schistosoma* spp.

1. **Which of the following statements are true for the causative agent of schistosomiasis?**
   A. It is a trematode or blood fluke.
      TRUE.
   B. It is restricted to North and South America.
      FALSE: schistosomiasis is found in South America, Africa, and parts of Asia.
   C. It lives in adult form in the human intestinal lumen.
      FALSE: the adults live in blood vessels around the intestine or the bladder.
   D. It reaches lengths of 20 cm in adult form.
      FALSE: in fact it reaches lengths of about 2 cm.
   E. It is associated with snail populations.
      TRUE: distinct aquatic (amphibious) snail species are associated with each schistosome species: *Biomphalaria*, *Bulinus*, and *Oncomelania*.

2. **Which of the following statements are true for the life cycle of *Schistosoma*?**
   A. There has to be human contact with fresh water.
      TRUE.
   B. Humans are infected by a larval form called a miracidium.
      FALSE: miracidia infect snails, cercariae infect humans.
   C. Within humans the first larval form is called the schistosomulum.
      TRUE.
   D. Adult male and female worms of *S. haematobium* normally live in blood vessels around the intestine.
      FALSE: *S. haematobium* lives around the bladder.
   E. Adult worms usually live for 3–5 months.
      FALSE: they usually live for 3–5 years, and sometimes longer (up to 30 years).

3. **Which of the following statements are true about schistosome eggs?**
   A. *S. mansoni* females lay about 3500 eggs per day.
      FALSE: in fact it is *S. japonicum* that lays about 3500 eggs per day.
   B. Eggs of *S. japonicum* normally pass through the bladder wall into urine.
      FALSE: *S. haematobium* eggs pass through the bladder wall.
   C. Granulomas form around eggs deposited in tissue.
      TRUE.
   D. *S. haematobium* eggs have a terminal spine.
      TRUE.
   E. *S. japonicum* eggs are larger than the other species.
      FALSE: they are smaller than *S. mansoni* and *S. haematobium*.

4. **Which of the following statements are true for the host immune response in schistosomiasis?**
   A. It is not thought to provide effective immunity to re-infection.
      FALSE: although the evidence is difficult to interpret there is a consensus that age-related acquired immunity develops.
   B. It is capable of killing schistosomula through IgE-mediated mechanisms.
      TRUE.
   C. It can lead to a hypersensitivity reaction with immune complex formation.
      TRUE: this being Katayama syndrome.
   D. It includes IgG_2- and IgG_4-mediated killing of adult worms.
      FALSE: the adult worms are not really subject to immune-mediated killing.
   E. It is responsible for the pathology of disease.
      TRUE: in that granulomas around eggs are the key to pathology.

5. **Which of the following are true for tissue granulomas in schistosomiasis?**
   A. They develop around schistosomula and adult worms.
      FALSE: tissue granulomas develop around eggs.
   B. They are purely due to a Th1 response in schistosomiasis.
      FALSE: in some granulomatous infections only a Th1 pattern is observed, but in the case of schistosomiasis both Th1 and Th2 are seen.
   C. Cytokines including IFN-γ and TNF-α are present.
      TRUE.
   D. They eventually dismantle and disappear without fibrosis.
      FALSE: in experimental animal models some granulomas seen in some infections do dismantle, but with schistosomiasis the usual fate is fibrosis and persistence.

E. They are associated with eosinophilia.
TRUE.

6. **Which of the following are clinical features of schistosomiasis?**
   A. A skin itch after swimming in water.
   TRUE.
   B. Katayama syndrome with dry cough and changes on chest X-ray.
   TRUE.
   C. Problems with adult worms blocking blood vessels.
   FALSE: adult worms themselves do not cause clinical features. Although adults are up to 2 cm long they do not block vessels to cause pathology.
   D. Nonspecific malaise and tiredness.
   TRUE.
   E. Blood in the urine.
   TRUE: for *S. haematobium*.

7. **Which of the following are true regarding the complications of schistosomiasis?**
   A. They are due to the inflammatory and fibrotic reaction around eggs.
   TRUE.
   B. They include scarring of the bladder wall and blockage of ureters.
   TRUE.
   C. They develop very soon after first infection.
   FALSE: problems take time to develop due to the accumulation of eggs.
   D. They include enlarged esophageal varices that arise from adult worms living in these blood vessels.
   FALSE: varices arise from liver fibrosis and not adult worms living in blood vessels.
   E. They include epilepsy.
   TRUE: occasionally with *S. japonicum*.

8. **Which of the following are true regarding the diagnosis of schistosomiasis?**
   A. It depends on finding characteristic eggs in a blood film.
   FALSE: eggs are found in urine, stool or tissue biopsy.
   B. It depends on finding eggs in urine usually provided first thing in the morning in the case of *S. haematobium*.
   FALSE: eggs are best seen in a terminal, mid-day sample of urine.
   C. It requires serological testing of returned travelers 10–14 days after water exposure.
   FALSE: for many infections convalescent serology at

10–14 days is a good time to check for antibodies. However, for schistosomiasis this is too early and testing after about 12 weeks is more sensitive.
   D. It is supported by the presence of eosinophilia.
   TRUE.
   E. It may need to be differentiated from Crohn's disease in the case of *S. mansoni*.
   TRUE: Crohn's disease can also cause bloody diarrhea with granulomas on tissue biopsy, although the latter will not contain eggs.

9. **Which of the following statements are true for the treatment of schistosomiasis?**
   A. A 6-week course of praziquantel is required.
   FALSE: treatment is complete in a single day.
   B. It depends on effective killing of eggs by praziquantel.
   FALSE: praziquantel kills adult worms and not eggs.
   C. It involves the use of praziquantel immediately after water exposure.
   FALSE: praziquantel is used to kill adult worms, and these are only present several weeks after water exposure.
   D. A combination of drugs is usually required.
   FALSE: usually treatment is with praziquantel monotherapy.
   E. It cures up to 90% of patients after the first treatment course.
   TRUE.

10. **Which of the following statements are true for the control and prevention of schistosomiasis?**
    A. It has resulted in elimination from South East Asia.
    FALSE.
    B. It is being achieved in some countries by population-wide mass chemotherapy.
    TRUE.
    C. It has been facilitated by the construction of dams and irrigation canals.
    FALSE: the situation has been made worse in most areas where water resources development projects have been implemented, as snail populations have flourished.
    D. It is not yet possible through vaccination.
    TRUE: although vaccine candidates are under investigation.
    E. It is easily achieved by killing intermediate host snail populations.
    FALSE: this is easier said than done, although there have been some successes in water-deprived areas such as around an oasis.

CASE 33: *Staphylococcus aureus*

1. **A young woman recently underwent rhinoplasty to correct a deviated septum and her nose was packed postoperatively. She experienced a 2–3-day period of low-grade fever, muscle aches, chills, and malaise. This was followed by the development of a sunburn-like rash that covered most of her body. The packing was removed and a nasal swab grew gram-positive cocci that were catalase-positive and produced colonies on mannitol-salt agar that were surrounded by a yellow halo. The bacteria clumped when mixed with rabbit plasma. What is your diagnosis?**
   A. Scalded skin syndrome.
   FALSE: the clinical presentation and age of the patient

are incorrect.
   B. Toxic shock syndrome.
   TRUE: the clinical presentation of TSS is an erythematous rash covering the whole body including the palms and the soles.
   C. Carbunculosis.
   FALSE: carbuncles present as a coalescence of furuncles and draining sinuses usually in the nape of the neck or the buttocks.
   D. Erysipelas.
   FALSE: erysipelas is an acute infection of the skin caused by *Streptococcus pyogenes*. The involved skin is inflamed and raised and clearly demarcated from the surrounding

unaffected skin.

E. Impetigo.
FALSE: the clinical presentation of impetigo is one of pus-filled vesicles on an erythematous base that occur primarily on the face and limbs. The vesicles burst and are covered with a honey-colored crust.

2. **Which one of the following mechanisms of action of toxic shock syndrome toxin-1 (TSST-1) produced by *Staphylococcus aureus* produces toxic shock syndrome?**
A. ADP-ribosylates a G protein of the adenylate cyclase complexes.
FALSE: this is the mode of action of pertussis toxin.
B. ADP-ribosylates elongation factor 2 (EF2) needed for protein synthesis.
FALSE: this s the mode of action of diphtheria toxin.
C. Binds to MHC class II molecules on antigen-presenting cells and to the CD4 T-cell receptor.
TRUE: TSST-1 functions as a superantigen.
D. Prevents release of glycine and other inhibitory neurotransmitters.
FALSE: this is the mode of action of tetanospasmin.
E. Membrane action results in host cell lysis.
FALSE: the membrane is the target for the *S. aureus* cytotoxins (α, β, γ, δ).

Refer to the case history below to answer questions 3–5.
A man presents to the emergency room with high fever and malaise. He claims to have felt this way for about 3 days, worse every day. He appears underweight and very ill. Physical examination reveals needle marks in both antecubital fossae. Listening for heart sounds, you hear a distinctive systolic heart murmur. You order blood cultures and make a presumptive diagnosis of acute bacterial endocarditis. Cultures confirm your initial diagnosis and the patient begins to respond to the drug regimen you ordered.

3. **Which one of the following antibiotics is most likely to be effective in this case?**
A. Nafcillin.
FALSE: a penicillinase-resistant penicillin not active against MRSA.
B. Methicillin.
FALSE: are resistant to methacillin.
C. Cephalosporin.
FALSE: these are susceptible to β-lactamase and MRSA is resistant to them.
D. Penicillin.
FALSE: susceptible to β-lactamase and MRSA is resistant.
E. Vancomycin.
TRUE: the antibiotic of choice, although resistance is beginning to appear.

4. **Which of the following molecules contribute to the virulence of this organism?**
A. M protein.
FALSE: M protein is the most important virulence determinant of *Streptococcus pyogenes*.
B. Polyglutamic acid capsule.
FALSE: polyglutamic capsule is a determinant of pathogenesis of *Bacillus anthracis*.
C. IgA1 protease.
FALSE: *S. aureus* does not possess an IgA1 protease.
D. Protein A.
TRUE: protein A contributes to the ability of *S. aureus* to avoid humoral immunity.
E. Exotoxin A.

FALSE: exotoxin A is a determinant of pathogenesis of *Pseudomonas aeruginosa* that inhibits protein synthesis, damages tissues, and is immunosuppressive.

5. **Which one of the following infections would MOST commonly be associated with the same organism that caused the man's infection?**
A. Community-acquired bacterial pneumonia.
FALSE: *S. aureus* is a rare cause of community-acquired pneumonia.
B. Bacterial meningitis.
FALSE: *S. aureus* does not generally cause meningitis.
C. Urinary tract infection.
FALSE: *S. aureus* is an uncommon cause of UTIs, although it may be seen with placement of urinary catheters. Bacteriuria may follow *S. aureus* bacteremia.
D. Toxic shock syndrome.
TRUE: *S. aureus* is an important cause of toxic shock syndrome.
E. Erysipelas.
FALSE: erysipelas is a skin infection caused by *Streptococcus pyogenes*.

Refer to the case history below to answer questions 6–8.
A 1-year-old boy is brought to his pediatrician with a severely inflamed cut on his knee. He has had four ear infections in the past year, and several additional wound infections that generally respond to topical therapy. Culture of the current wound reveals gram-positive cocci that are catalase-positive and coagulase-positive.

6. **Protein A, a determinant of pathogenesis for these bacteria is?**
A. Required for attachment to epithelial tissue.
FALSE: *S. aureus* binds to extracellular matrix molecules using surface components recognizing adhesive matrix molecules (MSCRAMMs).
B. A superantigen.
FALSE: protein A is not a superantigen.
C. Binds to the Fc portion of immunoglobulin G.
TRUE: protein A binds to the Fc region of IgG. As a consequence the bound immunoglobulin cannot act as an opsonin, cannot fix complement, and acts to mask the bacterium.
D. A cytotoxin.
FALSE: protein A is not a cytotoxin.
E. An exo-enzyme that aids in spreading of bacteria through tissue.
FALSE: protein A is not a spreading factor like hyaluronidase.

7. **Given the history, which of the following immune abnormalities would be MOST LIKELY?**
A. B-cell deficiency.
TRUE: a deficiency in opsonizing antibodies (and neutrophils) is associated with increased susceptibility to *S. aureus* infections.
B. T-cell deficiency.
FALSE: a difficult one because CD4 Th2 help B cells make antibody. However, the protective immune response against *S. aureus* is humoral and not cellular.
C. Complement deficiency.
FALSE: certainly complement pathways are involved in innate and acquired immunity to *S. aureus*, but complement deficiency has not been associated with increased susceptibility to *S. aureus* as is the case with pathogenic *Neisseria*.
D. Eosinophil deficiency.

FALSE: eosinophils are important in immunity to parasitic helminths and RNA viruses. A deficiency in these cells has not been reported to increase susceptibility to *S. aureus*.

E. Mast cell deficiency.
FALSE: a deficiency in these cells has not been reported to increase susceptibility to S. *aureus*.

8. **Scalded skin syndrome may develop in this child. Which of the following statements presents the MOST LIKELY reason why this may not occur?**

A. The bacterial strain causing this infection does not produce exfoliative toxin.
TRUE: not all strains of *S. aureus* produce exfoliative toxin.

B. The bacteria causing this infection are not the correct species to cause this disease.
FALSE: *S. aureus* is the only species of staphylococci that produces exfoliative toxin.

C. Underlying disease in this child will prevent the expression of the syndrome.
FALSE: underlying disease does not prevent the expression of the syndrome.

D. Development of the syndrome requires infection of a pre-existing burn, not just a cut.
FALSE: any breach of the epithelium that becomes infected with *S. aureus* can give rise to scalded skin syndrome.

E. Maternally derived IgG antibodies will neutralize the toxin.
FALSE: while serum IgG antibodies mitigate the activity of exfoliative toxin, maternal IgG antibodies would have decayed within 1 year post-partum.

9. **Six hours after consumption of potato salad at a picnic, several family members presented to the ER with the abrupt onset of nausea, vomiting, and diarrhea. Which one of the following is the MOST LIKELY causative pathogen?**

A. *Bacillus cereus*.
FALSE: there are two types of food poisoning caused by *B. cereus*, emetic and diarrheal. The emetic form results from consumption of rice contaminated with spores that germinate and produce enterotoxin following cooking. The course of the disease is rapid, like *S. aureus* food poisoning, with vomiting, nausea, and abdominal cramps but there is no diarrhea as there is with *S. aureus* food poisoning. The diarrheal form results from consumption of meats, vegetables or sauces that are contaminated with vegetative cells of *B. cereus*. After the bacteria are ingested they multiply in the gastrointestinal (GI) tract and produce enterotoxin. Thus, the incubation period is longer. The duration of the disease is about a day.

B. *Enterococcus faecalis*.
FALSE: *E. faecalis* does not cause food poisoning.

C. *Staphylococcus aureus*.
TRUE: the food poisoning caused by *S. aureus* fits the clinical scenario, with the consumption of potato salad and the abrupt onset of nausea, vomiting, and diarrhea.

D. *Escherichia coli*.
FALSE: *E. coli* is a common cause of gastroenteritis worldwide. Enteropathogenic (EPEC), enterotoxigenic (ETEC), enteroinvasive (EIEC), and enteroaggregative (EAEC) *E. coli* gastroenteritis are found in underdeveloped countries and the source is usually contaminated drinking water. There is watery diarrhea with or without blood depending on the type of *E. coli*. Enterohemorrhagic *E. coli* (EHEC) gastroenteritis is found in developed countries. The disease may range from mild diarrhea to hemorrhagic colitis. Undercooked ground beef, contaminated vegetables, fruit juices, and water are among the sources of contamination. The onset of disease is 3-4 days with diarrhea and abdominal pain. This progresses to a bloody diarrhea with severe abdominal pain. In most people the disease resolves within 4–10 days.

E. *Clostridium perfringens*.
FALSE: *C. perfringens* food poisoning is relatively common and is caused by meat products heavily contaminated with spores. Type A spores that survive the cooking process germinate if the food is held below 60°C after cooking. There needs to be in the order of 10^8–10^9 bacteria ingested to initiate the disease. In the GI tract the bacteria sporulate liberating enterotoxin. The incubation period is about 8–24 hours and the symptoms include abdominal cramps and watery diarrhea, but no nausea or vomiting. The clinical course lasts 1–2 days.

10. **An outbreak of sepsis caused by *Staphylococcus aureus* has occurred in the newborn nursery. You are asked to investigate. What is the most likely source of the organism?**

A. Scalp.
FALSE: colonized by CNS, but not commonly by *S. aureus*.

B. Nose.
TRUE: about 20% of healthy people are persistent carriers of *S. aureus* in the nose and this percentage increases in hospitalized patients and health-care personnel. About 10% of health-care personnel are colonized with MRSA.

C. Oropharynx.
FALSE: Less commonly colonized by *S. aureus*.

D. Vagina.
FALSE: may be colonized by *S. aureus* but an unlikely source of the bacterium in the newborn nursery.

E. Axilla.
FALSE: a common site of colonization with CNS, but not *S. aureus*.

CASE 34: *Streptococcus mitis*

1. **Which of the following distinguish the commensal viridans streptococci from the pneumococcus?**

A. Susceptibility to optochin.
TRUE: *S. mitis* and other commensal α-hemolytic streptococci are resistant to optochin, whereas the pneumococcus is sensitive.

B. Hemolysis.
FALSE: all are α-hemolytic.

C. Cellular morphology revealed by Gram staining.
TRUE: pneumococci are bullet-shaped diplococci whereas *S. mitis*, for example, grows in chains.

D. A and B.

FALSE: see answers to A and B.
E. A and C.
TRUE: see answers to A and C.

2. **Which one of the following bacteria is the most common cause of subacute bacterial endocarditis (SBE)?**
A. *Streptococcus pneumoniae*.
FALSE: rarely the cause of SBE.
B. *Streptococcus pyogenes*.
FALSE: rarely the cause of SBE.
C. *Streptococcus agalactiae*.
FALSE: rarely the cause of SBE.
D. *Streptococcus mitis*.
TRUE: an important cause of SBE, perhaps second to viridans streptococci.
E. *Enterococcus faecium*.
FALSE: overall enterococci account for 5–15% of cases of bacterial endocarditis.

3. **All of the following are characteristics of *Streptococcus mitis*, EXCEPT?**
A. Gram-positive.
TRUE: they are gram-positive.
B. α-Hemolytic.
TRUE: they are α-hemolytic.
C. Produce an IgA1 protease.
TRUE: the majority produce IgA1 protease.
D. Optochin-resistant.
TRUE: they are resistant to optochin.
E. Uniformly susceptible to penicillin.
FALSE: *S. mitis* is becoming increasingly resistant to penicillin.

4. ***Streptococcus mitis* is the most common cause of which one of the following infections that are caused by streptococci?**
A. Neonatal meningitis.
FALSE: they are not a cause of neonatal meningitis.
B. Necrotizing fasciitis.
FALSE: they do not cause necrotizing fasciitis.
C. Streptococcal toxic shock syndrome (STSS).
FALSE: although *S. mitis* has been reported to cause STSS, *S. pyogenes* is the most frequent cause.
D. Subacute bacterial endocarditis.
TRUE: viridans streptococci are the most common.
E. Lobar pneumonia.
FALSE: the pneumococcus is the most common cause of lobar pneumonia.

5. **All of the following statements about *S. mitis* or the disease it causes are correct, EXCEPT?**
A. It causes endogenous infections.
TRUE: it is an endogenous pathogen.
B. It is an opportunistic pathogen.
TRUE: it is an opportunistic pathogen.
C. It is a member of the oropharyngeal commensal microbiota.
TRUE: α-Hemolytic streptococci comprise a large fraction of the microbiotia in the oropharynx.
D. Disease results from horizontal transmission via respiratory droplets.
FALSE: disease does not result from horizontal transmission.
E. Most strains of *S. mitis* produce IgA1 protease.
TRUE: about two-thirds of isolates produce IgA1 protease.

6. ***S. mitis* has been reported to cause all of the following diseases, EXCEPT?**
A. Impetigo.
FALSE: *S. mitis* does not cause impetigo.
B. Subacute bacterial endocarditis (SBE).
TRUE: *S. mitis* and other viridans streptococci are the most frequent cause of SBE.
C. Intracerebral hemorrhage.
TRUE: if they seed the brain they may cause intracerebral hemorrhage.
D. Subarachnoid hemorrhage.
TRUE: if they seed the brain they may cause subarachnoid hemorrhage.
E. Toxic shock syndrome.
TRUE: *S. mitis* have been reported to cause toxic shock syndrome.

7. **Which of the following conditions predispose towards bacterial endocarditis caused by *S. mitis*?**
A. Intravenous drug use.
TRUE: intravenous drug users are at increased risk of bacterial endocarditis caused by *S. mitis*.
B. Rheumatic fever.
TRUE: persons who have had rheumatic fever are at increased risk of bacterial endocarditis caused by *S. mitis*.
C. Prosthetic heart valve.
TRUE: persons with prosthetic heart valves are at increased risk of bacterial endocarditis caused by *S. mitis*.
D. Congenital heart disease.
TRUE: persons with congenital heart disease are at increased risk of bacterial endocarditis caused by *S. mitis*.
E. All of the above.
TRUE: all are correct.

CASE 35: *Streptococcus pneumoniae*

1. **Which one of the following statements concerning the optochin disc test is correct?**
A. Used to determine whether a β-haemolytic streptococcus is *S. pyogenes*.
FALSE: optochin does not differentiate between β-haemolytic streptococci.
B. Inhibits viridans streptococci in respiratory specimens.
FALSE: optochin does not inhibit viridans streptococci other than *S. pneumoniae*.
C. Selects for *S. pneumoniae* in respiratory specimens.
FALSE: although *S. pneumoniae* is lysed by optochin, the disc test is not employed in this fashion.
D. Distinguishes *S. pneumoniae* from other species of viridans streptococci.
TRUE: this is the purpose of the test.
E. Tests the antibiotic sensitivity of gram-positive cocci.
FALSE: optochin is not a part of antibiotic sensitivity testing.

2. **Which one of the following statements concerning the pneumococcal conjugate vaccine is correct?**
A. Composed of capsular polysaccharide from 23 serotypes

of *S. pneumoniae.*
FALSE: the conjugate vaccine contains seven capsular polysachharides conjugated with a protein carrier.
B. Non-immunogenic in children under the age of 2 years.
FALSE: the conjugate vaccine is immunogenic in this age group.
C. Composed of seven capsular polysaccharide serotypes linked to a protein carrier.
TRUE: the conjugate vaccine contains seven capsular polysaccharides conjugated with a protein carrier.
D. A T-independent antigen.
FALSE: it is a T-dependent antigen since it is conjugated to a protein carrier.
E. A conjugate of pneumococcal surface protein A (PspA) and tetanus toxoid.
FALSE: the vaccine is capsular polysaccharide that may be conjugated to tetanus toxoid.

3. **All of the following statements concerning the capsule of pneumococci are correct, EXCEPT?**
A. It is an important virulence determinant.
TRUE: it is the principal virulence factor.
B. It is immunogenic in children over 2 years of age.
TRUE: children over age 2 years are able to respond to T-independent antigens.
C. It is anti-phagocytic.
TRUE: its principal role is to prevent phagocytosis.
D. It is antigenically diverse.
TRUE: there are over 90 serotypes.
E. All capsule types of pneumococci are equally invasive.
FALSE: certain capsule types, such as type III, are more invasive.

4. **All of the following statements concerning the envelope of *S. pneumoniae* are correct, EXCEPT?**
A. Autolyzes when treated with penicillin.
TRUE: penicillin activates the wall autolysins.
B. Pneumolysin is a principal cell wall protein.
FALSE: pneumolysin is located in the cytoplasm and released during cell lysis.
C. Fragments initiate inflammation.
TRUE: peptidoglycan fragments and teichoic acid fragments are pro-inflammatory.
D. Peptidoglycan is covalently linked to choline-containing teichoic acid.
TRUE: this is a feature of the architecture of the pneumococcal wall.
E. Contains several choline-binding adhesins.
TRUE: a family of choline-binding adhesins are found on the outer surface of the wall.

5. **Which one of the following bacteria is the most common cause of community-acquired pneumonia?**
A. *Streptococcus pneumoniae.*
TRUE: the pneumococcus is the most common cause.
B. *Legionella* species.
FALSE: the etiologic agent in about 15% of cases.
C. *Mycoplasma* species.
FALSE: the etiologic agent in about 15% of cases.
D. *Chlamydophila pneumoniae.*
FALSE: the etiologic agent in about 15% of cases.
E. *Pseudomonas aeruginosa.*
FALSE: but an important cause of nosocomial pneumonia.

6. **All of the following are characteristics of *Streptococcus pneumoniae*, EXCEPT?**

A. Gram-positive.
TRUE: the pneumococcus is a gram-positive diplococcus.
B. α-Hemolytic.
TRUE: the pneumococcus produces incomplete hemolysis.
C. Produce an IgA1 protease.
TRUE: the pneumococcus produces an IgA1 protease.
D. Bullet-shaped diplococci.
TRUE: the pneumococcus is a bullet-shaped diplococcus.
E. Uniformly susceptible to penicillin.
FALSE: penicillin resistance is a significant emerging problem.

7. **Streptococcus pneumoniae is the most common cause of bacterial meningitis in all of the following patients, EXCEPT?**
A. A 14-day-old neonate.
FALSE: transplacental anti-capsular IgG antibodies are protective in the post-partum period.
B. A 33-year-old male who developed penetrating head trauma in a motor vehicle accident.
TRUE: this is a predisposing condition.
C. A 45-year-old alcoholic.
TRUE: this is a predisposing condition.
D. A 64-year-old with diabetes mellitus.
TRUE: this is a predisposing condition.
E. A previously healthy 70-year-old with a recent bout of sinusitis.
TRUE: this is a predisposing condition.

8. **All of the following are features of the host–pneumococcus relationship, EXCEPT?**
A. Transmitted person to person by respiratory droplets.
TRUE: respiratory droplets are the principal mode of horizontal transmission.
B. Vertical transmission from mother to neonate during delivery.
FALSE: the pneumococcus does not inhabit the vaginal canal or cross the placenta so it is not transmitted vertically.
C. High rate of carriage in healthy children.
TRUE: there is a high carriage rate in children; less so in adults.
D. Repeat episodes of carriage are associated with different capsule serotypes.
TRUE: capsule serotypes are replaced probably as a result of the selective pressure of antibody.
E. Duration of carriage decreases with each successive colonization episode.
TRUE: duration of colonization decreases probably as a result of the selective pressure of antibody.

9. **The pneumococcus causes all of the following diseases, EXCEPT?**
A. Pneumonia.
FALSE: the pneumococcus is the most common cause of community-acquired pneumonia.
B. Otitis media.
FALSE: the pneumococcus is a common cause of middle ear infection.
C. Sinusitis.
FALSE: the pneumococcus is a common cause of sinusitis.
D. Meningitis.
FALSE: the pneumococcus is a common cause of bacterial meningitis.

E. Necrotizing fasciitis.
TRUE: the pneumococcus does not cause necrotizing fasciitis.

10. **Which of the following conditions predispose towards pneumococcal disease?**
A. Antecedent viral respiratory infection.
TRUE: antecedent viral respiratory infection is a predisposing factor.

B. Chronic pulmonary disease.
TRUE: chronic pulmonary disease is a predisposing factor.
C. Cigarette smoking.
TRUE: cigarette smoking is a predisposing factor.
D. Chronic renal disease.
TRUE: chronic renal disease is a predisposing factor.
E. All of the above.
TRUE: all are predisposing factors.

CASE 36: *Streptococcus pyogenes*

1. **All of the following are components of the GAS cell wall, EXCEPT?**
A. Thick peptidoglycan layer.
TRUE: all gram-positive bacteria have a thick peptidoglycan layer.
B. Teichoic acid
TRUE: teichoic acid is a component of almost all gram-positive bacteria.
C. Group-specific polysaccharide.
TRUE: this the Lancefield antigen.
D. Lipopolysaccharide.
FALSE: LPS is a component of the outer membrane of the gram-negative wall.
E. Lipoteichoic acid.
TRUE: lipoteichoic acid is a component of almost all gram-positive bacteria.

2. **All of the following are characteristics of GAS, EXCEPT?**
A. β-Hemolytic.
TRUE: GAS colonies are surrounded by a zone of complete hemolysis.
B. Resistant to bacitracin.
FALSE: GAS is susceptible to bacitracin.
C. Capnophilic i.e. grows best in the presence of CO_2.
TRUE: GAS grows best in the presence of 5% CO_2.
D. Catalase-negative.
TRUE: the genus *Streptococcus* does not produce the enzyme catalase.
E. PYR-positive.
TRUE: GAS is PYR-positive, which distinguishes it from Lancefield groups C and G.

3. **Which of the following laboratory tests are important in the diagnosis of GAS disease?**
A. Microscopy.
TRUE: the Gram stain shows gram-positive cocci growing in chains.
B. Culture.
TRUE: culture reveals the characteristic colony surrounded by a large zone of β-hemolysis.
C. Serology.
TRUE: anti-streptolysin O and anti-DNase B antibodies are elevated in patients with ARF and AGN, respectively.
D Antigen detection.
TRUE: immunological detection of the Lancefield antigen is the basis of the rapid test to detect GAS.
E. All of the above (A–D) are important.
TRUE: see answers A–D above.

4. **All of the following statements concerning the M protein of *Streptococcus pyogenes* are correct, EXCEPT?**
A. M protein is anti-phagocytic.
TRUE: this is a function of M protein.
B. Patients can have multiple episodes of *S. pyogenes* infection since there are multiple antigenic types of M protein.
TRUE: protective antibodies are M type-specific.
C. M protein is the most important virulence determinant.
TRUE: anti-M protein antibodies are protective.
D. All M types of *S. pyogenes* are equally virulent.
FALSE: certain M types like M1 are associated with invasive infections.
E. Epitopes on M protein cross-react with those on several human tissues.
TRUE: this is the basis for the autoimmune diseases ARF and AGN.

5. **Which one of the following statements concerning the capsule of *S. pyogenes* is true?**
A. It is associated with strains causing severe systemic infection.
TRUE: capsulate strains and those producing Spes are more invasive.
B. It is composed of hyaluronic acid.
TRUE: it simulates the mammalian intercellular matrix.
C. It masks the organism from the humoral immune system.
TRUE: the capsule is not seen as foreign by the immune system.
D. A and B.
FALSE: answer C is correct as well as answers A and B.
E. A, B, and C.
TRUE: answers A, B, and C are correct.

6. **All of the following statements about streptococcal pyrogenic exotoxins are correct, EXCEPT?**
A. They are superantigens.
TRUE: they function as superantigens.
B. They are encoded by bacteriophage.
TRUE: they are specified by bacteriophage.
C. They are responsible for the rash of scarlet fever.
TRUE: they cause the rash of scarlet fever.
D. They bind IgG via the Fc region.
FALSE: they do not bind immunoglobulin G.
E. They are associated with severe infections, including shock.
TRUE: M types producing Spes and capsule are associated with severe infections.

7. **All of the following are mechanisms by which GAS avoids host defenses, EXCEPT?**
   A. Inhibition of complement activation.
      TRUE: this is a major mechanism of immune evasion.
   B. Resist phagocytosis.
      TRUE: this is a major mechanism of immune evasion.
   C. Bind antibody via the Fc region.
      TRUE: the M-like proteins of GAS bind immunoglobulins via Fc.
   D. Cause unfocused activation of CD4 helper T cells.
      TRUE: the superantigens act in this way.
   B. Produce proteases that cleave antibody molecules.
      FALSE: GAS does not produce immunoglobulin proteases, although other streptococci do.

8. **All of the following may result from infection with GAS, EXCEPT?**
   A. Acute pharyngitis.
      TRUE: GAS is the most common bacterial cause of acute pharyngitis.
   B. Pyoderma.
      TRUE: GAS causes impetigo; often it is a mixed infection with *S. aureus*.
   C. Necrotizing fasciitis.
      TRUE: GAS causes necrotizing fasciitis, but so do other bacteria.
   D. Meningitis.
      FALSE: GAS does not cause bacterial meningitis.
   E. Cellulitis.
      TRUE: GAS is a common cause of skin and soft tissue infections.

9. **Prior infection with GAS can be demonstrated in patients with acute rheumatic fever by which one of the following?**
   A. Blood culture.
      FALSE: patients with ARF are not bacteremic.
   B. Skin culture.
      FALSE: even if GAS were to be isolated from the skin, skin infections do not lead to ARF.
   C. Culture of a heart valve in a patient with carditis.
      FALSE: ARF is an autoimmune disease and bacteria do not colonize the heart valves.
   D. A high titer of antibody against the hyaluronic acid capsule.
      FALSE: antibodies are not induced against the capsule.
   E. A high titer of antibody against streptolysin O.
      TRUE: a high anti-SLO titer is indicative of a recent GAS infection.

10. **All of the following statements about antibiotic treatment of GAS infections are correct, EXCEPT?**
    A. Penicillin is the antibiotic of choice unless the patient is allergic.
       TRUE: GAS is highly susceptible to penicillin.
    B. Penicillin need not be given to patients with GAS pharyngitis because the disease is self-limiting.
       FALSE: the purpose of penicillin therapy is prevent ARF rather than to resolve the pharyngitis.
    C. Penicillin is required for patients with GAS pharyngitis because it prevents the development of acute rheumatic fever.
       TRUE: penicillin given within 10 days after the onset of pharyngitis will protect against the development of ARF.
    D. Penicillin alone is inadequate treatment for necrotizing fasciitis.
       TRUE: surgical drainage and wound debridement are essential.
    E. Patients with a history of rheumatic fever require long-term antibiotic prophylaxis.
       TRUE: chronic antibiotic therapy must be given for several years to prevent recurrence of the disease.

CASE 37: *Toxoplasma gondii*

1. **Which of the following are true of the causative agent of toxoplasmosis?**
   A. *Toxoplasma gondii* is a protozoan intracellular pathogen.
      TRUE.
   B. The definitive host is the dog and other canids.
      FALSE: the cat and other felines.
   C. In highly endemic areas humans are a key reservoir for infection.
      FALSE: humans cannot pass infection on to others.
   D. The rapidly dividing form is called a tachyzoite.
      TRUE.
   E. Bradyzoites are the only extracellular stage.
      FALSE: bradyzoites are found in tissue cysts and tachyzoites are extracellular.

2. **Which of the following are true of the transmission of *T. gondii*?**
   A. Infection may be acquired by the ingestion of oocysts in the soil.
      TRUE.
   B. Vegetarians do not acquire toxoplasmosis.
      FALSE: vegetables can be contaminated by oocysts in the soil.
   C. Transmission through water supplies is possible.
      TRUE.
   D. Tissue cysts in meat are destroyed by thorough cooking, but not by freezing.
      FALSE: they are destroyed by freeze-thawing as well.
   E. Measures to reduce transmission are particularly important during pregnancy.
      TRUE.

3. **Which of the following are true of the host response to *T. gondii*?**
   A. Extracellular killing of tachyzoites is mediated by antibody and complement fixation.
      TRUE.
   B. Infected macrophages activate CD4+ T lymphocytes through antigen presented in the context of MHC class I molecules.
      FALSE: macrophages present via MHC class II molecules.
   C. Macrophages can kill phagocytosed tachyzoites by tryptophan depletion of the parasitophorous vacuole.
      TRUE.
   D. IFN-γ is produced by NK cells during infection.
      TRUE.
   E. TNF-α helps to activate infected macrophages.
      TRUE.

4. **Which of the following are true of the pathogenesis of toxoplasmosis?**
   A. Infection spreads through the body mostly after an immune response has been mounted.
   FALSE: most spread occurs before this.
   B. Highly infected tissues include cardiac muscle because it is an immunologically privileged site.
   FALSE: the eye and central nervous system are relatively privileged.
   C. Pathology is entirely due to tissue damage from dividing tachyzoites.
   FALSE: the host inflammatory response contributes to pathology.
   D. IL-10 promotes the tissue inflammatory response.
   FALSE: it does the opposite.
   E. Bradyzoites intermittently reactivate and cause disease when immune surveillance is compromised.
   TRUE.

5. **Which of the following are true of the clinical features of toxoplasmosis in immunocompetent individuals?**
   A. Infection in pregnancy is always symptomatic.
   FALSE.
   B. Fetal infection results in congenital abnormalities mainly when it occurs in the first trimester (0–13 weeks).
   FALSE: in fact the rate of congenital abnormality is highest between 10 and 24 weeks.
   C. The most common late manifestation of congenital toxoplasmosis is chorioretinitis.
   TRUE.
   D. Primary infection in adults is symptomatic in 80%.
   FALSE: in fact it is asymptomatic in 80%.
   E. Lymphadenopathy is the most common manifestation of primary infection in adults.
   TRUE.

6. **Which of the following are true of the clinical features of toxoplasmosis in immunocompromised individuals?**
   A. In HIV the most common site for reactivation is in the brain.
   TRUE.
   B. Toxoplasmic encephalitis is always manifest as a single focal lesion.
   FALSE: lesions may be multiple.
   C. *Toxoplasma* pneumonitis can occur in immunocompromised individuals.
   TRUE.
   D. Cardiac failure is possible in toxoplasmosis.
   TRUE.
   E. Congenital infection is more likely in the immunocompromised mother.
   TRUE: there may be a failure to kill circulating tachyzoites with an increased chance that these will cross the placenta.

7. **Which of the following are true of the diagnosis of toxoplasmosis?**
   A. Live tachyzoites are used in the Sabin-Feldman neutralization assay for antibodies.
   TRUE.
   B. Two negative serological tests are useful in ruling out infection.
   TRUE.
   C. A positive IgM assay indicates recent infection.
   FALSE: IgM may persist for a few years after primary infection.

D. PCR for *T. gondii* has high specificity, but variable sensitivity.
   TRUE.
   E. Histology of lymph nodes or infected tissue is always diagnostic.
   FALSE: sometimes characteristic tissue cysts and tachzoites or bradyzoites are seen, but not always.

8. **Which of the following are true of the differential diagnosis of toxoplasmosis?**
   A. Congenital *Toxoplasma* infection may have to be differentiated from cytomegalovirus (CMV) infection.
   TRUE.
   B. *Toxoplasma* cervical lymphadenopathy may have to be differentiated from glandular fever due to Epstein-Barr virus (EBV).
   TRUE.
   C. *Toxoplasma* chorioretinitis may have to be differentiated from sarcoidosis.
   TRUE.
   D. The radiological appearance of toxoplasmic encephalitis is characteristic, with few other possible alternatives.
   FALSE: there is a long list of possibilities, but this can be partly narrowed down by the context of the overall clinical picture.
   E. *T. gondii* is the most common of the pathogens responsible for the mononucleosis syndrome.
   FALSE: EBV is the most common.

9. **Which of the following are true of the treatment of toxoplasmosis?**
   A. Pyrimethamine and sulfadiazine are used to treat toxoplasmosis in immunocompetent hosts.
   FALSE: infections will usually self-limit.
   B. Pyrimethamine and sulfadiazine are used to treat all episodes of chorioretinitis.
   FALSE: some small lesions will have settled before a diagnosis is made.
   C. Pyrimethamine must be co-administered with folinic acid to prevent bone marrow suppression.
   TRUE.
   D. Spiramycin is used to treat fetal infection.
   FALSE: spiramycin does not cross the placenta.
   E. Treatment of fetal infection in the first trimester runs the risk of significant fetal malformation.
   TRUE.

10. **Which of the following are true of the control and prevention of toxoplasmosis?**
   A. Pets can be treated before purchase to prevent infection in the household.
   FALSE: this is not done and given the various sources and routes of infection may not be effective.
   B. Adequate filtration of water supplies is necessary to exclude oocysts.
   TRUE.
   C. Undercooked meats should not be eaten in pregnancy.
   TRUE.
   D. Gardens are safe for homes that do not have cats.
   FALSE: the neighbor's cats will shed oocysts wherever they roam and defecate.
   E. Prophylaxis with co-trimoxazole in immunocompromised individuals reduces the chances of reactivation of latent infection.
   TRUE.

CASE 38: *Trypanosoma* spp.

1. **Which of the following are true about *Trypanosoma*?**
   A. They are flagellated protozoan parasites.
      TRUE.
   B. *T. cruzi* is responsible for African trypanosomiasis.
      FALSE: *T. cruzi* is responsible for South American trypanosomiasis.
   C. They have animal reservoirs.
      TRUE.
   D. *T. brucei* species cause infections in Asia.
      FALSE: trypanosomes pathogenic to humans are found in Africa and South America.
   E. *T. cruzi* remains extracellular.
      FALSE: *T. cruzi* has a prolonged intracellular life stage.

2. **Which of the following are true about the life cycle of trypanosomiasis?**
   A. *T. brucei* species are transmitted by sandflies.
      FALSE: this species is transmitted by tsetse flies. Sandflies transmit leishmaniasis.
   B. *T. cruzi* is transmitted by triatomine bugs.
      TRUE.
   C. *T. cruzi* is inoculated into a new human host through saliva from the biting vector.
      FALSE: this is how tsetse flies transmit African trypanosomiasis. *T. cruzi* is excreted in the feces of biting triatomine bugs, and the feces are then rubbed into the bite wound by the host.
   D. Trypanosomiasis can be transmitted by blood transfusion.
      TRUE.
   E. Infections never pass from human to human.
      FALSE.

3. **Which of the following are true about the spread of trypanosomes?**
   A. They are inoculated directly into the bloodstream.
      FALSE: from local tissue they pass to lymph nodes and then enter the bloodstream.
   B. Invasion of the central nervous system occurs in the first few days.
      FALSE: for *T. brucei* species this takes a few weeks for *T. brucei rhodesiense* and several months for *T. brucei gambiense*.
   C. A key target for *T. cruzi* is the reticuloendothelial system including the liver, spleen, and bone marrow.
      FALSE: key targets are the heart and gastrointestinal tract.
   D. *T. cruzi* invades tissues and multiplies between cells rather than within cells.
      FALSE: multiplication is intracellular.
   E. *T. brucei gambiense* invades the heart tissue.
      FALSE: this species invades the central nervous system.

4. **Which of the following are true of the host response to *Trypanosoma brucei*?**
   A. Antibody and complement lyse *T. brucei* species.
      TRUE.
   B. CD4+ T lymphocytes help B lymphocytes to produce anti-trypanosomal antibodies.
      TRUE.
   C. Tumor necrosis factor-α inhibits trypanosomal growth.
      TRUE.

   D. *T. brucei* species must inhibit apolipoprotein L-1 to avoid lysosomal killing.
      TRUE.
   E. *T. brucei rhodesiense* evades host antibody responses by switching between about 100 variant surface glycoproteins.
      FALSE: in fact the number is about 1000.

5. **Which of the following are true of the host response to *Trypanosoma cruzi*?**
   A. Autoantibodies appear in the course of infection.
      TRUE.
   B. Antigenic variation is a clearly established immune evasion mechanism.
      FALSE: although it may happen, it is not established as an immune evasion mechanism.
   C. Tissue pseudocysts excite a tissue inflammatory reaction.
      FALSE: this only occurs when the cyst ruptures.
   D. CD8+ T lymphocytes can kill cells bearing *T. cruzi* antigen.
      TRUE.
   E. Macrophages are suppressed by *T. cruzi*.
      FALSE: macrophages are stimulated by *T. cruzi*-derived glycosylphosphatidylinositol (GPI).

6. **Which of the following statements are true about the clinical features of African trypanosomiasis?**
   A. A trypanosomal chancre may be seen in up to half of *T. brucei rhodesiense* infections.
      TRUE.
   B. Lymphadenopathy is not observed in *T. brucei gambiense* infection.
      FALSE.
   C. Fevers commence within a month of infection.
      TRUE.
   D. Death follows invasion of the central nervous system by trypanosomes.
      TRUE.
   E. Progression to death is more rapid with *T. brucei gambiense* than with *T. brucei rhodesiense*.
      FALSE.

7. **Which of the following are true of the clinical features of South American trypanosomiasis?**
   A. Inflammation at the site of inoculation is called a chagoma.
      TRUE.
   B. In the acute phase of infection the liver and spleen may be enlarged.
      TRUE.
   C. Long-term complications occur in all those infected.
      FALSE: they occur in about one-third.
   D. Infection of the heart results in thickening of heart muscle.
      FALSE: infection weakens and thins the muscle, which may become aneurysmal.
   E. Weakening of the gastrointestinal smooth muscle occurs with infection in all geographical areas.
      FALSE: this feature is mainly seen in Brazil.

8. **Which of the following are true of the diagnosis of trypanosomiasis?**
   A. South American trypanosomiasis, Chagas' disease, can always be diagnosed by characteristic clinical features.
   FALSE: although late features may be characteristic and highly suggestive, some infections remain subclinical.
   B. The card agglutination test for African trypanosomiasis tests for trypanosomal antigen in the bloodstream.
   FALSE: the card contains antigen to which antibody in the blood binds.
   C. The card agglutination test for trypanosomiasis has a low sensitivity, but a high specificity.
   FALSE: the converse is true, it has a high sensitivity but low specificity.
   D. For African trypanosomiasis diagnosis may be based on microscopy of blood or cerebrospinal fluid.
   TRUE.
   E. A serological test for antibody is available for Chagas' disease.
   TRUE.

9. **Which of the following are true for the treatment of trypanosomiasis?**
   A. The use of melarsoprol for African trypanosomiasis is associated with encephalopathy in about 20%.
   TRUE.
   B. Benznidazole is better tolerated by children than adults with *T. cruzi* infection.
   TRUE.

C. All individuals with *T. cruzi* infection should be treated to prevent complications.
FALSE: as some will not develop complications treatment may be more toxic than beneficial.
D. Pentamidine is used in the early stage of *T. brucei gambiense* infection.
TRUE.
E. Late stage African trypanosomiasis is treated with benznidazole.
FALSE: this drug is used for South American trypanosomiasis.

10. **Which of the following are true of the control of trypanosomiasis?**
   A. A vaccine is available.
   FALSE.
   B. Treatment of infected individuals is a key part in eliminating the reservoir of infection.
   FALSE: there is also an animal reservoir.
   C. Insecticide-treated traps are used to control tsetse fly populations.
   TRUE.
   D. Corrugated roofs are preferable to thatched roofs to limit triatomine bug populations.
   TRUE.
   E. Screening of blood before transfusion is ineffective in preventing transmission.
   FALSE: this is important and practiced in South America.

CASE 39: Varicella-zoster virus

1. **Which of the following are true of VZV?**
   A. It is an enveloped virus.
   TRUE: VZV is a herpesvirus and is therefore enveloped.
   B. It is a member of the *Poxviridae* (pox virus family).
   FALSE: VZV belongs to the family *Herpesviridae*.
   C. It carries a single-stranded DNA genome.
   FALSE: herpesviruses carry their genome in the form of double-stranded DNA.
   D. It undergoes latency in nerve cell bodies.
   TRUE: latency is a feature of all herpesviruses. Like HSV, the site of latency of VZV is the nerve cell body.
   E. It has a capsid of helical symmetry.
   FALSE: herpesviruses have an icosahedral capsid.

2. **Which of the following are true of VZV infection?**
   A. Infection is usually acquired via the gastrointestinal tract.
   FALSE: VZV infection is spread from person to person by direct contact, droplet, or air-borne spread of respiratory tract secretions.
   B. During the incubation period, VZV is spread around the body in the bloodstream.
   TRUE: viremic spread occurs about 10–12 days after infection, and accounts for the widespread skin rash all over the body.
   C. The rash of chickenpox is due to the deposition of immune complexes into the skin.
   FALSE: the vesicular rash of chickenpox arises through viral replication at that site.
   D. It is possible for a susceptible host to acquire infection by exposure to a patient with shingles (herpes zoster)
   TRUE: the vesicle fluid of herpes zoster contains viable virus that can be transmitted to susceptible individuals by

direct contact or droplets.
   E. The incubation period from infection to appearance of rash is of the order of 6–8 days.
   FALSE: the incubation period is of the order of 2–3 weeks, most usually 13–17 days.

3. **Which of the following statements are true?**
   A. The average age at primary infection is higher in tropical countries than in temperate ones.
   TRUE: thus, many more adults from tropical countries are susceptible to primary infection.
   B. In the UK, at least one-third of individuals greater than 20 years old have no evidence of prior infection with VZV infection.
   FALSE: a considerable majority of adults will have antibodies to VZV indicating prior infection, regardless of whether the infection was symptomatic or not.
   C. Hemorrhagic chickenpox usually arises as a result of a vigorous immune response to infection.
   FALSE: this is a rare complication of primary VZV infection, which usually occurs in an immunodeficient individual.
   D. The attack rate of infection in susceptible individuals following exposure to a case of chickenpox is less than 50%.
   FALSE: attack rates of over 70% are reported, indicating that a source patient is extremely infectious.
   E. During primary infection, virus goes latent in nerve cell bodies.
   TRUE.

4. **Which of the following statements concerning chickenpox are true?**
   A. Infection in an adult usually results in more severe disease than in a child.
   TRUE: infection in childhood is often asymptomatic, and even if a rash appears, systemic manifestations (fever, malaise) are usually mild. In adults, fever and constitutional disturbance may be severe.
   B. In adults, varicella encephalitis is the commonest life-threatening complication.
   FALSE: the most common life-threatening complication is varicella pneumonia.
   C. The absence of itch means that the rash is not likely to be due to VZV infection.
   FALSE: the occurrence of itch is idiosyncratic. The presence or absence of itch therefore does not impact on the differential diagnosis.
   D. New lesions may continue to appear for a few days after first appearance of the rash.
   TRUE: lesions commonly occur in successive crops such that several stages of lesion maturity are present at the same time.
   E. Lesions are usually more abundant at the extremities (i.e. scalp, hands, and feet) than over the rest of the body.
   FALSE: lesions are usually more abundant on covered rather than exposed parts of the body.

5. **With regard to chickenpox infection in pregnancy, which of the following are true?**
   A. Varicella embryopathy may arise from maternal chickenpox at any stage during the pregnancy.
   FALSE: cases of varicella embryopathy only arise from maternal chickenpox in the first 20 weeks of pregnancy.
   B. Over 10% of babies born to mothers who acquire chickenpox in pregnancy may suffer development defects.
   FALSE: the rate of embryopathy is 1–2% of pregnancies affected by maternal chickenpox in the first 20 weeks.
   C. Failure of limb bud development is one of the features of varicella embryopathy.
   TRUE: this was one of the striking features of the developmental defects of varicella embryopathy that led to its recognition as a complication of maternal chickenpox in pregnancy.
   D. Maternal shingles carries a significant risk that virus may cross the placenta and damage the developing fetus.
   FALSE: maternal shingles represents secondary reactivation of VZV in the mother, who will therefore already have antibodies that can cross the placenta and protect the fetus. No abnormalities arising from maternal shingles in pregnancy have been described.
   E. A baby whose mother develops chickenpox 2 days after delivery is not at risk of severe neonatal chickenpox.
   FALSE: the baby will not have protective antibodies (as its mother will have been antibody-negative during the pregnancy). VZV infection acquired at this age may be life-threatening.

6. **Which of the following factors increase the likelihood of reactivation of latent VZV infection?**
   A. Increasing age.
   TRUE: this reflects declining immune function with aging.
   B. HIV infection.
   TRUE: underlying immunodeficiency is an important predisposing factor to VZV reactivation. In HIV-infected individuals, this may happen before the patient is so immunodeficient that they present with an AIDS-defining illness.
   C. Antibody deficiency.
   FALSE: T-cell immunity is the dominant part of the immune response that protects against VZV reactivation.
   D. Hodgkin's lymphoma.
   TRUE: such patients have underlying T-cell dysfunction.
   E. Eczema.
   FALSE.

7. **Which of the following are recognized complications of herpes zoster (shingles) in an immunocompetent patient?**
   A. Ulceration of the cornea.
   TRUE: this is a severe complication of ophthalmic zoster, as virus may be present both above and below the eye, and may therefore be spread onto the cornea.
   B. Motor nerve paralysis.
   TRUE: a good example is zoster of the seventh cranial nerve, giving rise to facial paralysis (Bell's palsy).
   C. Persistence of pain after the rash has resolved.
   TRUE: post-herpetic neuralgia is very common, and can be very severe.
   D. Hepatitis.
   FALSE.
   E. Pneumonia.
   FALSE: this may arise as a consequence of primary infection in an adult, but not of reactivation in an otherwise immunocompetent individual.

8. **Which of the following statements regarding diagnosis of varicella-zoster virus infection are true?**
   A. Laboratory confirmation should be sought for all cases of clinically diagnosed chickenpox.
   FALSE: the clinical features of chickenpox are usually so clear-cut that a clinical diagnosis is accurate. Laboratory confirmation may be sought if there are atypical features or if there is any doubt.
   B. VZV can be isolated in tissue culture.
   TRUE: however, this takes days to weeks, and is not an ideal diagnostic approach.
   C. Antigen detection by immunofluorescence on cells scraped from the base of a vesicular lesion allows distinction between reactivated herpes simplex virus infection and reactivated VZV infection.
   TRUE: different 'spots' of cells would be stained with monoclonal antibodies against HSV-1, HSV-2, and VZV. Only one antibody would bind, depending on which is the causative agent.
   D. Electron microscopy of vesicle fluid allows distinction between reactivated herpes simplex virus infection and reactivated VZV infection.
   FALSE: electron microscopy of vesicle fluid may demonstrate the presence of herpesvirus particles, but both HSV and VZV are herpesviruses, and it is not possible to distinguish between them on the basis of EM morphology alone.
   E. Genome detection by polymerase chain reaction assay allows distinction between reactivated herpes simplex virus infection and reactivated VZV infection.
   TRUE: PCR requires the use of specific oligonucleotide primers, which will only result in amplification of that specific virus nucleic acid from which their sequence is derived.

9. **Which of the following patients should be offered antiviral therapy with aciclovir (or derivatives)?**
   A. 45-year-old man with two vesicular lesions on the left side of his forehead.
      YES: he has zoster in the territory of the ophthalmic branch of the trigeminal nerve, and ophthalmic zoster should always be treated as there is a danger of keratitis.
   B. 40-year-old woman with an uncomplicated shingles rash on her trunk which first appeared 36 hours ago.
      NO: as described, she does not possess any of the features that would suggest she is at risk of severe or complicated zoster.
   C. 28-year-old renal transplant recipient with three vesicular lesions just below his right nipple.
      YES: zoster in an immunosuppressed individual is potentially life-threatening, and should always be treated, no matter how mild it appears when first seen.
   D. 58-year-old woman with a painful vesicular rash on her left side that first appeared 5 days ago.
      NO: in order to be effective, antiviral therapy should be started within 72 hours of onset of the rash.
   E. 27-year-old male known to have HIV infection with a shingles rash in the distribution of the 10th thoracic nerve (i.e. at the level of his umbilicus).
      YES: he has HIV infection and therefore is potentially immunosuppressed. The same comment applies as for option C above.

10. **Which of the following statements regarding varicella vaccine are true?**
   A. The vaccine consists of a live attenuated virus.
      TRUE.
   B. It should not be given to children with leukemia.
      FALSE: the vaccine was actually first trialed in children in remission from their leukemia, in which it has an excellent safety and efficacy record.
   C. Susceptible health-care workers who work in areas where they may be exposed to VZV should be vaccinated.
      TRUE.
   D. Vaccinated patients may develop vaccine-associated shingles.
      TRUE: but unusual.
   E. Vaccine may be used as a routine vaccination in childhood.
      TRUE: it is licensed for such use in the USA (but not in the UK).

CASE 40: *Wuchereria bancrofti*

1. **Which of the following are true statements about *W. bancrofti*?**
   A. It is a tapeworm (cestode).
      FALSE: it is a round worm (nematode).
   B. Adult worms live in lymphatics.
      TRUE.
   C. Larvae (microfilariae) released from adult worms are ingested from lymphatics by mosquitoes.
      FALSE: larvae are ingested from the blood.
   D. Circulating microfilarial numbers are greatest around midday.
      FALSE: numbers are greatest around midnight.
   E. Adult worms have a lifespan that can reach 8 years.
      TRUE.

2. **Which of the following are true statements about the transmission of *W. bancrofti*?**
   A. There is an animal reservoir.
      FALSE: microfilaremic humans serve as the reservoir.
   B. Only female *Anopheles* mosquitoes are capable of transmission.
      FALSE: this only applies to malaria. Various species of mosquito can transmit filariasis.
   C. Microfilariae have to molt in the mosquito before transmission is possible.
      TRUE.
   D. Transmission from humans can only occur 8 months or longer after infection.
      TRUE: adult worms have to mature and mate first, before they release microfilariae.
   E. Microfilariae are inoculated into blood vessels when mosquitoes take a blood meal.
      FALSE: they enter the puncture wound left by the mosquito.

3. **Which of the following are true statements about the immune response in filariasis?**
   A. Repeated exposure to infectious mosquito bites dampens the immune response.
      FALSE.
   B. There is an absence of a cell-mediated immune response.
      FALSE: a cell-mediated response is present, but there is a dominance of a Th2 response over a Th1 response, especially in those with asymptomatic infection.
   C. High IgG4 levels may block potentially beneficial IgE-mediated degranulation of eosinophils and mast cells.
      TRUE.
   D. A weak immune response results in a failure to control infection and considerable symptoms and pathology.
      FALSE: infections can remain asymptomatic and an immune response contributes to pathology.
   E. Down-regulation of Toll-like receptors decreases stimulation of T lymphocytes.
      TRUE.

4. **Which of the following are true statements about protective immunity to filariasis?**
   A. Immunity to infection is directed against the L3 larvae.
      TRUE.
   B. Strong immune responses against adult worms can cause more pathology than protection to hosts.
      TRUE.
   C. Strong immune responses against microfilariae reduce the probability of mosquitoes ingesting microfilariae.
      TRUE: as circulating microfilarial numbers are very low.
   D. There is a genetic component to protective immunity.
      TRUE.
   E. Children born to microfilaremic mothers are protected against infection with L3 larvae.
      FALSE: the converse is TRUE.

5. **Which of the following are true about the clinical features of filariasis?**
   A. The death of adult worms results in resolution of lymphedema.
      FALSE: it persists or may be aggravated.

B. Adult worms cause lymphedema through inflammatory damage to lymphatics.
   TRUE.
C. Lymphedema is an inevitable consequence of all infections.
   FALSE: infections can be asymptomatic.
D. Chyluria occurs when lymph appears in the urine.
   TRUE.
E. Secondary bacterial infections can complicate lymphedema.
   TRUE.

6. **Which of the following are true of tropical pulmonary eosinophilia (TPE)?**
   A. There is a low titer of antifilarial antibodies.
      FALSE: titers are high.
   B. Fever, cough, and wheeze are accompanied by infiltrates on chest X-ray.
      TRUE.
   C. This manifestation is mainly seen in Asia.
      TRUE.
   D. During exacerbations high numbers of microfilariae are seen on a blood film.
      FALSE: the strong immune response clears microfilariae rapidly from the circulation.
   E. Long-term lung damage occurs with repeated bouts of TPE.
      TRUE.

7. **Which of the following are true of the diagnosis of filariasis?**
   A. The presence of microfilariae in a blood film is conclusive evidence that filariasis is responsible for a bout of febrile illness.
      FALSE: patients may have malaria or another febrile illness and an asymptomatic microfilaremia.
   B. Adult worms may be visualized in lymphatic vessels, especially in the scrotum.
      TRUE.
   C. Microfilariae are preferably found in blood films.
      TRUE.
   D. Diethylcarbamazine can provoke the appearance of microfilariae in the blood 30–60 minutes after administration.
      TRUE.
   E. Serological tests for antibody are highly sensitive and specific.
      FALSE: however, new filarial antigen tests are highly sensitive and specific.

8. **Which of the following are true of the differential diagnosis of filariasis?**
   A. A lymphedematous leg may also arise from walking barefoot in certain parts of Africa.
      TRUE: this being due to podoconiosis.
   B. In endemic areas gross lymphedema, elephantiasis, is invariably due to filariasis.
      TRUE.
   C. Intermittent cough and wheeze in TPE have to be differentiated from asthma.
      TRUE.
   D. Lymphedema due to filariasis can be distinguished from other causes because it is always unilateral.
      FALSE: it may be bilateral.
   E. Lymphadenopathy in the neck is usually due to other diagnoses.
      TRUE.

9. **In the treatment of filariasis which of the following are true?**
   A. Diethylcarbamazine does not kill microfilariae.
      FALSE: it kills about 70% of microfilariae.
   B. A single dose of diethylcarbamazine kills all adult worms.
      FALSE: it kills about 50% and annual cycles are used to try to clear adult worms.
   C. Albendazole acts synergistically with diethylcarbamazine to kill adult worms.
      TRUE.
   D. Ivermectin only kills microfilariae and reduces numbers in the short term.
      TRUE.
   E. Prompt treatment of secondary bacterial infections has little effect on long-term prognosis.
      FALSE: it limits progressive damage to lymphatics.

10. **Which of the following are true of the control of filariasis?**
    A. Mass chemotherapy is used for individuals who are microfilariae blood film positive.
       FALSE: blood films are not performed and all subjects in the community are controlled.
    B. In onchocerciasis-endemic areas diethylcarbamazine is preferred to ivermectin for mass chemotherapy.
       FALSE: the converse is TRUE.
    C. Mass chemotherapy is performed in repeated annual cycles.
       TRUE.
    D. Mosquito control measures must be combined with mass chemotherapy for maximum effect.
       TRUE.
    E. Doxycycline may now be used safely for future mass chemotherapy programs.
       FALSE: trials have not been performed for this purpose and there will be issues over the use of doxycycline in children and pregnant and breast-feeding mothers.

Figure Acknowledgments

Case 1. *Aspergillus fumigatus*

Figure 1. Reprint permission kindly granted by Science Photo Library. Additional photographic credit given as Sovereign, ISM and Science Photo Library (M108/626).

Figure 2. This figure is adapted from an image in the private library of the case study author, Dr. Nino Porakishvili.

Figure 3. Reprint permission kindly granted by Dr. David Ellis, University of Adelaide, Australia.

Figures 4–9. Reprint permission kindly granted by The Aspergillus Website, http://www.aspergillus.org.uk/index.html and Dr. Jennifer Bartholomew, School of Medicine, University of Manchester.

Case 2. *Borrelia burgdorferi* and related species

Table 1. Reprint permission kindly granted by *Eurosurveillance*, a publication of the European Centre for Disease Prevention and Control. Originally published in Volume 11, Issue 25, 22 June 2006.

Figure 1. Reprint permission kindly granted by Dr. Charles Goldberg, MD, and Regents of the University of California.

Figure 2. Reprint permission kindly given by the Centers for Disease Control, Atlanta, Georgia. Image is found in the Public Health Image Library #6631.

Figure 3. Image has been adapted from a Centers for Disease Control map in the public domain based on earlier map from Leningrad Nauka Publishers, 1985, a publisher which has subsequently been dissolved.

Figure 4. Reprint permission kindly given by Carolyn Klass, Cornell University. Image was originally published in "Integrated Pest Management for the Deer Tick" by Carolyn Klass, Cornell Cooperative Extension, 139IFS100.00, 12/93.

Figure 5. Reprint permission kindly given by the Centers for Disease Control, Division of Vector-Borne Infectious Disease, Fort Collins, Colorado.

Figure 6. Image permission kindly given by the Centers for Disease Control, Atlanta, Georgia. Image is found in the Public Health Image Library #9874. Additional photographic credit is given to Dr. James Gathany who took the photo in 2007.

Case 3. *Campylobacter jejuni*

Figure 1. Reprint permission kindly given by the Centers for Disease Control, Atlanta, Georgia. Image is found in the Public Health Image Library #6654. Additional photographic credit is given to Robert Weaver, PhD, who took the photo in 1980.

Figure 2. This figure is the creation of the case study author, Professor John Holton, and was produced specifically for this publication.

Case 4. *Chlamydia trachomatis*

Figure 1. Reprint permission kindly granted by the Seattle STD/HIV Prevention Training Center, University of Washington, Seattle.

Figure 2. Adapted with kind permission from Macmillan Publishers and Nature Publishing Group. Originally published in Nature Reviews Immunology 2005, "Immunology of Chlamydia infection: implications for a Chlamydia trachomatis vaccine" by Robert C. Brunham and José Rey-Ladino.

Figure 3. Adapted with kind permission from Macmillan Publishers and Nature Publishing Group. Originally published Nature Reviews Microbiology 2004, Volume 2, Number 7, pp 530 – 531 "Focus: Chlamydia" by Robert Belland, David M. Ojcius and Gerald I Byrne.

Figure 4. Reprint permission kindly granted by the Centers for Disease Control, Atlanta, Georgia. Image is found in the Public Health Image Library #4076. Additional photographic credit is given to Joe Miller who took the photo in 1976.

Figure 5. Reprint permission kindly granted by Elsevier Limited. Originally published in The Lancet, Volume 362, Issue 9379, page 7. "Trachoma" by David C W Mabey, Anthony W Solomon and Allen Foster.

Case 5. *Clostridium difficile*

Figure 1a. Reprint permission kindly granted by Science Photo Library. Additional photographic credit given as David M. Martin, MD and Science Photo Library (M130/559).

Figure 1b. Reprint permission kindly granted by Jae Lim , Calvin Oyer and Murray Resnick, editors and photographic contributors to the Digital Pathology website associated with Brown Medical School, Providence, Rhode Island.

Figure 2. Reprint permission kindly given by the Centers for Disease Control, Atlanta, Georgia. Image is found in the Public Health Image Library #3876. Additional photographic credit is given to Dr. Gilda Jones who took the photo in 1980.

Figure 3. Reprint permission kindly given by the Centers for Disease Control, Atlanta, Georgia. Image is found in the Public Health Image Library #3211. Additional photographic credit is given to Dr. Gilda Jones who took the photo in 1980.

Figure 4. This figure is the creation of the case study author, Professor John Holton, and was produced specifically for this publication.

Figure 5. Reprint permission kindly given by the Windeyer Institute of Medical Sciences, University College London, UK.

Figure 6. This figure is the creation of the case study author, Professor John Holton, and was produced specifically for this publication.

Case 6. *Coxiella burnetti*

Figure 1. Reprint permission kindly given by The European Association for Cardio-thoracic Surgery . Originally published in the volume Georghiou GP, Hirsch R, Vidne RA, Raanani E. *Coxiella burnetii* infection of an aortic graft: surgical view and a word of caution. Interactive Cardiovascular and Thoracic Surgery, 2004, 3: 333–335. URL http://icvts.ctsnetjournals.org/ Copyright © 2004 by The European Association for Cardio-thoracic Surgery.

Figure 2. Reprint permission kindly given by the photographer, Dennis Kunkel. Copyright Dennis Kunkel Microscopy, Inc.

Figure 3. Adapted with kind permission from Elsevier and The Lancet Infectious Diseases. The Lancet Infectious Diseases, Volume 3, Number 11, "Q Fever: A Biological Weapon in Your Backyard" by Miguel G Madariage, Katayoun Rezai, Gordon M. Trenholme and Robert A. Weinstein, November 2003.

Figure 4. Adapted with kind permission from Elsevier and The Lancet Infectious Diseases. The Lancet Infectious Diseases, Volume2, Number 11, "Q Fever in Children" by Helen C Maltezou and Didier Raoult, November 2002.

Figure 5. Adapted with kind permission from Elsevier and The Lancet Infectious Diseases. The Lancet Infectious Diseases, Volume 5, Number 4, "Natural History and Pathophysiology of Q Fever" by Didier Raoult, T J Marrie and J L Mege, April 2005.

Figure 6. Reprinted with kind permission from Dr. Yale Rosen and his website Atlas of Granuloma Diseases: http://granuloma.homestead.com/ .

Figure 7. Adapted with kind permission from Elsevier and The Lancet Infectious Diseases. The Lancet Infectious Diseases, Volume 5, Number 4, "Natural History and Pathophysiology of Q Fever" by Didier Raoult, T J Marrie and J L Mege, April 2005.

Figure 8. These two are in the public domain in the United States because it is a work of the United States Federal Government under the terms of 17 U.S.C. § 105. Combination of two x-rays found on the two websites http://www.fda.gov/cdrh/ct/what.html FDA website with normal chest x-ray http://www.cdc.gov/ncidod/eid/vol6no1/scrimgeourG2.htm CDC website documenting pneumonia All editing performed by me and released into public domain.

Figure 9. Adapted with kind permission from Elsevier and The Lancet Infectious Diseases. The Lancet Infectious Diseases, Volume 5, Number 4, "Natural History and Pathophysiology of Q Fever" by Didier Raoult, T J Marrie and J L Mege, April 2005.

Case 7. Coxsackie B virus

Figure 1. Reprint permission kindly granted by Science Photo Library (Image N532/015).

Figures 2 and 3. These figures are a component of the personal image library of the case study author, Professor Will Irving.

Case 8. *Echinococcus* spp.

Figure 1. Reprint permission kindly granted by Science Photo Library. Additional photographic credit given as Zephyr and Science Photo Library (M170/384).

Figure 2. This figure is the creation of the case study author, Dr. Prith Venkatesan, and was produced specifically for this publication.

Figure 3. Reprint permission kindly given by the Centers for Disease Control & Prevention, Atlanta, Georgia. Image is found in the Public Health Image Library #910. Additional photographic credit is given to Dr. Peter Schantz who took the photo in 1975.

Figure 4. Reprint permission kindly given by the Centers for Disease Control & Prevention, Atlanta, Georgia. Image is found in the Laboratory Identification of Parasites of Public Health Concern.

Case 9. Epstein-Barr virus

Figures 1and 2. Reprint permission kindly given by Elsevier LTD and the original author. This photo was published in "A Colour Atlas of Infectious Diseases" by Emond, Welsby and Rowland.

Figure 3. Adapted from "Principles and Practice of Clinical Virology", By Arie J. Zuckerman, Jangu E. Banatvala, J. R. Pattison, Paul Griffiths, Barry Schoub, 2004. Copyright John Wiley & Sons Limited. Reproduced with permission.

Figure 4. Asapted from the article "Cellular Responses to Viral Infection in Humans: Lessons from Epstein-Barr Virus", by Andrew D. Hislop, Graham S. Taylor, Delphine Sauce and Alan B. Rickinson.Reprint permission kindly given by the *Annual Review of Immunology*, Volume 25 © by Annual Reviews www.annualreviews.org. Additional reprint permission given by the lead author, Alan B. Rickinson.

Figure 5a and 5b. Reprint permission kindly given by John Wiley & Sons, Ltd. Originally published in the Encyclopedia of Life Sciences, *Infectious Mononucleosis*, by M. A. Epstein, University of Oxford, DOI: 10.1038/npg.els.0002318. Article Online Posting Date: January 29, 2003. Copyright © 2003 John Wiley & Sons, Ltd. All rights reserved.

Case 10. *Escherichia coli*

Figure 1. This photo was taken by the case study author, Professor John Holton and the copyright is in his possession.

Figure 2. Reprint permission kindly given by the Centers for Disease Control, Atlanta, Georgia. Image is found in the Public Health Image Library #3211. The photo was taken in 1979.

Figure 3. Reprint permission kindly given by the photographer, Dennis Kunkel. Copyright Dennis Kunkel Microscopy, Inc.

Figure 4. This figure is the creation of the case study author, Professor John Holton, and was produced specifically for this publication.

Case 11. *Giardia lamblia*

Figure 1. Adapted with permission kindly given by the Centers for Disease Control & Prevention, Atlanta, Georgia. Original image is found in the Public Health Image Library #3394. Additional photographic credit is given to Alexander J. da Silva, PhD, and Melanie Moser who created the image in 2002.

Figure 2. Reprint permission kindly granted by Science Photo Library. Additional photographic credit given as CNRI and Science Photo Library (Z100/059).

Figure 3. Reprint permission kindly given by the Centers for Disease Control & Prevention, Atlanta, Georgia. Image is found in the Public Health Image Library #3741. Additional photographic credit is given to Dr. Mae Melvin who took the photo in 1977.

Figure 4. Reprint permission kindly given by the Centers for Disease Control & Prevention, Atlanta, Georgia. Image is found in the Public Health Image Library #7833. Additional photographic credit is given to DPDX/ Melanie Moser who created the original image.

Figure 5. Reprint permission kindly given by the Centers for Disease Control & Prevention, Atlanta, Georgia. Image is found in the Laboratory Identification of Parasites of Public Health Concern.

Case 12. *Helicobacter pylori*

Figure 1. Reprint permission kindly given by Professor DinoVaira, First Medical Clinic University of Bologna, Italy.

Figure 2. Adapted with kind permission from a map created by the Helicobacter Foundation and originally published at the following web address: http://www.helico.com/h_epidemiology.html.

Figure 3. Adapted with kind permission from an image created by The Nobel Committee for Physiology or Medicine and originally published at the following web address: http://nobelprize.org/nobel_prizes/medicine/laureates/2005/press.html.

Figures 4–6. These figures are the creation of the case study author, Professor John Holton, and were produced specifically for this publication.

Figure 7. Reprint permission kindly given by Dr. Lorraine Racusen, Johns Hopkins Medical School, Johns Hopkins University, Baltimore, Maryland.

Figures 8–10. These figures are the creation of the case study author, Professor John Holton, and were produced specifically for this publication.

Case 13. Hepatitis B Virus

Figure 1. Reprint permission kindly given by the Centers for Disease Control & Prevention, Atlanta, Georgia. Image is found in the Public Health Image Library #2860. Additional photographic credit is given to Thomas F. Sellers and Emory University who took the photo in 1963.

Figure 2. This figure is the creation of the case study author, Professor Will Irving, and was produced specifically for this publication.

Figure 3. Adapted with kind permission from a map created by the United States Centers for Disease Control and Prevention.

Figure 4. This figure is the creation of the case study author, Professor Will Irving, and was produced specifically for this publication.

Case 14. Herpes simplex virus 1 (HSV-1)

Figure 1. Reprint permission kindly granted by Science Photo Library. Additional photographic credit given as St Bartholomew's Hospital, London and Science Photo Library (M150/026).

Figure 2. Adapted with kind permission from the Wellcome Trust.

Figure 3. This figure is the property of the case study author, Professor Will Irving, and is located within his personal library.

Figure 4. Reprint permission kindly granted by Oxford University Press. Originally published in Humphreys and Irving, page 191, Figure 42.1.

Case 15. Herpes simplex virus 2 (HSV-2)

Figures 1 and 2. These figures are the property of the case study author, Professor Will Irving, and are located within his personal library.

Figure 3. Reprint permission kindly given by the Centers for Disease Control & Prevention, Atlanta, Georgia. Image is found in the Public Health Image Library #6510. Additional photographic credit is given to Judith Faulk who took the photo in 1973.

Figure 4. Reprint permission kindly given by Oxford University Press. Image originally published in Antimicrobial Chemotherapy, editor Greenwood et al. 2007. figure 6-2 p 99, 5th edition.

Case 16. *Histoplasma capsulatum*

Figure 1. Reprint permission kindly given by the Centers for Disease Control & Prevention, Atlanta, Georgia. Image is found in the Public Health Image Library #3954. Additional photographic credit is given to M. Renz who took the photo in 1963.

Figure 2. Reprint permission kindly given by the Centers for Disease Control & Prevention, Atlanta, Georgia. Image is found in the Public Health Image Library #3191. Additional photographic credit is given to Dr. William Kaplan who took the photo in 1969.

Figure 3a. Reprint permission kindly given by the Centers for Disease Control & Prevention, Atlanta, Georgia. Image is found in the Public Health Image Library #4023. Additional photographic credit is given to Dr. Libero Ajello who took the photo in 1968.

Figure 3b. Reprint permission kindly given by the Centers for Disease Control & Prevention, Atlanta, Georgia. Image is found in the Public Health Image Library #4022. Additional photographic credit is given to Dr. Libero Ajello who took the photo in 1968.

Figure 4. Reprint permission kindly given by Professor Michael R. McGinnis of DoctorFungus.org. www.doctorfungus.org.

Figure 5. Adapted with permission kindly given by Professor Michael R. McGinnis of DoctorFungus.org. www.doctorfungus.org.

Figure 6. Reprint permission kindly given by the Centers for Disease Control & Prevention, Atlanta, Georgia. Image is found in the Public Health Image Library #4223. Additional photographic credit is given to Dr. Libero Ajello who took the photo in 1972.

Figure 7. Reprint permission kindly given by the Centers for Disease Control & Prevention, Atlanta, Georgia. Image is found in the Public Health Image Library #6840. Additional photographic credit is given to Susan Lindsley, VD, who took the photo in 1973.

Case 17. Human immunodeficiency virus (HIV)

Figure 1. Reprint permission kindly given by Elsevier Health Sciences Publishing. Image originally published in "Immunology 7th edition" by David K. Mail, Jonathan Brostroff, Ivan Maurice Roitt, and David Roth, 2006.

Figure 2. Reprint permission kindly given by the Centers for Disease Control & Prevention, Atlanta, Georgia. Image is found in the Public Health Image Library #960. Additional photographic credit is given to Dr. Edwin P. Ewing who took the photo in 1984.

Figure 3. Reprint permission kindly given by the Centers for Disease Control & Prevention, Atlanta, Georgia. Image is found in the Public Health Image Library #6070. Additional photographic credit is given to Sol Silverman, Jr., D.D.S., University of California, San Francisco who took the image in 1987.

Figure 4. Reprint permission kindly given by Elsevier Health Sciences Publishing. Image originally published in "Immunology 7th edition" by David K. Mail, Jonathan Brostroff, Ivan Maurice Roitt, and David Roth, 2006.

Figure 5. Reprint permission kindly given by Elsevier Health Sciences Publishing. Image originally published in "Immunology 7th edition" by David K. Mail, Jonathan Brostroff, Ivan Maurice Roitt, and David Roth, 2006.

Figure 6. Reprint permission kindly given by Professor Leonard Poulter, Professor Emeritus, University College London.

Figure 7. Reprint permission kindly given by Elsevier Limited Publishing. Image originally published in journal *Placenta*, Volume 15, Number 1, 1994.

Figure 8. Adapted with kind permission of UNAIDS, www.unairds.org.

Figure 9. Reprint permission kindly given by Elsevier Health Sciences Publishing. Image originally published in "Immunology 7th edition" by David K. Mail, Jonathan Brostroff, Ivan Maurice Roitt, and David Roth, 2006.

Figure 10. Originally published in Garland Science title "Instant Notes in Immunology", 2nd edition, page 191.

Case 18. Influenza virus

Figure 1. This figure was produced specifically for this publication.
Figure 2. Adapted with kind permission from Oxford University Press. Originally published in Humphreys and Irving, page 72, Figure 17.5.

Figure 3. This figure was produced specifically for this publication.

Figure 4. This figure is the creation of the case study author, Professor Will Irving, and was produced specifically for this publication.

Figure 5. Adapted with kind permission from Oxford University Press. Originally published in Humphreys and Irving, page 80, Figure 21.2.

Figure 6. Adapted with kind permission from the New England Journal of Medicine Volume 353: 1363 – 1373, Page 1364, Figure 1. © 2005 Massachusetts Medical Society.

Case 19. *Leishmania* spp.

Figure 1. Reprint permission kindly given by the World Health Organization, Special Programme for Research and Training in Tropical Diseases, http://www.who.int/tdr/index.html image #9706290. Additional photographic credit is given to Andy Crump who took the photograph in 1997 in Sudan.

Figure 2. Reprint permission kindly given by the Centers for Disease Control & Prevention, Atlanta, Georgia. Image is found in the Public Health Image Library #331. Additional photographic credit is given to Dr. Martin D. Hicklin who created the image in 1964.

Figure 3. Reprint permission kindly given by the Centers for Disease Control & Prevention, Atlanta, Georgia. Image is found in the Public Health Image Library #544.

Figure 4. Adapted with kind permission from the Centers for Disease Control & Prevention, Atlanta, Georgia. Image is found in the Public Health Image Library #3400. Additional photographic credit is given to Alexander J. da Silva, PhD, and Melanie Moser who created the image in 2002.

Figure 5. Reprint permission kindly given by the Centers for Disease Control & Prevention, Atlanta, Georgia. Image is found in the Public Health Image Library #6274. Additional photographic credit is given as follows: World Health Organization (WHO), Geneva, Switzerland.

Figure 6. Reprint permission kindly given by the Centers for Disease Control & Prevention, Atlanta, Georgia. Image is found in the Public Health Image Library #352. Additional photographic credit is given to Dr. D. S. Martin.

Figure 7. Reprint permission kindly given by the World Health Organization, Special Programme for Research and Training in Tropical Diseases, http://www.who.int/tdr/index.html image #9106015. Additional photographic data indicates that the photo was taken in Switzerland in 1990.

Figure 8. Reprint permission kindly given by the Centers for Disease Control & Prevention, Atlanta, Georgia. Image is found in the Public Health Image Library #30. Additional photographic credit is given to Dr. Francis W. Chandler who created the image in 1979.

Case 20. *Leptospira* spp.

Figure 1. Reprint permission kindly given by the Centers for Disease Control & Prevention, Atlanta, Georgia. Image is found in the Public Health Image Library #1220. Additional photographic credit is given to Janice Haney Carr who took the image and the CDC NCID and Rob Weyant who provided the image for CDC PHIL website.

Figure 2. Adapted with kind permission from Dr. Samuel Baron and the University of Texas Medical Branch at Galveston, Department of Microbiolgy and Immunology.

Case 21. *Listeria monocytogenes*

Figure 1. Reprint permission kindly granted by Science Photo Library. Additional photographic credit given as CC STUDIO / SCIENCE PHOTO LIBRARY (M874/658).

Figure 2. Reprint permission kindly granted by Science Photo Library. Additional photographic credit given as A.B. DOWSETT / SCIENCE PHOTO LIBRARY (B220/273).

Figure 3. Reprint permission kindly granted by Science Photo Library. Additional photographic credit given as MAXIMILIAN

STOCK LTD / SCIENCE PHOTO LIBRARY (H110/1266).

Figure 4. This figure is the creation of the case study author, Professor John Holton, and was produced specifically for this publication.

Figure 5. Reprint permission kindly granted by Barrow Neurological Institute, Barrow Quarterly, 19 (4): 20–24, Figure 3, 2003. © Barrow Neurological Institute.

Figure 6. Reprint permission kindly granted by Becton, Dickinson and Company, http://www.bd.com/ © Becton, Dickinson and Company.

Case 22. *Mycobacterium lepra*

Figures 1 and 2. Reprint permission kindly granted by the U.S. Department of Health and Human Services.

Figure 3. This figure was produced specifically for this publication based on a variety of data sources.

Figure 4. Adapted with kind permission from Elsevier and The Lancet. The Lancet, Volume 363, Number 9416, "Leprosy" by Warwick J. Britton and Diana Lockwood, 10 April 2004.

Case 23. *Mycobacterium tuberculosis*

Figures 1 and 2. These figures are the property of the case study author, Professor John Holton, and are located within his personal library.

Figure 2. This figure is the property of the case study author, Professor John Holton, and is located within his personal library.

Figure 3. This figure was produced specifically for this publication based on a variety of data sources.

Figure 4. This figure is the property of the case study author, Professor John Holton, and is located within his personal library.

Figure 5. Adapted from *Instant Notes in Immunology*, Lydyard, Whelan Fanger, 2004, Garland Science, Taylor and Francis Publishing.

Figure 6. Adapted from *Janeway's Immunology*, Garland Science, Taylor and Francis Publishing.

Figure 7. This figure is the property of the case study author, Professor John Holton, and is located within his personal library.

Case 24. *Neisseria gonorrhoeae*

Figure 1. Reprint permission kindly given by the Centers for Disease Control & Prevention, Atlanta, Georgia. Image is found in the Public Health Image Library 4065.

Figure 2. Reprint permission kindly given by the Centers for Disease Control & Prevention, Atlanta, Georgia. Image is found in the Public Health Image Library #3694. Additional photographic credit is given to Joe Miller who took the photo in 1979.

Figure 3. Reprint permission kindly given by the Centers for Disease Control & Prevention, Atlanta, Georgia. Image is found in the Public Health Image Library #4087.

Figure 4. Reprint permission kindly given by the Centers for Disease Control & Prevention, Atlanta, Georgia. Image is found in the Public Health Image Library #5516. Additional photographic credit is given to Joe Miller who took the photo in 1975.

Figure 5. Reprint permission kindly given by the Centers for Disease Control & Prevention, Atlanta, Georgia. Image is found in the Public Health Image Library #6384 Additional photographic credit is given to Dr. Weisner who took the photo in 1972.

Figure 6a. Reprint permission kindly given by Science Photo Library. Additional photographic credit given as Science Photo Library (M874/670). Additional photographic credit given as CC STUDIO / SCIENCE PHOTO LIBRARY.

Figure 6b. Reprint permission kindly given by Science Photo Library. Additional photographic credit given as Science Photo Library (M874/548). Additional photographic credit given as CNRI / SCIENCE PHOTO LIBRARY.

Figure 7. Reprint permission kindly given by the Centers for Disease Control & Prevention, Atlanta, Georgia. Image is found in the Public Health Image Library #6505. Additional photographic credit is given to Renelle Woodall who took the photo in 1969.

Figure 8. Reprint permission kindly given by the Centers for Disease Control & Prevention, Atlanta, Georgia. Image is found in the Public Health Image Library #3798. Additional photographic credit is given to Joe Miller who took the photo in 1980.

Figure 9. Reprint permission kindly given by Gary E. Kaiser, Professor of Microbiology, Biology Department, Community College of Baltimore County, Cantonsville Campus, Baltimore, Maryland.

Case 25. *Neisseria meningitidis*

Figure 1. Reprint permission kindly given by Science Photo Library. Additional photographic credit given as Science Photo Library (M210/018). Additional photographic credit given as JOHN RADCLIFFE HOSPITAL/ SCIENCE PHOTO LIBRARY.

Figure 2. Reprint permission kindly given by the Centers for Disease Control & Prevention, Atlanta, Georgia. Image is found in the Public Health Image Library #6423. Additional photographic credit is given to Dr. Brodsky who took the photo in 1966.

Figure 3. Reprint permission kindly given for image in *Laboratory Methods for the Diagnosis of meningitis Caused by Neisseria meningitidis, Streptococcus pneumoniae, and Haemophilus influenzae* CDC, Centre for Disease Control and Prevention August 1998, Figure 9a, page 26.

Figure 4. Reprint permission kindly given for image in *Laboratory Methods for the Diagnosis of meningitis Caused by Neisseria meningitidis, Streptococcus pneumoniae, and Haemophilus influenzae* CDC, Centre for Disease Control and Prevention August 1998, Figure 16, page 37.

Figure 5. Adapted with kind permission by Wolters Kluwer Health. Originally published in Color Atlas and Textbook of Diagnostic Microbiology, 5th Edition, Koneman, Allen, Janda, Schreckenberger, Winn Jr; Lippincott, Color Plate 10-1 Page 495.

Figure 6. Reprint permission kindly given by the Department of Pathology, University of Pittsburgh School of Medicine.

Figure 7. Reprint permission kindly given for image in *Laboratory Methods for the Diagnosis of meningitis Caused by Neisseria meningitidis, Streptococcus pneumoniae, and Haemophilus influenzae* CDC, Centre for Disease Control and Prevention August 1998, Figure 6, page 21.

Figure 8. Reprint permission kindly given for image in *Laboratory Methods for the Diagnosis of meningitis Caused by Neisseria meningitidis, Streptococcus pneumoniae, and Haemophilus influenzae* CDC, Centre for Disease Control and Prevention August 1998, Figure 111, page 31.

Figure 9. Reprint permission kindly given for image in *Laboratory Methods for the Diagnosis of meningitis Caused by Neisseria meningitidis, Streptococcus pneumoniae, and Haemophilus influenzae* CDC, Centre for Disease Control and Prevention August 1998, Figure 15, Page 36.

Figure 10. Reprint permission kindly given for image in *Laboratory Methods for the Diagnosis of meningitis Caused by Neisseria meningitidis, Streptococcus pneumoniae, and Haemophilus influenzae* CDC, Centre for Disease Control and Prevention August 1998, Figure 18, page 41.

Case 26. Norovirus

Figure 1. This figure was produced specifically for this publication.

Figure 2. Reprint permission kindly given for image by Hodder Education. Originally published in *Topley & Wilson Microbiology and Microbial Infection*, 10*th* edition CD Set, 2005, Figure 42.7, Brian E. J. Mahey et al. Reproduced by permission of Edward Arnold (Publishers) Ltd.

Figure 3. This figure was produced specifically for this publication.

Figure 4. This figure was produced specifically for this publication.

Figure 5a. Reprint permission kindly given by Dr. James Gray of the Health Protection Agency (originator of the image) and by Elsevier Publishing. Originally due to be published as Figure 23.3 in *Notes on Medical Microbiology* by Dr. Kate Ward.

Figure 5b. Reprint permission kindly given by Dr. James Gray of the Health Protection Agency (originator of the image) and by Elsevier Publishing. Originally due to be published as Figure 23.3 in *Notes on Medical Microbiology* by Dr. Kate Ward.

Case 27. Parvovirus

Figure 1. This figure is the property of the case study author, Professor Will Irving, and is located within his personal library.

Figure 2. Reprint permission kindly granted by Oxford University Press. Originally published in Humphreys and Irving, page 322, Figure 67.1.

Figure 3. Reprint permission kindly given by the Centers for Disease Control & Prevention, Atlanta, Georgia. Image is found in the Public Health Image Library 4508.

Case 28. *Plasmodium* spp.

Figure 1. Reprint permission kindly given by the Centers for Disease Control & Prevention, Atlanta, Georgia. Image is found in the Public Health Image Library #7192. Additional photographic credit is given to James Gathany, Dr. Frank Collins and the University of Notre Dame. The image was taken in 2005 by James Gathany.

Figure 2. Reprint permission kindly given by the Centers for Disease Control & Prevention, Atlanta, Georgia. Image is found in the Public Health Image Library #3405. Additional photographic credit is given to Alexander J. da Silva, PhD and Melanie Moser. The image was created in 2002.

Figure 3. Adapted with kind permission from Elsevier's publication *Mim'sMedical Microbiology*, 4th edition, 2008, Figure 27.11.

Figure 4. Adapted with kind permission from Carlos Guerra and the Malaria Atlas Project. First published under Creative Commons License Agreement in the publication: "The Limits and Intensity of Plasmodium falciparum Transmission: Implications for Malaria Control and Elimination Worldwide." http://www.map.ox.ac.uk/.

Figure 5. Reprint permission kindly given by the Centers for Disease Control & Prevention, Atlanta, Georgia. Image is found in the Public Health Image Library #4884. Additional photographic credit is given to Dr. Mae Martin who took the photo in 1971.

Case 29. Respiratory syncytial virus (RSV)

Figure 1. Reprinted under the terms of the Creative Commons Attribution License (http://creativecommons.org/licenses/by/2.0), which permits unrestricted use, distribution, and reproduction in any medium, provided the original work is properly cited. Originally published in the *Journal of Medical Case Reports*, 2008, 2:212, doi:10.1186/1752-1947-2-212. The electronic version of this article is the complete one and can be found online at: http://www.jmedical-casereports.com/content/2/1/212.

Figure 2. Reprinted with kind permission from Dr. Jeremy A. Garson and the Centre for Virology, Department of Infection, Royal Free and University College Medical School, Windeyer Building, London.

Figure 3. Reprinted with kind permission from the American Society for Microbiology. Originally published in the *Clinical Microbiology Reviews*, April 01, 1999, Volume 12, Issue 2, "Respiratory Syncytial Virus Infection: Immune Reposne, Immunopathogenesis, and Treatment" by Joseph B. Domachowske and Helene F. Rosenberg.

Figure 4. This figure is the creation of the case study author, Kate Ward, and was produced specifically for this publication.

Figure 5. Adapted with kind permission from the Centre for Infections, Health Protection Agency, England.

Case 30. *Rickettsia* spp.

Figure 1. Reprint permission kindly given by Science Photo Library. Additional photographic credit given as Science Photo Library (M320/295). Additional photographic credit given as DR P. MARAZZI / SCIENCE PHOTO LIBRARY.

Figure 2. This figure is the creation of the case study author, Professor John Holton, and was produced specifically for this publication.

Figure 3. Reprint permission kindly given by the Centers for Disease Control & Prevention, Atlanta, Georgia. Image is found in the Public Health Image Library 5977.

Figure 4. Reprint permission kindly given by the Centers for Disease Control & Prevention, Atlanta, Georgia. Image is found in the Public Health Image Library 5636. Additional photographic credit is given to CDC/ DVBID, BZB, Entomology and Ecology Activity, Vector Ecology & Control Laboratory, Fort Collins, CO. and was created in 2004.

Figure 5. Reprint permission kindly given by the Centers for Disease Control & Prevention, Atlanta, Georgia. Image is found in the Public Health Image Library 9217. Additional photographic credit is given to CDC/ Frank Collins, Ph.D. and was created in 2006.

Figure 6. Reprint permission kindly given by the Centers for Disease Control & Prevention, Atlanta, Georgia. Image is found in the Public Health Image Library 5447.

Figure 7. This figure is the creation of the case study author, Professor John Holton, and was produced specifically for this publication.

Case 31. *Salmonella typhi*

Figure 1. Reprint permission kindly given by the Centers for Disease Control & Prevention, Atlanta, Georgia. Image is found in the Public Health Image Library 2114.

Figure 2. This figure is the creation of the case study author, Professor John Holton, and was produced specifically for this publication.

Figure 3. Reprint permission kindly given by the Centers for Disease Control & Prevention, Atlanta, Georgia. Image is found in the Public Health Image Library 2215. Additional photographic credit is given to the Armed Forces Institute of Pathology and Charles N. Farmer who created the image in 1964.

Figure 4. Reprint permission kindly given by the Centers for Disease Control & Prevention, Atlanta, Georgia. Image is found in the Public Health Image Library 6619.

Case 32. *Schistosoma* spp.

Figure 1. Reprinted with kind permission from Dr. Alan Mills and the Virginia Commonwealth University Department of Pathology.

Figure 2. Reprint permission kindly given by the Centers for Disease Control & Prevention, Atlanta, Georgia. Image is found in the Public Health Image Library #5252.

Figure 3. Adapted with kind permission from the Centers for Disease Control & Prevention, Atlanta, Georgia. Image is found in the Public Health Image Library #3417. Additional photographic credit is given to Alexander J. da Silva, PhD, and Melanie Moser who created the image in 2002.

Figure 4. Reprint permission kindly given by the Centers for Disease Control & Prevention, Atlanta, Georgia. Image is found in the Public Health Image Library #8556. Additional photographic credit is given to Minnesota Department of Health, R.N. Barr Library; Librarians Melissa Rethlefsen and Marie Jones, Prof. William A. Riley and the photo was created in 1942.

Figure 5a. Reprint permission kindly given by the Centers for Disease Control & Prevention, Atlanta, Georgia. Image is found in the Public Health Image Library #4843.

Figure 5b. Reprint permission kindly given by the Centers for Disease Control & Prevention, Atlanta, Georgia. Image is found in the Public Health Image Library #4841.

Figure 5c. Reprint permission kindly given by the Centers for Disease Control & Prevention, Atlanta, Georgia. Image is found in the Public Health Image Library #4842.

Figure 6. Reprinted with kind permission from Dr. Yale Rosen and his website Atlas of Granuloma Diseases: http://granuloma.homestead.com/.

Figure 7. Reprint permission kindly given by the Centers for Disease Control & Prevention, Atlanta, Georgia. Image is found in the Public Health Image Library #35. Additional photographic credit is given Dr. Edwin P. Ewing, Jr. who took the photo in 1973.

Figure 8. Reprint permission kindly given by the Centers for Disease Control & Prevention, Atlanta, Georgia. Image is found in the Public Health Image Library #5250.

Case 33. *Staphylococcus aureus*

Figure 1. Adapted with permission kindly given by Elsevier Publishing, New York. Image originally published in Medical Microbiology 5th edition, by Murray, Rosenthal, and Pfaller, Figure 22.2, page 233.

Figure 2. Reprint permission kindly given by DermAtlas.org © Bernard Cohen - Contributor, Dermatlas; http://www.dermatlas.org.

Figure 3. Reprint permission kindly given by Science Photo Library. Additional photographic credit given as Science Photo Library (M120/077).

Figure 4. Reprint permission kindly given by Science Photo Library. Additional photographic credit given as Science Photo Library (M160/044).

Figure 5. Reprint permission kindly granted by Science Photo Library. Additional photographic credit given as Science Photo Library (M130/226).

Figure 6. Reprint permission kindly given by the Centers for Disease Control & Prevention, Atlanta, Georgia. Image is found in the Public Health Image Library #5113.

Figure 7. Reprinted by permission from Macmillan Publishers Ltd: *Journal of Investigative Dermatology* Volume 124: Pages 700 – 703, Figure S1, 2005.

Figure 8. Reprint permission kindly given by the Centers for Disease Control & Prevention, Atlanta, Georgia. Image is found in the Public Health Image Library #2297. Additional photographic credit is given Dr. Richard Facklin who took the photo in 1973.

Figure 9. Reprint permission kindly given by Professor A. M. Sefton, Institute of Cell and Molecular Science Centre for Infectious Disease, Barts and the London School of Medicine and Dentistry.

Figure 10. Reprint permission kindly given by the American Society for Microbiology, MicrobeLibrary Visual Resources Licensing.

Figures 11a and 11b. Reprint permission kindly given by Jay Hardy, President of Hary Diagnostics, www.Hardy.Diagnostics.com.

Figure 12. Adapted with permission kindly given by Dr. Celeste Chong-Cerrillo, Instructional Laboratory Manager, Department of Biological Sciences, College of Letters, Arts, and Sciences, University of Southern California.

Case 34. *Streptococcus mitis*

Figure 1a. Reprint permission kindly given by Nature Publishing Company and Macmillan Publishers, Ltd. Originally published in journal *Eye*, Volume 19, Issue 5, 2004. doi:10.1038/sj.eye.6701530.

Figure 1b. Reprint permission kindly granted by Science Photo Library (Image M155/186).

Figure 1c. Reprint permission kindly granted by Science Photo Library (Image M172/281). Additional photographic credit as Dr. P Marazzi.

Figure 2. Reprint permission kindly given by the Centers for Disease Control & Prevention, Atlanta, Georgia. Image is found in the Public Health Image Library #2897. Additional photographic credit is given to Dr. Mike Miller who took the photo in 1978.

Figure 3. Reprint permission kindly given by the American Society for Microbiology, MicrobeLibrary Visual Resources Licensing.

Figure 4. Reprint permission kindly given for image in *Laboratory Methods for the Diagnosis of meningitis Caused by Neisseria meningitidis, Streptococcus pneumoniae, and Haemophilus influenzae* CDC, Centre for Disease Control and Prevention August 1998, Figure 20, Page 46.

Case 35. *Streptococcus pneumoniae*

Figure 1. Reprint permission kindly given by the Centers for Disease Control & Prevention, Atlanta, Georgia. Image is found in the Public Health Image Library #2897. Additional photographic credit is given to Dr. Thomas Hooten who took the photo in 1978.

Figure 2. Reprint permission kindly granted by Science Photo Library (Image B236/056). Additional photographic credit given as LEBEAU / CUSTOM MEDICAL STOCK PHOTO / SCIENCE PHOTO LIBRARY.

Figure 3. Reprint permission kindly granted by Science Photo Library (Image B236/145). Additional photographic credit CNRI / SCIENCE PHOTO LIBRARY.

Figure 4. Reprint permission kindly given by the American Society for Microbiology, MicrobeLibrary Visual Resources Licensing.

Figure 5. Adapted with permission kindly given by Nature Publishing Company and Macmillan Publishers, Ltd. Originally published in journal *EMBO Reports*, Volume 3, Issue 8, 2002.

Figure 6a. Reprint permission kindly given for image in *Laboratory*

Methods for the Diagnosis of meningitis Caused by Neisseria meningitidis, Streptococcus pneumoniae, and Haemophilus influenzae CDC, Centre for Disease Control and Prevention August 1998, Figure 7, Page 22.

Figure 6b. Reprint permission kindly given for image in *Laboratory Methods for the Diagnosis of meningitis Caused by Neisseria meningitidis, Streptococcus pneumoniae, and Haemophilus influenzae* CDC, Centre for Disease Control and Prevention August 1998, Figure 12, Page 32.

Figure 7. Reprint permission kindly given for image in *Laboratory Methods for the Diagnosis of meningitis Caused by Neisseria meningitidis, Streptococcus pneumoniae, and Haemophilus influenzae* CDC, Centre for Disease Control and Prevention August 1998, Figure 20, Page 46.

Case 36. *Streptococcus pyogenes*

Figure 1. Reprint permission kindly given by the Centers for Disease Control & Prevention, Atlanta, Georgia. Image is found in the Public Health Image Library #3185. Additional photographic credit is given to Dr. Heinz F. Eichenwald who took the photo in 1958.

Figure 2. Reprint permission kindly granted by Science Photo Library (Image B236/157). Additional photographic credit EYE OF SCIENCE / SCIENCE PHOTO LIBRARY.

Figure 3. Reprint permission kindly given by the Centers for Disease Control & Prevention, Atlanta, Georgia. Image is found in the Public Health Image Library #8170. Additional photographic credit is given to Richard R. Facklam, Ph.D., who took the photo in 1977.

Figure 4. Adapted with permission kindly given by Elsevier. Originally published in *Medical Microbiology 5th edition*, by Murray, Rosenthal, and Pfaller, 2005.

Figure 5a. Reprint permission kindly granted by Science Photo Library (Image M260/209). Additional photographic credit BIO-PHOTO ASSOCIATES / SCIENCE PHOTO LIBRARY.

Figure 5b–5d. Reprint permission kindly given by Elsevier. Originally published in *Clinical dermatology: A color guide to diagnosis & therapy*, by Thomas P. Habif, Fourth edition, 2004.

Figure 6. Reprint permission kindly given by the Centers for Disease Control & Prevention, Atlanta, Georgia. Image is found in the Public Health Image Library #2874. Additional photographic credit is given to Dr. Thomas F. Sellers and Emory University who took the photo in 1963.

Figure 7. Reprint permission kindly granted by Science Photo Library (Image M130/994). Additional photographic credit DR. P. MARAZZI / SCIENCE PHOTO LIBRARY.

Figure 8. Reprint permission kindly given by Elsevier. Originally published in *Diseases of the Skin 2nd edition*, found at the following website: http://www.merckmedicus.com/ppdocs/us/hcp/content/white/chapters/white-ch-024-s002.htm.

Figure 9. Reprint permission kindly granted by Science Photo Library (Image M874/036). Additional photographic credit VAN BUCHER / SCIENCE PHOTO LIBRARY.

Figure 10. Reprint permission kindly given by Jay Hardy, President of Hary Diagnostics, www.Hardy.Diagnostics.com.

Figure 11. Reprint permission kindly given by Mark Reed and Pro-Lab Diagnostics, www.pro-lab.com.

Case 37. *Toxoplasma gondii*

Figure 1. This figure is the property of the case study author, Professor Will Irving, and is located within his personal library.

Figure 2. Reprint permission kindly given by the Centers for Disease Control & Prevention, Atlanta, Georgia. Image is found in the entry for Toxoplasmosis in the DPDx Parasite Image Library at the following website address: http://www.dpd.cdc.gov/dpdx/HTML/Image_Library.htm.

Figure 3. Reprint permission kindly given by the Centers for Disease Control & Prevention, Atlanta, Georgia. Image is found in the Public Health Image Library #3421. Additional photographic credit is given to Alexander J. da Silva, PhD, and Melanie Moser who created the image in 2002.

Figure 4. Reprint permission kindly given by the Centers for Disease Control & Prevention, Atlanta, Georgia. Image is found in the entry for Toxoplasmosis in the DPDx Parasite Image Library at the following website address: http://www.dpd.cdc.gov/dpdx/HTML/Image_Library.htm.

Figure 5. Reprint permission kindly given by the Centers for Disease Control & Prevention, Atlanta, Georgia. Image is found in the Public Health Image Library #966. Additional photographic credit is given to Dr. Edwin P. Ewing who created the image in 1984.

Figures 6a and 6b. Reprint permission kindly given by the Centers for Disease Control & Prevention, Atlanta, Georgia. Image is found in the entry for Toxoplasmosis in the DPDx Parasite Image Library at the following website address: http://www.dpd.cdc.gov/dpdx/HTML/Image_Library.htm.

Case 38. *Trypanosoma* spp.

Figure 1. Reprint permission kindly given by the World Health Organization, Special Programme for Research and Training in Tropical Diseases, http://www.who.int/tdr/index.html image #9905349. Additional photographic credit is given to Andy Crump who took the photograph in 1999 in Argentina.

Figure 2. Reprint permission kindly given by the Centers for Disease Control & Prevention, Atlanta, Georgia. Image is found in the Public Health Image Library #613. Additional photographic credit is given to Dr. Myron G. Schultz who took the photo in 1970.

Figure 3. Reprint permission kindly given by the Centers for Disease Control & Prevention, Atlanta, Georgia. Image is found in the Public Health Image Library #543.

Figure 4. Reprint permission kindly given by the Centers for Disease Control & Prevention, Atlanta, Georgia. Image is found in the Public Health Image Library #613. Additional photographic credit notes that the images was donated by the World Health Organization, Geneva, Switzerland, in 1976.

Figure 5. Reprint permission kindly granted by Science Photo Library (Image Z340/031). Additional photographic credit attributed to Martin Dohrn.

Figure 6. Adapted with kind permission from the Centers for Disease Control & Prevention, Atlanta, Georgia. Image is found in the Public Health Image Library #3418. Additional photographic credit is given to Alexander J. da Silva, PhD, and Melanie Moser who created the image in 2003.

Figure 7. Adapted with kind permission from the Centers for Disease Control & Prevention, Atlanta, Georgia. Image is found in the Public Health Image Library #3384. Additional photographic credit is given to Alexander J. da Silva, PhD, and Melanie Moser who created the image in 2002.

Figure 8. This figure was produced specifically for this publication.

Figure 9. Reprint permission kindly given by the Centers for Disease Control & Prevention, Atlanta, Georgia. Image is found in the Public Health Image Library #2617. Additional photographic credit notes that the photo was taken by Dr. Mae Melvin in 1962.

Figure 10. Reprint permission kindly given by the World Health Organization, Special Programme for Research and Training in Tropical Diseases, http://www.who.int/tdr/index.html image #9105027. Additional photographic notes indicate the image was taken in Brazil in 1990.

Figure 11. Reprint permission kindly given by the World Health Organization, Special Programme for Research and Training in Tropical Diseases, http://www.who.int/tdr/index.html image #9604658. Additional photographic credit is given to Andy Crump who took the photograph in 1996 in Uganda.

Case 39. Varicella-zoster virus

Figure 1. Reprint permission kindly given by the Centers for Disease Control & Prevention, Atlanta, Georgia. Image is found in the Public Health Image Library #10169. Additional photographic credit notes that the photo was taken by Dr. Alexander D. Langmuir in 1960.

Figure 2. This figure is the property of the case study author, Professor Will Irving, and is located within his personal library.

Figure 3. Adapted with kind permission from The Wellcome Trust, London, UK.

Figure 4. Reprint permission kindly given by the Centers for Disease Control & Prevention, Atlanta, Georgia. Image is found in the Public Health Image Library #6121. Additional photographic credit notes that the photo was taken by Dr. Alexander D. Langmuir in 1960.

Case 40. *Wuchereria bancrofti*

Figure 1. Reprint permission kindly given by the Centers for Disease Control & Prevention, Atlanta, Georgia. Image is found in the Public Health Image Library #373.

Figure 2. Reprint permission kindly given by the Centers for Disease Control & Prevention, Atlanta, Georgia. Image is found in the Public Health Image Library #3008. Additional photographic credit is given to Dr. Mae Melvine who took the photo in 1978.

Figure 3. Adapted with kind permission from the Centers for Disease Control & Prevention, Atlanta, Georgia. Image is found in the Public Health Image Library #3425. Additional photographic credit is given to Alexander J. da Silva, PhD, and Melanie Moser who created the image in 2003.

Figure 4. Reprint permission kindly given by the World Health Organization, Special Programme for Research and Training in Tropical Diseases, http://www.who.int/tdr/index.html image #01021639. Additional photographic credit is given to Andy Crump who took the photograph in 2001 in India.

Figure 5. Reprint permission kindly given by the World Health Organization, Special Programme for Research and Training in Tropical Diseases, http://www.who.int/tdr/index.html image #01021471. Additional photographic credit is given to Andy Crump who took the photograph in 2001 in Haiti.

Index